MONTGOMERY COLLEGE LIBRARY
ROCKVILLE CAMPUS

THE EFFECTS OF TROPICAL STORM AGNES ON THE CHESAPEAKE BAY ESTUARINE SYSTEM

THE EFFECTS OF TROPICAL STORM AGNES ON THE CHESAPEAKE BAY ESTUARINE SYSTEM

THE CHESAPEAKE RESEARCH CONSORTIUM, INC.

The Johns Hopkins University
Smithsonian Institution
University of Maryland
Virginia Institute of Marine Science

Project Coordinator, Jackson Davis (VIMS)
Volume Coordinator, Beverly Laird (VIMS)

Section Editors

Evon P. Ruzecki, Hydrological Effects (VIMS)
J. R. Schubel, Geological Effects (JHU)
Robert J. Huggett, Water Quality Effects (VIMS)
Aven M. Anderson, Biological Effects, Commercial (U.Md.)
Marvin L. Wass, Biological Effects, Non-Commercial (VIMS)
Richard J. Marasco, Economic Impacts (U.Md.)
M. P. Lynch, Public Health Impacts (VIMS)

November 1976

CRC Publication No. 54

*Published for The Chesapeake Research Consortium, Inc.,
by The Johns Hopkins University Press, Baltimore and London*

Copyright © 1977 by The Chesapeake Research Consortium, Inc.

All rights reserved. No part of this book may be
reproduced or transmitted in any form or by any means,
electronic or mechanical, including photocopying,
recording, xerography, or any information storage and
retrieval system, without permission in writing
from the publisher.

Manufactured in the United States of America

The Johns Hopkins University Press, Baltimore, Maryland 21218
The Johns Hopkins Press Ltd., London

Library of Congress Catalog Card Number 76-47392
ISBN 0-8018-1945-8

Library of Congress Cataloging in Publication data
will be found on the last printed page of this book.

Preface

During June 1972 Tropical Storm Agnes released record amounts of rainfall on the watersheds of most of the major tributaries of Chesapeake Bay. The resulting floods, categorized as a once-in-100-to-200-year occurrence, caused perturbations of the environment in Chesapeake Bay, the nation's greatest estuary.

This volume is an attempt to bring together analyses of the effects of this exceptional natural event on the hydrology, geology, water quality, and biology of Chesapeake Bay and to consider the impact of these effects on the economy of the Tidewater Region and on public health.

It is to be hoped that these analyses of the event will usefully serve government agencies and private sectors of society in their planning and evaluation of measures to cope with and ameliorate damage from estuarine flooding. It is also to be hoped that the scientific and technical sectors of society will gain a better understanding of the fundamental nature of the myriad and interrelated phenomena that is the Chesapeake Bay ecosystem. Presumably much of what was learned about Chesapeake Bay will be applicable to estuarine systems elsewhere in the world. Most of the papers comprising this volume were presented at a symposium held May 6-7, 1974, at College Park, Maryland, under the sponsorship of the Chesapeake Research Consortium, Inc., with support from the Baltimore District, U.S. Army Corps of Engineers (Contract No. DACW 31-73-C-0189). An early and necessarily incomplete assessment, *The Effects of Hurricane Agnes on the Environment and Organisms of Chesapeake Bay* was prepared by personnel from the Chesapeake Bay Institute (CBI), the Chesapeake Biological Laboratory (CBL), and the Virginia Institute of Marine Science (VIMS) for the Philadelphia District, U.S. Army Corps of Engineers. Most of the scientists who contributed to the early report conducted further analyses and wrote papers forming a part of this report on the effects of Agnes. Additional contributions have been prepared by other scientists, most notably in the fields of biological effects and economics.

The report represents an attempt to bring together all data, no matter how fragmentary, relating to the topic. The authors are to be congratulated for the generally high quality of their work. Those who might question, in parts of the purse, the fineness of the silk must keep in mind the nature of the sow's ears from which it was spun. This is not to disparage the effort, but only to recognize that the data were collected under circumstances which at best were less than ideal. When the flood waters surged into the Bay there was no time for painstaking experimental design. There were not enough instruments to take as many measurements as the investigators would have desired. There were not enough containers to obtain the needed samples or enough reagents to analyze them. There were not enough technicians and clerks to collect and tabulate the data. While the days seemed far too short to accomplish the job at hand, they undoubtedly seemed far too long to the beleaguered field parties, vessel crews, laboratory technicians, and scientists who worked double shifts regularly and around the clock on many occasions. To these dedicated men and women, whose quality of performance and perseverance under trying circumstances were outstanding, society owes an especial debt of gratitude.

It should be noted that the Chesapeake Bay Institute, the Chesapeake Biological Laboratory, and the Virginia Institute of Marine Science, the three major laboratories doing research on Chesapeake Bay, undertook extensive data-gathering programs, requiring sizable commitments of personnel and equipment, without assurance that financial support would be provided. The emergency existed, and the scientists recognized both an obligation to assist in ameliorating its destructive effects and a rare scientific opportunity to better understand the ecosystem. They proceeded to organize a coordinated program in the hope that financial arrangements could be worked out later. Fortunately, their hopes proved well founded. Financial and logistic assistance was provided by a large number of agencies that recognized the seriousness and uniqueness of the Agnes phenomenon. A list of those who aided is appended. Their support is gratefully acknowledged.

This document consists of a series of detailed technical reports preceded by a summary. The summary emphasizes effects having social or economic impact. The authors of each of the technical reports are indicated. To these scientists, the editors extend thanks and commendations for their painstaking work.

Several members of the staff of the Baltimore District, U.S. Army Corps of Engineers, worked with the editors on this contract. We gratefully acknowledge the helpful assistance of Mr. Noel E. Beegle, Chief, Study Coordination and Evaluation Section, who served as Study Manager; Dr. James H. McKay, Chief, Technical Studies and Data Development Section; and Mr. Alfred E. Robinson, Jr., Chief of the Chesapeake Bay Study Group.

The editors are also grateful to Vickie Krahn for typing the Technical Reports and to Alice Lee Tillage and Barbara Crewe for typing the Summary.

The Summary was compiled from summaries of each section prepared by the section editors. I fear that it is too much to hope that, in my attempts to distill the voluminous, detailed, and well-prepared papers and section summaries, I have not distorted meanings, excluded useful information or overextended conclusions. For whatever shortcomings and inaccuracies that exist in the Summary, I offer my apologies.

Jackson Davis
Project Coordinator

Acknowledgements

The Chesapeake Research Consortium, Inc. is indebted to the following groups for their logistic and/or financial aid to one or more of the consortium institutions in support of investigations into the effects of Tropical Storm Agnes.

U. S. Army
-- Corps of Engineers, Baltimore District
-- Corps of Engineers, Norfolk District
-- Corps of Engineers, Philadelphia District
-- Transportation Corps, Fort Eustis, Virginia

U. S. Navy
-- Naval Ordnance Laboratory
-- Coastal River Squadron Two, Little Creek, Virginia
-- Assault Creek Unit Two, Little Creek, Virginia
-- Explosive Ordnance Disposal Unit Two, Fort Story, Virginia
-- Naval Ordnance Laboratory, White Oak, Maryland

U. S. Coast Guard
-- Reserve Training Center
-- Coast Guard Station, Little Creek, Virginia
-- Portsmouth Supply Depot
-- Light Towers (Diamond Shoal, Five Fathom Bank, and Chesapeake)

National Oceanic and Atmospheric Administration
-- National Marine Fisheries Service (Woods Hole, Massachusetts and Sandy Hook, New Jersey)

The National Science Foundation

Food and Drug Administration

Environmental Protection Agency

U. S. Office of Emergency Preparedness

State of Maryland, Department of Natural Resources

Commonwealth of Virginia, Office of Emergency Preparedness

TABLE OF CONTENTS

Preface.	v
Acknowledgements	vii
SUMMARY.	1
Hydrological Effects	1
The Storm and Resulting Flood.	1
Effects of Flood Waters on the Salinity Distribution in Chesapeake Bay, Its Major Tributaries and Contiguous Continental Shelf.	6
Effects of Agnes Flooding on Smaller Tributaries to Chesapeake Bay	12
Geological Effects	12
Water Quality Effects.	15
Biological Effects	16
Shellfishes.	16
Fishes	18
Blue Crabs	19
Aquatic Plants	19
Jellyfish.	19
Plankton and Benthos	19
Economic Impact.	20
Shellfish and Finfish Industries	21
Economic Impact on Recreation Industries and Users . . .	23
Other Impacts.	27
Public Health Impacts.	27
Shellfish Closings	28
Water Contact Closings	28
Shellfish Contamination.	28
Waterborne Pathogens	29
Miscellaneous Hazards.	29

APPENDICES: TECHNICAL REPORTS

A. Hydrological Effects

    Effects of Agnes on the distribution of salinity along the main axis of the bay and in contiguous shelf waters, *Schubel, Carter, Cronin*. 33

Changes in salinity structure of the James, York and
Rappahannock estuaries resulting from the effects
of Tropical Storm Agnes, *Hyer, Ruzecki* 66

The effects of the Agnes flood in the salinity structure
of the lower Chesapeake Bay and contiguous waters,
Kuo, Ruzecki, Fang 81

Flood wave-tide interaction on the James River during
the Agnes flood, *Jacobson, Fang* 104

Daily rainfall over the Chesapeake Bay drainage basin
from Tropical Storm Agnes, *Astling* 118

The effect of Tropical Storm Agnes on the salinity
distribution in the Chesapeake and Delaware Canal,
Gardner 130

Agnes impact on an eastern shore tributary: Chester
River, Maryland, *Tzou, Palmer* 136

Tributary embayment response to Tropical Storm Agnes:
Rhode and West rivers, *Han* 149

Rhode River water quality and Tropical Storm Agnes,
Cory, Redding 168

B. Geological Effects

Effects of Agnes on the suspended sediment of the Chesapeake
Bay and contiguous shelf waters, *Schubel* 179

Response and recovery to sediment influx in the Rappahannock
estuary: a summary, *Nichols, Thompson, Nelson* 201

The effects of Tropical Storm Agnes on the copper and zinc
budgets of the Rappahannock River, *Huggett, Bender* 205

Agnes in Maryland: Shoreline recession and landslides,
McMullan 216

Chester River sedimentation and erosion: equivocal evidence,
Palmer 223

Effect of Tropical Storm Agnes on the beach and nearshore
profile, *Kerhin* 227

Agnes in the geological record of the upper Chesapeake Bay,
Schubel, Zabawa 240

C. Water Quality Effects

Some effects of Tropical Storm Agnes on water quality in the Patuxent River estuary, *Flemer, Ulanowicz, Taylor* 251

Indirect effects of Tropical Storm Agnes upon the Rhode River, *Correll* 288

Effects of Tropical Storm Agnes on nutrient flux and distribution in lower Chesapeake Bay, *Smith, MacIntyre, Lake, Windsor* 299

Effects of Agnes on the distribution of nutrients in upper Chesapeake Bay, *Schubel, Taylor, Grant, Cronin, Glendening* 311

The effect of Tropical Storm Agnes on heavy metal and pesticide residues in the eastern oyster from southern Chesapeake Bay, *Bender, Huggett* 320

Effects of Agnes on the distribution of dissolved oxygen along the main axis of the bay, *Schubel, Cronin* 335

Observations on dissolved oxygen conditions in three Virginia estuaries after Tropical Storm Agnes (summer 1972), *Jordan* 348

The effect of Tropical Storm Agnes as reflected in chlorophyll $\underline{a}$ and heterotrophic potential of the lower Chesapeake Bay, *Zubkoff, Warinner* 368

Calvert Cliffs sediment radioactivities before and after Tropical Storm Agnes, *Cressy* 389

D. Biological Effects

Distribution and abundance of aquatic vegetation in the upper Chesapeake Bay, 1971-1974, *Kerwin, Munro, Peterson* 393

Effects of Tropical Storm Agnes and dredge spoils on benthic macroinvertebrates at a site in the upper Chesapeake Bay, *Pearson, Bender* 401

Some effects of Tropical Storm Agnes on the sea nettle population in the Chesapeake Bay, *Cargo* 417

Effects of Tropical Storm Agnes on Zooplankton in the
lower Chesapeake Bay, *Grant, Bryan, Jacobs, Olney*. 425

Effects of Tropical Storm Agnes on standing crops
and age structure of zooplankton in middle Chesapeake
Bay, *Heinle, Millsaps, Millsaps*. 443

Short-term Response of Fish to Tropical Storm Agnes in
mid-Chesapeake Bay, *Ritchie*. 460

The effects of Tropical Storm Agnes on fishes in the
James, York, and Rapahannock rivers of Virginia,
Hoagman, Wilson. 464

Mortalities caused by Tropical Storm Agnes to clams and
and oysters in the Rhode River area of Chesapeake Bay,
Cory, Redding. 478

The effect of Tropical Storm Agnes on oysters, hard clams,
soft clams, and oyster drills in Virginia, *Haven, Hargis,
Loesch, Whitcomb* 488

A comparative study of primary production and standing
crops of phytoplankton in a portion of the upper
Chesapeake Bay subsequent to Tropical Storm Agnes,
Loftus, Seliger. 509

Effects of Tropical Storm Agnes on the bacterial flora
of Chesapeake Bay, *Nelson, Colwell* 522

Species diversity among sarcodine protozoa from Rhode
River, Maryland, following Tropical Storm Agnes, *Sawyer,
Maclean, Coats, Hilfiker, Riordan, Small* 531

The impact of Tropical Storm Agnes on mid-bay infauna,
Hamilton 544

Patterns of distribution of estuarine organisms and their
response to a catastrophic decrease in salinity,
Larsen 555

The effect of Tropical Storm Agnes on the benthic fauna of
eelgrass, <u>Zostera marina</u>, in the lower Chesapeake Bay,
Orth 566

Effect of Tropical Storm Agnes on setting of shipworms
at Gloucester Point, Virginia, *Wass*. 584

The displacement and loss of larval fishes from the
  Rappahannock and James rivers, Virginia, following
  a major tropical storm, *Hoagman, Merriner* 591

  Effect of Agnes on jellyfish in southern Chesapeake Bay,
  Morales-Alamo, Haven 594

E. Economic Impacts

  Economic impacts of Tropical Storm Agnes in Virginia,
  Garrett, Schifrin 597

  The Maryland commercial and recreational fishing industries:
  an assessment of the economic impact of Tropical Storm
  Agnes, *Smith, Marasco* 611

F. Public Health Impacts

  Public health aspects of Tropical Storm Agnes in Virginia's
  portion of Chesapeake Bay and its tributaries, *Lynch,
  Jones* 625

  Public health aspects of Tropical Storm Agnes in Maryland's
  portion of Chesapeake Bay, *Andersen* 636

THE EFFECTS OF TROPICAL STORM AGNES ON THE CHESAPEAKE BAY ESTUARINE SYSTEM

THE EFFECTS OF TROPICAL STORM AGNES
ON THE CHESAPEAKE BAY ESTUARINE SYSTEM

SUMMARY REPORT

HYDROLOGICAL EFFECTS

The Storm and Resulting Flood

The Tropical Storm named Agnes started on the Yucatan Peninsula as a tropical disturbance on 14 June 1972. Within four days it developed to hurricane intensity in the Gulf of Mexico northwest of Cuba. The path of Agnes from tropical depression through hurricane and finally to extratropical stage is shown in Figure 1. While Agnes was in her infancy on the Yucatan Peninsula, a weak cold front moved through the region which serves as the watershed to Chesapeake Bay and its tributary rivers. This cold front was responsible for rainfall in the Chesapeake watershed in amounts from one to three inches with isolated stations in Maryland reporting up to six inches. This rainfall, although not exceptionally strong for the region, served to further saturate a watershed which was already suffering from an unusually wet winter and spring.

Rains directly attributable to Agnes reached the watershed on 21 June and lasted through 23 June. Figure 2, a grouping of radar echo pictures, shows the movement of precipitation echoes through the watershed from southwest to northeast on 21 and 22 June. The entire watershed was subjected to measured rainfall in excess of five inches with approximately one third of the region receiving more than twelve inches of water and isolated locations recording eighteen inches in the three day period. Other storms (e.g. Camille) have produced greater amounts of rainfall at individual stations, but none in the period of record has produced such heavy rainfall over as large a region as Agnes. The two largest contributors of fresh water, the Susquehanna River and the Potomac River, each had 3-day accumulations in excess of 8 inches (8.28 Potomac, 8.09 Susquehanna).

This deluge, on a saturated watershed, resulted in immediate flooding of the major tributaries to Chesapeake Bay. Major tributaries to Chesapeake Bay are the Susquehanna, Potomac, Rappahannock, York and James rivers. Their relative positions are shown in Figure 3. Most rivers crested at levels higher than previously recorded. Table 1 lists average flows for the month of June as well as average daily flows and instantaneous peak discharges for major tributaries to Chesapeake Bay for the period 20 to 27 June 1972. These flows were measured (or estimated) at the furthest downstream gauging station in each river (usually just upstream of the region of tidal influence).

From 21 to 30 June 1972, the Susquehanna River, usually responsible for 61% of the fresh water contributed to Chesapeake Bay in June, had flows averaging 15.5 times greater than normal. This river accounted for 64% of the fresh inflow to the Bay for the ten-day period. This volume of water from the major tributary resulted in a 30 nautical mile translation of fresh water down the Bay (based on the movement of the 5 ppt isohaline). Had Chesapeake Bay been a reservoir (i.e. if a dam existed between the Virginia Capes), the water level in the Bay and all its tidal tributaries would have been increased by approximately two feet from the ten-day Agnes-induced flooding of all major tributaries to the Bay.

Summary

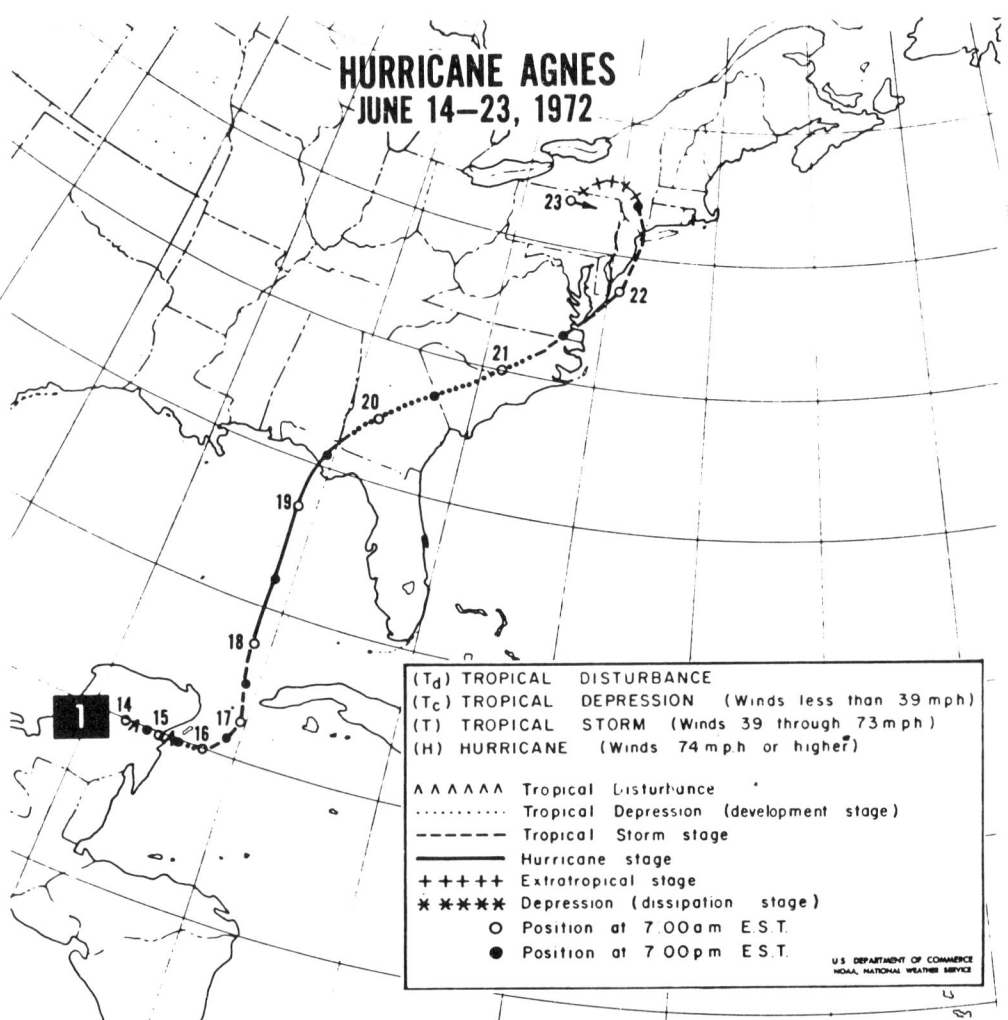

Figure 1. Track of the center of Hurricane Agnes. Rains and winds extended out from the storm center over long distances. From DeAngelis and Hodge 1972. Preliminary Climatic Data Report: Hurricane Agnes, June 14-23, 1972. U. S. Dept. of Commerce, NOAA, Techn. Mem. EDS NCC-1, iv + 62 p.

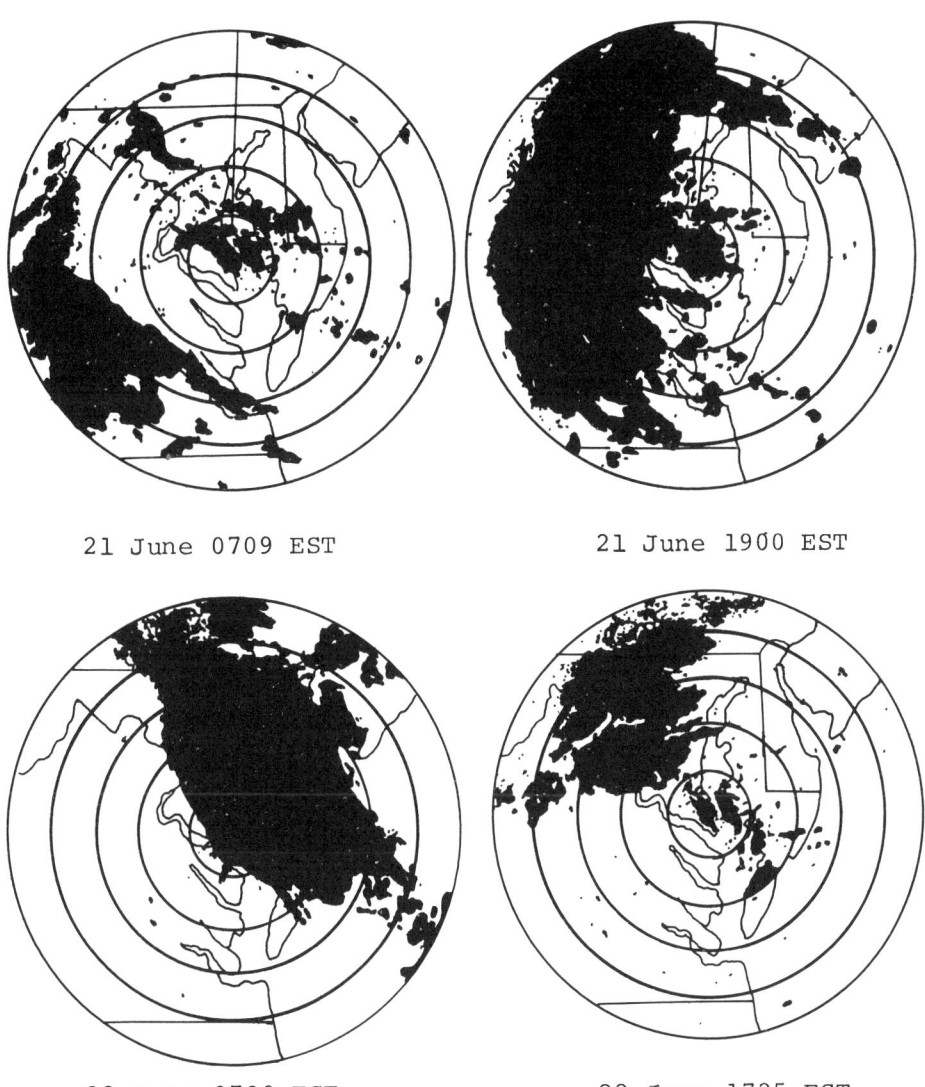

Figure 2. Precipitation echoes over the Chesapeake Bay watershed observed by the Patuxent weather station.

4   Summary

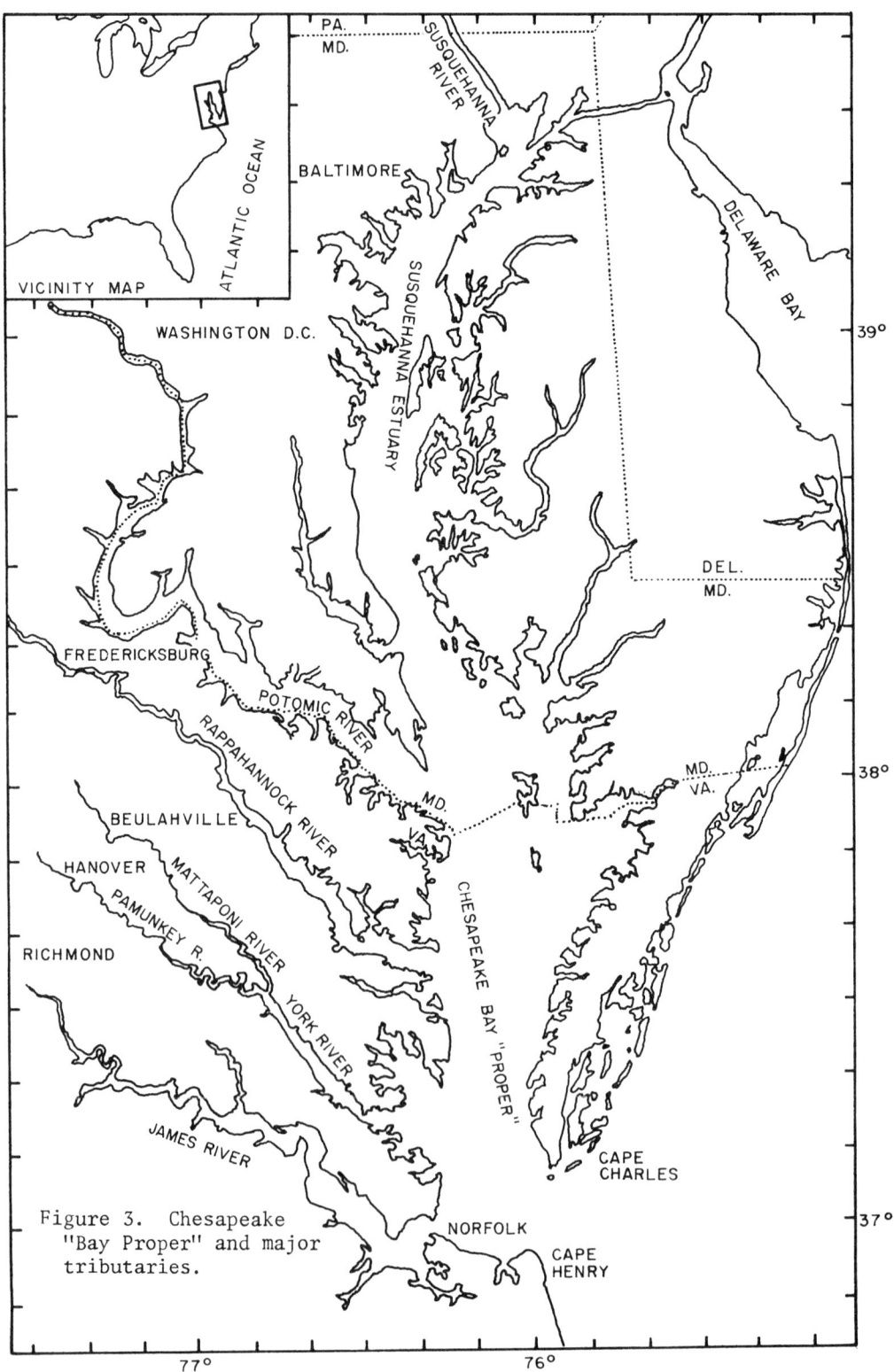

Figure 3. Chesapeake "Bay Proper" and major tributaries.

Table 1. Gauged flows for major tributaries to Chesapeake Bay during flooding from Tropical Storm Agnes. (Numbers in parenthesis indicate instantaneous peak flows).

River (Gauging Station)	Drainage Area Above Fall Line (Sq. Mi.)[3]	Normal Av. June Flows (cfs)[1]	Average daily flows (cfs)[1] for June 1972							
			20	21	22	23	24	25	26	27
Susquehanna (Conowingo, Md.)	27,089	35,000*	54,200	50,400	445,000	1,040,000	1,120,000 (1,130,000)	1,010,000	696,000	418,000
Potomac (Washington, D.C.)	11,560	8,020	7,160	12,000	172,000	268,000	334,000 (359,000)	200,000	79,800	35,600
Rappahannock (Fredericksburg, Va.)	1,599	1,240	1,090	6,380	84,200 (107,000)	44,500	18,200	7,410	5,100	4,600
York (Beulahville, Va.) Hanover, Va.)	1,691	859	1,793	3,091	10,460	30,730 (35,630)[2]	34,300	32,500 (33,200)[2]	28,700	21,480
James (Richmond, Va.)	6,757	4,650	9,220	14,300	136,000	296,000 (313,000)	210,000	88,700	28,700	19,900

Note:
1- cfs = cubic feet per second.
2- The York River has two major tributaries where furthest downstream gauging stations are located: On the Mattaponi River, the gauging station at Beulahville, Va. recorded a peak instantaneous flow of 16,900 cfs on 25 June; on the Pamunkey River the gauging station near Hanover, Va. recorded a peak instantaneous flow of 29,900 cfs on 23 June. These instantaneous peak flows, combined with average daily flows in the associated York tributary gave combined peak flows into the York system as shown.
3- Drainage area statistics from Seitz, R. C. Drainage Area Statistics for Chesapeake Bay Freshwater Drainage Basin. CBI Special Report 19, Feb. 1971.
*- Based on USGS info available in this office this value appears high by $\pm$ 8000 cfs.

The relative effect of Agnes flooding on each major tributary becomes apparent when flows are normalized to average June flows. Flows are normalized by dividing each daily flow by the normal daily June flow for each river. Figure 4 shows daily normalized flows for the major tributaries to the Bay during the period 20 June - 5 July 1972. Normalized peak instantaneous flows are also shown. From this figure, it is apparent that two forms of flooding occurred: (1) an abrupt flow increase in excess of 60 times normal, followed by an equally abrupt decrease in flow back to approximately six times normal as is illustrated by the James and Rappahannock rivers, and (2) a somewhat slower increase in flow to 30 or 40 times normal followed by a decrease in flow which took twice as long as the increase as is illustrated by the Susquehanna and York rivers. The normalized record for the Potomac River falls somewhere between these two. Table 2 illustrates that for normalized flows, flooding in the York River was most severe. When actual volumes of water are considered, however, flooding in the York was less severe because of the low flows usually experienced in June (see Table 1). Additionally, peak flooding in the Mattaponi and Pamunkey rivers, the major tributaries to the York, occurred at different times (see Table 1). Of all major tributaries to the Bay, the York was least affected by Agnes-induced flooding.

Table 2. Normalized flows for major tributaries to Chesapeake Bay for various periods of flooding due to Tropical Storm Agnes.

River	Normalized Flows* (1972)		
	7 days 21 to 27 June	10 days 21 to 30 June	15 days 21 June to 5 July
Susquehanna	19.5	15.5	11.6
Potomac	19.7	15.4	9.2
Rappahannock	19.6	**15.2**	11.3
York	25.3	21.2	15.2
James	24.4	18.0	12.3
Total for all rivers shown	22.2	17.3	12.4

$$* \text{ Normalized flow} = \frac{\text{average flow per period}}{\text{normal average June flow}}$$

Effects of Flood Waters on the Salinity Distribution in Chesapeake Bay, Its Major Tributaries and Contiguous Continental Shelf

Prior to the Agnes flood, Chesapeake Bay was in an unusual hydrologic condition. Whereas water temperature was similar to that expected in late spring or early summer the salinity distribution was most akin to that expected in mid-spring owing to greater than average flows during the preceding winter and spring. Hence salinity was depressed more than in June of a year of more normal rainfall.

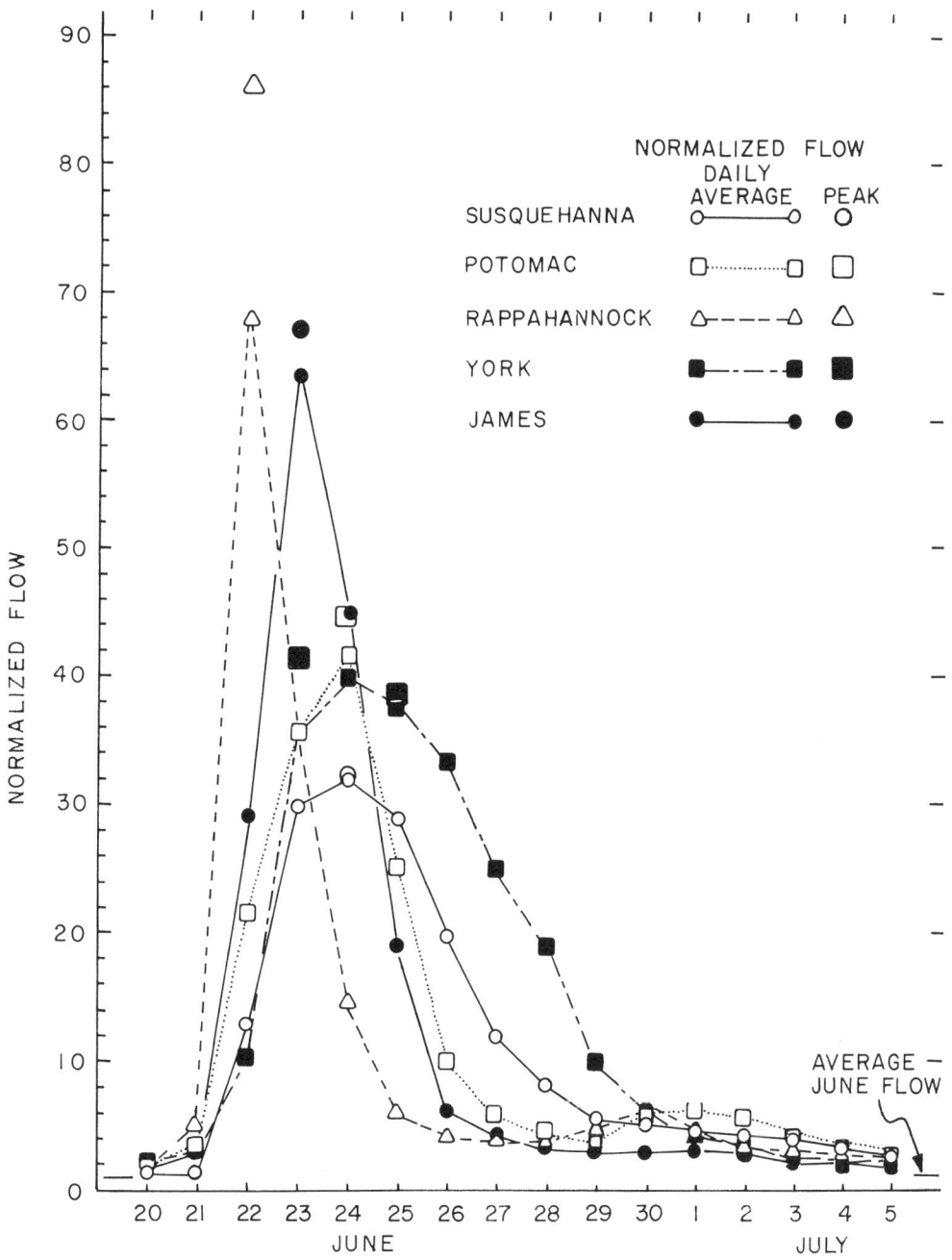

Figure 4. Normalized daily flows for major tributaries to Chesapeake Bay during flooding from Tropical Storm Agnes, June-July 1972.

Chesapeake Bay and each of its major tributaries showed similar reactions to Agnes flooding. Generally, four stages were observed (Fig. 5).

1) Initially, flood waters forced surface salinities downstream several miles while bottom salinities remained somewhat constant, producing highly stratified estuaries. Distance and duration of the displacement were dependent on the dimensions of each particular basin and the magnitude of flooding within that basin.

2) The second stage of reaction to the flood was similar to the first but operated on bottom rather than surface waters shifting them downstream. This resulted in vertically homogeneous estuaries of very low salinity.

3) The third stage was essentially a reaction to the first two and is presumed to be the result of gravitational circulation. During this stage, there was a net transport of salt up the estuaries. This transport started in the lower layers, eventually acted on surface water and, particularly in the lower layers, moved salt water upstream substantially beyond the pre-Agnes position.

4) The final stage was vertical mixing between surface and bottom water which resulted in salinity structure similar to that expected during a "normal" summer. This final stage of the reaction to Agnes-induced flooding was generally underway, for the Chesapeake Bay system, by the end of September, approximately 100 days after the flood waters crested at the fall line.

Chesapeake Bay and its major tributaries can be segmented into two categories: The "Bay Proper" and the tidal-estuarine portions of the major tributaries as shown in Figure 3. If we consider the northern portion of the Bay (north of the Potomac River) as the estuarine portion of the Susquehanna River, then the southern portion of the Bay is the "Bay Proper". The generalized sequence of events described above was evident to some extent throughout the estuarine system but was most pronounced in the James and York rivers and the "Bay Proper". The remaining major tributaries showed the first two (downstream directed) stages but were subjected to bay-tributary interactions during the third (upstream directed) stage. At the time the Potomac and Rappahannock rivers went through the third stage, the up-bay encroachment of high salinity water had not reached their mouths. The result was an upstream movement of slightly salty water into these rivers from the northern-most portion of the "Bay Proper". This situation did not occur in the James and York rivers because of their proximity to the ocean.

The lower Bay was subjected to a cascade of flooding from the major Virginia tributaries as well as the initial flooding from the Susquehanna. The early effect is shown in Figure 6 which illustrates surface salinities for the period 29 June to 3 July. Flood waters from the Susquehanna and Potomac coursed down the center of the Bay bypassing pockets of higher salinity water in smaller tributaries on either side. Approximately seven days later (Fig. 7) large patches of freshened water from the Potomac and Rappahannock had progressed some distance downstream from their mouths.

Analyses of over 120 sampling runs conducted on the James, York, and Rappahannock rivers to measure temperature, salinity, dissolved oxygen and

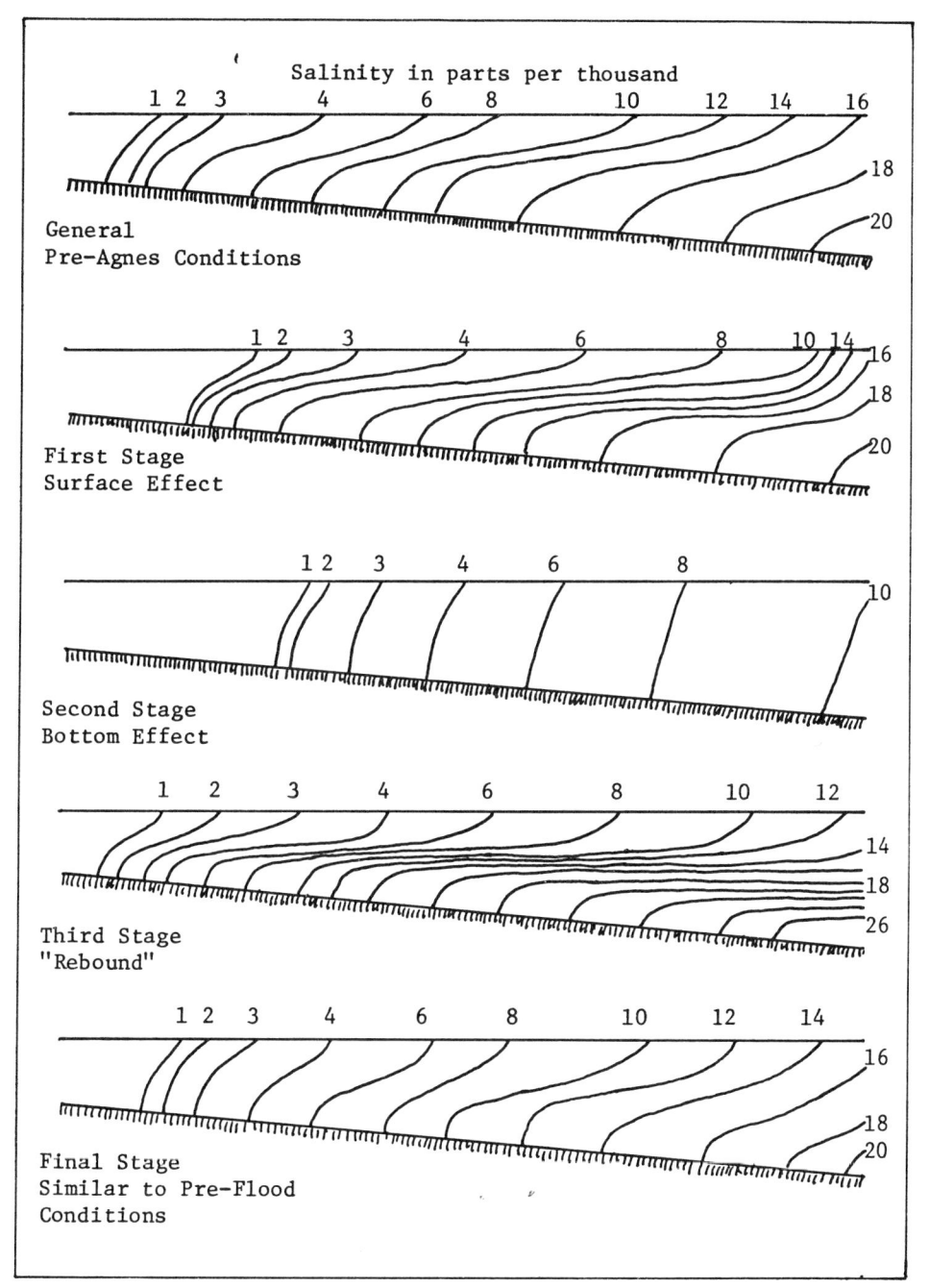

Figure 5. Schematic representation of reaction to Agnes flooding.

10 Summary

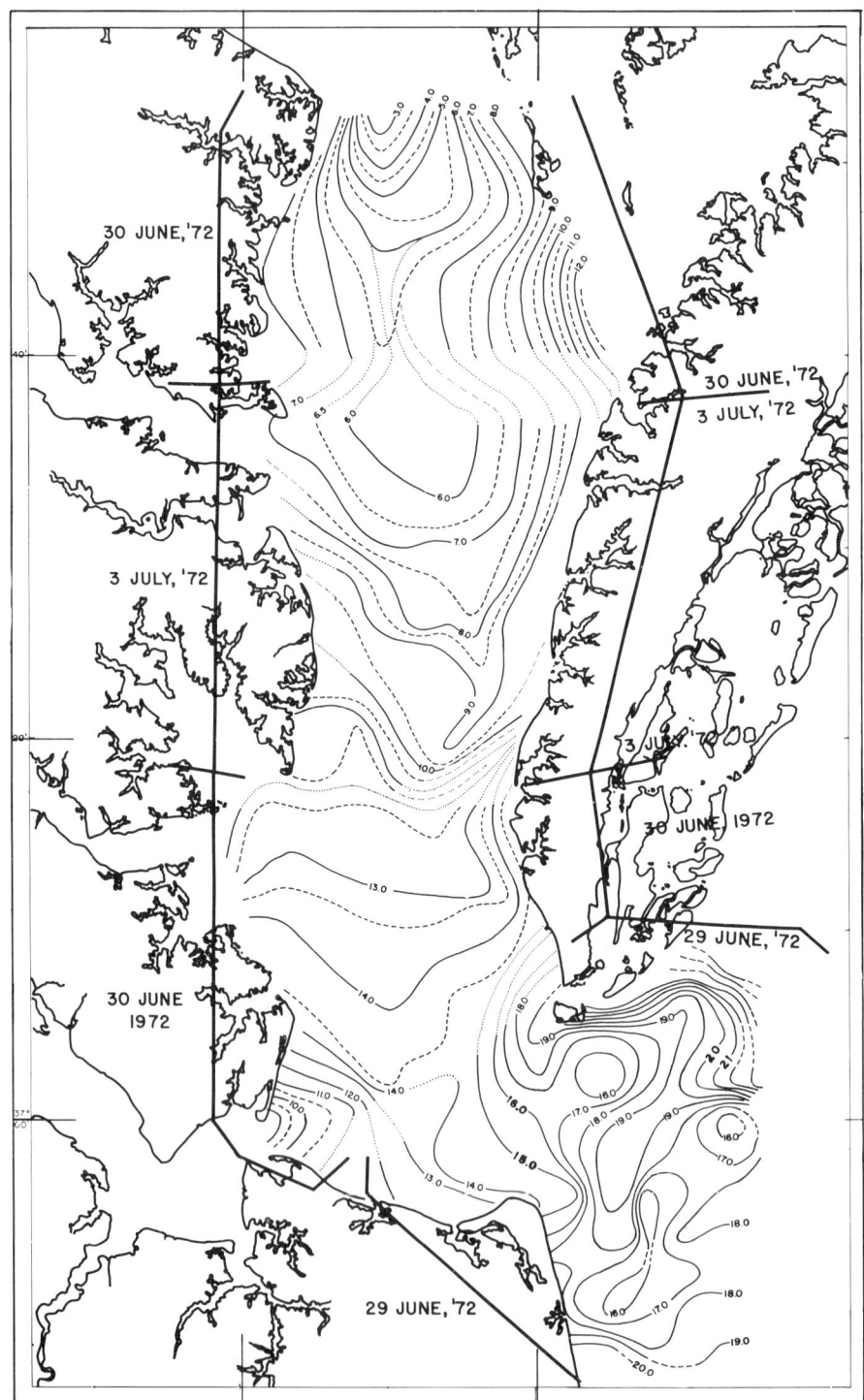

Figure 6. Surface salinities for lower Chesapeake Bay during the period 29 June - 3 July 1972.

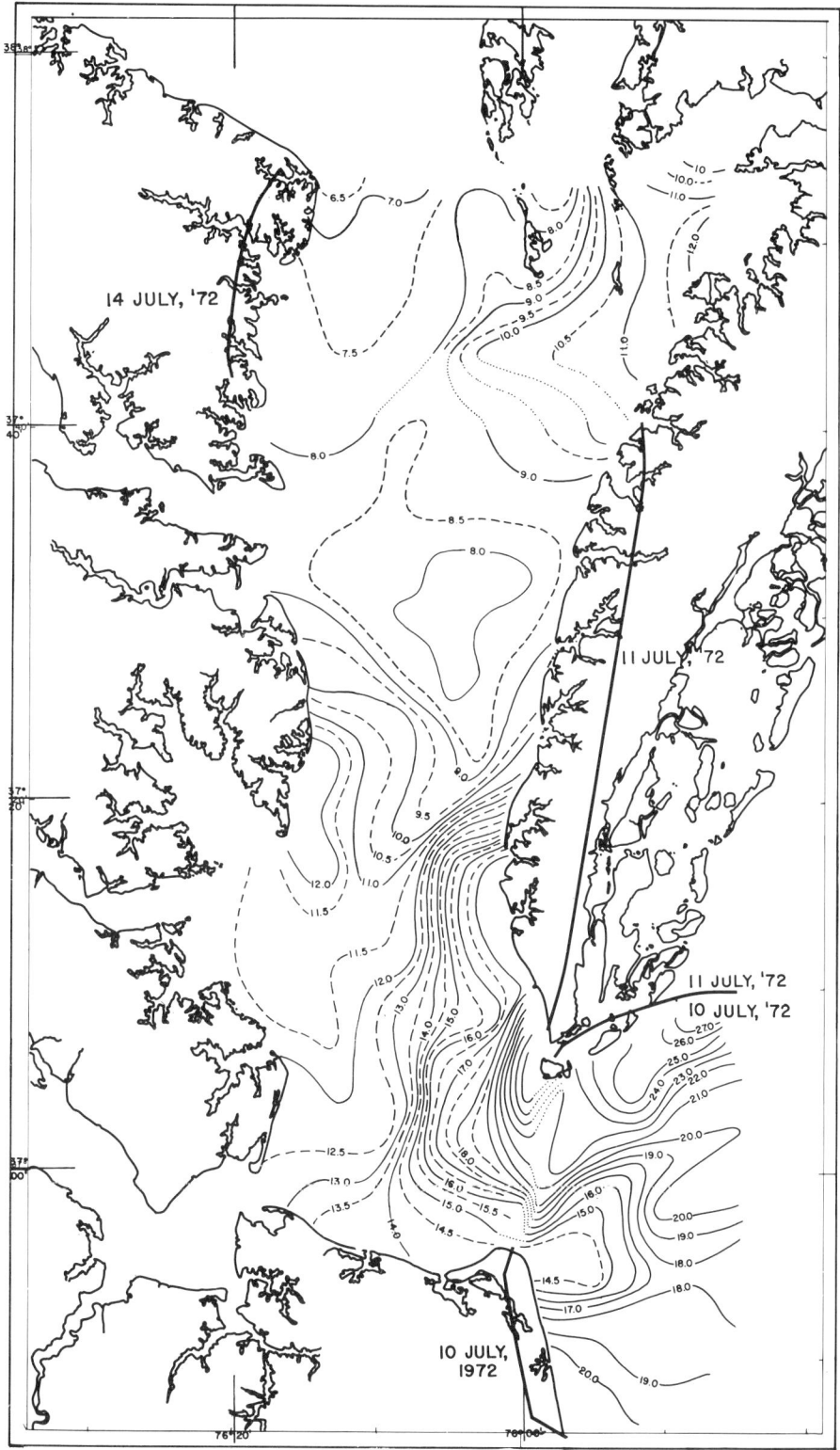

Figure 7. Surface salinities for lower Chesapeake Bay during the period 10 - 14 July 1972.

suspended sediments indicate that these tributaries to the Bay were subjected to internal seiches (essentially resonating waves within a particular basin) which were generated by the flood shock. These internal oscillations with periods from four to fifteen days helped to mix these estuaries vertically.

Analyses of current and tide data in the James River indicate the following:

1) Rise in water level was slight indeed in the tidal rivers in comparison with that experienced above the Fall Line. Water level elevations of approximately 6 feet above predicted tidal levels occurred in the upper portions of the tidal rivers, but no change was discernible at the mouths. Passage of the storm's low pressure center caused an increase in water level of a few inches.

2) The normal tidal current pattern was disrupted, there being a continuously ebbing current for several days as far downstream as the zone of transition from fresh to salt water. Downstream of that zone, surface waters ebbed continuously for three days, but lower layers showed normal ebb and flood current oscillations.

Movement of Agnes flood waters onto the continental shelf was traced with one cruise by the Chesapeake Bay Institute and five cruises by the Virginia Institute of Marine Science. Flood waters leaving the mouth of the Bay moved southward along the coast. Freshened water remained in the upper 10 meters (33 feet) of the water column and was broken into large patches by tidal motion at the Bay mouth. Mixing of the patches of fresh water with ocean water was most prominent on their eastern boundaries, there being little vertical or north-south mixing.

Effects of Agnes Flooding on Smaller Tributaries to Chesapeake Bay

In general, small tributaries to the Bay became reverse estuaries after the passage of the Agnes flood. Their normal source of salt water, the Bay, became substantially fresher than these small rivers. Fresh water moved upstream from the mouth in the surface layers and saltier water moved from the upstream reaches toward the mouth in the lower layers. These conditions persisted for varying lengths of time depending on the recovery of adjacent portions of the Bay and especially on the vertical structure of the water column in the adjacent Bay in relation to the depth of the mouth of the subestuary or bay. Seiche conditions similar to those observed in the large tributaries were evident as surface phenomena and were attributed to wind set-up rather than freshwater flooding.

GEOLOGICAL EFFECTS

Tropical Storm Agnes clearly constituted a major event in the history of the Chesapeake Bay estuary. The Chesapeake Bay, like other estuaries, is an ephemeral feature on a geological time scale. The sediments that are rapidly filling the Bay basin will gradually expel the intruding sea that formed the estuary and convert the Bay back into a river valley system. If sea level remains relatively constant, this process will take at most a few tens of thousands of years to complete. If sea level rises, the life of the

estuary will be prolonged. If sea level falls, its lifetime will be shor
The rate of filling of the basin is generally greatest in the upper reaches
of the estuaries because of the proximity to the fluvial sources, and because
of the dynamic processes in the estuary.

The sedimentological processes that characterize the Bay under "normal"
or "average" conditions have been described by a number of investigators,
but there was, until Agnes, a dearth of observations of the effects of
episodic events such as floods and hurricanes on the geological history of the
Bay. Sampling during episodes is often difficult and the infrequency of such
events makes the likelihood of fortuitous observations very small. Tropical
Storm Agnes presented scientists with an unusual opportunity to document the
impact of a major flood on the Chesapeake Bay estuarine system. Since there
was little wind associated with Agnes by the time she reached the Bay,
sampling in the Bay was not hampered by high seas. The heavy rains stripped
large quantities of soil from throughout much of the drainage basin, and the
swollen streams and rivers produced significant bank erosion in upland areas.
Much of the fine-grained sediment was carried into the Chesapeake Bay estuarine
system. The primary mode of sediment transport both into and within the Bay
was as suspended load.

In a ten-day period (21-30 June 1972) the Susquehanna River -- the long-
term supplier of approximately half of the total fresh water input to the
Chesapeake Bay estuarine system-- discharged more sediment than during the
preceding ten years, and probably during the previous quarter of a century,
and perhaps even longer. The suspended sediment discharge at Conowingo
during this ten day period was estimated to be more than 31 million metric
tons, compared to an average annual input during "normal" years of only
one-half to one million metric tons. The Agnes sediment was very similar
in texture to that normally discharged. It was primarily silt and clay,
but a significant amount of fine sand was also discharged.

Sediment discharges of the other major western shore tributaries were
also anomalously high, but there is a paucity of suspended sediment data
for these tributaries, except the Rappahannock River. The Rappahannock
probably discharged more than a million metric tons of sediment during Agnes.
Since the Eastern Shore (the Delmarva Peninsula) did not receive the heavy
rainfall associated with Agnes, increases in the water and sediment discharges of the Eastern Shore tributaries were relatively small.

The large influxes of suspended sediment by the major tributaries raised
concentrations of suspended sediment throughout much of the Chesapeake Bay
estuarine system to levels higher than any previously reported. In the
upper Bay concentrations exceeded 1000 mg/l in the surface waters for a few
days, and in the upper reaches of the Potomac, Rappahannock, James and other
major estuaries, concentrations of several hundred mg/l were observed.
Normally concentrations are near 10 mg/l in surface waters. In the upper
"Bay Proper", and in the upper reaches of the tributaries, there were marked
downstream gradients of the concentration of suspended sediment during the
period of peak flow. The downstream decreases were produced primarily by
the settling out of sediment, dilution of riverwater by Bay water being a
minor factor. The bulk of the sediment discharged by the Susquehanna, probably
more than 75% of it, was deposited in the upper reaches of the Bay -- upstream from Tolchester. The bulk of the sediment discharge by the other rivers
was trapped in a similar fashion in the upper reaches of the estuaries of
those rivers, and relatively little reached the "Bay Proper". For example,

:harged by the Rappahannock River was deposited within
y, upstream from the juncture of that tributary with
y.

.nna is the only river that discharges directly into
tne m... 	ay without an intervening estuary, its effect on the
distribution of suspended sediment in the "Bay Proper" was most apparent.
This input, combined with the suspended sediment that escaped the tributary
estuaries sent concentrations of suspended sediment in the middle and lower
reaches of the Bay during late June and July to levels two-to-threefold higher
than "normal" for that time of year. The outflow from the Bay could be traced
as a low salinity, turbid band of water that turned south after leaving the
Virginia Capes, and flowed along the Carolina coast. The concentrations of
suspended sediment in this nearshore band were higher than any previously
reported for this segment of the shelf.

As riverflows subsided, the normal two-layered estuarine circulation
patterns were re-establshed in the upper Bay, and in the upper reaches of the
tributary estuaries. As salt water was advected upstream in the lower layer
there was a net upstream movement of fine sediment suspended in the lower
layer. Sediment previously carried downstream and deposited by the Agnes floodwaters was resuspended by tidal currents and gradually transported back up the
estuary. The routes of sediment dispersal are clear, but the rates of movement
are obscure in nearly all of the Bay and its tributaries. The available data
do not permit reliable estimates of sediment transport, particularly during
the recovery period. Studies of the longitudinal distributions of transition
metals in the bottom sediments suggest the importance of this upstream
movement during the recovery period.

Sedimentation rates were greatest in the upper reaches of the Bay and its
tributaries, and decreased markedly in a downstream direction. In large
areas of the upper Bay between Turkey Point and Tolchester 15-25 cm of new
sediment was deposited by Agnes. In the shipping channel in this same region
as much as one meter of new material was deposited. According to CBI observations, farther upstream on the Susquehanna Flats approximately ten acres of
new islands and several hundred acres of new inter-tidal areas were formed.
In addition, there was appreciable fill in several stretches of the shipping
channel that extends from Havre de Grace to the head of the Bay.

In the middle and upper Rappahannock estuary between 2 and 7.5 mm of new
sediment was deposited during Agnes. This corresponds to approximately one-
third the average annual accumulation of sediment in this stretch of the
estuary. The depositional pattern of the Agnes sediment was indicated by the
distribution of "extractable" copper in the bottom sediments before and after
Agnes. The inorganic fraction of the copper content of the Rappahannock sediments increased by a factor of 2 to 3 during Agnes, but returned to "normal"
levels within one year. The return to normal levels probably resulted from
a combination of processes: upstream transport of copper-rich Agnes sediments
in the lower layer, resuspension and seaward transport of Agnes sediments
in the upper layer, mixing of Agnes sediments with underlying sediments by
burrowing organisms, and chemical re-equilibrium.

Since there was little wind associated with Agnes when she reached the
Bay, seas were not unusually high and evidence of abnormal shore erosion by
wind waves was equivocal. Significant erosion did occur along some stretches

of the shoreline where relief is high and banks are comprised of normally erosion resistant clayey sediments. Banks comprised of apparently more easily erodible, unconsolidated Pleistocene sediments suffered less erosion. Erosion was most pronounced in the non-marine Cretaceous clays of the Potomac Group and in which the erosion was produced primarily by a combination of torrential rains, movement of groundwater, and failures (slippages) of water saturated sediments on unstable slopes. It is likely that these processes account for much of the erosion of the margins of the Bay even during "normal" years.

Floods of the magnitude of Agnes have a recurrence interval of about 100 to 200 years; floods of lesser magnitude occur even more frequently. Comparison of sedimentological data collected during and immediately following Agnes with data collected during "average" conditions over many years indicates that episodic events probably have a much greater and more persistent impact on the sedimentary history of the Chesapeake Bay than do average conditions.

The impact of Agnes on the Bay was clearly depositional and not erosional; the erosion occurred farther upstream in the drainage basin. Some parts of the Bay aged, geologically, by more than a decade in a week. It appears unlikely however, that the sediments deposited during Agnes will be preserved in the geological record of the Bay as a discrete sedimentary layer covering large areas of the Bay and its tributaries. Repeated corings during the year subsequent to Agnes revealed that the distinguishing features of the Agnes layer were slowly being obliterated by burrowing organisms. Furthermore, variability among the cores was so great at some stations that strata could not be correlated in consecutive cores taken at a given time from an anchored vessel.

WATER QUALITY EFFECTS

The large amount of runoff from land resulted in unseasonably high concentrations of dissolved inorganic nitrogen. Concentrations of nitrate and nitrite were some two to three times greater than normal in the northern half of Chesapeake Bay, but were only slightly elevated at the Bay mouth. In contrast to nitrogen, phosphate concentration remained essentially normal in the Bay despite elevated concentrations measured in some tributaries. Although several investigators measured nutrients at different places, comparison among them is equivocal because different techniques were applied or because different components were measured, such as dissolved vs. particulate. The flood waters also contributed a pulse of dissolved organic matter.

The nutrients were rapidly lost to the sediments. Measurement of nutrient flux at the Bay mouth indicated that little of the input was exported to the ocean. Release of nutrients from the sediments in the summer one year after Agnes resulted in abnormally extensive blooms of algae.

Dissolved oxygen concentrations were depressed both in Chesapeake Bay and in the tributaries. Deep waters in the lower portions of the tributaries normally become oxygen-deficient during summer, but the post-Agnes deficiencies extended over greater area.

Trace metal and pesticide budgets of the Bay changed, but not drastically. The metals cadmium, copper and zinc in oysters from the lower Bay were examined and compared with quantities present in 1971. The flood effected shifts in concentration gradients of these metals but not to an extent which caused public health hazards or toxicity to the animals. The pesticide analyses (chlorinated hydrocarbons) showed no addition of these compounds due to Agnes and decreases of pesticide concentrations in oysters from some areas relative to before the storm were noted.

As was the case with salinity, water quality in the small subestuaries and bays was more affected by input from Chesapeake Bay than from their watersheds, which are small and wholly within the coastal plain.

BIOLOGICAL EFFECTS

The effects of Tropical Storm Agnes were generally minor and temporary on finfish, blue crabs, and hard clams. Soft-shell clams and oysters, however, suffered heavy mortalities. Successful spawnings by the surviving soft-shell clams led to a rapid recovery but the surviving oysters produced no significant set in 1972 and 1973. The entire biological community was disrupted to some degree. Although effects remained discernible two years after the advent of Agnes, the Chesapeake Bay ecosystem has demonstrated great resiliency.

Shellfishes

No group of organisms was more strikingly influenced than the molluscs, or shellfishes. The influx of fresh water exposed extensive beds of shellfish to low salinities for periods longer than the shellfish could endure. The severity of mortality depended on the kind of shellfish involved, the duration of exposure to subminimal salinity, the temperature, and in some instances, on dissolved oxygen levels.

Oysters are active only when salinity is approximately 4 ppt or higher. At lower salinity they neither feed nor pump water over their gills for respiration. The length of time oysters can survive salinity less than 4 ppt depends largely on temperature. In winter oysters can survive unfavorable salinity for two to three months but in summer at temperatures of 70 to 80° F survival longer than three weeks cannot be expected. Similarly, in summer hard clams will survive salinity as low as 10 ppt for only one to two weeks, and soft clams are killed if salinity persists below 2.5 ppt for one to two weeks.

Oysters

Chesapeake Bay oysters suffered massive mortalities from the prolonged low salinities caused by Tropical Storm Agnes. The mortalities were greatest in the upper reaches of the estuaries and in the upper Bay, and least along the eastern shore and at the lower ends of estuaries where salinities remained fairly high.

Oysters north of the Chesapeake Bay Bridge, west of Cobb Island in the Potomac River, and in the upper ends of many of the tributaries suffered nearly 100% mortalities. Many of these oysters were already stressed from the low salinities that preceded Agnes. Heavy mortalities (greater than 25%) also occurred among oysters living along the western shore from Bay Bridge 15 miles down bay to Herring Bay (especially among oysters living in less than 20 feet of water), in the upper Patuxent, and in the middle parts of the estuaries of the Potomac. Light mortalities occurred to oysters along the eastern shore of Maryland, and in the lower end of the Potomac River. The Maryland Department of Natural Resources estimates that more than two million bushels of market-sized oysters died on Maryland oyster bars because of Agnes.

In Virginia mortality of mature oysters on public oyster bars was as follows: James River 5%, Rappahannock River 2%, Potomac River tributaries 50%, York River none. Grounds leased for private culture of oysters, being generally less favorably situated to withstand freshets, sustained greater mortality as follows: James River 10%, Rappahannock River 50%, Potomac River tributaries 70%, York River 2%. Economic impact of shellfish losses is considered in a subsequent section.

Somewhat surprisingly, the total harvest of oysters from Chesapeake Bay during the 1972-1973 season was excellent. Maryland oystermen, for example, landed some 3.1 million bushels--the highest total in 31 years. Landings, however, were down in the Potomac and in Virginia.

Reproduction by oysters was a disastrous failure in 1972. Spatfall, production of young oysters, approached the normal level only in the tributaries of Mobjack Bay. Elsewhere it was essentially nil with the exception of a very slight set late in the season in the James River. Abnormally low salinity during the breeding season prevented successful reproduction. In 1973 spatfall was again abnormally low, especially so in the upper Bay and Potomac River. Thus few oysters have been added to replace those killed and seed for rehabilitation of decimated bars has been scarce.

Of some benefit to the shellfishing industry was the destruction of oyster drills, snails which prey on oysters. Drills are killed in three weeks or less by salinity below about 9 ppt. The Rappahannock River and the Piankatank River were essentially freed of drills and extent of the area infested was reduced in the York River, James River and Tangier Sound. Drills are slow to repopulate an area, therefore oyster growers can expect a few years of reduced predation in those areas in which drills were killed.

Rehabilitation of oyster populations was attempted by the Maryland Department of Natural Resources, the Virginia Marine Resources Commission and the Potomac River Fisheries Commission. Hatchery-reared seed planted on a pilot scale in the Potomac River and its tributaries survived and grew satisfactorily. Brood stock was moved into depleted areas, and cultch was added both by exposing shell already on the bottom and by adding large quantities of new shell. Rehabilitation efforts were hampered by the scarcity of seed oysters.

Even with the rehabilitation efforts, the future of the Chesapeake Bay oyster industry appears gloomy. Heavy mortalities followed by large harvests and poor recruitment indicate that harvests will drop and that recovery from Agnes will take a long time.

Summary

Soft Shell Clams

Over 90% of the soft-shell clams died from the combined stresses of low salinity and high water temperature. The amount of productive clam bottom in Maryland, for example, decreased from about 26,600 acres three months before Agnes to only 500 acres after Agnes. Careful sampling in the Rhode and South rivers shortly after Agnes, yielded not one live clam. Soft-shell clams were destroyed in the Rappahannock River, but survived in the York River and in Chesapeake Bay between these two rivers and also on the eastern shore of Maryland (Eastern Bay, Wye River, and Miles River), where the Bay waters retained salinities greater than 4 ppt. Fortunately, the surviving clams spawned successfully in the fall of 1972. Seed clams were found throughout the area, except in the Potomac River.

In 1973 the Maryland Department of Health and Mental Hygiene found high levels of bacteria in soft-shell clams throughout most of the waters under their jurisdiction and, therefore, closed the fishery. It is unclear if the high level of bacteria was a direct result of Agnes or if it merely occurred at the same time. The result was the complete loss of clam production for 1972 and for much of 1973.

Hard Clams

Hard clams occur primarily in the York River, the James River and the intervening section of Chesapeake Bay. Losses in natural populations occurred only in the York River and these were slight. However an estimated 50% of the clams transplanted from the James River to shallow bays in the York River and nearby areas died from exposure to low salinity.

Fishes

Adults and juvenile fish weathered the flood and reduced salinities well, although most moved downstream as much as 10 miles or into deeper water where the salinity approximated that of their normal habitats. Freshwater fishes also moved downstream into normally saline areas, but not so far as the marine forms. With the return of normal salinities the fishes returned to their normal habitats. Displacement of fishes disrupted both the recreational and commercial fisheries to a minor extent.

Many fish eggs and larvae were washed out of the nursery areas. But because information is lacking on the normal yearly fluctuations in spawning success, the relationship between spawning success and future adults, and many other factors, the effect these losses might have on future fish abundance is in question.

Plankton samples collected at the mouth of the James and Rappahannock rivers during the Agnes flood indicated that numerous fish larvae were being washed out of these rivers, as many as 6.5 million larvae per hour out of the Rappahannock during the peak of the flood. Most were goby and anchovy larvae, but other species, including shad and river herring were also captured. The loss of larvae from the Rappahannock and James is probably indictative of the losses from other nursery areas.

Once out of the nurseries the larvae were probably killed by unfavorable salinities and temperatures, predators, or the lack of proper food. At any rate, the loss of larvae from the nursery grounds might seriously affect the future abundance and distribution of fish in Chesapeake Bay. The effect of these losses on the fishery will first be felt in 1976 when maturing shad and river herring will return to spawn.

Blue Crabs

Effects of Tropical Storm Agnes on blue crab stocks appear to have been limited to 1) widespread and immediate but temporary downriver displacement of crabs, and 2) sporadic instances of crab deaths involving small numbers of crabs within two weeks following the storm. Crab fishing effort was measurably reduced after the storm, as some fishermen lost crab pots and others caught too few crabs to justify the cost and effort of baiting and resetting pots. Landings of hard crabs were about equal to those reported for the same months the two previous years despite the disruption of the fishery.

If Agnes affected reproduction, the impact was not great inasmuch as the 1972 year class fell within the normal range of variation.

Aquatic Plants

Perhaps, next to oysters and soft clams, the ecological group most depressed was submerged aquatic plants, especially in the Susquehanna Flats and along the western shore. Plants were sampled in August-September of 1971, '72, and '73 to determine trends in abundance, composition and distribution in the Bay, and major environmental factors, e.g., salinity and turbidity, effecting changes.

Widgeongrass, Eurasian watermilfoil, wildcelery, naiads and macrophytic algae all decreased significantly through 1973. However, eelgrass decreased the most, 89%. For all species combined, the decrease was 67%. Only sago pondweed and horned pondweed increased during this period. The reduction in aquatic plants was due to a loss in area covered. In 1971, plants occurred on 29% of the area censused; in 1973 only 10% was covered. No significant change in standing crop per unit area occurred. Lowered salinity and increased turbidity apparently effected this loss. Aquatic plants are an important food for waterfowl and are the basis of a community which serves as a nursery for fishes and crabs.

Jellyfish

The population of jellyfish or sea nettles had dwindled considerably in the fall of 1971 and very few were present in 1972. Flood waters further reduced the population of medusae to one-tenth or less of normal abundance. There was minor recovery of the population late in 1972. Three years after the storm the population appeared to be on the verge of full recovery.

Plankton and Benthos

The plankton and the benthos are complex biological communities consisting of a myriad of kinds of organisms, many of which are minute and distinguishable only by specialists. Individual species comprising these communities were, as was the case with fishes, crabs and shellfishes, differently affected by Tropical Storm Agnes. These details are, however, left to the technical reports. Only general trends are treated in this summary. Some dominant species, although not immediately affected, suffered reproductive failures after the storm. Perturbations of the communities allowed the eruption of opportunistic species. Such reverberations to the initial perturbation of Agnes likely will continue at a diminishing magnitude, gradually becomming a part of the background "noise" caused by each year's various deviations from the ecological norm.

In general the effects of the flood on benthos and plankton were greatest in the higher salinity (greater than about 15 ppt) portion of the system. This estuarine segment, termed the polyhaline zone, is inhabited by typically marine organisms which are relatively intolerant of low salinity. Many of these were extirpated from much of their normal range. Included were many tunicates, echinoderms, molluscs, crustaceans, and algae. In contrast, organisms inhabiting the mesohaline (5-18 ppt) and oligohaline (0.5 - 5 ppt) zones are tolerant of a much wider range of salinity, longer periods of depressed salinity, and more rapid salinity changes. Although mortalities of species in zones of lower salinity occurred (e.g. oysters, soft clams, barnacles and hooked mussels), none of these were eradicated. The most striking faunal change in the mesohaline zone was an influx of organisms displaced from areas upstream. The fauna of oligohaline and tidal freshwater was not changed in species composition, though some individuals moved downstream to waters approximating their normal salinity range.

Phytoplankton increased rapidly, nourished by the organic material and other nutrients furnished by the flood waters. The algal bloom continued through the fall of 1972 and in 1973 phytoplankton was denser than usual. Neither species composition nor seasonality of the blooms was the same as the normal plankton blooms. Red or mahogany algae were unusually prevalent. Some zooplankton species likewise experienced unusually large blooms at abnormal times of the year while other species were decimated.

With the exception of some minor impacts in the fisheries, these changes in the benthos and plankton have not influenced man's day-to-day interactions with Chesapeake Bay. They are, however, of significance to those investigating the impact of man-induced changes because they have given some indication of the resilience of the natural communities and their responses to severe perturbation.

ECONOMIC IMPACT

Tropical Storm Agnes set records for physical and economic damage to the Eastern Seaboard. Accordingly, assessment was made of the economic losses suffered by the commercial fisheries, the recreation industry, recreators, and the cost of debris removal and repair to navigation works. The assessment was limited to the first round or primary effects. No attempt was made to isolate multiplier effects; that is economic repercussions experienced by other sectors of the Maryland and Virginia economies. Total losses in current dollars in these sectors of the economy were $43 million. The impact was unevenly distributed through the economy, and geographically, being most severe in the shellfishing, tourism, and recreation industries and less severe in other industries. The impact was greater in general on the western shore than on the eastern. Losses were both immediate and long term, especially in the shellfishery. Table 3 summarizes the economic losses from Tropical Storm Agnes.

Table 3. Total economic losses in Maryland and Virginia attributable to Tropical Storm Agnes.

Item	Loss	
	Current Dollars	Constant Dollars*
		(1967=100)
Shellfish and Finfish Industries	33,628,200	24,848,200
Recreation Industries	7,498,700	5,984,597
Channel Dredging, and Boat and Ship Damages	1,615,000	1,288,907
TOTAL	42,741,900	32,121,704

*To convert to constant dollars the current dollar losses were divided by the appropriate consumer price index. The Survey of Current Business, U. S. Department of Commerce, Social and Economic Statistics Administration, Bureau of Economic Analysis, February 1973, Vol. 53, #2, and July 1974, Vol. 54, #7 were the sources of the consumer price indexes.

Shellfish and Finfish Industries

Oysters

Tropical Storm Agnes caused large oyster mortalities in the Chesapeake Bay and its tributaries during the summer of 1972. Resulting loss in current dollars to the harvesting sector was $13.8 million and to the processing sector $8.4 million. In order to determine the economic impact of Agnes, for the years 1972 and 1973, it was necessary to estimate what landings and revenues would have been without Agnes. Estimation of losses for 1974 required estimates for both actual landings and revenues given the occurrence of Agnes and potential landings and revenues which would have been realized if Agnes had not affected the Bay. These are summarized in Table 4.

Table 4. Estimated current dollar loss in Maryland and Virginia oyster landings and value added in the oyster processing industry attributable to Tropical Storm Agnes.

Year	Pounds Lost (lbs.)	Dollar Loss in Landings Maryland	Dollar Loss in Landings Virginia	Value Added Loss in Processing	Total
1972	6,355,900	$1,602,000	$2,241,800	$3,178,000	$7,021,800
1973	4,935,800	0	2,220,800	2,467,900	4,688,700
1974	5,585,500	5,300,000	2,440,000	2,792,800	10,532,800

A major effect of Agnes on oyster production may not, however, be felt until the late 1970's or later and probably can never be accurately documented. The aftermath of Agnes created water conditions that virtually eliminated oyster spawn and set during 1972 in both Maryland and Virginia waters. While the short run reduction in oyster production that will occur in 1975 and to a lesser extent in 1976 represents a direct effect, a more serious long range effect can result from a semi-permanent reduction in the stock of oysters producing spawn and set. Failure to replenish the oyster stock needed to provide seed oysters could cause a much longer recovery period to reach some "normal" or "desired" level of production.

The decreased volume of oysters landed also had a fairly large impact on the oyster processing industry in the Bay region (Table 4). The normal markup by oyster processors is about 78 percent of the exvessel price. This suggested that the processors' markup was about 50 cents per pound in 1971. Assuming that the processors' markup remained at 50 cents per pound, the value-added loss to Maryland and Virginia in 1972-73 was $5.6 million, bringing the total loss accruing to the harvesting and processing sectors, for the two years, to $11.7 million (Table 4). Extimated total losses for 1974 were projected to be $10.5 million (Table 4).

Soft Clams

In the summer of 1972 the soft clam industry of Chesapeake Bay was centered in Maryland, there being essentially no active harvesting operations in Virginia waters at that time. Biologic surveys taken in Maryland indicated that market-sized soft shell clams suffered over 90% mortality and that about 95% of the surviving commercially harvestable soft shell clams were concentrated in less than 5,000 acres (Maryland State Department of Health and Mental Hygiene). This decimation of adult soft clams led the Maryland State Department of Natural Resources to impose a ban on clamming in September 1972. Attributing all the decrease in landings to Agnes resulted in estimates of 1972-73 losses at the fishing and processing levels of $5,456,300 and $2,334,000 respectively. Losses for 1974 were projected to be $2,094,300 at the fishermen's level, and $1,457,900 at the processors'

level. Total direct losses in current dollars attributable to Agnes were
$11.3 million.

Hard Clams

Agnes produced a relatively minor economic impact on the hard clam
industry in Virginia and none in Maryland. Hard clams survived the fresh
waters of Agnes with one prominent exception. Clams that had been relocated
from the James River to the York River and its tributaries in order to
undergo a 3-week cleansing period suffered almost 50 percent mortality.
Normally, minimal loss is observed in relocated clams, and since only relocated clams were affected, the combination of stress and low salinity
levels were apparently responsible. Using data from the Virginia Marine
Resources Commission it was estimated that 7,035 bushels of relocated clams
were affected by Agnes producing a direct loss of $42,410.

Crabs and Finfish

Revenues from the production of crabs and finfish indicate that Agnes
did not significantly reduce catches in 1972 and 1973. The crab catch
continued to conform to biologists' expectations and catches of menhaden,
the most significant contributor to the dollar value of Virginia finfish,
increased approximately one-third in 1972. Fishing was disrupted for periods
ranging up to three weeks, but no estimates of the costs of these disruptions
have been made. Agnes probably reduced the reproductive success of some
fishes, most notably shad and river herring. The economic impact will not
be felt until 1976 when the 1972 yearclass enters the fishery.

Economic Impact on Recreation Industries and Users

The recreation activities most predominant on the Chesapeake Bay
historically have been fishing, hunting, crabbing, boating, swimming and
sightseeing. Many of these activities were curtailed by Agnes. Boating
and therefore fishing were made difficult by floating debris in many areas
of the Bay. Beaches were closed because of potential or confirmed health
hazard for varying periods of time. Sportcrabbing and sportfishing were
less productive than normal. Businesses which served the recreationists -
motels, restaurants, marinas, bait and boat sales places - lost revenue
because of the decrease in recreation activity. Losses totaled $7.5 million
in the two states in those segments of the tourism and recreation industries
for which a basis of estimating existed. This is considered a minimal estimate of the loss to the economy of the area inasmuch as no data base
exists from which to estimate some losses. In the following discussion,
losses in each state are discussed separately because the investigators had
available to them different data bases and used somewhat different approaches.

Maryland

Motels. In order to measure the economic impact of Agnes on the Maryland
motel industry, motels in all of the counties which touch on the Bay were
surveyed. According to survey response, the total decrease in business
which motel operators attribute to Agnes' effect on the Bay amounted to

around $181,000[1]. For some motels business decreased only during the immediate period of the storm, but for others, business was poor all summer.

Other Recreation Industries. There was no direct information on the impact of Agnes on the Maryland restaurant business. However, the motel survey found that 7,726 motel customer days were lost because of Agnes. Motel customers make heavy use of the restaurant industry. An A. D. Little study of tourism in Maryland in 1972 found that motel-staying boaters spend about 80% as much at eating and drinking places as at motels. This figure was applied to the motel loss to obtain an estimate of the loss to Bay area restaurants, or $145,000. This was considered a minimum estimate of losses because it only measures the loss of customers who stayed in motels. Potential restaurant customers who live in the area or make day trips to boat, fish or sightsee, were also discouraged from these activities because of Agnes and its aftermath.

The A. D. Little study also reported that a motel-staying recreationist would probably spend an additional $6 to $8 in the area, per day. Using the same method as above, at least $54,000 in gross revenue was lost to retail establishments, such as gas stations, because of Agnes. State parks also serve the recreationist. Four state parks along the Bay attributed to Agnes decreases in income amounting to $70,000.

Marinas. In 1972 there were approximately 320 marinas (places which charged for rental of 10 or more slips and moorings) along the Chesapeake Bay and its estuarine tributaries. Information on the impact of Agnes on marinas was derived from Economic Analysis of Marinas in Maryland, The Boating Almanac, and a small telephone survey conducted in March 1974.

Had Agnes not occurred, the marina business would have grossed close to $9 million not including restaurant business. This figure was arrived at by updating the average revenue per marina in Economic Analysis of Marinas in Maryland by use of the average increase in slip rental price, in gas prices, and in the consumer price index, and multiplying by the number of marinas operating on the Bay in 1972. Marina operators on average believed their loss of revenue was about 13% of their business, or a gross dollar loss of 1.1 million.

Impact on Recreation Users. The use of the Bay for swimming, fishing, boating, and other recreation activities affords utility to large members of recreators. The travel-cost method was used to estimate the value of the Bay to sport-fishermen. There have been extensive creel censuses of the Bay which provided information on how many anglers came from different distance zones in order to fish on the Chesapeake Bay in the summer. The total value to fishermen of their use of the upper Bay proper for the summer season, the area under the resources demand curve, was estimated to be approximately $1,378,000 in 1972.

---

[1] From the sample it was estimated that 7,726 visitor days were lost. This was multiplied by a per person room rate of $9 and inflated by 2.6 to represent all motels.

After Agnes, portions of the Bay were closed to boating. Even when boating was allowed, it was more hazardous than normal because of floating debris. Pier fishing was also interrupted, due to inclement weather and lack of fish. In any case, pier fishing accounts for less than 1% of the angler days on the Bay proper. On average, fishing from boats was not possible for three weeks of usually prime fishing time. According to the 1962 Elser creel census, 16% of the season's angler-days would have occurred during this three week period. The value lost to fishing recreationists then was about $220,500. The loss estimated above was the value lost to fishermen only. Non-fishing boat-owners were similarly affected. National and state boat-owner surveys have found that only 40% of all boat owners use their boats for fishing. This indicates that there were 1.5 times as many non-fishing boat owners as fishing ones. Assuming that they use their boats for pleasure cruising or water skiing with the same frequency as fishermen, and their demand curve for the Bay was similar to that of the sport fishing boat owners, an additional $330,700 in recreational value was estimated to have been lost by non-fishing recreationists because of Agnes.

The closure of the upper part of the Bay to water contact sports, which lasted until August 12, 1972 caused a loss to large number of people. However, because of the lack of information, it was impossible to estimate the loss to recreationists who would have participated in water contact sports.

Virginia

Tourism. While precise data were not available on tourism expenditures *per se*, the absolute dollar expenditures on three tourist-oriented industries were compared, by quarters, from 1970 through 1973 for counties and cities in Tidewater and Eastern Shore to determine the effect of Agnes. The data were obtained from the Department of Taxation, Taxable Sales Annual Reports and cover expenditures for "motels and hotels," "restaurants" and "sports and hobbies".

In order to determine the direct effects of Agnes on tourism, expected revenue was calculated for the spring quarter of 1972 for the three industry groups, and subtracted from the actual spring quarter revenue. Expected revenue was determined by a simple average of the revenue for the spring quarters of 1971 and 1973 and checked with the growth rate for 1970-73 for consistency. A check for inordinately large increases between 1972-73 during the summer quarter was made in order to ensure that the increase was not overly influenced by additions to facilities between 1972 and 1973.

The total direct loss in tourism revenues attributed to Agnes was estimated to be $3.5 million, with $2.4 million of that loss in the Norfolk and Virginia Beach areas. This was considered a conservative estimate since the assessment included only those industries primarily tourist oriented; i.e., purchases of all other goods and services by tourists, such as gasoline, groceries, theater admissions, etc., have not been included.

Sportfishing. The economic impact of Agnes fell also on the many business firms and individuals who derive part or all of their income from the very sizable sportfishing industry in Virginia.

There are really two separate sectors of the sportfishing industry, one of which might be called the "commercial" sportfishing industry, the other the "private" sportfishing industry. Of these, the commercial sector, which includes charter and head-boats, fishing piers, and boat rental businesses, is the more visible, and it did suffer some losses from Agnes. But much larger losses befell the "private" sector, which represents those many businesses that sell the variety of goods and services used by private boat owners in operating their own craft for their own enjoyment.

1) *Commercial Sportfishing*. The revenue losses accruing to charter and head-boat fishing because of Agnes were confined to the inshore boats, which are fairly few in number, usually lower in prices, and see less intensive normal utilization than the offshore boats. Charter and head-boat fishing involve large outlays by passengers, on the order of up to $165 per day for a six-passenger chartered boat and $8-20 per day per passenger for head-boats carrying as many as 40 passengers, and Virginia apparently has many such boats. Losses were minimized, however, because of the interplay of two factors: one, the impact on ocean (as contrasted with Bay) fishing from Agnes was virtually nil and, secondly, the large majority of Virginia boats either are exclusively for offshore use or have the size, range and location that enable them to head offshore as an alternative to fishing in or around the Bay. The largest charter and head-boat fleets in Virginia are either already ocean-based, such as those at Wachapreague on the Eastern Shore or at Rudee Inlet in Virginia Beach, or have easy access to the ocean, such as those based in the Norfolk area from Willoughby Bay to Little Creek, and on the North shore of Virginia Beach in Lynnhaven Inlet.

Thus, to whatever extent charter and head-boat fishing revenues were affected by Agnes, losses were confined to the inshore boats, which are fairly few in number, usually lower priced, and see less intensive normal utilization than the offshore boats. Estimates of even these losses are difficult to come by, because these boats are largely decentralized in location and do not use central booking services. However, a total loss to them within the $25,000-$50,000 range is estimated to have occurred, and the mid-point of that range $37,500, is accepted as a reasonably accurate loss figure.

While charter boat fishing is by far the dominant fraction of the commercial sportfishing industry in Virginia and losses to it related to Agnes were not large, there are other elements of commercial sportfishing which, though much smaller in size, suffered proportionately more damages. In this regard, particularly, are the fishing pier and small-boat rental businesses.

There are nine commerical fishing piers in Virginia, only two of which extended into the ocean; the remaining seven are all located at the mouth of or inside the Bay. While precise damage figures to these seven are not available on an individual basis, our survey of pier owners and operators indicates that the average losses to the four larger piers in the lower Bay attributable to Agnes were approximately $15,000, representing more the continuing impact throughout the summer than the absence of business during the storm period. Losses to other piers totalled no more than $10,000. Thus, Agnes-related revenue losses to fishing pier operators are estimated at $70,000.

Another facet of this economic sector is small boat rentals. A survey of these small boats, salt water rental firms indicated that total losses to them from Agnes probably came to only $10,000.

Thus, the total Agnes-related losses to all elements of the commercial sportfishing industry in Virginia total approximately $117,500.

2) *Private Sportfishing*. But as commercial sportfishing losses were small, the impact of the decrease in private sportfishing was not. The analysis indicated that losses in this sector were the product of two factors: the total number of boating days (average boats per day multiplied by the number of days affected) lost because of Agnes and the typical monetary outlay per boat day. About $1.78 million dollars were assumed not spent in the lower Bay area by private sportfishermen on gas, bait, food and beverages.

Other Impacts

Tropical Storm Agnes caused economic losses other than those relating to commercial and private sportfishing. Among the other losses are those that concern channel maintenance, debris removal, and boat and marine equipment damages.

However, of these, only the first looms large as an out-of-pocket loss. Reports for the U. S. Army Corps of Engineers, Norfolk District, indicated that thus far the flooding of 1972 in the tidewater estuaries of Chesapeake Bay has caused $695,000 in dredging expenses, above normal maintenance costs.

Extensive interviews support the opinion that debris removal was largely left to nature itself, or was done by marina and waterfront property owners and boaters themselves on a non-commercial basis. Furthermore, it also appeared that many boat damages were to the small boat owner and to a considerable extent were of small individual magnitude. Many of these damages were hull scrapes and gouges that were repaired by the owners themselves.

In regard to damages to larger boats which tend to be more costly per incident, data provided by the United States Coast Guard Station in Norfolk indicated that there were Agnes-connected incidents of damage to five barges and one Navy vessel, also totaling an estimated $10,000 in damages. The reporting requirements for marine casualty damages specify at least a $1,500 damage level, and thus all losses above that amount presumably were reported.

Accordingly, the economic damages from Agnes relating to channel clearance, debris removal, and boat damages totaled $1.6 million in Chesapeake Bay and its tidal tributaries of which $715,000 was expended in Virginia and $900,000 in Maryland.[1]

PUBLIC HEALTH IMPACTS

Prompt action by the State Health Departments in Maryland and Virginia in closing the waters of the Bay to shellfish harvesting for direct human

---

[1] The U. S. Army Corps of Engineers served as the major source of these data.

consumption undoubtedly prevented marketing of contaminated shellfish. The higher bacterial counts in the Virginia portion of the Bay appeared to be primarily associated with runoff from the initial rains. In Maryland, the bacterial levels were to a greater extent attributed to material entering with the flood waters. The overall public health impact of Agnes on Chesapeake Bay was minimal with the possible exception of economic dislocations caused by shellfish and water contact recreation closings.

Shellfish Closings

On 23 June 1972, the waters of Chesapeake Bay were closed for the taking of shellfish for direct consumption by the Maryland Department of Health and Mental Hygiene (MDH) and the Virginia Department of Health (VDH) in waters under their respective jurisdiction. On 10 July the MDH reopened the first of the areas closed due to Agnes on the eastern side of the Bay. By 1 August most of the Eastern Shore waters in Maryland were reopened.

In Virginia, the "Bay Proper" below New Point Comfort was opened on 20 July. The York River was opened on 1 August followed by the James on 3 August and the Rappahannock on 9 August. By 5 October 1972 all remaining areas in Virginia closed due to Agnes had been reopened.

Reopenings in Maryland were complicated by a Maryland Department of Natural Resources (DNR) ban on all soft shell clam harvesting to protect remaining clam populations. On 18 September 1972 the Maryland DNR opened the oyster season.

Water Contact Closings

On 30 June 1972, the MDH closed the Maryland portion of the Bay to water contact recreation. No comparable closing was made for Virginia waters. Since the Potomac River to the Virginia shore is within Maryland jurisdiction, this closing did impact the water related recreation industry within these Virginia localities with shorelines abutting on the Potomac River.

On 7 July the ban was lifted for Baltimore, Ann Arundel and Calvert Counties. By 15 July the ban was lifted for all but the Maryland portion of the Susquehanna River, the upper Choptank River, the Patuxent River and the Potomac River above the Potomac River Bridge. On 28 July the ban was lifted on the remaining areas.

In the Virginia portions of Chesapeake Bay, water contact bans imposed by the Norfolk, Virginia Health Department and the U. S. Navy at beaches under their jurisdiction on 14 July and 19 July respectively were initially attributed in part to Agnes flood waters. Subsequent investigation, however, attributed these closings to a faulty sewer main in the Norfolk system.

Shellfish Contamination

Analysis of pesticide and heavy metal accumulation in shellfish indicate that there was no accumulation of these materials sufficient to cause a public health hazard.

Waterborne Pathogens

No evidence of increased incidence of infection by waterborn pathogens as a result of Agnes was found. Fifty-six percent fewer cases of infectious hepatitis were reported in June, July and August 1972 in Virginia than in the comparable period in 1971.

Miscellaneous Hazards

No incidents of injuries or health impairment due to hazardous substances entering the estuarine portions of the Bay region with flood waters were reported.

HYDROLOGICAL EFFECTS

Evon P. Ruzecki, Editor

EFFECTS OF AGNES ON THE DISTRIBUTION OF
SALINITY ALONG THE MAIN AXIS OF THE BAY
AND IN CONTIGUOUS SHELF WATERS[1]

J. R. Schubel[2]
H. H. Carter[3]
W. B. Cronin[3]

ABSTRACT

The passage of Tropical Storm Agnes through the drainage basin of the Chesapeake Bay in June 1972 resulted in record, or near-record, flooding of most of the major tributaries. The tidal reaches of the Susquehanna River were extended nearly to the Bay Bridge at Annapolis, nearly 35 km farther seaward than previously reported. The high discharges sent salinities in most of the Chesapeake Bay estuarine system to levels lower than any previously reported.

Minimum salinities in the surface layer of the Maryland portion of the Bay were reached within a few days of peak riverflow, and the zone of surface salinities less than 1 ppt extended seaward to the mouth of the Little Choptank. Minimum salinities in the bottom waters of the upper Bay were not reached in some areas until the middle of July, and minimum salinities were not observed in the mouth of the Bay until about a month after peak flooding.

Surface salinities remained depressed throughout the summer, while bottom salinities increased to "normal" levels over this same time interval by the upstream movement of salty water in the lower layer. The large freshwater inputs and the compensatory upstream flow of salty water in the lower layer produced large vertical salinity gradients. Even in early fall vertical gradients were typical of spring conditions.

The outflow of water from the Bay could be traced as a band of low salinity water leaving the Bay, turning south, and moving along the Virginia and North Carolina coasts.

INTRODUCTION

The Chesapeake Bay, like other estuaries, is a dynamic environment characterized by marked natural fluctuations of many of its physical and chemical properties. In addition to the "normal" year-to-year variations, marked fluctuations can result from catastrophic events such as floods or hurricanes. The distribution of salinity is one of the most important and characteristic features of the Chesapeake Bay and of other estuaries. Salinity varies from near zero at the head of the Bay to nearly that of full sea water at its mouth. Dynamically, gravity acting upon the density difference between the fresh water and the sea water gives rise to the characteristic two-layer estuarine circulation pattern. The distribution of salinity is also important because of its

---

[1]Contribution No. 230, Chesapeake Bay Institute, The Johns Hopkins University.
[2]Marine Sciences Research Center, State University of New York, Stony Brook, N. Y. 11794
[3]Chesapeake Bay Institute, The John Hopkins University, Baltimore, Md. 21318

effect on a number of important physico-chemico properties such as the rates of uptake and release of materials adsorbed to fine-grained sediments. Salinity is also obviously very important biologically. It exerts a marked influence on the distribution and activity of many organisms that inhabit the Chesapeake Bay estuarine system.

The distributions of salinity in the Bay and its major tributaries have been studied by the Chesapeake Bay Institute for more than 25 years. The results have been presented in a series of graphical summary reports (Whaley & Hopkins 1952; Stroup & Lynn 1963; Seitz 1971). In the reports of Whaley and Hopkins (1952) and Stroup and Lynn (1963), the distributions were compiled from data collected on the same cruise, but at different phases of the tide. In 1968, the Chesapeake Bay Institute initiated a program of regular "same slack" cruises to delineate the distributions of temperature and salinity along the main axis of the Bay at the same phase of the tide--slack before flood. Since the phase velocity of the flooding tidal wave between the mouth of the Bay and Baltimore is very nearly 7.7 m/sec (15 knots), a vessel traveling at this velocity will experience the "same" tidal phase throughout its trip. From April 1968 through October 1973, 91 same slack cruises were made; results from the first 25 have been described by Seitz (1971).

All of these data were collected during "normal" conditions. There was, until Agnes, a dearth of direct observations of the effects of a major flood on the distribution of salinity not only in the Chesapeake Bay but in the entire estuarine environment. Sampling during episodes is frequently difficult, and the infrequency of such occurrences makes the probability of fortuitous observations very small. Tropical Storm Agnes presented Bay scientists with an unusual opportunity to document the impact of a major storm on the Chesapeake Bay. There was little wind associated with Agnes when she reached this area, but torrential rains sent flows of the major tributaries to record or near-record levels. The massive inputs of fresh water were clearly manifested in the distributions of salinity.

The Chesapeake Bay Institute made extensive observations to document the impact of Agnes not only on the distribution of salinity, but also on the distributions of temperature, dissolved oxygen, nutrients, and suspended sediment. Salinity determinations were made in the Bay proper, in the larger estuaries tributary to the Maryland portion of the Bay, and in the waters overlying the continental shelf off the mouth of the Bay. Sampling was initiated during the period of peak flooding and was extended over approximately one year to chronicle the impact of Agnes, and the subsequent recovery to "normal" conditions.

The primary purpose of this report is to assess the impact of Agnes on the distribution of salinity along the main axis of the Bay, and in the contiguous shelf waters. The emphasis will be on the Maryland portion of the Bay proper since this is where we have the most observations. A secondary purpose is to use the Agnes data to test the ability of Boicourt's (1969) numerical model of the upper Bay to reproduce the time-varying salinity distribution associated with such an unusual event as Agnes.

RAINFALL AND RUNOFF

Although these topics are covered elsewhere in this volume, some additional information is required for the ensuing discussion of the distribution of salinity following Agnes.

The late winter of 1971 and the early spring of 1972 was an unusually "wet" period with riverflow to the Bay well above the long-term average. For the 10-month period August 1971 through May 1972, the discharges of Maryland's two major rivers, the Susquehanna and Potomac, exceeded their long-term mean flows for the same 10-month period by about 50%. With the passage of Agnes, torrential rains fell throughout most of the drainage basin of the Chesapeake Bay estuarine system. Total rainfall from Agnes ranged from 20 to more than 45 cm over the drainage basin of the Susquehanna. Comparable precipitation fell over most of the remainder of the drainage basins of the western shore tributaries. Rainfall over the Eastern Shore, the Delmarva Peninsula, however, was considerably less, averaging less than 10 cm. The torrential rains produced disastrous flash flooding of record or near-record proportions of virtually all the western shore tributaries.

With a long-term average discharge of about 985 m^3/sec, the Susquehanna River accounts for nearly 50% of the total freshwater input to the Chesapeake Bay estuarine system, and for more than 85% of the total freshwater input to the Bay above the mouth of the Potomac. The characteristic seasonal variation of riverflow, high discharge in spring followed by low to moderate flow throughout the summer and most of the fall is revealed by an ensemble average by month and is typical of mid-latitude rivers (Fig. 1). Following Agnes there was record flooding of the Susquehanna. The day the River crested, 24 June 1972, the average daily flow exceeded 27,750 m^3/sec--the highest average daily flow ever recorded, exceeding the previous daily average high by approximately 33%. The instantaneous peak flow on 24 June 1972 of more than 32,000 m^3/sec was the highest instantaneous flow reported over the 185 years of record. The daily average discharge of the Susquehanna River during 1972 at Conowingo Md. is plotted in Fig. 2. Conowingo is approximately 15 km upstream from the River's mouth at Havre de Grace. The *monthly* average discharge of the Susquehanna of about 5100 m^3/sec for June 1972 was the highest average discharge for *any* month over the past 185 years, and was more than nine times the average June discharge over this same interval. A comparison of the monthly average discharge of the Susquehanna during 1972 and the ensemble monthly average discharge over the period 1929-1966 clearly shows the departure of the 1972 June flow from the long-term average June flow, Fig. 1.

The Potomac, with a long-term average discharge of about 310 m^3/sec, is the second largest river debouching into the Bay, accounting for approximately 19% of the total freshwater input. The Potomac also had very high flows following Agnes. On 24 June 1972, the day the River crested, the daily average discharge at Washington, D. C. reached approximately 10,080 m^3/sec--the fourth highest average discharge in the 83 years of record. The monthly average discharge of the Potomac for June 1972 of 1330 m^3/sec was more than six times the median June flow over the past 30 years.

METHODS

Temperature and salinity determinations were made at a number of stations along the axis of the Bay. The most frequently made stations are plotted in Fig. 3. Data on sampling frequencies in time and space are summarized in Table 1. The sampling period considered in this report extends from 5 June 1972 (pre-Agnes) to approximately one year after Agnes, 29 June 1973. All measurements were made with an induction-conductivity-temperature device. The basic measurements have an accuracy of ± 0.02°C for temperature, and ± 0.03 millimhos per cm for conductivity.

Table 1. Summary Data of the Sampling Frequencies in Time and Space.

DATE	37°10'/37°00'	37°20'/37°10'	37°30'/37°20'	37°40'/37°30'	37°50'/37°40'	38°00'/37°50'	38°10'/38°00'	38°20'/38°10'	38°30'/38°20'	38°40'/38°30'	38°50'/38°40'	39°00'/38°50'	39°10'/39°00'	39°20'/39°10'	39°30'/39°20'
							NUMBER OF STATIONS								
June 26-30 72				1			2	2		2	2	5	6	4	4
July 1-5				1							3	3	9	8	6
July 6-10	8	1		1		1	5			2	2	3	9	5	4
July 11-15				5		2	1	1		2	1	10	8	6	11
July 16-20				1		1	1	1	1	1		3	7	4	4
July 21-25	6	1		1		1	2			2	1	3	6	3	2
July 26-31				1		1	1					1	3	2	2
Aug 1-15	7	1		1		1	4			3	1	4	7	6	5
Aug 16-31	9	2		2		4	4			2	5	6	9	6	7
Sept 1-15															
Sept 16-30	4	2		2		2	2				5	4	7	4	5
Oct 1-15	2	1		1		2	2			2	3	3	6	4	5
Oct 16-31	7	1		1		1	4		4	5	2	3	3	4	6
Nov 1-15											1	1	3		2
Nov 16-30	2	1		1		1	1				2	2	4	2	2
Dec 1-15	2	1		1		1	1				1	1	3	2	2
Dec 16-31															
Jan 1-15 73															
Jan 16-31	9	2		2		2	3			2	1	1	3	2	1
Feb 1-15											2	2	5	4	2
Feb 16-28											1	1	3	2	3
Mar 1-15	10	2		2		1	1			1	1	1	4	3	5
Mar 16-31	9	2		2		2	5			2	2	2	5	4	5
Apr 1-15															
Apr 16-30	9	2		2		2	3			2	2	2	5	4	3
May 1-15															
May 16-31	9	2		2		2	2			2	2	1	3	2	2
June 1-15											1	1	3	2	2
June 16-30	9	2		2		2	2			2	2	2	5	3	4

SALINITY DISTRIBUTIONS

Data from selected cruises are plotted in Figs. 4-22. Some of these data were previously described in a report to the U. S. Army Corps of Engineers, Philadelphia District, by the Chesapeake Bay Research Council (1973).

The Upper Bay

For the purposes of this discussion, the upper Bay is considered to be that part of the main body of the Bay above the mouth of the Potomac. With the flooding from Agnes, salinities in the upper Chesapeake Bay fell sharply; seaward of about Station 913R salinities reached values less than any previously reported for those positions. On 26 June 1972 the front separating the fresh river water was more than 35 km farther seaward than previously reported (Fig. 4).

Variations in the surface and bottom salinity along the axis of the upper Bay on several specific dates in June and July of 1972 are plotted in Fig. 5. The dates are 6-7 June (prior to Agnes), 26-29 June (a few days after the rivers crested), 6-7 July, and 14-15 July. Similar distributions for a comparable time, July 11, during a "normal" year, 1968, are given for reference. It is apparent from these data that even before Agnes the salinities were depressed, and that with the massive influx of Agnes floodwaters salinities fell precipitously. The lag between the time of maximum freshwater discharge and the time of minimum salinity varied, of course, with position and depth. Because the discharges of all of the major rivers, and of many of the smaller streams, were record or near-record highs, minimum salinities were reached almost everywhere in the surface layers of the Maryland portion of the Bay, and its western shore tributaries within the period 26-29 June. At this time surface salinities were less than 1 ppt all the way from the head of the Bay to the mouth of the Little Choptank. Three weeks earlier on 6-7 June 1972, the zone of the Bay proper with surface salinities less than 1 ppt extended only to Tolchester, and the surface salinity off the mouth of the Choptank was about 8.5 ppt. In some areas of the upper Bay minimum salinities in the bottom waters were not reached until about 15 July.

Fig. 6 is similar to Fig. 5 except that data from later cruises are plotted. The dates covered for 1972 are 22-24 July, 5 August, 16-17 August, and for reference, 9 August 1968. It is apparent from these data that even three weeks after Agnes, surface salinities throughout the Maryland portion of the Bay proper were depresssed below "normal" levels as observed in August 1968. In the bottom waters, however, salinities were near normal by the middle of August. Data from other selected cruises are plotted in Figs. 7-8.

In assessing the biological effects of Agnes on the Bay, one must consider the distributions of salinity in both space and time; the effects on organisms of exposure to low salinities depend both upon the actual salinity values, and upon the duration of exposure. An interesting, and useful, way of presenting some of the salinity data previously described is shown in Fig. 9. This figure depicts the temporal variations of the downstream positions of the 1 ppt and 5 ppt isohalines at depths of 2 and 6 m. The positions are relative to the mouth of the Susquehanna at Havre de Grace, Md. From this figure it can be seen, for example, that at the Chesapeake Bay Bridge the salinity remained below 5 ppt for about 6 weeks at all depths less than 6 m.

Lower Bay and Continental Shelf

For the purposes of this report, the lower Chesapeake Bay is considered to be that segment of the Bay proper extending from the mouth of the Potomac estuary to the mouth of the Bay at the Virginia Capes. During and following the passage of Agnes, the lower Chesapeake Bay was subjected to several pulses of fresh water that entered the Bay at different times and locations.

The results of a series of same-slack cruises made over the entire length of the Bay are presented in Figs. 10-21. The figures show that minimum salinities were not observed in the mouth of the Bay until about a month after the time of peak riverflow.

Reestablishment of the "normal" salinity distribution is effected by the flow of more saline waters up the estuary in the lower layer and subsequent slow vertical mixing of the lower and upper layers primarily by tidal energy. The combination of large freshwater inputs accompanying Agnes, and the compensating upstream flow of salty water in the lower layer produced large vertical salinity gradients, gradients larger than any previously reported throughout much of the Chesapeake Bay estuarine system. Abnormally large gradients persisted throughout the summer. Even in early autumn the vertical salinity gradients were more typical of spring conditions than those characteristic of the fall season.

CBI made one cruise on the continental shelf from the mouth of the Bay to about 35°40'N between 28 July and 3 August 1972. The surface data presented in Fig. 22 clearly reveal a band of low salinity water flowing out of the Bay, turning south, and moving along the Virginia and North Carolina coasts. The band extended beyond Oregon Inlet, and probably became entrained in the Gulf Stream. The flood-derived waters were largely confined to the upper 10 m of the water column. This band of low salinity water also contained relatively high concentrations of suspended sediment (Schubel, this volume). Boicourt (1973) suggested that the secondary minimum in salinity (<28 ppt) observed on the outer portion of the shelf might be due to outflow from the Delaware Bay.

USE OF AGNES DATA FOR MODEL VERIFICATION

The studies of the effects of Agnes on the Chesapeake Bay provide the first comprehensive chronicling of the impact of a major flood on the Bay and the subsequent recovery to "normal" conditions. These data will be useful in verifying the hydraulic model of the Bay now under construction. Since data collected under more normal conditions will be used to adjust the hydraulic model, verification that the model could reproduce the time-varying salinity distribution observed under such unusual conditions as those associated with Agnes would greatly increase confidence in the model's ability to predict consequences of man's activities that fall outside the range of conditions used in the adjustment of the model. The data will also be valuable in verifying numerical models of the time-varying salinity distribution in the Bay and its tributaries. Although the Chesapeake Bay hydraulic model has not been completed as of this date, there is in existence a numerical model of the upper Bay which relates the salinity distribution to the Susquehanna river flow. This model was constructed by William Boicourt in 1969 and has been used by D. W. Pritchard to predict the effect of enlarging the Chesapeake and Delaware Canal on the diversion of fresh water from Chesapeake Bay to the Delaware River. It was felt, therefore, that a comparison of predicted cross-sectional mean salinities from Boicourt's model with measured values for the period before, during, and after the passage of Agnes at three or four upper bay locations would be interesting even though few felt that the model would be able to deal properly with events as catastrophic as Agnes.

Boicourt's model is time-dependent and one-dimensional in that variations in salinity are related only to position along the estuary and time, and is based on the concepts of salt and water continuity. The resulting differential equations were converted to finite difference equations and solved numerically. Aside from the geometry of the upper bay and the flow of the Susquehanna River, the model requires knowledge of the diffusion coefficient $K_{x,\sigma}$, as a function of position in the estuary and river flow, and the boundary conditions on salinity for the two ends of the longitudinal axis, i.e., the mouth of the Susquehanna River and a position 5 km below the Bay Bridge.

Since the salinity at the downstream boundary is variable, a separate model was constructed by Boicourt to predict the downstream boundary salinity. For this purpose, Boicourt found it necessary to use three separate polynomial predictor equations since the downstream boundary salinity depends not only on river flow but also on its time history, the appropriate length being dependent in turn on the absolute value of the river flow. The three predictor equations essentially parameterized the effect of river flow history into high, moderate, and low river flow conditions with the computation programs being written to select the proper predictor equation.

The product of σ, the cross-sectional area and the diffusion coefficient, $K_{x,\sigma}$, was considered by Boicourt to be a function of river flow and longitudinal position in the estuary. Based on 160 CBI hydrographic stations occupied between March 1966 and March 1967, he developed exponential relationships between $\sigma K_{x,\sigma}$ and downstream distance, x, in the form

$$\sigma K = \exp(C_0 + C_1 x)$$

with C_0 and C_1 depending on the river flow. For additional details concerning the model, the reader is referred to Boicourt (1969).

The model was run with appropriate values (weekly averages) of the Susquehanna River flow for the year 1972. For comparison with salinity measurements made in connection with the Agnes program, predictions for the period between 5 June 1972 and 28 August 1972 for Stations 914S (41.2 km from the origin), 909 (55.7 km from the origin), and 858C (73.5 km from the origin) are plotted on Figs. 23 through 25 together with measured salinities (appropriately adjusted to cross-sectional mean values).

It is clear from an examination of the figures that the model failed to predict sufficient initial freshening associated with the peak flow at Conowingo which occurred on 24 June 1972. In addition, the measured values of salinity suggest that this initial freshening preceded first a rise and then a fall in salinity followed by a slow rise to "normal" conditions. Neither the peak nor the secondary minimum in salinity was predicted by the model.

In connection with the first deficiency, it should be noted from the figures that the model predicts salinities that are too high by about 2 ppt just prior to Agnes. Since the model essentially predicts changes in salinity, the absolute value of the first salinity minimum would not be correctly predicted. Note, however, that the change in salinity predicted by the model between the week ending 12 June and the week ending 26 June is in good agreement with the measured change. The failure of the model to predict pre-Agnes conditions is probably a result of deficiencies in the predictor equations for the downstream boundary salinity.

The absence of the salinity peak and secondary minimum in the model predictions also suggest shortcomings in the downstream boundary value predictions since the predictor equations do not contain explicitly any downstream effects, such as modifications in the lower layer source salinity by the Potomac and/or

James Rivers. The salinity maximum appears to be a normal response of the lower layer to increased upstream river flow with the subsequent minimum being caused by a freshening of the lower layer downstream from the lower boundary by the influx of fresh water from the Potomac which, of course, would not be predicted by the model. On the other hand, the predicted time for the upper Bay to return to "normal" conditions seems to be in excellent agreement with the measured salinities.

On balance, the authors feel that the results of the comparison increase our confidence in application of Boicourt's model to other practical problems, and suggest several ways in which the model might be improved.

ACKNOWLEDGEMENTS

We thank C. F. Zabawa, T. W. Kana, R. C. Seitz, and M. Glendening, for their assistance in making the measurements described in this report, and D. W. Pritchard for his helpful suggestions. The figures were drawn by D. Pendleton, and photographed by J. Sullivan. The manuscript was typed by A. Sullivan. This research was supported in part by the U. S. Army Corps of Engineers (Baltimore and Philadelphia Districts); in part by the Oceanography Section, National Science Foundation, NSF Grant GA-36091; and in part by a project jointly funded by the Fish and Wildlife Administration, State of Maryland, and the Bureau of Sport Fisheries and Wildlife, U. S. Department of the Interior, using Dingell-Johnson Funds.

LITERATURE CITED

Boicourt, W. C. 1969. A numerical model of the salinity distribution in upper Chesapeake Bay. Chesapeake Bay Inst., Johns Hopkins Univ. Tech. Rep. 54, Ref. 69-7, 59 p. + Appendices.

Boicourt, W. C. 1973. The circulation of water on the Continental Shelf from Chesapeake Bay to Cape Hatteras. Johns Hopkins Univ. Doctoral Dissertation, 183 p.

Chesapeake Bay Research Council. 1973. The effects of Hurricane Agnes on the environment and organisms of Chesapeake Bay. A Report to the Philadelphia District U. S. Army Corps of Engrs. (Aven M. Andersen, Coordinator), Chesapeake Bay Inst. Contrib. 187, Nat. Resour. Inst. Contrib. 529, Va. Inst. Mar. Sci. Spec. Rep. Appl. Mar. Sci. & Ocean Engr. 29, 172 p.

Seitz, R. C. 1971. Temperature and salinity distributions in vertical sections along the longitudinal axis and across the entrance of the Chesapeake Bay (April 1968 to March 1969). Chesapeake Bay Inst., Johns Hopkins Univ. Graph. Sum. Rep. 5, Ref. 71-7, 99 p.

Stroup, E. D. and R. J. Lynn. 1963. Atlas of salinity and temperature distributions in Chesapeake Bay, 1952-1961 and seasonal averages 1949-1961. Chesapeake Bay Inst., Johns Hopkins Univ. Graph. Sum. Rep. 2, Ref. 63-1, 410 p.

Whaley, H. H. and T. C. Hopkins. 1952. Atlas of the salinity and temperature distribution of Chesapeake Bay, 1949-51. Chesapeake Bay Inst., Johns Hopkins Univ. Graph. Sum. Rep. 1, Ref. 52-3.

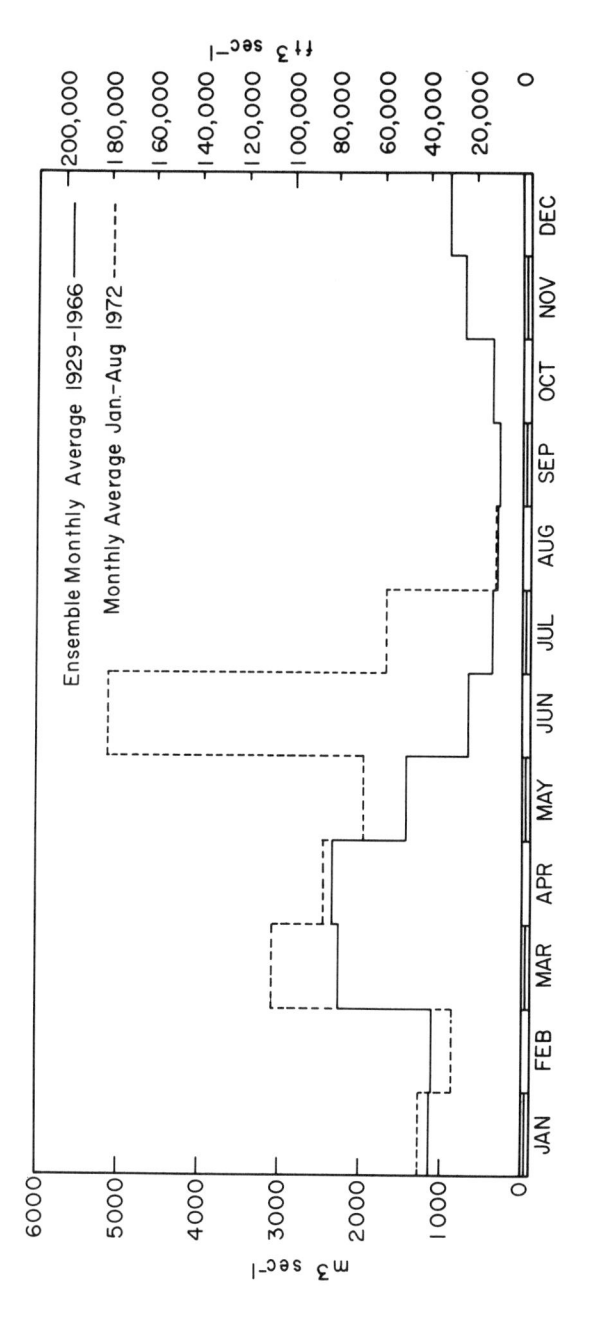

Figure 1. Ensemble monthly average riverflow of the Susquehanna River at Conowingo, Md., over the period 1929-1966; and monthly average discharge for 1972.

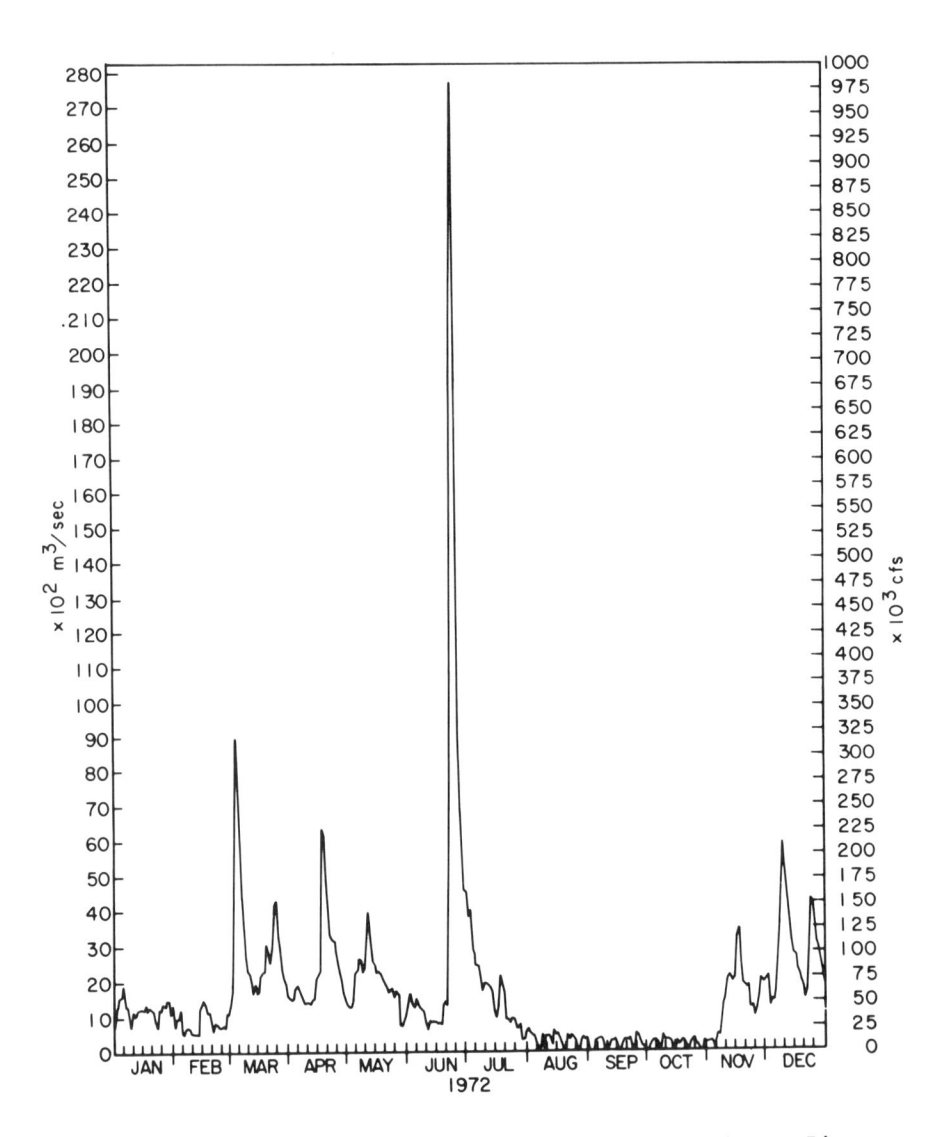

Figure 2. Daily average discharge of the Susquehanna River at Conowingo, Md., during 1972.

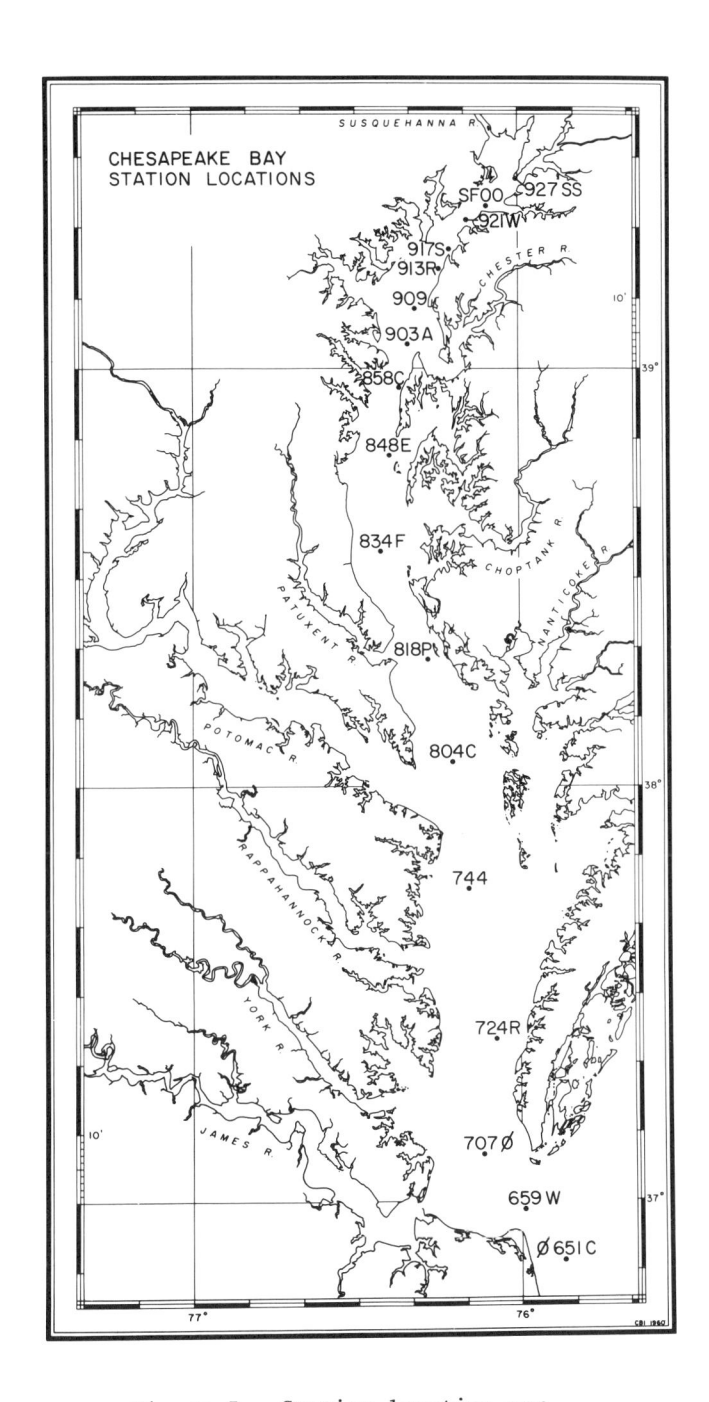

Figure 3. Station location map.

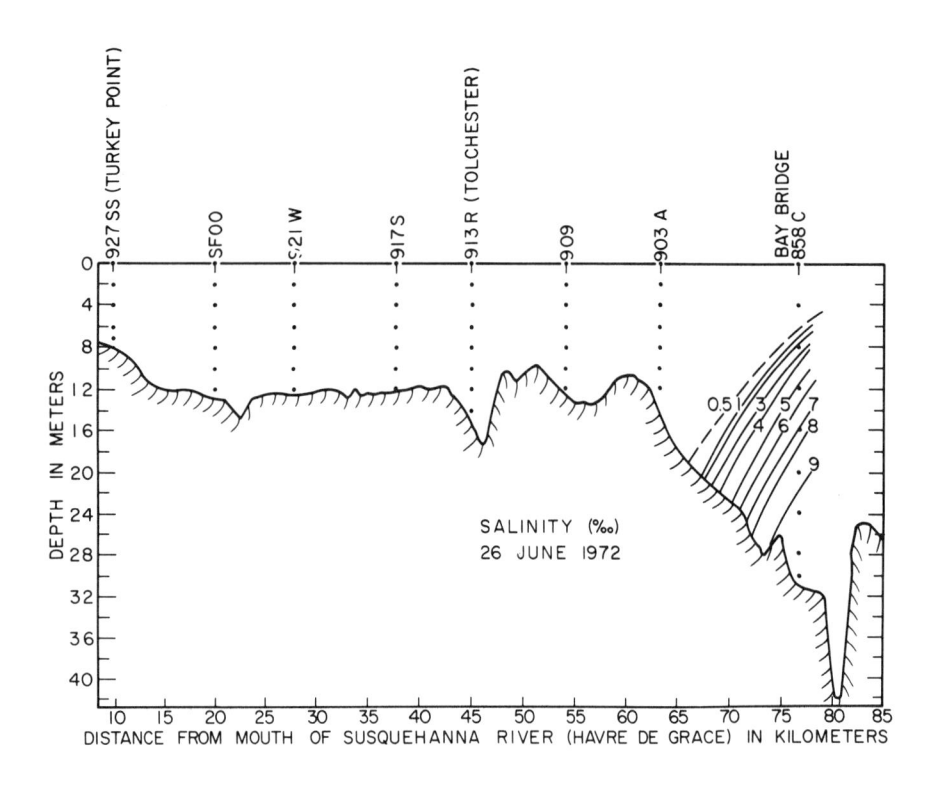

Figure 4. Longitudinal salinity distribution along the axis of the upper Bay on 26 June 1972, two days after the Susquehanna crested.

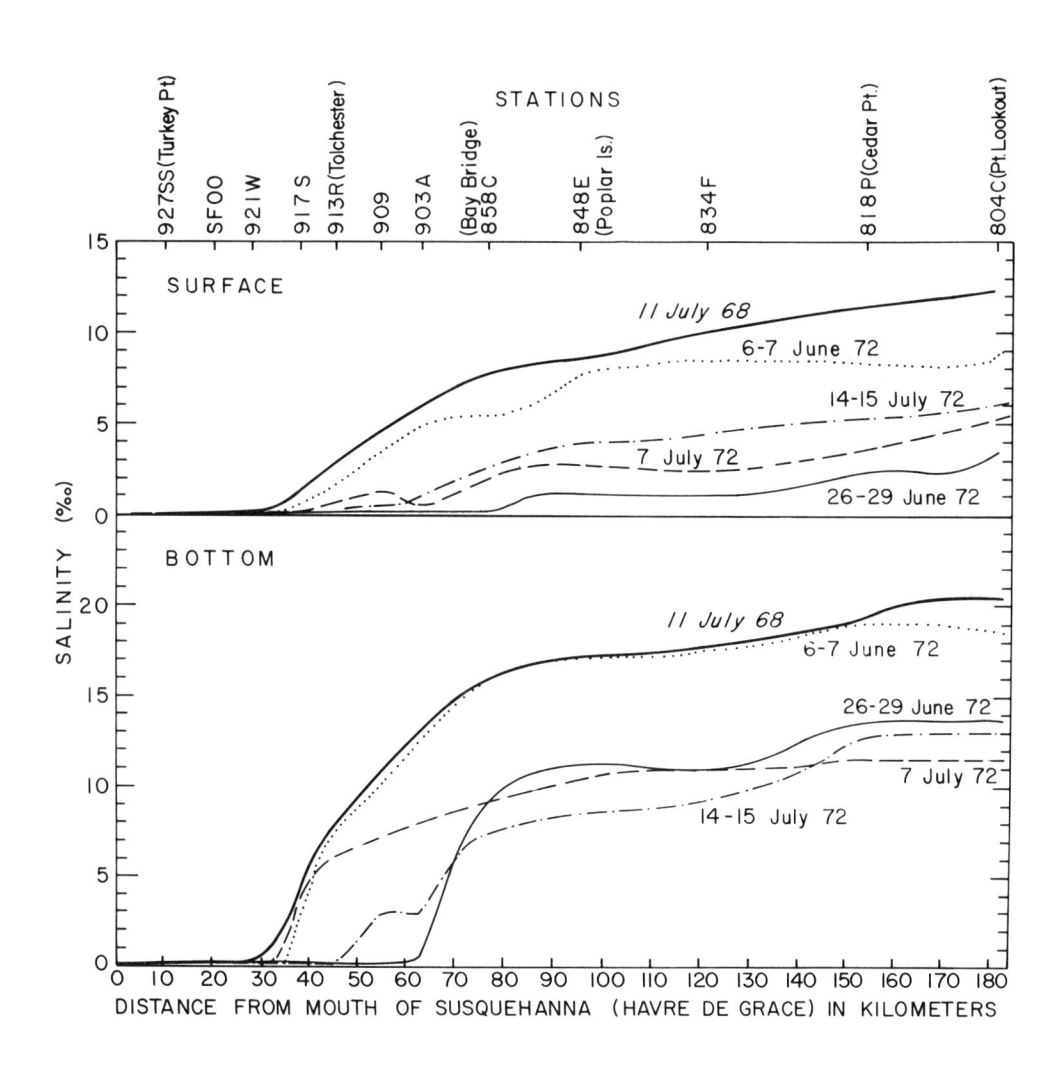

Figure 5. Variations in the surface and bottom salinities along the axis of the upper Bay on specific dates in June and July 1972. Data from a more normal year, 1968, are also given for comparison.

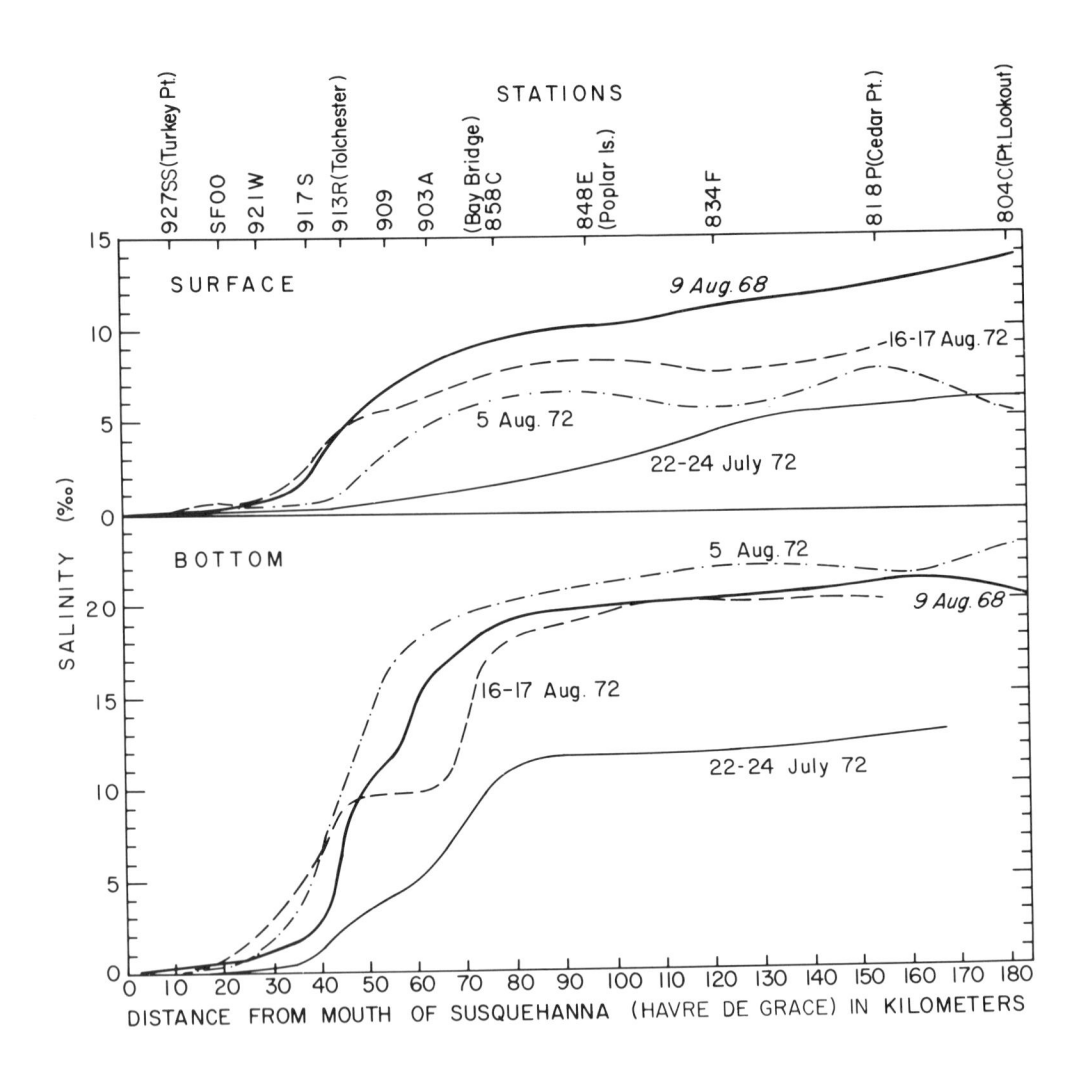

Figure 6. Variations in the surface and bottom salinities along the axis of the upper Bay on specific dates in July and August 1972. Data from a more normal year, 1968, are also given for comparison.

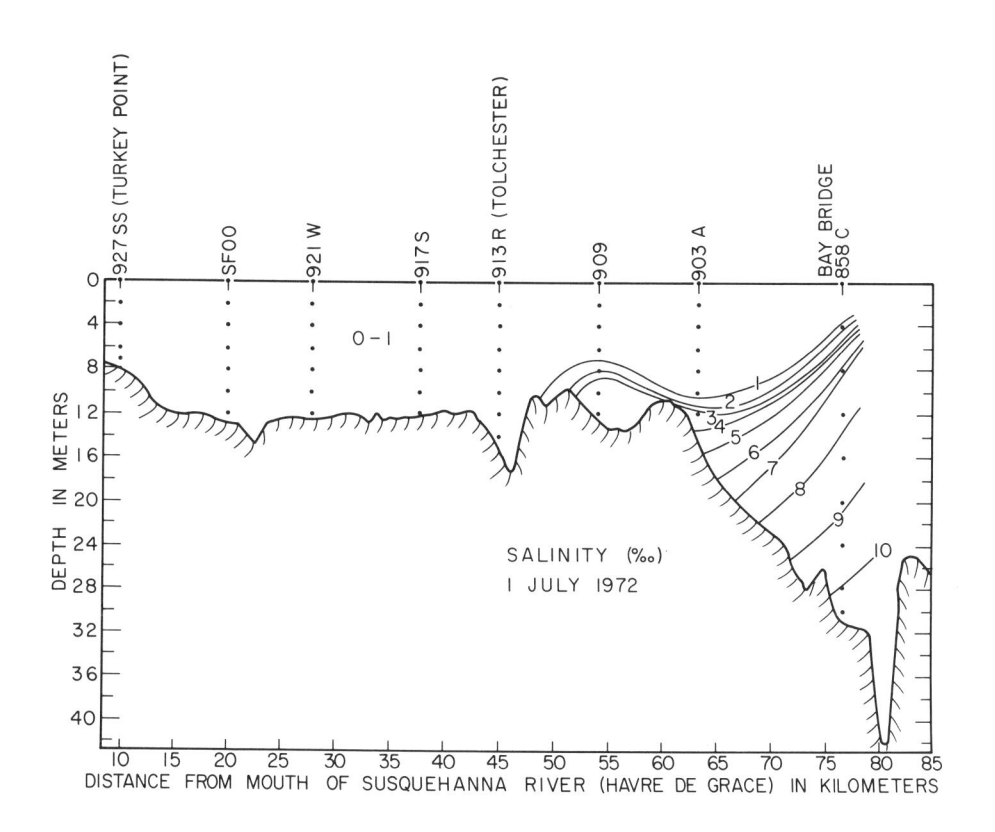

Figure 7. Longitudinal salinity distribution along the axis of the upper Bay on 1 July 1972.

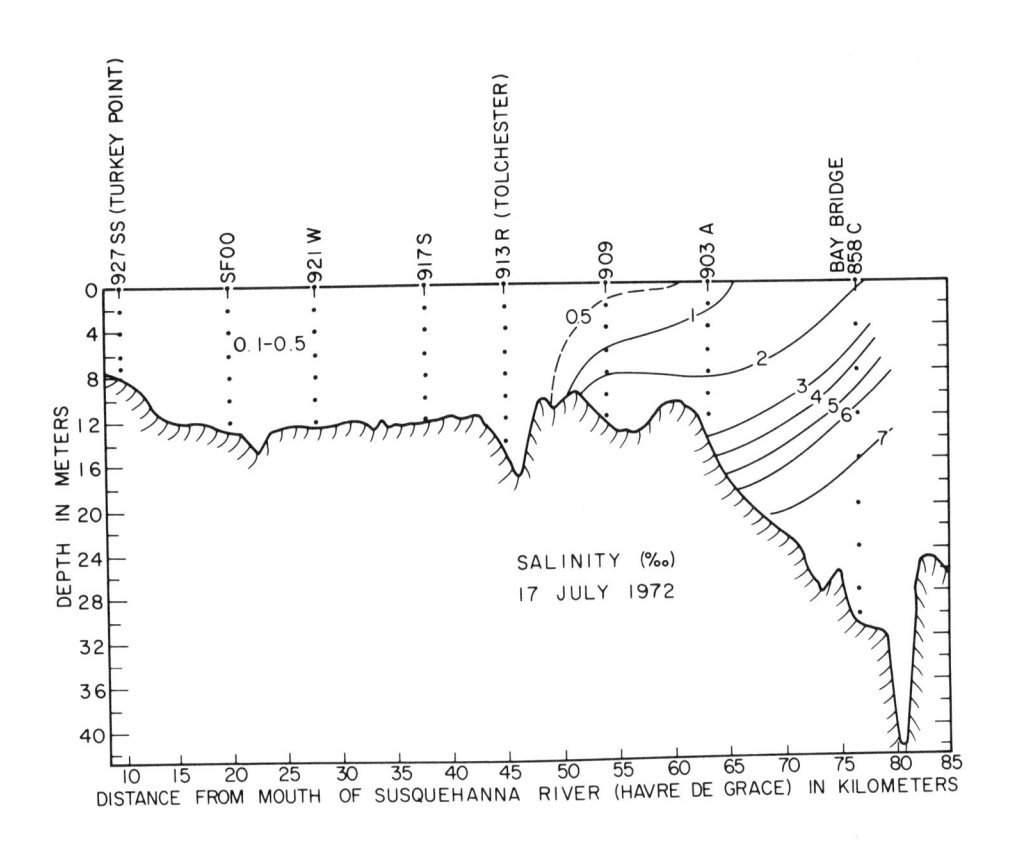

Figure 8. Longitudinal salinity distribution along the axis of the upper Bay on 17 July 1972.

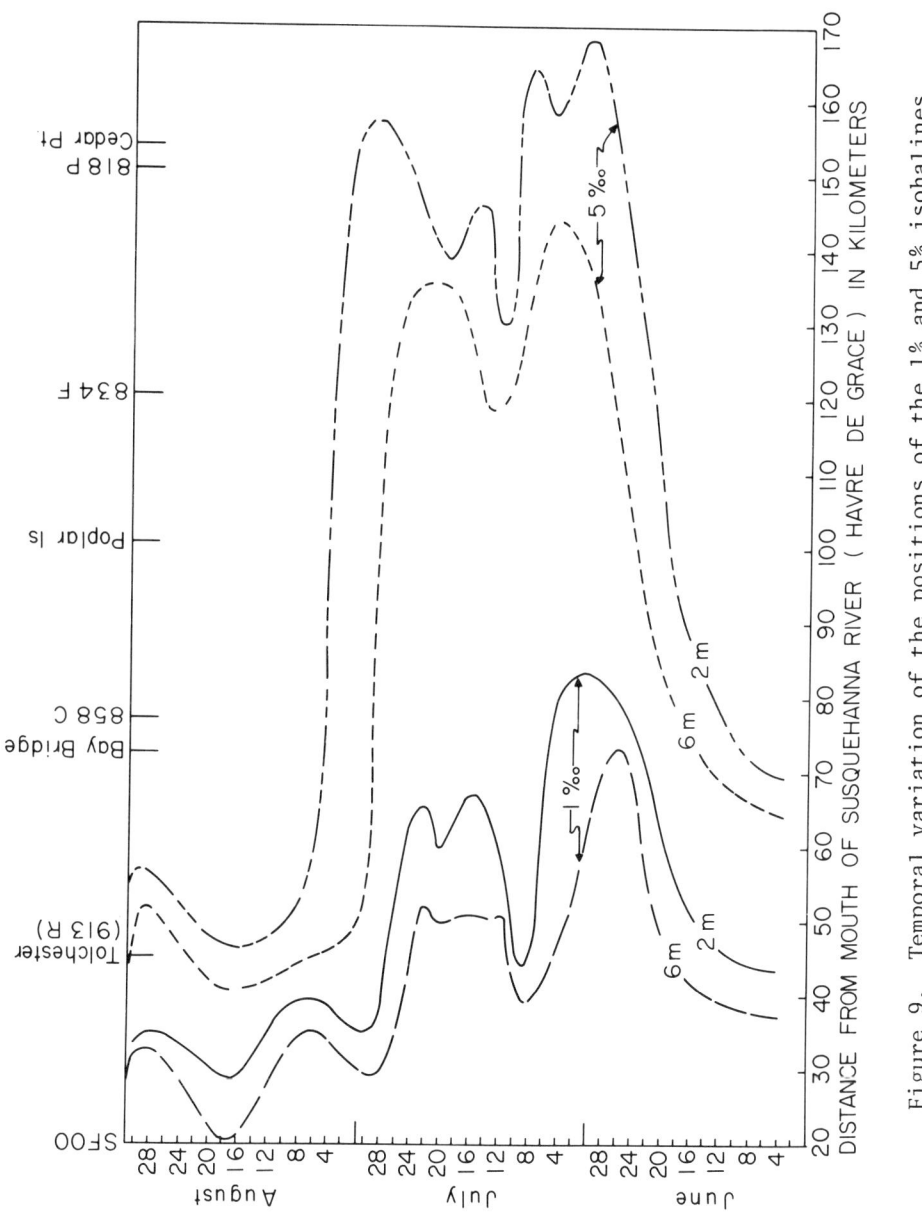

Figure 9. Temporal variation of the positions of the 1‰ and 5‰ isohalines at the 2 m and 6 m levels along the axis of the Bay following Agnes.

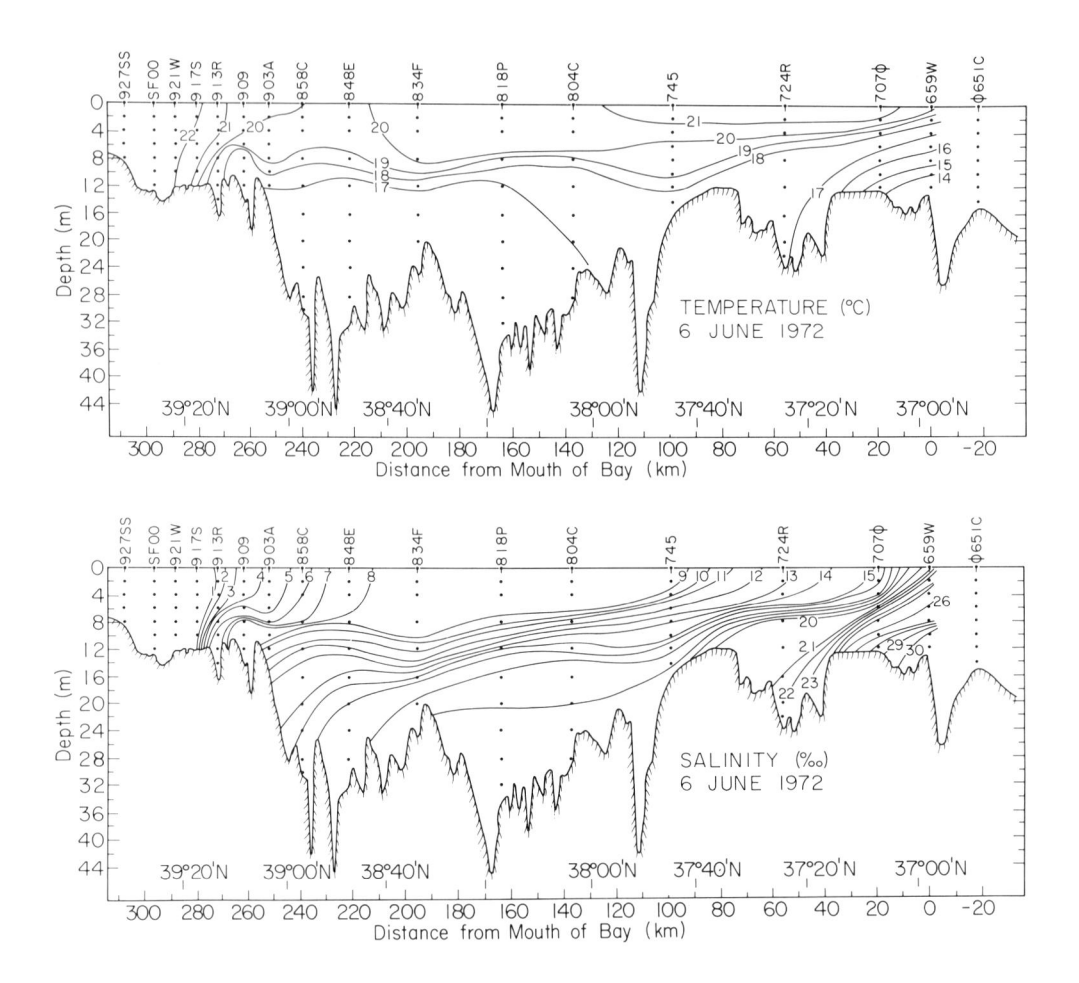

Figure 10. Longitudinal distributions of temperature and salinity along the axis of the Bay on 6 June 1972.

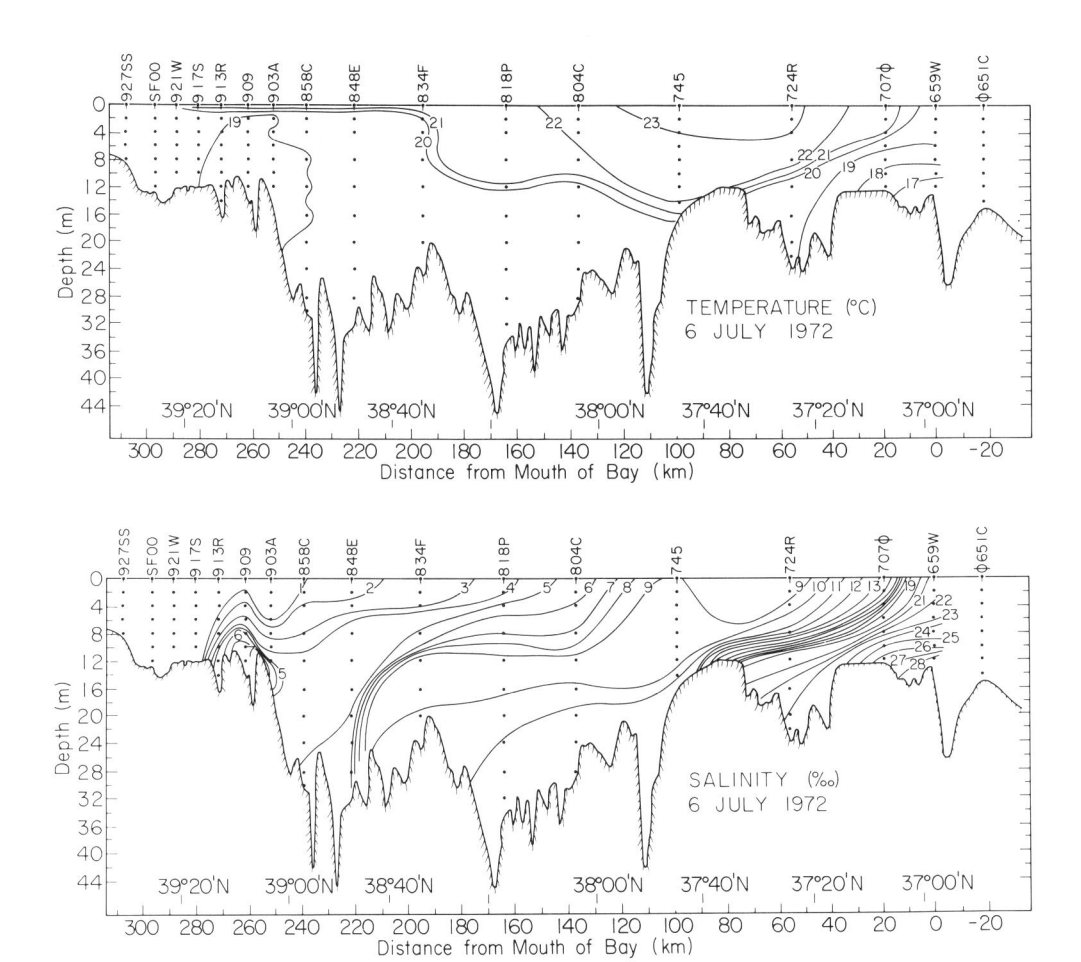

Figure 11. Longitudinal distributions of temperature and salinity along the axis of the Bay on 6 July 1972.

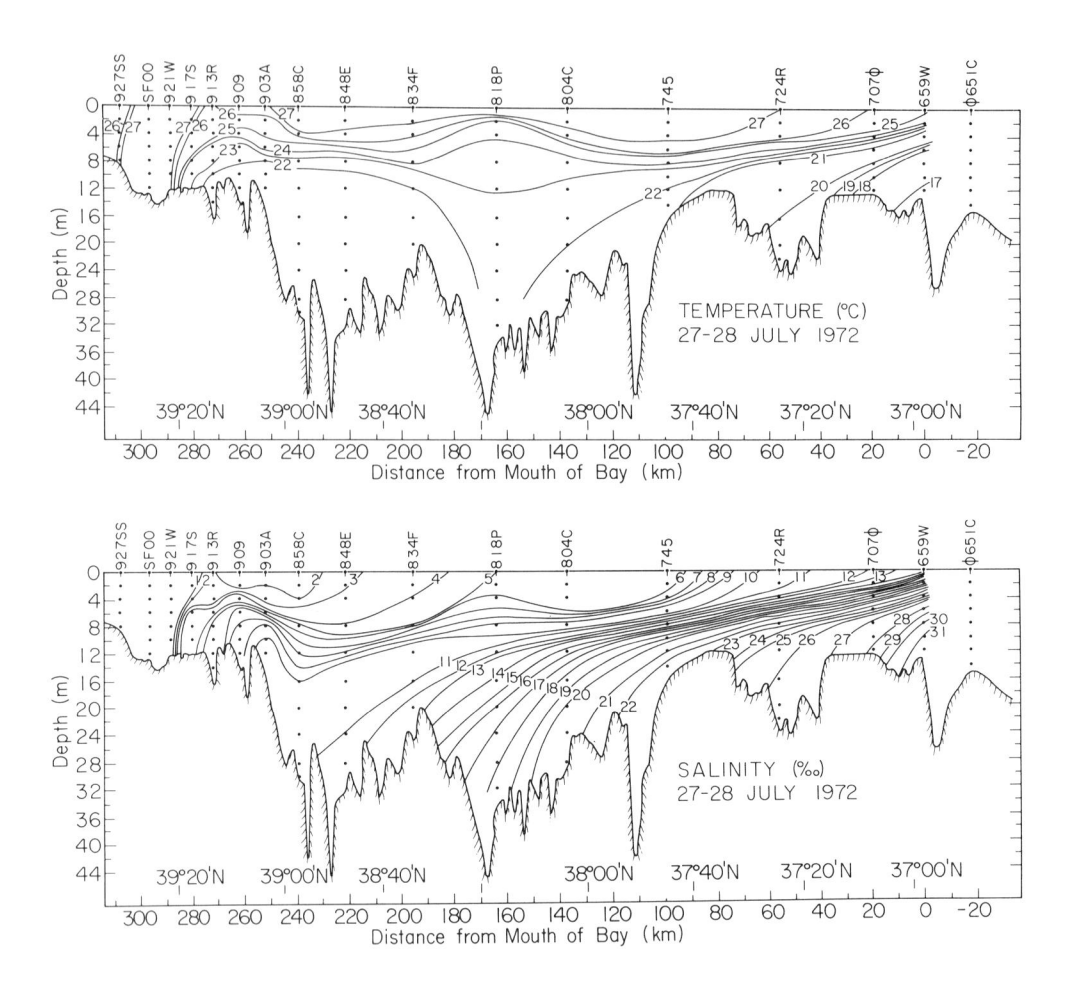

Figure 12. Longitudinal distributions of temperature and salinity along the axis of the Bay on 27-28 July 1972.

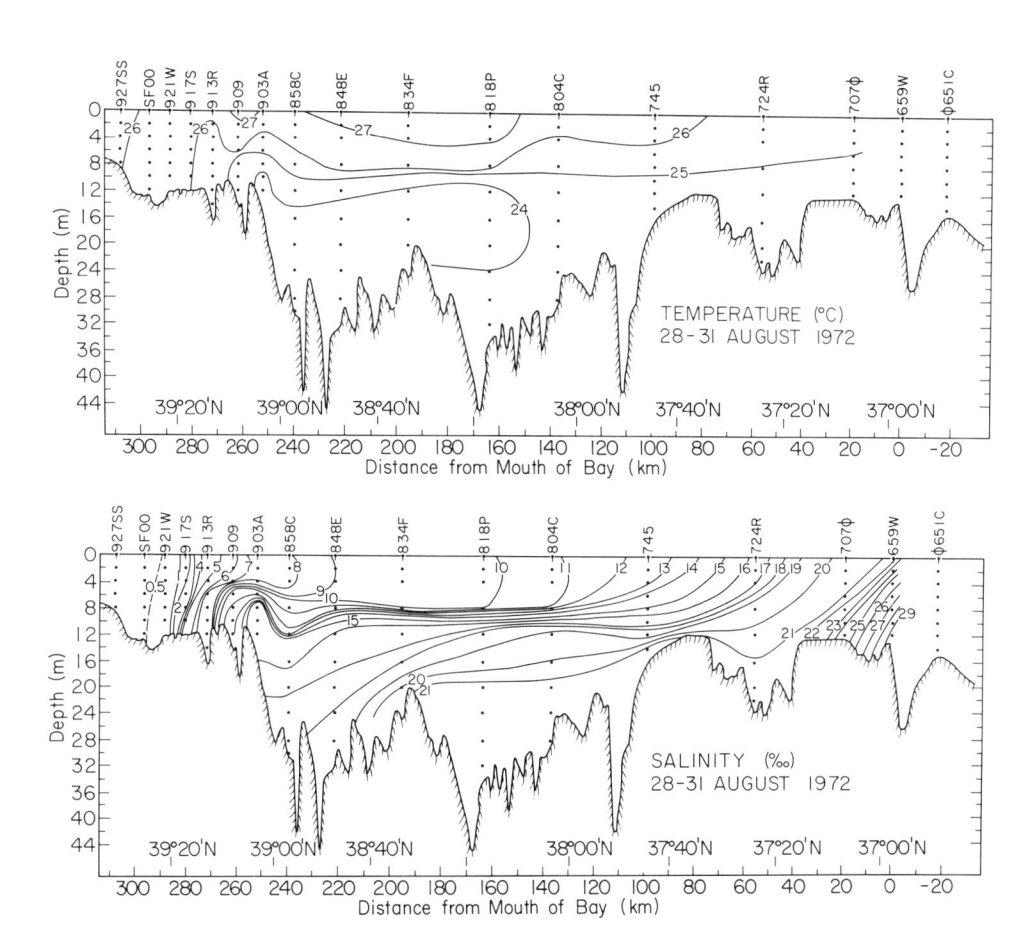

Figure 13. Longitudinal distributions of temperature and salinity along the axis of the Bay on 28-31 August 1972.

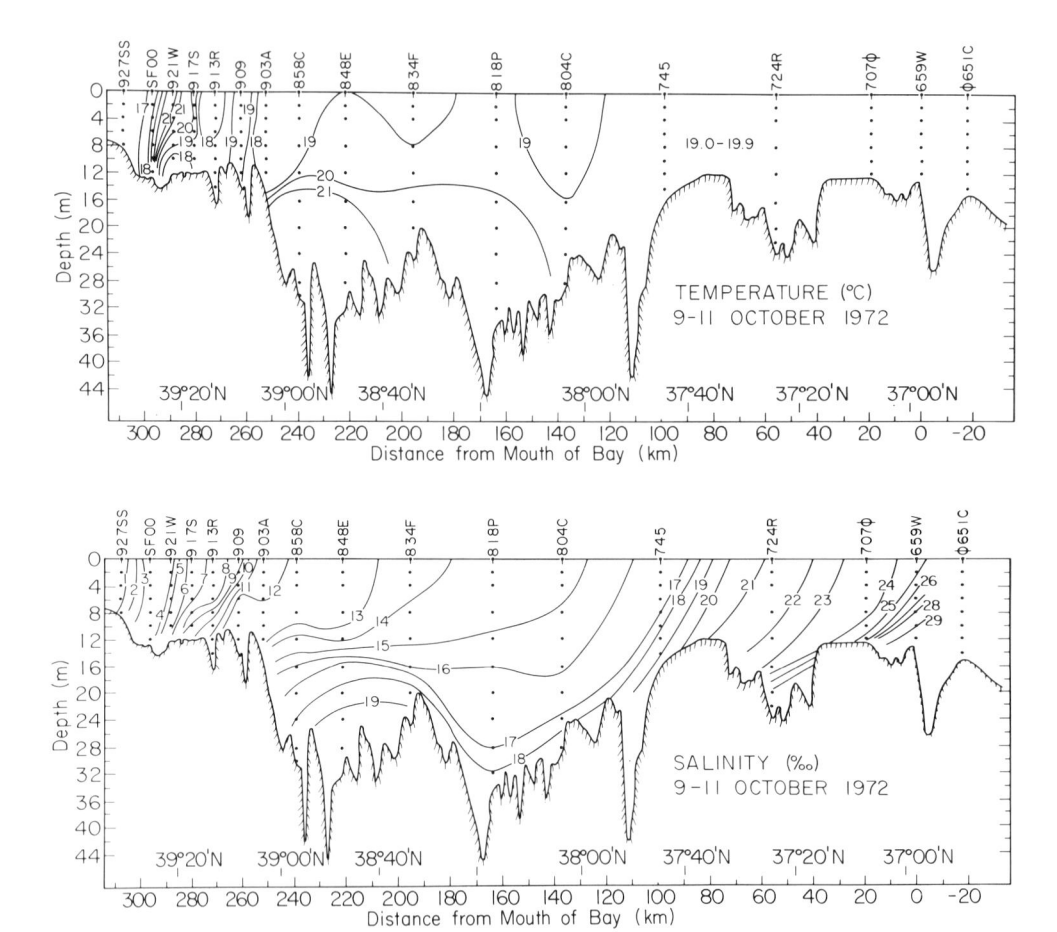

Figure 14. Longitudinal distributions of temperature and salinity along the axis of the Bay on 9-11 October 1972.

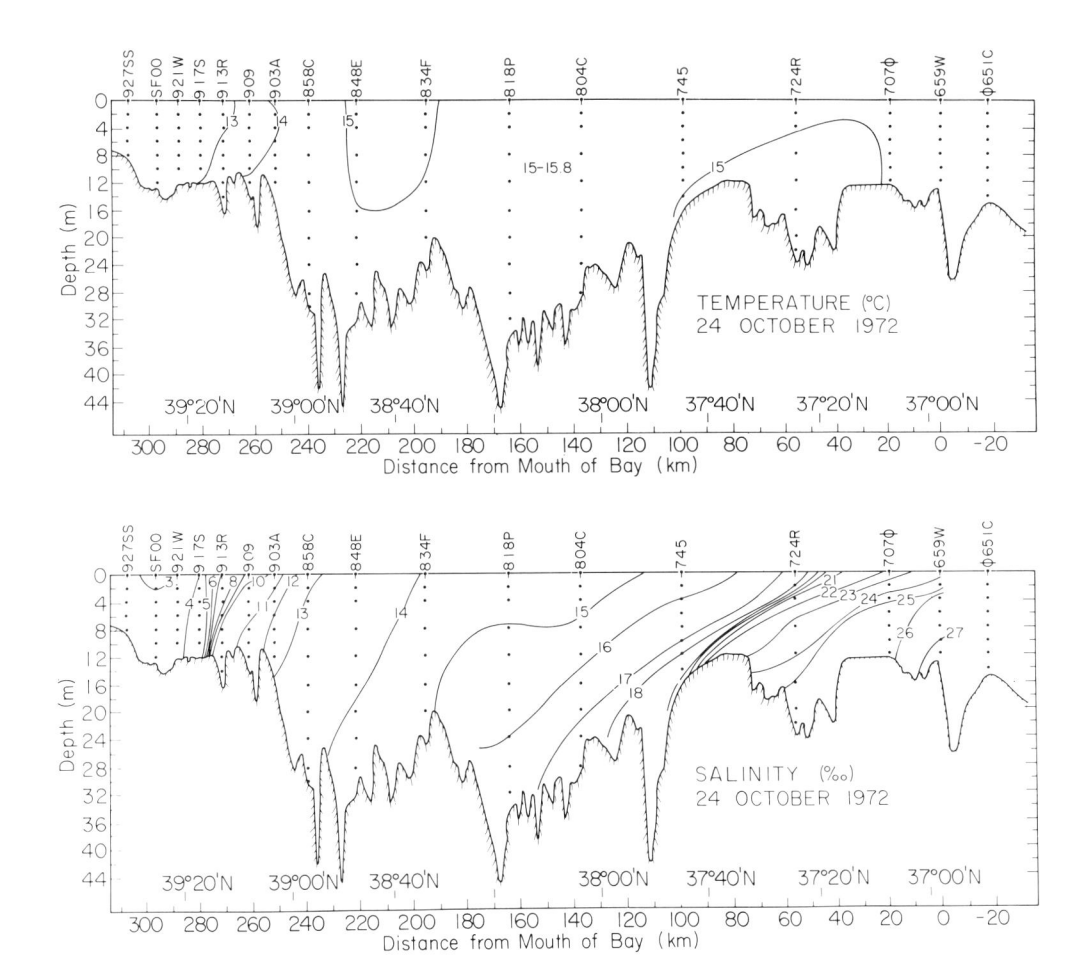

Figure 15. Longitudinal distributions of temperature and salinity along the axis of the Bay on 24 October 1972.

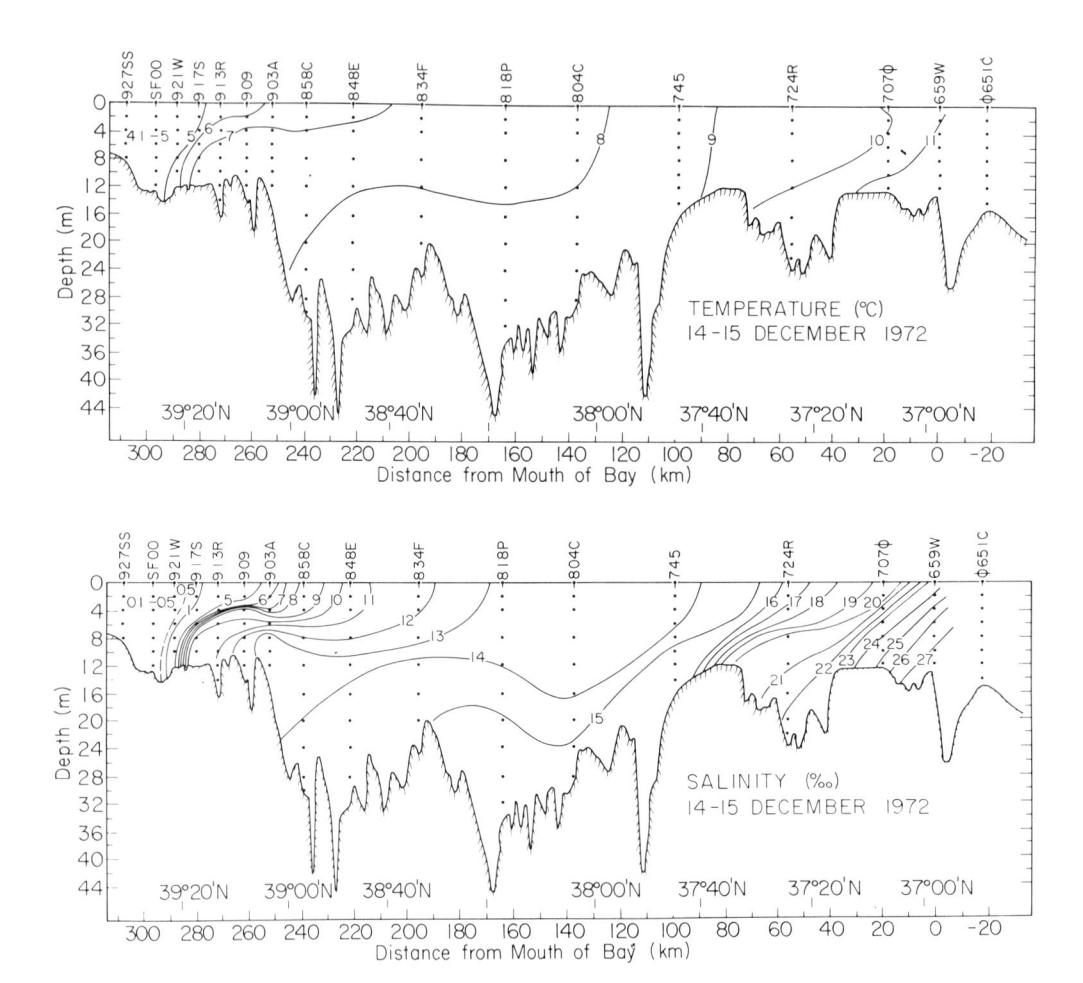

Figure 16. Longitudinal distributions of temperature and salinity along the axis of the Bay on 14-15 December 1972.

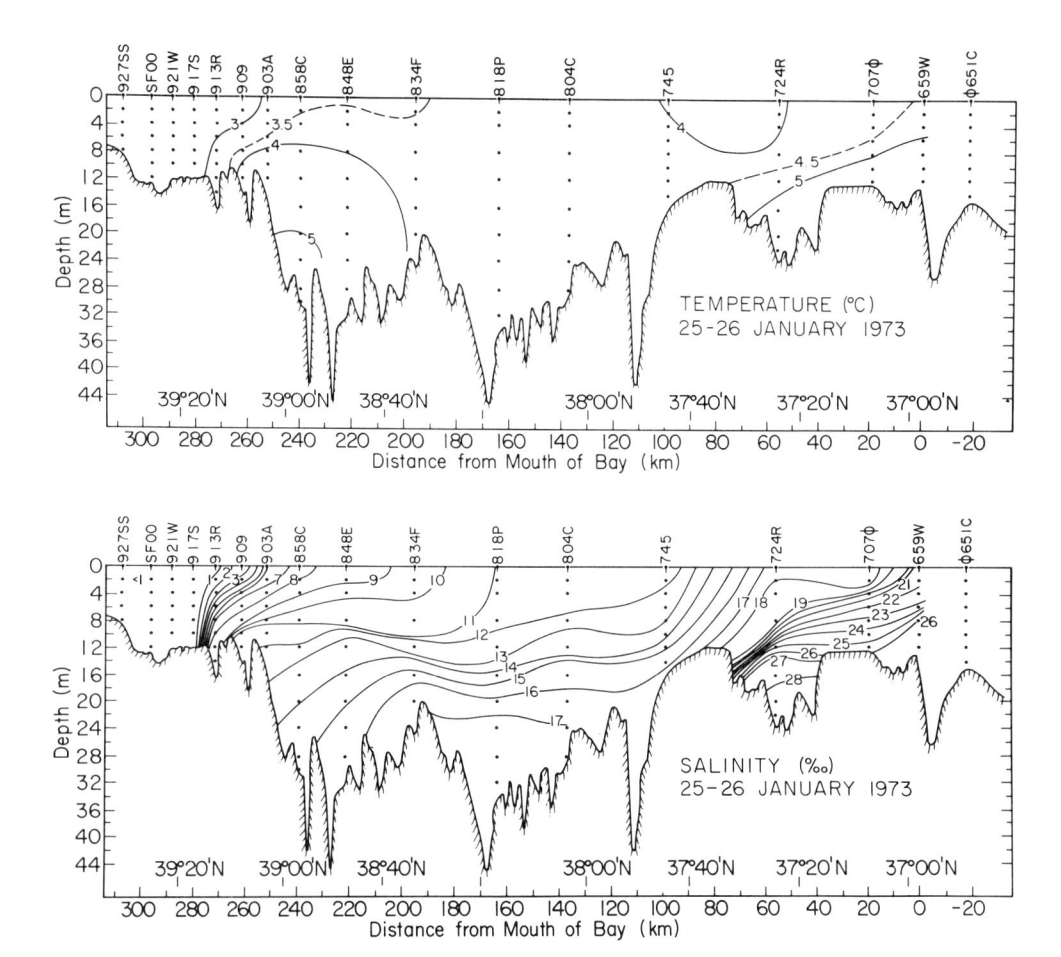

Figure 17. Longitudinal distributions of temperature and salinity along the axis of the Bay on 25-26 January 1973.

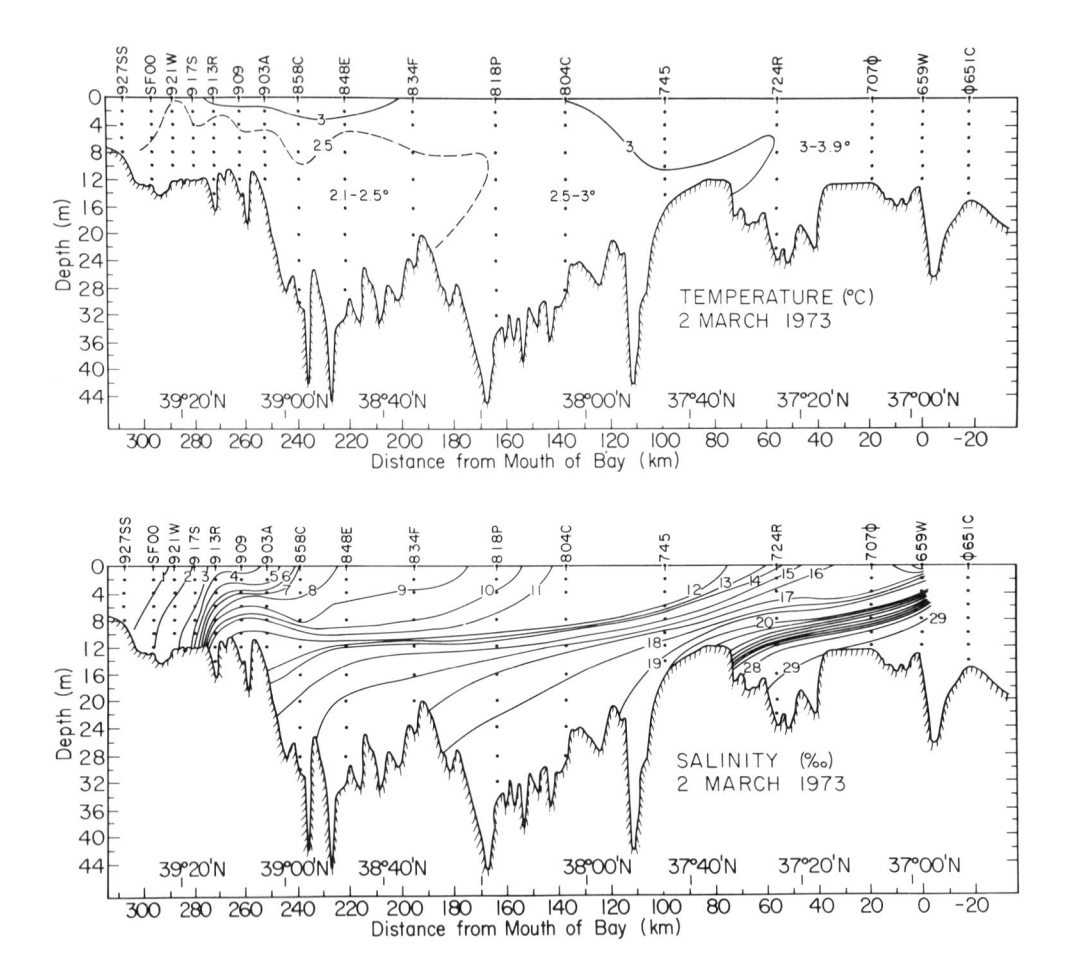

Figure 18. Longitudinal distributions of temperature and salinity along the axis of the Bay on 2 March 1973.

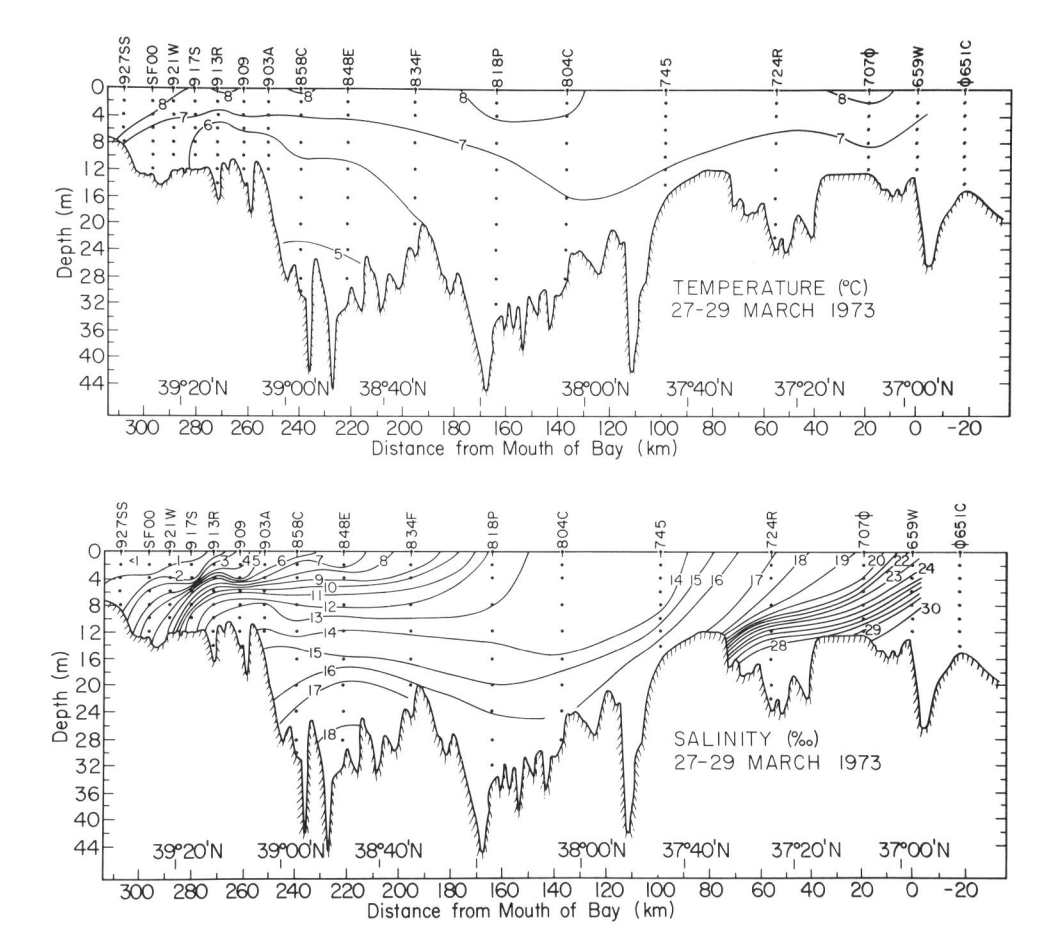

Figure 19. Longitudinal distributions of temperature and salinity along the axis of the Bay 27-29 March 1973.

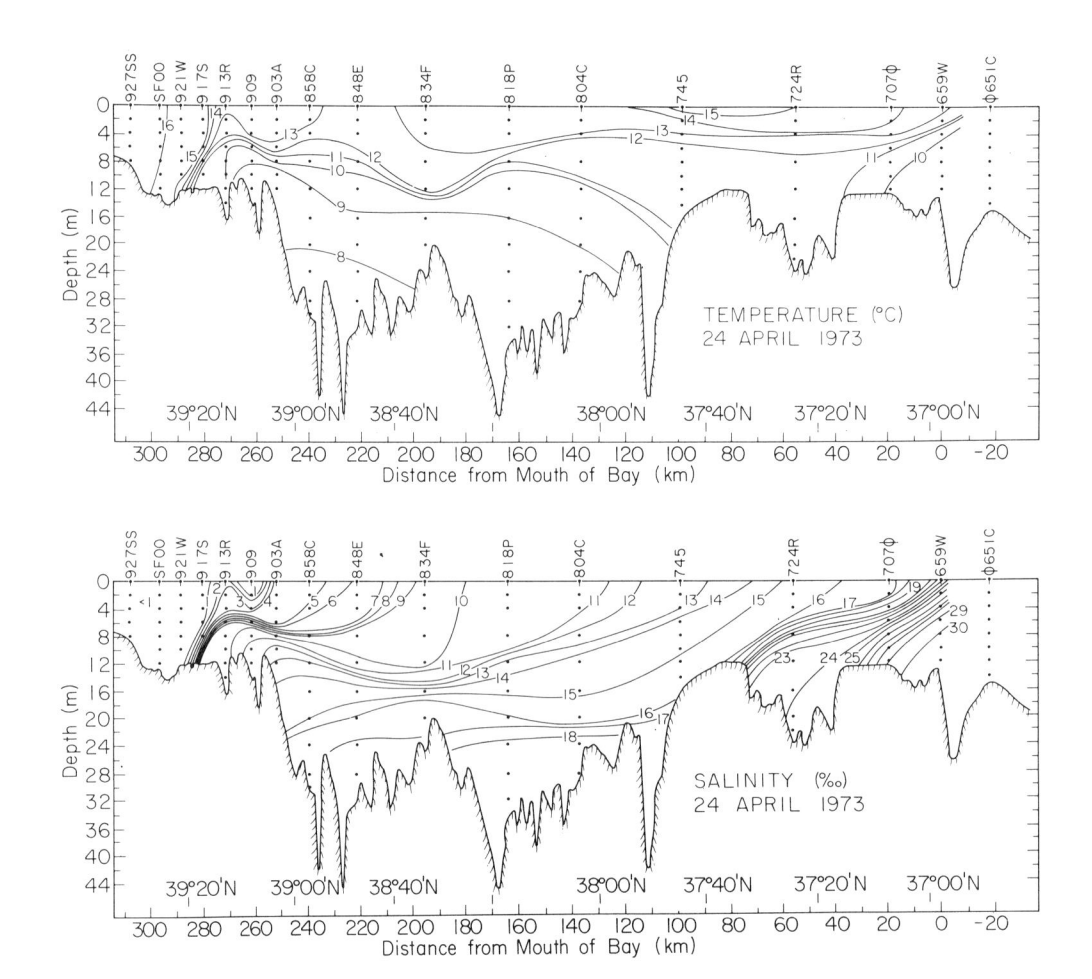

Figure 20. Longitudinal distributions of temperature and salinity along the axis of the Bay on 24 April 1973.

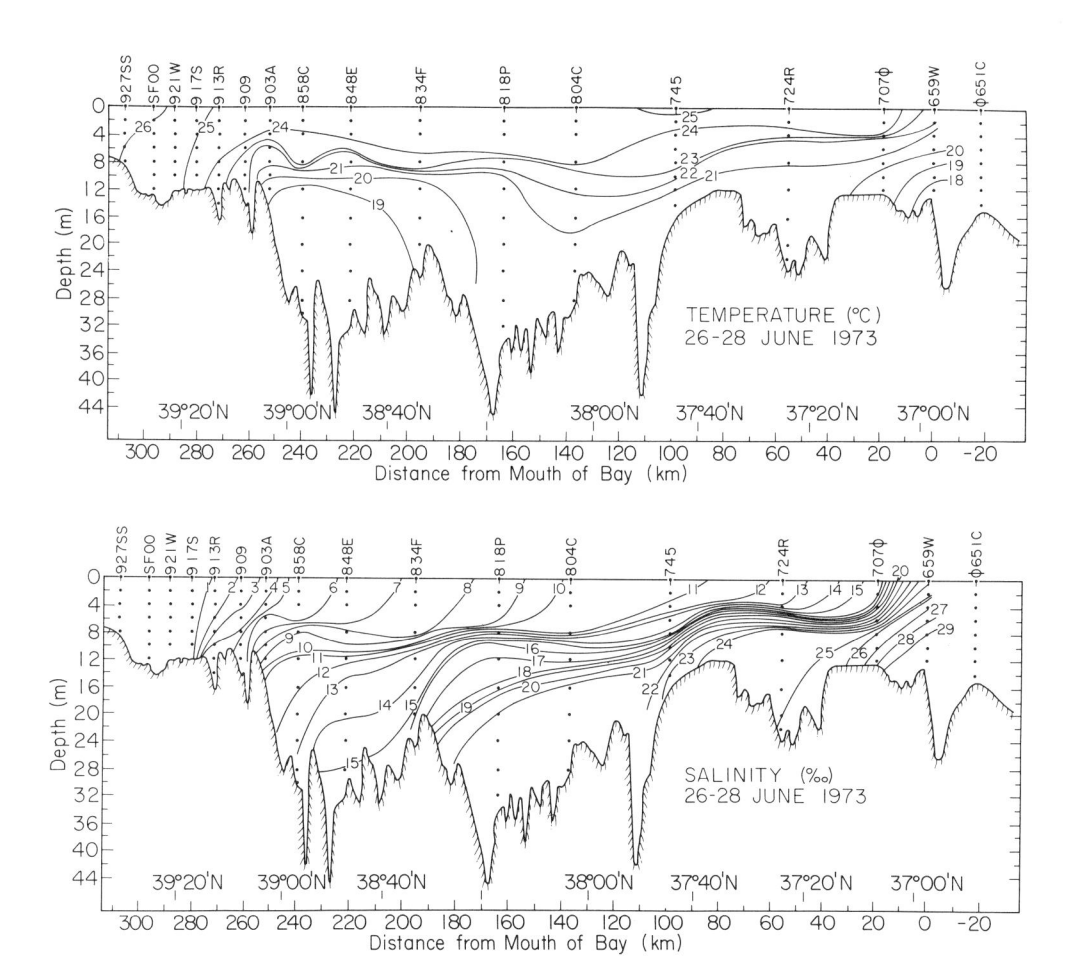

Figure 21. Longitudinal distributions of temperature and
salinity along the axis of the Bay on 26-28
June 1973.

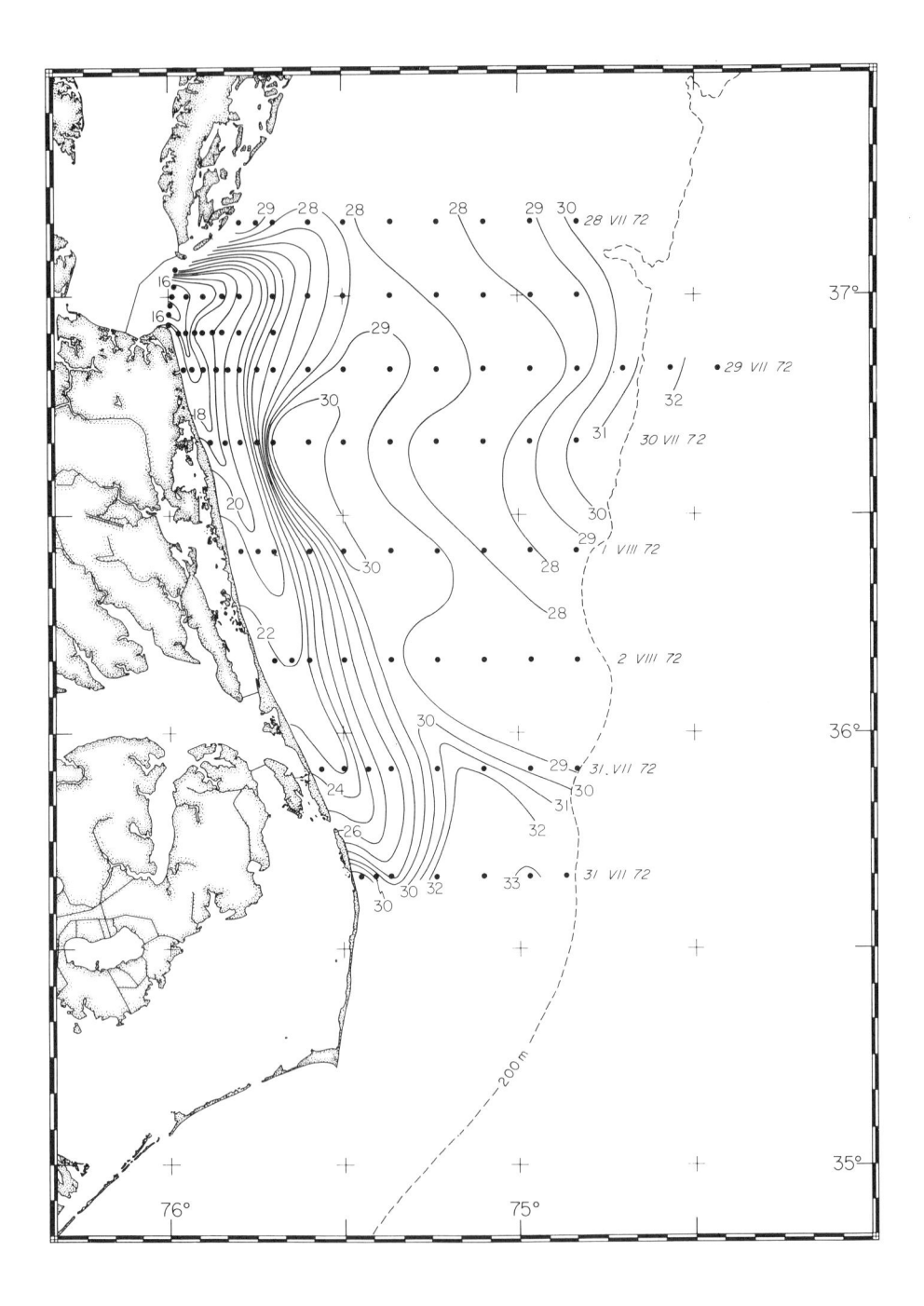

Figure 22. Distribution of surface salinity in the waters overlying the continental shelf off the mouth of Chesapeake Bay, 28 July-2 August 1972.

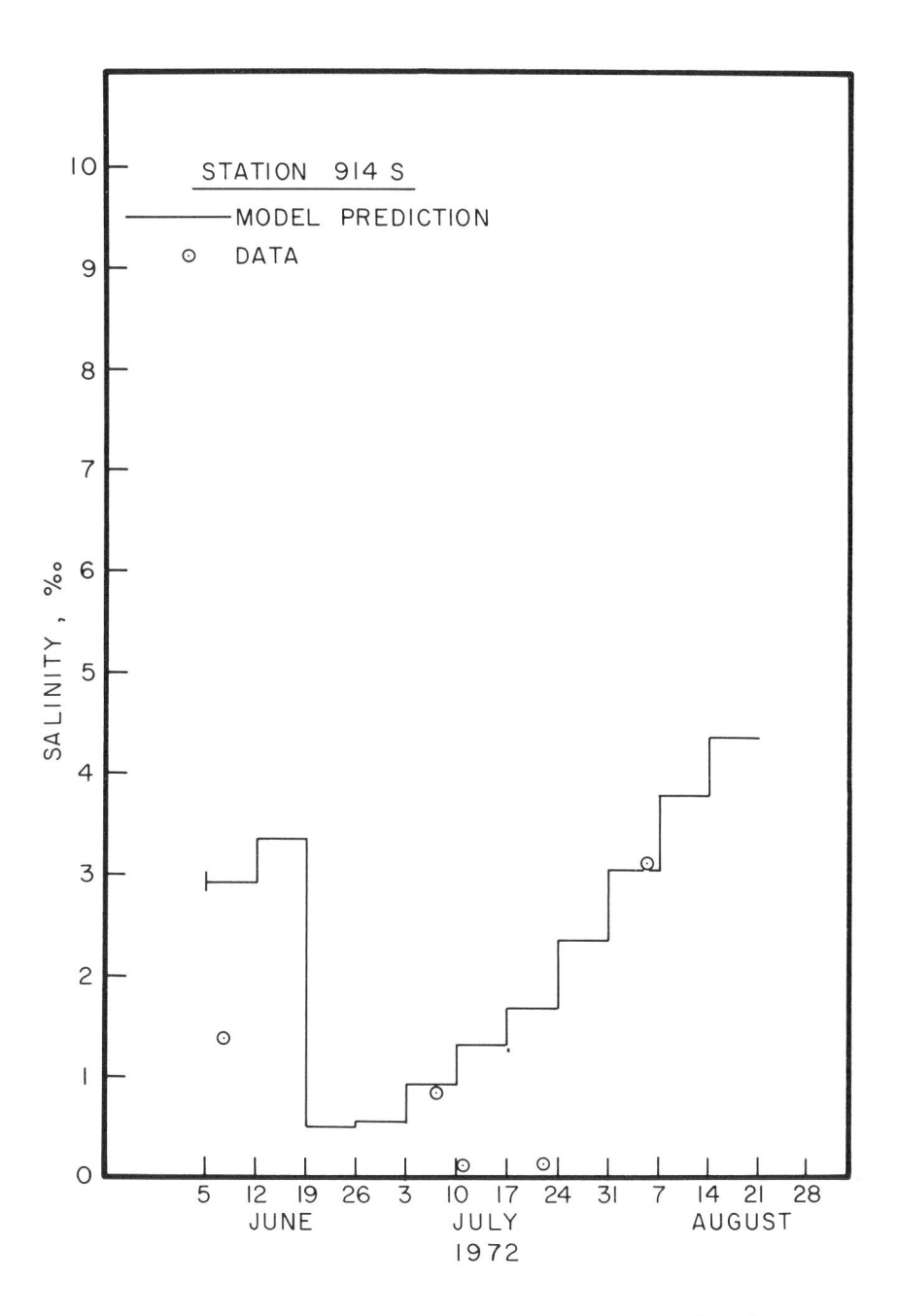

Figure 23. Comparison of Boicourt model prediction (solid line) with measured values of cross-sectional mean salinities (☉) at Station 914S between 5 June and 28 August 1972.

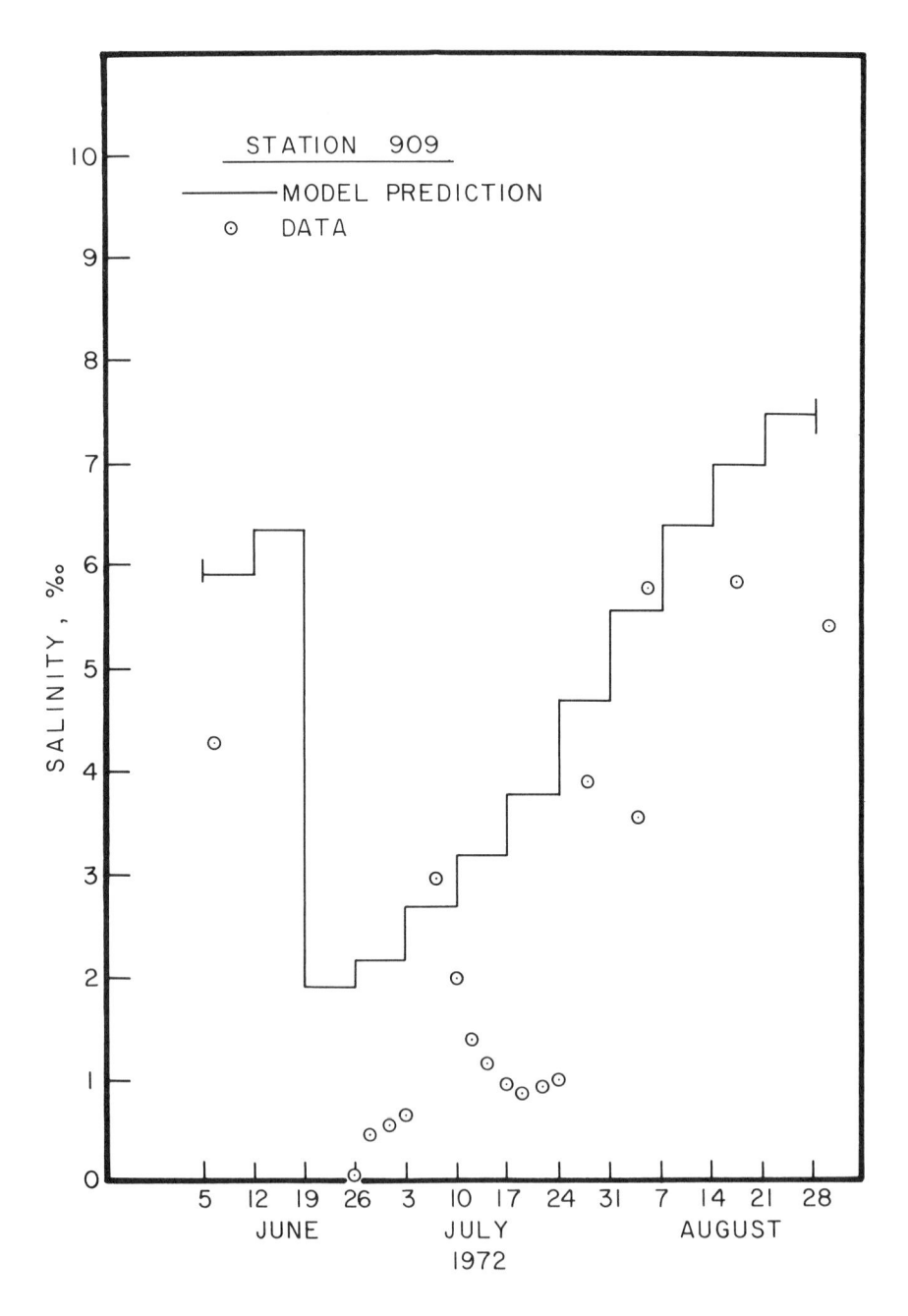

Figure 24. Comparison of Boicourt model prediction (solid line) with measured values of cross-sectional mean salinities (☉) at Station 909 between 5 June and 28 August 1972.

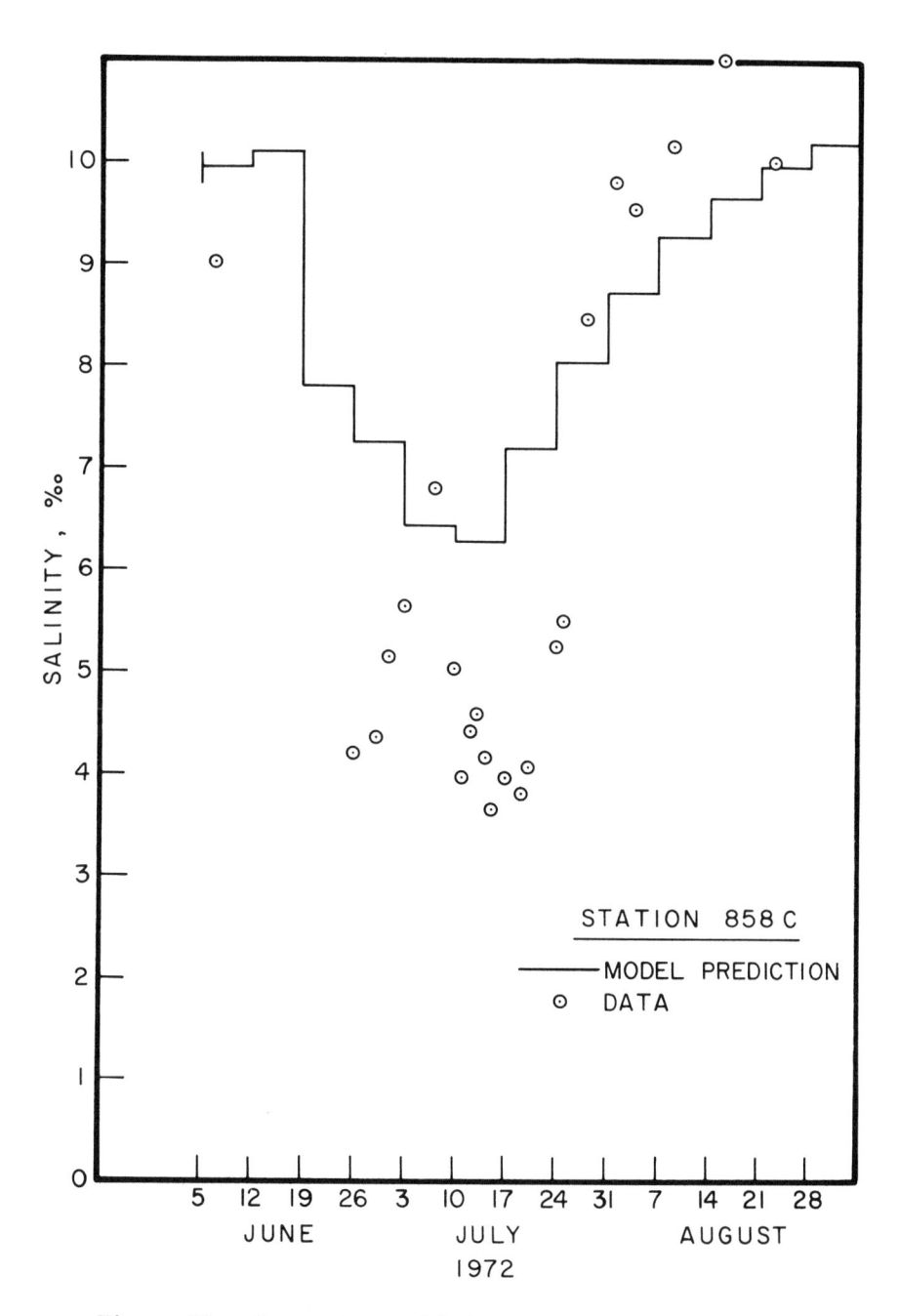

Figure 25. Comparison of Boicourt model prediction
(solid line) with measured values of cross-
sectional mean salinities (☉) at Station
858C between 5 June and 28 August 1972.

CHANGES IN SALINITY STRUCTURE OF THE JAMES,
YORK AND RAPPAHANNOCK ESTUARIES RESULTING
FROM THE EFFECTS OF TROPICAL STORM AGNES[1]

Paul V. Hyer[2]
Evon P. Ruzecki[2]

ABSTRACT

The peak effect of the flood waters produced by Tropical Storm Agnes was seen on June 25 in the James, June 26 in the Rappahannock, and June 30 in the York. Recovery toward normal salinity conditions after the high runoffs proceeded discontinuously, with alternating periods of vertical stratification and destratification. During strongly stratified stages, saline water advanced upstream along the bottom. In the York and James Rivers, the most dramatic stratification occurred about July 20-25. This event resulted in bottom salinity values exceeding normal ambient values and, at the river mouths, reaching values hitherto unobserved. This event was apparently controlled by the salinity distribution in the Bay. Less pronounced stratification maxima occurred in the James about July 6 and August 18 and in the York during August. These events do not appear to be correlated with stream gauge flow records or local precipitation. These events are possible instances of overshooting of equilibrium by the intruding salt water near the bottom.

INTRODUCTION

Virginia's three major estuaries emptying into the Chesapeake Bay were all affected by flood runoff from the rains of Agnes in June 1972. Instead of three special cases, however, they constitute three parts of a larger system which includes the Bay itself and all its tributaries. Fig. 1 shows the Chesapeake Bay drainage system. The stations sampled in the three major estuaries are shown. Recovery of salinity structure in each individual estuary depends therefore in part on the boundary condition at that estuary's mouth, as controlled by the Bay.

Qualitative summaries of the salinity distribution in each of the three estuaries will be presented for the major recovery period, beginning at the time of passage of Agnes and extending to about the end of August.

OBSERVATIONS

James River

Figs. 2 and 3 summarize the salinity distribution throughout the recovery period. These figures are derived from twenty-eight slack water runs performed in the two-month period following Agnes. Fig. 2 shows the recovery process at the river mouth, while Fig. 3 illustrates the recovery process by showing the progress upstream of selected isohalines. In Fig. 2 and its counterparts for the Rappahannock and York, the stage of the tide is distinguished, although

---
[1]Contribution No. 756, Virginia Institute of Marine Science.
[2]Virginia Institute of Marine Science, Gloucester Point, Va. 23062

there seems to be little systematic difference in the salinity structure when measured at high slack and low slack. Beginning June 24, strong stratification is evidenced by the difference in salinity between the surface and bottom, while the maximum depression of surface salinity occurred on 25 June. Bottom salinity, however, reached a minimum on June 27, a time when surface salinity had begun to recover. A period of slowly increasing salinity followed, accompanied by oscillations with a period of about four days, evident both in the salinity at the river mouth and the salinity intrusion upstream. Vertical stratification also decreased until about July 7, and, near the mouth, until July 14. However, from July 14 until July 21, there was a dramatic increase in vertical stratification at the river mouth, with a characteristic pattern of an initial decrease at the surface, followed by a considerable increase at the bottom and an increase at the surface. At the same time, salinity began to move upstream, but reached its maximum excursion several days after salinity at the river mouth reached its maximum value. This "overshoot" of equilibrium was apparently initiated by conditions in the Chesapeake Bay. Bottom salinity at the mouth of the James reached a value of 30.03 ppt on July 21, the highest value hitherto recorded for that point (Fig. 4). The station numbers shown in this figure and the other salinity contour plots represent the distance from the river mouth in nautical miles. Both this maximum value of salinity and the subsequent maximum upstream penetration of salt water were observed at low water slack. The subsequent retreat of salt water appears to be characterized by overall vertical mixing, but with a spike about two weeks after the late July event. The system finally reached a well-mixed state on August 25. Slack water data from September 1971 are included in Figs. 2 and 3 for comparison. Background data from 1973 are also included.

Rappahannock River

The freshening effect of the flood waters was evident in the Rappahannock on the first date sampled, but surface salinity at the mouth reached a minimum on June 26, accompanied by prominent vertical stratification (Figs. 5 and 6). There were thirty-three slack water runs made in the Rappahannock in the summer of 1972. By June 30, bottom salinity at the mouth had achieved a local minimum in time, but not an absolute minimum over the history of the flood. Vertical stratification was strong upstream, as evidenced by the fact that the 5 ppt isohaline met the surface. The density structure at the mouth at this time corresponded to a reverse three-layer system, with a stratified river connected to a well-mixed Bay (Fig. 7). In such a system, there is net flow inward at mid-depth and outflow at surface and bottom (Pritchard & Carpenter 1960). From the beginning of July until about July 10, salt water penetrated upstream, as a result of gravitational circulation. At the same time, however, salinity at the river mouth was decreasing due to flushing of the Bay. Bottom salinity at the river mouth reached a minimum value of 7.2 ppt. The result was a well-mixed, low-salinity, over-extended system. Salt water retreated rapidly until mid-July, with the system staying well mixed. Surface salinity at the mouth was notably constant for this period. From July 15 to July 25, an increase in bottom salinity at the river mouth set in motion an upstream transport of salt, with characteristic strong vertical stratification. The balance of July was a period of vertical mixing as can be seen by the fact that the 5 ppt isohaline became relatively steep. In early August, bottom salinity at the mouth increased rapidly, (due to recovery within the Bay) reaching a value greater than 20 ppt, while the surface salinity remained below 8 ppt (Fig. 6). The system soon reverted to a well-mixed state. However, the salt supplied by this event made possible an increase in salinity throughout the estuary.

York River

York River mouth salinity and saline intrusion history are shown in Figs. 8 and 9 respectively. A total of 38 slack water runs are summarized. Of the three major tributaries of the lower Chesapeake Bay, the York seems to have been the least affected and had the slowest response to Agnes-caused flooding. This was probably due to two main factors: the relatively small size of the York watershed and the fact that a new reservoir on the North Anna River (a secondary tributary to the York) served as a catchment basin for a good portion of the flood waters. Minimum surface salinity at the river mouth occurred on 30 June, with bottom salinity reaching a minimum two days later. Over the first ten days of July there was a trend toward recovery, albeit with fluctuations. The mouth of the York became well-mixed by 13 July. Then, as in the James, there occurred a strong perturbation originating in the bay. Bottom salinities increased dramatically, causing an increase in vertical stratification throughout the estuary. Fig. 10 shows the situation on 26 July. Bottom salinities at the mouth of the York reached the highest values hitherto observed, as was the case in the James. A period of salinity advance and a trend toward mixing but with a spike (indicating a short-period increase in salinity) occurring a week after maximum salinity was observed. The salinity advance upstream was rapid, as can be seen by the progress of the 5 ppt isohaline (Fig. 9). By August 29 the estuary was well mixed.

Another interesting feature of the York River is the high salinity lens observed on several occasions at low water slack, between Tue Marsh Light and Gloucester Point (Fig. 11). This feature probably results from upwelling of the ebbing water as it is forced out of the narrow, deep channel between Gloucester Point and Yorktown and into a reach with shallower, broader profile. A lens in the James, on the other hand, was seen on only one occasion and probably reveals transient response to Agnes flood waters.

DISCUSSION

The fluctuations observed in the salinity structure of the James, York, and Rappahannock Rivers indicate oscillatory motion and sensitivity to a time-dependent boundary condition, i.e., the salinity in the Chesapeake Bay. All three systems received an initial perturbation from the flood. The York and James received a second disturbance in the period July 20-25. The Rappahannock apparently received a second disturbance about August 7. Both disturbances occurring in the James and the late July disturbance in the York seemed to be followed by an oscillatory approach to equilibrium. Apparently the returning salt water in the lower depths gained sufficient momentum to overshoot equilibrium, so that an oscillation resulted. This behavior was not observed in the Rappahannock estuary, which is extremely long compared to the others. The initial impulse delivered to the York was apparently too gradual to cause any oscillation.

CONCLUSIONS

The James, Rappahannock, and York estuaries were greatly disturbed first by the flood runoff from Agnes and again, over a month later, by dramatic changes of the salinity structure in Chesapeake Bay. Following these events, the James and York showed an oscillatory behavior suggesting an internal freely-oscillating seiche. Salinity conditions in the York and Rappahannock in late September were not greatly different from those a year earlier.

LITERATURE CITED

Pritchard, D. W. and J. H. Carpenter. 1960. Measurements of turbulent diffusion in estuarine and inshore waters. Bull. Int. Assoc. Sci. Hydrol. 20:37-50.

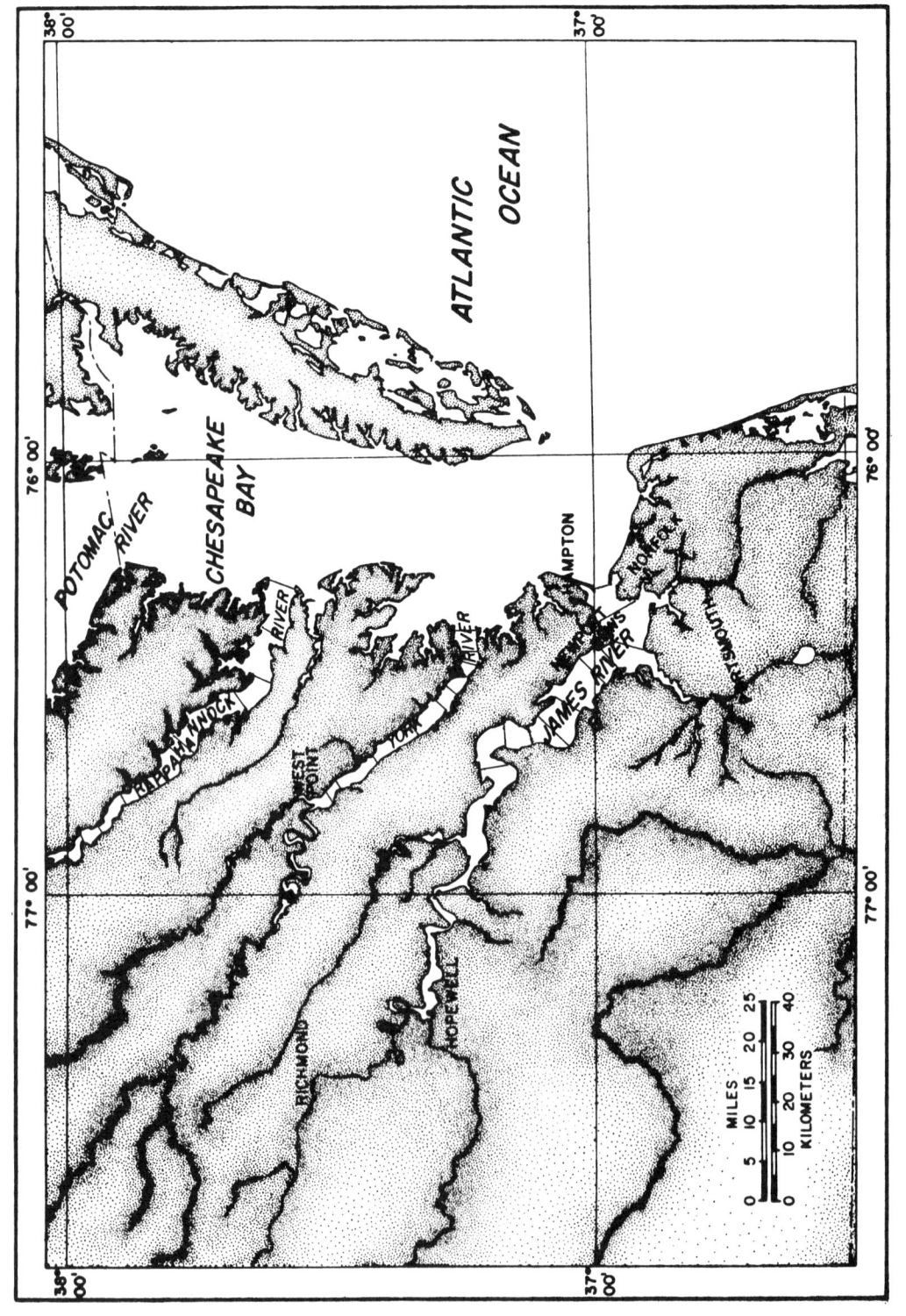

Figure 1. Post-Agnes sampling points in Virginia estuaries.

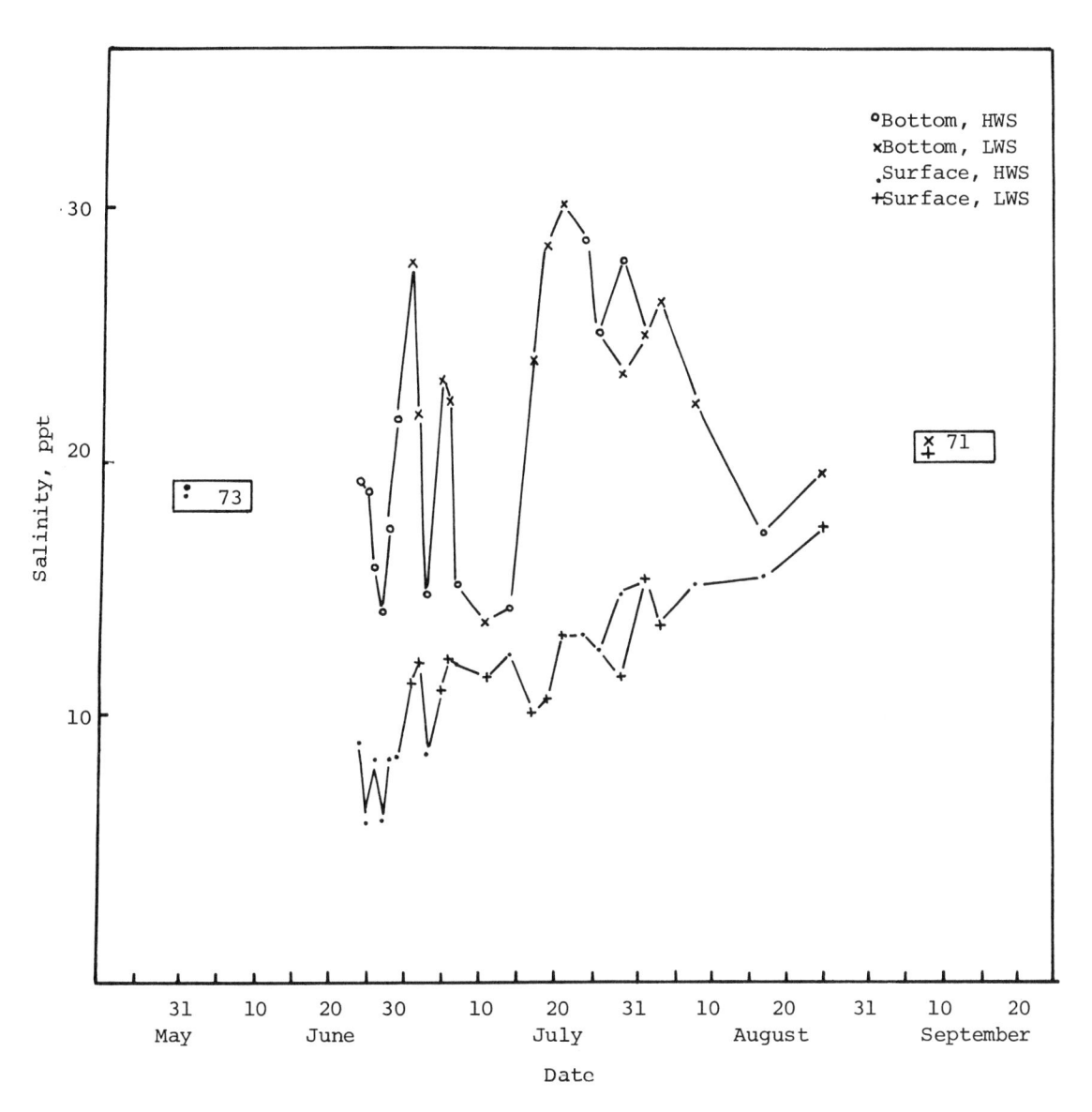

Figure 2. Surface and bottom salinity at the James River mouth as a function of time.

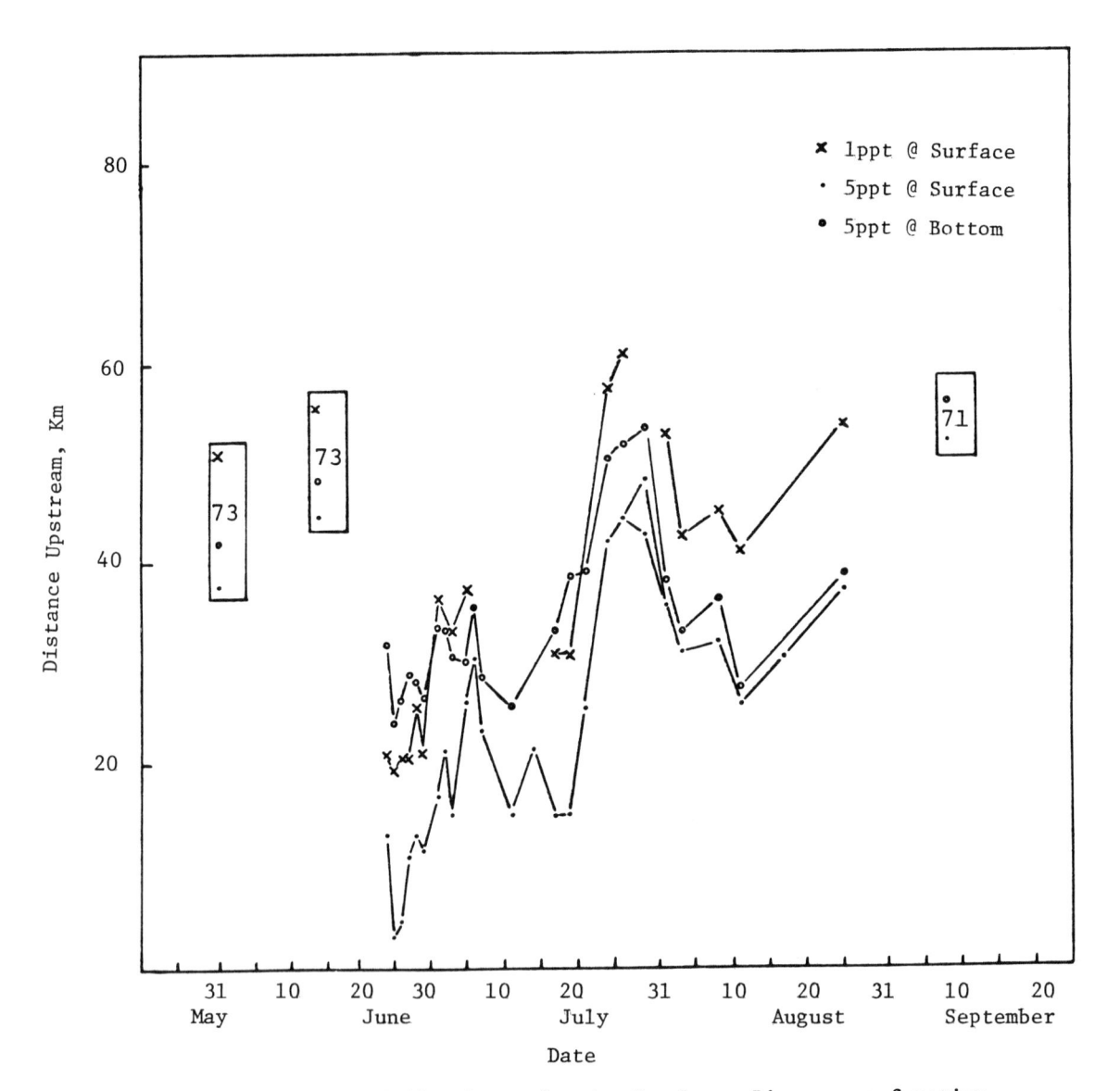

Figure 3. Saline intrusion in the James River as a function of time.

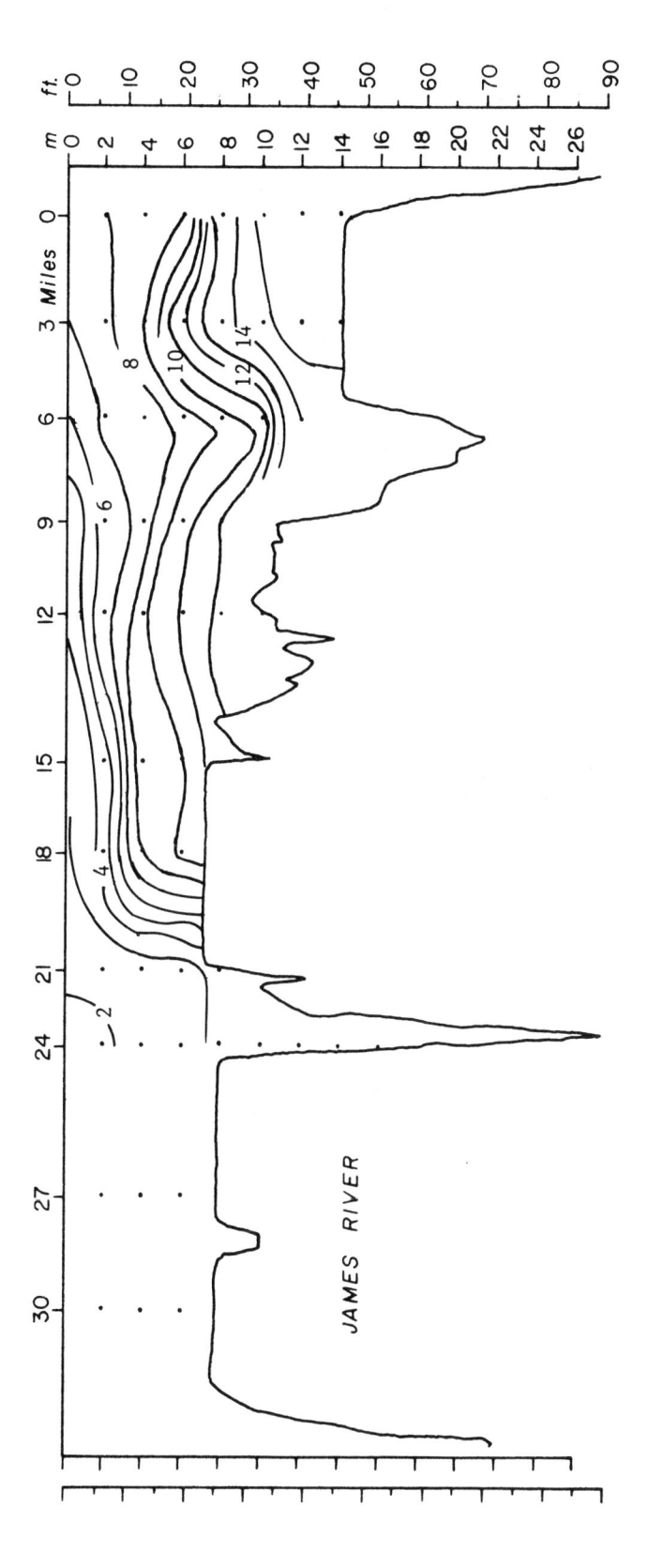

Figure 4. Salinity at low water slack, July 21, 1972.

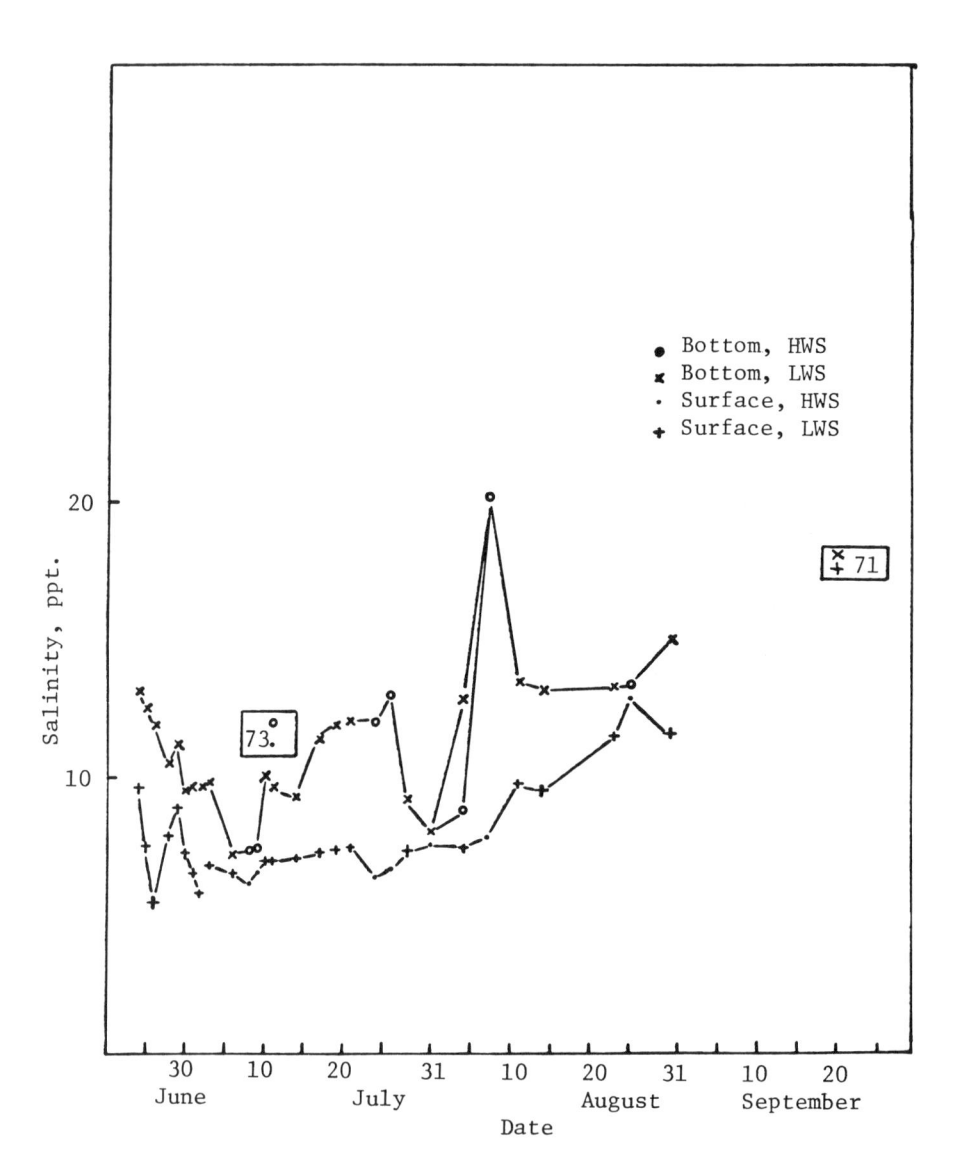

Figure 5. Surface and bottom salinity at the Rappahannock River mouth as a function of time.

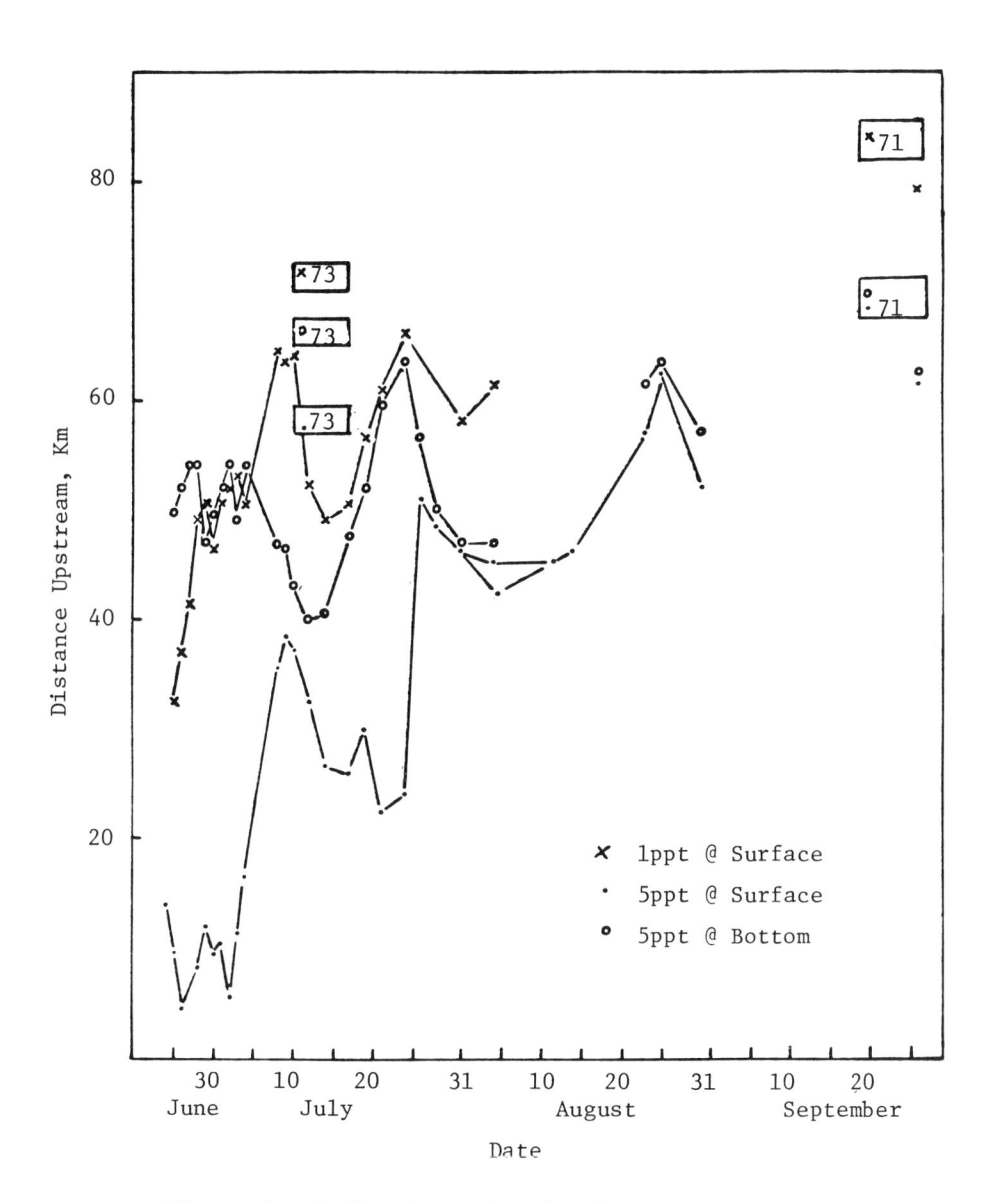

Figure 6. Saline intrusion in the Rappahannock River as a function of time.

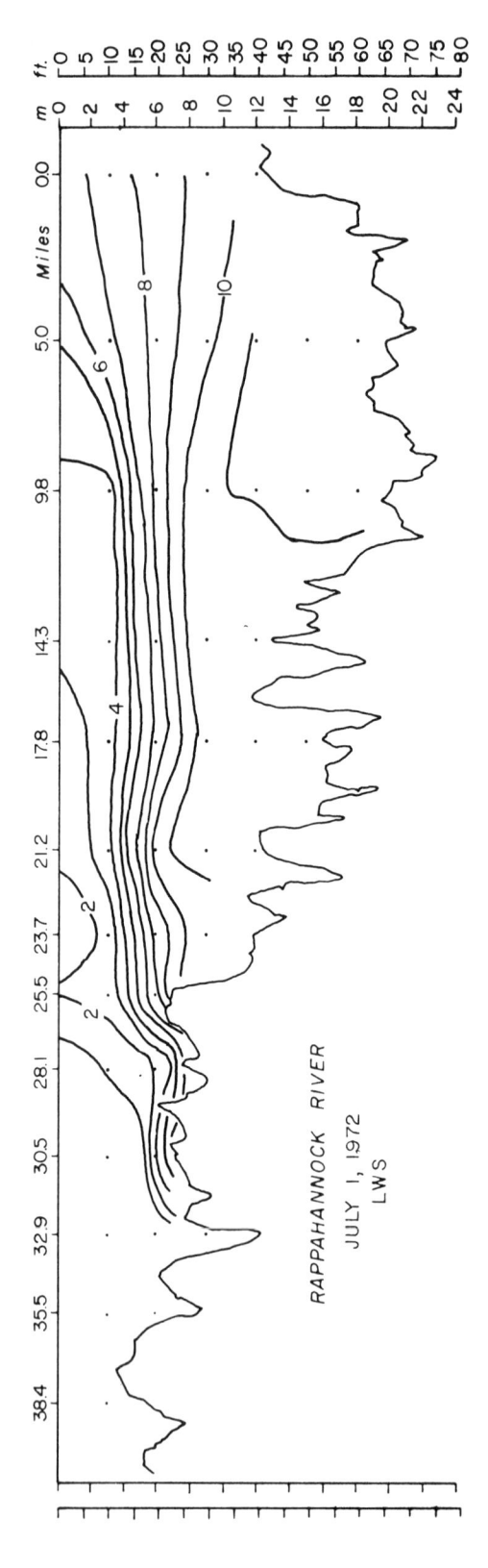

Figure 7. Salinity at low slack water, Rappahannock River, on July 1, 1972.

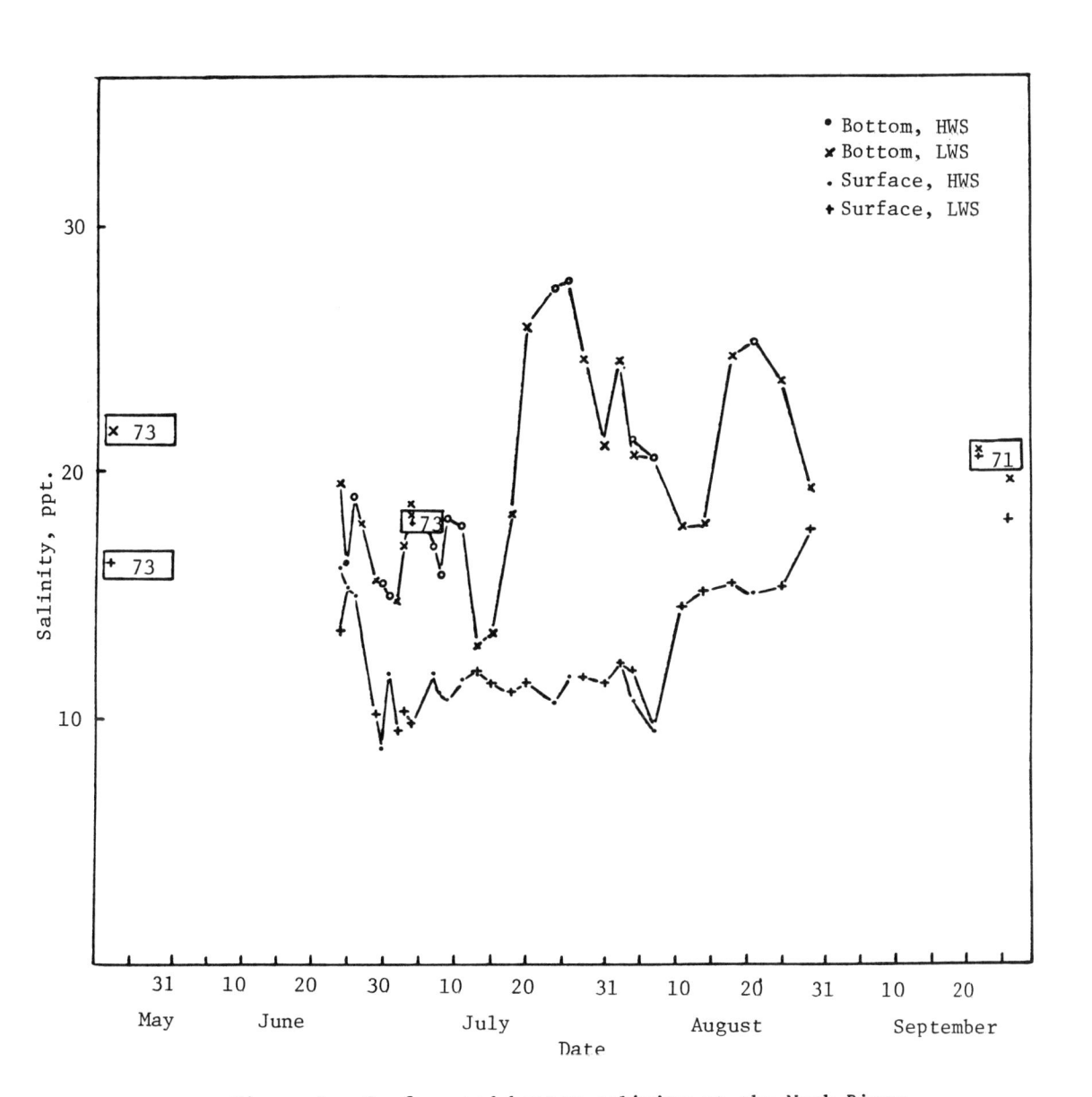

Figure 8. Surface and bottom salinity at the York River mouth as a function of time.

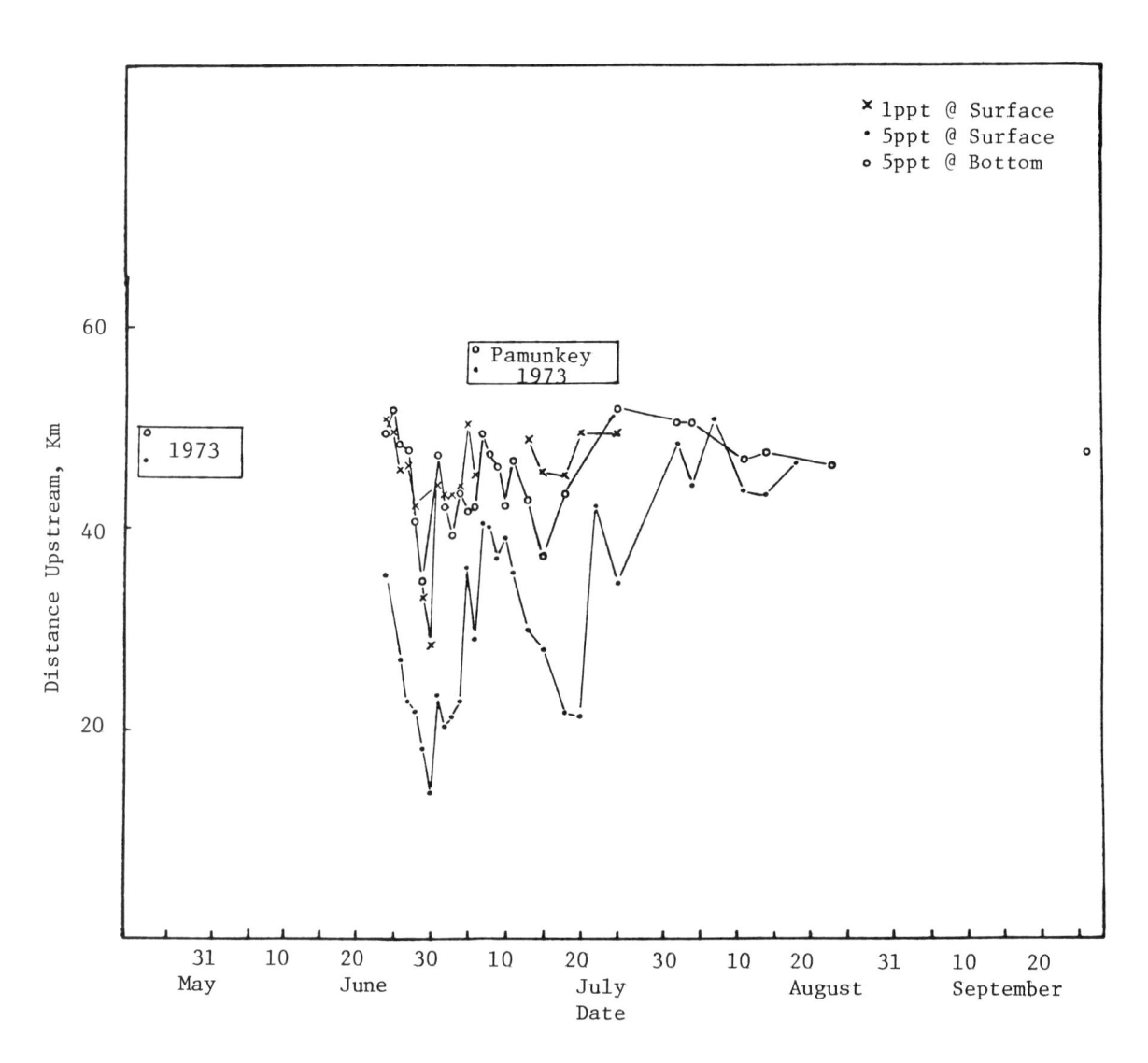

Figure 9. Saline intrusion in the York River as a function of time.

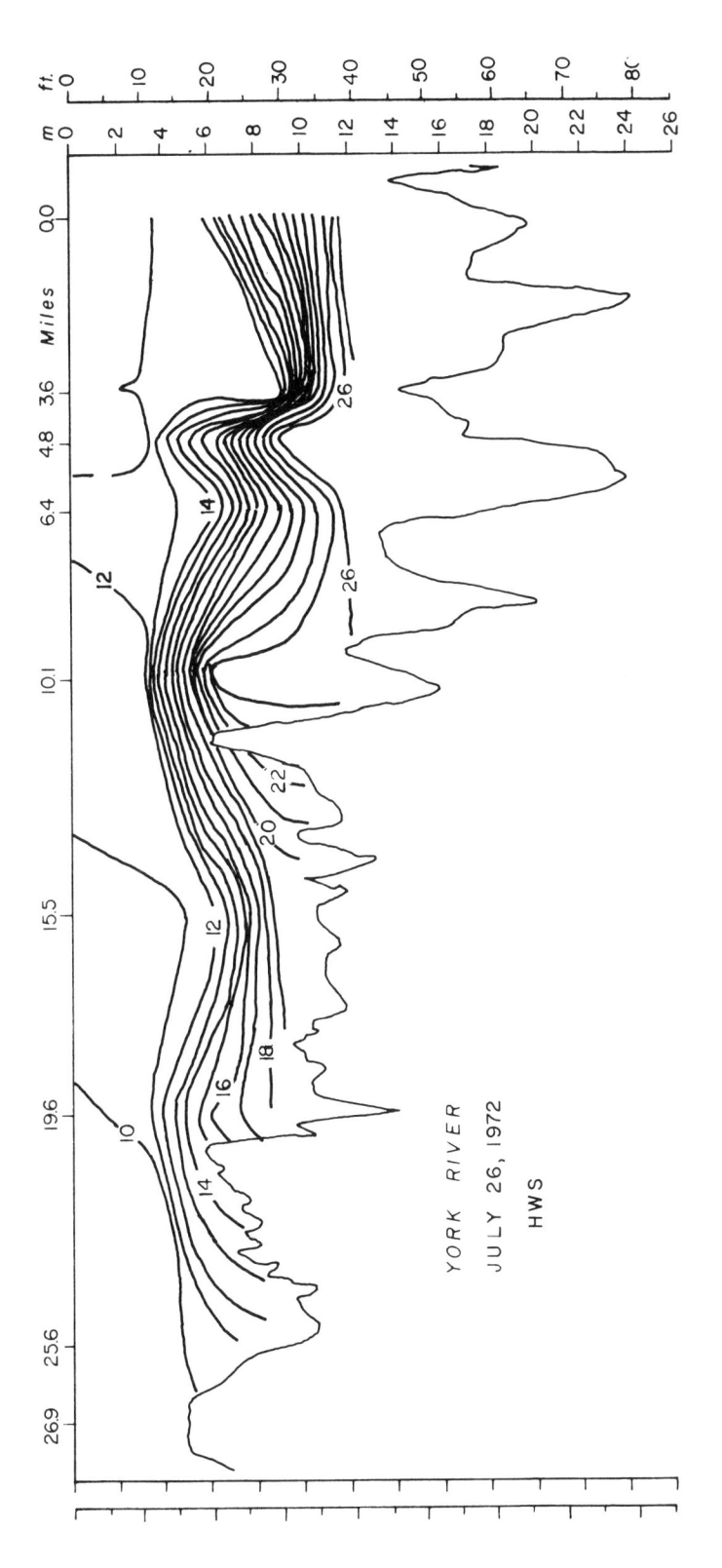

Figure 10. Salinity at high water slack, York River, July 26, 1972.

Figure 11. Salinity at low water slack, York River, July 4, 1972.

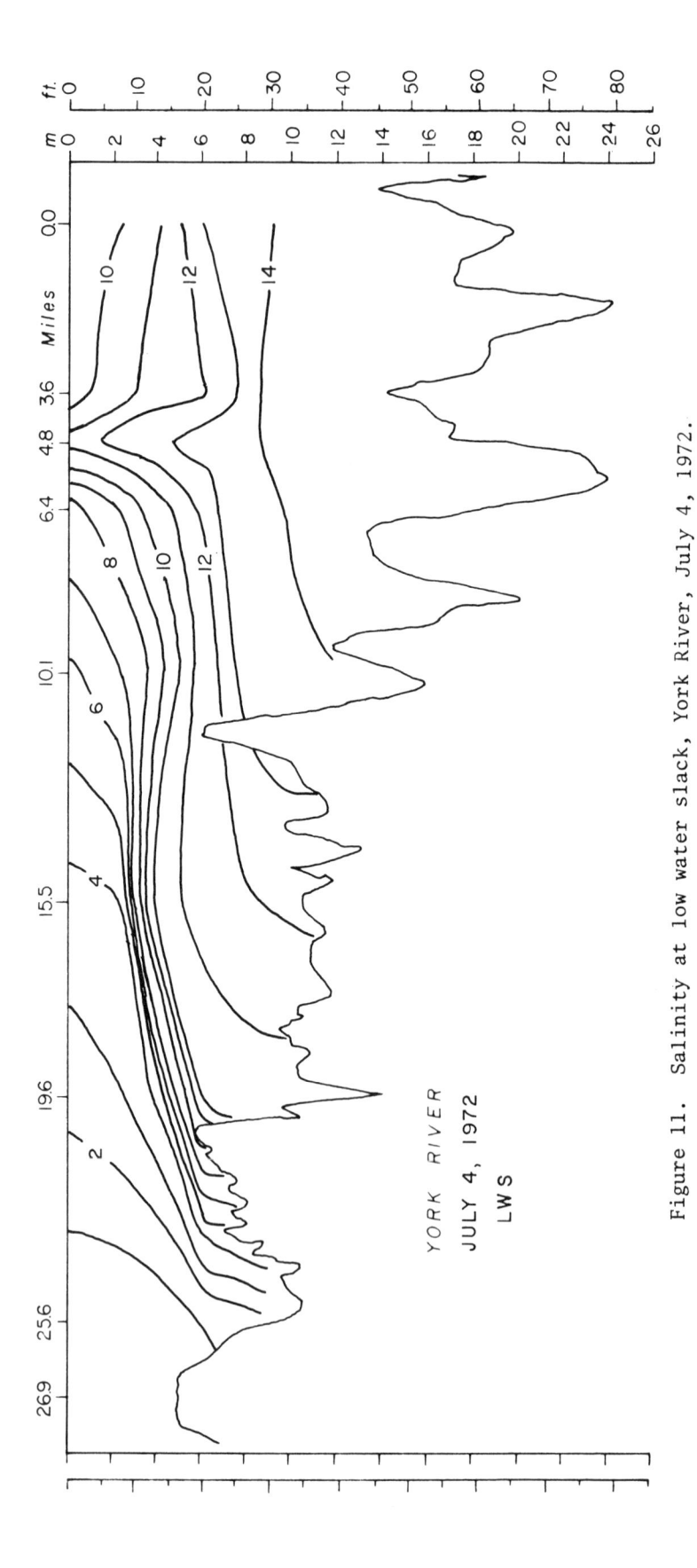

THE EFFECTS OF THE AGNES FLOOD ON THE
SALINITY STRUCTURE OF THE LOWER CHESAPEAKE BAY
AND CONTIGUOUS WATERS[1]

A. Y. Kuo[2]
E. P. Ruzecki[2]
C. S. Fang[2]

ABSTRACT

The transient response of salinity distribution in lower Chesapeake Bay to flood waters from Tropical Storm Agnes is studied in terms of a two-layered, partially mixed estuary. Prior to 30 June 1972, surface salinities were well depressed throughout the Bay while those at the bottom near the Bay mouth were not depressed by 5 July. This resulted in a highly stratified situation normally found in the spring of the year. Stratification decreased when bottom waters were flushed down-bay by the flood (on 5 to 10 July for the region south of New Point Comfort). The "rebound" of salinity structure started immediately after the passage of the flood water which otherwise retarded up-bay movement of bottom waters. This "rebound" began on 13 July near the Bay mouth and progressed up-bay reaching the mouth of the Potomac River by 20 July. During this period, surface salinity remained low, resulting in strong stratification again. The recovery of surface salinity by tidal mixing finally weakened stratification to a near "normal" salinity structure by the end of August. The large mass of flood water leaving the Bay mouth is treated as a natural tracer release. The distribution of flood water on the continental shelf indicates that pulses of freshened surface water left the Bay on ebb tide and were separated from one another by intrusion of saltier shelf water on flood tide. During the period when the wind speed was below 4 m/s, the flood water remained in the upper 10 meters of the water column and traveled southward with a speed of 80 cm/sec.

INTRODUCTION

In July and August of 1972, the two-month period after Agnes brought a record-breaking amount of rainfall to the Chesapeake Bay drainage basin, the Virginia Institute of Marine Science conducted twelve slack water runs along the axis of the lower Bay and five cruises on the continental shelf off the Bay mouth. Salinity and temperature were the two physical parameters measured in each of these field surveys. Because of the wide range of salinity variation, the temperature effect on water density was negligible in comparison with salinity effect and relative density changes reflected the salinity structure in most instances; hence, only results of salinity data are presented and discussed in this paper. The salinity structure is used to infer the path of flood water as it passed down Chesapeake Bay and out into the continental shelf.

---

[1]Contribution No. 592, Virginia Institute of Marine Science
[2]Virginia Institute of Marine Science, Gloucester Point, Va. 23062

SALINITY DISTRIBUTION IN THE LOWER CHESAPEAKE BAY

Field Surveys and Results

For the purpose of studying the effects of the Agnes flood, eleven sampling stations (Fig. 1) were set up along the axis of the lower Chesapeake Bay from the Virginia Capes to the mouth of the Potomac River. Each survey was started from the station at the Bay mouth and proceeded up-bay following the tidal wave propagation. At each of the sampling stations, data were collected every 2 to 3 meters in depth.

Samples of the results from field surveys are shown in Figs. 3-8 as the longitudinal distribution of salinity along the Bay axis. To indicate the tidal phase at which the field surveys were conducted, both the time span of field surveys and the predicted time of slack water at the mouth of Chesapeake Bay (NOAA 1972) are indicated. A period of approximately four hours is required for a tidal wave to travel from the Virginia Capes region to the mouth of the Potomac River.

Fig. 2 shows a typical July salinity distribution in lower Chesapeake Bay (Seitz 1971). Comparison between Figs. 2 and 3 shows salinities were well depressed by flood waters everywhere in the Bay except at the bottom near the Bay mouth. Fig. 3 additionally shows a compression of isohalines at the surface layer near the Bay mouth resulting in a high longitudinal salinity gradient. The magnitude of flood effect may be indicated by the displacement of a particular isohaline. In July 1968, the 10 ppt isohaline was located at 180 km (from the Bay mouth) at the surface and 250 km at the bottom, more than 100 km up-bay from its location on 5 July 1972 (Fig. 3).

Transient Response of Salinity Structure

Vertical distribution of salinity at stations B702 and B734 is shown in Fig. 9. On 5 July the salinity at station B702 was about 18 ppt throughout the upper 6 meters and increased sharply to 30 ppt at 14 meters. The structure of the lower 60 km of the Bay was therefore considered to be similar to that of a two-layered coastal plain estuary (Pritchard 1952). Between 5 and 11 July, surface salinities remained unchanged while the bottom salinities decreased from 30 ppt to 22 ppt. The resulting reduction in stratification is illustrated in Fig. 4. From 11 to 20 July, surface salinities remained little changed while bottom salinities increased from 22 ppt to 30 ppt resulting, once more, in a highly stratified water column (Figs. 5 & 6). From 20 July to 31 August, the surface salinity increased from 18 ppt to 24 ppt and was accompanied by a slight decrease in bottom salinity. This resulted in a moderately stratified water column (Figs. 7 & 8), similar to more normal conditions illustrated by Fig. 2.

Vertical distribution of salinity at other stations shows a similar two-layered response sequence. As also shown in Fig. 9, both the surface and bottom salinities had been depressed by 5 July 1972 further up the Bay (station B734). Bottom salinities then increased, and finally, surface salinities increased resulting in more normal conditions toward the end of August.

Based on the above discussion, the transient response of the salinity structure to a sudden, large injection of fresh water may be separated into four stages. Identification of the principal mechanisms which bring about each stage of response will aid in tracing the flood water down Chesapeake Bay.

To show the time sequence of salinity response, the salinities of surface and bottom layers at stations B702 and B734 were plotted as a function of time in Figs. 10 and 11 respectively. Some of the points were inferred from data collected by the Chesapeake Bay Institute and reported earlier (CBRC 1973).

The first stage of the response was the suppression of surface salinity due to the flushing and dilution of surface layer by flood water. Figs. 10 and 11 show that this first stage of the response occurred throughout the Bay before 5 July, the date of first slack water run. Results of a helicopter survey of surface salinity (Fig. 12) indicate that the surface salinity was well suppressed throughout the lower Bay by 30 June and that freshening was most prominent throughout the central portion of the Bay with pockets of more saline water bypassed, particularly in the region of Tangier and Pocomoke Sounds.

The second stage of the response was the subsequent suppression of bottom salinity, which, at station B702 was not completed until 13 July. Two possible mechanisms may be responsible for this stage of the response: flushing of the bottom layer by flood water from the upper Bay or, the dilution, by vertical mixing with the low-salinity upper layer. Most likely, vertical mixing is not the principal mechanism for the following reasons:
1) To reduce the salinity of bottom water at station B702 from 30 ppt (on 5 July) to 20 ppt (on 13 July) by mixing it with 18 ppt surface water would require replacing 83% of the bottom water with surface water. This is highly improbable because the strong vertical stratification would prevent any large amount of turbulent mixing.
2) If vertical mixing is the responsible mechanism, the decrease of bottom salinities should continue beyond 13 July when the vertical stratification is weakened. The data of station B702 show otherwise.

The third stage of the response is the rebound of salinity in the lower layer due to gravitational circulation which moved the saline ocean water up-bay in this layer. This stage of the response started at the Bay mouth and progressed up-bay. The salinity rebound commenced on 13 July at station B702, reaching the station at the mouth of the Potomac River between 20 and 25 July. This implies a velocity of gravitational circulation on the order of 10 to 17 cm/s. It should be pointed out that the 20 July field survey was conducted at slack before ebb while most of the other surveys were conducted around slack before flood. Therefore, there should be a downward adjustment of 20 July data. The rebound of bottom salinity had a tendency to overshoot to a higher than normal value before the transient response ceased. This resulted in a highly stratified estuary as shown in Figs. 5 and 6. This feature of "overshoot" was also observed in the Virginia estuarine rivers (Hyer & Ruzecki, this volume).

The fourth stage of the response was the recovery of salinities in the upper layer and re-adjustment of bottom-layer salinities by tidal mixing. This resulted in a weakly stratified estuary normally existing at the end of the summer.

FLOOD WATER ON THE CONTINENTAL SHELF

Five cruises on the continental shelf were conducted in July and August 1972. Salinity and temperature data were collected throughout the water column at each of the selected sampling stations, which varied from cruise to cruise.

To study the path of flood water on the continental shelf, the large mass of flood water leaving the Bay mouth was treated as a natural "dye release", which provided an opportunity to observe the fate of fresh water from the Bay drainage system. The "concentration" of flood water, or fraction of flood water, at any given point was calculated from salinity data with a formula used by Ketchum and Keen (1955):

$$f = \frac{S_o - S}{S_o}$$

where
 f = fraction of flood water, or concentration of flood water,
 s = salinity at the point in question
 S_o = background salinity.
Background salinity was taken as 32 ppt, allowing for some fresh water normally present in the coastal waters.

Figs. 13, 14, and 15 show the sampling stations and dates of three of the five shelf cruises. The other two cruises were conducted on 11-12 August and 29 August, with sampling stations directly due east of the Bay mouth. Contours of constant flood water concentration at the surface were also plotted in Figs. 13 and 14. These contours are equivalent to isohalines, but give a more direct indication of the degree of dilution of flood water by shelf water caused by mixing and entrainment. Figs. 13 and 14 show that the flood water generally went south along the coast after it left the Bay mouth.

The distribution of flood water in vertical cross-sections is shown in Figs. 16 to 30. Except for those of the 3-4 August cruise, vertical sections are east-west with longitude shown in the figures as abscissa. Data of the 3-4 August cruise were plotted in the vertical sections along lines A, B and C as shown in Fig. 15.

On 6-8 July, the flood water was well mixed within a 10-meter surface layer before its concentration decreased sharply with depth (Figs. 16 to 20). This situation is better demonstrated in Figs. 19 and 20, where the contours of constant flood water concentration are vertical down to the 10 meter depth, and then run horizontally. In Figs. 21 to 28 most of the constant concentration contours are horizontal which indicates strong stratification and weak vertical mixing. The 1% constant concentration contour stays at about the 10 meter level in most cases implying the flood water was largely confined in the upper 10 meters. The difference between the vertical distributions of flood water on 6-8 July and those on the other dates was attributed to the turbulent mixing induced by wind. Average wind speeds recorded at Chesapeake Light Station were 5.6 m/s for 6-7 July, 3.3 m/s for 19-20 August and 2.9 m/s for 3-4 August.

Fig. 25 shows the distribution of flood water in the vertical section along line A (Fig. 15) on 3-4 August, during which the cruise went as far south as Oregon Inlet, N. C. It is seen that there was no vertical mixing when the flood water moved south, since the 0.01, 0.05 and 0.10 contours stayed at a constant depth. Three separated regions of high flood water concentration are identifiable and are spaced at approximately 56 km intervals. These low salinity cells might result from pulses of fresher flood water leaving the Bay on successive ebb tides and were separated from one another by intrusions of saltier shelf water on flood tides. Adjusting the time lag for the research vessel cruise speed, the speed of flood water was calculated to be of the order of 80 cm/sec. Interpreting drogued buoy data for the same region, Welch (1974) has reported that the southward movement of water along the coast from the Chesapeake Bay mouth to Cape Hatteras occasionally reaches 100 cm/sec.

The maximum flood water concentration in the successive pulses was reduced from 0.60 to 0.35, and 0.30. This gives a rate concentration reduction slower than $-\frac{1}{2}$ power of distance, a characteristic closer to a 2-dimensional plane jet than to a 3-dimensional radial jet. This is supported by the fact that vertical mixing is negligible along the course of travel of the flood water. Furthermore, because of the coast line boundary on the west, large scale mixing and entrainment are active only on the ocean side.

ACKNOWLEDGEMENTS

Field work, from which the data of this paper were derived, was a coordinated effort involving several departments at VIMS. We would like to express our thanks to all personnel involved. Special thanks are due to Mr. J. Norcross for his assistance in the offshore cruises. We are also indebted to Dr. John Zeigler for his suggestion to treat the flood water as a natural tracer on the continental shelf. Analysis of data was supported in part by the Oceanography Section, National Science Foundation NSF Grant GA-35446; in part by the U. S. Army Corps of Engineers Contract #DACW-73-C-0189 and, in part, by the Commonwealth of Virginia.

LITERATURE CITED

Chesapeake Bay Research Council. 1973. The effects of Hurricane Agnes on the environment and organisms of Chesapeake Bay. A Report to the Philadelphia District U. S. Army Corps of Engrs. (Aven M. Andersen, Coordinator), Chesapeake Bay Inst. Contrib. 187, Nat. Resour. Inst. Contrib. 529, Va. Inst. Mar. Sci. Spec. Rep. Appl. Mar. Sci. & Ocean Engr. 29, 172 p.

Ketchum, B. H. and D. J. Keen. 1955. The accumulation of river water over the continental shelf between Cape Cod and Chesapeake Bay. Deep-Sea Res. 3 (Suppl.):346-357.

National Oceanic and Atmospheric Administration. 1972. Tidal current tables; Atlantic coast of North America.

Pritchard, D. W. 1952. Salinity distribution and circulation in the Chesapeake Bay estuarine system. J. Mar. Res. 11(2):106-123.

Seitz, R. C. 1971. Temperature and salinity distributions in vertical sections along the longitudinal axis and across the entrance of the Chesapeake Bay (April 1968 to March 1969). Chesapeake Bay Inst., Johns Hopkins Univ. Graph. Sum. Rep. 5, Ref. 71-7, 99 p.

Welch, C. 1974. VIMS-NASA LRC EOLE buoy program. AIAA Symp. on Free Drifting Buoys.

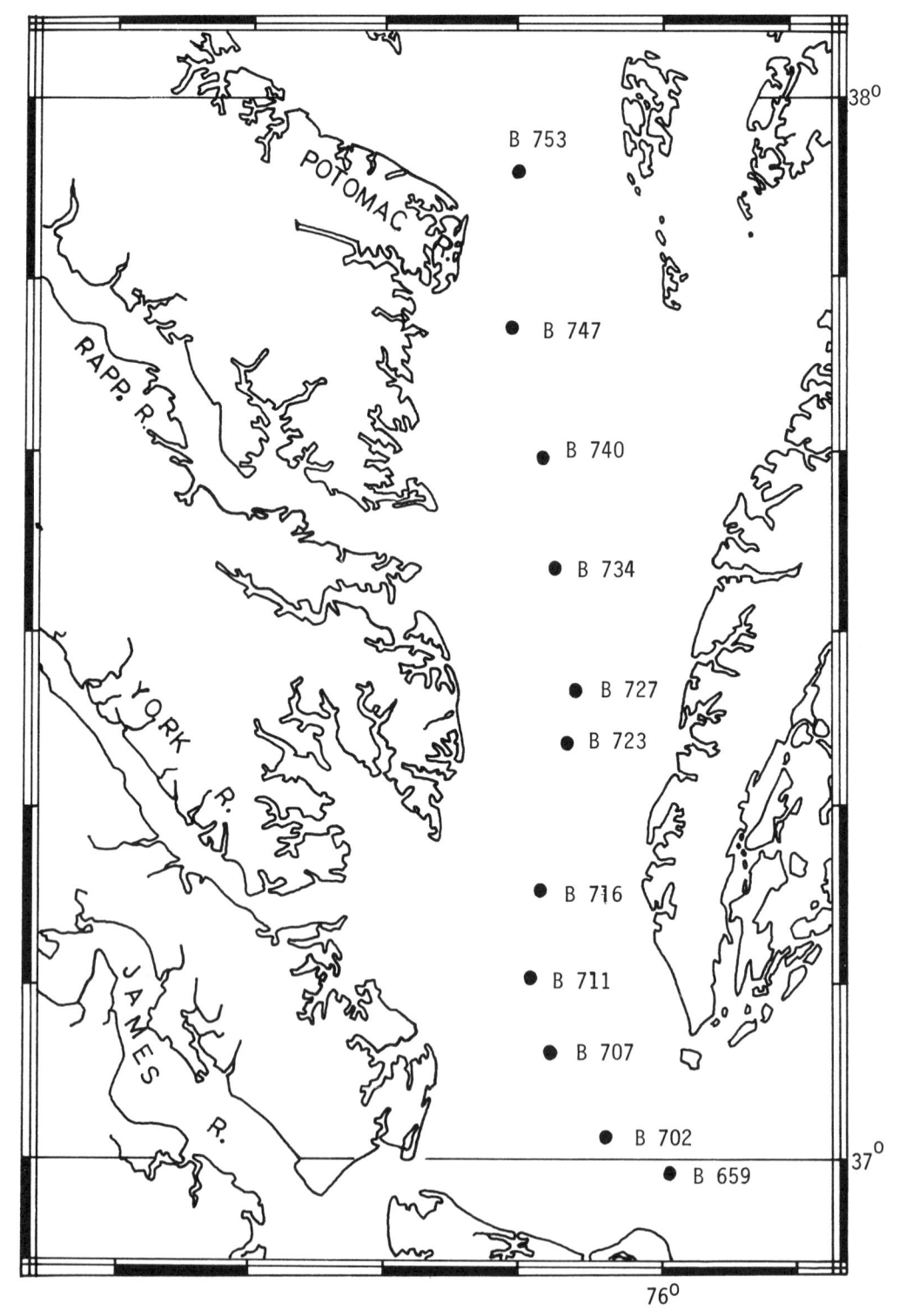

Figure 1. Map of the lower Chesapeake Bay showing the locations of sampling stations.

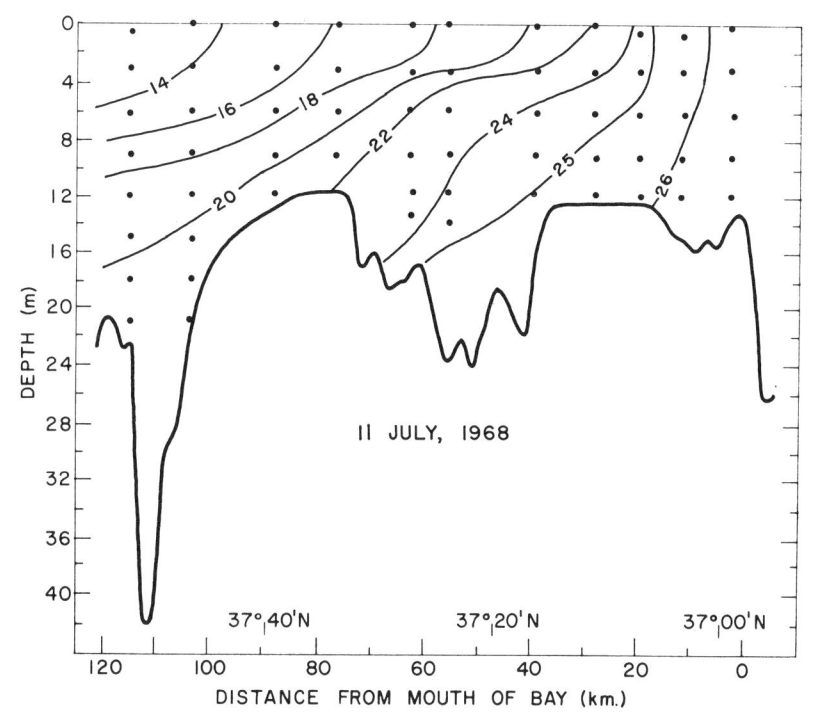

Figure 2. Longitudinal distribution of salinity along the axis of the lower Bay on 11 July 1968 (Seitz 1971).

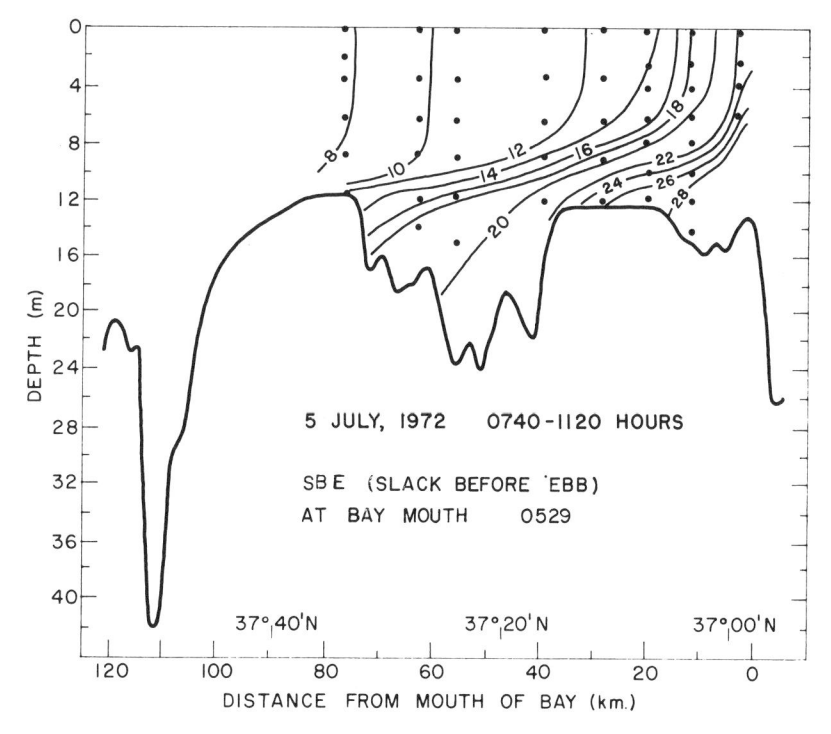

Figure 3. Longitudinal distribution of salinity along the axis of the lower Bay on 5 July 1972.

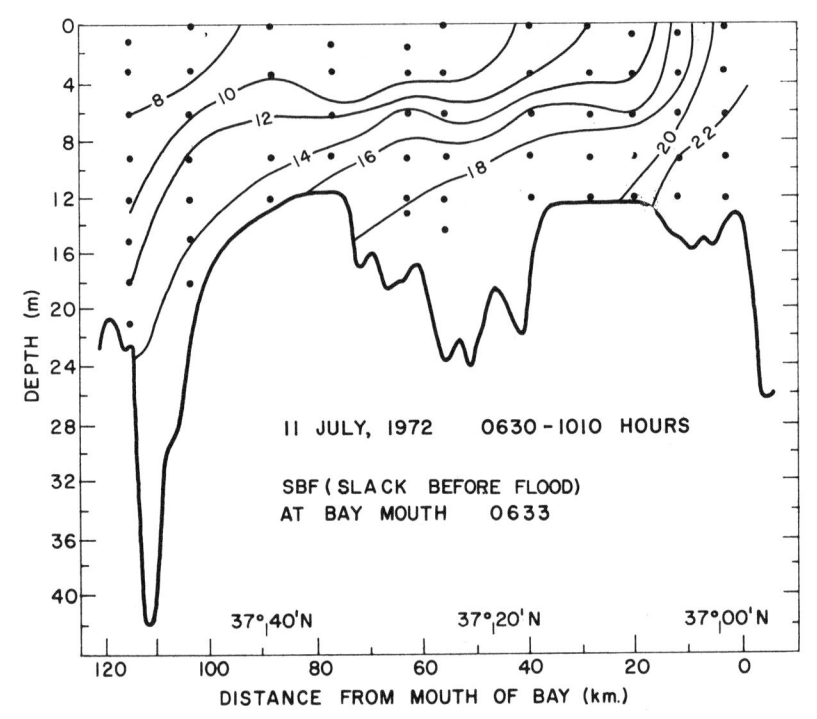

Figure 4. Longitudinal distribution of salinity along the axis of the lower Bay on 11 July 1972.

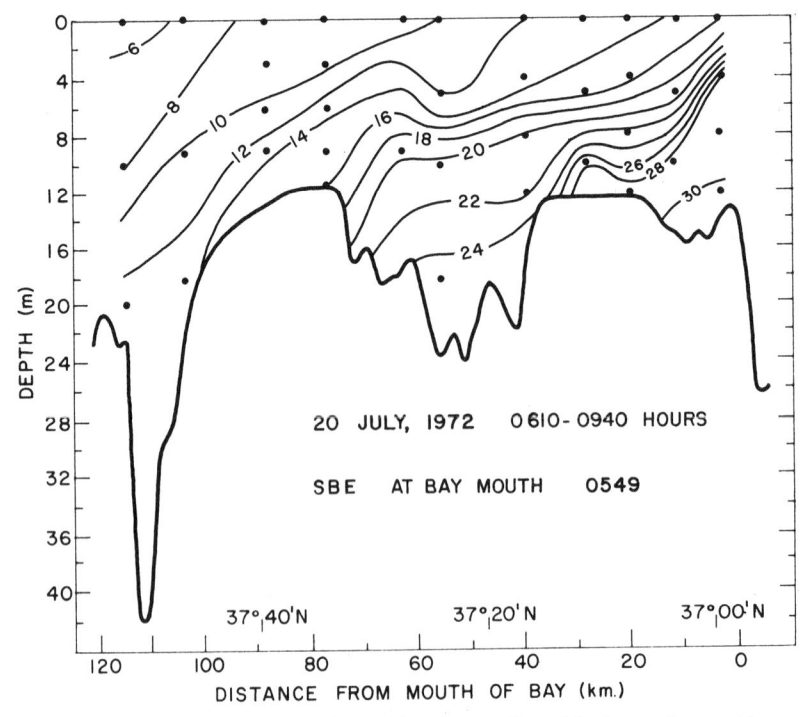

Figure 5. Longitudinal distribution of salinity along the axis of the lower Bay on 20 July 1972.

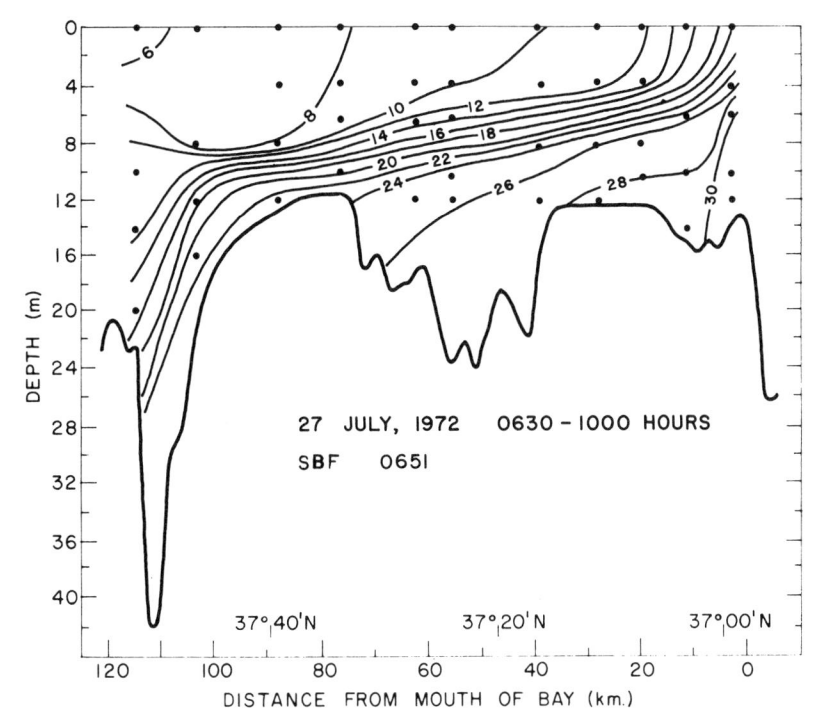

Figure 6. Longitudinal distribution of salinity along the axis of the lower Bay on 27 July 1972.

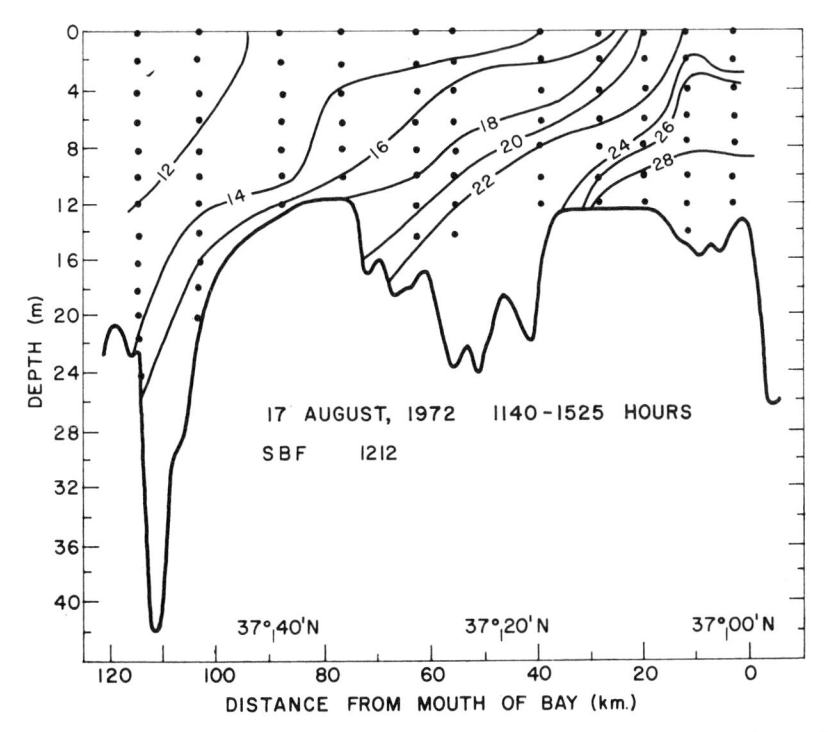

Figure 7. Longitudinal distribution of salinity along the axis of the lower Bay on 17 August 1972.

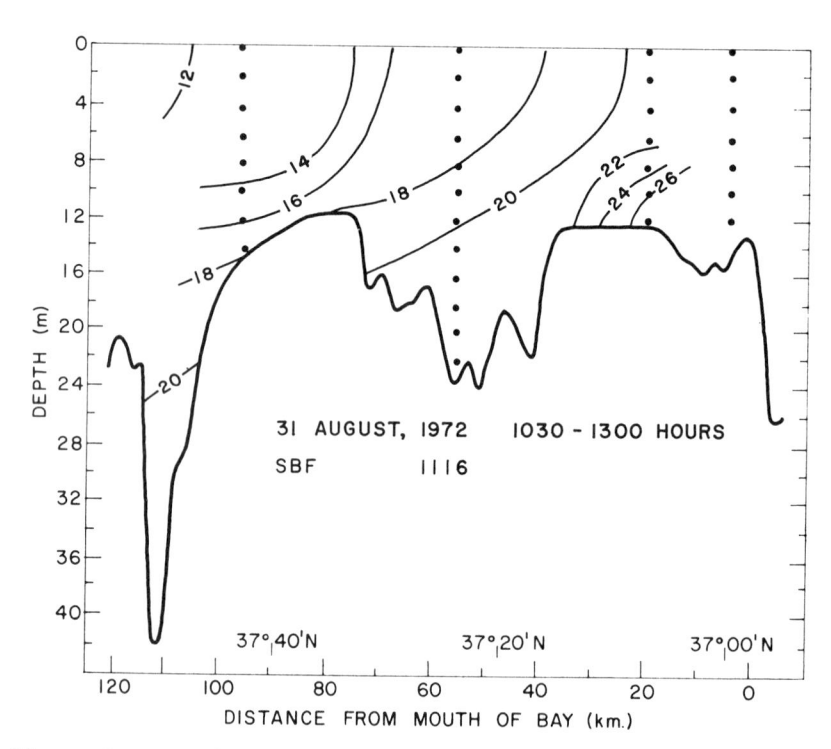

Figure 8. Longitudinal distribution of salinity along the axis of the lower Bay on 31 August 1972.

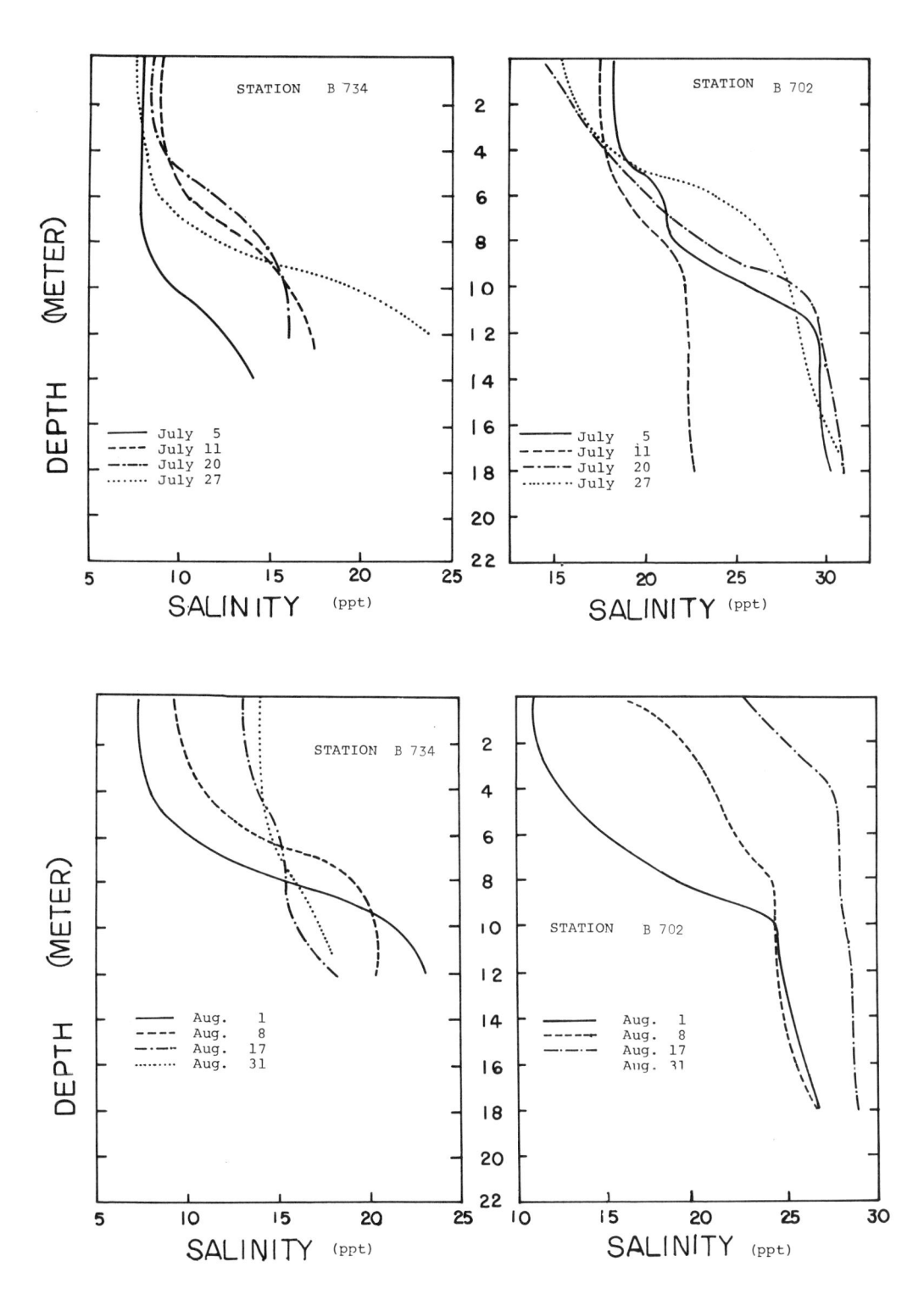

Figure 9. Vertical distributions of salinity at stations B702 and B734.

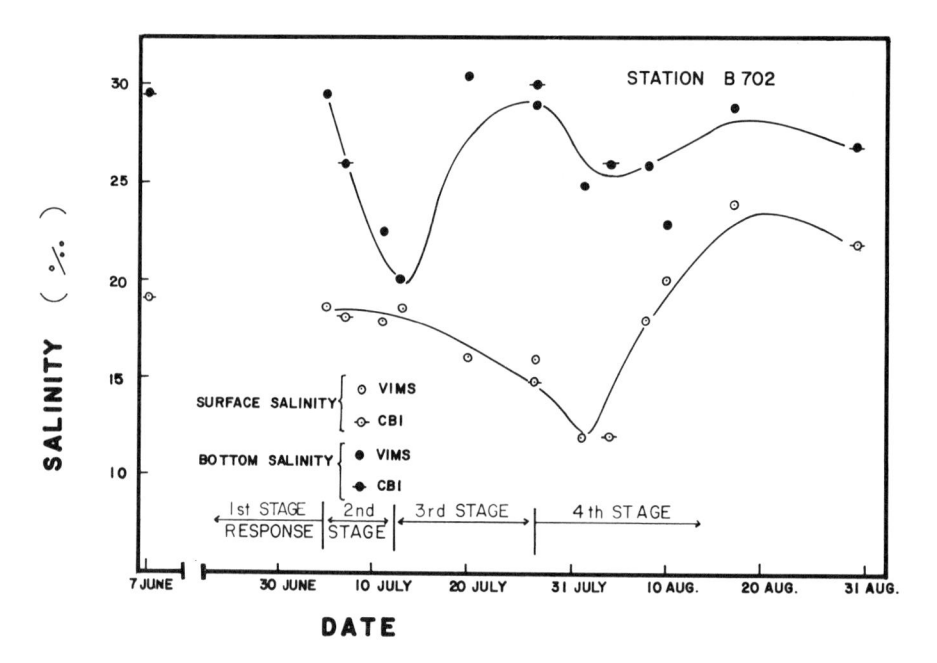

Figure 10. Variations of surface and bottom salinities with time at station B702.

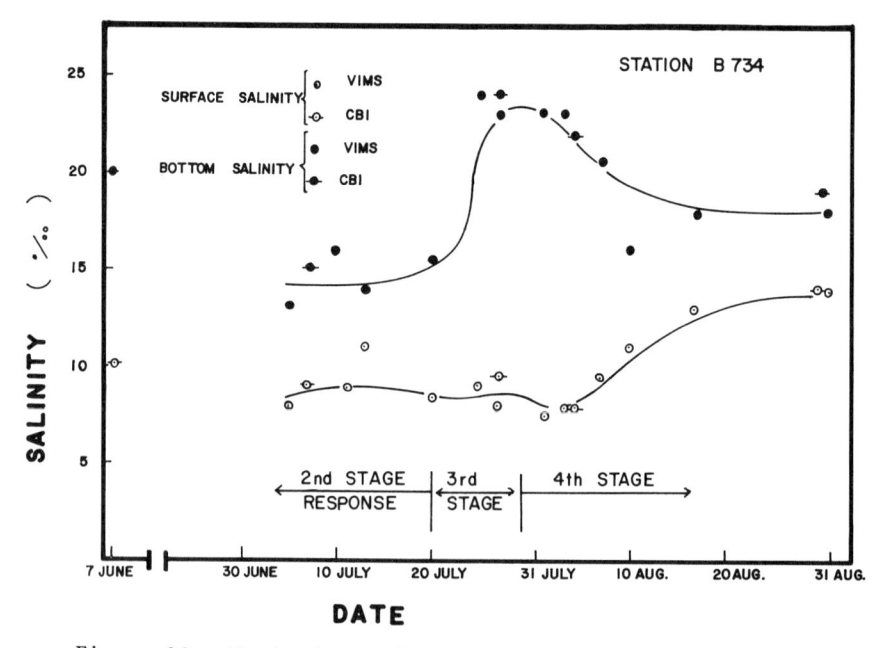

Figure 11. Variations of surface and bottom salinities with time at station B734.

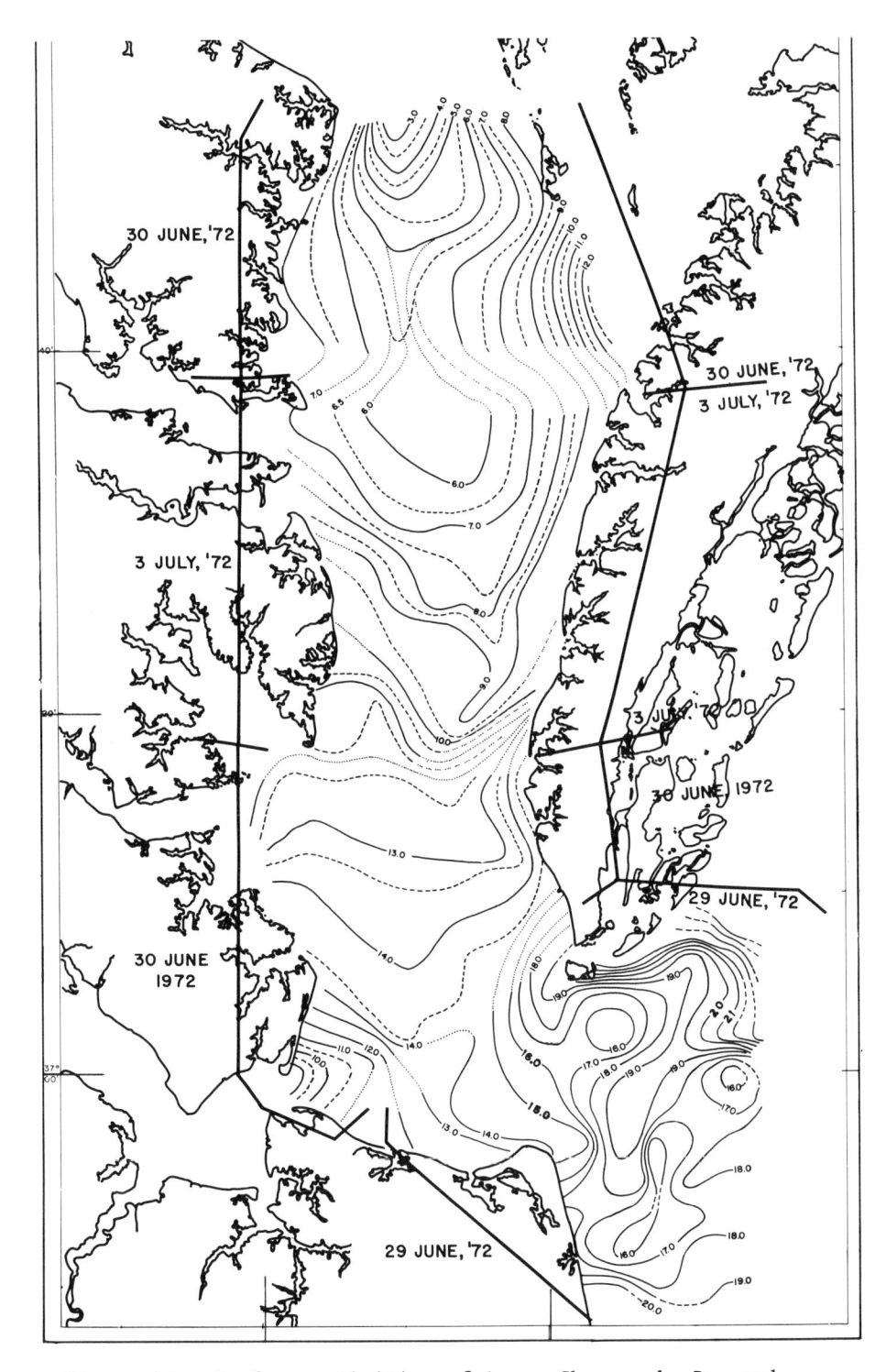

Figure 12. Surface salinities of lower Chesapeake Bay and nearshore shelf region based on samples collected on 29, 30 June and 3 July 1972.

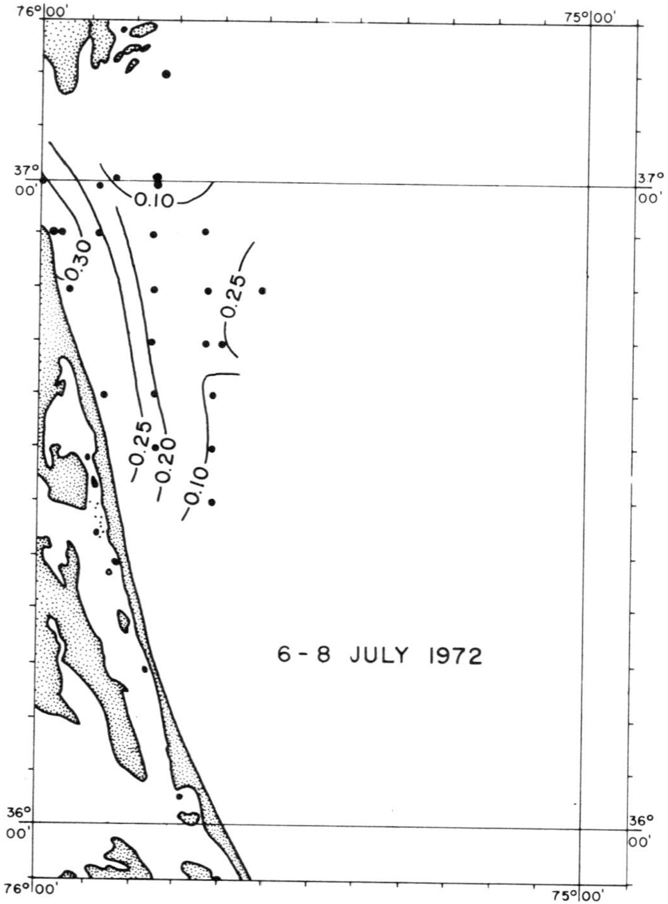

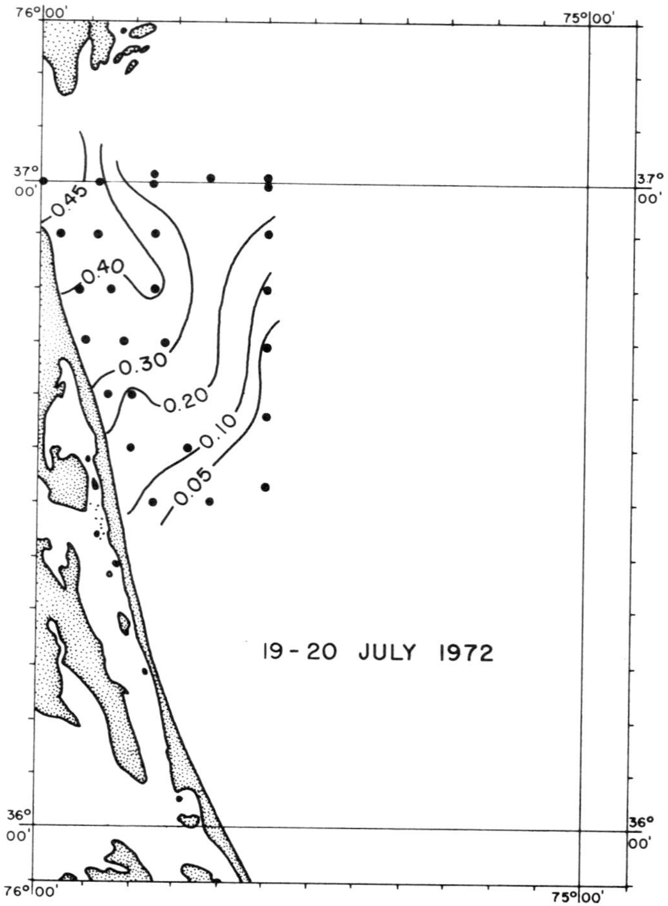

Figure 14. The concentration of flood water in the surface water overlying the continental shelf off the Bay mouth, 19-20 July 1972.

Figure 13. The concentration of flood water in the surface water overlying the continental shelf off the Bay mouth, 7-8 July 1972.

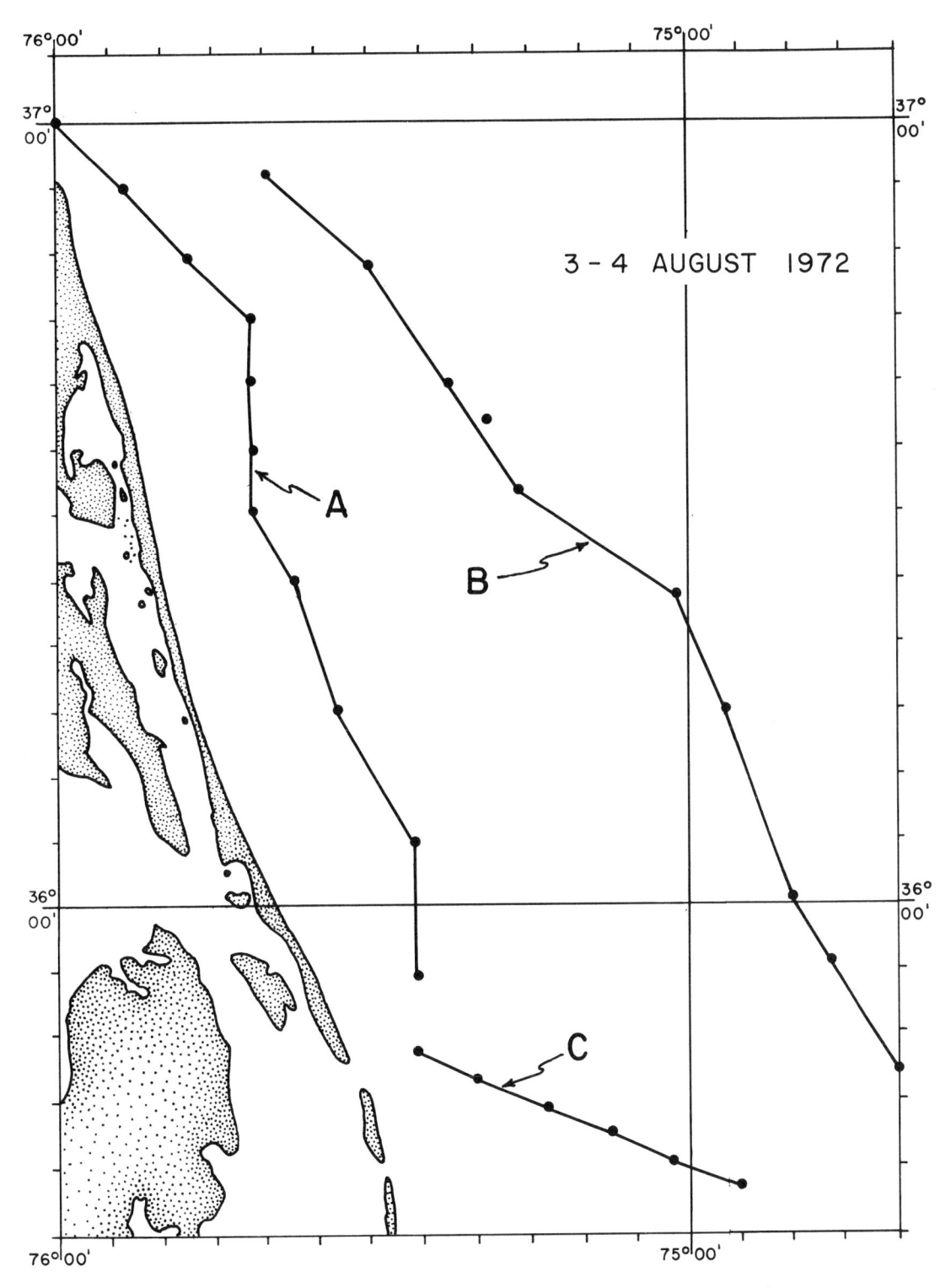

Figure 15. Map of the continental shelf showing the sampling stations of the cruise on 3-4 August 1972.

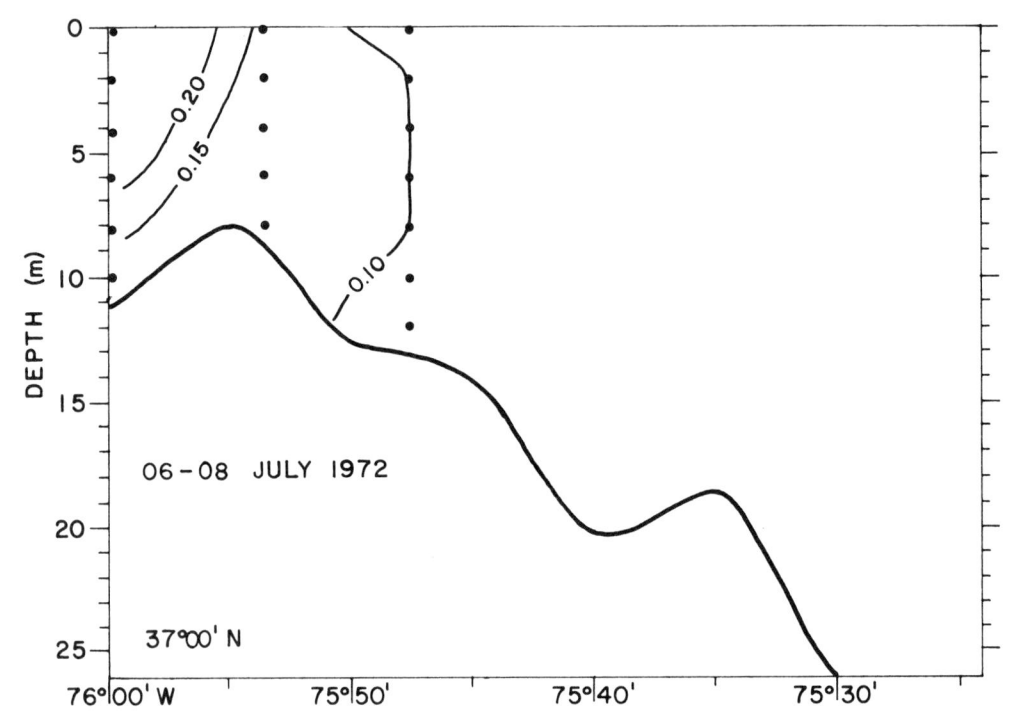

Figure 16. The concentration of flood water in the vertical section along 37°00'N on 6-8 July 1972.

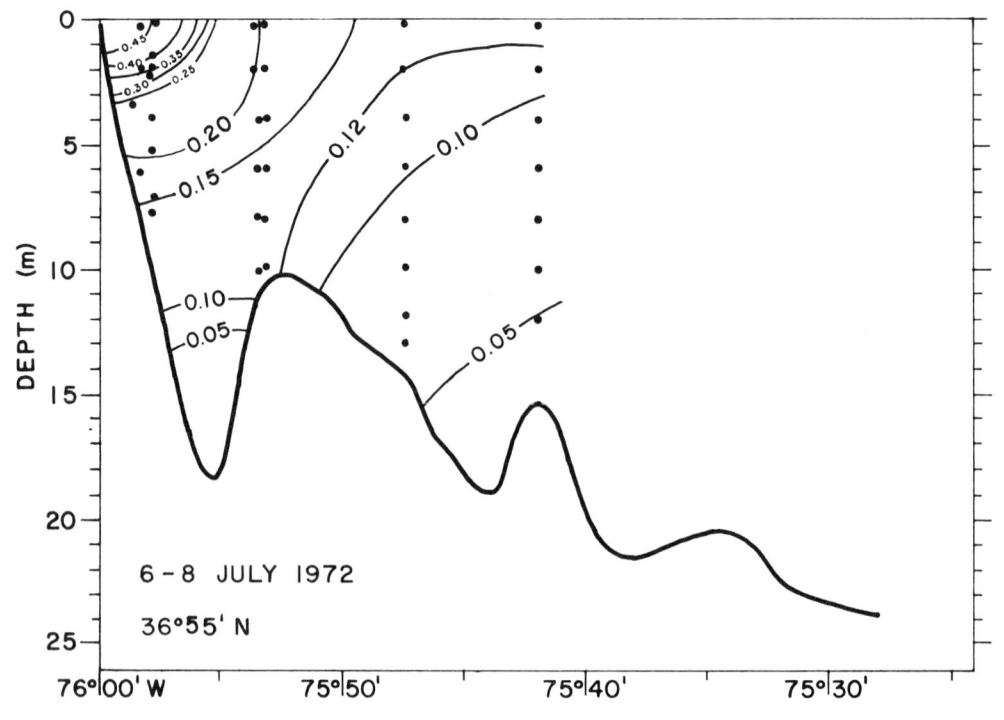

Figure 17. The concentration of flood water in the vertical section along 36°55'N on 6-8 July 1972.

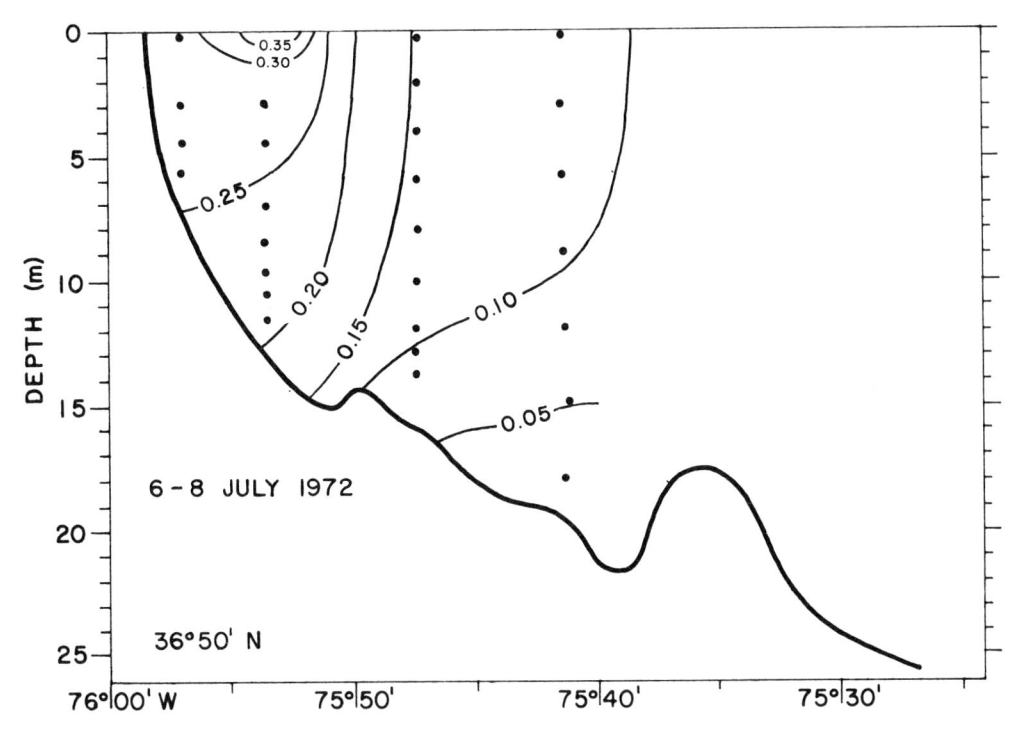

Figure 18. The concentration of flood water in the vertical section along 36°50'N on 6-8 July 1972.

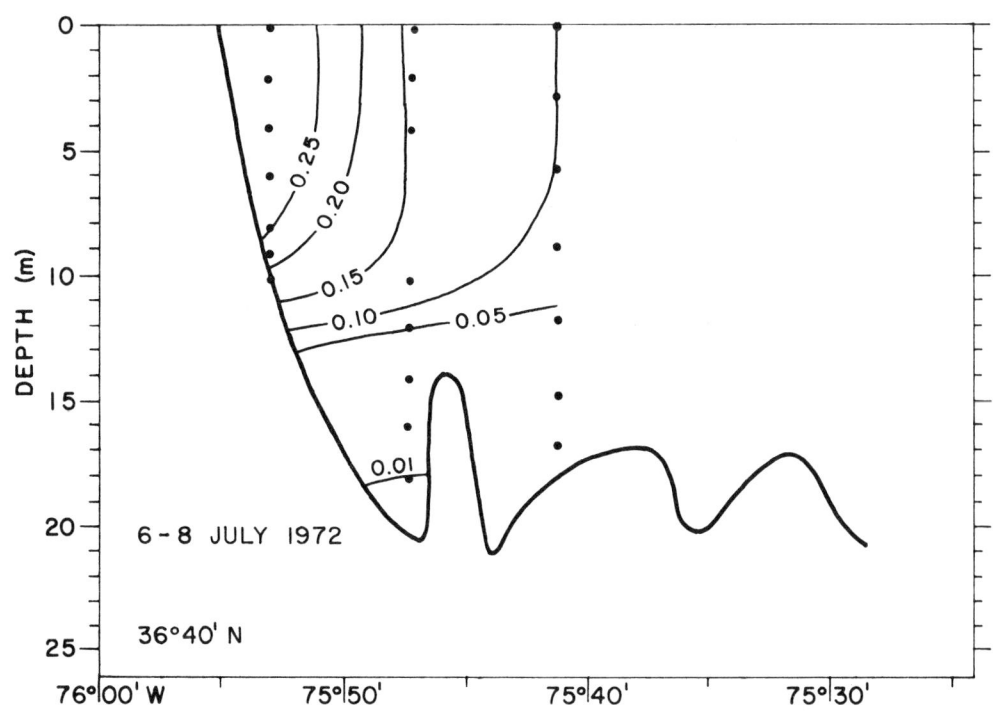

Figure 19. The concentration of flood water in the vertical section along 36°40'N on 6-8 July 1972.

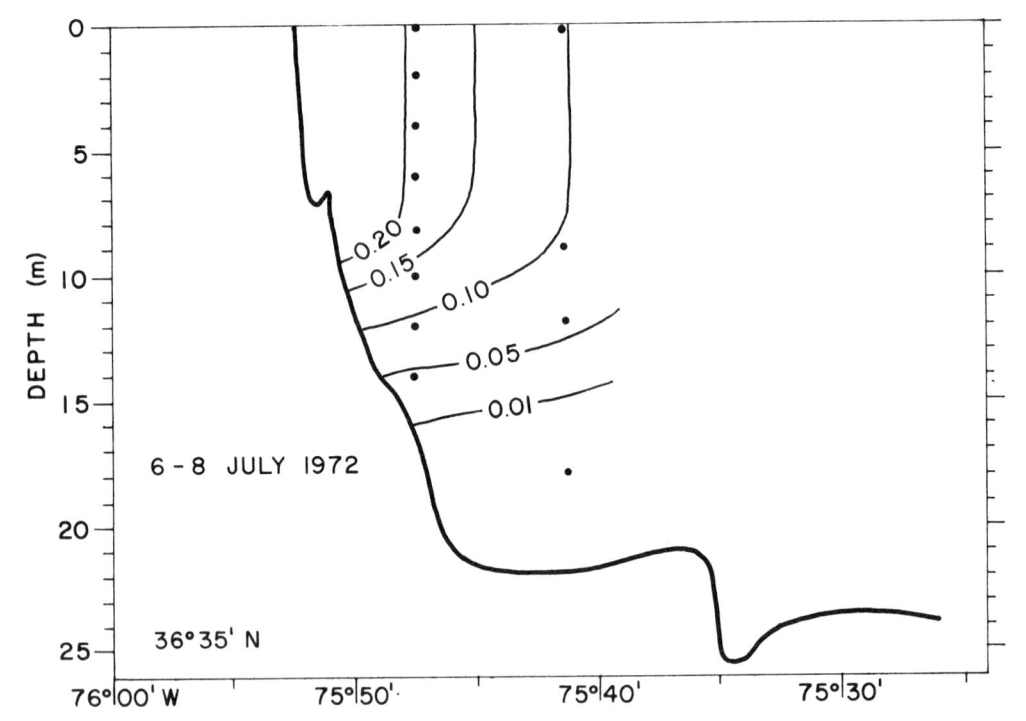

Figure 20. The concentration of flood water in the vertical section along 36°35'N on 6-8 July 1972.

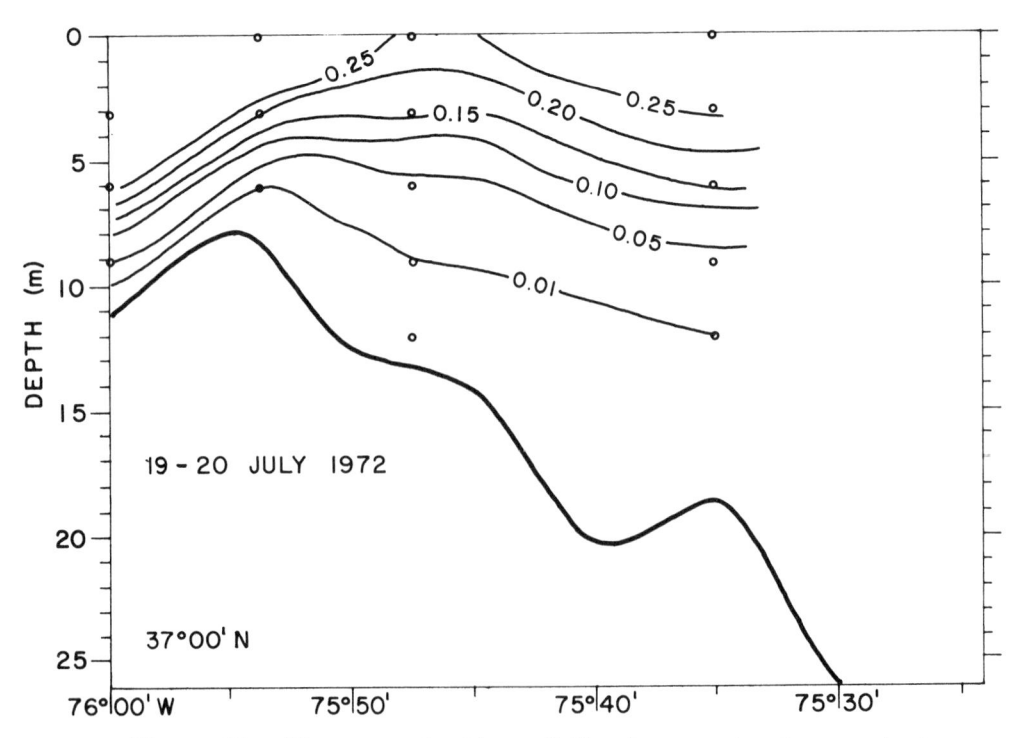

Figure 21. The concentration of flood water in the vertical section along 37°00'N on 19-20 July 1972.

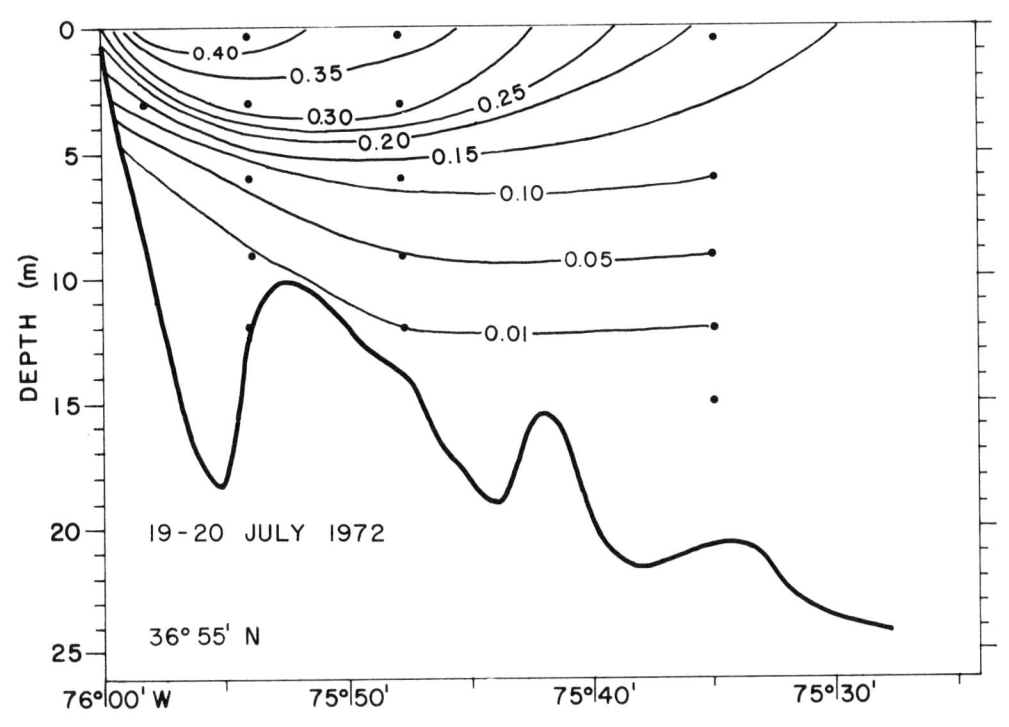

Figure 22. The concentration of flood water in the vertical section along 36°55'N on 19-20 July 1972.

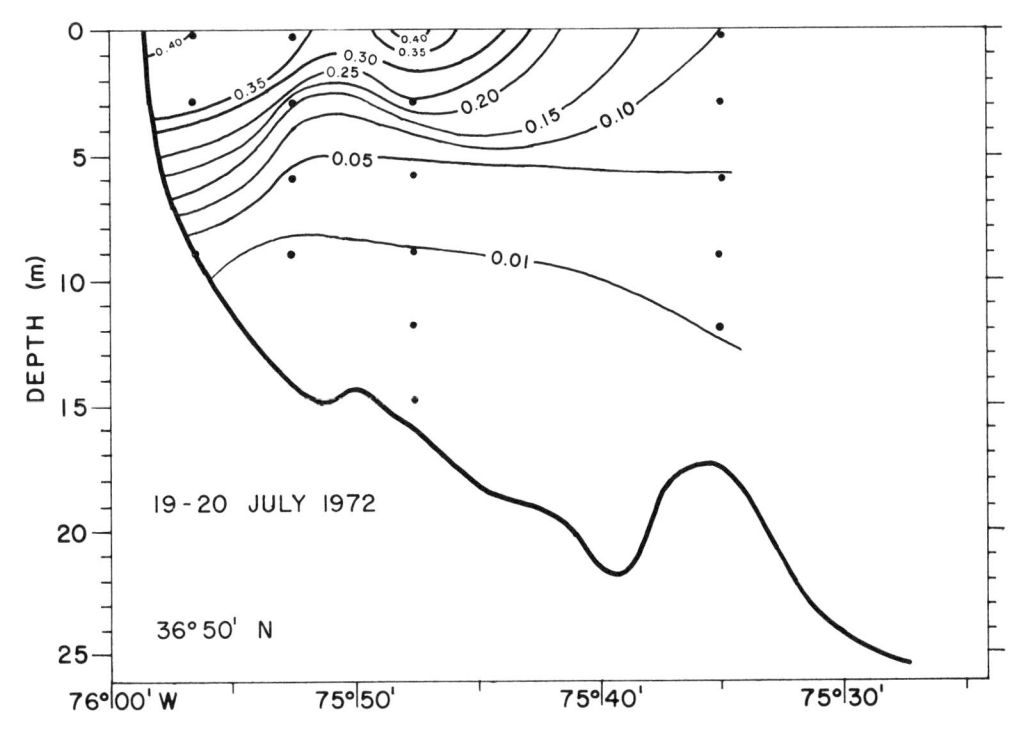

Figure 23. The concentration of flood water in the vertical section along 36°50'N on 19-20 July 1972.

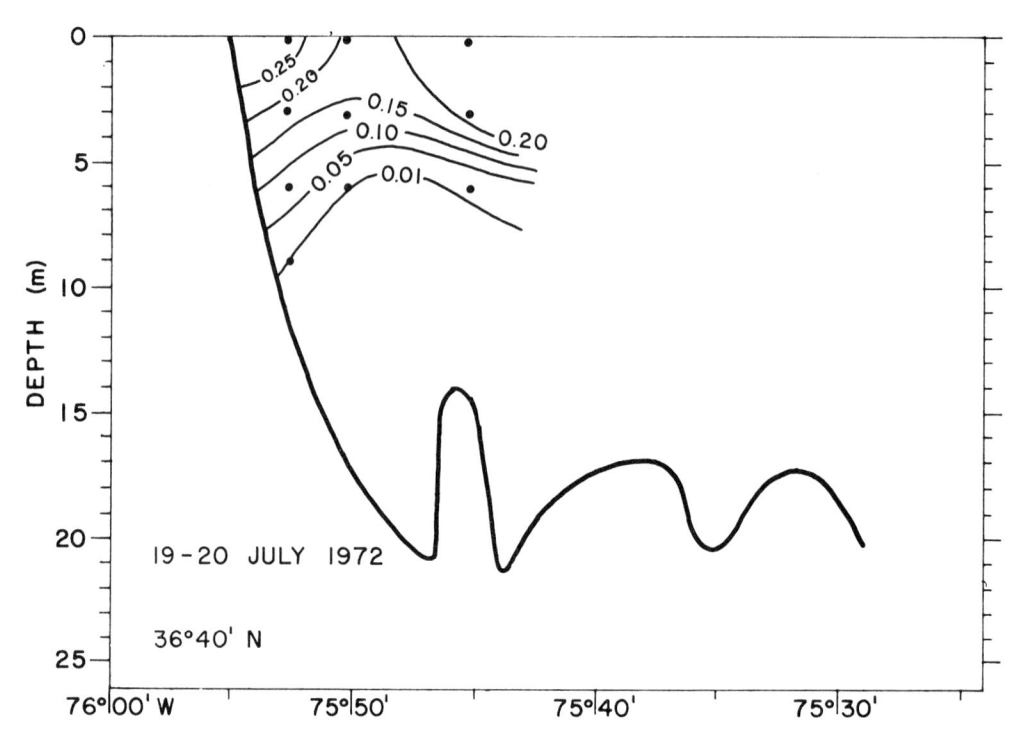

Figure 24. The concentration of flood water in the vertical section along 36°40'N on 19-20 July 1972.

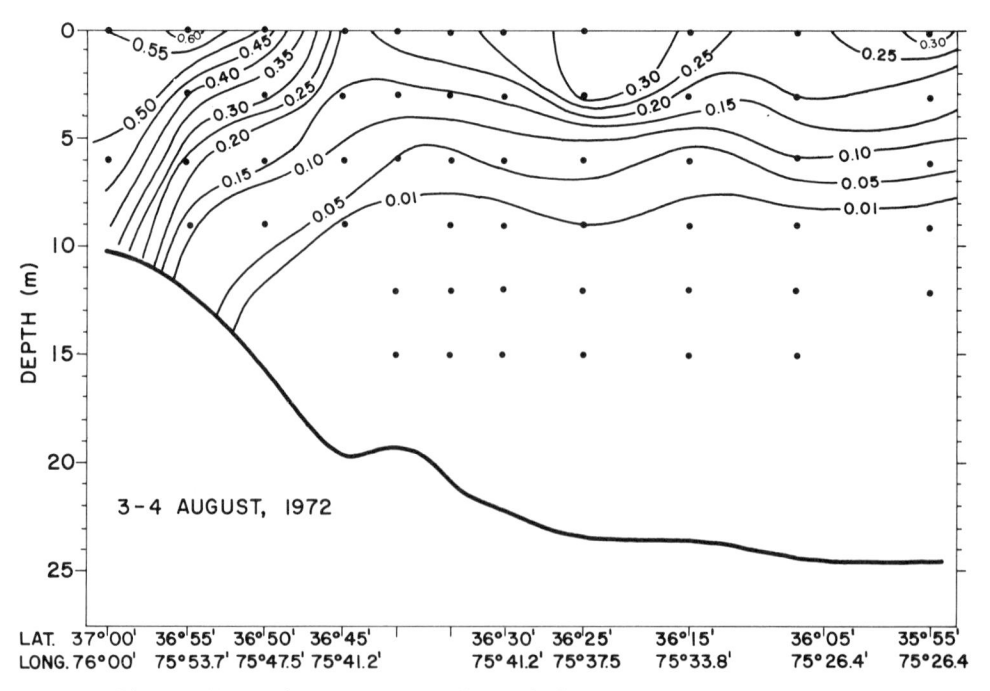

Figure 25. The concentration of flood water in the vertical section along line A of Figure 15 on 3-4 August 1972.

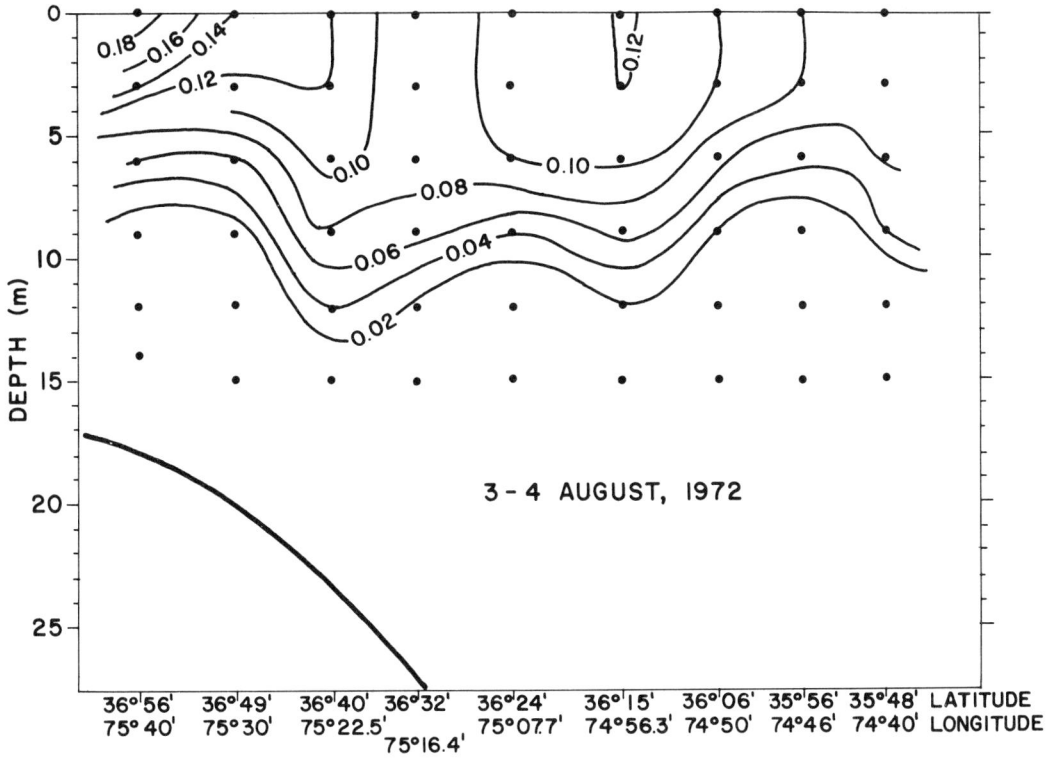

Figure 26. The concentration of flood water in the vertical section along line B of Figure 15 on 3-4 August 1972.

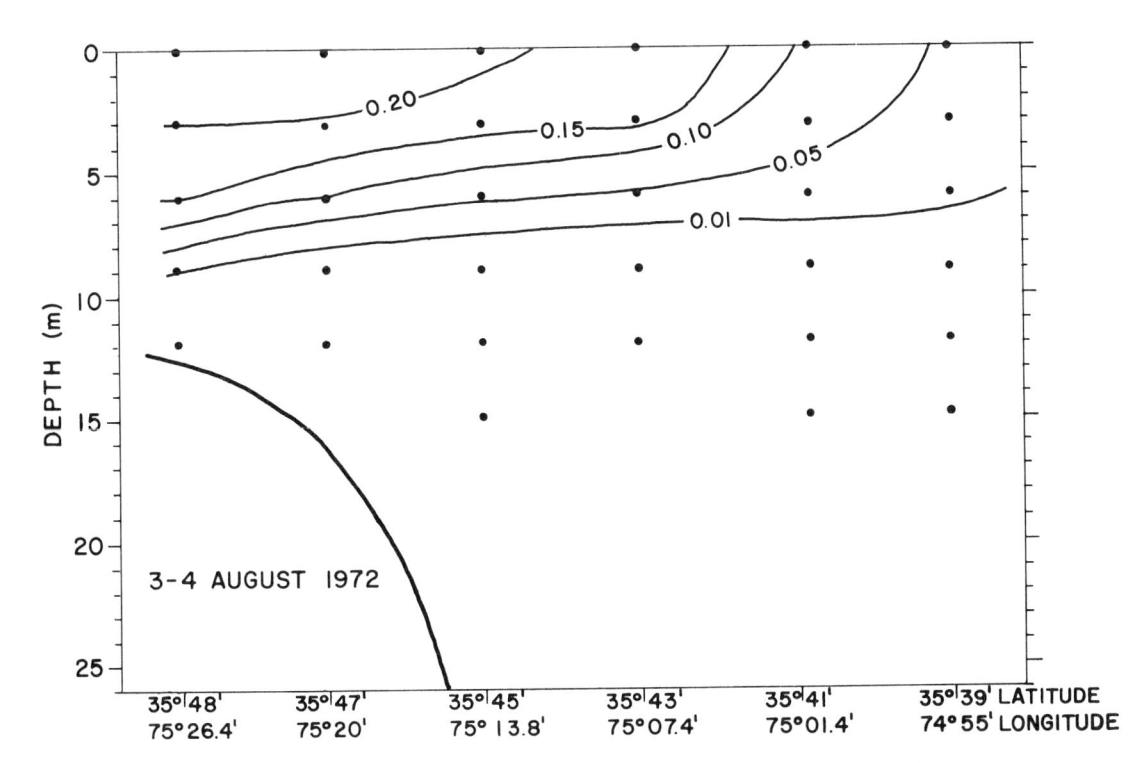

Figure 27. The concentration of flood water in the vertical section along line C of Figure 15 on 3-4 August 1972.

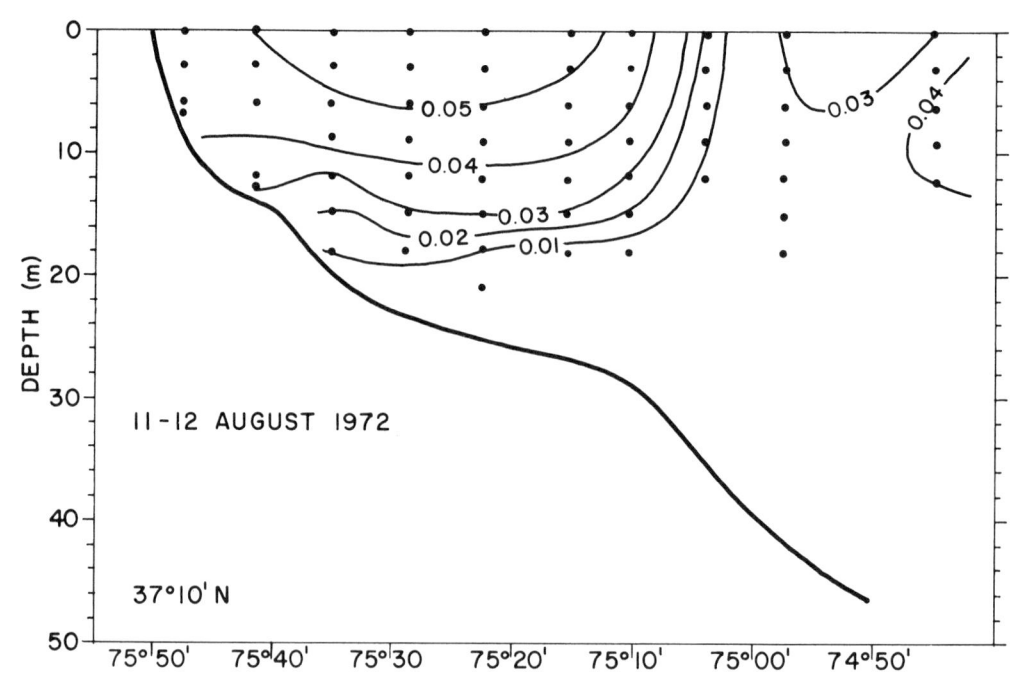

Figure 28. The concentration of flood water in the vertical section along 37°10'N on 11-12 August 1972.

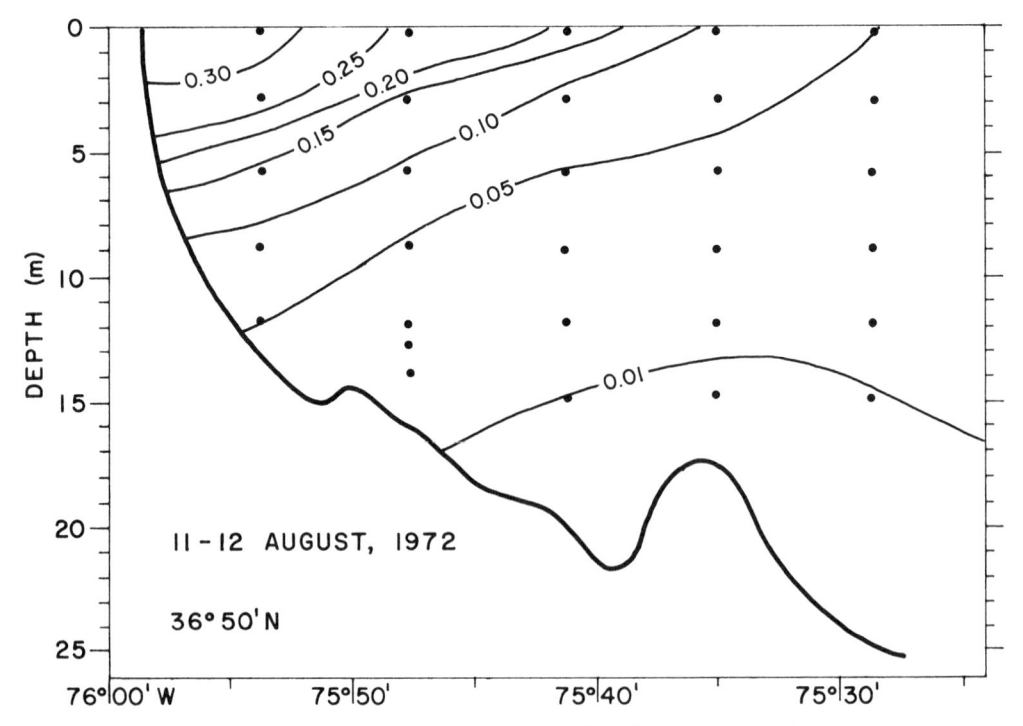

Figure 29. The concentration of flood water in the vertical section along 36°50'N on 11-12 August 1972.

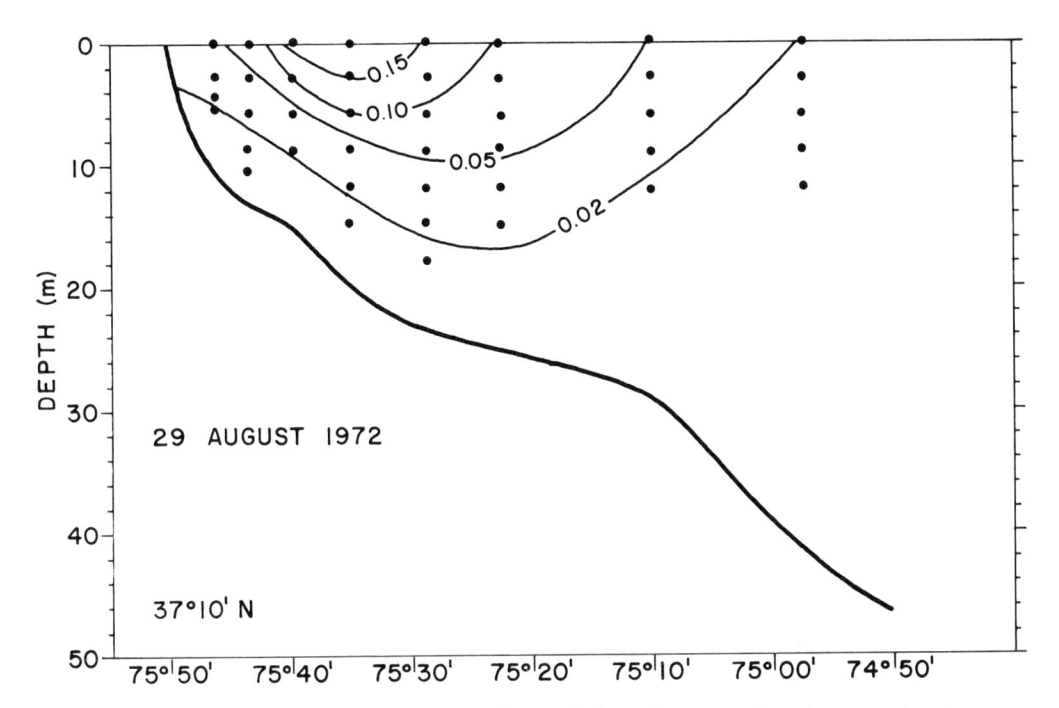

Figure 30. The concentration of flood water in the vertical section along 37°10'N on 29 August 1972.

FLOOD WAVE-TIDE WAVE INTERACTION ON THE JAMES RIVER
DURING THE AGNES FLOOD[1]

John P. Jacobson[2]
C. S. Fang[2]

ABSTRACT

During the Agnes flood hourly tidal height data were collected at seven locations along the tidal James River and currents were measured at two transects in the lower James. A comparison between actual tides and currents and the predicted tidal features as given by the tide and tidal current tables of NOAA was made. Results of this comparison show that Agnes did significantly affect water levels in the upper portion of the tidal James, especially near Richmond. However in the lower portion of the James no discernible rise was evident due to the passage of the flood crest. A small storm surge (<2 feet) was noted on the day of the passage of Agnes, 21 June, throughout the tidal James. A phase shift in times of high and low water due to the interaction of the two wave systems was not observed. In the freshwater portion of the tidal James currents continually ebbed during the passage of the flood crest. In the saline portion of the system, the flood effect on the currents was limited to the surface portion of the channel.

INTRODUCTION

The James River, the largest river in Virginia, is approximately 400 miles in length with a drainage area of over 10,000 square miles. It is tidal from its mouth at Hampton Roads to Richmond, a distance of about 100 miles with an average tidal range of 3 feet. The tides at Richmond lag those at the mouth by 5 hours, the time it takes the shallow water wave to travel up the river. The average depth of the tidal portion of the James is approximately 20 feet. The average saltwater intrusion reaches to Jamestown Island, 40 miles upstream. The flood wave generated by Agnes also was a shallow water wave. The interaction of these two wave systems, the tide wave traveling upstream and the flood wave moving downstream, is studied using tide gage data, current meter data and the tide and tidal current prediction tables of NOAA.

METHODS

During the Agnes flood, 7 tide gages on the James remained operational. These data were provided to the Virginia Institute of Marine Science (VIMS) by the Baltimore District Corps of Engineers. In addition, current velocity and direction were measured at two transects in the lower James using savonius rotor current meters (Fig. 1). These meters were placed in the river on 24 June 1972 and remained operational for over one month. By comparing the recorded data

---

[1]Contribution No. 757, Virginia Institute of Marine Science
[2]Virginia Institute of Marine Science, Gloucester Point, Va. 23062

against the predicted tides and currents from the NOAA tide tables, the wave interaction was determined.

WATER LEVELS AT RICHMOND

Records available at Richmond indicate that the James started rising on the afternoon of 21 June 1972 (Fig. 2). The observed crest was 36.5 feet, 27.5 feet above flood stage, which occurred at 4PM on 23 June 1972. By 27 June the river at Richmond had fallen to normal water levels. The crest travelled downstream as a kinematic wave at a speed of somewhat less than 5 MPH. Thus the maximum effect of the flood wave on the tidal James should be seen during the period 23-25 June. Flows on the tributaries to the James below Richmond were less than 1/10 those at Richmond and are therefore neglected in the analysis.

WATER LEVELS

At Hopewell, river mile 76, the tide gage record (Fig. 3) shows two distinct peaks in the difference between the predicted and actual tide, one on 21 June when Agnes passed and the other on 24 June due to the passage of the flood wave. The peak on the 21st is probably due to the combined effects of low barometric pressure, wind and local rainfall and runoff causing a small storm surge (<2 feet) which was observed throughout the tidal portion of the river. The major feature, on 24 June, due to the flood, shows large oscillations (>3 feet) in the difference between actual and predicted tides. Looking at the actual water surface there is very little tidal fluctuation (<0.5 foot) whereas the tide range is usually about 3 feet. The flood wave is interacting with the tide wave almost totally washing out the tidal oscillation. Thus when a difference is taken between the observed and predicted tidal heights the normal tidal oscillation is misleadingly emphasized. The maximum difference of 5 feet was observed on 24 June. A phase shift due to the interaction of the wave systems was not obvious.

Further downstream, at Wilcox Wharf, (Fig. 4) river mile 69, a similar situation exists. Two peaks are again observed, on the 21st and on the 24th. However the anomaly on the 24th is less severe than at Hopewell, approximately 3 feet, due to the larger cross-sectional area and thus a larger volume in this segment of river to accept the flood waters. No phase shift in time of high and low water was noted.

At Claremont, (Fig. 5) river mile 52, two peaks are still discernible, but now they are equal in height, approximately one foot. No phase shift occurred. At Scotland, (Fig. 6) by Jamestown Island at river mile 42, two distinct peaks of about one foot remain discernible and again occur on the 21st and 24th. No phase shift was detectable at this location. This tide gage is located near the normal limit of saltwater intrusion. However, the flood waters of Agnes moved this intrusion limit 20 miles downstream. Current measurements were taken near this point which will be commented on later.

The flood crest was not discernible from tide records at the lower 3 stations, Holliday Point (Fig. 7) on the Nansemond River (river mile 10) Sewells Point (Fig. 8, river mile 3), and Old Point Comfort (Fig. 9) at the mouth. The storm surge on the 21st is again noticeable as it was throughout the tidal portion of the James. Thus a rare hydrological event such as the flood of Agnes, the worst flood on record at Richmond, had no appreciable effect on the water surface elevation of the lower tidal James. It appears that water levels in the lower tidal James are much more affected by storm surge associated with hurricanes than the flood water that may be released.

CURRENTS

Current meters were set by VIMS at two transects in the James on 24 June 1972 shortly after Agnes left the area. One transect was located at Jamestown Island, river mile 40, and the other was off Newport News just below the James River Bridge at river mile 11. The current meters were set just prior to the passage of the Agnes flood crest.

At the Jamestown transect two current meter stations were occupied. On the south side, in the flats part of the river, flood current was recorded for only one hour during each of the first two tidal cycles after emplacement (top, Fig. 10). This is due to the passage of the flood crest. On the afternoon of 25 June the current returned to a more normal duration of ebb and flood. A phase shift may be noticed here but this probably is due to increased friction and decreased wave speed in the shallows rather than wave interaction. In the channel (bottom, Fig. 10), the current at all depths ebbed continuously for the one full tidal cycle recorded on the 24th. On 25 June the current returns to near predicted values. Thus, at this freshwater transect currents due to the flood wave were large enough to keep the tide from flooding over the whole cross-section for at least one full tidal cycle. The effect of the flood wave is evident longer in the shallows due to a slower time of travel caused by the lesser depth. No phase shift is evident in the main channel.

Near the mouth of the James, a different situation prevailed. In the flats area on the south side of the river, the predicted and recorded currents coincided closely (top, Fig. 11). This is despite the fact that salinity recorded in conjunction with the current measurements indicate that the water is less than 2 ppt (usually greater than 15 ppt) at that station during 24-26 June. In the main channel, near the surface (3 feet), the tide does not flood for the first two tidal cycles recorded, on the 24th and 25th (bottom, Fig. 11). However at the lower depths, 15 feet and 26 feet, the tide floods and ebbs as predicted. Thus in this saline portion of the river, the influence of the flood wave is found mainly in the channel and only in the essentially fresh surface layer that is overriding the salt water. Currents return quickly to correspond with predictions after the passage of the flood. No phase shift was evident.

SUMMARY AND CONCLUSIONS

Agnes did significantly affect water levels in the upper portion of the tidal James, especially near Richmond. In the lower portion of the James no discernible rise was evident due to the passage of the flood wave. A small storm surge (<2 feet) was evident throughout the tidal James on 21 June, the day Agnes passed through this area. A phase shift due to the interaction of the two wave systems was not obvious, probably because the frequencies of the two wave systems were significantly different. In the freshwater portion of the tidal James, currents continually ebbed during passage of the flood crest. In the saline portion the flood effect on the currents was limited to the surface portion of the channel with the fresh water essentially overriding the salty water.

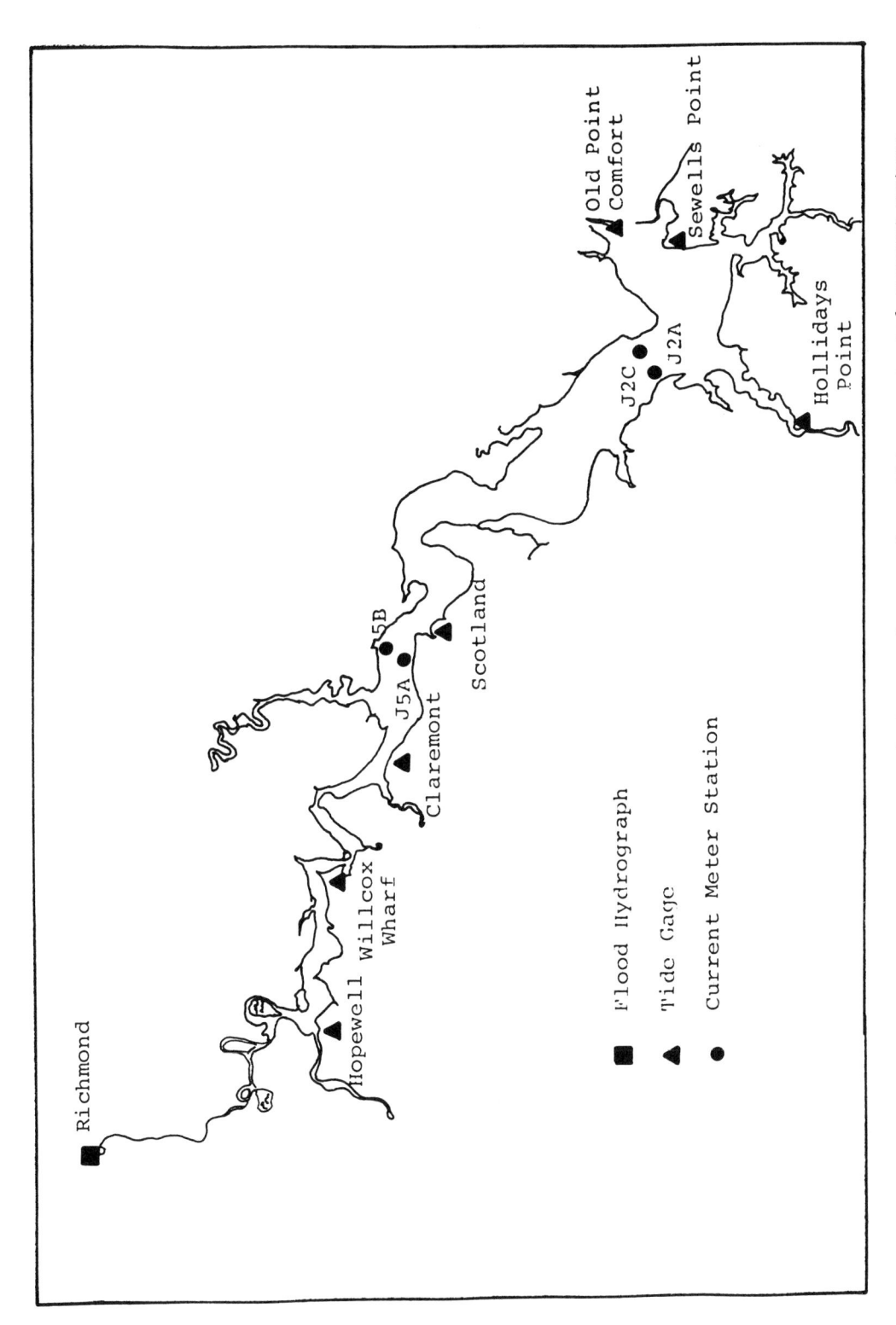

Figure 1. James River showing locations of the gauging station, tide gages, and current meters.

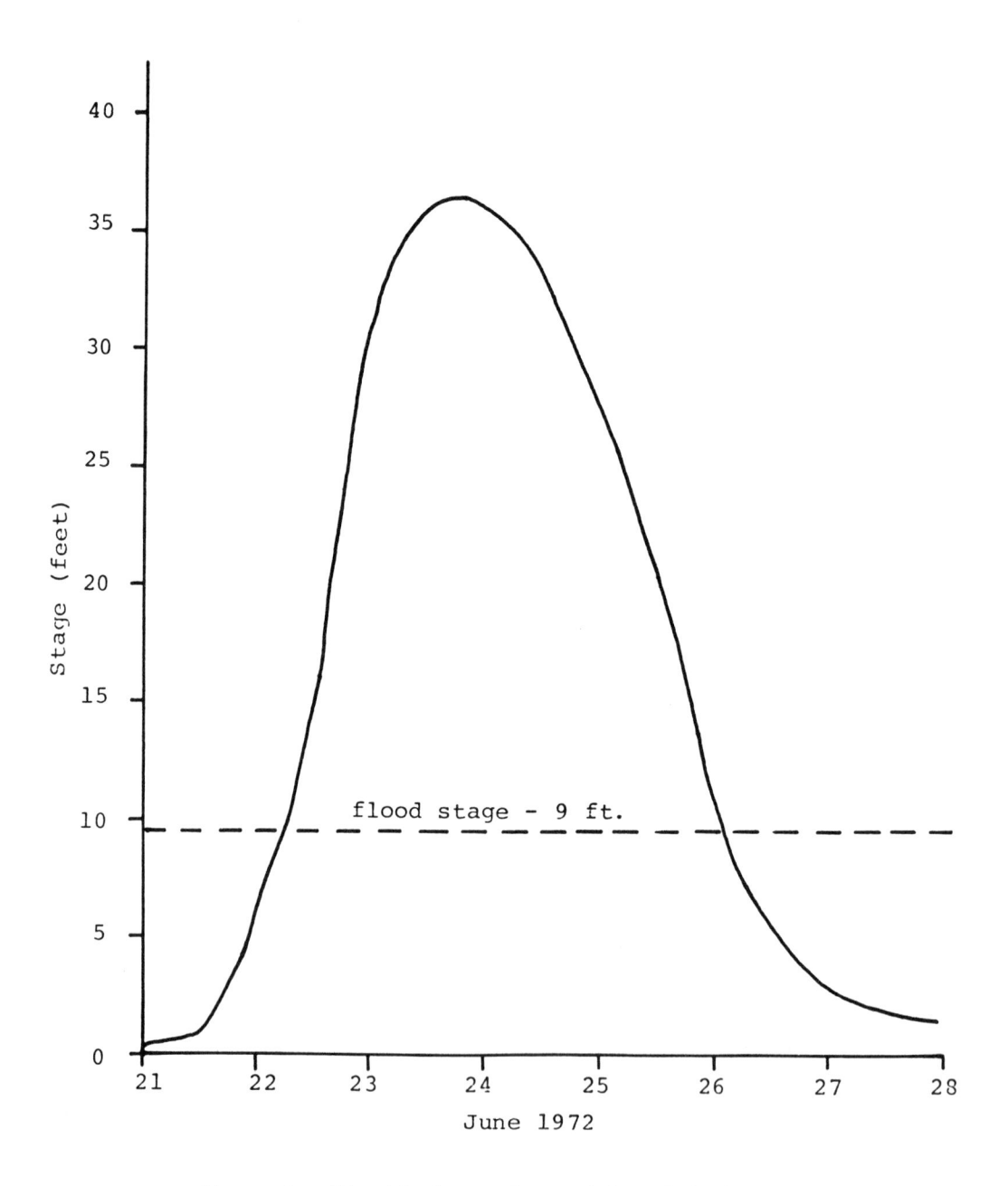

Figure 2. Flood hydrograph at Richmond during Agnes.

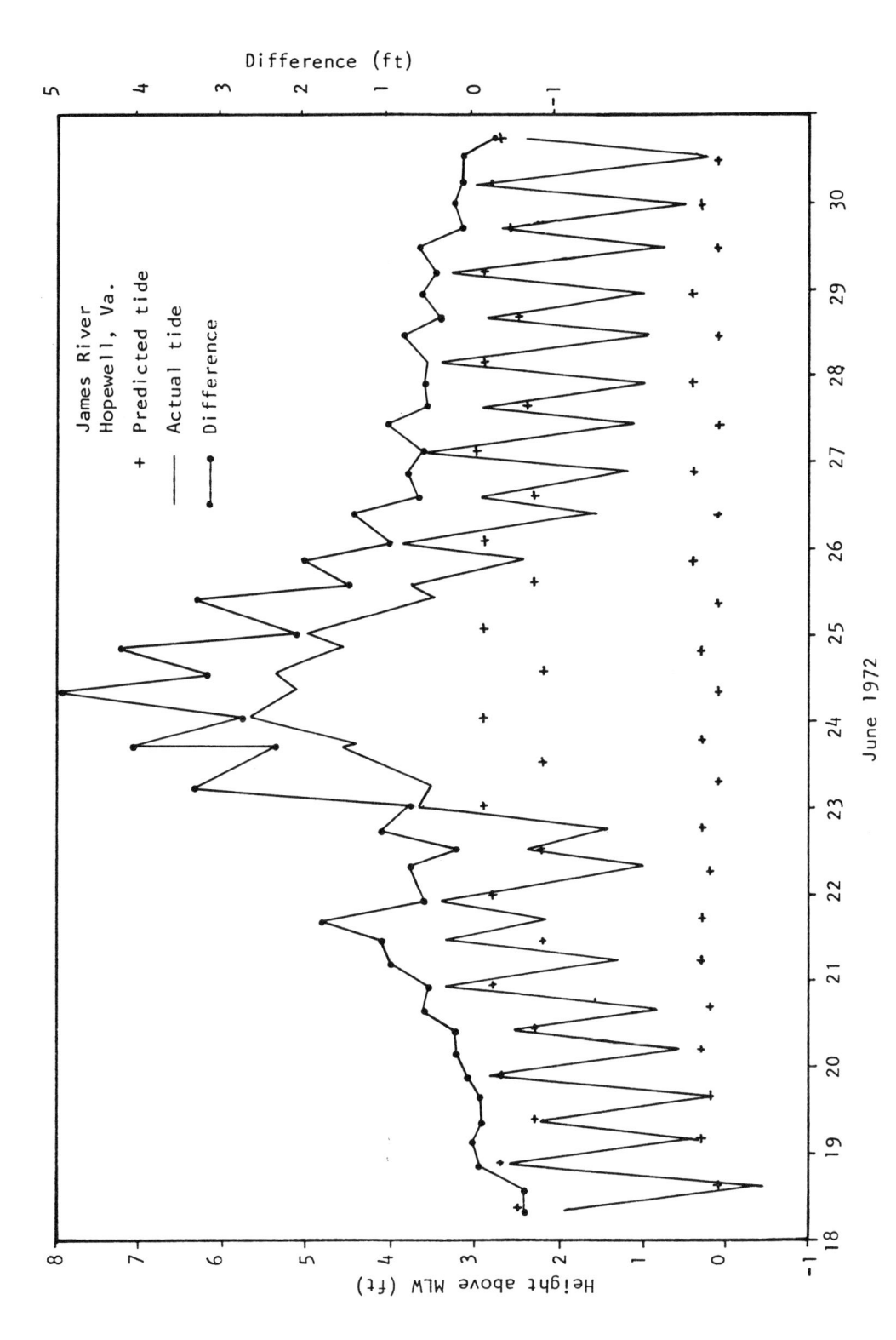

Figure 3. Predicted, recorded, and difference between recorded and predicted tidal heights at Hopewell, Va. during the Agnes flood.

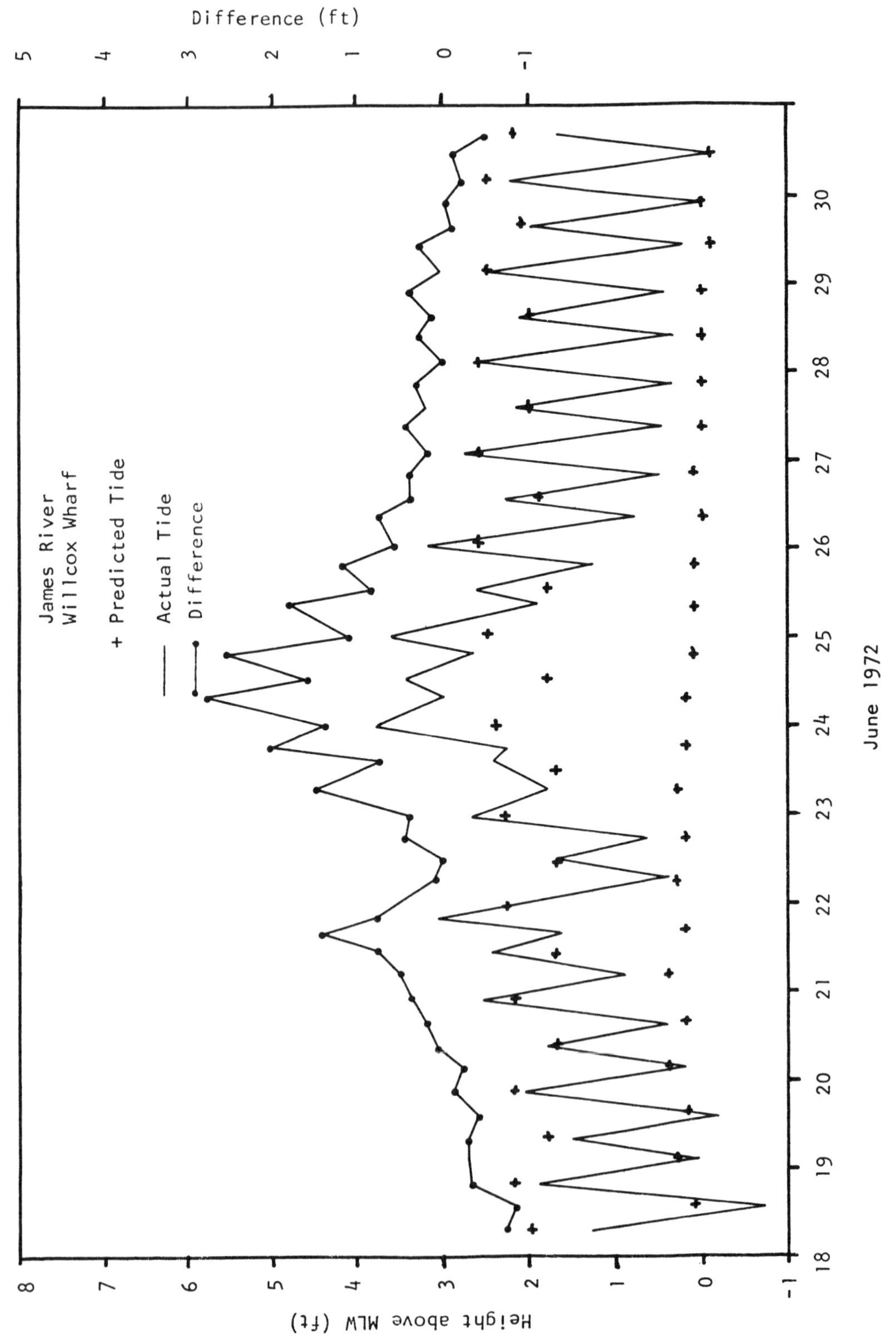

Figure 4. Predicted, recorded, and difference between recorded and predicted tidal heights at Willcox Wharf, Va. during the Agnes flood.

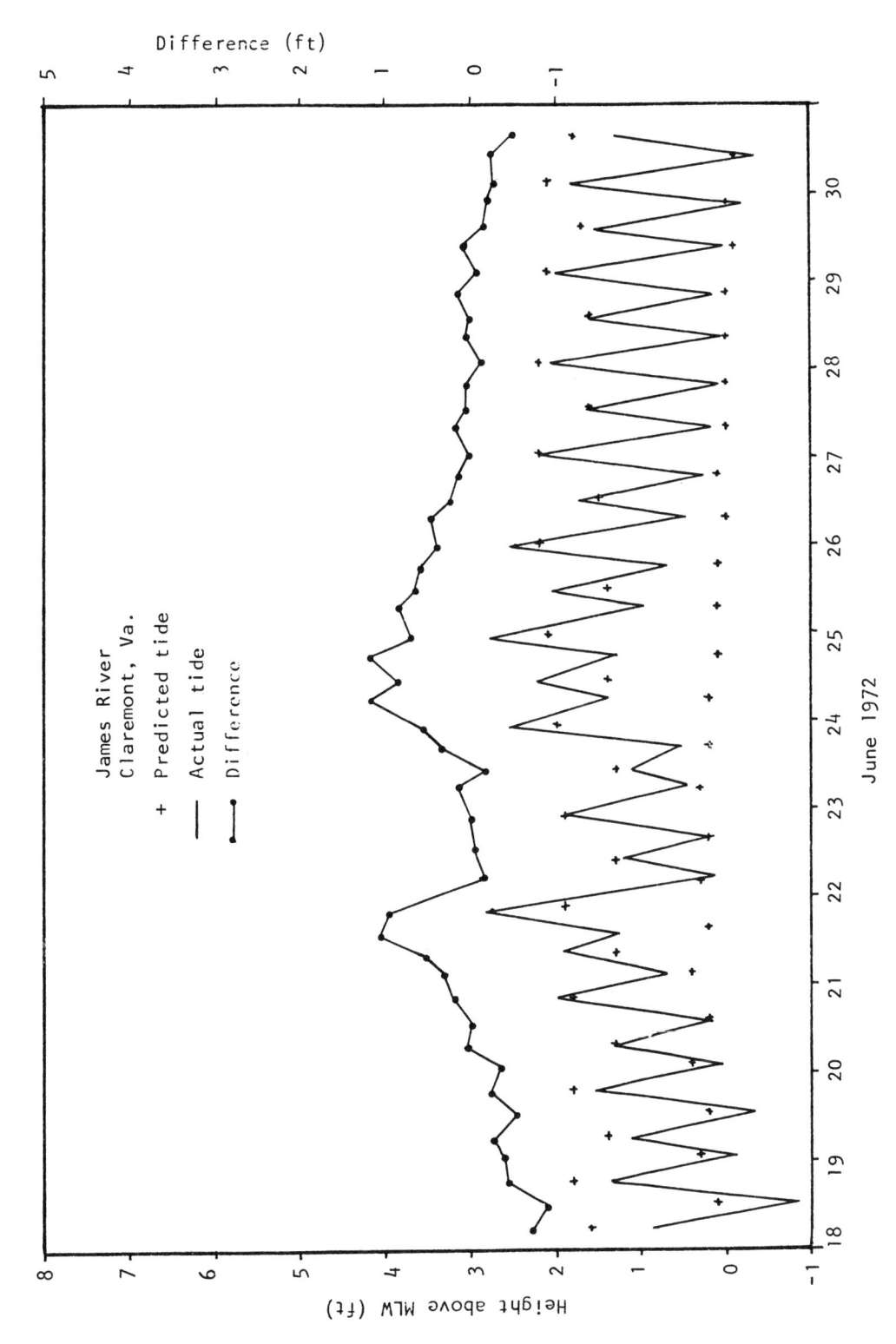

Figure 5. Predicted, recorded, and difference between recorded and predicted tidal heights at Claremont, Va. during the Agnes flood.

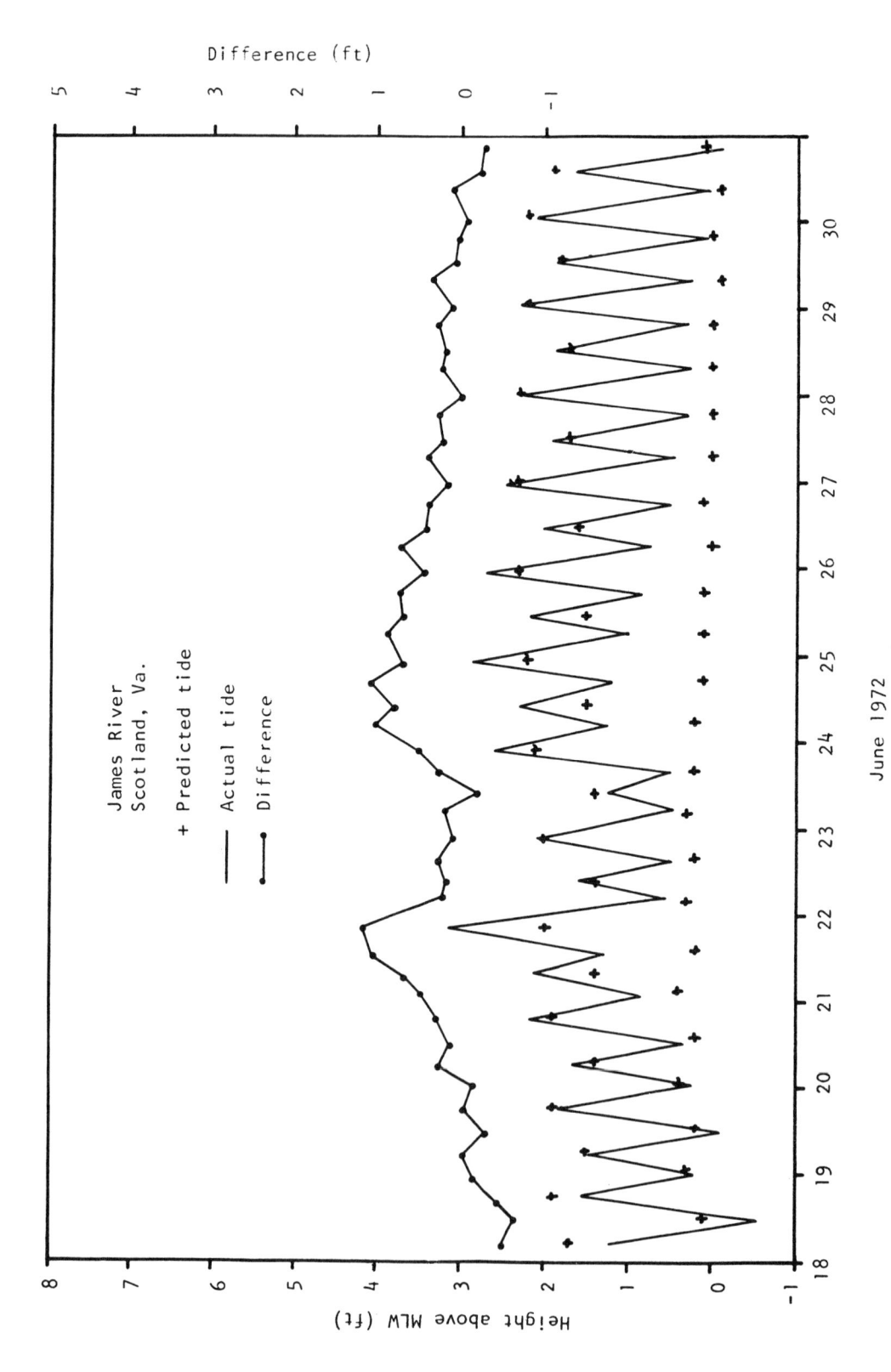

Figure 6. Predicted, recorded, and difference between recorded and predicted tidal heights at Scotland, Va. during the Agnes flood.

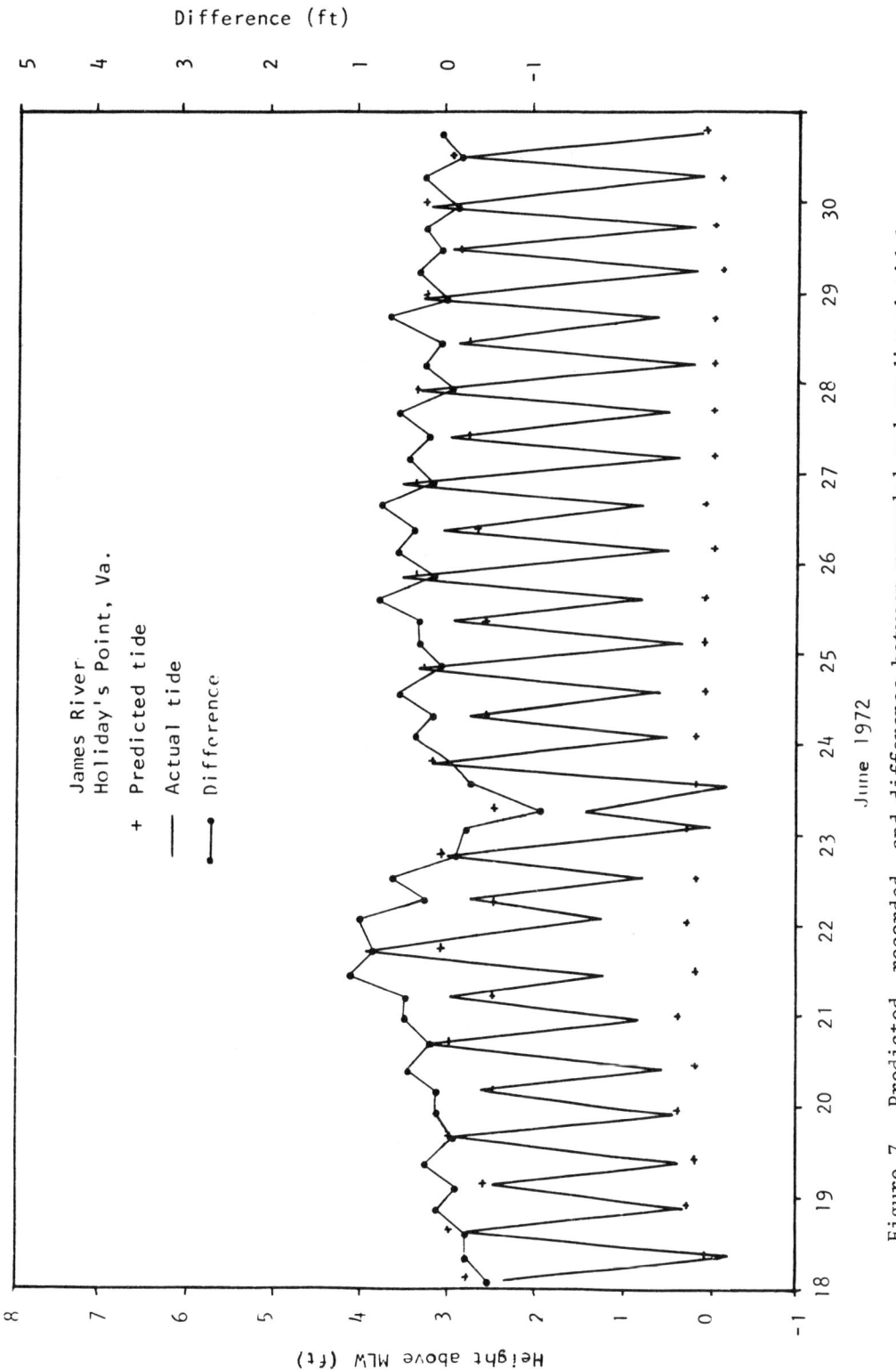

Figure 7. Predicted, recorded, and difference between recorded and predicted tidal heights at Holiday's Point, Va. during the Agnes flood.

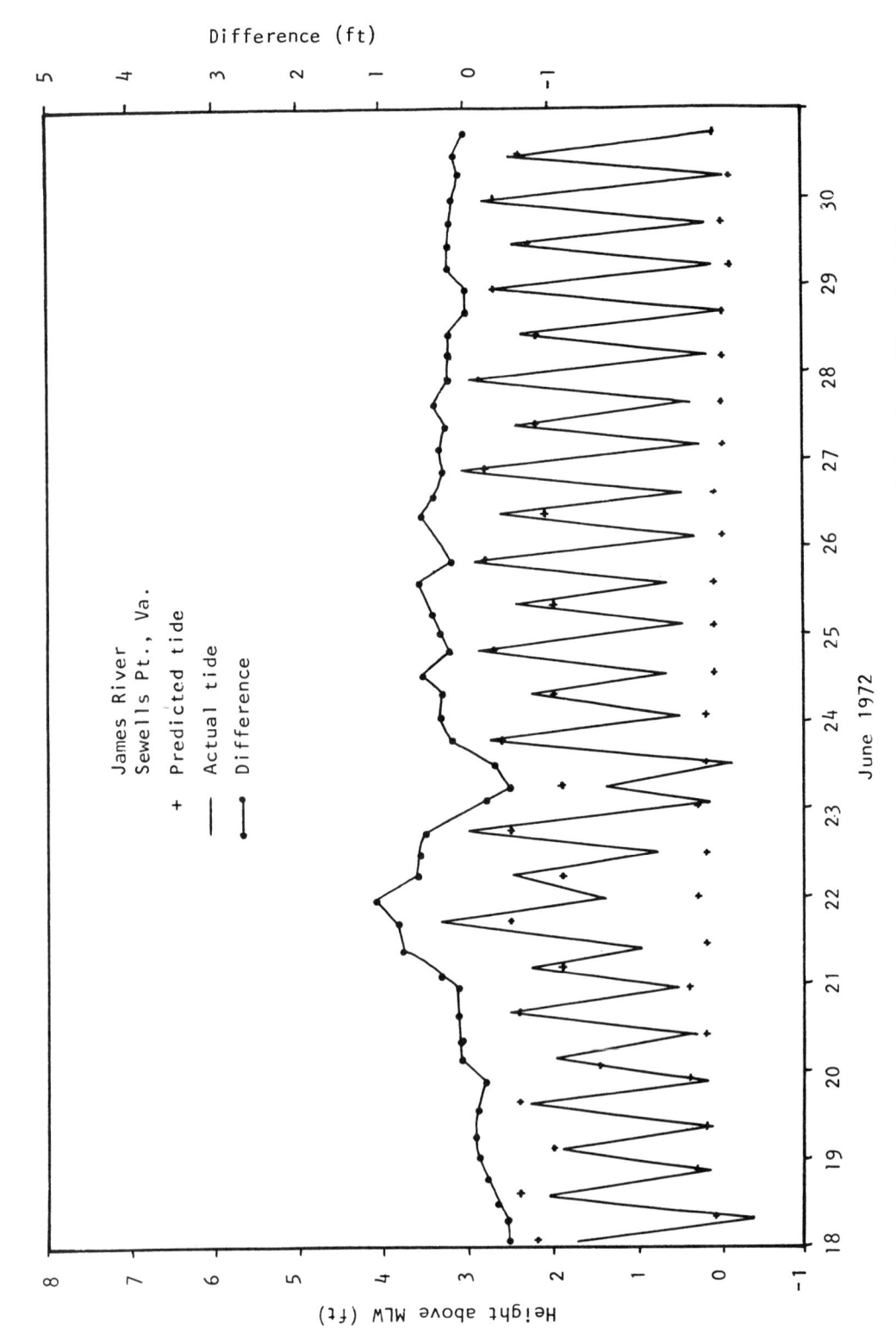

Figure 8. Predicted, recorded, and difference between recorded and predicted tidal heights at Sewells Pt., Va. during the Agnes flood.

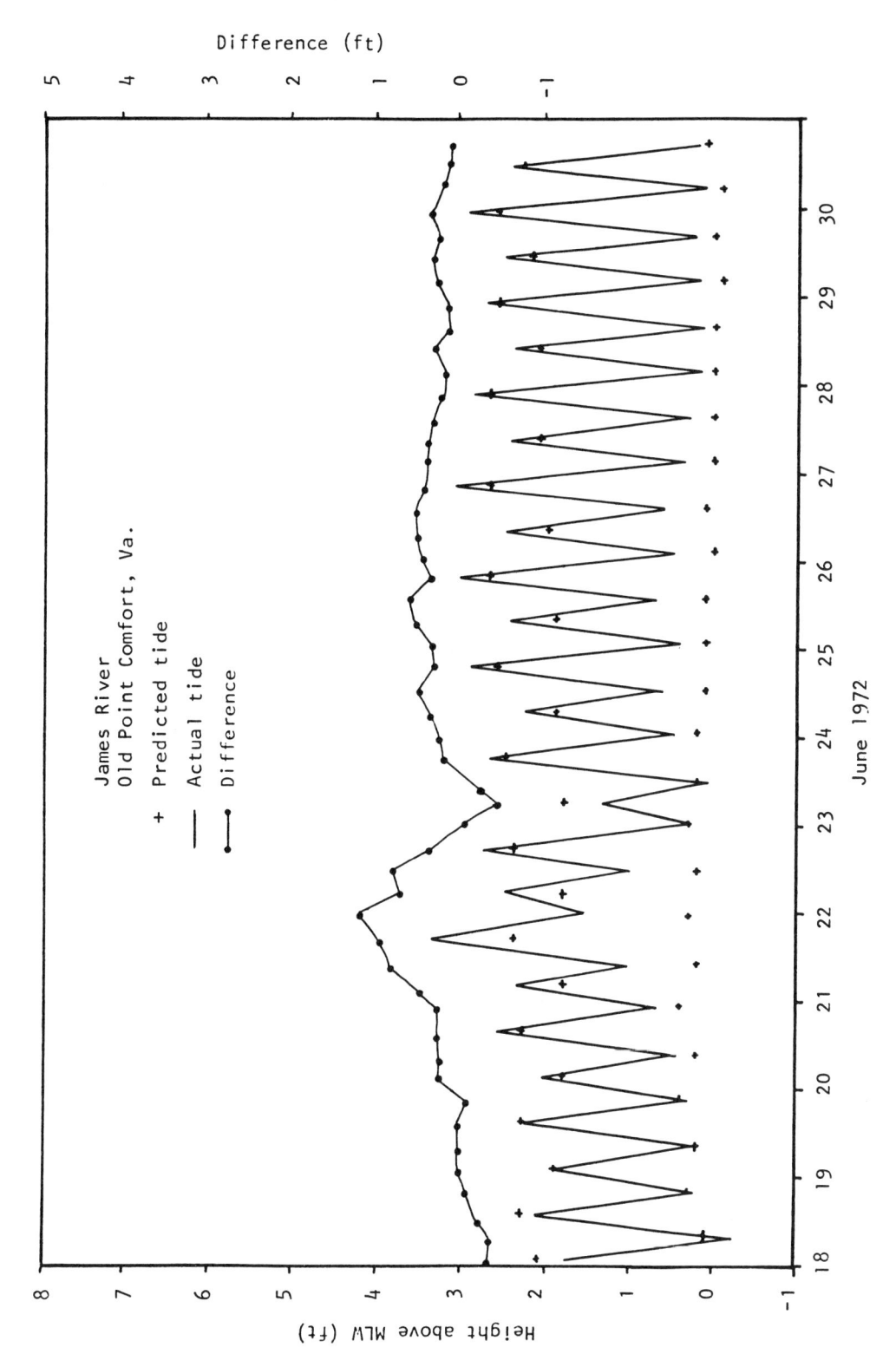

Figure 9. Predicted, recorded, and difference between recorded and predicted tidal heights at Old Point Comfort, Va. during the Agnes flood.

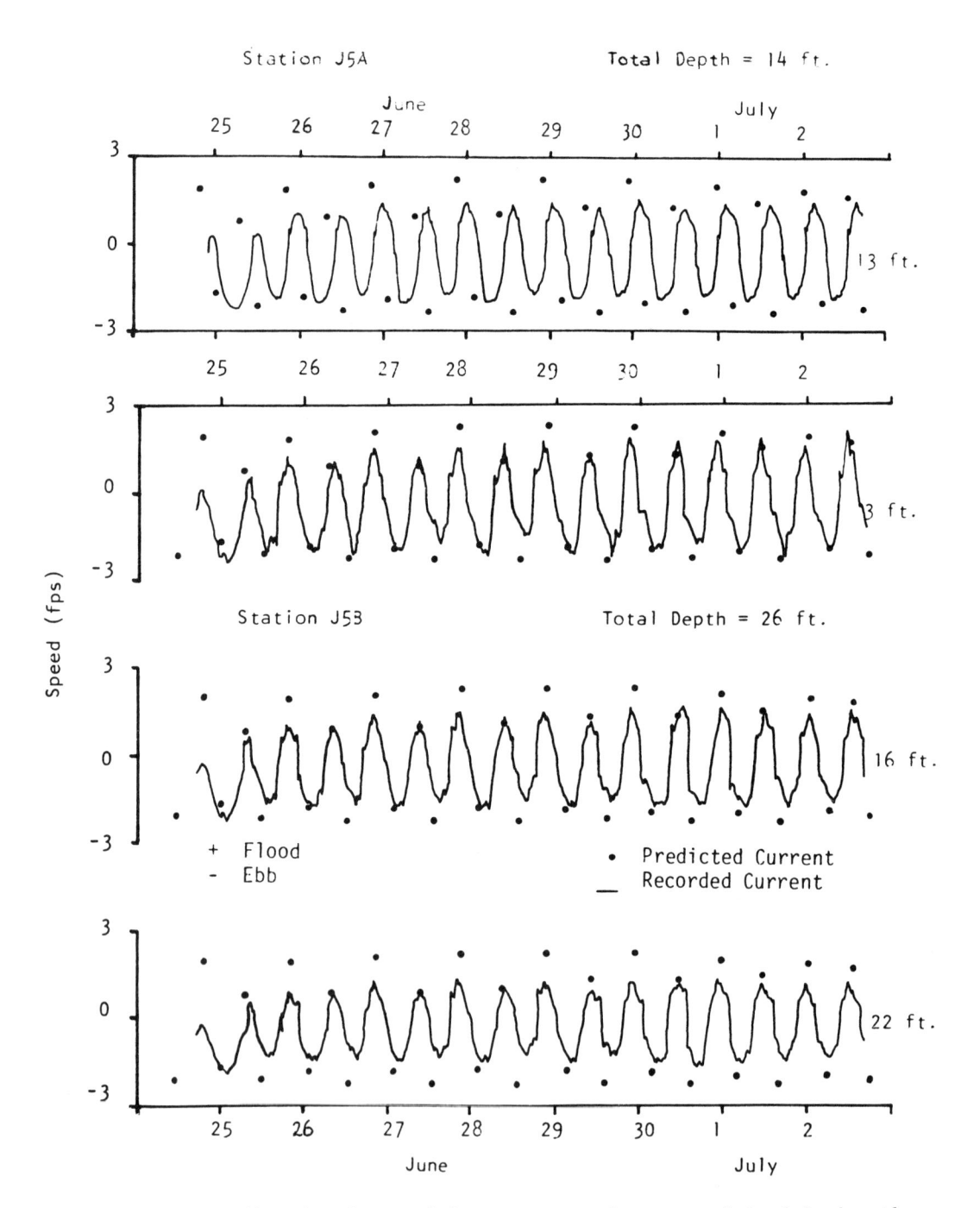

Figure 10. Predicted and recorded currents at Jamestown Island during the Agnes flood.

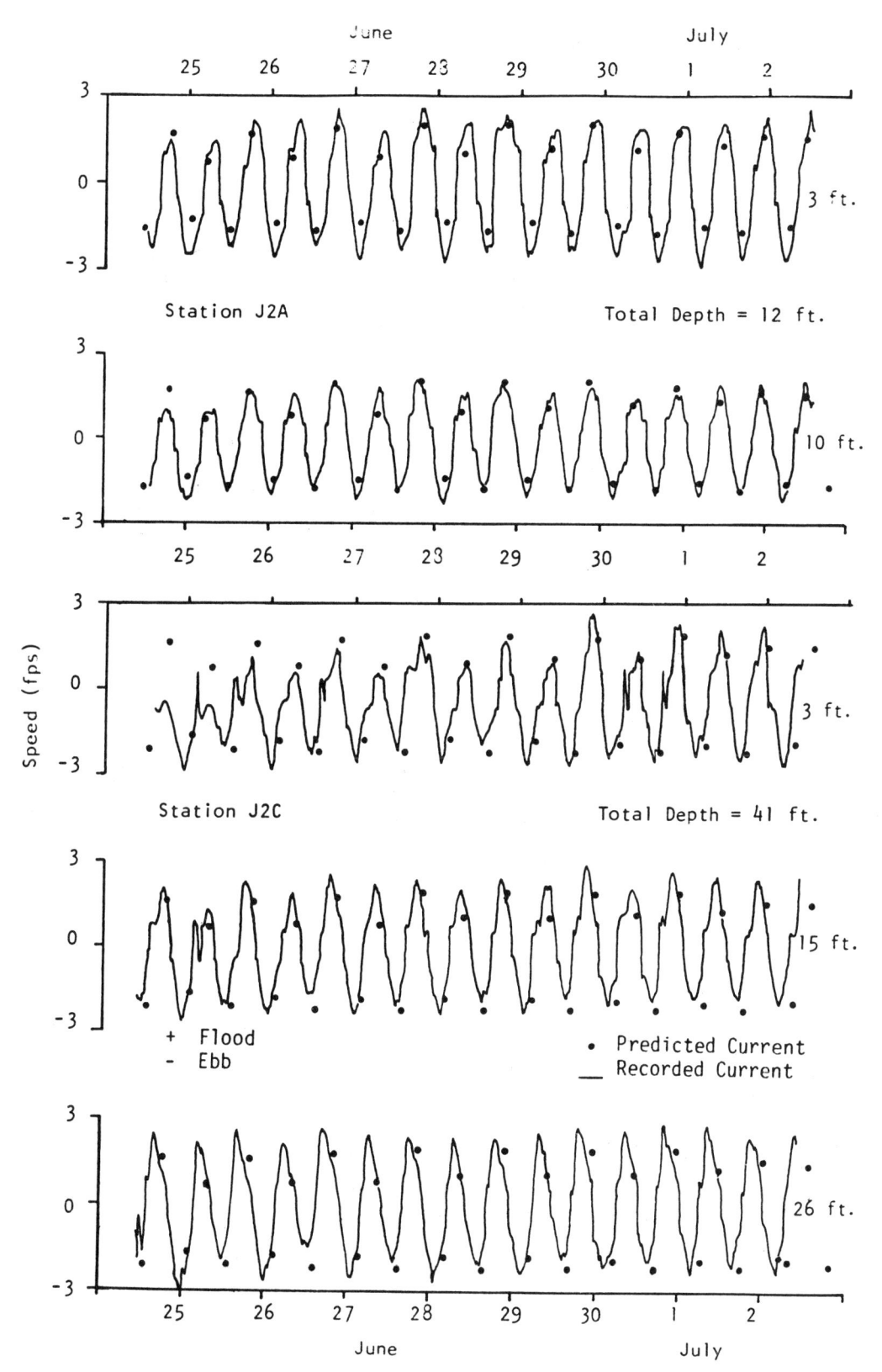

Figure 11. Predicted and recorded currents off Newport News during the Agnes flood.

DAILY RAINFALL OVER THE CHESAPEAKE BAY DRAINAGE BASIN FROM TROPICAL STORM AGNES

E. G. Astling[1]

ABSTRACT

Weather satellite pictures and infrared data were combined with conventional weather observations to investigate the extent and characteristics of Tropical Storm Agnes as the storm passed over Chesapeake Bay and its drainage basins. Radar imagery provided a means to describe subsynoptic scale rain areas. Precipitation observations were employed to map daily rainfall distribution.

Rainfall produced by Agnes was most intense over the Chesapeake Bay environs when the storm moved across the region between June 21 and 23, 1972. Although rainfall amounts reported by individual stations were greater during Camille's passage over Virginia in August 1969, the precipitation associated with Agnes was more extensive and was centered over the drainage basins flowing into the Chesapeake Bay to produce greater flooding.

The effects of the storm were enhanced by heavy rain showers that preceded Agnes when a weak cold front moved across the mid-Atlantic region on June 17 and 18. At this time, Agnes was a hurricane deriving its moisture source from the eastern Gulf of Mexico and producing a rain shield that spread inland ahead of the storm. After the hurricane reached landfall, it weakened to a tropical depression and continued to move northward (DeAngelis & Hodge 1972).

This paper presents 24-hour precipitation accumulations for the 3-day period from June 21 to 23, 1972. Analysis of the precipitation data was based on approximately 200 station observations located in the drainage basins of the Chesapeake Bay. Radar echoes of precipitation and weather satellite cloud pictures supplemented the conventional data to describe the spatial and temporal variations of precipitation.

JUNE 21, 1972

The weather over the Chesapeake Bay was first influenced by Agnes when the depression moved into North Carolina. Low level easterly flow around the north side of its center provided an atmospheric moisture source from the Atlantic Ocean for rain that was deposited over the Piedmont Plateau and along the eastern side of the Blue Ridge Mountains.

On July 21, 1972, daily rainfall amounts exceeded 2 inches over large portions of the upper James, York and Rappahannock River basins as depicted in Fig. 1a. Several stations reported 24-hour accumulations in excess of 6 inches and one station (Dulles Airport) reported 10.67 inches in the Potomac River basin (Environmental Data Service 1972). In general, the Chesapeake Bay estuarine system received less than 1 inch of rain during this first day.

---

[1] Old Dominion University, Norfolk, Va. 23508

Fig. 1b shows 24-hour rainfall amounts over Pennsylvania for June 21. Only isolated thunderstorms had occurred in the central portion of this region and consequently only three small areas were reported where there were precipitation amounts in excess of 2 inches. Over the Allegheny Plateau a broad area of precipitation was associated with an approaching trough system from the eastern Great Lakes. This area is outside the drainage basins flowing into the Chesapeake Bay; however, this heavy rainfall area moved eastward and merged with Agnes on the next day.

The mesoscale structure of the precipitation areas was evident in weather radar and satellite cloud imagery. Fig. 2 shows precipitation echoes that were observed within a 125-mile radius of the Patuxent radar station. These data together with weather satellite cloud pictures for June 21 revealed clusters or groups of thunderstorm complexes which tended to be oriented in bands. These bands moved towards the northwest from the Atlantic Ocean around the north side of the low center. Large rainfall amounts associated with these echoes were attributed to frictional convergence along the coastal region and orographic lifting further inland.

JUNE 22, 1972

On the second day, Agnes moved across the southern portion of the Chesapeake Bay and into the Atlantic Ocean and intensified into a tropical storm. The winds changed from an easterly to a northwesterly direction over the Chesapeake Bay as the storm center moved eastward. Further inland, heavy rain continued and the extensive precipitation area moved around the west side of the low pressure center.

Fig. 3 shows the 24-hour rainfall amounts for June 22. Amounts in excess of 2 inches covered almost the entire map. In the upper drainage basins over the Piedmont Plateau, amounts exceeded 4 inches, and along the Blue Ridge Mountains, 24-hour rainfall exceeded 6 inches. In Maryland and the upper Potomac River basin, several stations reported amounts greater than 10 inches. Heaviest amounts were reported in the lower portion of the Susquehanna River basin.

The Patuxent radar echoes from late on June 21 to June 22 (Figs. 4 & 5) show the extensive characteristics of the broad area of heavy precipitation. A complete overcast of echoes on the west side of the Chesapeake Bay extended from central Virginia to Pennsylvania. This is in contrast to the previous day when lines of thundershowers and isolated radar cells were present. Late in the afternoon on June 22, the clouds and precipitation dissipated over southern and central Virginia. By this time, cold air from mid-latitudes had fed into the low center and Agnes began to acquire extratropical characteristics.

JUNE 23, 1972

The storm center moved inland over Long Island and into western Pennsylvania on June 23. The precipitation area spread westward over the Susquehanna River basin as Agnes moved inland. Fig. 6b shows that most of Pennsylvania received more than 2 inches of rain, and a very broad area of amounts exceeding 4 inches occurred along the Susquehanna River. The greatest 24-hour rainfall amounts during the entire 3-day period were reported in the middle Susquehanna Valley (Bear Gap) with 13.96 inches. To the south, the rain had subsided and only extreme northern Virginia, West Virginia, and Maryland observed precipitation on June 23.

Agnes weakened further and moved northward. By the next day, it was no longer depositing heavy rainfall amounts in the Chesapeake Bay drainage basins.

SUMMARY

Agnes left in its wake, some of the most extensive 24-hour rainfall amounts to occur in the past 30 years over the Chesapeake Bay drainage basins. The average 24-hour precipitation amounts over the five major river basins that drain into the Chesapeake Bay are summarized in Table 1. The two largest basins, the Potomac and Susquehanna, had the largest amounts, exceeding 8 inches for the 3-day period. Other storms have produced greater rainfall at individual stations (i.e., Camile, in Thompson 1969) but none have had such heavy rainfall amounts extend over as large a region as Agnes.

Table 1. Average 24-hour precipitation amounts for June 21 to 23, 1972 for the major river basin areas which drain into the Chesapeake Bay. Units are inches.

River Basin	June 21	June 22	June 23	Total Accumulation
James	1.35	4.04	0.01	5.40
York	1.28	3.98	0.00	5.26
Rappahannock	1.36	3.53	0.13	5.02
Potomac	2.69	4.55	1.04	8.28
Susquehanna	0.28	4.16	3.65	8.09

LITERATURE CITED

DeAngelis, R. M. and W. T. Hodge. 1972. Preliminary climatic data report, Hurricane Agnes, June 14-23, 1972. NOAA Tech. Memo. EDS NCC-1, Asheville, N. C.

Environmental Data Service. 1972. Climatological data for Virginia, Maryland, and Pennsylvania. NOAA Publication.

Thompson, H. J. 1969. The James River flood of August 1969 in Virginia. Weatherwise 22:180-183.

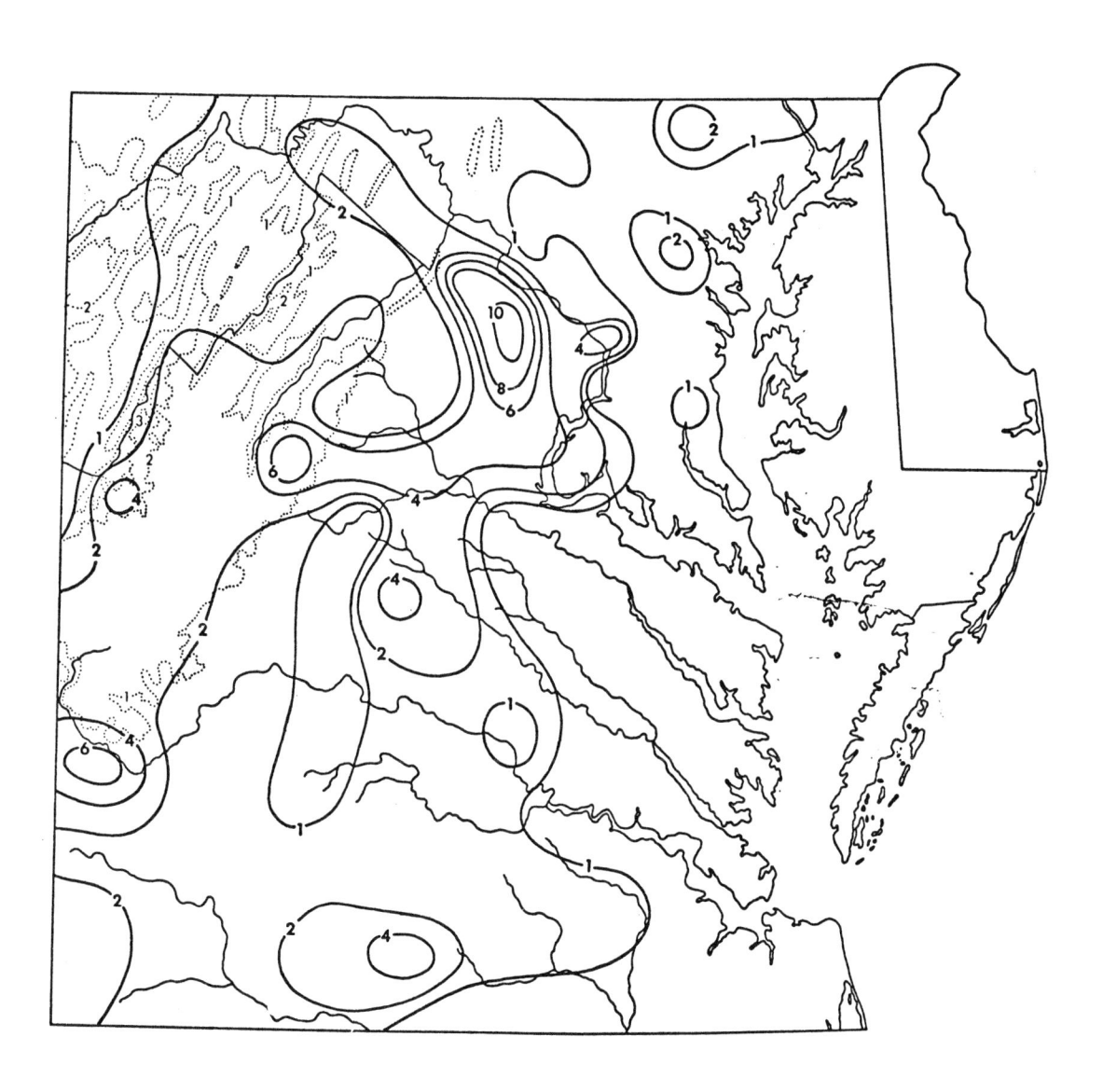

Figure 1a. Daily rainfall amounts in inches for 21 June 1972 over Virginia, Maryland, and Delaware. Isohyets are indicated by heavy solid lines and terrain heights for 1000 ft intervals by light dotted lines.

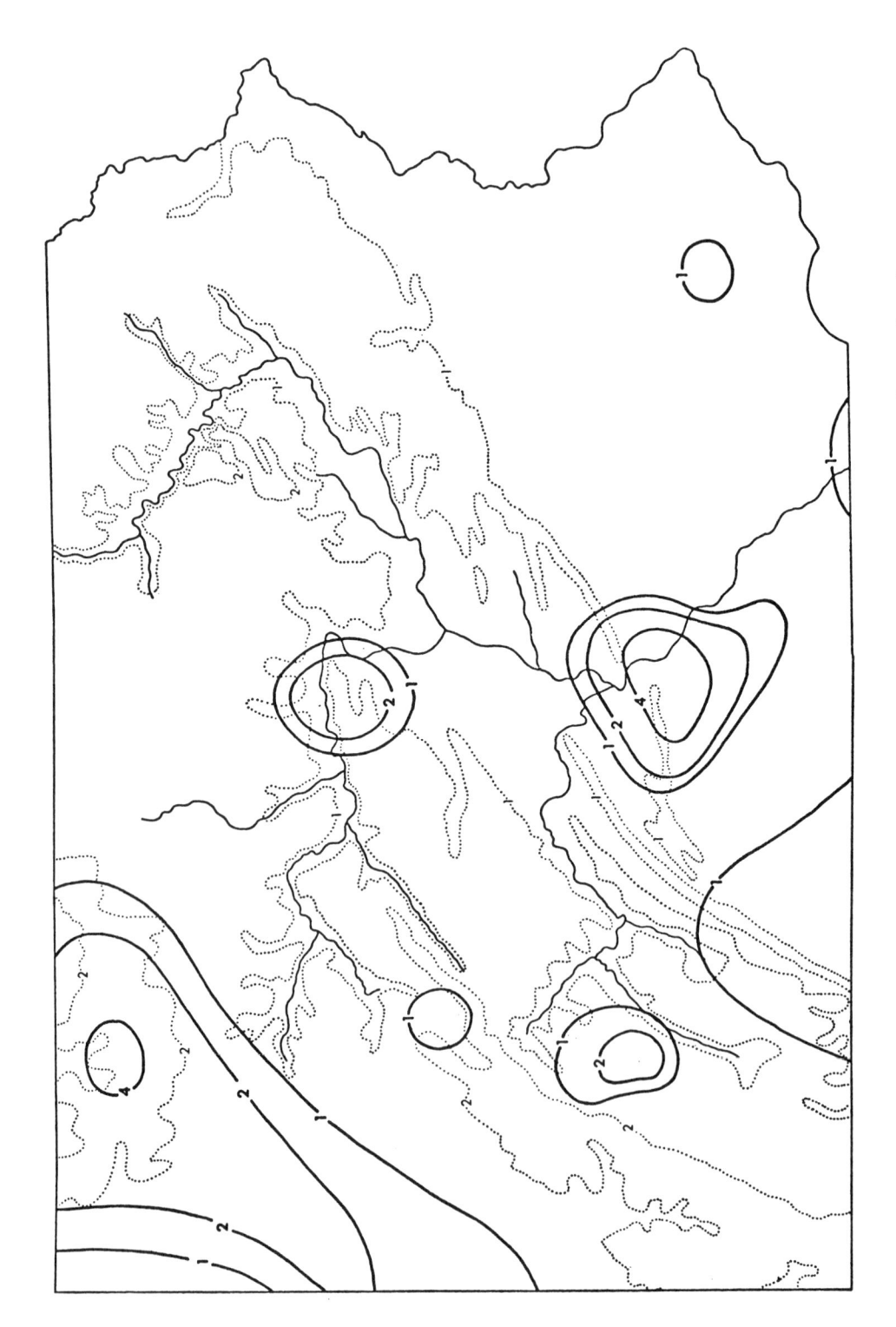

Figure 1b. Daily rainfall amounts in inches for 21 June 1972 over Pennsylvania.

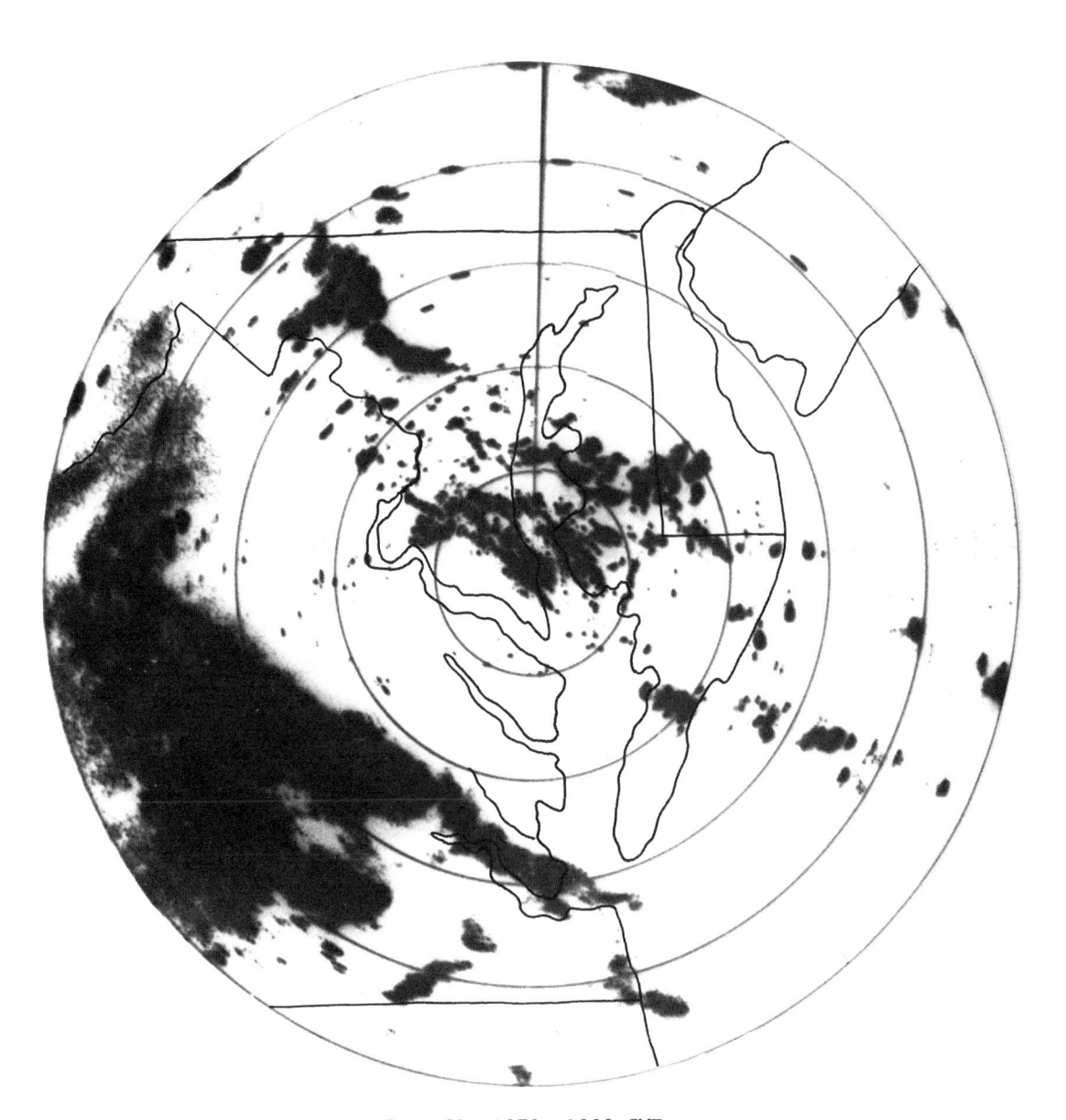

June 21, 1972   1209 GMT

Figure 2. Precipitation echoes observed by the Patuxent weather
radar station for 21 June 1972 at 0709 EST.

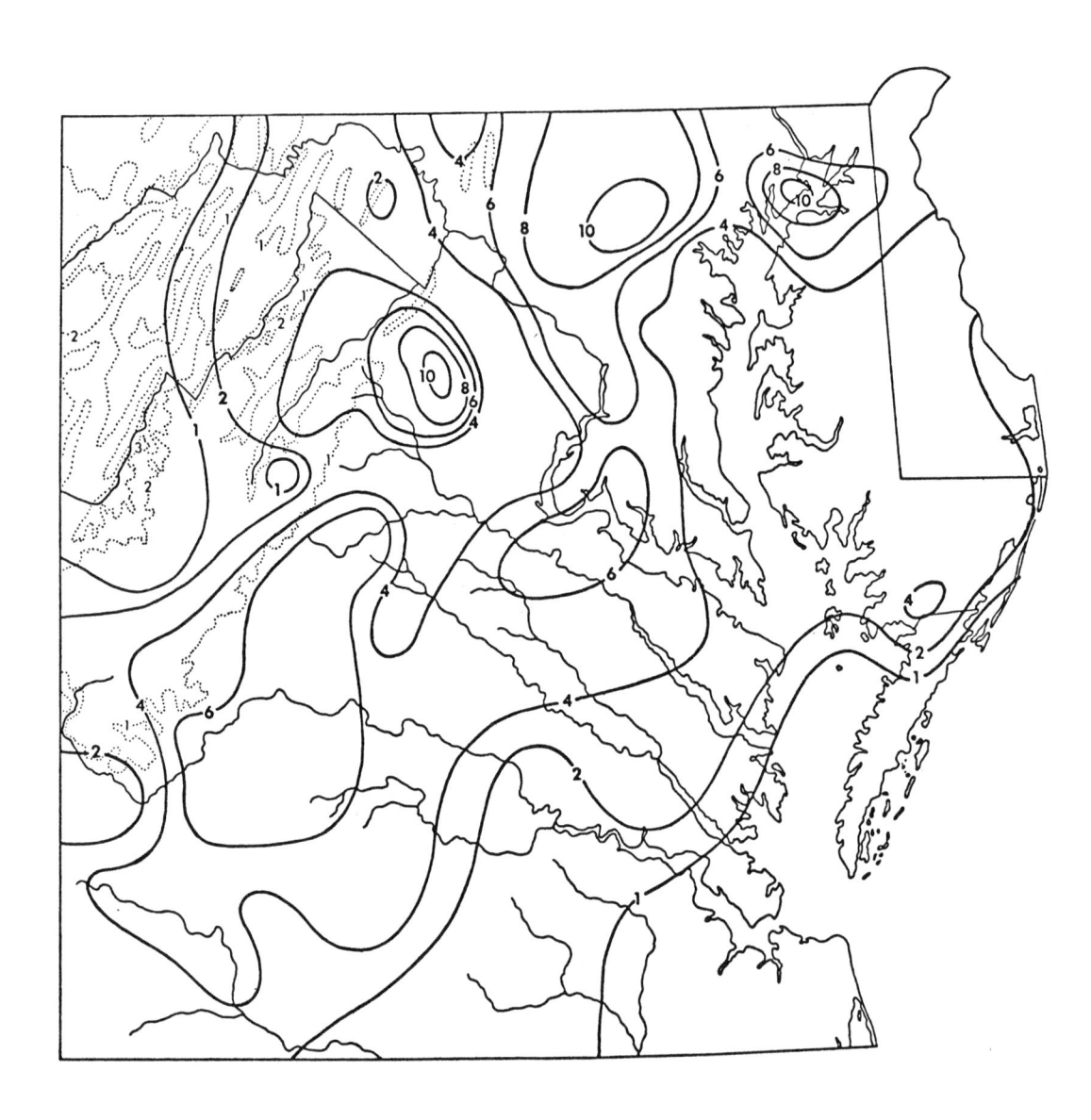

Figure 3a. Daily rainfall amounts in inches for 22 June 1972 over Virginia, Maryland, and Delaware.

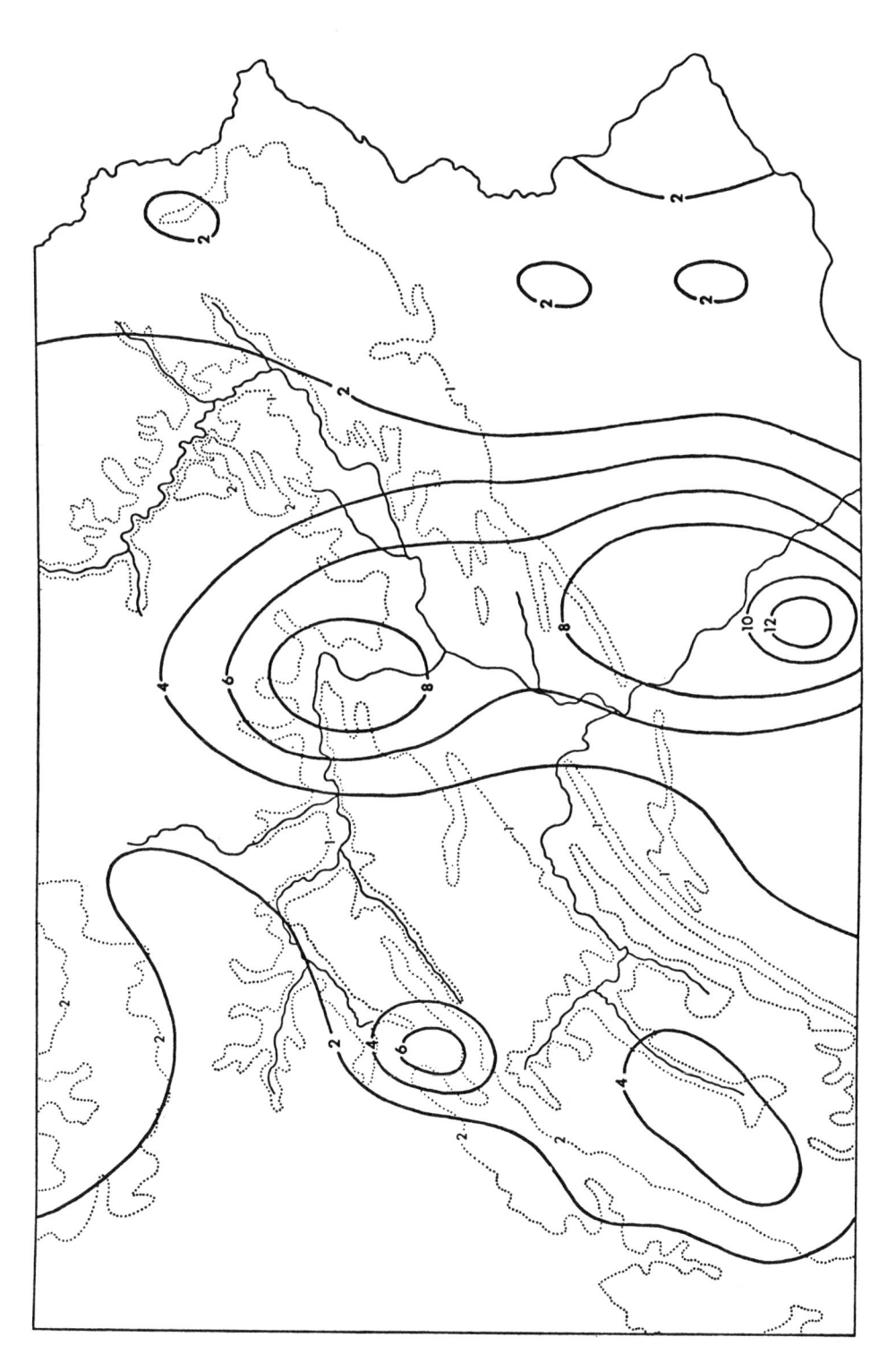

Figure 3b. Daily rainfall amounts in inches for 22 June 1972 over Pennsylvania.

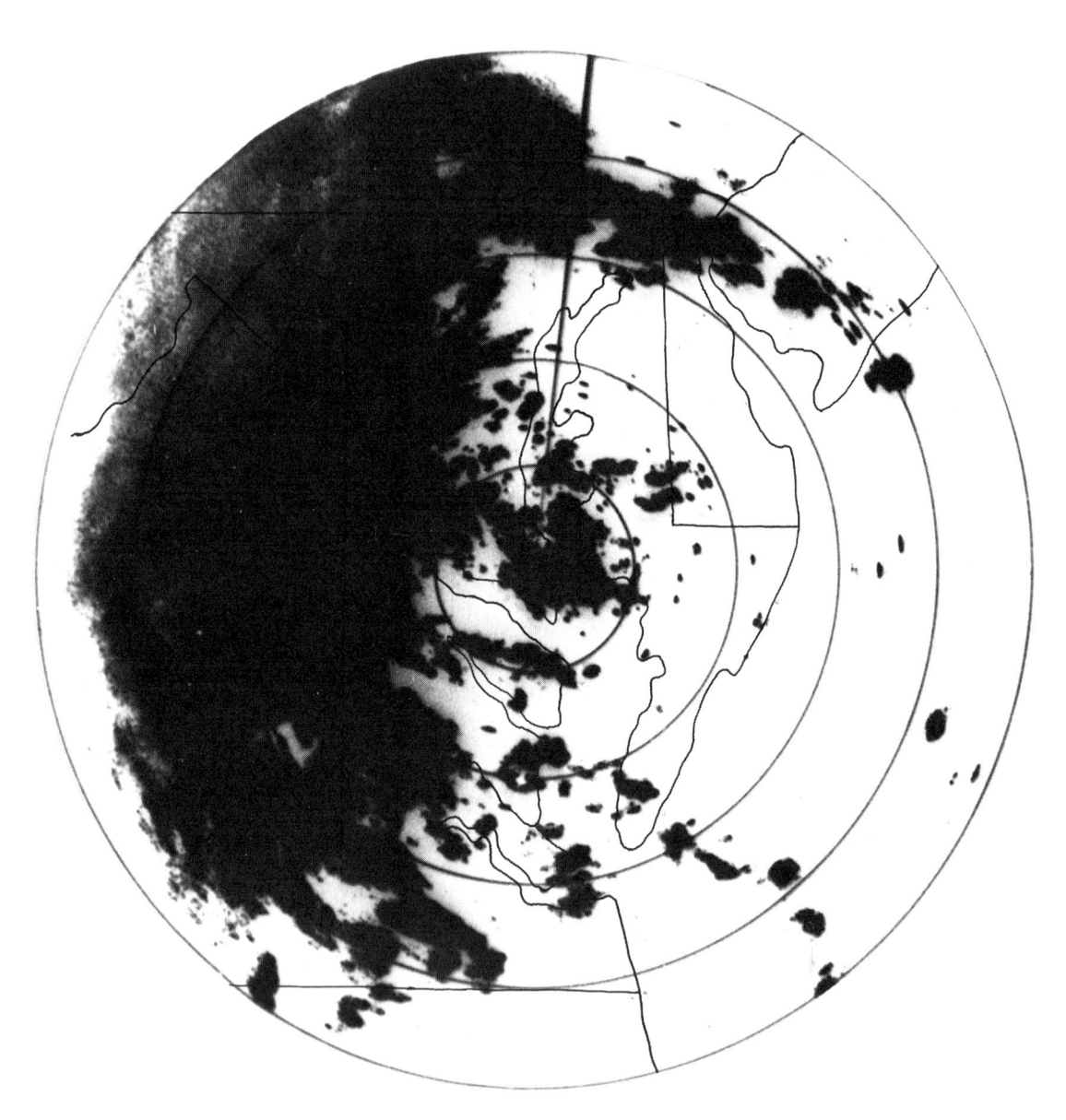

June 22, 1972 0000 GMT

Figure 4. Precipitation echoes observed by the Patuxent weather radar station for 21 June 1972 at 1900 EST.

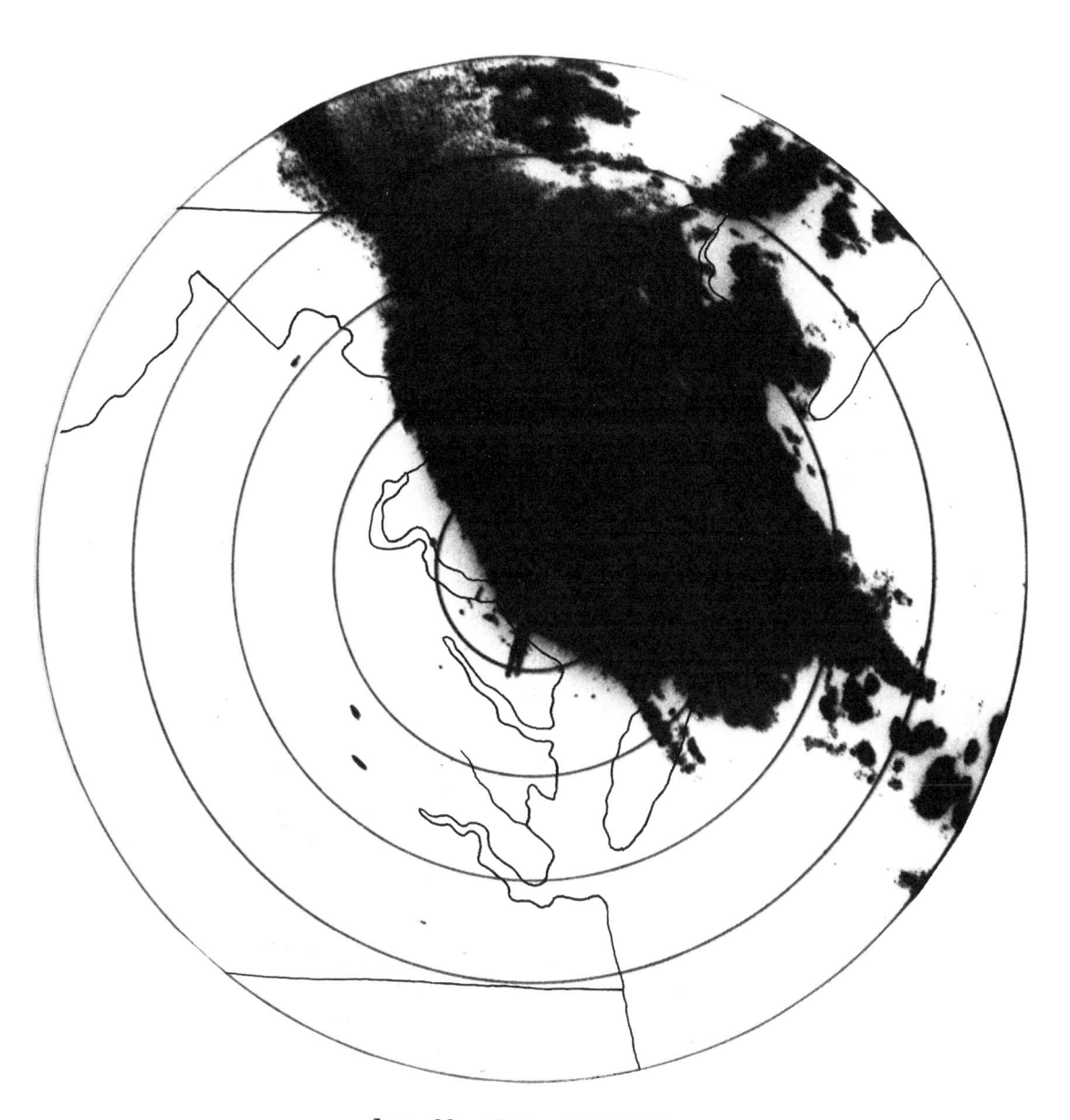

June 22, 1972   1200 GMT

Figure 5. Precipitation echoes observed by Patuxent weather radar station for 22 June 1972 at 0700 EST.

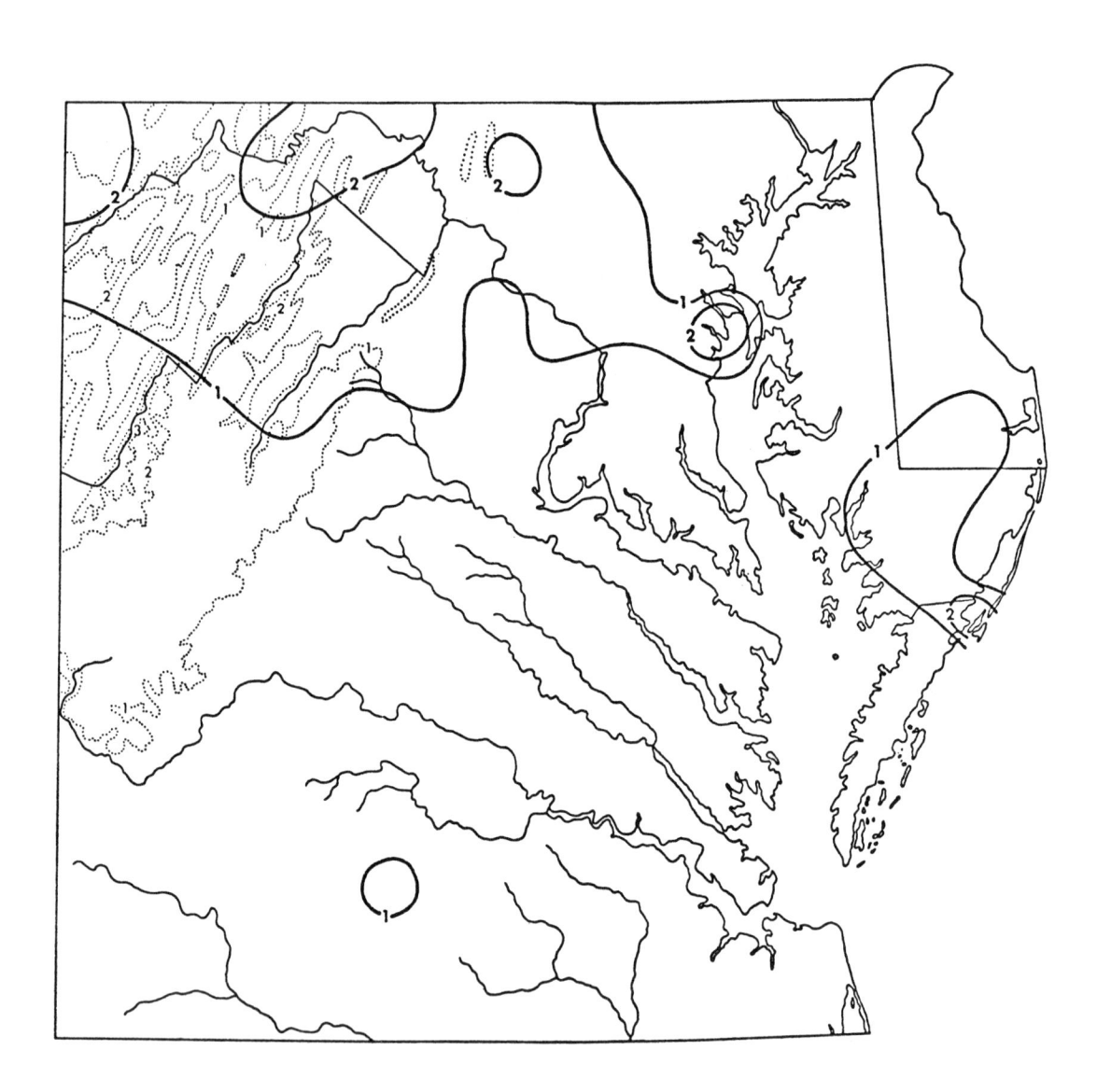

Figure 6a. Daily rainfall amounts in inches for 23 June 1972 over Virginia, Maryland, and Delaware.

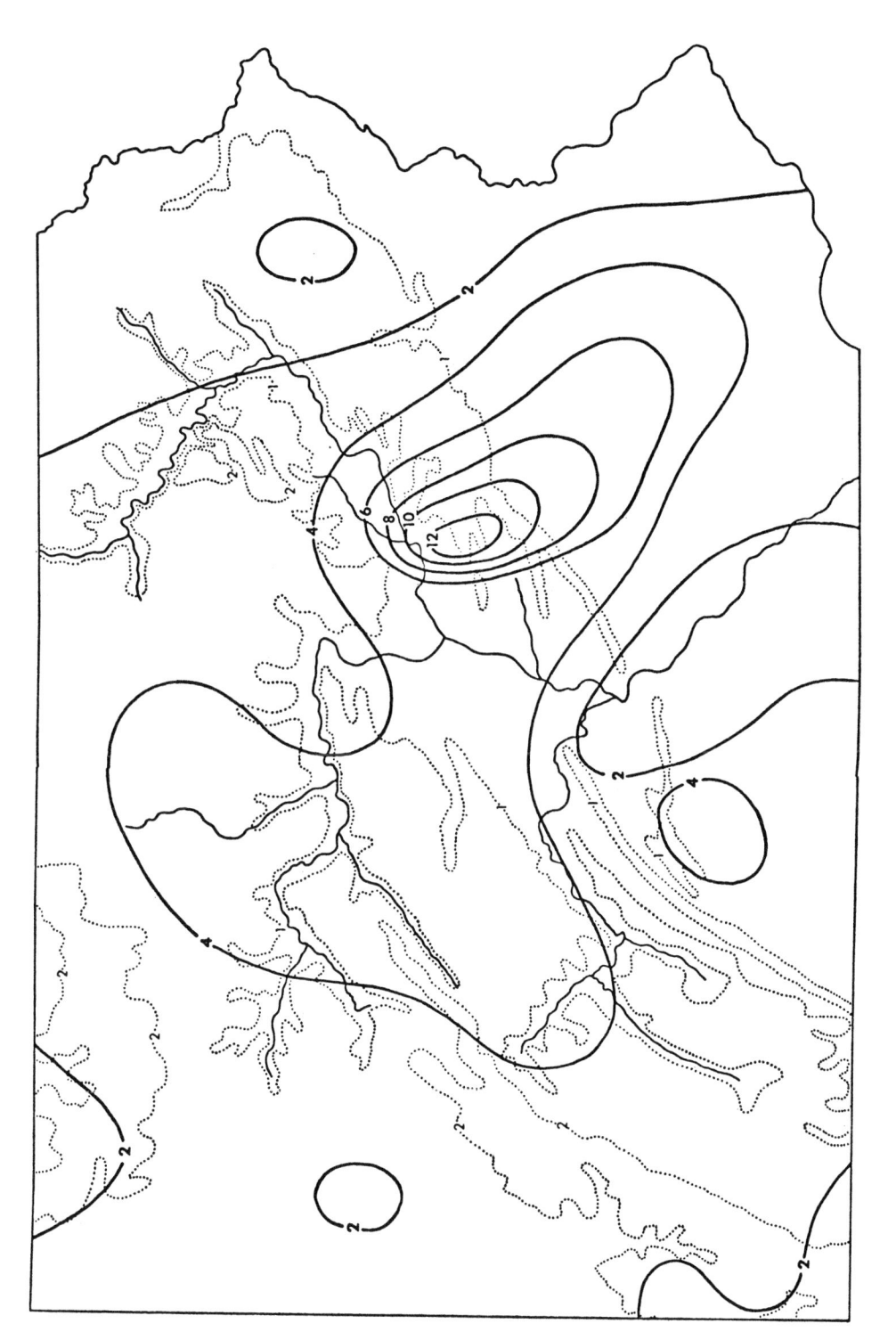

Figure 6b. Daily rainfall amounts in inches for 23 June 1972 over Pennsylvania.

THE EFFECT OF TROPICAL STORM AGNES
ON THE SALINITY DISTRIBUTION IN THE
CHESAPEAKE AND DELAWARE CANAL[1]

G. B. Gardner[2]

ABSTRACT

Salinity measurements in the Chesapeake and Delaware Canal made by the Chesapeake Bay Institute at approximately monthly intervals since May 1969 have provided a good indication of the salinity distribution in the Canal. A survey made on 11-12 July 1972, about two weeks after the storm, indicated that the water was essentially fresh throughout the Canal. A survey made on 8-9 August 1972 showed that the salinity distribution had returned to near normal conditions.

INTRODUCTION

The Chesapeake and Delaware Canal is a sea level, man-made waterway connecting the Chesapeake Bay to the Delaware River. As seen in Fig. 1, the connection to the Chesapeake Bay is through the Elk River, near the head of the Bay.

At the time of Agnes the Chesapeake Bay Institute was conducting a study of the hydrography of the Chesapeake and Delaware Canal. In connection with this study, monthly surveys of temperature and salinity had been made beginning in May 1969. These surveys provide an indication of the annual variation of salinity distribution in the Canal and of the effect of Agnes on that distribution.

The salinity distribution in the Canal is controlled by the salinity at the ends of the Canal and the flow through the Canal. The flow of fresh water from the Susquehanna River is the dominant factor in the salinity regime in the upper Chesapeake Bay, and thus largely determines the salinity at the western end of the Canal. The net transport through the Canal is difficult to define precisely, due to its highly variable nature. Averages over periods of approximately three weeks have shown the normal net flow to be about 2000 cfs toward the Delaware, with a maximum of about 3500 cfs during the spring maximum in Susquehanna flow. The tidal transport in the Canal is much larger than the net transport. The peak transport is generally between 60,000 cfs and 80,000 cfs on both the ebb (westward flow) and flood (eastward flow). Because of the net easterly flow, the salinity, at least in the western portion of the Canal, is largely determined by the Susquehanna River flow. A more complete description of the hydrography of the Canal is given in Pritchard and Gardner (1974).

The net flow through the Canal is related to a higher mean tide level in the upper Chesapeake Bay than in the Delaware River at the eastern end of the Canal. Since Agnes had a larger effect on the Susquehanna than on the Delaware, it is reasonable to assume that the level difference was increased, producing an increase in the net transport.

---

[1]Contribution No. 233, Chesapeake Bay Institute, The Johns Hopkins University
[2]Chesapeake Bay Institute, The Johns Hopkins University, Baltimore, Md. 21218

COMPARISON OF TYPICAL AND AGNES INDUCED SALINITY DISTRIBUTIONS

In view of the preceding discussion it is not surprising that the Canal was essentially fresh two weeks after Agnes. Fig. 2 shows examples of typical springtime minimum salinity conditions--typical July salinity conditions in the Canal. The sections extend from Grove Pt. (924QQ) through the Elk River and Canal to the Delaware River (DR2). A survey of salinity made on 11-12 July 1972 showed no water above 0.2 ppt. In fact, little salt was found north of Tolchester.

The next survey of the Canal was made on 8-9 August. At that time the salinity distribution had returned to near normal conditions. Normal conditions are hard to define for an area as variable as the Chesapeake and Delaware Canal. Fig. 3 provides an indication of the mean annual salinity variations in the Canal (solid line) and the standard deviation about the mean (dotted line), based on data taken from May 1969 to February 1973. The points plotted in Fig. 3 are for year 1972. All values plotted in Fig. 3 represent the surface salinity averaged over the central area of the Canal. The extreme low value for July 1972 is evident here. Also note that by August, salinity had increased to within one standard deviation of the mean.

While the salinity pattern in the Canal itself had returned to near normal by 8-9 August, the upper Chesapeake Bay had recovered less. Fig. 4 compares sections made on 9 August 1972 and 14 August 1970. Note that the salinity distribution in the Canal is very similar; the salinity in the vicinity of Turkey Point is still about 0.2 ppt on 9 August 1972, compared with 1.0 ppt on 14 August 1970. Presumably the salt in the Canal on 9 August 1972 came primarily from the Delaware River. It has been found that salt is frequently advected into the Canal from the Delaware due to the phase relation between the tides in the Delaware River and the tides in the Canal (Pritchard & Gardner 1974). This effect accounts for the salinity maximum at the eastern end of the Canal in Figs. 2 and 4.

CONCLUSION

Agnes resulted in the only instance in nearly four years of measurement when the Chesapeake and Delaware Canal was essentially fresh throughout its length. Within six weeks of the storm the Canal itself had returned to essentially normal conditions while the salinity at the mouth of the Elk River was still well below normal. Recovery in the Canal was apparently speeded by its connection to the Delaware River.

ACKNOWLEDGEMENTS

The data described in this paper were gathered under Contract No. DACW G1-71-C-0062 with the Philadelphia District, U. S. Army Corps of Engineers. The author wishes to express his thanks to Ms. Arlene Sullivan for typing the original manuscript, to Ms. Linda Jensen for typing the final manuscript, and to Mrs. Dean Pendleton for her assistance in the preparation of figures.

LITERATURE CITED

Pritchard, D. W. and G. B. Gardner. 1974. Hydrographic and ecological effects of enlargement of the Chesapeake and Delaware Canal, Final Report, Philadelphia District, U. S. Army Corps of Engineers; Appen. XIV. Chesapeake Bay Inst., Johns Hopkins Univ. Tech. Rep. 85, Ref. 74-1.

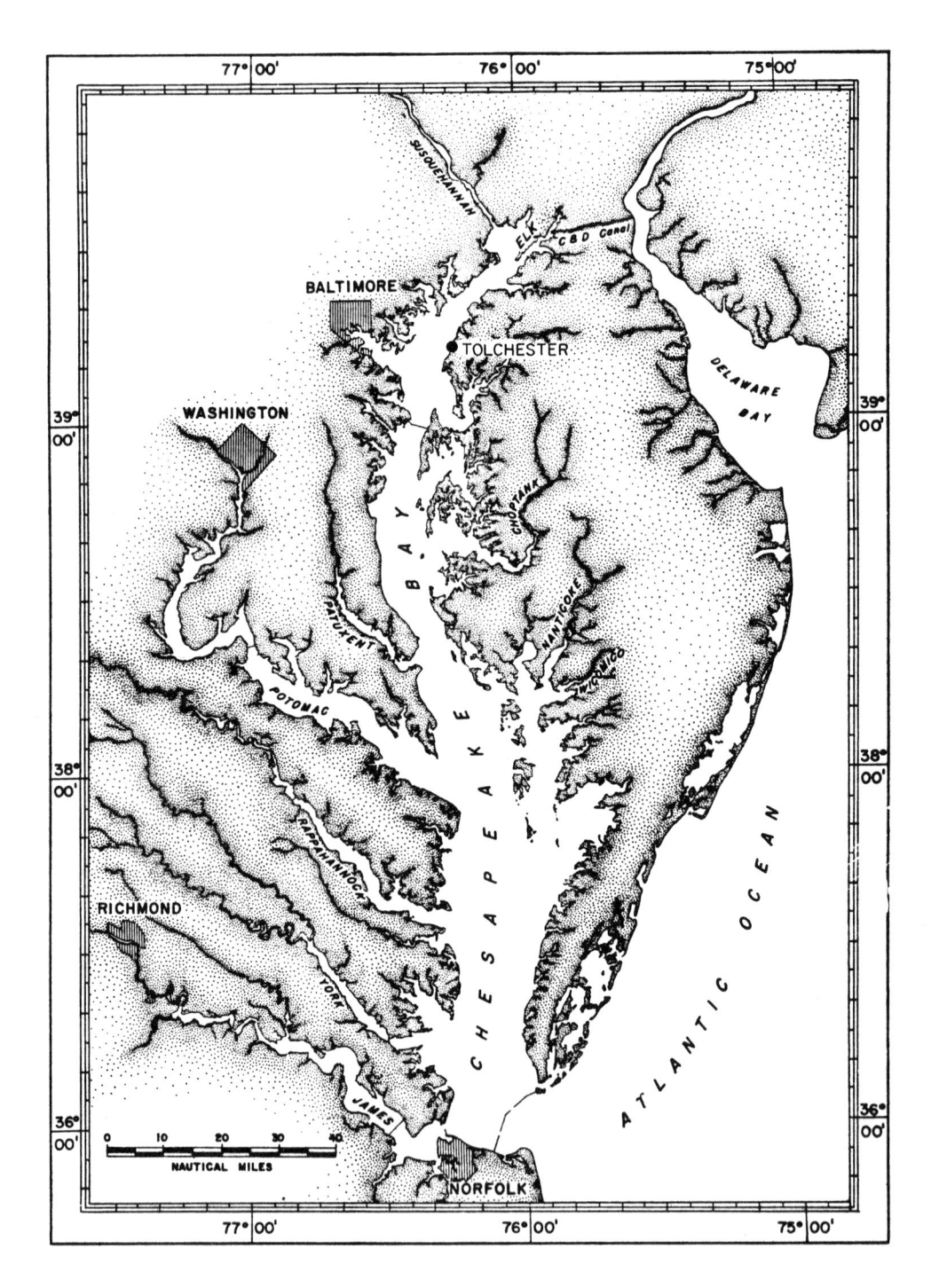

Figure 1. Location map showing relation of C & D Canal to Chesapeake Bay and Delaware Bay.

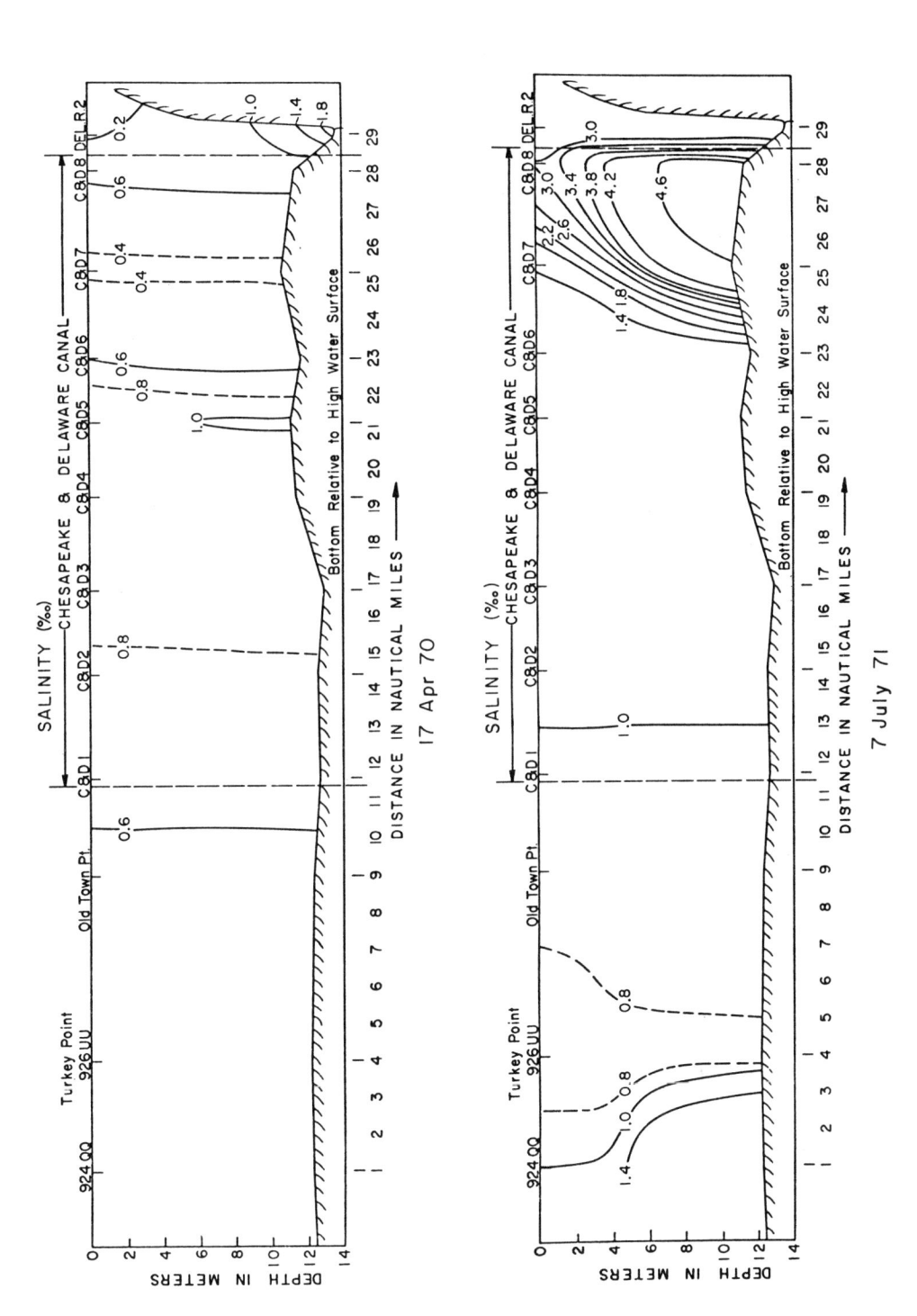

Figure 2. Longitudinal salinity sections showing typical conditions in springtime and in July.

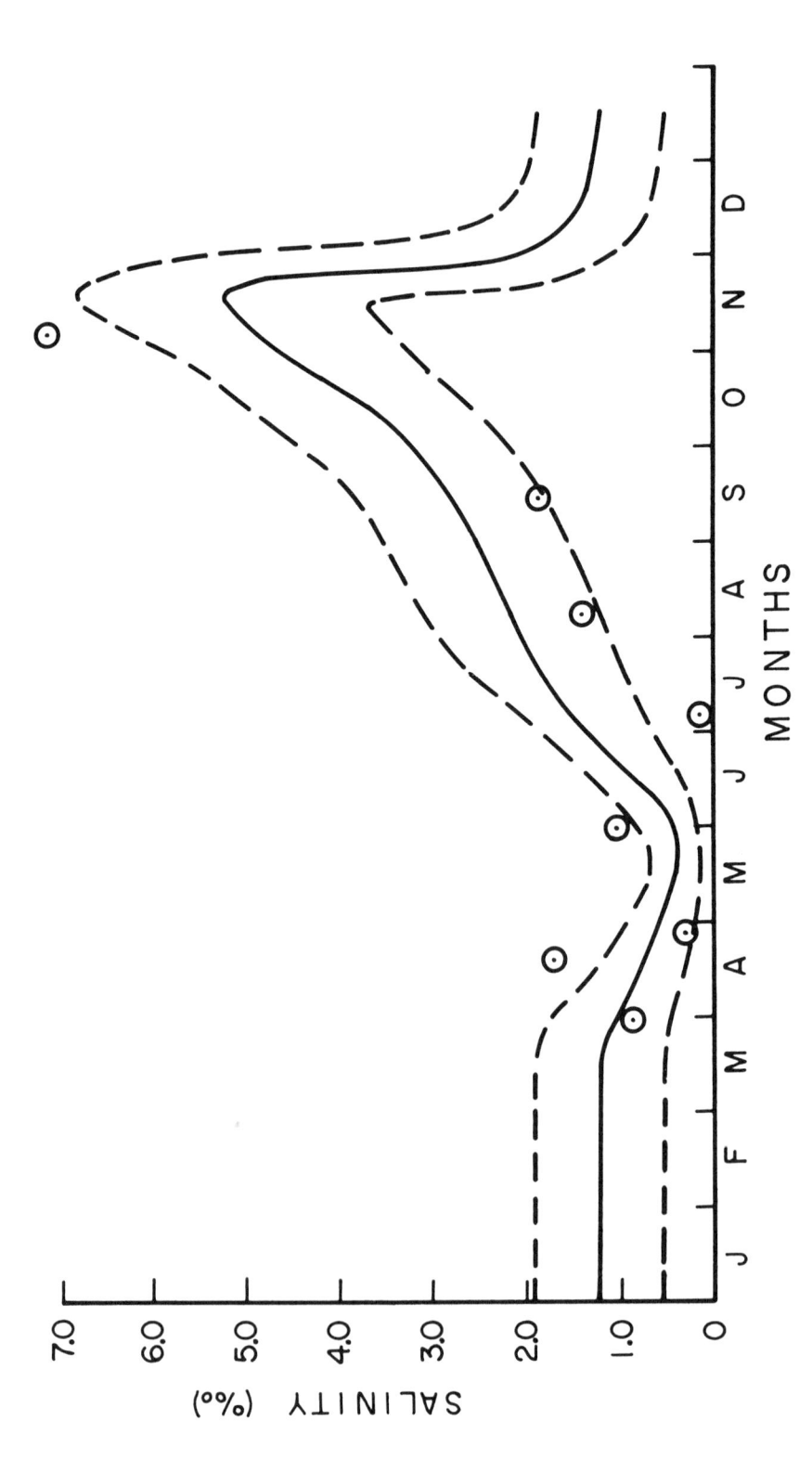

Figure 3. Average annual salinity variation. Solid line represents mean, dotted line standard deviation, and points the values for 1972.

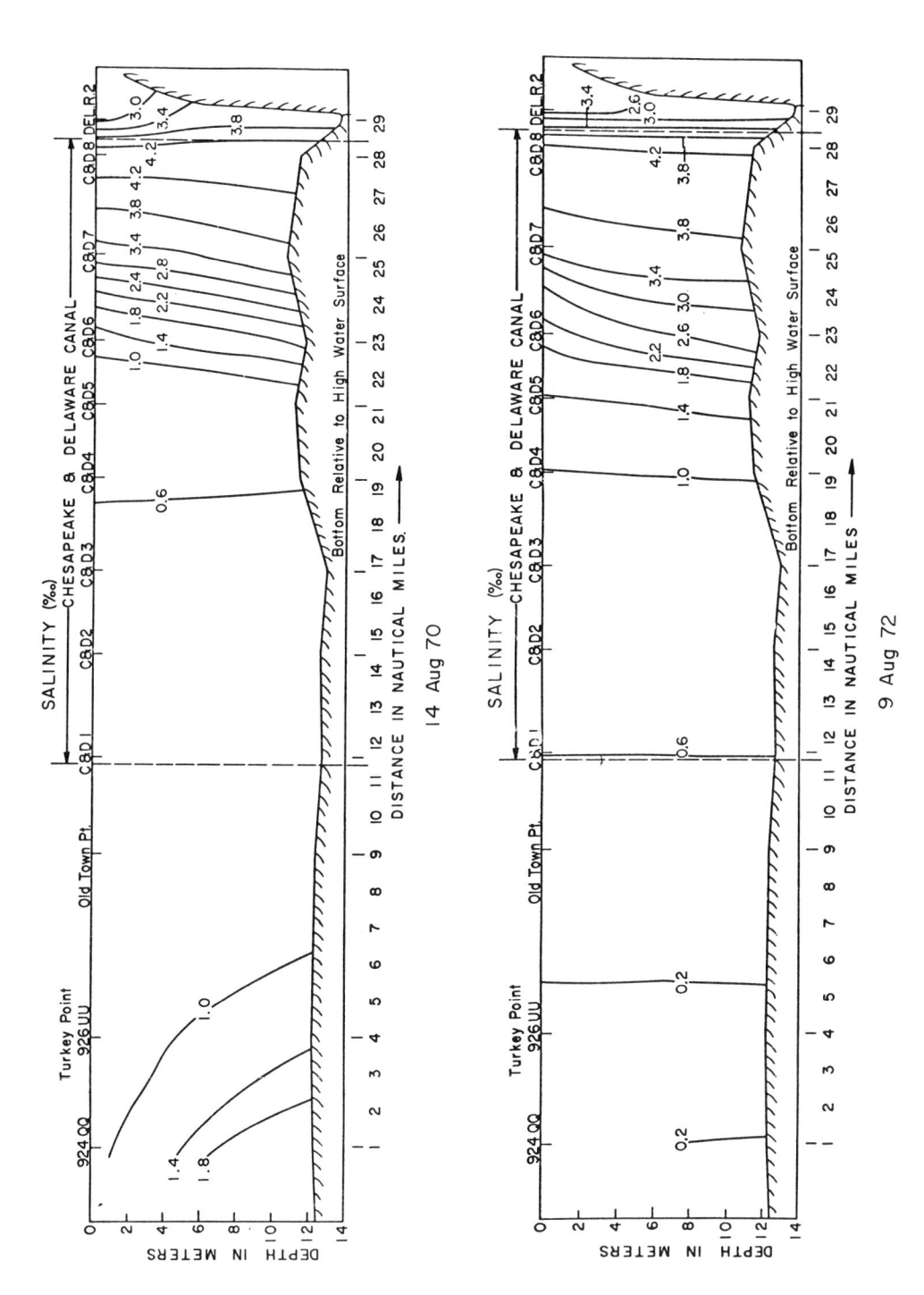

Figure 4. Longitudinal salinity sections comparing conditions in August 1972 with those in August 1970.

AGNES IMPACT ON AN EASTERN SHORE TRIBUTARY:
CHESTER RIVER, MARYLAND

K. T. S. Tzou[1]
H. D. Palmer[1]

ABSTRACT

The impact of Tropical Storm Agnes was not as severe on the Eastern Shore of Maryland as it was in the north central portion of the state. The Chester River Basin received only about six inches (15 cm) of rainfall, while to the west and north, twice this amount was recorded. Although some instantaneous stream discharges increased by a factor of 1000 over average flows for this period, the total runoff from the Chester River was insignificant in comparison to the mass influx from the Susquehanna and other rivers entering Chesapeake Bay from the west.

Water temperature at the lower reaches of the river reverted to a winter-type profile in that a thermal inversion in the upper half of the water column occurred as a result of the influx of colder, fresher Bay water. The salinity dropped to a record low of 0.60 ppt and as a result of this freshening, a great loss of shellfish, particularly the soft-shell clam, occurred within the river system. Surface salinity and temperature generally increased in an upstream direction, a direct effect of the predominance of the Bay water intrusion into the river. High winds accompanying the passage of the storm were responsible for much damage, and as a result of peak winds and high water, a minor seiche was set in motion at the mouth of the river during the most intense phase of the storm.

INTRODUCTION

The Chester River system is typical of the drainage basins on the Eastern Shore of the Chesapeake Bay (Fig. 1). It has an area of approximately 440 square miles (1140 square kilometers). The length of the main course of the river is about 51 miles (82 km) and the contained volume of water at low tide is calculated to be 1.8×10^{10} cubic feet (5.1×10^8 cubic meters) (Webb & Heidel 1970). In the main channel of the river, the boundary between tidal and non-tidal water is considered to lie near Millington, Md., some 42 miles (68 km) upstream from the mouth of the river. Thus, most of the river course is under tidal influence.

EXTREME METEROLOGICAL AND HYDROLOGICAL CONDITIONS

Agnes moved across the Chester River Basin in the early morning of 22 June 1972. The extremes in meteorological and hydrological conditions in this region during the storm are summarized as follows:
- Greatest Daily Precipitation
  6.28 inches (15.9 cm) at Chestertown
- Strongest Wind Velocity
  41 knots (76 km/hr) from North West at Love Point
- Peak Stream Flow
  7500 cfs (212 cms) for a drainage area of 10.5 sq. mi. at Morgan Creek

---
[1]Westinghouse Ocean Research Laboratory, Annapolis, Maryland

- o Highest Water Level
     4.54 feet (1.39 m) above MLW at Chestertown
- o Storm Surge and Seiche
     2.0 feet (0.61 m) of storm surge with a seiche of
     0.6 feet (18 cm) in amplitude and 15 minutes in p
- o Lowest Salinity
     0.6 ppt at the surface and 2.0 ppt on the bottom
     Love Point

STORM EFFECTS ON THE CHESTER RIVER SYSTEM

Agnes did not strike the Eastern Shore area as heavily as other parts of Maryland. The Chester River Basin received about six inches (15 cm) of rainfall which was less than one half of the rainfall in the north central portion of the state. Mean daily discharge from Unicorn Branch and Morgan Creek for June of 1972 is presented in Fig. 2. In one instance, on 22 June, peak flow in Morgan Creek reached 7,500 cfs (about 1,000 times the long term average flow during June). The recurrence frequency of that peak flow was about once in 95 years based on the available data from October 1952 up to December 1972. Fig. 3 shows wind fields during the storm. A dramatic change in wind speed and direction occurred shortly after 1:00 AM on 22 June. The strong effects of the meteorological conditions on the hydrological regime of the river system are illustrated by Figs. 4 and 5. Because of the high wind speed and the sudden change in its direction, a storm surge of 2.0 feet (0.61 m) in height with a seiche of 0.6 foot (18 cm) in amplitude and 15 minutes in period was created at the mouth of the river during the storm.

Comparisons of the vertical distributions of temperature, salinity, dissolved oxygen (DO), and pH along the central channel of river before and after the flood are presented in Figs. 6 and 9. As a result of the storm and floodwater run-off from the Susquehanna River, the salinity at Love Point (Station 5) dropped from 7.07 ppt to 0.58 ppt at the surface and from 7.38 ppt to 4.50 ppt near the bottom. At Station 2, the drop in bottom salinity was even greater, to 1.98 ppt. This freshwater inflow also changed the pH of the surface water from 7.80 to 7.10 while DO increased from 7.55 ppm to 8.35 ppm. Because of the large volume of relatively cold, fresh water entering the Bay from the Susquehanna River, the vertical temperature distribution in the upper half of the water column at Love Point became inverted, a situation similar to winter temperature distributions. The lower half of the water column retained the normal (summer) thermal structure where temperature decreases with depth. The temperature and salinity distributions in Fig. 10 are data taken one week after the storm. Surface temperature and salinity generally increased in the upstream direction from 20.0° C and 0.6 ppt at Love Point to around 25.0° C and 2.9 ppt in the Queenstown to Tilghman Neck area, giving the impression that a mass of water was effectively blocked off in that portion of the river by the flood waters entering the Bay from the Susquehanna River. The same situation is reflected in the deeper waters, indicating that flow through the lower Chester River was severely restricted during this period. Even though a marked drop in salinity occurred over most of the water column in the Queenstown to Tilghman Neck area, it was not as drastic as in Chesapeake Bay proper and the temperature profile in this portion of the river remained more typical of summer conditions.

ACKNOWLEDGEMENTS

This paper is part of an interdisciplinary study, a jointly funded program of the State of Maryland and Westinghouse Electric Corporation. The writers also wish to express their appreciation to Mr. W. J. Moyer of NOAA, former Climatologist of the State of Maryland and Delaware for providing information on Agnes, to Mr. P. Pfannebecker of Water Resources Division of the U. S. Geological Survey for furnishing stream flow data, to Mr. S. C. Berkman of National Ocean Survey of NOAA for supplying tide data.

LITERATURE CITED

Webb, W. E. and S. G. Heidel. 1970. Extent of brackish water in the tidal rivers of Maryland. Maryland Geological Survey Rep. No. 13.

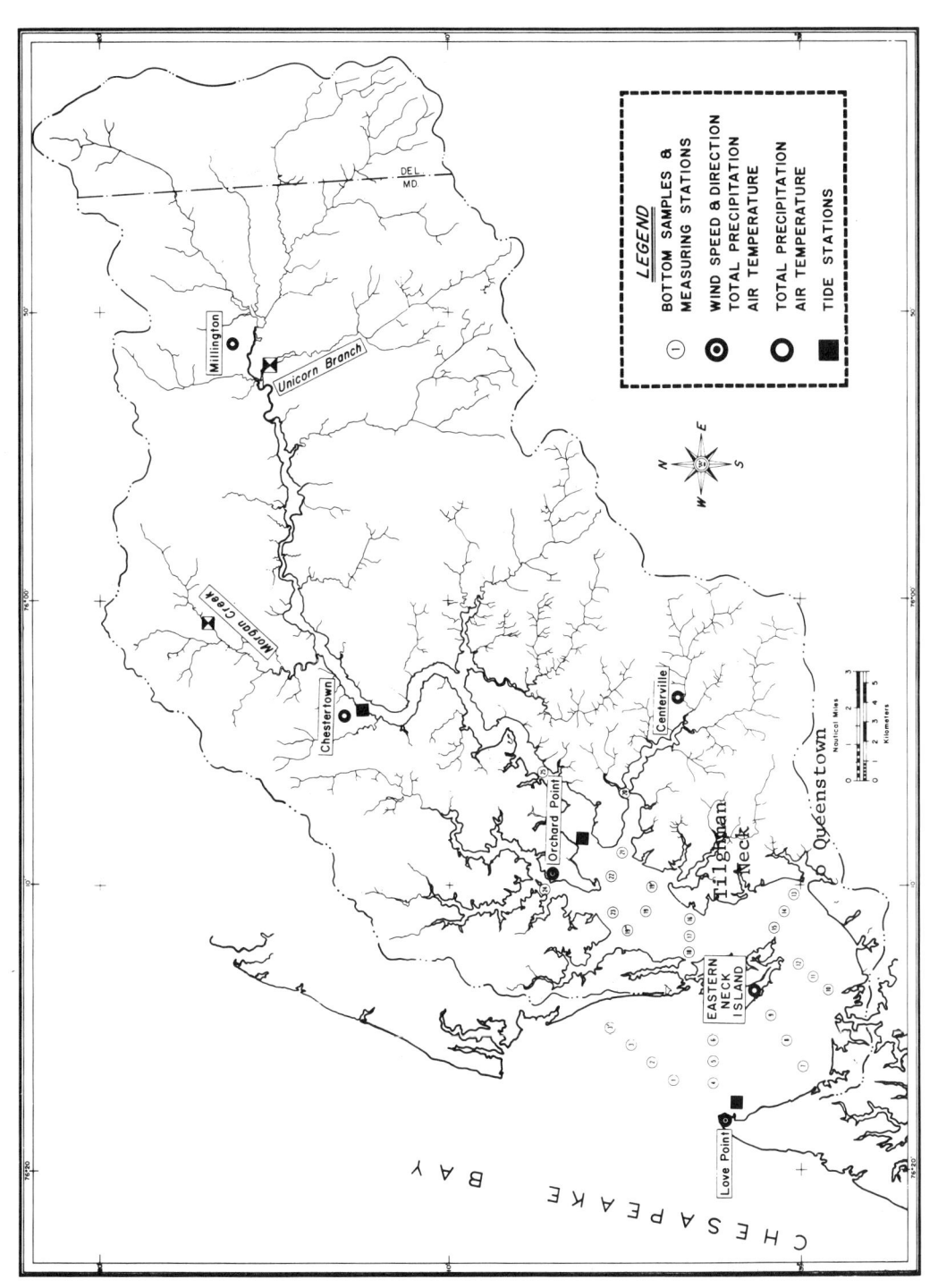

Figure 1. Chester River drainage basin.

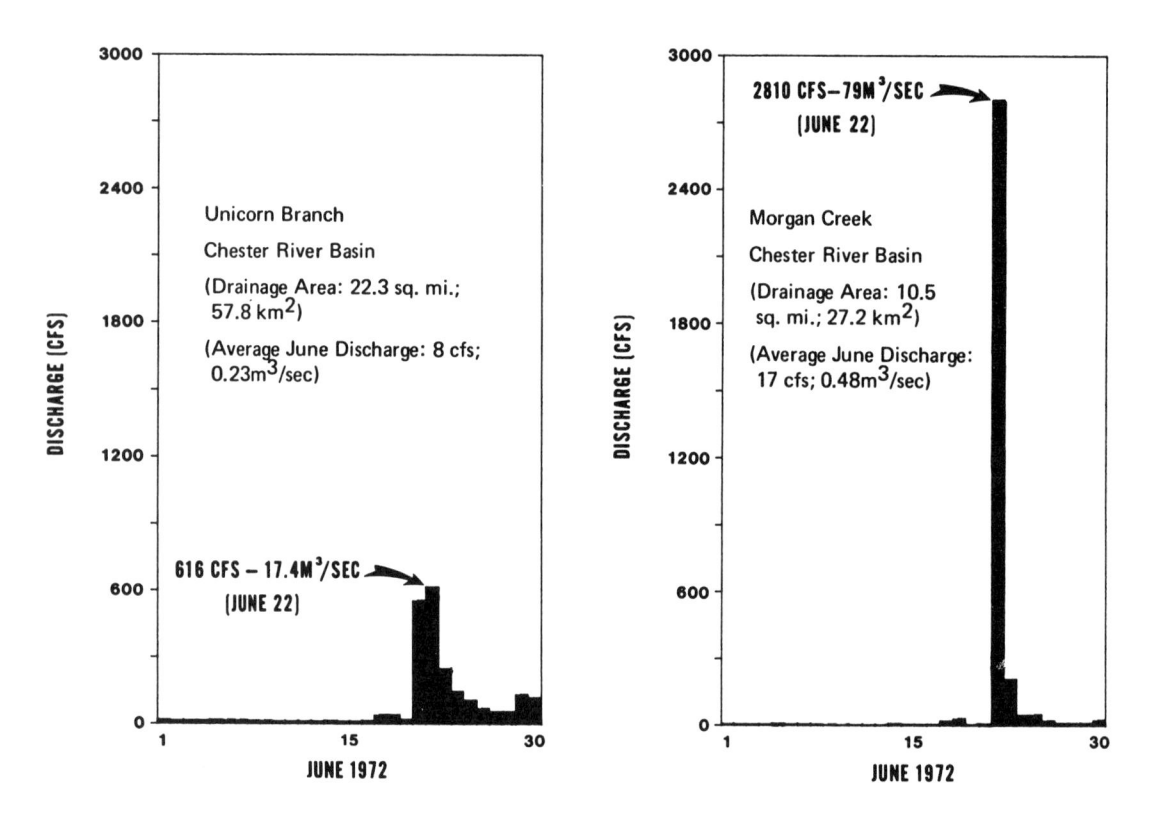

Figure 2. Mean daily discharge from Unicorn Branch and Morgan Creek before and after Agnes.

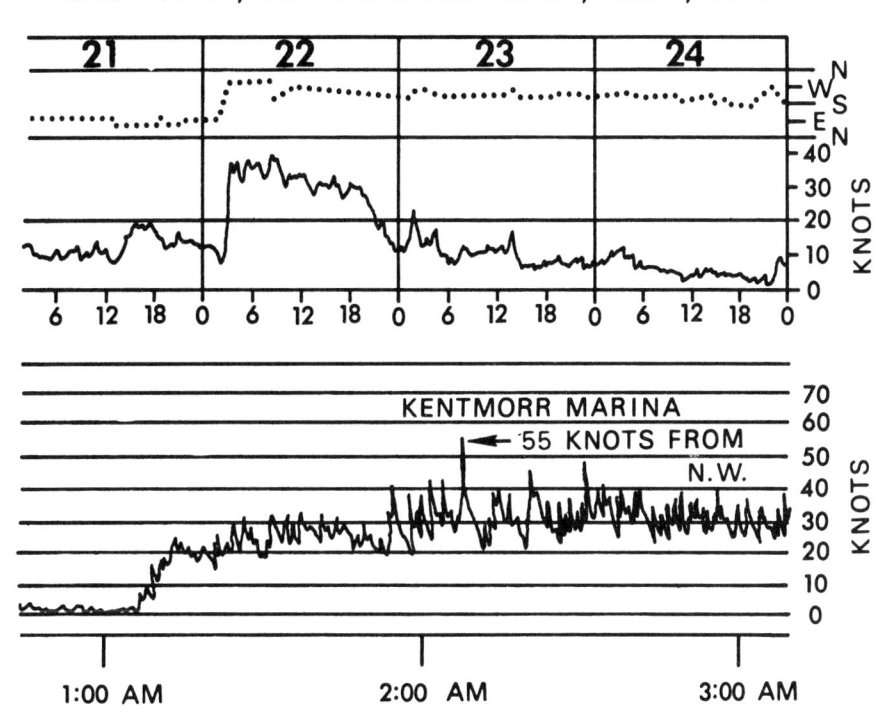

Figure 3. Wind speed and direction at Love Point and Kentmorr Marina during Agnes. Kentmorr Marina records with 55 knots (28.2 m/sec) peak are for early morning, 22 June 1972.

142 Tzou, Palmer

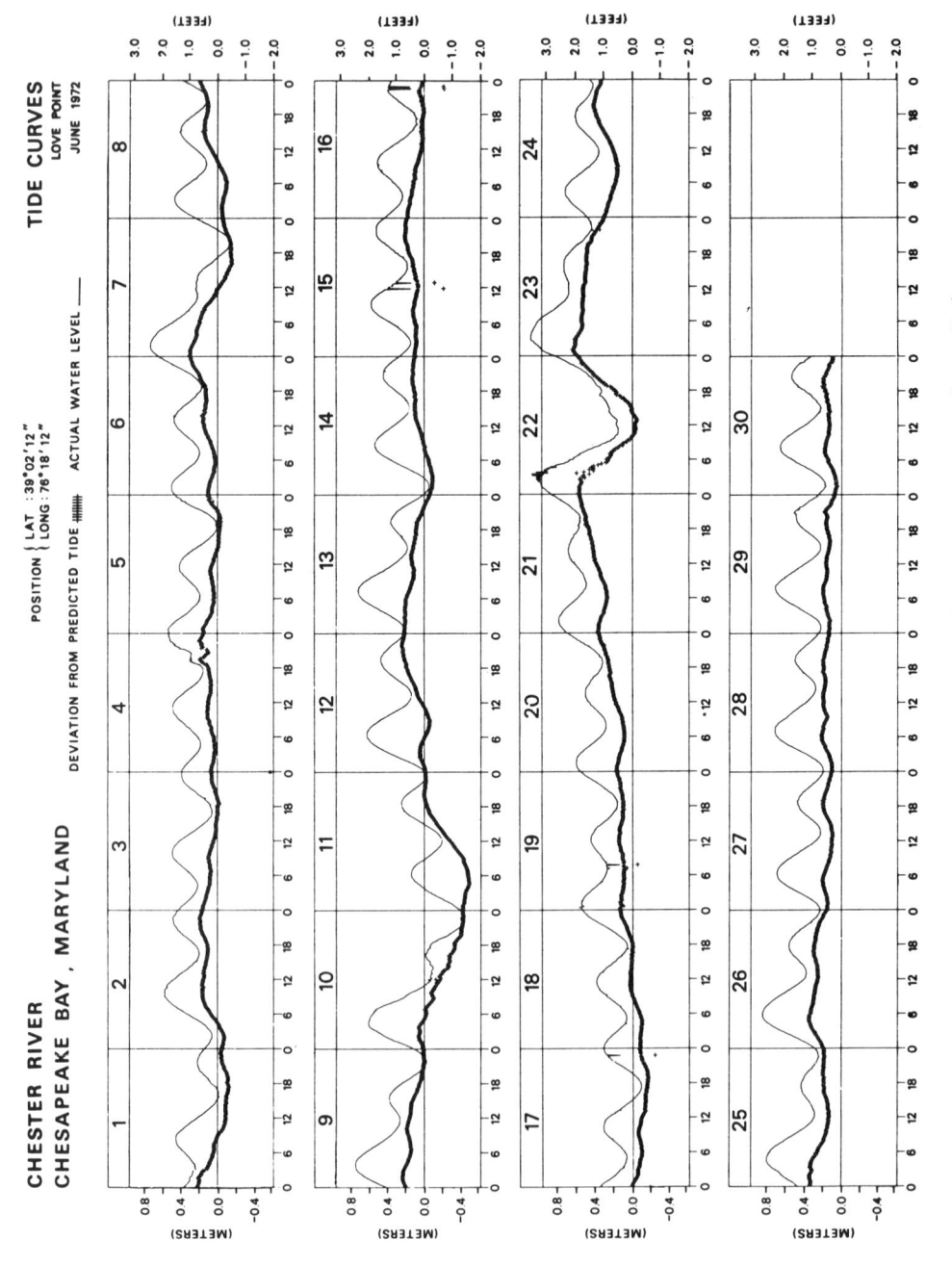

Figure 4. Actual water level and storm surge at Love Point.

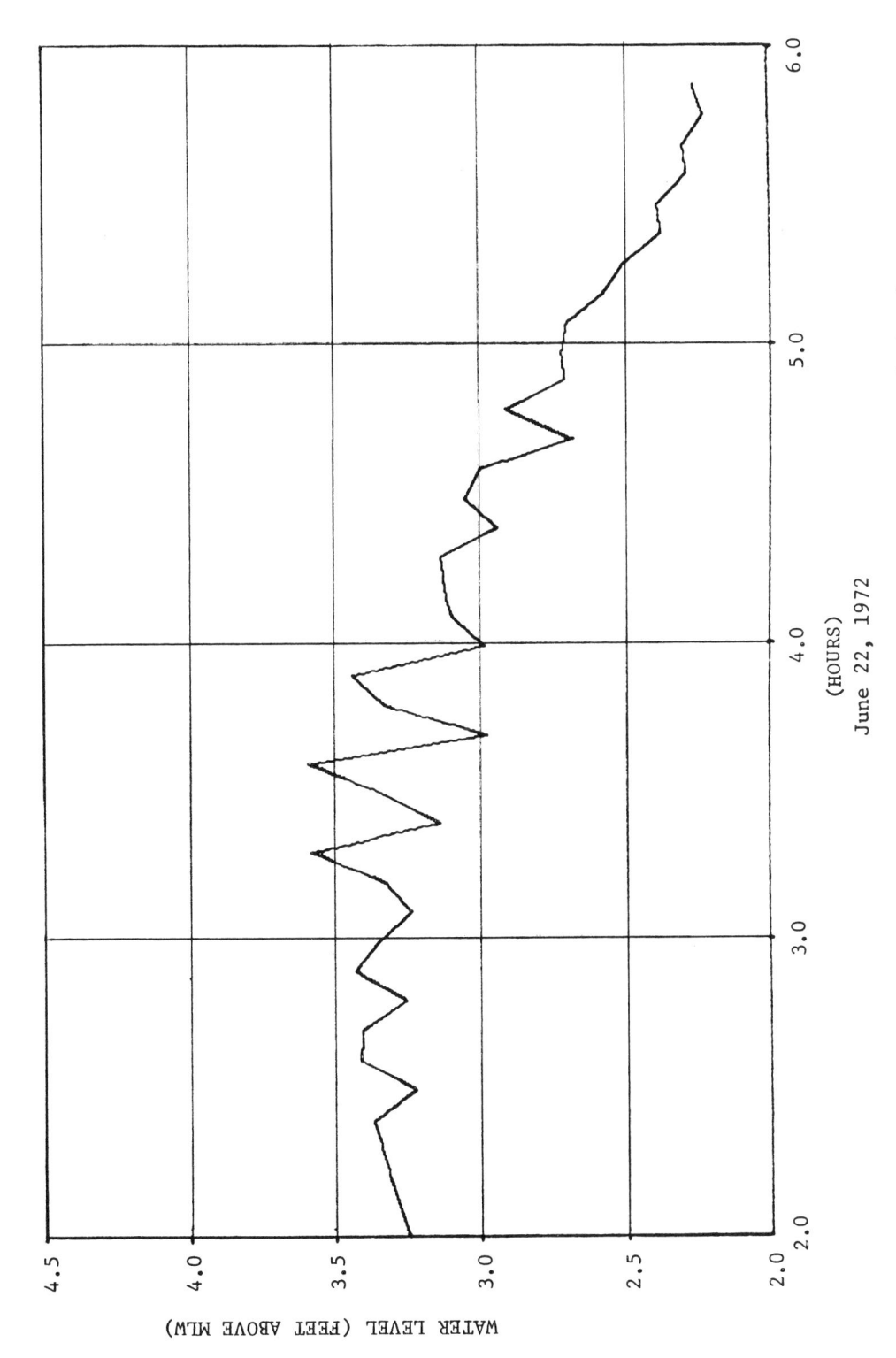

Figure 5. Seiche at Love Point of the Chester River during Agnes.

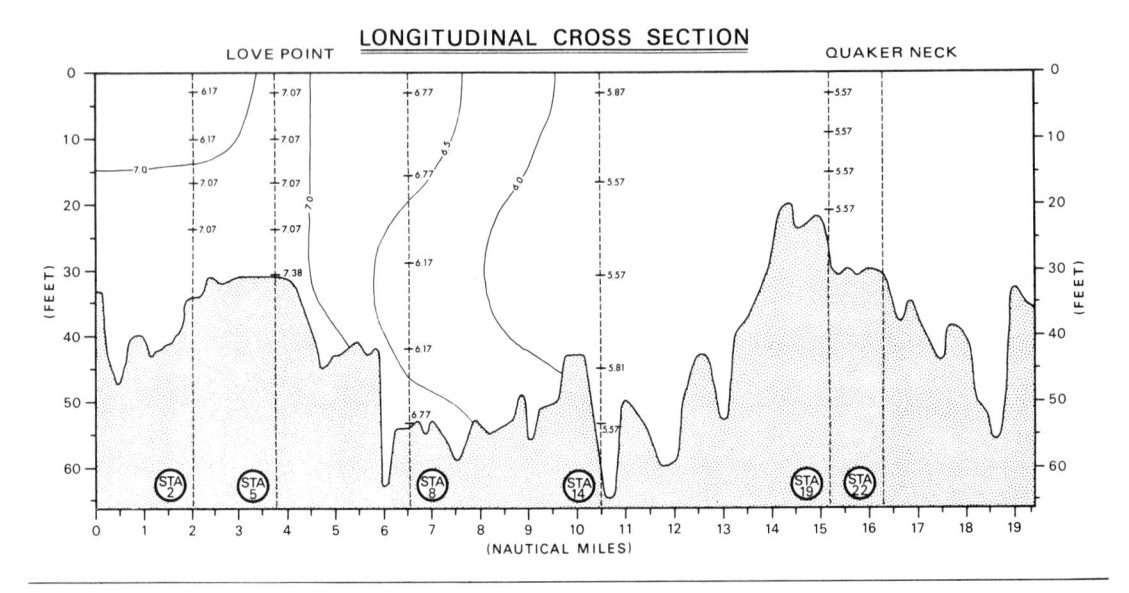

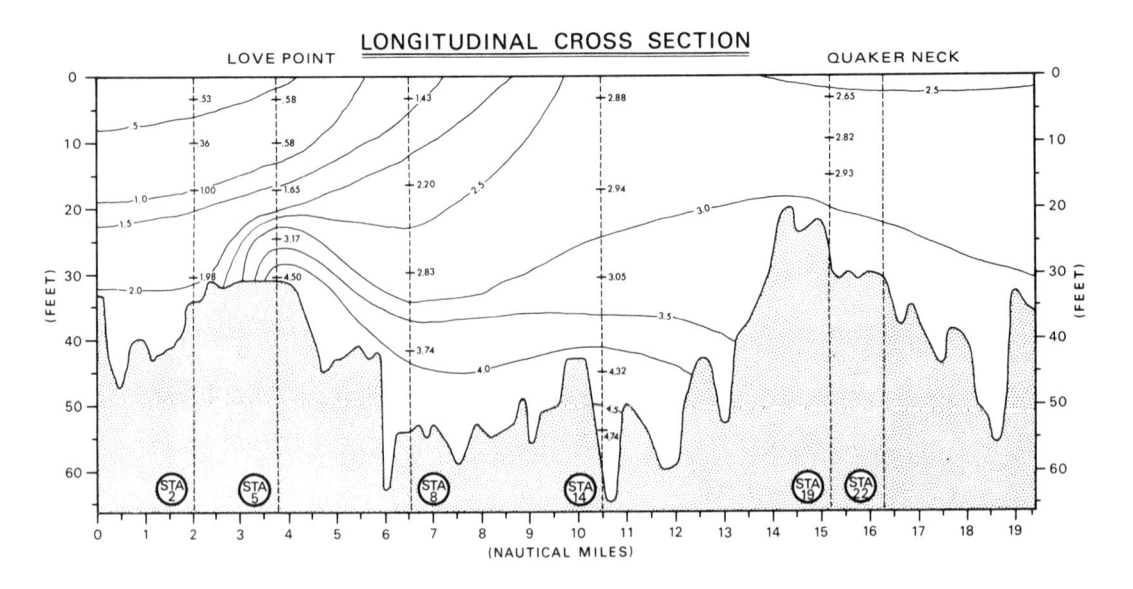

Figure 6. Variations in salinity along the main channel of the Chester River before and after Agnes.

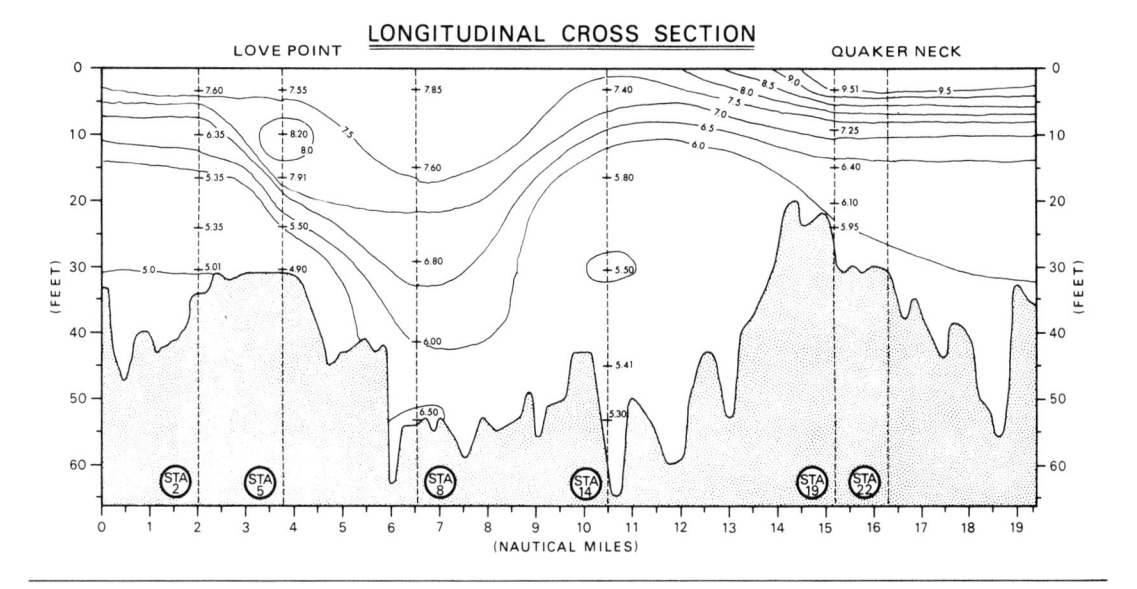

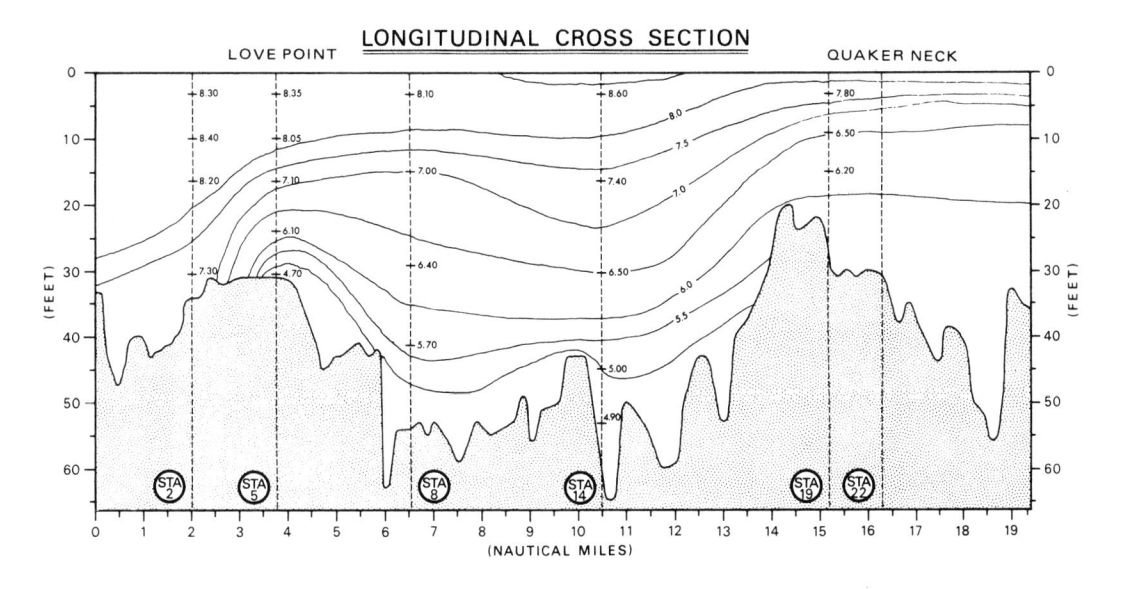

Figure 7. Variations in dissolved oxygen along the main channel of the Chester River before and after Agnes.

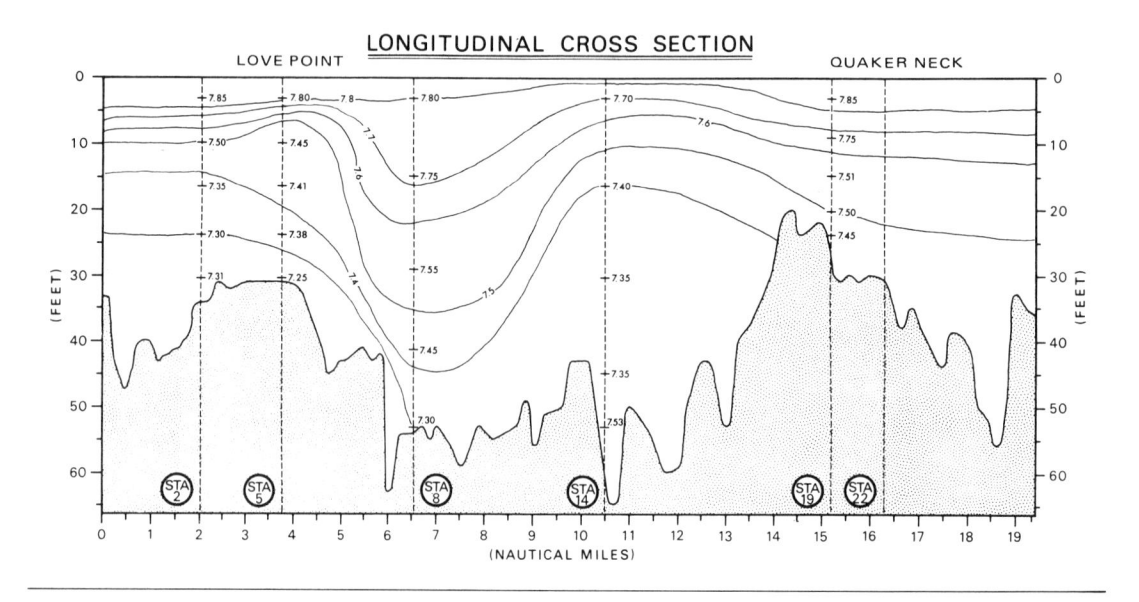

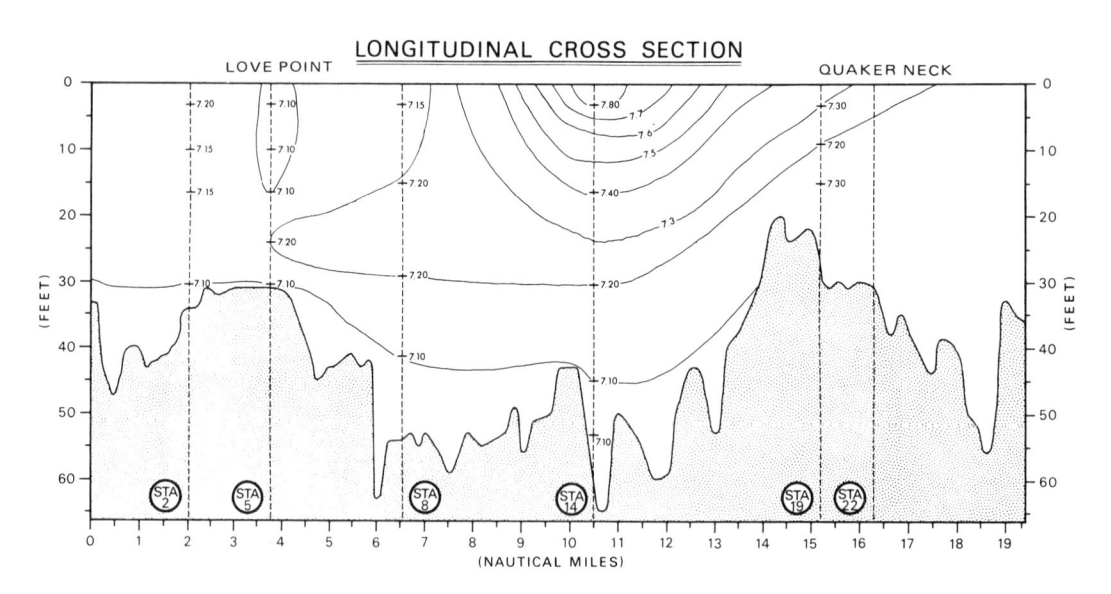

Figure 8. Variations in pH along the main channel of the Chester River before and after Agnes.

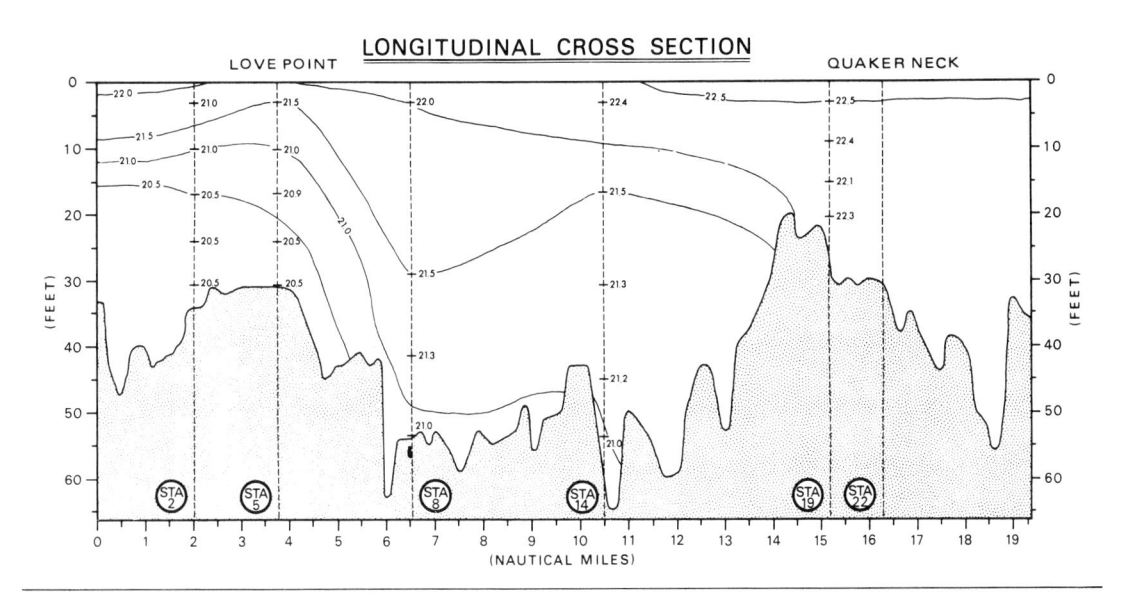

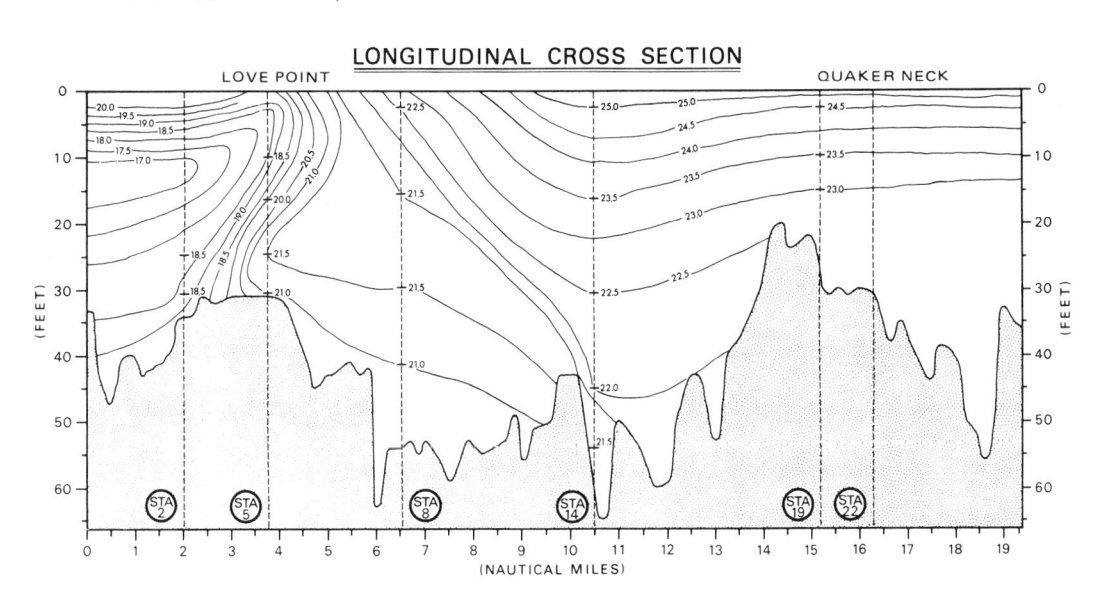

Figure 9. Variations in water temperature along the main channel of the Chester River before and after Agnes.

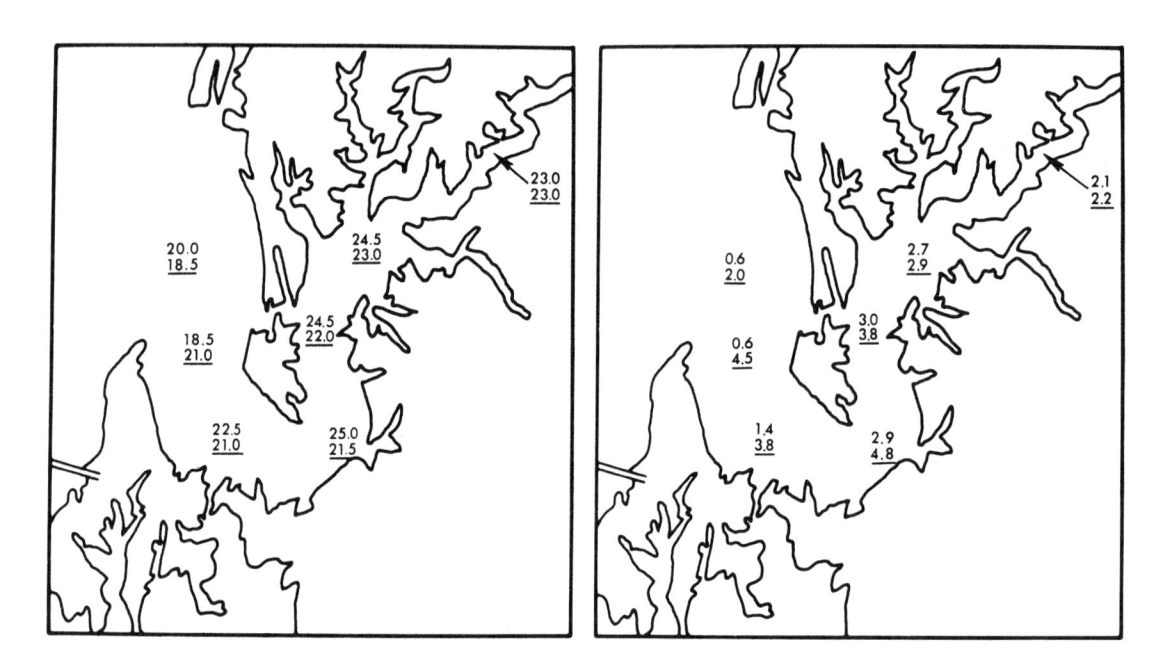

Figure 10. Temperature and salinity distributions in the Chester River one week after Agnes (30 June 1972). Underlined values are bottom measurements; others are surface measurements.

TRIBUTARY EMBAYMENT RESPONSE TO TROPICAL STORM AGNES:
RHODE AND WEST RIVERS[1]

Gregory C. Han[2]

ABSTRACT

Tributary embayments are little affected by direct rainfall and the freshwater runoff from their small drainage basins due to the large effect of temporal changes of salinity in the main estuarine system. Salinity observations in the Rhode and West Rivers from 14 June - 14 August 1972 documented the embayment's response to the changing salinity structure in the bay from Tropical Storm Agnes. By 3 July an influx of fresh water (1 ppt) from the bay had displaced and mixed with the water resident in the embayment and produced water with an average salinity of 2.2 ppt. Throughout July the embayment responded to the slow increase in salinity of the layer of the Bay above the halocline. In early August more saline water (7-9 ppt) from below the sharp halocline in the bay reached the depth of the mouth of the embayment (4 m), causing rapid inflow of higher salinity water along the bottom and producing a 0.8 ppt/day increase in salinity and a 3 ppt/meter vertical salinity gradient in the inner reaches of the embayment. Results from a dye tracing experiment during the highly stratified period confirmed the presence of a strong two-layered flow and reduced vertical mixing across the density gradient.

INTRODUCTION

Tributary embayments are estuaries whose salt balance is primarily dependent upon salinity variations in the bay to which they are tributary rather than upon the variation in freshwater flow from their drainage basins. The Rhode and West Rivers, which are located in Fig. 1, are two tributary embayments of the Chesapeake Bay. Salinity observations made from 14 June - 16 August 1972, as discussed herein, document the response of the embayments to salinity changes in the bay due to Agnes and show how that response is a function of not only the salinity in the bay but also its vertical structure. The study will concentrate on the Rhode River, but measurements on the West River will also be presented and comparisons made between the two. The effect of the increased runoff from the Susquehanna River upon the salinity in the bay is discussed by Schubel, Carter and Cronin (this volume).

EQUIPMENT AND PROCEDURE

Temperature and conductivity measurements were made at the stations shown in Fig. 2 at approximately weekly intervals from 14 June to 2 August 1972 and at one to three day intervals from 2 August to 16 August 1972. The first observation

---
[1]Contribution No. 207, Chesapeake Bay Institute, The Johns Hopkins University.
[2]Present address: NOAA, Atlantic Oceanographic and Meteorological Laboratory, 15 Rickenbacker Causeway, Miami, Fla. 33149

after Agnes was not made until 27 June. Measurements were taken within two hours of high water slack in most cases, but were always within the interval for the August intensive study period. Station spacing is approximately 1 km longitudinally, 300 to 700 meters laterally and 1 meter vertically to the bottom. The vertical spacing was reduced in regions of sharp haloclines.

Measurements of temperature and conductivity representative of the region of interest in the bay were taken by Schubel et al. (this volume) at Station 858C (Fig. 1) which is 5.7 km northeast of the study area in the main channel of the bay. Their observations were not all at the same tidal phase however.

All measurements were made with an Interocean temperature and conductivity sensor using a display unit built by the Chesapeake Bay Institute. Accuracy is ± .02°C for temperature and ± .03 mho/cm for conductivity at normal salinities but the instrument is probably less accurate at the low salinities encountered (<1 ppt).

A dye tracing experiment was also conducted in the Rhode River during the study period. A flourescent dye (Rhodamine WT 30%) was metered continuously at 3.2 ml/min from a moored dinghy at a station midway between stations RR4A and RR3A at 0.5 meters beneath the surface. Dye tracing was conducted with a Turner Model III fluorometer with a high volume flow cell, using standard fluorometric techniques (Carter 1972). Dye observations were made by making lateral transects while pumping water from a 0.5 meter depth and continuously recording the fluorometer readings. Vertical profiles were also made at all regular temperature and conductivity stations. These stations did not necessarily correspond to the position of maximum concentration found on the lateral transects but distinct plumes of dye were not usually present away from the dye source. Only the vertical profiles of dye are discussed here. The dye concentrations are plotted in arbitrary units (1 unit = 0.143 ppb dye). The fluorometer has a detection limit of 0.1 relative units.

OBSERVATIONS AT BAY STATION 858C

The salinity structure with time at Station 858C is shown on Fig. 3. The salinity distribution at 858C is taken as indicative of the bay salinity near the Rhode and West Rivers since conditions at this point are free of any influence from tributaries and are a consequence of the response of the bay to the Susquehanna outflow.

Two major events can be seen in the record at 858C, each of which affect the salinity distribution in the Rhode and West Rivers. The two major events were the rapid decrease in salinity due to Agnes and the rapid increase in early August. These events were separated by a 10-day period with no halocline present which allowed vertical mixing to raise the surface salinity. Following Agnes, a sharp drop in the vertically averaged salinity is recorded with the presence of a 1 ppt/m halocline at 5-10 meters. Minimum surface salinities (0.31 ppt) were found on 1 July 1972. In early July, conditions produced a salinity maximum at the surface (3.73 ppt) with no halocline present followed by a second decrease in mid-July to a relative salinity minimum (1.55 ppt). In late July and early August, salinities rose sharply; and by 5 August, a 2.8 ppt/m halocline had developed. In mid-August, the bay developed a two-layered structure with homogeneous top and bottom layers separated by a 0.9 ppt/m halocline.

OBSERVATION IN RHODE AND WEST RIVERS

The results of the measurement program are shown primarily as selected longitudinal sections of salinity. Lateral variations were generally not important. Temperature measurements are not shown because the unusual effects of Agnes were confined to the salinity field only.

The influx of fresh water from the bay after Agnes was first observed on 27 June. No observations were made between 14 and 27 June but Cory and Redding (this volume) observed the initial effect of Agnes as a decrease in salinity above station RR4A on 22-23 June from upland runoff followed on 24 June by a salinity increase to values comparable to pre-Agnes salinities as runoff diminished. They first observed decreasing salinities due to the influence from the bay on 25 June.

The observations on 27 June (Figs. 4, 5, 6, 7) show a well developed circulation with water of less than 2 ppt salinity moving in at the surface, displacing and mixing only slightly with water resident in the embayments. Charts of surface and 3-meter salinities (Figs. 4 & 5) show the striking nature of the circulation as saline water moved out through the deeper channel. Bear Neck Creek, a deep tributary of Rhode River, also showed this two layer structure while the other shallower tributaries had already freshened. By 29 June (Fig. 8), most of the more saline water had been displaced from the Rhode River except for a mass of 2.5 - 3.5 ppt water at the bottom near the mouth; but by 3 July, the entire embayment was less than 2.5 ppt.

Minimum volume averaged salinities for the period (2 ppt) were not recorded until 7 July (Fig. 9). Normal one-layered exchange occurred throughout the remainder of July (Fig. 10). On 2 August (Fig. 11), salinities in the bay began to increase; by 7 August (Figs. 13 & 14), a halocline began to appear in the bay; and by 8 August (Fig. 15), a two-layered advective flow became established in the embayment. The fact that a strong two-layered flow occurred on 8 August can be checked by observation of the dye distribution. On 7 August when the dye release began, measurements showed that the dye mixed rapidly into the water column (4-4.2 ppt) and a salinity intrusion from the bay had just begun. By 8 August, 6-7.5 ppt salinity water had moved into the lower layer displacing upwards the 4 ppt water which was tagged with dye. The dye distribution (Fig. 16) shows this upward displacement and the lack of mixing with water from the bay. On 9 August (Fig. 17), conditions remained much the same but the halocline in the bay had disappeared and the stratification in the embayment began to disperse. The dye measurements (Fig. 18) showed that vertical mixing between the upper and lower layers was still restricted on 9 August; but mixing on 10 August, probably from the 7-19 knot NW winds, caused the halocline to disappear, and both the salinity and dye distributions on 11 August appeared as vertically homogeneous (Figs. 19 & 20). On 14 August (Figs. 21 & 22), the salinity patterns had regained a more normal magnitude and vertical structure. Salinity continued to increase at an average rate through November 1972. For hydrographic purposes, the unusual events associated with Agnes which had begun on 25 June were finished by 14 August, though a sharp halocline persisted in the bay beyond that date.

DISCUSSION

The response of the embayments to the bay is seen in Fig. 23a as the 0-4 meter average salinity at Station 858C and the volume average of segment B1 at the mouth of the two embayments. The response of the interiors of the embayments to changes at the mouth is seen in Fig. 23b as the volume averaged salinity versus time of two comparable interior segments of the Rhode and West Rivers. Salinities in segment B1 are a filtered response to conditions in the bay with both reduced amplitude and a phase lag, which can also be said of the salinity response of interior segment WR2 and RR2 to segment B1.

The rapid changes in bay salinities, i.e., the decrease in late June and the increase in early August, were quickly duplicated in the embayments; whereas, during the more gradual changes in bay salinity in July, the exchange with the embayments was much slower. The relative salinity maximum in the bay on 10 July increased the salinity in the embayments, but a less well-defined maximum was achieved due to the slower response of the embayments. During the times when rapid exchange was occurring, the salinity gradient between the upper and lower layers of segment B1 (Fig. 23c) was large (0.85 ppt/m). These large gradients in salinity are a result of the salinity stratification in the bay causing a two layer advective exchange. On 7-9 August, the embayments were exchanging rapidly, not only because the salinities in the bay were increasing, but largely because the halocline in the bay was shallow enough to allow water in the halocline to intrude directly into the embayments. When there was little stratification in the bay as in mid-July, or when the halocline was deep enough so the embayment could only communicate with the upper layer of the bay as on 14 August, the exchange of the embayments was much slower. During the time when there is little stratification in the bay, the embayments lend themselves more readily to modeling in one dimension with all advective fluxes except river flow treated as diffusion.

The depth of the halocline at 858C is always greater than that observed at the mouths of the embayments. This difference can be due to a mechanism such as the tilting of the halocline across the bay due to the wind. No data are available to further illustrate the details of the position or movement of the halocline during this period.

The West River has not been discussed directly because better salinity data and dye study results are available from the Rhode River and the West River is very similar to the Rhode River in morphology and circulation patterns. The chief difference between the two embayments is that West River has a shorter exchange time. This can be seen qualitatively in Figs. 7, 12, & 14 by comparing the corresponding figures from the Rhode River. In each case, water from the bay has moved further into the West River and changed the salinity faster than in the Rhode River. The freshwater intrusion on 27 June caused the salinity to drop to less than 3 ppt throughout a large portion of the West River while a major portion of the Rhode River had salinities greater than 4 ppt. Fig. 23b shows how segment WR2 always leads segment RR2 during any salinity change because of West River's more rapid response.

The effect of the circulations described could be important to organisms in the embayments. The delayed response of the embayments to changes in the bay directly following Agnes prevented the salinities in the embayments from decreasing below 2 ppt when bay salinities were less than 1 ppt. In August, the high salinity waters in the halocline were also high in nutrients (Correll, personal communication). Intrusion of these waters into the embayments and subsequent vertical mixing brought these nutrients into the euphotic zone. The same waters in the bay were prevented from mixing into the surface layer by the strong halocline which persisted into the fall.

CONCLUSIONS

The freshwater runoff from the Susquehanna River due to Agnes produced unusual salinity patterns in the Chesapeake Bay off the mouth of the Rhode and West Rivers. The observations made in these two embayments illustrated the effect of the rapid change in salinities and the importance of the depth of the halocline in determining the rate of exchange between the embayments and the bay.

Circulation patterns and rapid changes in salinity similar to those in the Rhode and West Rivers can be expected to have occurred in other tributary embayments throughout the bay. The controlling factor in the time and speed of salinity recovery in other embayments would depend upon their depth and the behavior of the halocline in the adjacent section of the bay. The eastern shore and the southern part of the western shore which had little rainfall in their local drainage basins could be expected to have had events similar to those observed in the Rhode and West Rivers.

ACKNOWLEDGMENTS

The author wishes to thank Mr. H. H. Whaley, Mr. R. C. Whaley and the staff of the Chesapeake Bay Institute and Dr. F. Williamson and the staff of the Chesapeake Bay Center for Environmental studies for use of their field facilities and their cooperation throughout the study. Mr. H. Carter reviewed the manuscript and made helpful suggestions. Mrs. Dean Pendleton prepared the figures and Miss Kay Snyder typed the manuscript. This work was supported under NSF/RANN Grant GI 34869.

LITERATURE CITED

Carter, H. H. The measurement of Rhodamine tracers in natural systems by fluorescence. Presented at Symposium on the physical processes responsible for the dispersal of pollutants in the sea, 1972. Arhus, Denmark (in press).

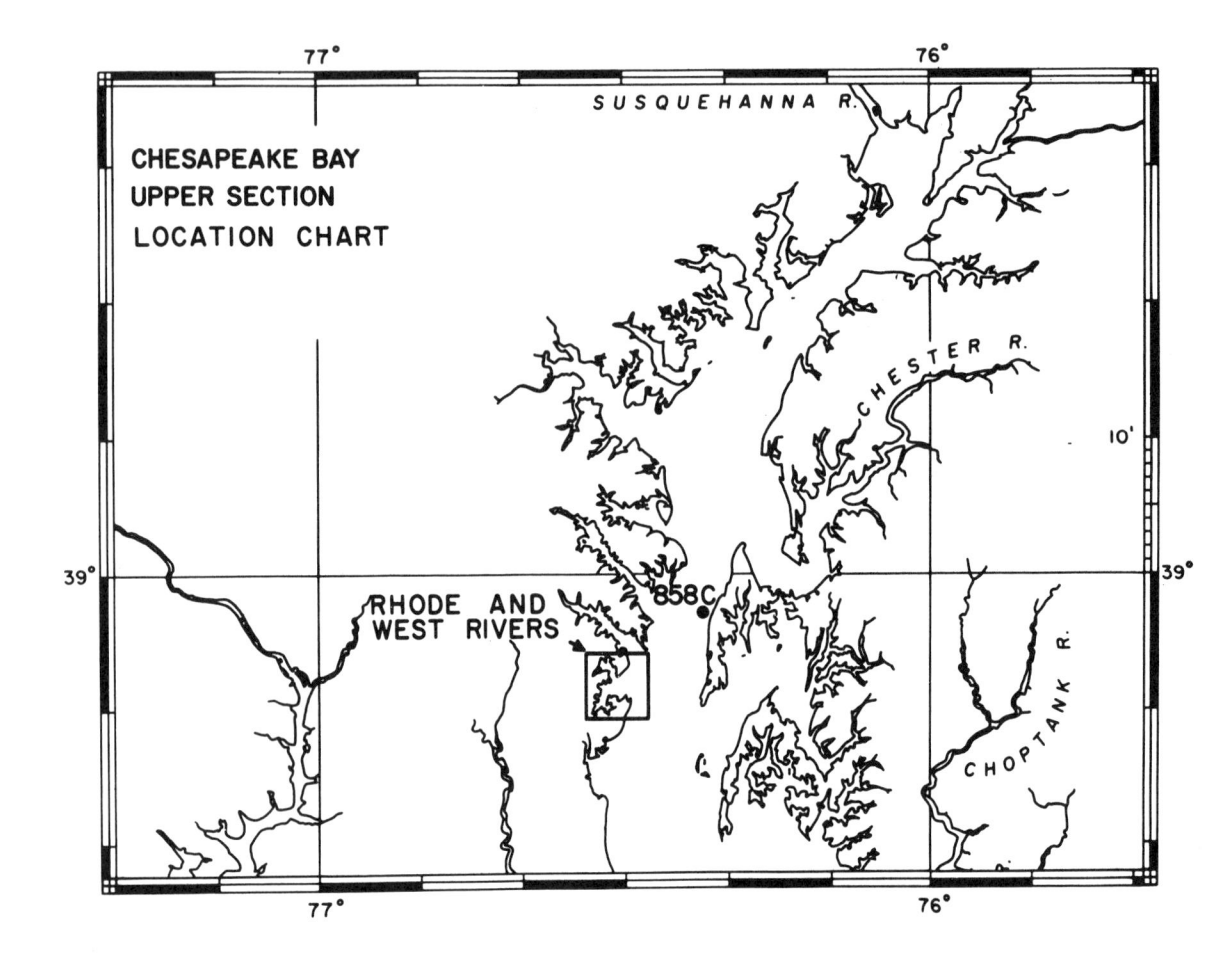

Figure 1. Location of Rhode and West Rivers and Station 858C in upper section of Chesapeake Bay.

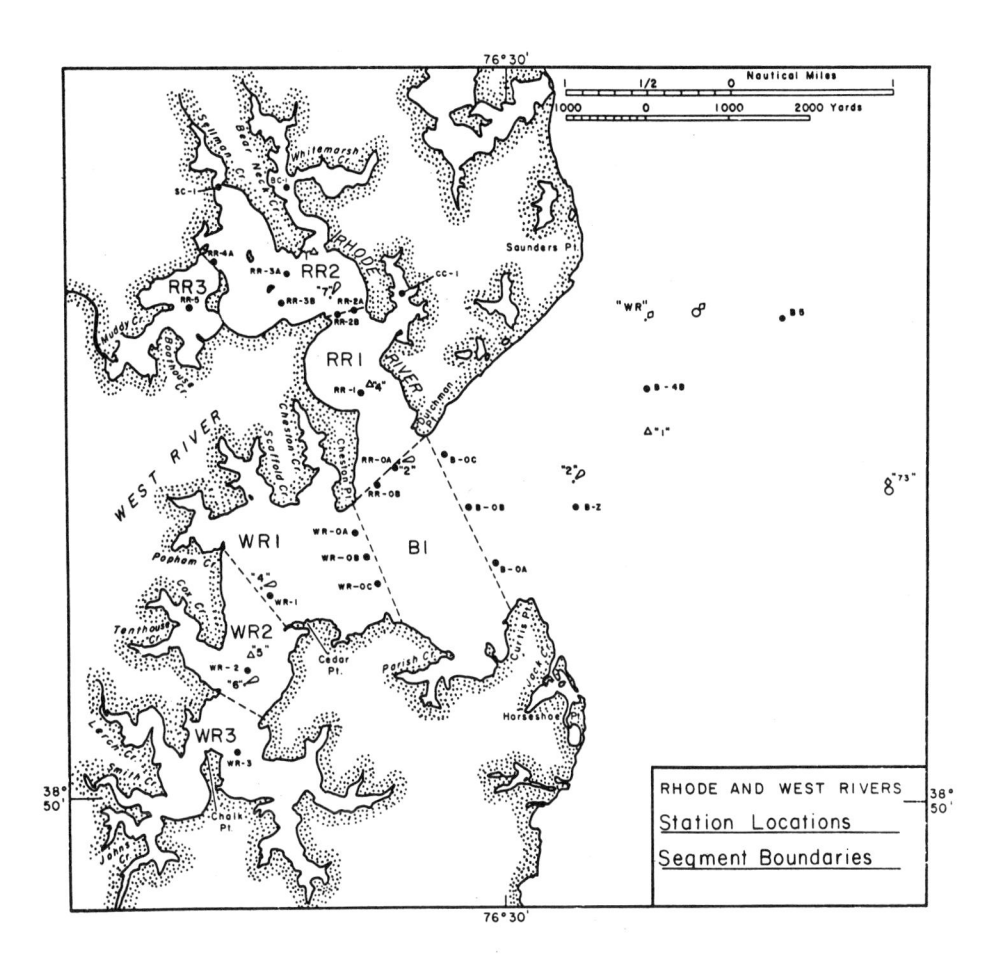

Figure 2. Station locations and segment boundaries.

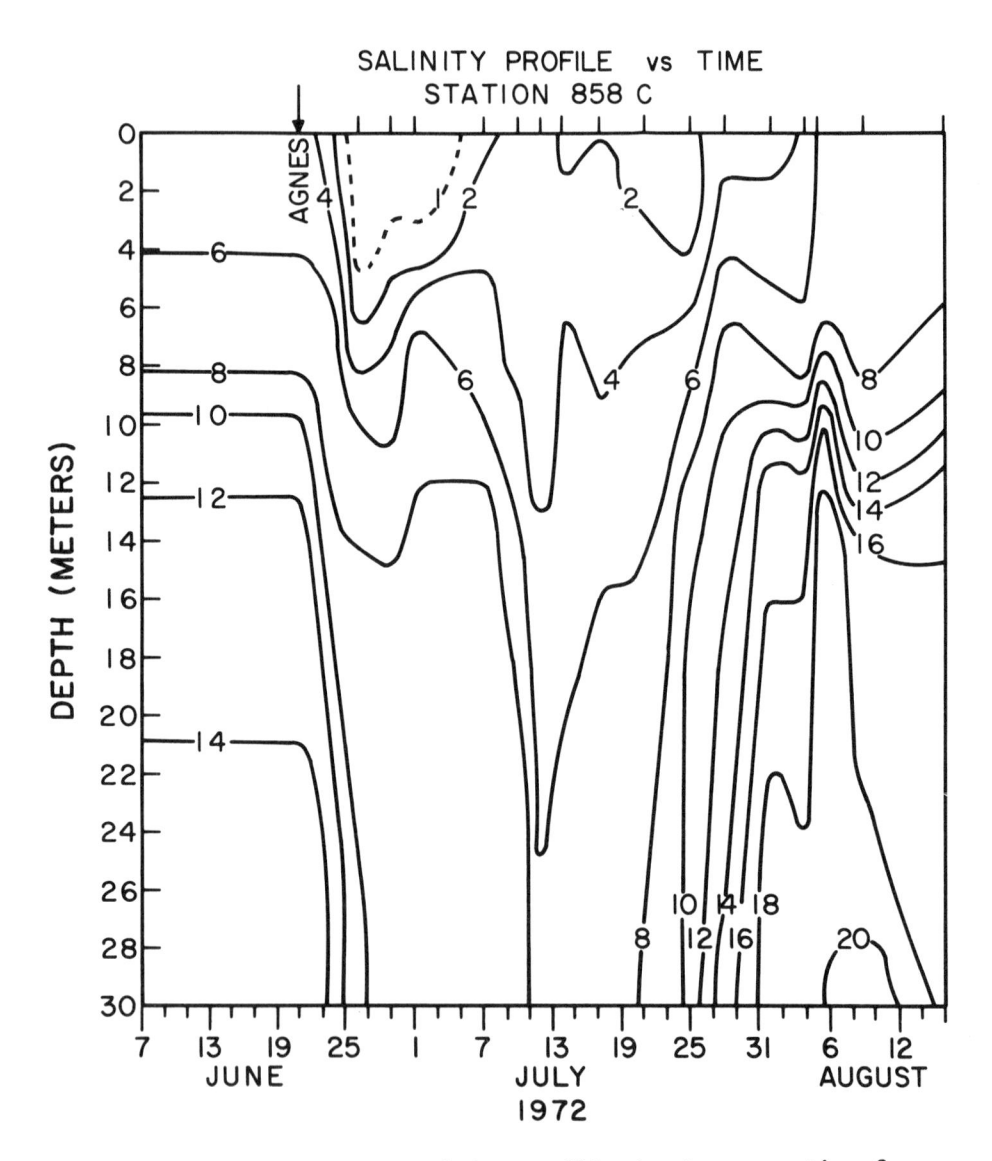

Figure 3. Vertical salinity profile (ppt) versus time for Station 858C (after Schubel, Carter, and Cronin, this volume).

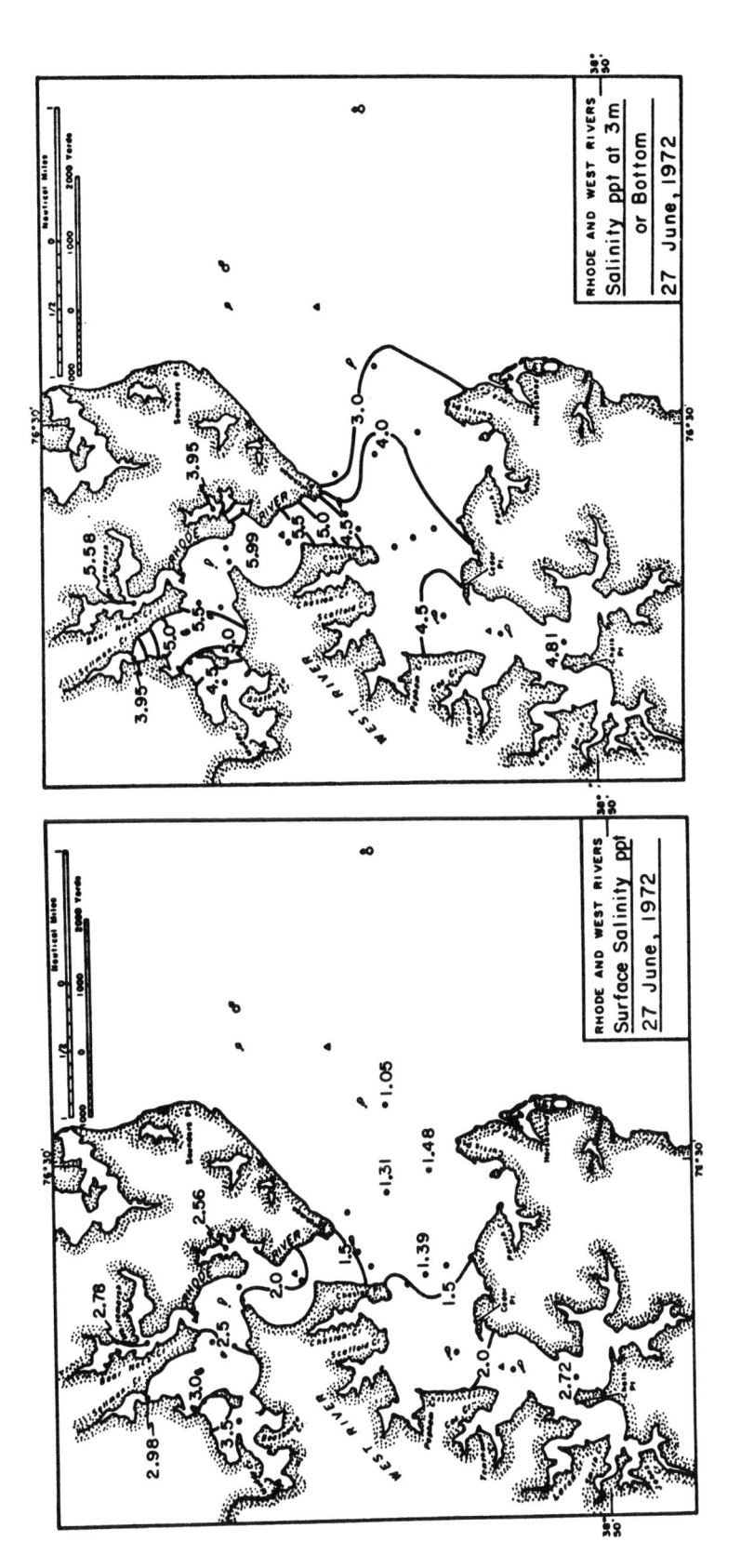

Figure 4. Surface salinity on 27 June 1972, Rhode and West Rivers.

Figure 5. Salinity at 3 m or bottom on 27 June 1972, Rhode and West Rivers.

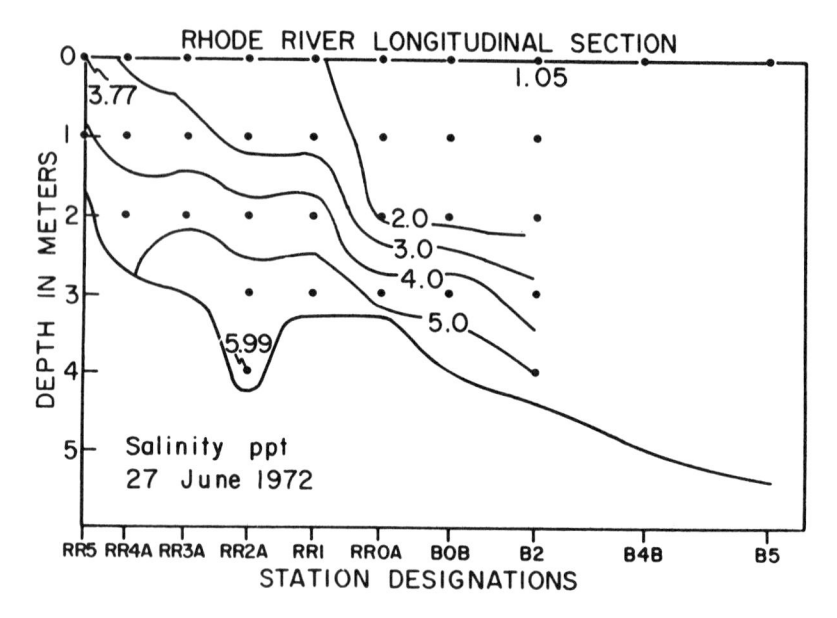

Figure 6. Rhode River salinity, 27 June 1972.

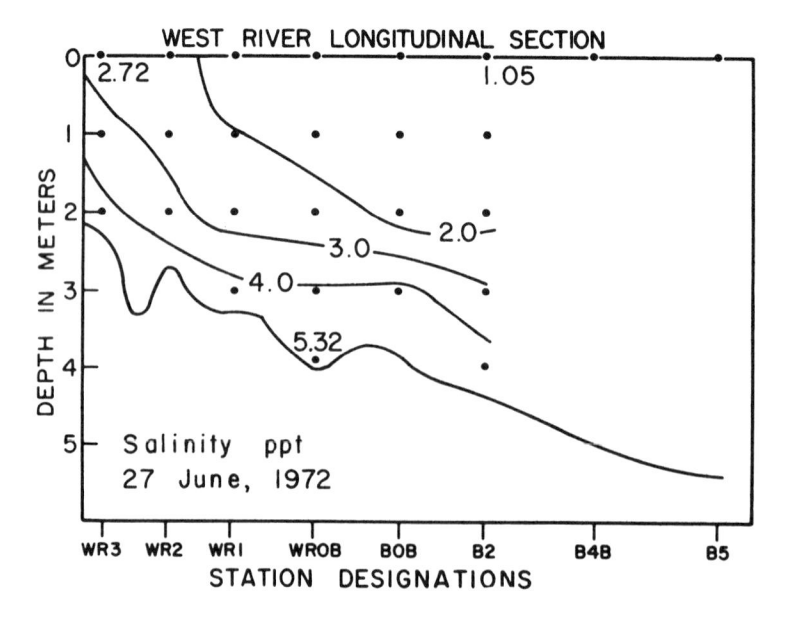

Figure 7. West River salinity, 27 June 1972.

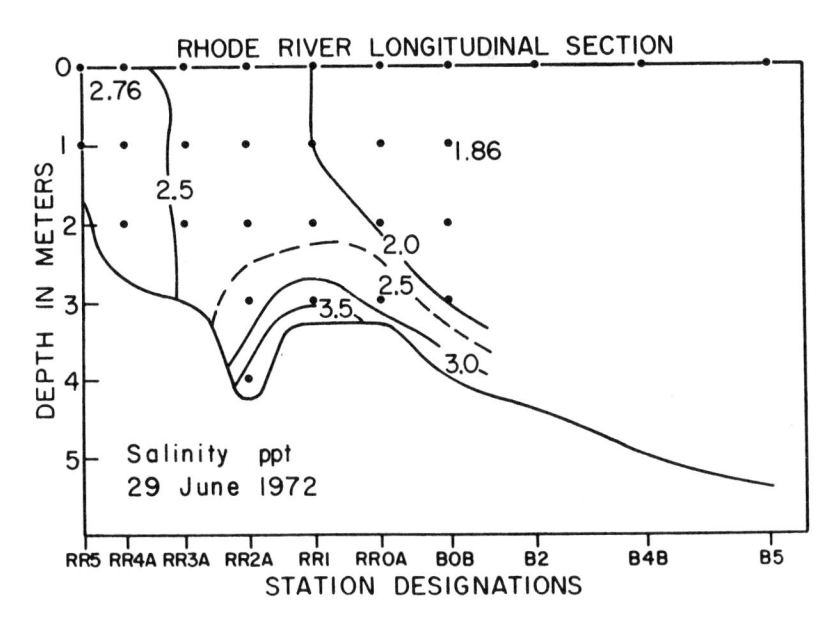

Figure 8. Rhode River salinity, 29 June 1972.

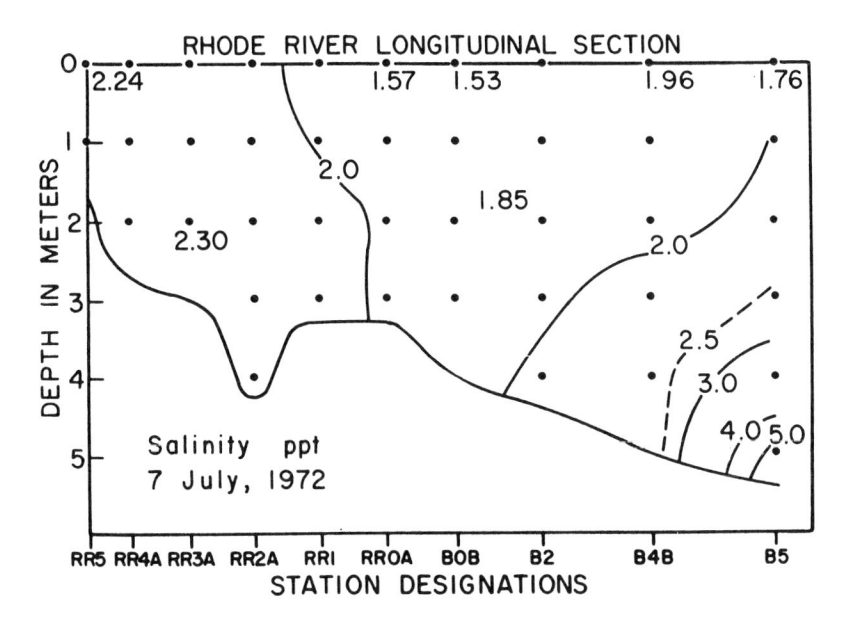

Figure 9. Rhode River salinity, 7 July 1972.

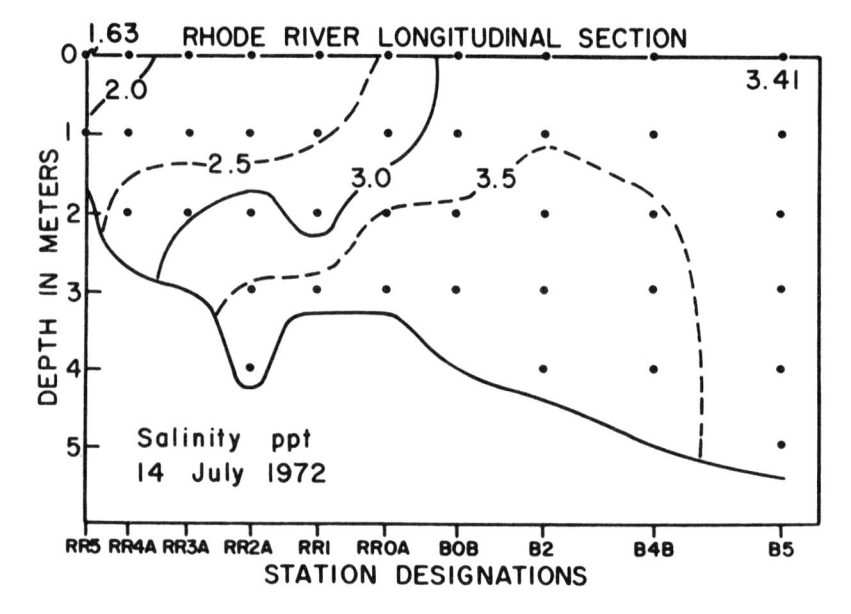

Figure 10. Rhode River salinity, 14 July 1972.

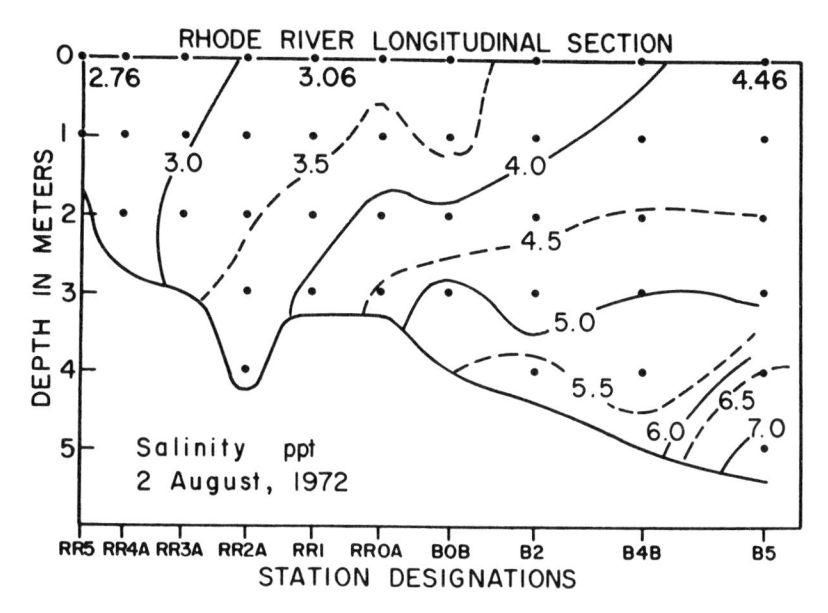

Figure 11. Rhode River salinity, 2 August 1972.

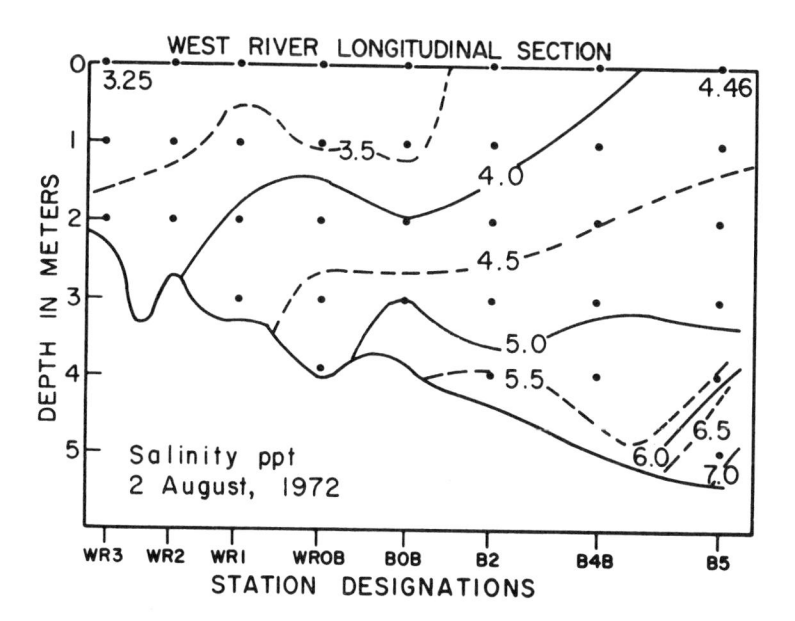

Figure 12. West River salinity, 2 August 1972.

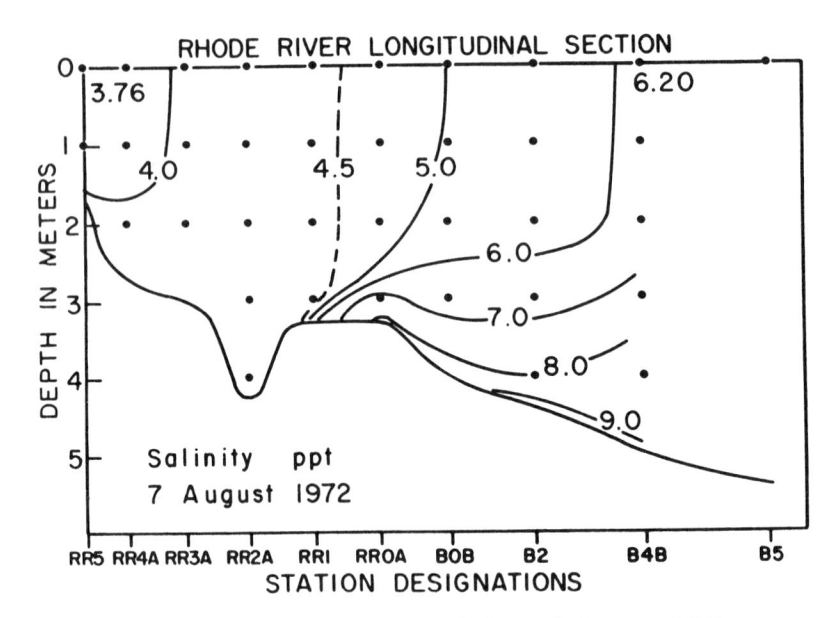

Figure 13. Rhode River salinity, 7 August 1972.

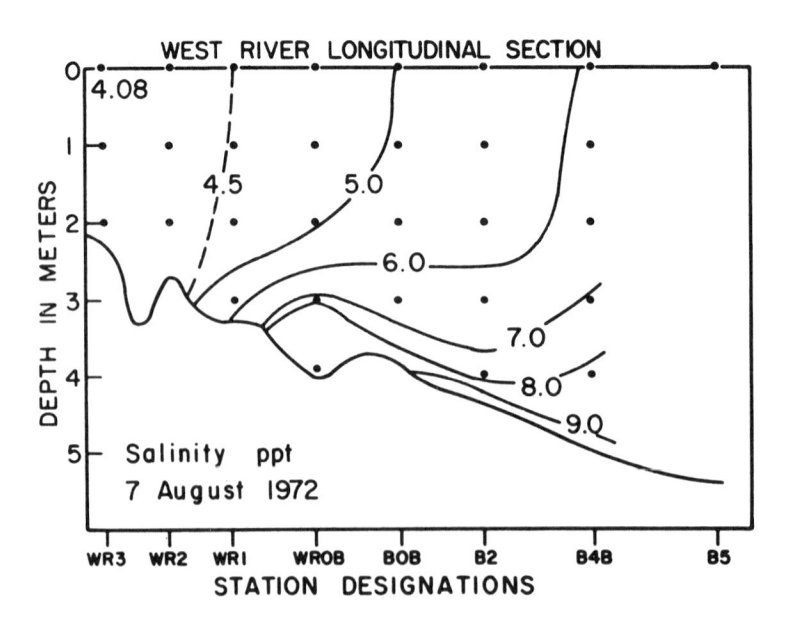

Figure 14. West River salinity, 7 August 1972.

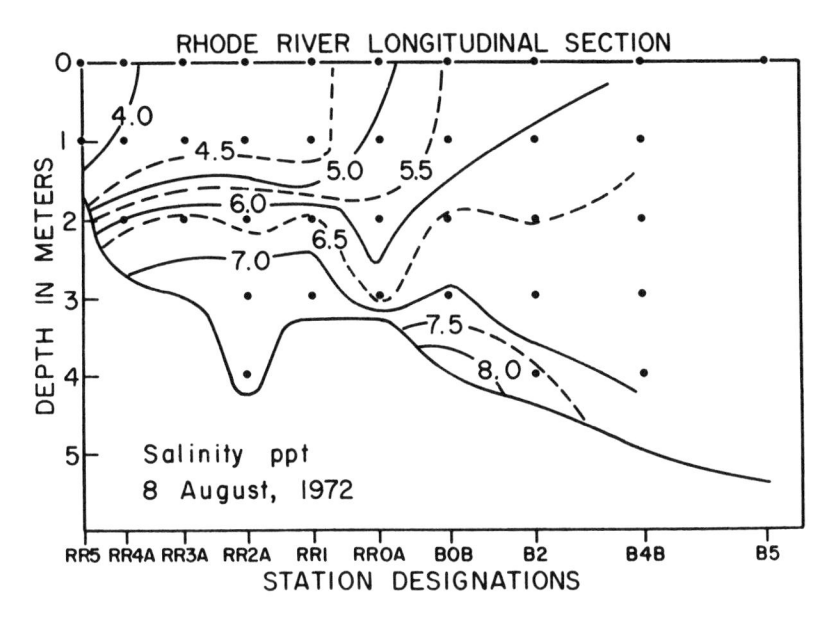

Figure 15. Rhode River salinity, 8 August 1972.

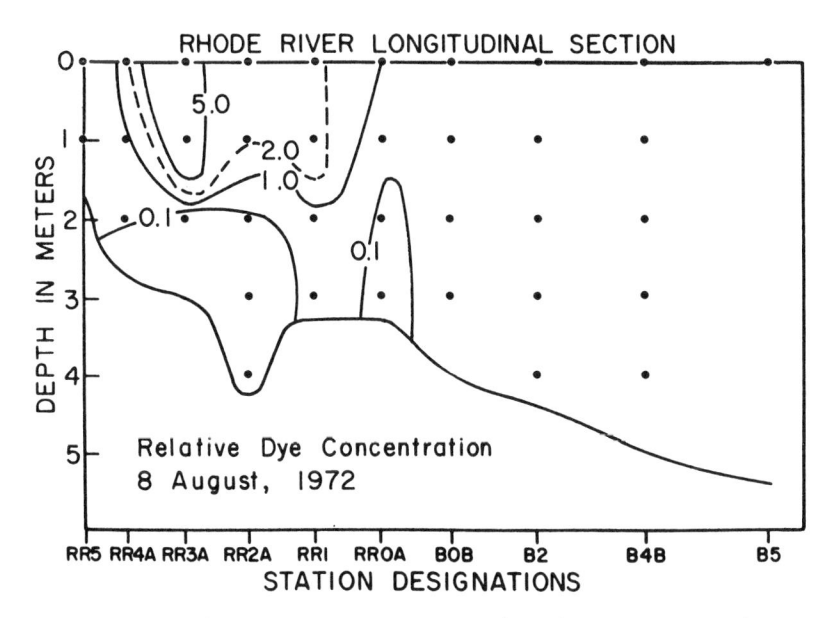

Figure 16. Rhode River, relative dye concentration, 8 August 1972.

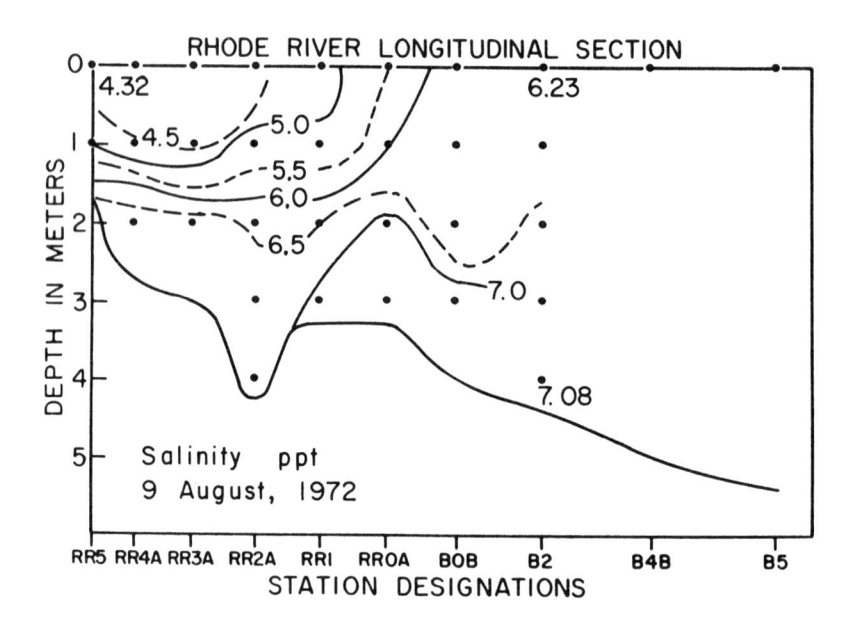

Figure 17. Rhode River salinity, 9 August 1972.

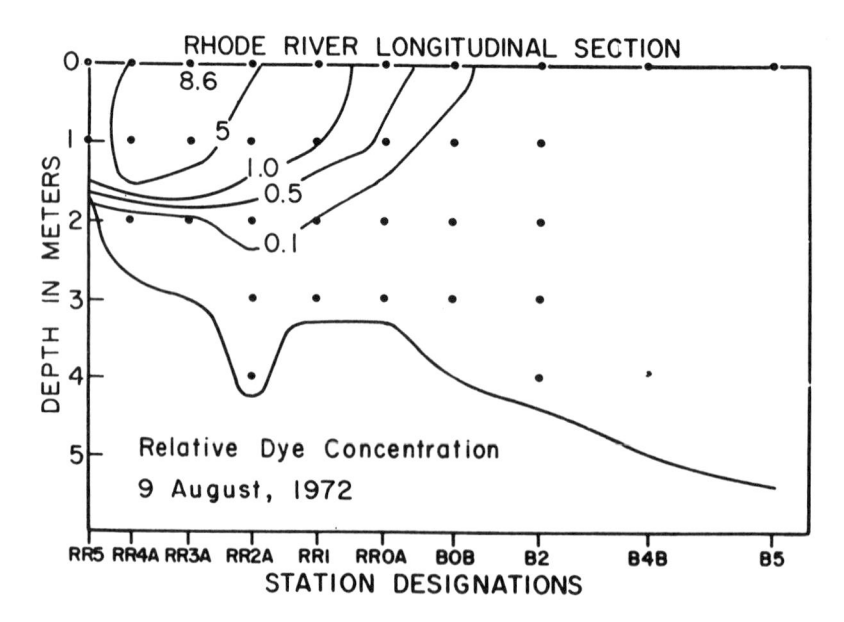

Figure 18. Rhode River, relative dye concentration, 9 August 1972.

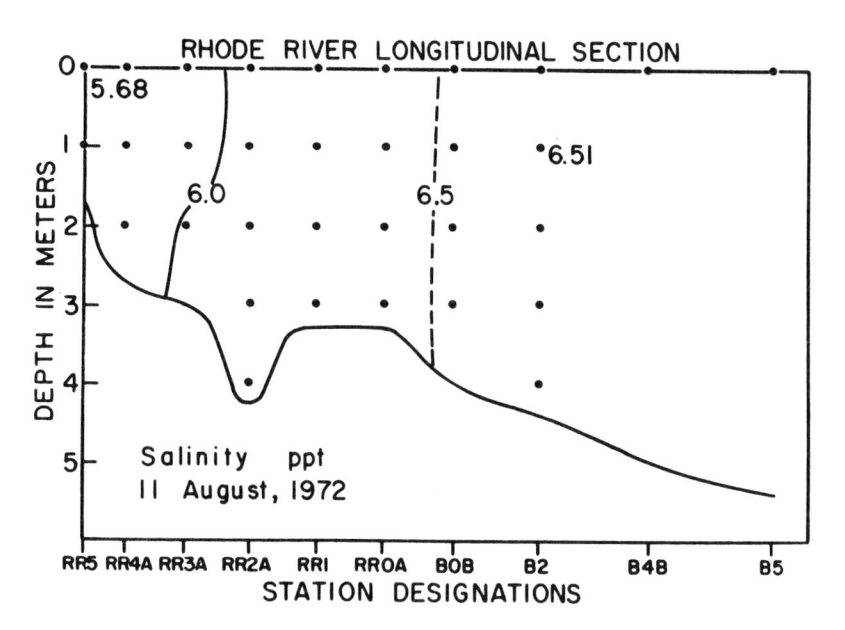

Figure 19. Rhode River salinity, 11 August 1972.

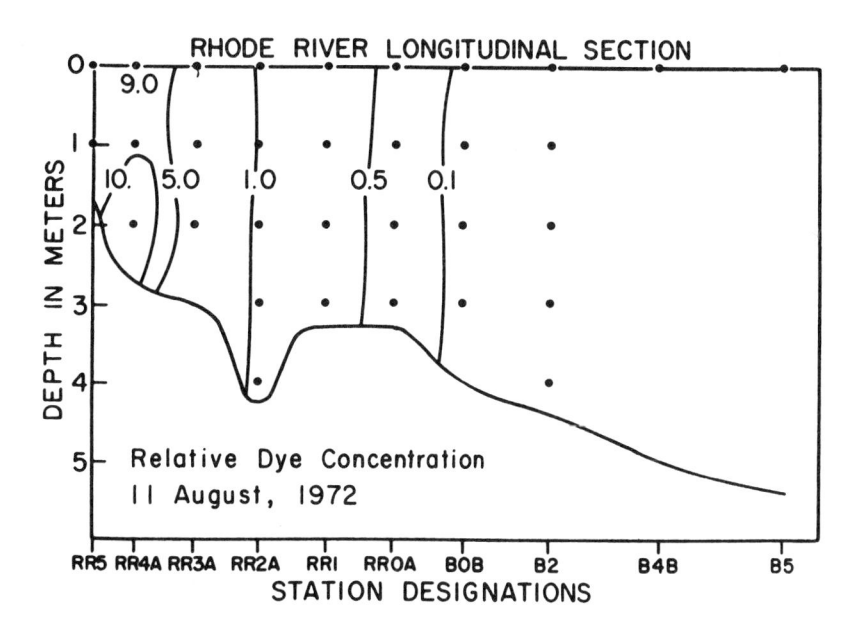

Figure 20. Rhode River, relative dye concentration, 11 August 1972.

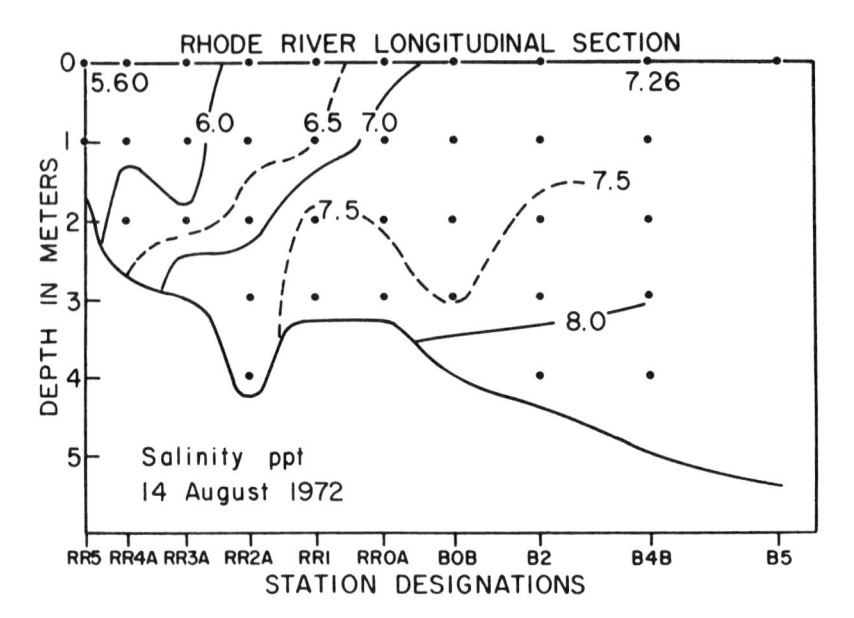

Figure 21. Rhode River salinity, 14 August 1972.

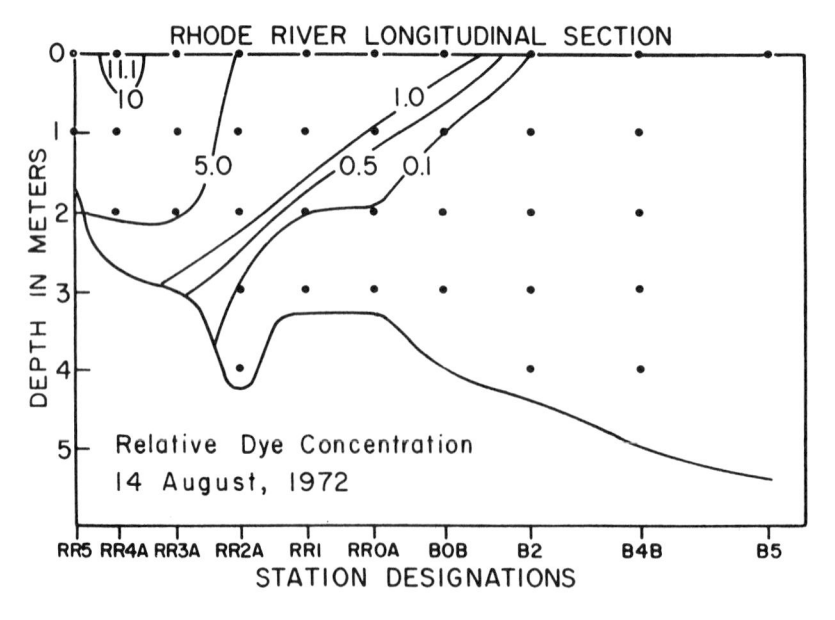

Figure 22. Rhode River, relative dye concentration, 14 August 1972.

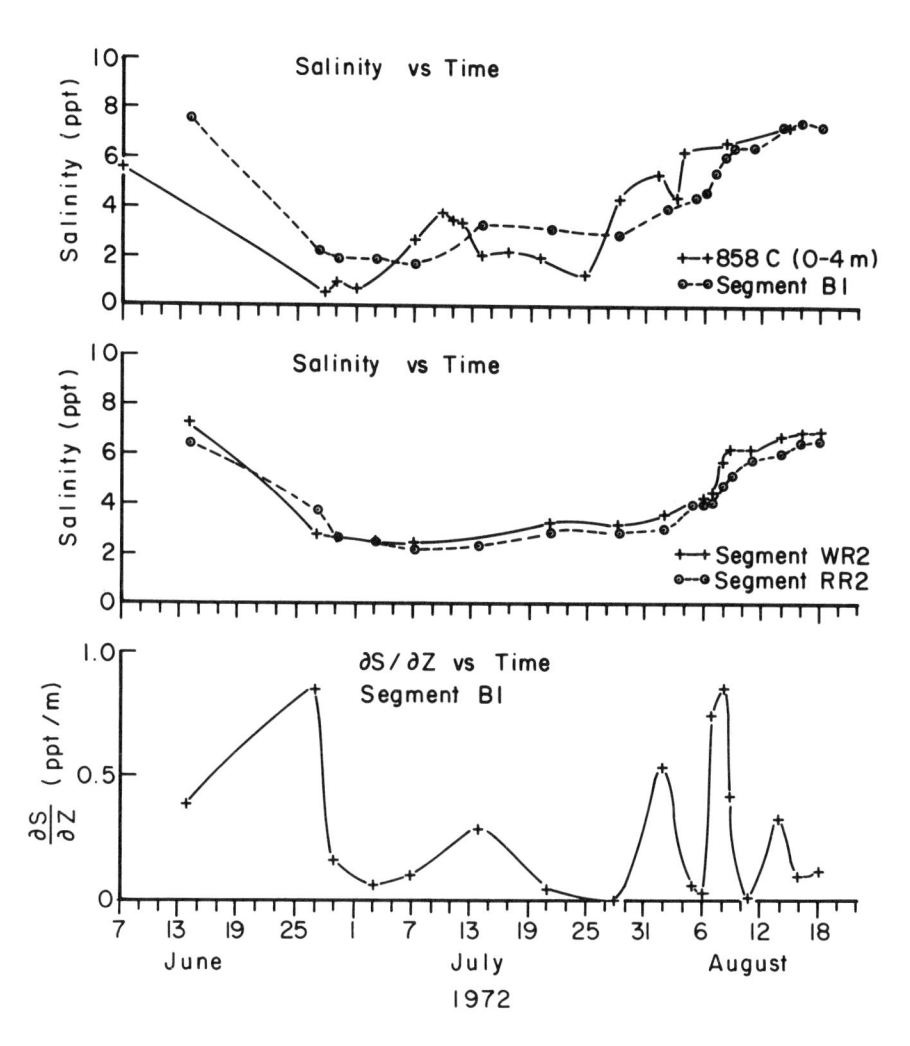

Figure 23. Volume averaged salinity versus time:
a. Station 858C and Segment B1;
b. Segments WR1 and RR1;
c. Vertical gradient of volume averaged salinities for Segment B1.

RHODE RIVER WATER QUALITY AND TROPICAL STORM AGNES[1]

Robert L. Cory[2]
J. Michael Redding[3]

ABSTRACT

Water quality data from the Rhode River, Md. were utilized to determine the effects of Tropical Storm Agnes. Immediate storm influence was first seen in tidal record irregularities when a series of seiches were superimposed on tidal oscillations that were larger than usual. Short period changes in temperature, salinity, dissolved oxygen (DO), and pH also indicated the storm's passage. A hydroclimagraph of temperature vs. salinity clearly indicates a period of midsummer stress in 1972 compared to similar data from a "typical" year.

Changes in daily oxygen and pH values were indicators of change in biotic metabolism as well as changes due to influx of cooler, fresher water from the bay. A comparison of daily oxygen changes in 1972 with a previous yearly record indicates metabolism was higher during the summer period following Agnes than it had been the previous year.

INTRODUCTION

When Agnes passed to the west of the upper Chesapeake area it was scarcely noticed in the immediate vicinity of the Rhode River. The local effects were a few windy, rainy days with higher than usual tides. Its greatest impact was in the deluge of fresh water that was deposited on the western portion of the bay's vast drainage basin, resulting in excessive discharge in the major tributary rivers of the western shore from James River in the South to the Susquehanna in the north. Total average monthly stream flow into the bay for the month of June was 9,204 cubic meters per second (m^3/s) about 2,832 m^3/s more than had ever been recorded in any month for the previous 21 years. About 60% of this flow (5,267 m^3/s) was delivered by the Susquehanna River. It was the Susquehanna's discharge that caused the major effect on the upper bay and sub-estuaries like the Rhode and West Rivers. To document the storm's effects in the Rhode and West Rivers an analysis of water quality data from a continuous recording monitor and an observation on changes in total oxygen budget are discussed here.

METHODS

In April 1970 a U. S. Geological Survey continuous recording water quality system was installed at the seaward end of the Smithsonian Institution's Pier. The pier is located just beyond the mouth of Muddy Creek, the principal stream of the Rhode River watershed. Water temperature, DO, turbidity, conductivity, and hydrogen ion (pH) measurements are made on water pumped into the monitor's sampling chamber. The pump intake is suspended from a float at a depth of one meter from the surface in water that averages about two meters in depth. Solar

---

[1] Publication authorized by the Director of the U. S. Geological Survey.
[2] U. S. Geological Survey, Chesapeake Bay Center for Environmental Studies, Smithsonian Institution, Rt. 4, Box 622, Edgewater, Md. 21037
[3] School of Graduate Studies, Oregon State University, 644 N. W. 28th St., Corvallis, Ore. 97330.

radiation, wind direction, wind velocity and tide stage complete the nine variables recorded. The collected data furnish baseline information needed for ecosystem studies at the Smithsonian's Chesapeake Bay Center, furnish information on open water metabolism, are used to detect subtle long period changes in the variables measured, and also to assess short term disturbances such as Agnes.

Vertical profiles of DO were obtained with a portable, battery operated, polaragraphic oxygen sensor calibrated against the Carpenter-modified Winkler analysis (Carpenter 1965).

RESULTS

Water Level

From June 19 to 1700 June 20 water levels at the S. I. Pier face fluctuated in a normal, lunar-dominated fashion ranging from 1.79 to 2.19 meters (5.9 to 7.2 ft.) (Fig. 1). Irregular fluctuations recorded on June 20 and 21 and a gradual increase in water level are first evidences of the approaching storm. The data for June 22 and 23 (Fig. 1) indicate the full impact of the storm's meteorological effects on local water levels, when variations over a 15-hour period ranged from 1.68 to 2.56 meters (5.5 to 8.4 ft.), nearly double the normal .46 meters (1.5 ft.) range.

Of particular interest were the seiches or standing waves which developed as a result of the storm. Though causes of such oscillations are not fully understood it is known that only a small amount of energy is needed to produce and maintain them. Periodic fluctuations in wind or barometric pressure are the usual causes (Sverdrup, Johnson & Fleming 1942).

The Rhode River seiches, recorded from 0800 to 1700 hours June 22 (Fig. 1) had an average nodal period (T.) of 68 minutes with periods ranging from 48 to 87 minutes. The greatest amplitude followed by diminishing amplitudes was 13.7 cm (5.4 in.). Similar seiches associated with meteorological disturbances are often observed at this site.

Water Temperature

During June 19 and 20, water temperatures ranged from 23.2°C to 26.6°C (Fig. 1). A slight decrease occurred on June 21 followed by a rapid cooling on the 22nd. Cooling continued at a slower rate through June 24 when a low of 19.6°C was observed (Fig. 1). Temperatures were generally depressed for the remainder of June, averaging about 2°C lower than recorded during June of 1970 and 1971. Though not shown here, a prolonged heat wave in July, when daily air temperatures exceeded 32°C for 12 consecutive days, resulted in a period of rapid rise in water temperatures. By late July average water temperature at the 1 meter depth was 31°C with daytime peaks up to 32.5°C. Shallow water temperatures probably exceeded these values and together with unseasonally low salinity stressed some of the estuarine fauna during this time (see the biological portion of these reports).

Specific Conductance (Salinity)

During June 20 to 24 a total of 4 inches of rainfall was recorded at the

Chesapeake Bay Center with about 90% of it falling on June 21 and 22. Stream flow in Muddy Creek, the principal tributary of the Rhode estuary, rose from a low of .01 m^3/s on the 20th to a peak flow of .35 m^3/s on the 22nd then dropped to .05 m^3/s by 25 June. On the 22nd specific conductance at 25°C at the pier site, showed a rapid decrease from 11,000 to 8,000 micromhos per centimeter (μmho/cm), (6.4-4.5 ppt salinity) (Fig. 1). Fluctuations corresponding with the observed seiches are apparent. Conductivity remained at about 9,000 μmhos/cm (5.1 ppt sal.) until 2000 hours on the 23rd when it increased to 12,400 μmhos/cm (7.0 ppt sal.). By the end of June 24 a steady decrease had begun which continued to mid-July 1972 when a minimum value of 3,800 μmhos/cm (2.0 ppt sal.) was recorded. The first short period depression in conductance-salinity observed during the storm was caused by the high streamflow in Muddy Creek; the subsequent lowering was caused by the influx of Susquehanna River water via Chesapeake Bay. A detailed description of this phenomena is given in this symposium by Greg Han of Chesapeake Bay Institute, Johns Hopkins University.

To compare the year of Agnes with a typical year a hydroclimagraph was prepared (Fig. 2). This figure was constructed by plotting bi-weekly averages of temperature against salinity taken from the U. S. G. S. monitor record for the period April 1970 to April 1971, which depicts a "typical" year (solid line Fig. 2) and January through December 1972 (broken line Fig. 2). The envelopes described by these plots show the annual ranges in these variables to which the estuarine biota are subject. The large decrease in salinity together with the rapid increase in temperature which occurred post-Agnes in 1972 is depicted in Fig. 2 by the pronounced deflection from a normally circular shape.

Dissolved Oxygen and pH

Changes in DO and pH during the period of storm passage reflect both biotic metabolism and water mass variations as the tidal currents move the water past the pump intake (Fig. 3).

On a daily basis the large changes observed are most often attributed to biological metabolism, whereby plants utilizing water dissolved CO_2 bind carbon and produce an excess of oxygen during the day while at night oxygen is lost due to plant and animal respiration. The pH values reflect the changes in dissolved carbon dioxide and verify the oxygen data. At the 1 meter sampling depth longitudinal changes over the length of a tidal excursion are usually slight even though strong vertical gradients may exist.

Prior to June 19, the weather was partly cloudy with very little wind and rather uniform daily temperatures. During this period respiration appeared to exceed daily production and there was a net loss of oxygen and lowering of pH. The 19th, a day of full sunshine, resulted in high primary productivity indicated by rapid increases in DO and pH. From the 20th through the 22nd reduced metabolic activity caused by the low light levels of these rainy, windy, high turbidity days was reflected in the small daily changes in DO and pH and overall reduction in both values. On the 23rd and 24th and for 2 days after, large, short period changes in both values are evident. These can be attributed to (1) discontinuities in the water mass associated with the influx of creek water and (2) to the rapid development of phytoplankton blooms observed during this period. Patches of reddish brown water were readily observed and the short period high values of oxygen and pH attested to the high state of phytoplankton activity in them. Though not documented here, phytoplankton samples and cell counts made by Leslie Kauffman and Daniel Davies from Johns Hopkins Biology Department (personal communication) showed a marked decrease in phytoplankton the day after the storm.

On the 24th patches of high phytoplankton cell counts were observed and a few days later a massive bloom occupied the entire area. By June 26th the water appeared uniformly brownish-red in color, 24 hour changes in oxygen and pH were large, and the frequency of short term changes in oxygen and pH diminished.

A histogram of weekly averages of daily oxygen changes for the years 1970-1971 and 1972 was prepared (Fig. 4). As stated above, these daily changes reflect variations in biotic metabolism. Seasonal and yearly differences are apparent with 1972 indicating a higher summer activity. The average daily change for each of the years plotted was 2.96 $mgO_2 \ell^{-1}$ for 1971 and 3.48 for 1972. Increases in the 1972 late summer metabolism also resulted in lower daily minimum oxygens (Fig. 4).

Agnes affected the total oxygen budget of the Rhode and West Rivers. Midday vertical profiles of DO were taken June 14, before and June 27, after the storm's passage. The Rhode and West Rivers were subdivided and volume averages of oxygen for each day measured are shown in each subdivision (Fig. 5) The averaged values of oxygen were multiplied by the volumes and data from each box were totaled. The total DO in this system had increased by 48% from 206 to 301 metric tons. A similar increase can be seen in the plot of average weekly minimum oxygen values shown in the lower half of Fig. 4.

The increase in DO can be attributed to the influx of cooler less saline bay water displacing and mixing with the warmer more saline water of the Rhode and West River subestuaries. Although metabolism increased during the post-Agnes period the net result was a loss of oxygen shown by the weekly minimums of late August and early September (Fig. 4).

CONCLUSIONS

Water quality monitoring at a point in the Rhode estuary has furnished useful information on both short term and long period effects of a tropical storm.

A combination of both historical and storm-affected data on the water quality variables of temperature and salinity enables one to define a prolonged period of environmental aberration. Similar comparison of DO and pH data indicates increases in daily changes of these variables during the year of the storm. These increases are indicators of probable increases in biotic metabolism.

LITERATURE CITED

Carpenter, J. H. 1966. New measurements of oxygen solubility in pure and natural water. Limnol. Oceanogr. 11(2):264-277.

Sverdrup, H. U., M. W. Johnson, and R. H. Fleming. 1957. The Oceans. Prentice-Hall, Inc., Englewood Cliffs, N. J., pp. 538-542.

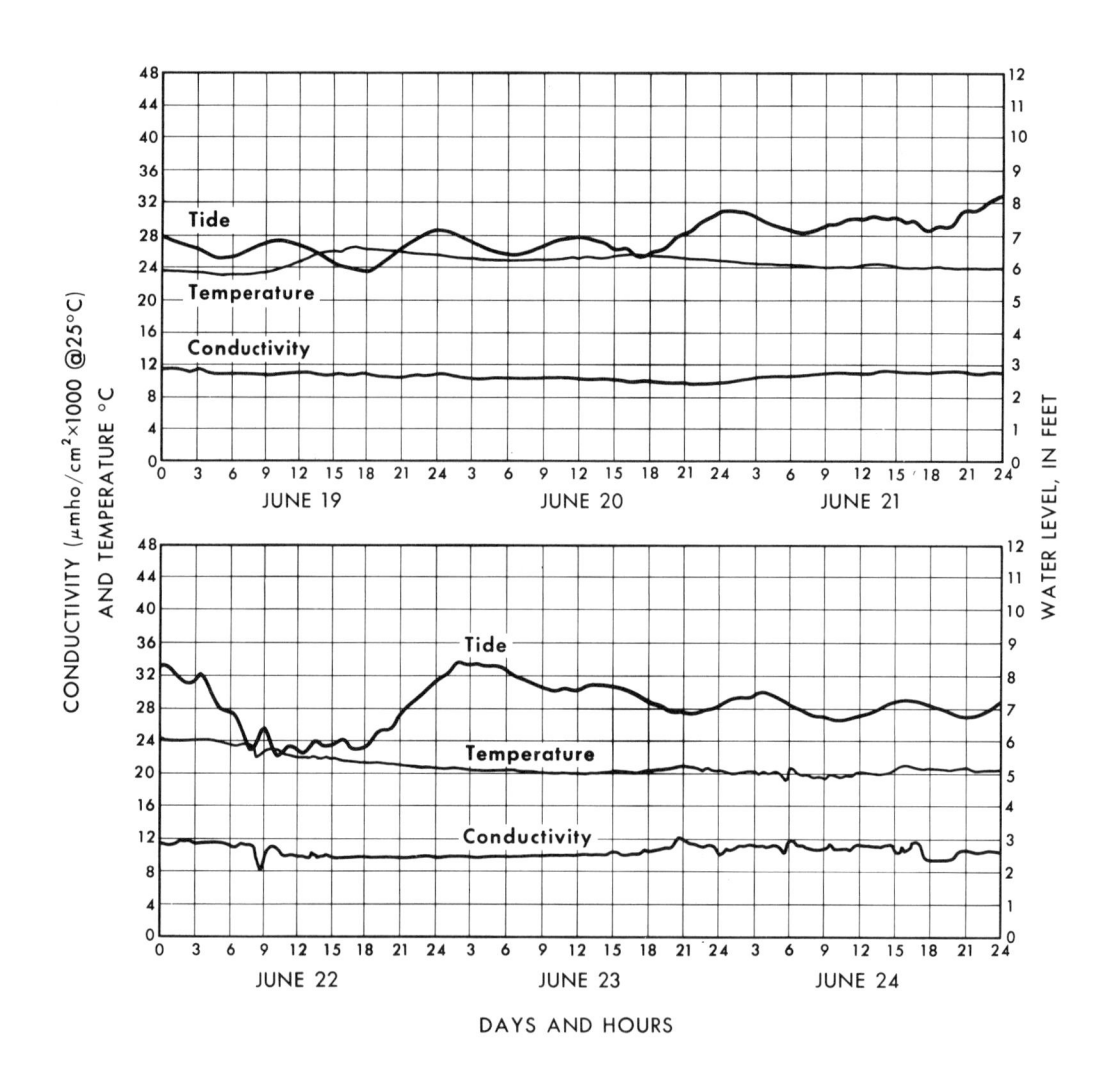

Figure 1. Water quality at Smithsonian Institution's pier, Rhode River, Md., June 1972.

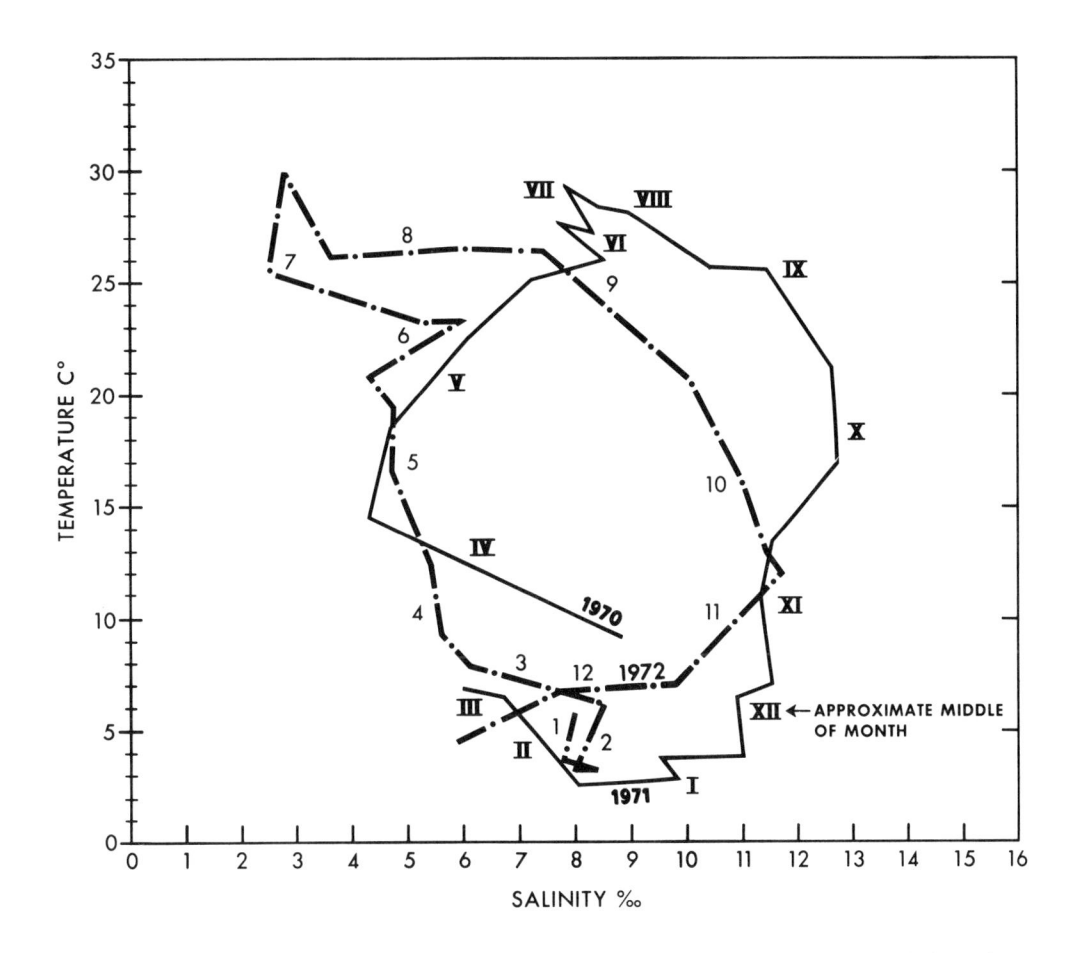

Figure 2. Hydroclimagraph of temperature and salinity at Smithsonian pier, Rhode River, Md.

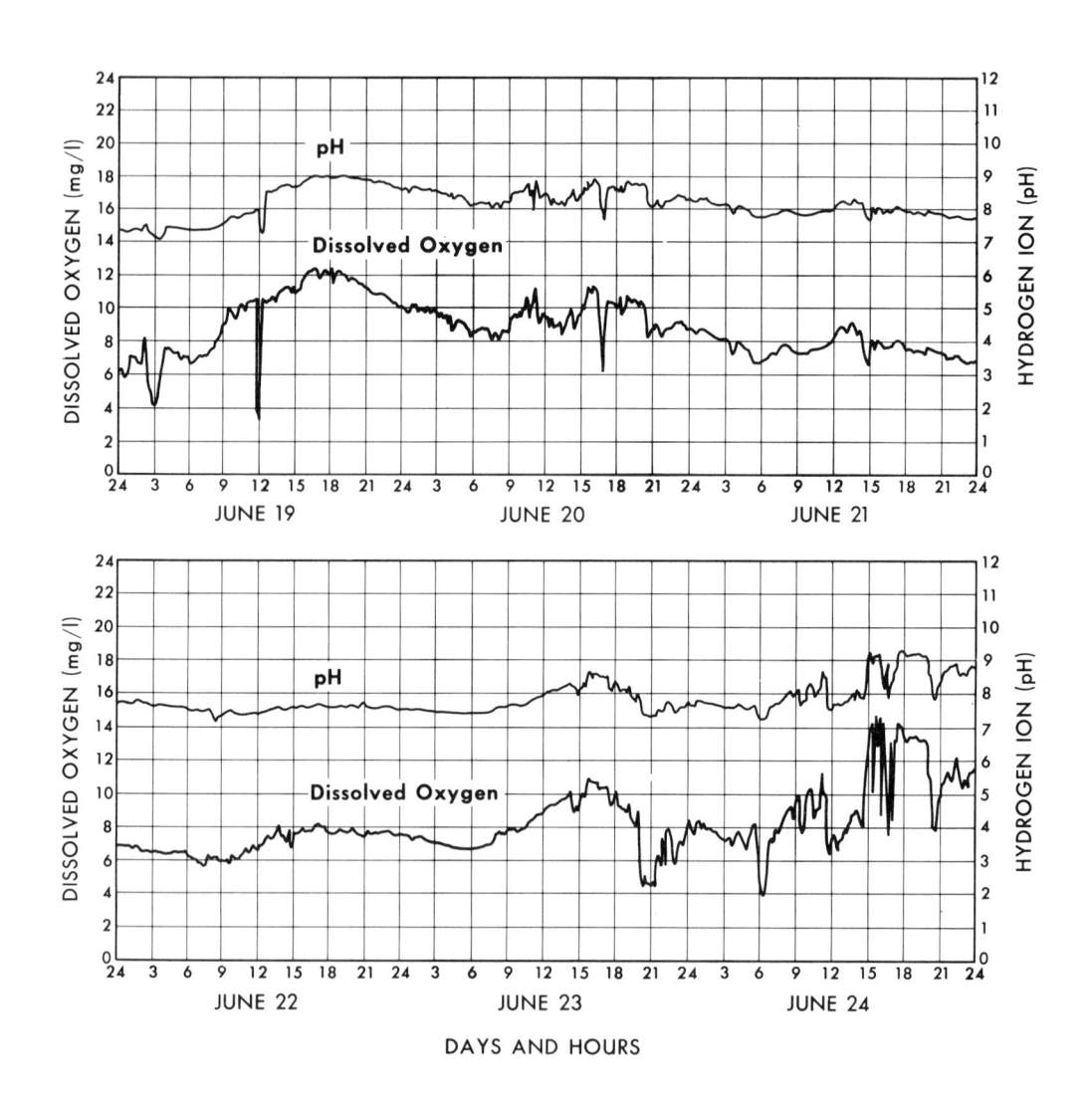

Figure 3. Water quality at Smithsonian's pier, Rhode River, June 1972.

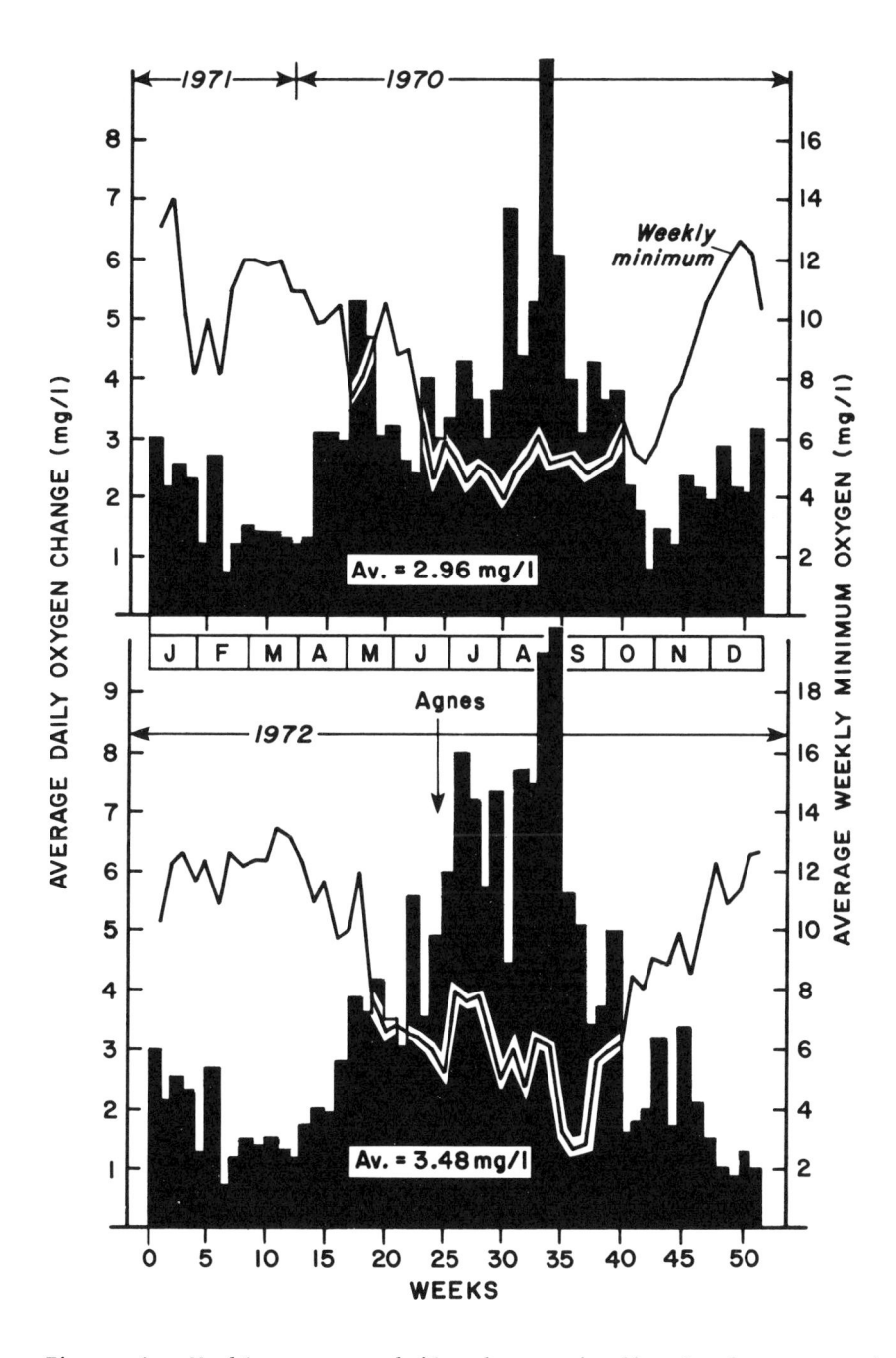

Figure 4. Weekly average daily changes in dissolved oxygen and weekly average minimum dissolved oxygen at Smithsonian pier, Rhode River.

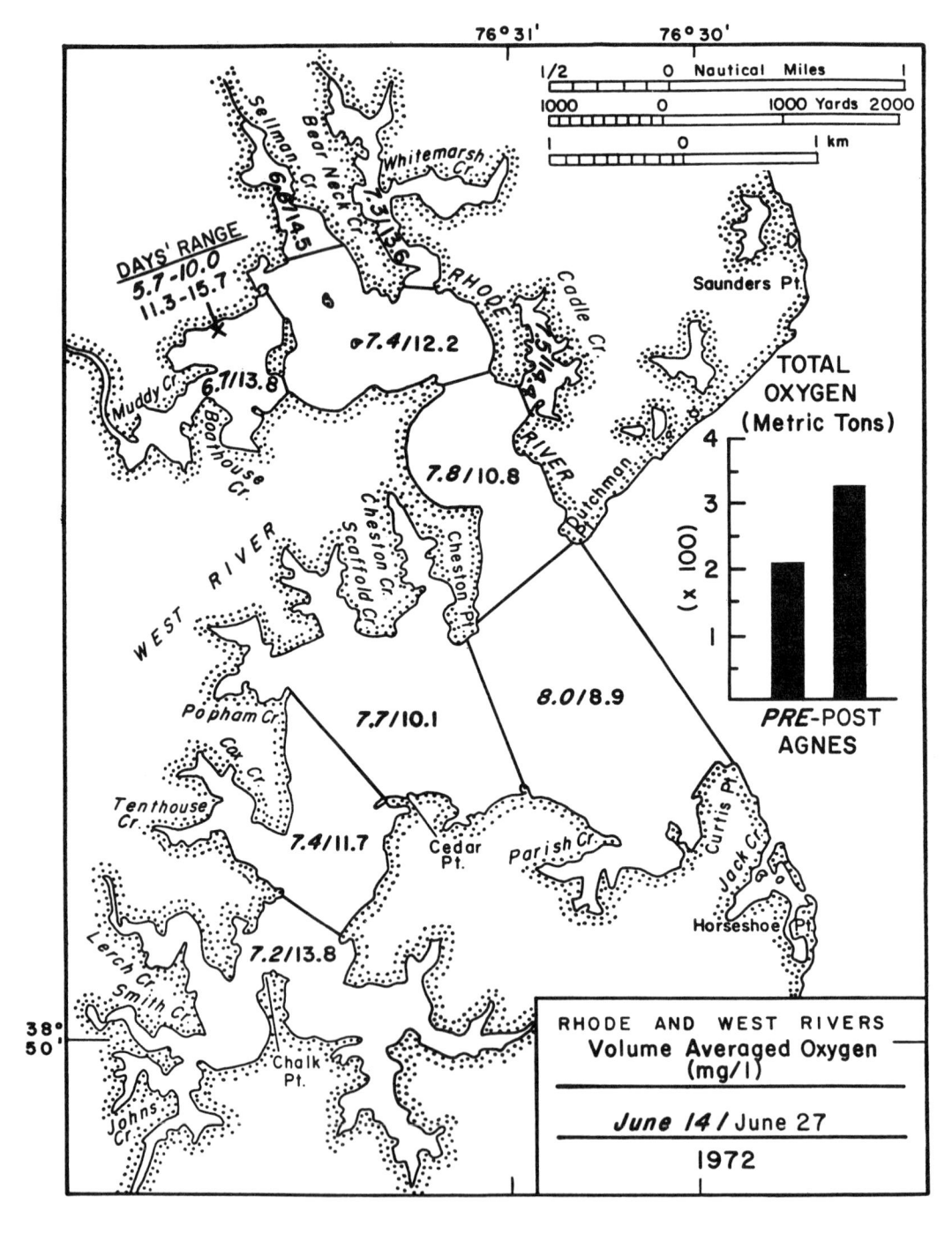

Figure 5. Volume averaged dissolved oxygen budget in the Rhode and West Rivers before-after Agnes.

GEOLOGICAL EFFECTS

J. R. Schubel, Editor

EFFECTS OF AGNES ON THE SUSPENDED SEDIMENT OF THE CHESAPEAKE BAY AND CONTIGUOUS SHELF WATERS[1]

J. R. Schubel[2]

ABSTRACT

The flooding rivers discharged massive amounts of suspended sediment into the Chesapeake Bay estuarine system. In the 10-day period, 21-30 June 1972, the Susquehanna discharged more than 31×10^6 metric tons of suspended sediment into the Bay. Its annual input during most years is only $0.5-1.0 \times 10^6$ metric tons. Concentrations of suspended sediment throughout much of the Chesapeake Bay estuarine system were higher than any previously reported. During the period of peak riverflow the upper Bay was characterized by a marked longitudinal gradient of suspended sediment. On 26 June 1972, two days after the Susquehanna crested, the concentration of suspended sediment in the surface waters dropped from more than 700 mg/ℓ at the head of the Bay (Turkey Point) to 400 mg/ℓ at Tolchester, and to 175 mg/ℓ at the Annapolis Bay Bridge. Concentrations of suspended sediment in the upper Bay remained anomalously high for about a month after Agnes.

In the middle and lower reaches of the Bay concentrations of suspended sediment were 2-3x higher than "normal," and the outflow from the Bay could be traced as a band of low salinity, turbid water that turned south and flowed along the Virginia and Carolina coasts.

INTRODUCTION

The Chesapeake Bay, like other estuaries is an ephemeral feature on a geologic time scale. It is being rapidly filled with sediments; sediments from rivers, from shore erosion, from the remains of organisms that inhabit it, and from the sea. The sources are thus external, marginal, and internal. On a system wide basis, the external sources are predominant, and the rivers account for the vast majority of this input. The characteristic mode of sediment transport both into and within the estuarine portions of the Bay is as suspended load. The distribution and transportation of suspended sediment[3] in the main body of the Bay have been discussed by Schubel (1968a,b,c, 1969, 1972), Schubel and Biggs (1969), and by Biggs (1970). Studies of the distribution and transportation of suspended sediment in the Rappahannock and James River estuaries have been reported by Nichols (1972, 1974); there are few published data for any of the other estuaries tributary to the Bay.

The investigations to date have described "typical" or "average" conditions. But in addition to these "normal" variations, marked fluctuations can result from catastrophic events such as floods or hurricanes. There was, until Agnes, a dearth of direct observations of the effects of such "rare" events on the suspended solids population not only of the Chesapeake Bay, but of the entire coastal environment. Sampling during episodes is frequently difficult, and the infrequency of such occurrences makes the likelihood of fortuitous observations very

---

[1]Contribution No.234, Chesapeake Bay Institute, The Johns Hopkins University.
[2]Chesapeake Bay Institute, The Johns Hopkins University, Baltimore, Md. 21218
[3]The phrases suspended sediment and total suspended solids are used interchangeably throughout the text.

small. As a result, the sedimentary impact of severe storms has much more commonly been inferred from a "post-mortem" examination of the "record" than from direct observations of sediment dispersal.

Agnes presented scientists with an unusual opportunity to document the impact of a major storm on the Chesapeake Bay. There was little wind associated with Agnes when she reached this area, but torrential rains sent flows of the major tributaries to record or near-record levels. The heavy rains stripped large quantities of soil from throughout much of the drainage basin, and the flooding rivers carried significant quantities of this sediment into the Chesapeake Bay estuarine system.

The Chesapeake Bay Institute made extensive observations to document the impact of Tropical Storm Agnes on the distributions not only of suspended sediment but also of salinity, temperature, dissolved oxygen, and nutrients in the Bay proper, in the Potomac estuary, and in the shelf waters off the mouth of the Bay. Sampling was initiated during the period of peak flooding and was extended over approximately one year to chronicle the impact of Agnes, and the subsequent recovery of the Bay to "normal" conditions. The primary purpose of this paper is to summarize the suspended sediment data from the main body of the Bay and contiguous shelf waters. The emphasis will be on the upper Chesapeake Bay since this is the segment of the Bay proper where the effects of Agnes were greatest, and where we have the most observations. To assess the impact of Agnes, these observations must be compared with data from more "normal" years to determine whether they fall outside of the range of "normal" year to year variations.

METHODS

The Agnes stations are plotted in Fig. 1, and data on sampling frequency are summarized in Table 1. The concentrations of suspended sediment were determined gravimetrically. In the Bay, water samples were collected with a submersible pump and filtered through pre-weighed 47 mm, 0.6 μ APD Nuclepore membrane filters. The filters and sediment were washed several times with distilled water to remove any salt, then placed in individual desiccators. The filters were desiccated at ambient temperature over silica gel for at least 72 hours, re-weighed, and concentration of total suspended solids (suspended sediment) calculated.

Determinations of the concentration of suspended sediment were also made in the Susquehanna River at the Conowingo Hydroelectric Plant to estimate the suspended sediment discharge of the Susquehanna. Water samples were collected from the tailrace with a weighted plastic bottle and processed in the manner described above. All weighings were made to ± 0.02 mg.

Concentrations of suspended sediment in shelf waters were determined by filtering 2 to 4ℓ water samples through pre-weighed, 90 mm, 0.4 μ APD Nuclepore membrane filters using special filtering water bottles (Schubel et al. 1972).

OBSERVATIONS

Input by the Susquehanna

Nineteen seventy-two started out not very unlike most years, although it was somewhat wetter. During the spring freshet in March, flow of the Susquehanna was

Table 1. Summary of suspended sediment sampling locations and dates. See Fig. 1 for station locations.

Date	927SS	SF00	921W	917S	913R	909	903A	858C	848E	834F	818P	804C	745	724R	707∅	659W
25 June 72				x												
26 June 72		x	x	x	x	x	x	x	x							
27 June 72		x	x	x	x	x	x	x	x							
29 June 72		x	x	x	x	x	x	x	x	x						
1 July 72		x	x	x	x	x	x	x	x							
3 July 72		x	x	x	x	x	x	x	x	x						
5-6 July 72		x	x	x	x	x	x	x	x	x			x	x	x	x
7-8 July 72		x	x	x	x	x	x	x	x	x	x	x				
10 July 72		x	x	x	x	x	x	x								
12-13 July 72		x	x	x	x	x	x	x	x	x	x	x	x	x		
14 July 72		x	x	x	x	x	x	x								
17 July 72		x	x	x	x	x	x	x	x							
19-20 July 72		x	x	x	x	x	x	x	x	x	x	x	x	x		
24-25 July 72		x	x	x	x	x	x	x	x	x	x	x	x	x	x	x
28 July-1 Aug 72		x	x	x	x	x	x	x	x	x	x	x	x	x		
4 Aug 72		x	x	x	x	x	x	x	x							
9 Aug 72		x	x	x	x	x	x	x	x							
15-17 Aug 72		x	x	x	x	x	x	x	x	x	x	x	x			
28-31 Aug 72		x	x	x	x	x	x	x	x	x	x	x	x	x	x	x
16-19 Sept 72		x	x	x	x	x	x	x	x	x	x	x	x	x	x	x
25-26 Sept 72		x	x	x	x	x	x	x	x							
9-10 Oct 72		x	x	x	x	x	x	x	x	x	x	x	x	x	x	x
17-18 Oct 72		x	x	x	x	x	x	x	x	x	x	x				
7 Nov 72		x	x	x	x	x	x	x	x							
16 Nov 72					x	x	x	x	x	x	x	x	x	x	x	x
27 Nov 72		x	x	x	x	x	x	x	x							
11-15 Dec 72		x	x	x	x	x	x	x	x	x	x	x	x	x	x	x

fairly high, exceeding 8900 m³/sec, and the concentration of total suspended solids in the Susquehanna River at Conowingo, Md., on one day reached 190 mg/ℓ. The Conowingo Hydroelectric Plant is located approximately 15 km upstream from the River's "mouth" at Havre de Grace, Md., and is the last dam before the River enters the Bay. Between 1 January 1972 and 21 June 1972 the concentration of total suspended solids at Conowingo exceeded 100 mg/ℓ on only four days--not unlike most years. During May and the first 20 days of June of 1972 the concentration of total suspended solids, which was generally between 10-25 mg/ℓ was somewhat higher than average for that time of the year, but was not really "abnormal." Then Agnes entered the area. Torrential rains fell throughout most of the drainage basin of the Susquehanna producing record flooding. The day the Susquehanna crested, 24 June 1972, the average daily flow exceeded 27,750 m³/sec-- the highest average daily flow ever recorded, exceeding the previous daily average high by approximately 33%. The instantaneous peak flow on 24 June 1972 of more than 32,000 m³/sec was the highest instantaneous flow reported over the 185 years of record, Fig. 2. The *monthly* average discharge of the Susquehanna of approximately 5100 m³/sec for June 1972 was the highest average discharge for *any* month over the past 185 years, and was more than nine times the average June discharge over this same interval. It was also the first time on record that the monthly maximum occurred in the month of June. Comparison of the monthly average discharge of the Susquehanna during 1972 with the ensemble monthly average over the period 1929-1966 clearly shows the departure of the 1972 June flow from the long-term average June flow, Fig. 3.

Even before Agnes, 1972 had been a "wet" year, Fig. 3., and the concentration of total suspended solids in the Susquehanna and in the upper Bay were somewhat higher than normal for that time of year. Then on 22 June 1972 riverflow increased rapidly as a result of heavy rainfall accompanying Agnes, and the concentration of suspended solids at Conowingo reached 400 mg/ℓ, Fig. 4. On 23 June 1972 riverflow exceeded 24,400 m^3/sec, and the concentration of suspended solids jumped to more than 10,000 mg/ℓ--a concentration more than 40 times greater than any previously reported for the lower Susquehanna. This concentration was determined on the downstream side of the dam at Conowingo. Unfortunately, no sediment sample was collected on 24 June 1972, the day the Susquehanna crested, because the dam had to be evacuated for safety reasons. The average daily flow on June 24 was more than 27,750 m^3/sec. By June 25 the riverflow had decreased to about 23,100 m^3/sec, and the concentration of suspended solids to approximately 1,450 mg/ℓ. By 30 June riverflow had subsided to about 4,600 m^3/sec, and the concentration of suspended solids to about 70 mg/ℓ.

Except near the bottom, concentrations of total suspended solids greater than 100 mg/ℓ are relatively uncommon in the northern Chesapeake Bay, even in the mouth of the Susquehanna. In 1967 the maximum concentration in the mouth of the Susquehanna at Havre de Grace, Md., was 140 mg/ℓ (Schubel 1968b). In 1969 the maximum concentration of suspended sediment at Conowingo was only about 57 mg/ℓ (Schubel 1972). Since 1969 samples of suspended solids have been collected on the downstream side of the Conowingo Dam on nearly a daily basis. In 1970 the maximum concentration at Conowingo was 253 mg/ℓ, but it exceeded 100 mg/ℓ on only five days during the entire year. In 1971 the peak concentration at Conowingo was 142 mg/ℓ, and it exceeded 100 mg/ℓ on only four days during the year. This sediment is nearly all silt and clay; the coarser particles are trapped upstream in the reservoirs along the lower reaches of the River (Schubel 1968a,b; 1969).

During the ten-day period, 21-30 June 1972, the Susquehanna probably discharged more than 31 x 10^6 metric tons of suspended sediment into the upper Chesapeake Bay. The daily water and suspended sediment discharges for this ten-day period are tabulated in Table 2. The water discharges were obtained from the records of the Conowingo Hydroelectric Plant. The suspended sediment discharge for each of the days, except June 24, was calculated using the water discharge and the concentration of suspended sediment in a single sample of water collected sometime during that day. The suspended sediment samples were obtained by filtration of water samples collected from the tailrace with a weighted plastic bottle. Since the suspended sediment discharge for each day is based on a single determination of the concentration of suspended sediment, the implicit assumption is that that concentration is representative of the entire day. The extremely high suspended sediment discharge estimated for June 23 may have resulted from a spurious determination of the concentration of suspended sediment. Because of the very turbulent discharge and the method of sampling, this investigator expects that the concentration would, if anything, be an underestimate of the average concentration of suspended sediment at the time of sampling. It is however, conceivable that a relatively high density turbidity current was generated by erosion of the bottom of the Conowingo reservoir by the rampaging River, that the sample was taken during this period, and that this concentration is not representative of the entire day. It may also be that although the concentration is very high, it is a realistic representation of the average concentration of suspended sediment on 23 June 1972. It is unlikely that this question will be resolved.

Table 2. Susquehanna River discharges of water and suspended sediment past Harrisburg (Pa.) and Conowingo (Md.) from 21-30 June 1972.

DATE (June 1972)	CONOWINGO, MD Water Discharge (m³/sec)	Suspended Sediment Discharge (metric tons/day)	HARRISBURG, PA* Water Discharge (m³/sec)	Suspended Sediment Discharge (metric tons/day)
21	1,316	2,287	1,215	9,350
22	10,795	372,636	9,373	751,000
23	24,419	21,353,311	25,202	2,080,000
24	27,795	5,000,000**	27,014	2,210,000
25	23,071	2,901,856	20,388	1,470,000
26	16,150	1,418,846	13,677	725,000
27	10,350	413,489	8,637	147,000
28	6,816	99,346	6,116	63,100
29	5,697	49,682	4,899	40,700
30	4,600	29,628	4,474	27,700
TOTALS		31,641,081		7,523,850

*A. B. Commings (personal communication, 1974)
**Estimated from Fig. 5.

Since no water sample was collected on June 24, the suspended sediment discharge on that day had to be estimated by other means. A plot was made of the daily water and suspended sediment discharges at Conowingo for each sampling day in 1972 (Fig. 5); the suspended sediment discharge for 24 June 1972 was then estimated from this graph. Because of the appreciable scatter in these data, only a visual estimate was made. This may be off by a factor or two, but it appears likely that the estimate is on the low side. Table 2 also contains estimates of the water and suspended sediment discharges of the Susquehanna past Harrisburg, Pa., which is approximately 100 km upstream from Conowingo, Md. These estimates were made by the U. S. Geological Survey and are based on standard gauge height measurements and on vertically integrated samples of suspended sediment (A. B. Commings, personal communication, 1974). Suspended sediment samples were taken at least daily, and on a few days two or three samples were collected. The Table indicates that on June 21-22 the total suspended sediment discharge past Harrisburg (760,350 metric tons) was nearly twice that discharged past Conowingo (374,923 metric tons), even though the average water discharge rate past Conowingo for this two day period (6,056 m³/sec) was greater than that past Harrisburg (5,294 m³/sec). For the remainder of this ten-day period however, the daily suspended sediment discharge past Conowingo appears to have been consistently, and considerably, greater than that past Harrisburg. Over the ten-day period, 21-30 June 1972 the total estimated suspended sediment discharge past Conowingo was more than 31.6×10^6 metric tons, while that past Harrisburg was only 7.5×10^6 metric tons. Over this same period the average riverflow increased by less than 10% between Harrisburg and Conowingo. Previously, Schubel (1974) suggested that the apparent marked increase in the suspended sediment discharge between Harrisburg and Conowingo may have been due in large part to erosion of sediment from the bottom of Conowingo reservoir and perhaps from one or more of the other reservoirs between Conowingo, Md., and Harrisburg, Pa. A bathymetric survey was made of the Conowingo reservoir in 1967 and again late in the summer of 1972 following Agnes. A detailed examination of these data indicates that over this five year period there was some erosion in the upstream part of the Conowingo reservoir, but that in the middle and lower parts of the reservoir there was deposition, and that the volume of material depositied was much greater than could be accounted for by erosion farther upstream in the reservoir. This does not mean, of course, there there was not erosion of the bottom of the reservoir during Agnes.

Despite the uncertainties, it appears clear that in the 10-day period from 21-30 June 1972 more suspended sediment was discharged into the upper Chesapeake Bay by the Susquehanna River than during the previous decade, and perhaps during the previous quarter of a century, or even longer. In most years, the Susquehanna probably discharges between 0.5-1.0×10^6 metric tons of suspended sediment into the upper Chesapeake Bay (Schubel 1968a, b, 1969; Biggs 1970). The bulk of the sediment discharged during Agnes was silt and clay. Pipette analyses of sediment samples from cores taken throughout the upper Bay failed to show any significant textural differences between the Agnes and pre-Agnes sediments (Schubel and Zabawa, this volume).

A significant, but undetermined, amount of the sediment discharged by the Susquehanna past Conowingo during Agnes was deposited below Havre de Grace where the Susquehanna opens into the broad, shoal region known as the Susquehanna Flats. Approximately 4×10^4 m^2 (10 acres) of new islands and several hundred thousand m^2 of new inter-tidal areas were formed during Agnes. In addition, more than 38,000 m^3 of sediment had to be dredged from one section of the main shipping channel just below Havre de Grace to restore it to its original project depth, and there was appreciable shoaling in other stretches of the channel farther downstream on the Flats. The U. S. Army Corps of Engineers made several bathymetric transects on the Flats following Agnes, but the results have not been analyzed (J. McKay, personal communication, 1974)

Distribution of Suspended Sediment

The large influxes of suspended sediment by the Susquehanna and by the other tributary rivers produced anomalously high concentrations of suspended sediment throughout much of the Chesapeake Bay estuarine system. Since the Susquehanna is the only *river* that debouches directly into the main body of the Bay, its effect on the distribution of suspended sediment was most apparent, particularly in the upper Bay where concentrations soared to levels higher than any previously reported. Most of the sediment discharged by the Susquehanna was deposited in the upper reaches of the Bay, upstream from Station 909. Most of the sediment discharged by the other rivers was apparently deposited in the upper reaches of their estuaries and did not reach the Bay proper (see, for example, Nichols et al. this volume).

On 25 June 1972 the concentrations of total suspended solids at Station 917S exceeded 800 mg/ℓ at the surface, and 1100 mg/ℓ at mid-depth (6 m). Near the time of peak flooding there was a marked longitudinal gradient of the concentration of suspended solids in the upper Bay. On 26 June 1972, two days after the Susquehanna crested, the concentration of total suspended solids at the surface dropped from more than 700 mg/ℓ off Turkey Point at the head of the Bay to about 400 mg/ℓ at Tolchester (30 km farther seaward), and to approximately 175 mg/ℓ at the Bay Bridge near Annapolis, Fig. 6. The concentration of suspended solids at mid-depth in the upper reaches of the Bay showed a similar distribution pattern although the concentrations were generally greater than near the surface. Seaward of Station 903A, however, there was an abrupt decrease of the concentration of total suspended solids below a depth of about 10 m. This distribution resulted from the over-riding of relatively "clean" estuary water by the sediment-laden Susquehanna River water.

The marked downstream decrease in the concentration of total suspended solids in the upper Bay clearly resulted almost entirely from the deposition of material by settling; there was little dilution of Susquehanna River water by Bay water in this segment of the Bay. Riverflow was so great that the tidal reaches of the Susquehanna were pushed seaward more than 80 km from the mouth of the River at Havre de Grace--nearly to the Chesapeake Bay Bridge at Annapolis, Fig. 7. The front separating the fresh river water from the salty estuary water was more than 35 km farther seaward than ever previously reported.

The distribution of suspended sediment in the upper Bay on 27 June 1972 is depicted in Fig. 8. By 29 June 1972, the concentrations of suspended sediment had decreased significantly throughout the upper Bay, Fig. 9. Maximum concentrations at that time were observed between Stations 917S and 909, and did not exceed 300 mg/ℓ. The concentration of suspended sediment decreased both upstream and downstream of this approximately 20 km long segment. The longitudinal gradient of suspended sediment that had characterized the upper Bay on June 26 and 27 had disappeared. Longitudinal distributions of total suspended solids in the upper Bay during the week following Agnes, Figs. 6, 8, 9, and 10, show that the concentrations dropped quickly following peak discharge, and that the bulk of the material discharged into the main body of the Bay at Turkey Point was deposited above Station 903A. Concentrations of suspended solids were relatively high, however, over all of the Maryland portion of the Bay proper.

The concentrations of total suspended solids remained anomalously high throughout most of the upper Bay for about a month. Figs. 11 and 12 depict the longitudinal variations of the surface and mid-depth concentrations of suspended sediment in the upper Bay on specific dates following Agnes. The distribution observed on 8 July in 1968--a more "normal" year--is plotted for reference, Fig. 12. The effects of Agnes on the distribution of suspended sediment in the Bay proper were not only greatest in the upper reaches of the Bay, but were also more persistent in that segment of the Bay than in more seaward segments. By the end of July the distribution of total suspended solids in the upper Bay was near "normal" for that time of year.

As the normal two-layered circulation pattern was reestablished throughout the upper reaches of the Bay, there was a net upstream movement of sediment suspended in the lower layer. Sediment previously carried downstream and deposited by the flooding Agnes waters was resuspended by tidal currents and gradually transported back up the estuary. The routes of sediment dispersal are clear, but the rates of movement are obscure. The data do not permit reliable estimates of the rates of sediment transport, particularly during the recovery period.

Data from selected cruises that extended over the middle and lower reaches of the Bay are summarized in Figs. 13-16. Comparison of these data with the data from more "normal" years indicates that concentrations of total suspended solids in the middle and lower Bay were, throughout July and August 1972, 2-3x higher than average for this time of year. Seaward of Station 858C concentrations in July and August did not exceed 10 mg/ℓ except near the bottom. It was previously reported by VIMS (Chesapeake Bay Research Council 1973) that on 24 July 1972 the thirteen-hour average "turbidity" at a section across the Bay off Smith's Point at the mouth of the Potomac was 17.6 mg/ℓ. On this same date we occupied stations along the axis of the Bay north (Station 818P) and south (Station 745) of this section. Our data, Fig. 14, indicate that the concentration of total suspended solids at these stations did not exceed 10 mg/ℓ, and that it exceeded 5 mg/ℓ only near the bottom. Since both of our stations were located on the axis of the Bay, the concentrations may not be representative of the entire cross-section. Concentrations of suspended sediment may have been higher on the western side of the Bay because of outflow from the Potomac.

One cruise was made on the continental shelf between 28 July and 2 August 1972. The outflow from the Bay at that time could be traced as a band of low salinity water that turned south after leaving the Bay and moved along the Virginia and Carolina coasts (Schubel, Carter, Cronin, this volume). This band was also characterized by relatively high concentrations of suspended sediment (Fig. 17)--concentrations higher than any previously reported for this segment of the shelf.

SUMMARY AND CONCLUSIONS

The floodwaters from Agnes discharged massive amounts of suspended sediment into the Chesapeake Bay estuarine system. In the ten-day period from 21-30 June 1972 the Susquehanna discharged more suspended sediment into the Bay than it had during the previous decade, and probably during the previous quarter of a century, or perhaps even longer. Our data indicate that during this ten-day period the suspended sediment discharge at Conowingo, Md. was more than 31×10^6 metric tons. This influx sent concentrations of suspended sediment throughout the upper Bay soaring to levels higher than any previously reported. The marked longitudinal gradient of suspended sediment that characterized the upper Bay during the period of peak discharge resulted largely from deposition; there was little dilution of river water by Bay water in the upper Bay. Concentrations of suspended sediment remained anomalously high for about a month after Agnes. Sedimentation rates in the upper reaches of the Bay were appreciably higher than farther seaward (Schubel and Zabawa, this volume).

In the middle and lower reaches of the Bay concentrations of suspended sediment following Agnes were 2-3x higher than "normal" for that time of year. The outflow from the Bay could be traced as a low salinity, turbid band of water that turned south and flowed along the Virginia and Carolina coasts. Concentrations of suspended sediment in this band were higher than any previously reported for this segment of the shelf.

Floods of the magnitude of Agnes have a recurrence interval of about 100 years; floods of lesser magnitude occur even more frequently. Comparison of our sedimentological data collected during Agnes with data collected during "average" conditions over many years indicates that episodic events may have a much greater and more persistent impact on the sedimentary history of the Chesapeake Bay estuarine system than do "chronic" or "average" conditions.

ACKNOWLEDGEMENTS

I am indebted to C. F. Zabawa, W. B. Cronin, T. W. Kana, M. Glendening, C. H. Morrow, and L. Smith for their help both in the field and in the laboratory. The figures were drafted by D. Pendleton and photographed by J. Sullivan. The manuscript was typed by A. Sullivan.

This research was supported in part by the Oceanography Section, National Science Foundation, NSF Grant GA-36091; in part by a project jointly funded by the Fish and Wildlife Administration, State of Maryland, and the Bureau of Sport Fisheries and Wildlife, U. S. Department of Interior, using Dingell-Johnson Funds; and in part by the U. S. Army Corps of Engineers (Baltimore and Philadelphia Districts).

LITERATURE CITED

Biggs, R. B. 1970. Sources and distribution of suspended sediment in northern Chesapeake Bay. Mar. Geol. 9(1970):187-201.

Chesapeake Bay Research Council. 1973. Effects of Hurricane Agnes on the environment and organisms of Chesapeake Bay. A Report to the Philadelphia District U. S. Army Corps of Engrs. (Aven M. Andersen, Coordinator), Chesapeake Bay Inst. Contrib. 187, Nat. Resour. Inst. Contrib. 529, Va. Inst. Mar. Sci. Spec. Rep. Appl. Mar. Sci. & Ocean Engr. 29, 172 p.

Nichols, M. 1972. Sediments of the James River estuary. In: Nelson, B. E., ed., Environmental Framework of Coastal Plain Estuaries. Geol. Soc. Am. Memoir 133.

Nichols, M. 1974. Development of the turbidity maximum in the Rappahannock estuary, summary. Proc. Int. Conf. on Estuary-Shelf Interrelationships, in press.

Schubel, J. R. 1968a. Suspended sediment of the northern Chesapeake Bay. Chesapeake Bay Inst., Johns Hopkins Univ. Tech. Rep. 35, Ref. 68-2, 264 p.

Schubel, J. R. 1968b. Suspended sediment discharge of the Susquehanna River at Havre de Grace, Md., during the period 1 April 1966 through 31 March 1967. Chesapeake Sci. 9(2):131-135.

Schubel, J. R. 1968c. The turbidity maximum of the Chesapeake Bay. Science 161(3845):1013-1015.

Schubel, J. R. 1969. Size distributions of the suspended particles of the Chesapeake Bay turbidity maximum. Neth. J. Sea Res. 4(3):283-309.

Schubel, J. R. 1972. Suspended sediment discharge of the Susquehanna River at Conowingo, Md. during 1969. Chesapeake Sci. 13(1):53-58.

Schubel, J. R. and R. B. Biggs. 1969. Distribution of seston in upper Chesapeake Bay. Chesapeake Sci. 10(1):18-23.

Schubel, J. R., E. W. Schiemer, T. W. Kana, and G. M. Schmidt. 1972. A new filtering water bottle. J. Sediment. Petrol. 42(2):482-484.

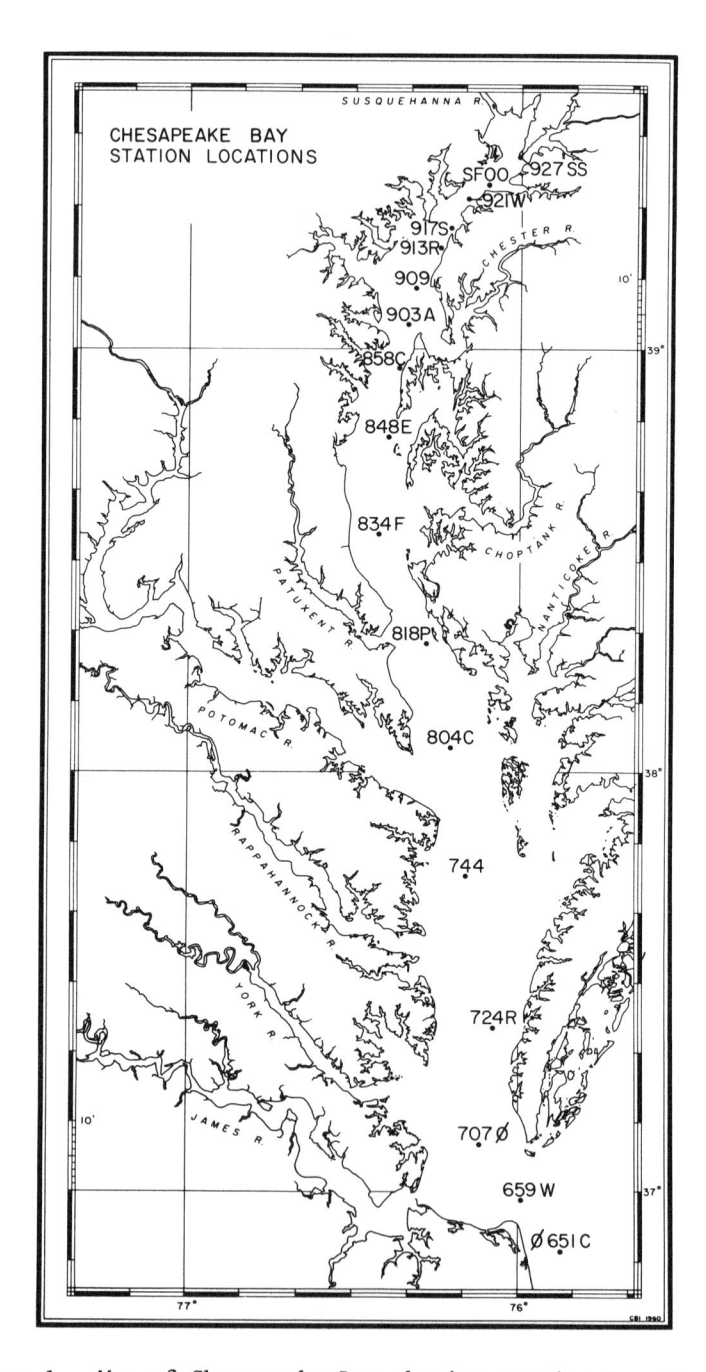

Figure 1. Map of Chesapeake Bay showing station locations.

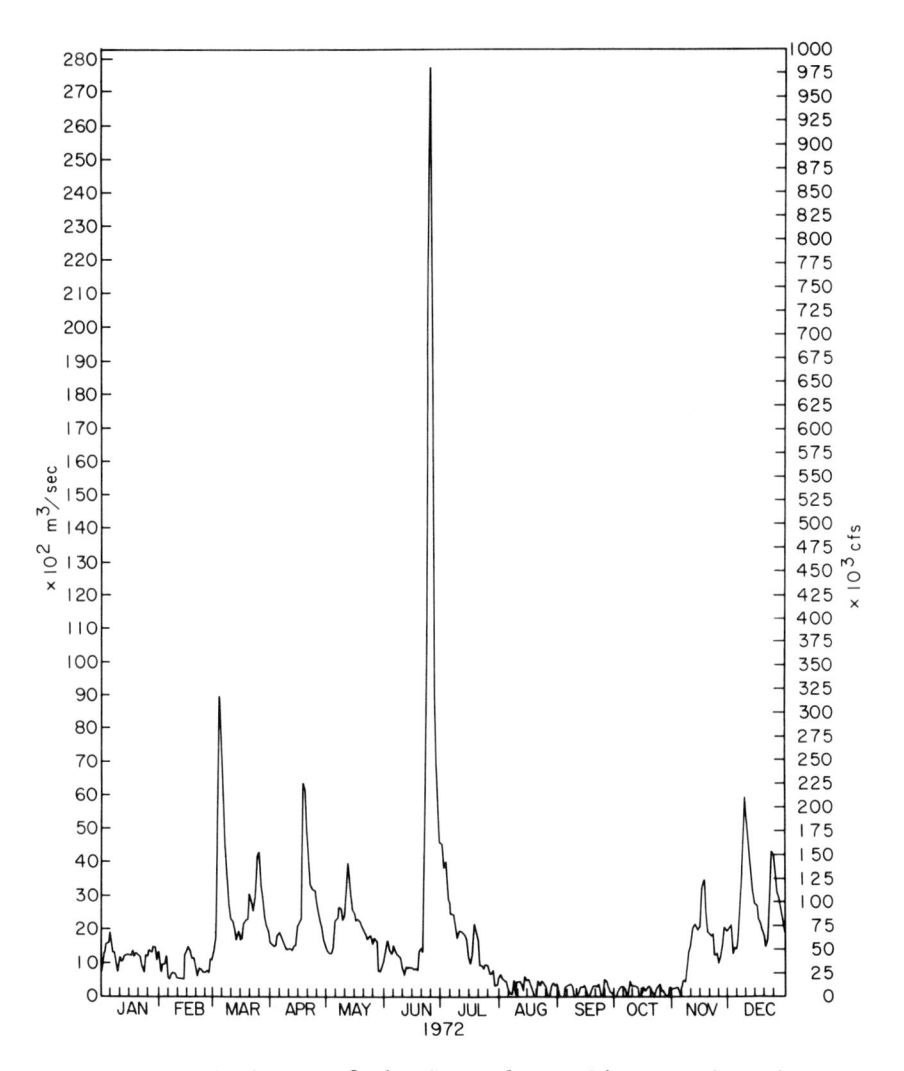

Figure 2. Discharge of the Susquehanna River at Conowingo, Md., during 1972.

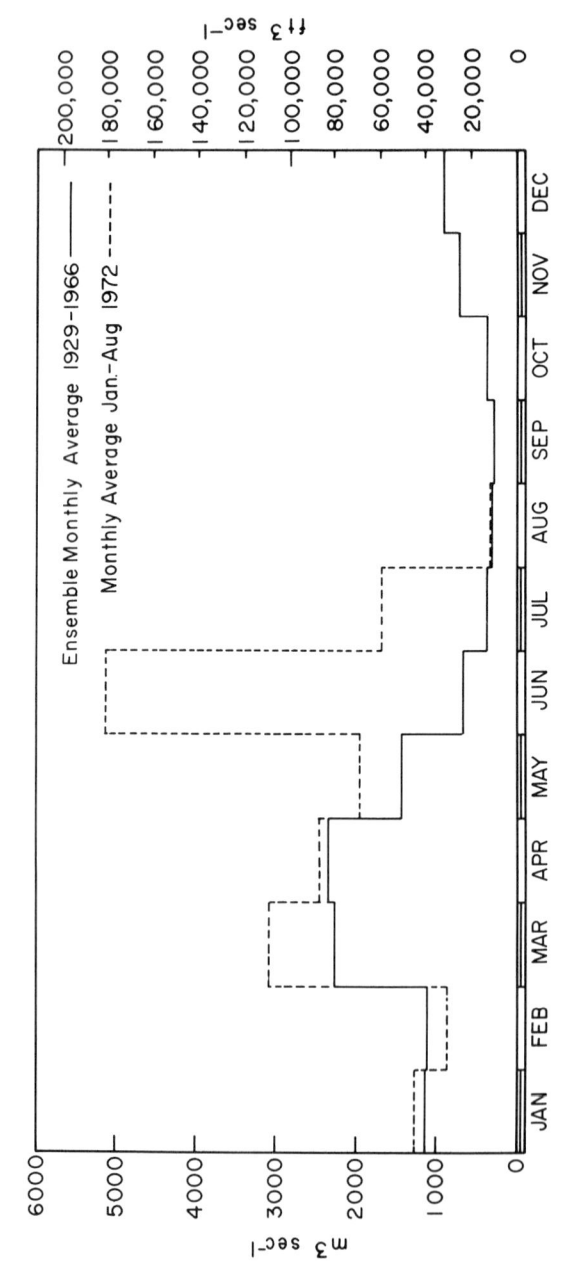

Figure 3. Ensemble monthly average river flow at Conowingo, Md., over the period 1929-1966, and the monthly average discharge during 1972.

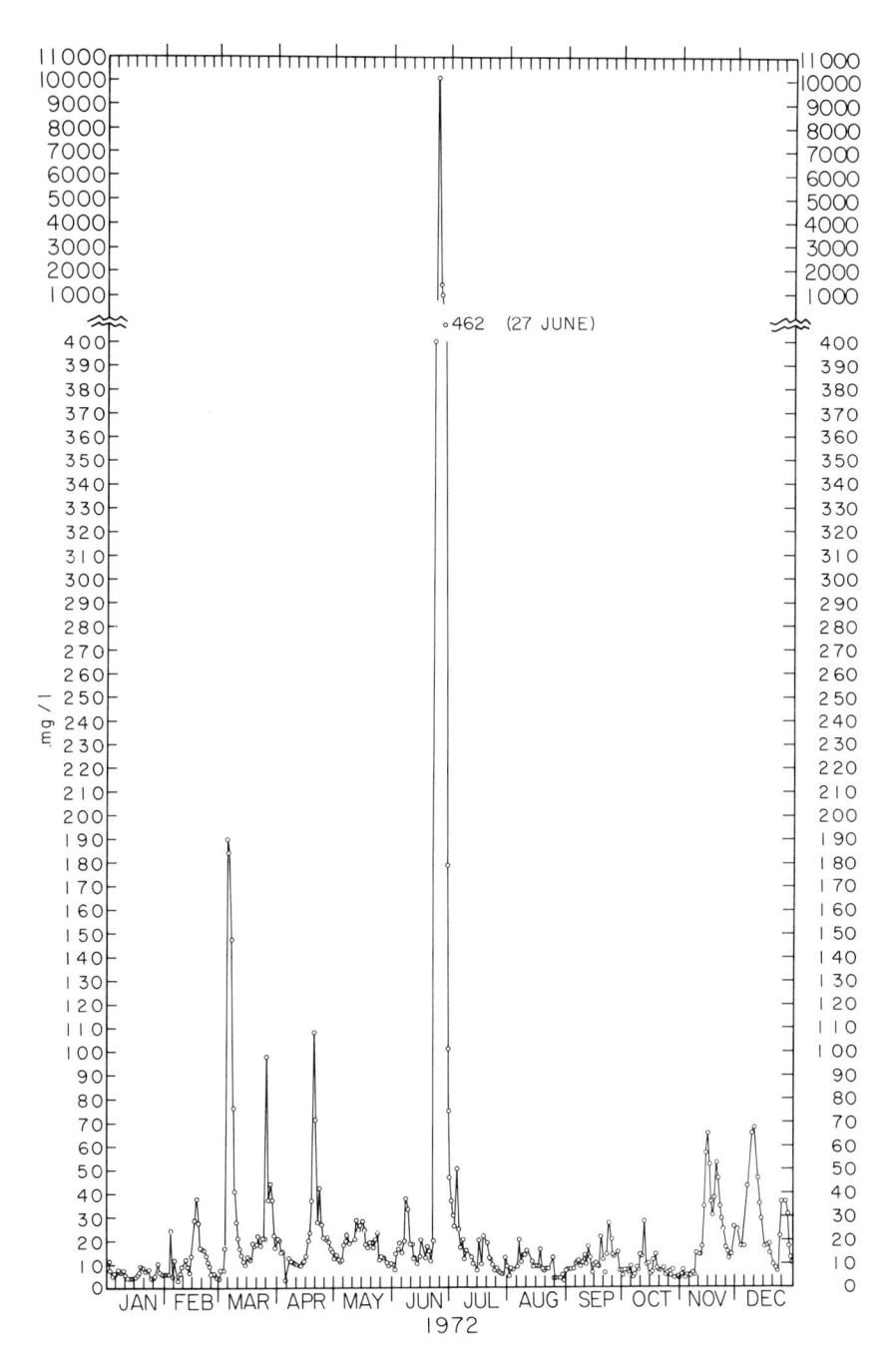

Figure 4. Concentration of total suspended solids (mg/l) in the Susquehanna River at Conowingo, Md., during 1972.

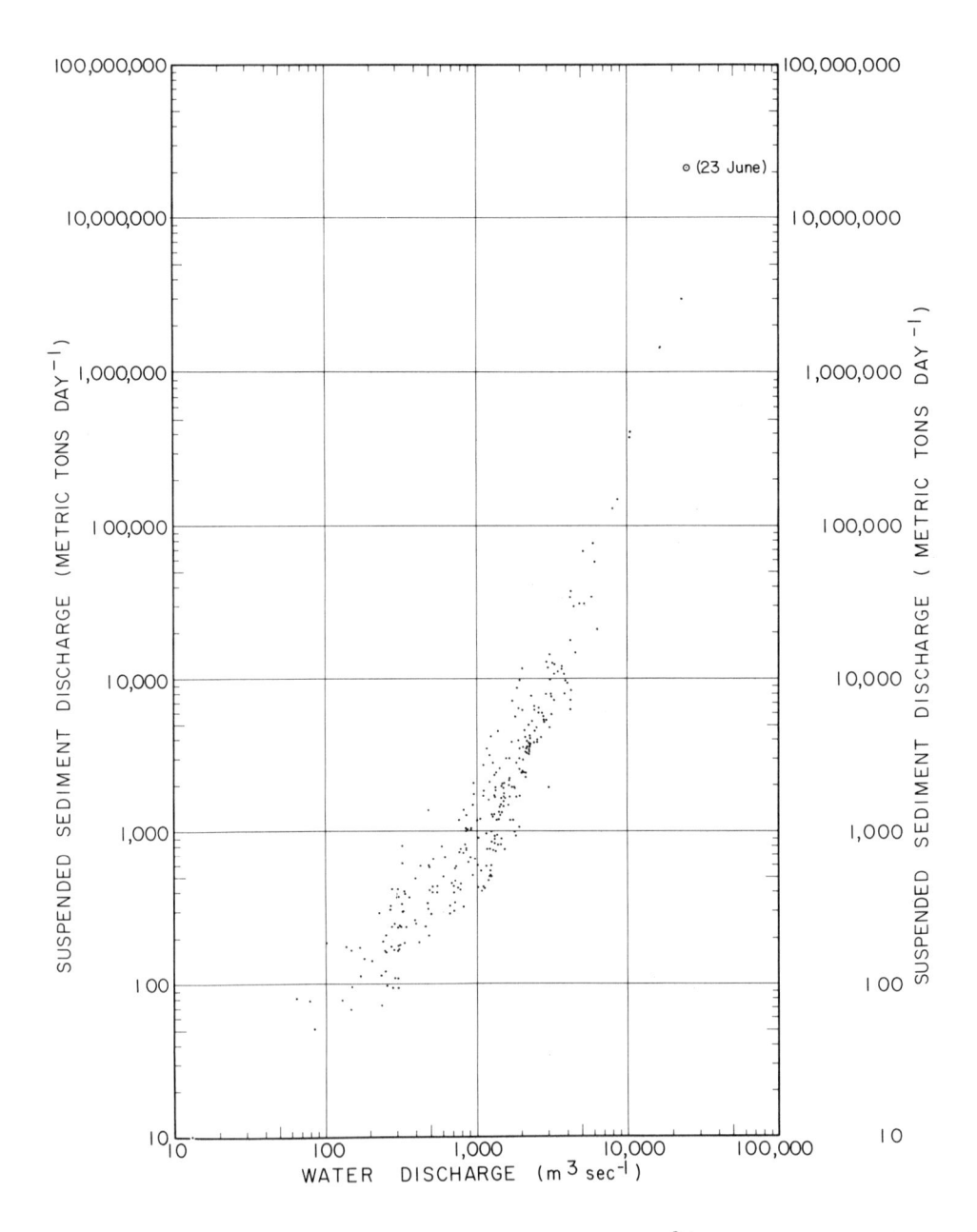

Figure 5. Plot of daily average water discharge (m³/sec) vs. suspended sediment discharge (metric tons/day) at Conowingo, Md., during 1972.

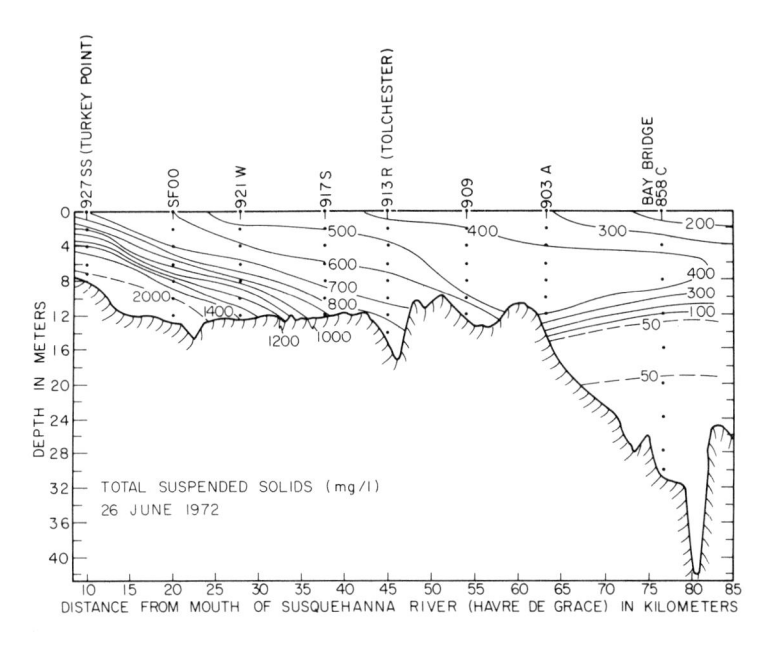

Figure 6. Distribution of total suspended solids along the axis of the upper Bay on 26 June 1972.

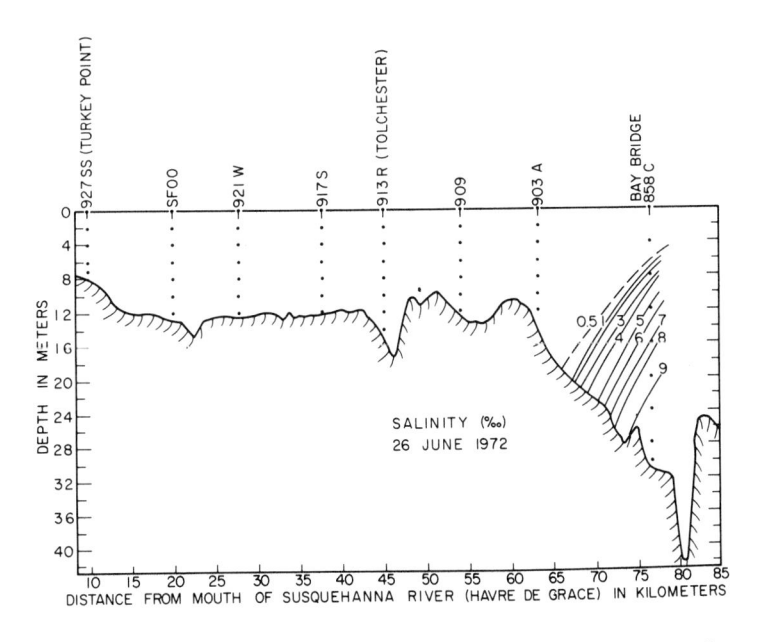

Figure 7. Distribution of salinity along the axis of the upper Bay on 26 June 1972.

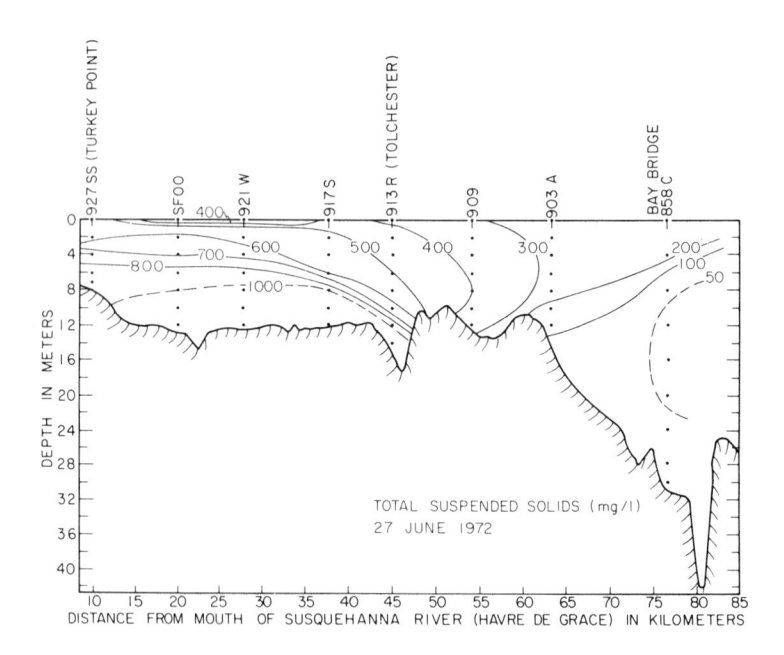

Figure 8. Distribution of total suspended solids along the axis of the upper Bay on 27 June 1972.

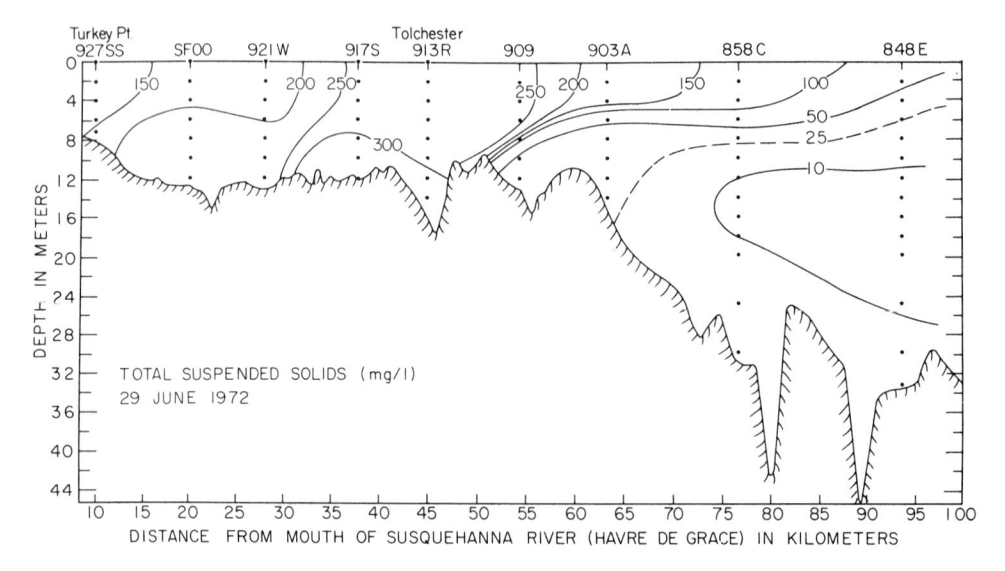

Figure 9. Distribution of total suspended solids along the axis of the upper Bay on 29 June 1972.

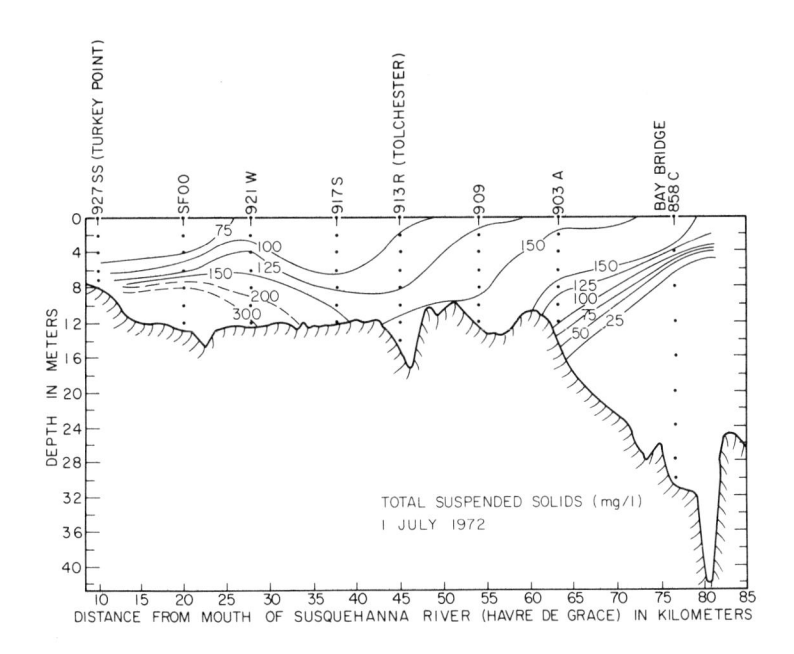

Figure 10. Distribution of total suspended solids along the upper Bay on 1 July 1972.

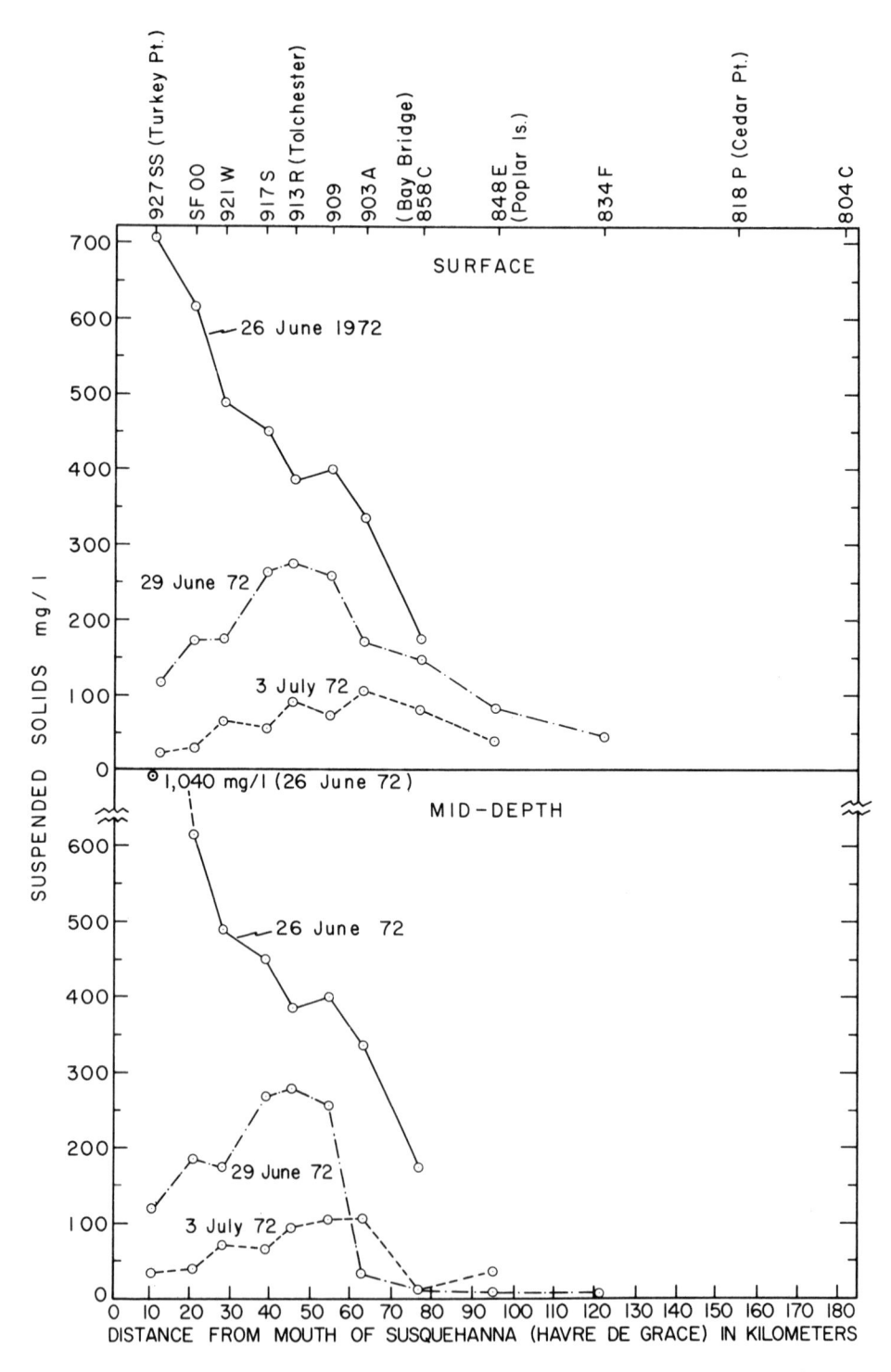

Figure 11. Variation of surface and mid-depth concentrations of total suspended solids along the axis of the upper Bay on specific dates following Agnes.

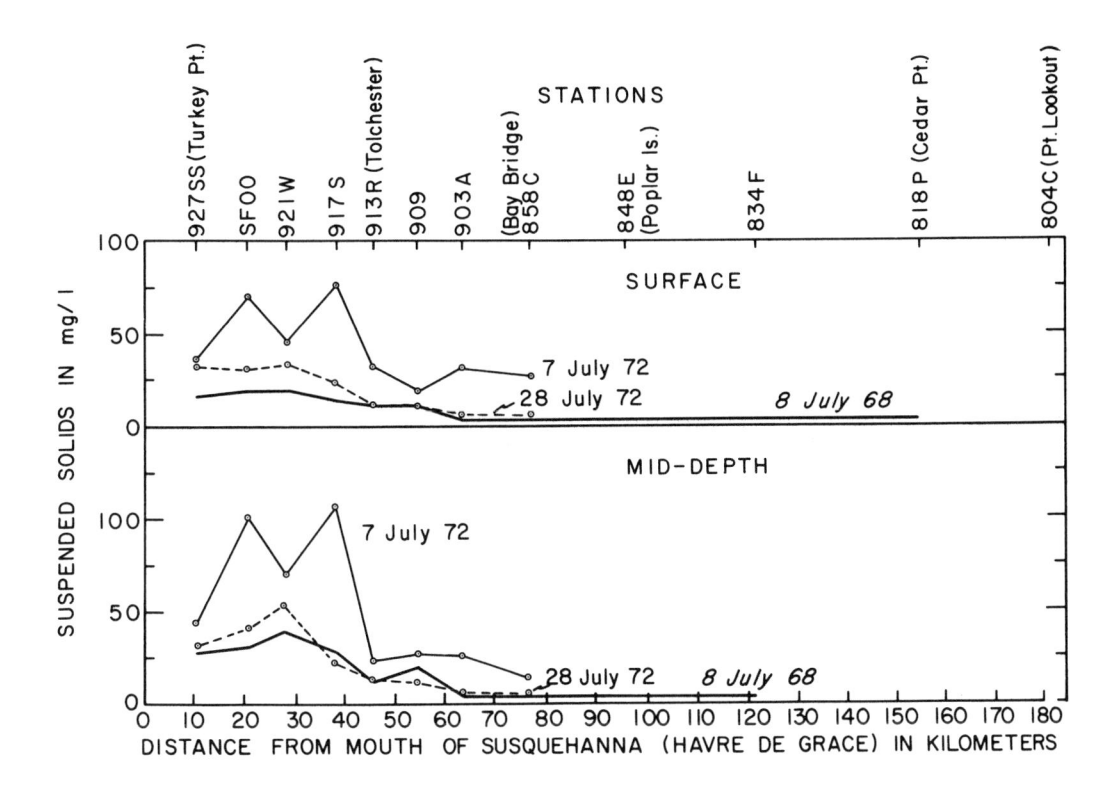

Figure 12. Variation of surface and mid-depth concentrations of total suspended solids along the axis of the upper Bay on specific dates following Agnes.

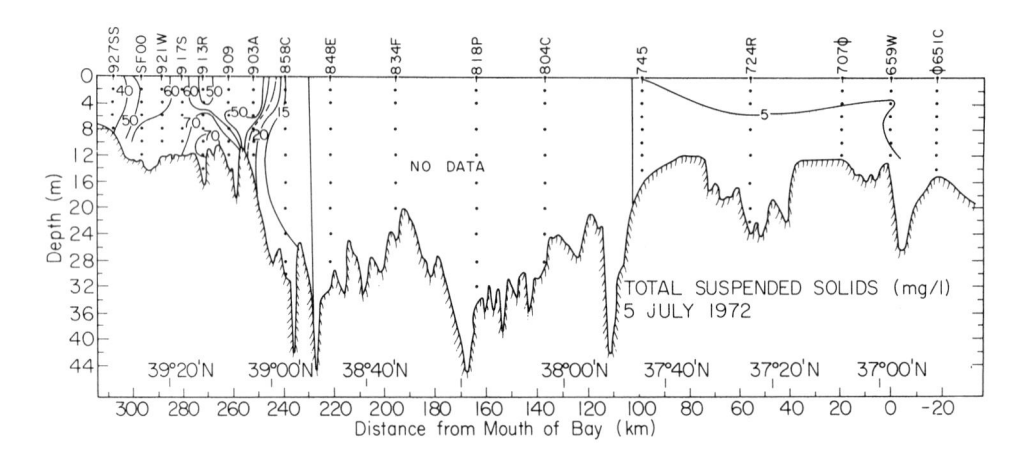

Figure 13. Distribution of total suspended solids along the axis of the Bay on 5 July 1972.

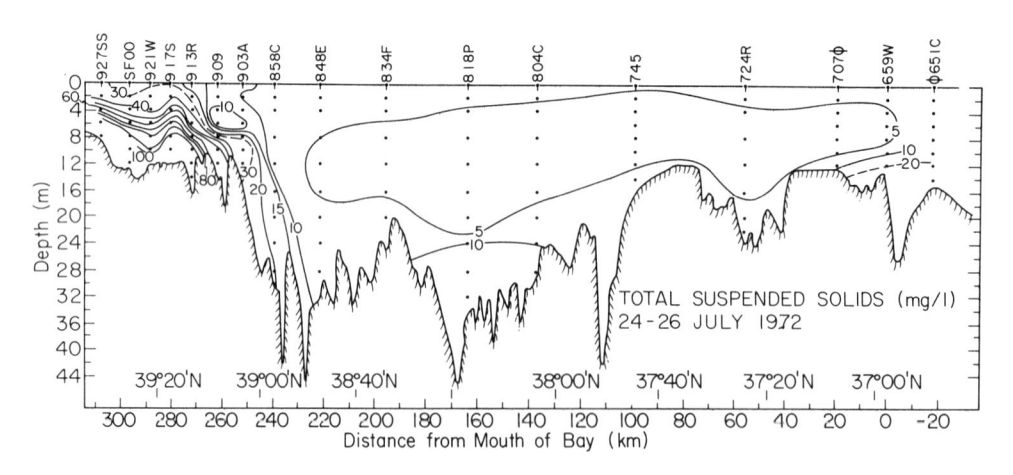

Figure 14. Distribution of total suspended solids along the axis of the Bay on 24-26 July 1972.

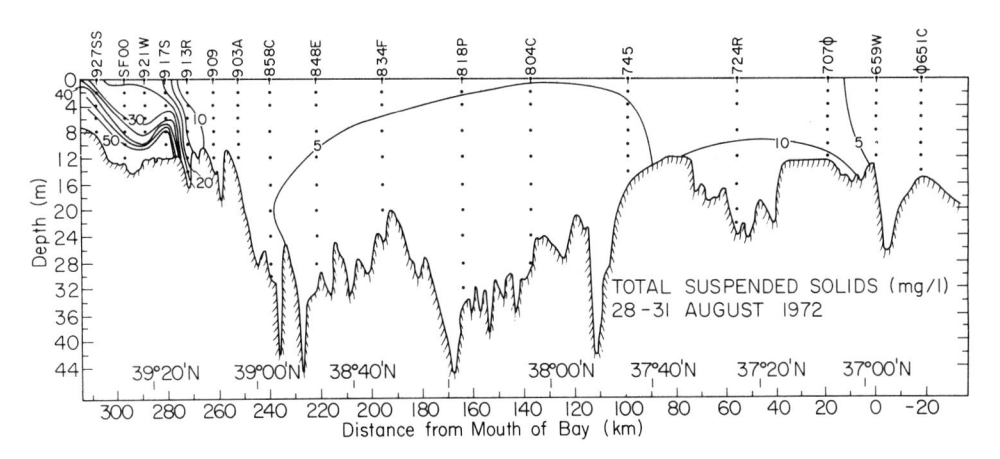

Figure 15. Distribution of total suspended solids along the axis of the Bay on 28-31 August 1972.

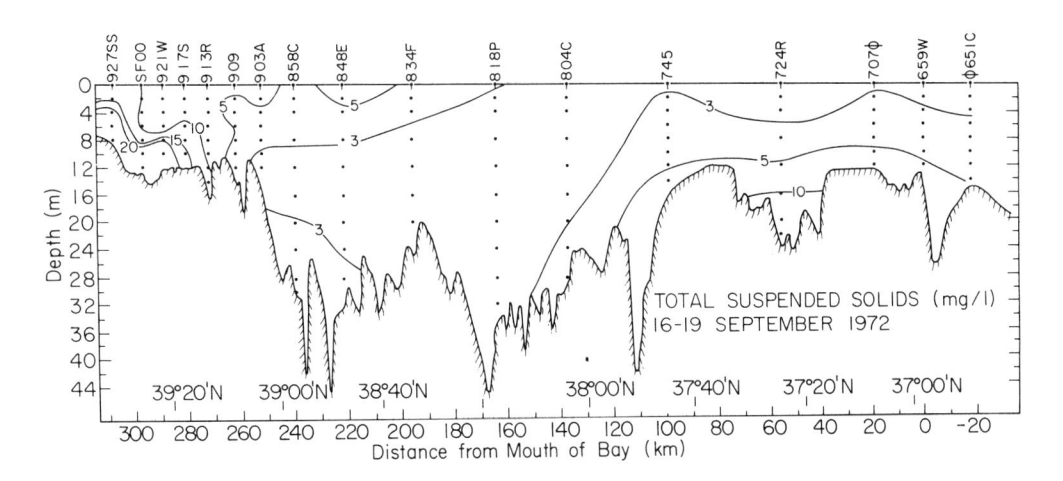

Figure 16. Distribution of total suspended solids along the axis of the Bay on 16-19 September 1972.

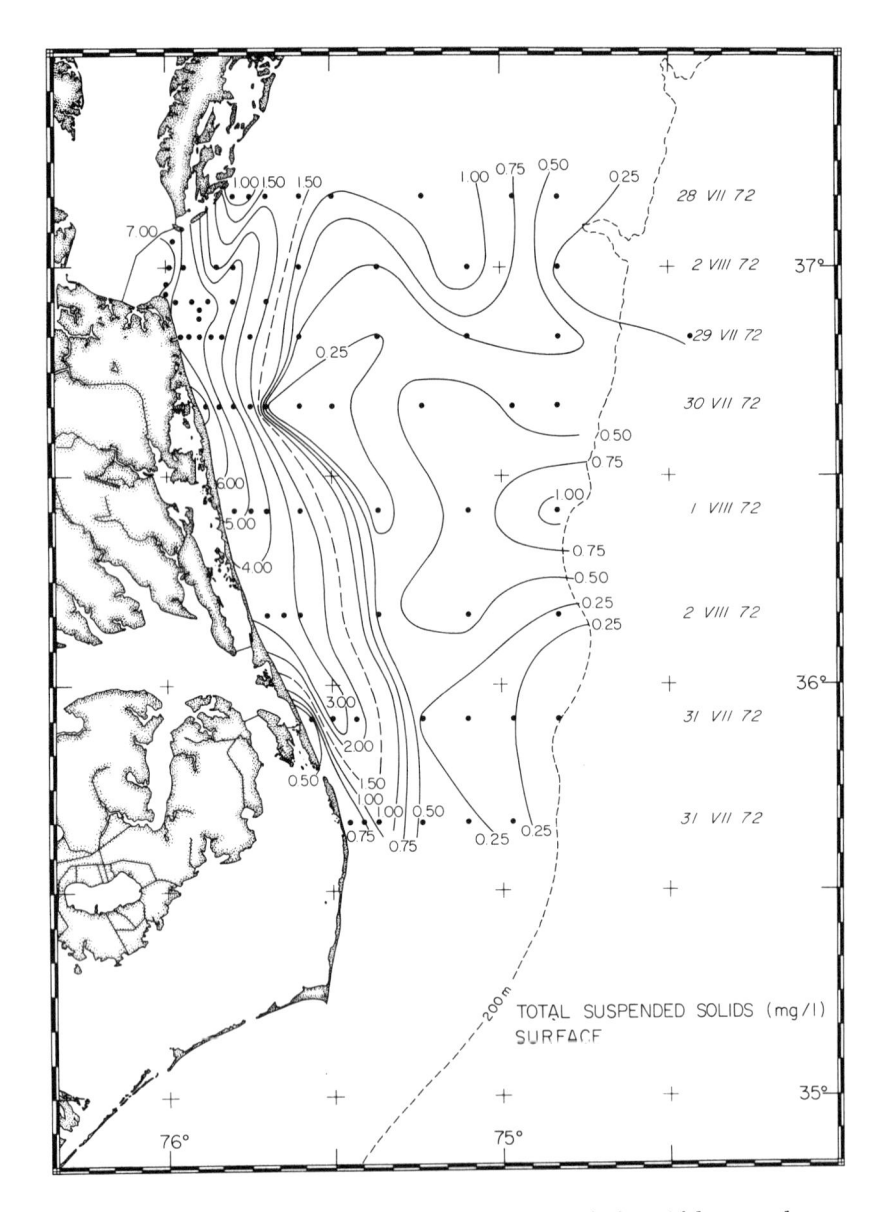

Figure 17. Distribution of total suspended solids at the surface of the waters overlying the continental shelf off the mouth of the Chesapeake Bay, 28 July-2 August 1972.

RESPONSE AND RECOVERY TO SEDIMENT INFLUX
IN THE RAPPAHANNOCK ESTUARY:
A SUMMARY[1]

Maynard M. Nichols[2]
Galen Thompson[2]
Bruce Nelson[3]

ABSTRACT

Flooding from Tropical Storm Agnes produced unique hydrographic conditions for transport and dispersal of sediment in the Rappahannock and James estuaries. Analyses indicate two cycles of response and recovery to the shock of extreme freshwater and sediment influx; one cycle in response to Rappahannock inflow; the other to intense mixing within the estuary. Important stages in the sequence consist of: (1) an initial response and seaward surge of river water and sediment; (2) shock with downstream translation of the salt intrusion head with a near-bottom salinity front and high turbidity in surface and in bottom water; (3) rebound with intense stratification and formation of an enriched turbidity maximum; (4) partial recovery with salinity intrusion strengthened by upstream flow along the bottom; landward migration of the maximum; (5) full recovery and return to partly-mixed state with decay of turbidity maximum over a broad zone 30 days after flooding. Sediment was derived initially from lateral tributaries and then from the main river. The bulk of the load sedimented above the salt intrusion during the first three days of flooding. Sediment dispersed into the estuarine circulation system later was effectively trapped by upstream flow along the bottom. Over the entire event, 91% of the sediment load was trapped.

INTRODUCTION

Flooding caused a record influx of 0.1 megatons of sediment in the Rappahannock. Sediment was derived largely from the main river drainage basin. Clay minerals from the estuary, illite, chlorite and muscovite, indicate a Piedmont source. The chief question is, where does the sediment go? Is it either flushed through the estuary, trapped in suspension, or deposited on the channel floor?

The data analyzed consist of: (1) longitudinal sections of salinity and suspended sediment concentration observed daily near slack water along the estuary length; and (2) time distributions of salinity, current velocity, and suspended sediment concentration observed hourly at 3-4 depths from an anchor station in the lower estuary. The mass of data was reduced and evaluated to determine how the estuary responded to the 3-fold stress of freshwater inflow, sediment influx, and tide.

---

[1]Contribution No. 758, Virginia Institute of Marine Science.
[2]Virginia Institute of Marine Science, Gloucester Point, Va. 23062
[3]University of South Carolina, Columbia, S. C. 29208

STRESS AND RESPONSE

Freshwater inflow, which reached a peak of 2,382 m^3 per sec, June 22, 1972 at Fredericksburg on the fall line, receded gradually over a 15-day period from June 22 to July 6, 1972, except for a slight increase on June 30. Sediment influx reached an estimated 56,680 tons per day at Fredericksburg and, except for the "second surge", diminished with recession of inflow. Tidal heights at Tappahannock reached a maximum during late stages of the storm, June 21, and high waters gradually diminished over the 15-day period. At the same time the tide range increased from neap to spring, thus acting to increase tidal currents and haline mixing during the progress of flooding.

Suspended sediment concentrations in surface water of the lower estuary reached a maximum 5 days after peak flooding on the fall line. As shown in Fig. 1, the magnitude and direction of net velocity varied through a sequence of phases from "initial response" to "shock", "rebound", "reversal", "second shock", "homogeneity", and "recovery". Despite fluctuating sediment loads and diminishing concentrations after the 6th day, sediment transport follows the time-trends of net velocity. A similar sequence of changes occurred at depth in the lower layer. However, they are smaller than in the upper layer and the net current is directed mainly upstream except for a short period of reversal. Thus, the two-layered estuarine circulation responded quickly to flooding and prevented escape of most sediment from the estuary.

The complete sequence of stress and response consists of 7 stages (grouped chronologically by days after peak flooding on the fall line):

1) *Initial response* (+1 to +2 day). Flooding began with increasing inflow and sediment influx from lower tributary creeks. Storm tides flooded and the combined force of storm surge, wind-waves and tidal currents intensely stirred the estuary floor. Freshwater inflow pushed the salt intrusion downstream and transformed it into a salt wedge. Therefore, currents in the upper estuary reversed from upstream to downstream and allowed river-borne suspended sediment to pass into the upper estuary. The high sediment influx overwhelmed the turbidity maximum that normally resides in the upper estuary (+2 day).

2) *Shock* (+3 to +4 day). As the main surge from mainstream drainage entered the estuary it freshened near-surface water, lowered the halocline and created a salt wedge with high stratification. The high hydrostatic head created a near-surface gravitational current that carried part of the sediment load seaward through the upper layer into the lower estuary. But seaward transport below mid-depth was arrested by the salt wedge. The main load of sediment accumulated landward of the convergence and created a high longitudinal concentration gradient (+4 day).

3) *Rebound* (+4 to +5 day). As the main surge diminished the salt wedge penetrated landward into the upper estuary channel (+5 day). Stratification became intense and the convergence strengthened. Enriched turbid aureoles, a turbidity maximum, formed just landward of the near-bottom convergence. At the same time the estuarine circulation was maintained in the lower estuary and net currents returned to near-normal speeds. The suspended sediment "minimum" at mid-depth disintegrated as tidal resuspension diminished and as sediment settled from the upper layer.

4) *Reversal* (+6 to +7 day). Freshening of Chesapeake Bay temporarily created an inverse salinity gradient in the lower estuary and resulted in a reversal of the estuarine circulation. At the same time the salt wedge penetrated farther landward and the turbidity maximum shifted upstream.

5) *Second shock* (+8 through +11 day). A second surge of mainstream flooding again depressed the salt wedge. It freshened the upper layer, increased stratification and strengthened the estuarine circulation. The turbidity maximum became larger and shifted downstream. The second shock was less intense than the first shock but it persisted longer.

6) *Haline homogeneity* (+12 to +22 day). As the second surge diminished, bottom water continued to freshen as spring tides weakened stratification and mixed near-surface water downward. The estuarine circulation persisted but net current was very weak. Sediment influx from the bay and intense mixing maintained the turbidity maximum.

7) *Recovery* (+22 to +60 day). As river inflow subsided, and as Bay water regained normal salinity, the estuary returned to its normal partly-mixed state. The salt intrusion stabilized in the upper estuary and the turbidity maximum decayed over a broad zone (e.g. +30 day).

DISPERSAL AND DEPOSITION

Sediment was initially transported through the main channel of the upper estuary and later through the upper layer of the salt intrusion in both the channel and over bordering shoals. However, the sediment transport budget indicates 91% of the total river-borne input was trapped. It was partly retained for a while in the turbidity maximum upstream of the salt intrusion and gradually deposited as the maximum shifted landward during recovery. For another part, suspended sediment was progressively diluted with distance downstream and deposited by simple gravitational settling rather than by flocculation. Thus the flood-borne load was spread out on the channel floor of the upper and middle estuary from 7.5 to 2mm thick (estimated by Huggett, this volume). This deposition, which amounts to one-third of the annual average deposition, is part of a long-term trend of sedimentary and seaward shifting of the locus of shoaling. Unless existing channel depths are maintained by dredging, shoaling will shift the head of navigation seaward, reduce salinity in the estuary, and shift the circulation pattern from a type B to a type C (Pritchard 1955). Thus, deposition will change the hydraulic regime from a trapping mode to an escape mode.

LITERATURE CITED

Pritchard, D. W. 1955. Estuarine circulation patterns. Am. Soc. Civil Engineers, Proc. 81:1-11.

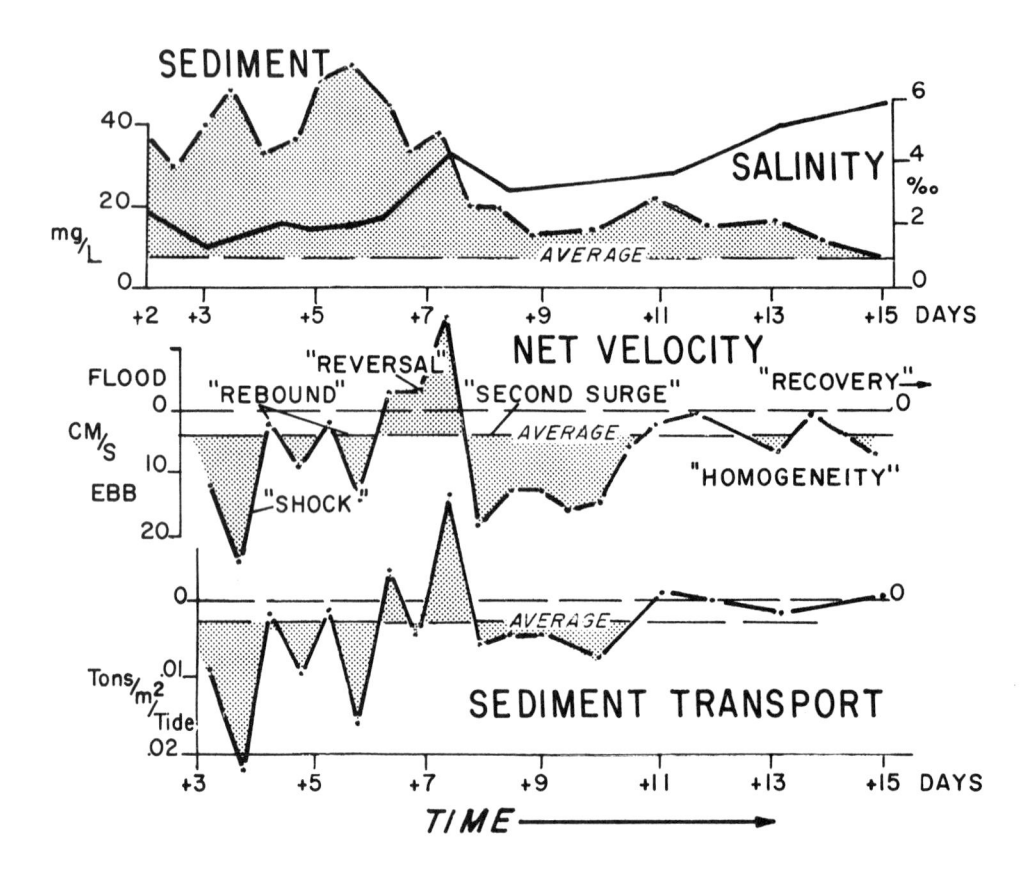

Figure 1. Time-variations of total suspended matter (sediment), salinity (top), net velocity (middle), and sediment transport (bottom) in surface water of the lower Rappahannock Estuary near Urbanna between June 24 (+ 2 day) and July 7 (+ 15 day) 1972. Sequence of events, defined mainly by net velocity response, are labeled "shock", "rebound", etc. Deviation from average values in this reach of the estuary, shaded.

THE EFFECTS OF TROPICAL STORM AGNES ON THE COPPER AND ZINC BUDGETS OF THE RAPPAHANNOCK RIVER[1]

Robert J. Huggett[2]
Michael E. Bender[2]

ABSTRACT

The metals copper and zinc were analyzed in bottom sediments (top 1 cm) from the Rappahannock River before and after Tropical Storm Agnes. By extracting the sediments with various techniques (HNO_3, HCl) the nature of the metal speciation can be estimated. Data show that the inorganic copper was increased by a factor of 2 to 3 in the normally saline portion of the river as a result of Agnes but returned to before-Agnes levels within one year.

Metal analyses of suspended sediments collected during the Agnes flooding allows an estimate of sedimentation indicating at least 7.5 mm of new sediments at mile 40, decreasing nearly linearly to 1 mm at mile 15.

INTRODUCTION

The Rappahannock River is a coastal plane estuary located on the Chesapeake Bay (Fig. 1). It is tidal for approximately 100 nautical miles (185 kilometers) with the first 45 miles (80 kilometers) at normal river flows being estuarine having salinities greater than 0.4 ppt. This system is relatively pristine in nature in its estuarine portion with occasional agricultural development along its banks. To the authors' knowledge there are no man-induced trace metal sources in the river with the possible exception of drainage and sewage from Fredericksburg, Virginia, located at the fall line, 185 kilometers upstream.

In an attempt to define and understand the trace metal budgets of large coastal plane estuaries, the Rappahannock River was extensively sampled in 1972 and 1973. During this period three major sampling runs were conducted: one in January 1972, one in October 1972, and one in June 1973. The first sampling was approximately six months before Agnes passed over the system, the second was two months, and the third was 12 months after.

The work reported here was originally intended to describe the background levels of copper and zinc in the top 1 cm of Rappahannock River bottom sediments and to correlate the concentrations found to the normally analyzed estuarine variables of the pH and salinity. Six months after such a background study, Agnes passed through the system.

To ascertain the effects of this deluge on the sediment metals budget on the Rappahannock River for which we had good background data, the system was resampled and analyzed. Since some changes were noted, the system was sampled again 12 months after the storm to note recovery if any.

---

[1] Contribution No.759, Virginia Institute of Marine Science.
[2] Virginia Institute of Marine Science, Gloucester Point, Va. 23062

METHODS AND PROCEDURES

The original sampling was in January 1972 and consisted of samples taken in the channel at 0.5 nautical mile intervals from the mouth to 20 miles above the freshwater-saltwater interface. In all, 63 miles of the stream were sampled, yielding a total of 126 samples. After Agnes, in October 1972, the samples were taken at 2-mile intervals from the mouth to approximately 30 miles upstream and then at 1-mile intervals up to mile 60. The last sampling was conducted in June 1973, approximately one year after Agnes, and consisted of samples taken at 5-mile intervals from the mouth to mile 35 and then at 2-mile intervals up to mile 63. The sampling intervals increased with each subsequent sampling because we were initially unaware of the natural variations; therefore as many samples as we could analyze were taken. As more was learned about the system, fewer samples were taken.

The samples were collected with a ponar grab sampler which was lowered slowly to the bottom. When tension was released on the wire, the sampler closed and was returned to the boat, opened and the top 1 centimeter of sediment was removed, being careful not to collect material which had come in contact with the sides of the sampler.

Each sample was wet sieved and only the less-than or equal-to 63 micron portion was saved for analysis. Since the concentrations of metals sorbed or coated to sediment grains is a function of the surface area per unit mass of the grains, the sieving was necessary to help normalize the samples.

After the $\geq 63\mu$ fractions were obtained from the samples, they were dried first in air and then at $105^\circ C$. Each dried sample was then split and one portion was extracted with 0.1N HCl at room temperature for one hour and another portion was extracted with fuming (not boiling) concentrated HNO_3. The exact details of the extractions are given by Huggett, Bender, and Slone (1972).

The various methods of extraction yield two distinct metal fractions: the HCl should release non-crystalline metals which are bound to the sediments by absorption, precipitation and co-precipitation reactions. The HNO_3 extraction should release these metals as well as those bound within organic matrices. The difference between the HNO_3 and HCl yields should approximate the organic metals. The extracts were all analyzed by standard Atomic Absorption techniques.

Suspended sediments were obtained from Dr. B. Nelson (University of South Carolina and Dr. M. Nichols (Virginia Institute of Marine Science) who collected them from the Rappahannock River during the Agnes flooding. The suspended sediments were separated on 0.45μ membrane filters by filtration.

The suspended matter was scraped from the filters with a glass rod and extracted for copper by the previously mentioned HNO_3 procedure. Since the suspended sediments had been stored approximately 18 months before analysis, it was feared that the samples may have lost their integrity with respect to organic-inorganic copper phases. Therefore the HNO_3 extract, which should extract both, was used.

RESULTS AND DISCUSSION

The precipitated-co-precipitated zinc data are graphically displayed in Fig. 2. The concentrations at a station are nearly the same for all three sampling periods. In the normally freshwater portion of the river (above mile 45), the concentrations are relatively constant at approximately 18 ppm. From mile 45 downstream to mile 10 there is an increase to between 50 and 60 ppm. This increase may be due to either an increase in pore-water pH towards the mouth of the estuary (Nelson 1973) or an increase in surface area of the downstream sediment grains. If such a sediment grain surface area increase is true, it must be for particles below 1µ since nearly the same size distribution for particles greater than 1µ exist in the estuarine portion of the stream (Nelson & Nichols, unpublished) From mile 10 to the mouth of the estuary the concentrations vary between 50 and 60 ppm.

The organic zinc concentrations are given in Fig. 3. As in the case of the precipitated-co-precipitated zinc fraction, the levels of organic zinc are nearly the same for all three sampling periods. From mile 45 to mile 60, the values range between 40 and 50 ppm but decrease from about 45 ppm beginning at mile 45 to 20 ppm at the mouth of the estuary at mile 0. This decrease is gradual but quite linear and opposite the trend observed for the precipitated-co-precipitated zinc fraction. This suggests that the metal bound inorganically is not available for organic reactions.

The precipitated-co-precipitated copper data are presented in Fig. 4. These data clearly show that shortly after Agnes passed over the Rappahannock River, the precipitated-co-precipitated copper was a factor of 2 to 3 times higher in the normally saltwater portion of the river than either six months before or one year after. In the normally freshwater section, the values did not significantly change during the eighteen months of study. It is the authors' belief that the increase was due directly to Agnes. The estuarine section of the river was turned nearly fresh by the deluge and since this section showed elevated copper but the normally freshwater portion did not, a salinity controlled reaction for the precipitated-co-precipitated copper appears possible. However, the investigators did not note such a phenomenon even after subjecting Rappahannock River sediments to various salinities and dissolved copper concentrations in the laboratory. Another possibility is that elevated dissolved copper was brought into the system from upstream runoff. This appears unlikely since the concentrations did not change at the upstream stations. It also appears unlikely that the copper was transported into the estuary from the Chesapeake Bay since the net flow of the stream was into the Bay during this period.

The final and most likely explanation of the increase is that new sediments high in copper were transported from the land to the river during the storm's rain and runoff and were deposited in the estuary. This hypothesis is substantiated by the studies of Nichols, Nelson, and Thompson (this volume) which showed massive amounts of erosional products being swept into the Rappahannock estuary by Agnes runoff. Further substantiation will be presented later in the discussion of copper analysis of suspended sediments collected after Agnes in the Rappahannock River.

The organic copper concentrations are given in Fig. 5. The levels are nearly the same for all sampling periods, ranging between 10 and 15 ppm. The elevated precipitated-co-precipitated copper previously mentioned was not evident in the organic fraction suggesting that the inorganically bound metals are not readily available for organic reactions.

The suspended sediments collected during the Agnes runoff period were analyzed for organic copper (Table 1). Due to the extremely small sample sizes only this fraction could be extracted and analyzed. The data in Table 1 show that the copper concentrations tend to increase for samples taken closer to the estuary's mouth. This is explained by the fact that the suspended sediments should have a greater surface area the further downstream they travel since larger particles would be settling out.

Table 1. Copper in Suspended Sediments Collected During the Agnes Rappahannock River Flood

River Miles From Mouth	Date	Depth	HNO$_3$ ppm Cu
14.8	6/24/72	Bottom	180
14.8	6/27/72	Bottom	370
18	6/29/72	Surface	220
21	6/29/72	Surface	370
21	6/29/72	Bottom	280
27	6/29/72	Surface	230
27	6/29/72	Bottom	190
29	6/29/72	Surface	93
29	6/29/72	Bottom	400
31	6/29/72	Surface	100
31	6/29/72	Bottom	130
33	6/29/72	Surface	130
33	6/28/72	Bottom	57
36	6/28/72	Bottom	70
36	6/29/72	Surface	100
36	6/29/72	Bottom	95
39	6/29/72	Surface	84

If the hypothesis that the elevated copper concentrations (Fig. 4) were due to Agnes-induced new sediments is true, then by comparing the pre-Agnes sediment copper concentrations with the after-Agnes values and the suspended sediment copper levels, an estimate of sedimentation can be obtained. In order to do this several assumptions must be made:

1) That all the suspended materials were of the same or similar origin with respect to their precipitated-co-precipitated copper concentrations.

2) That the copper concentrations did not significantly change from time of deposition until sampling.

3) That the suspended sediments collected at any one place were similar with respect to copper as those deposited at that point.

4) That the new sediments were not mixed below 1 cm by either biological or physical factors.

The first assumption appears valid since samples collected on different days from the same locations had similar copper concentrations. The second assumption may not be entirely valid. Since samples collected one year after the storm showed copper levels to have returned to normal, it is logical that the re-equilibration started soon after the waters returned to normal (∼1 month before the October sampling). This would result in the sedimentation estimates being low. The third assumption is probably valid since the river is tidal and therefore the suspended sediments move up and downstream depending on the tide stage. This assumption would probably not be true if the system were non-tidal. The fourth assumption, if not true, would again result in a lower estimate of sedimentation. The authors know of no way to check this assumption.

With these assumptions, the before-Agnes and after-Agnes bottom sediment copper data, and the suspended sediment copper values, the percent of the top 1 cm of bottom sediment due to Agnes can be calculated at each location for which all these values are known by the following formula:

$$X = \frac{a-c}{a-b} \cdot 10$$

It must be noted this formula is valid only if the top 1 cm or bottom sediment is sampled. In this equation "x" is the millimeters of new sediment in the top 1 cm of the bottom material after-Agnes; "a" is before-Agnes sediment concentration; "b" is the suspended sediment concentration, and "c" is the after-Agnes sediment concentration.

The data used for these calculations were the raw sediment analyses rather than the moving averages presented in the previous figures. The moving average technique was used to smooth out "noise" in the data but still show trends. The un-averaged data must be used in the sedimentation calculations to assure accurate estimates at each location. The results are presented in Fig. 6. In this figure the range of values as well as the means are given for each location in which there were suspended sediment samples. The data show that at least 7.5 millimeters of new sediment were deposited in the channel at mile 39. The amount of new material decreases nearly linearly to about 1 millimeter at mile 15. This trend is logical and may be thought of as a proof of the calculations because more sediments should have been deposited upstream since these areas are closer to the source of the suspended sediments.

A year after the storm the sediment copper levels were back to normal. This could be due to: migration of the sediments upstream on the estuarine salt wedge; bottom sediments being resuspended and carried seaward in the surface waters; mixing of the new sediments with old underlying material by burrowing animals; or chemical re-equilibration of copper to normal with the return of stable salinity and pH structure. The authors do not know the exact mechanism; perhaps a combination of all.

CONCLUSIONS

Agnes caused a 2- to 3-fold increase in the precipitated-co-precipitated copper content of the estuarine surface sediments of the Rappahannock River. The sediments did, however, return to "normal" within one year after the storm. The organic copper and zinc and the precipitated-co-precipitated zinc levels were not affected by the storm.

A calculation based on the deposition of suspended material, high in precipitated-co-precipitated copper on material relatively low in this copper phase, resulting in a sediment with a copper content between the two, shows that at least 7.5 millimeters of new sediment were deposited at mile 39 with amounts decreasing downstream to about 1 millimeter at mile 15. This technique appears extremely sensitive to small sedimentation amounts and may prove useful to other investigators.

ACKNOWLEDGEMENTS

The authors wish to acknowledge the assistance of H. D. Slone and J. Lunz of the Virginia Institute of Marine Science in this project and Dr. B. Nelson of the University of South Carolina and Dr. M. Nichols of the Virginia Institute of Marine Science who graciously supplied suspended sediment samples for this project.

This project was funded through the Chesapeake Research Consortium by the National Science Foundation - RANN Program Grant G. I. 34869.

LITERATURE CITED

Huggett, R. J., M. E. Bender, and H. Slone. 1972. Sediment heavy metal relationships in the Rappahannock River sediments. In: Annual Report, June 1971-May 1972, Chesapeake Research Consortium, Inc., Baltimore, Maryland.

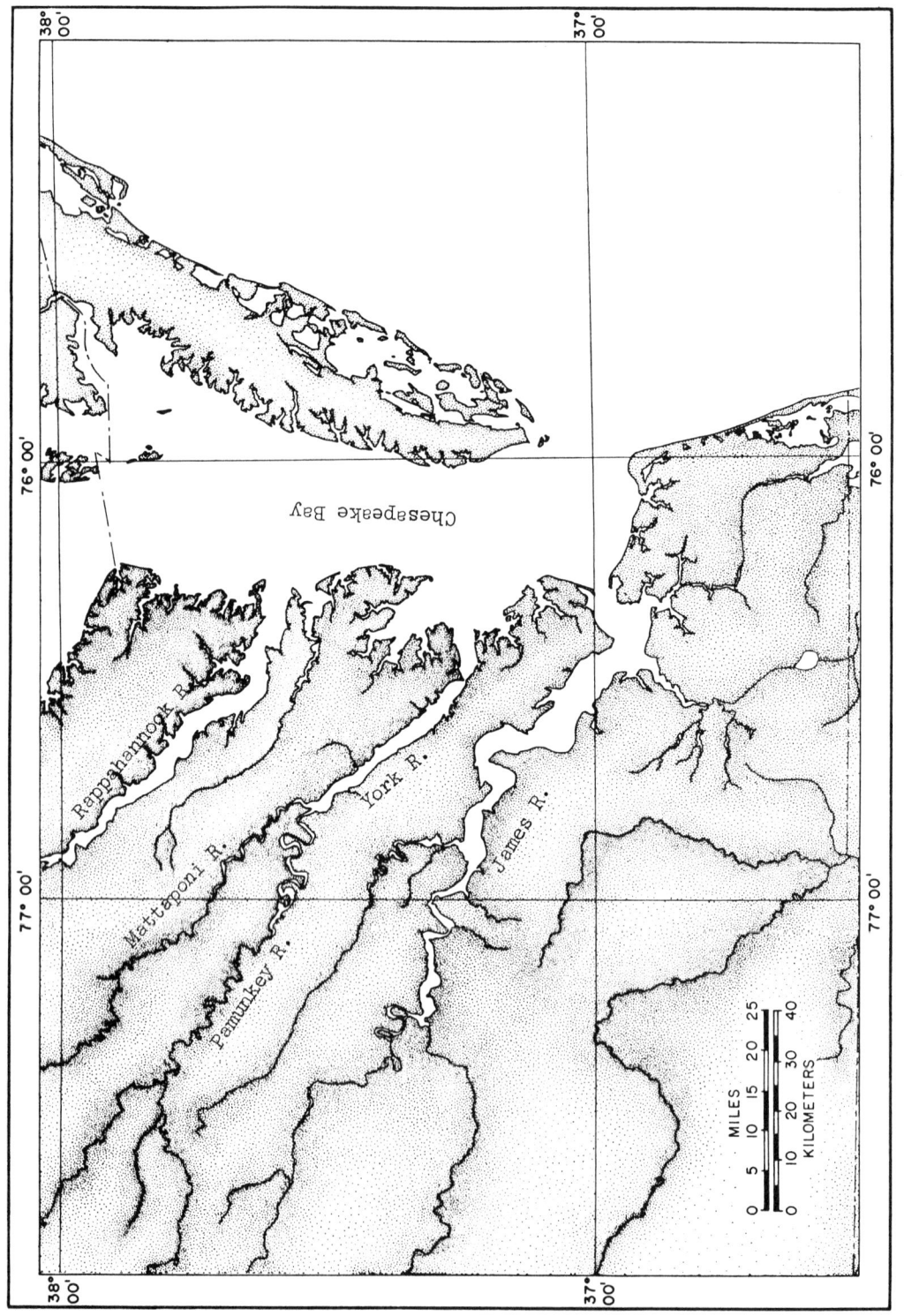

Figure 1. Map of the Chesapeake Bay.

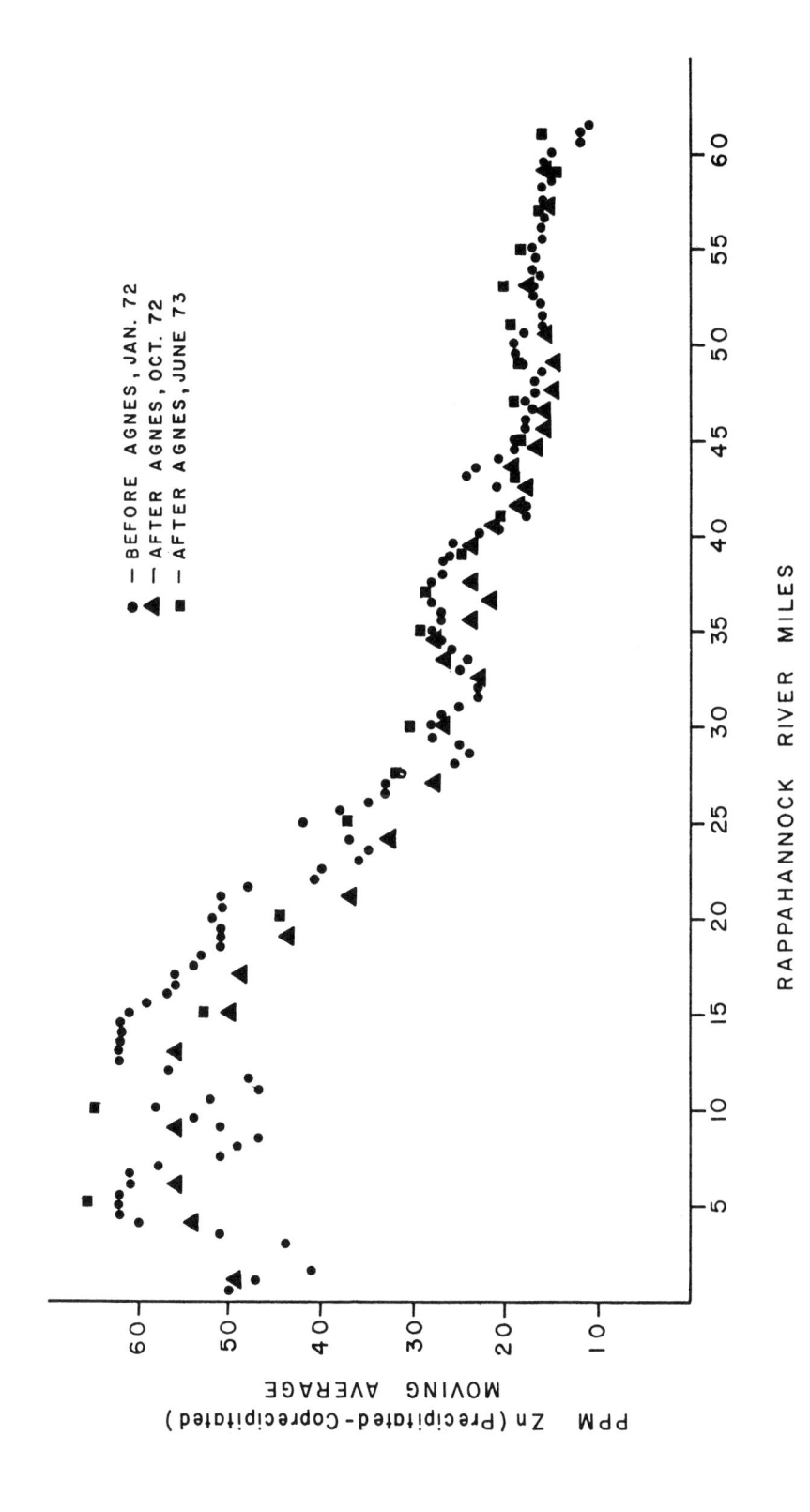

Figure 2. Precipitated-co-precipitated zinc in Rappahannock River sediments.

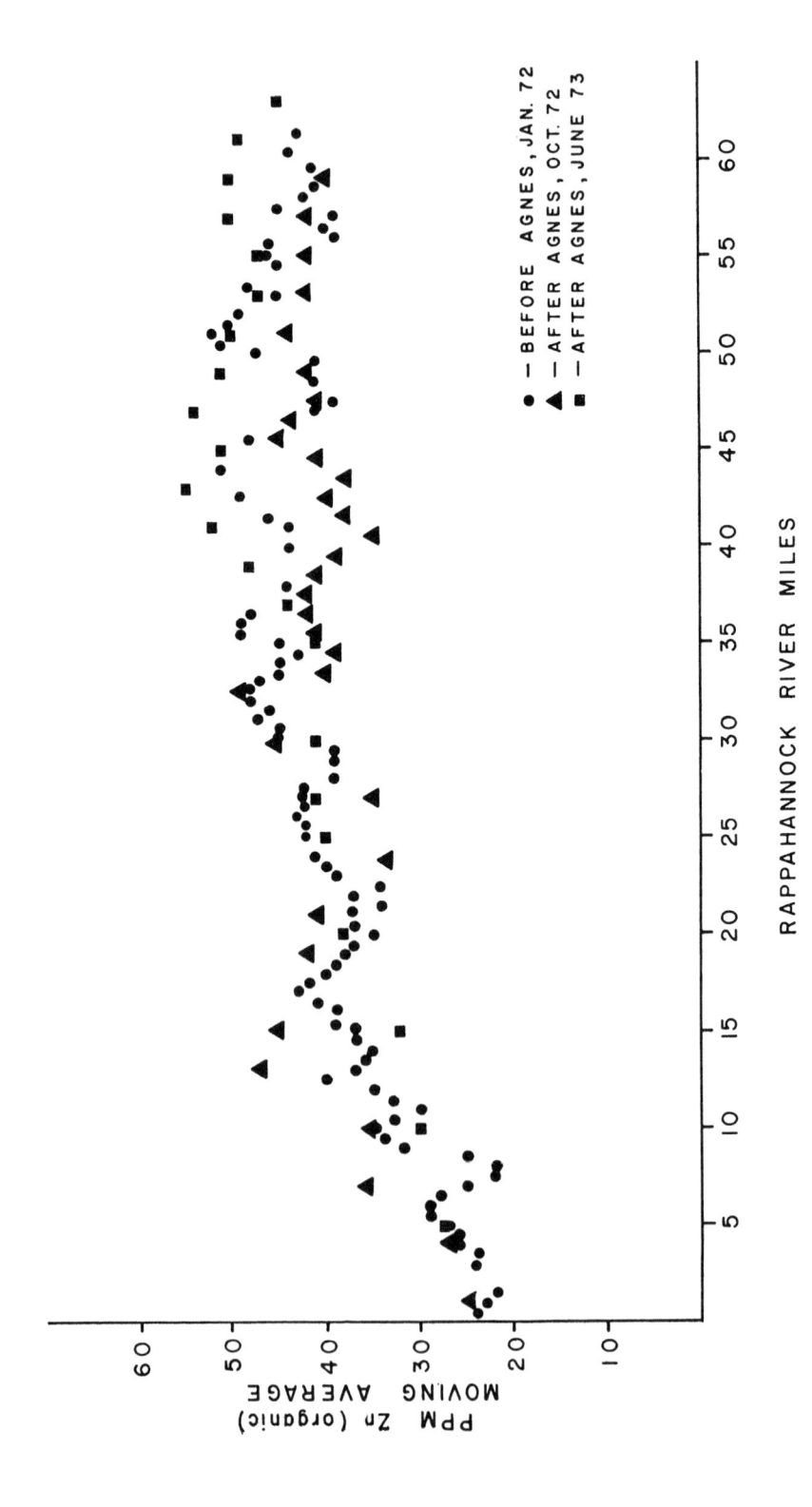

Figure 3. Organic zinc in Rappahannock River sediments.

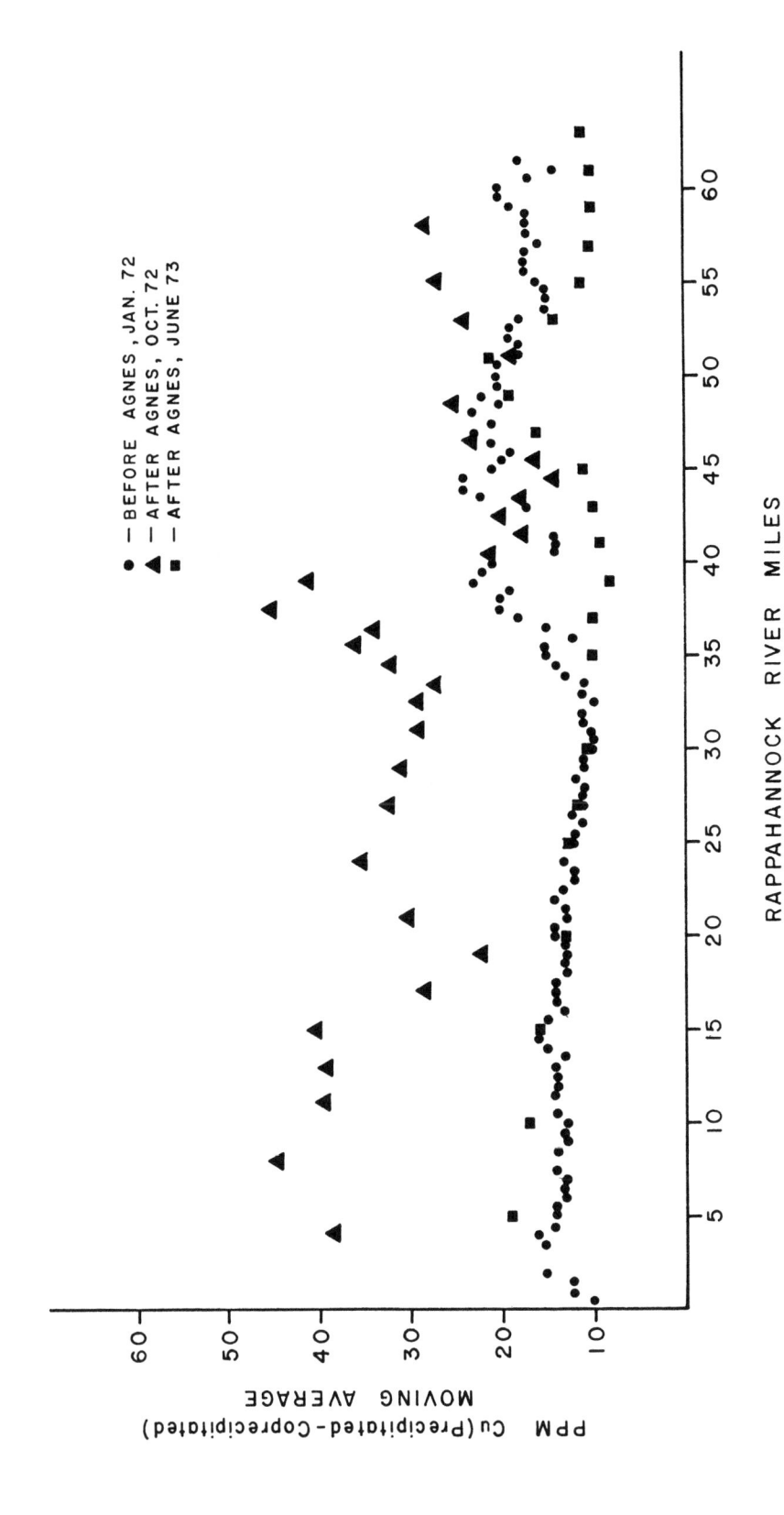

Figure 4. Precipitated-co-precipitated copper in Rappahannock River sediments.

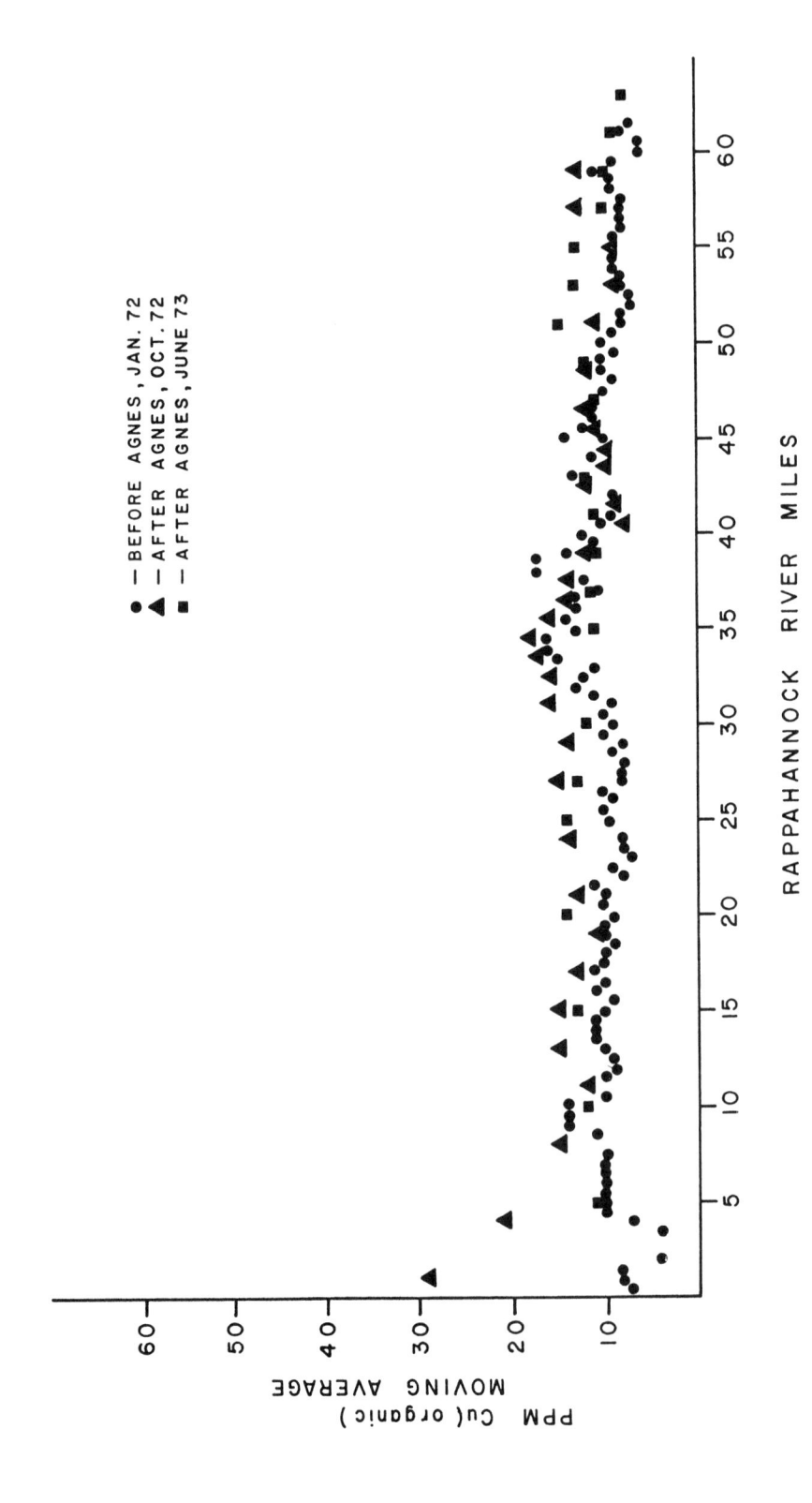

Figure 5. Organic copper in Rappahannock River sediments.

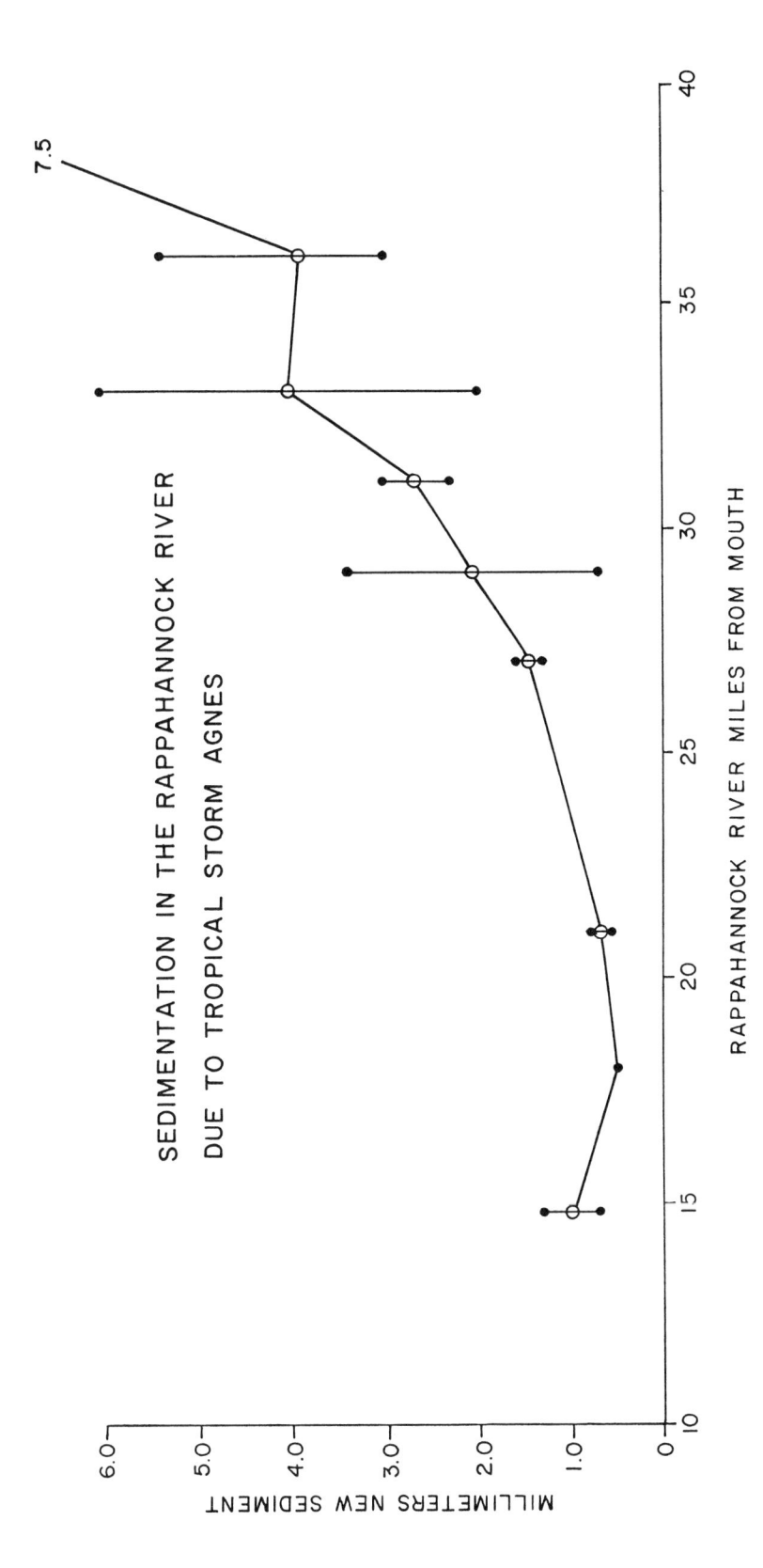

Figure 6. Sedimentation in the Rappahannock River due to Agnes.

AGNES IN MARYLAND: SHORELINE RECESSION AND LANDSLIDES

Barry G. McMullan[1]

ABSTRACT

The mean yearly rate and the summer seasonal rate of coastal recession were compared with a short term rapid erosion rate resulting from the high energy regime of Tropical Storm Agnes. Data from more than two hundred monitoring stations around the Chesapeake Bay disclosed variable increases in the short term erosion over the normal summer seasonal rate. These increases are attributable to abnormal conditions of tidal range, precipitation and wind velocity in association with local variations in the geology. A comparison of the rainfall distribution, the locations of greatest coastal recession, and the local geologic formations produced evidence that the most severe erosion occurred on non-marine Cretaceous clays of the Potomac Group subjected to conditions of torrential rainfall. These same clays are better known inland for their failures along cut slopes due to slippage along fracture systems. To further document this correlation, additional observations of Cretaceous clay bank failures were made at supplemental locations by field inspections and by utilizing before and after ground photography and NASA-ERTS underflight photography.

INTRODUCTION

The influence of hurricanes on coastal processes in the Chesapeake Bay has been noted in the past (Singewald & Slaughter 1949). Earlier reports have indicated that linear recession of exposed shorelines has been as great as thirty to forty feet as a result of high tides and large waves from one storm. Prior hurricanes that have caused extreme erosion were the storms of August 1933, Hazel-October 1954, and the twin hurricanes Connie and Diane of August 1955. The influence of Tropical Storm Agnes (June 1973) on Maryland shores has been evaluated through the collection of shore erosion measurement data. A basis for comparison was available from shore erosion data collected shortly before and immediately after the occurrence of Agnes.

IMPACT OF A TROPICAL STORM ON THE POTOMAC GROUP CLAYS

Shore erosion data are collected periodically during the year throughout the Maryland portion of the Chesapeake Bay. Mean yearly and summer seasonal rates of coastal erosion were compared with the short term erosion rate caused by Agnes. Minor increases in erosion rates were noted which may be attributed to the effect of above-normal tides, precipitation, and wind velocity on localized geology. An overlay of lines of equal rainfall in inches on a map of Maryland shows that areas suffering severe erosion during Agnes were located within the six, eight, and ten inch isohyets. Fig. 1, a superposition overlay of isohyets on the geologic map of

---

[1]Maryland Geological Survey, The Johns Hopkins University, Baltimore, Md. 21218

the State of Maryland, further reveals that the outcrop belt of the lower Cretaceous Formations in Maryland lies within the six, eight, and ten inch isohyets. Of particular interest are the non-marine clays of the Potomac Group, composed of the following formations: the Patapsco Formation, the Arundel Clay, and the Patuxent Formation. The Potomac Group clays when saturated are well known for their slope instability (Miller 1967; Smith 1972 & 1973; Withington 1967).

Two examples of slope failure in Potomac Group clay illustrate the problem of slope instability. A cut slope failure at the Kline's Department Store in Greenbelt, Md. resulted in the death of five and the injury of eleven construction men on December 28, 1962. (Withington 1962; Engineering News Records 1963). A $30,000 home was completely destroyed as a result of slope failure in suburban Washington D. C. The entire saturated upper layer of the hillside slid down slope, seriously damaging ten other homes as well (Klingebiel 1967). These Potomac Group clays, where disturbed by cut slopes on highways, such as the interstate system, are sites of frequent failures caused by rain and groundwater. Most of the shoreline outcrops of Potomac Group clays are steep or vertical banks which are fairly resistant to wave erosion along the Bay coast (Slaughter, pers. comm.). During Agnes, the normally erosion resistant clay banks eroded at a greater rate than the highly erodible unconsolidated Pleistocene sands, clays and gravels of the lower Eastern Shore of the Chesapeake Bay. Therefore, it seems reasonable to assume that increased erosion rates of Potomac Group clay banks during Agnes were caused by surface runoff and groundwater creating an unstable slope condition.

The following areas experienced above normal erosion rates on bluffs of Potomac Group clay exposed to the intense rains of Agnes. Cecil County shorelines at Red Point, Mauldin Mountain, and Turkey Point experienced slips along shear planes in clays of the Patapsco Formation. Grove Point also had slope failures in the form of sand flows in Pleistocene sands. In Kent County along the Chesapeake Bay Shore, Worton Point suffered slope failures from shear slips in clays of the Patapsco Formation. In the area north and south of Fairlee Creek, soil flows of an unusual shape occurred along the shore. These soil flows showed conical shapes similar to sinkholes with vertical openings along the face of the bank. These flows may have resulted from unusual soil conditions, causing ponding of water, thus saturating the soil near cliff faces. A few Pleistocene sand flows occurred at Howell Point. At Indian Head in Charles County on the Potomac River, slope failures occurred as shear slips in clay bluffs of the Patapsco Formation. Along the Bay shore at Bodkin Point in Anne Arundel County, shear slips were evident in slope failures of the Patapsco Formation clay. In Baltimore and Harford Counties slip failures occurred in the Potomac Group clay at Clay Bank Point and on the east side of Plum Point.

THE PHYSICAL CHARACTERISTICS, MINERALOGY, DISTRIBUTION AND ORIGIN OF THE POTOMAC GROUP CLAYS

The Potomac Group clays are commonly variegated or mottled red to white silty clays. They appear to be internally massive and stratification is difficult to recognize. The clay when damp is firm to sticky and exhibits some swell-shrink characteristics. It has vertical slickensided fractures forming polygons which may have resulted from: dessication and shrinkage before burial, subaerial shrinkage, destruction by action of plant roots, unloading of over-consolidated sediments, gravity induced stress along natural or manmade slopes or freezing and thawing action. Individual bodies of Potomac Group clays may be one hundred to one hundred and fifty feet in thickness, dip southeastward sixty to one hundred and fifty feet per mile and outcrop along a five to twenty mile wide belt from Virginia to New Jersey.

The Potomac Group clays were deposited in a complex of fluvial and deltaic environments. The slickensided fractures are typical of flood plain deposits in general. Sedimentary structures and textural character suggests deposition on a low deltaic plain by sluggish, low gradient, meandering rivers (Glaser 1969). The predominance of fine-grained sediments indicated deposition in flood basins and backswamp areas. Studies have shown these non-marine clays to be dominantly kaolinitic with no montmorillonite whereas marine clays generally contain montmorillonite (Groot & Glass 1958)

The clay mineralogy of 12 random samples of the Potomac Group clays in the Maryland area reveals the following percentages of clay minerals in the less than two micron size fraction (Knechtel et al. 1961).

Mineral	Average %	Maximum %	Minimum %
Kaolinite	59.2	76	41
Illite	23.7	48	10
Mixed Layer	16.3	29	3

The cohesiveness of the clay is weakened after a period of physical and chemical weathering. After weathering, slope failures in the Potomac Group clay are generally believed to be caused by groundwater movement in the vertical fractures which expands the fractures, saturates the clays and lubricates the faces of the fractures. It has been estimated that most highway slope failures in Patasco Formation clays occur after several years, possibly resulting from weathering which breaks down the cohesion (Barber, undated; Sallberg (Pers. Comm.)

CAUSES OF SLOPE FAILURE

The groundwater movement which causes failures is normally directly resultant from heavy rainfall which raises the local water table for short durations, compounding the instability problem by increasing the vertical load on the clays and increasing the hydraulic gradient.

Though slope failures result from water problems the primary cause of failure is slope angle. Steep slopes are caused by removal of the slope toe by such processes as river erosion, coastal wave action and onland construction. When slopes in Potomac Group clay reach an angle of repose greater than two and one-half to one, they are in danger of failure (Barber, undated).

All of the clay slope failures encountered along the Bay shores were slopes of steepness (45°-80°), near vertical (80°-90°) or vertical (90°). A two to one slope generally results in clay failures along Maryland highways. These slopes place the clays in a vulnerable position which when combined with a water problem results in slope failures.

The following is a list of factors which contribute to failures on slopes of Potomac Group clay (Eckel 1958): External-Higher Stress--1) Increased slope-less lateral support, 2) Increased overburden-larger load, 3) Water/ice pressure in vertical fractures, 4) Water lubrication of cracks-less friction, 5) Vibration (a) Highway vehicles (b) Water waves, 6) Permeable overburden-rapid saturation, 7) Removal of vegetation-faster saturation. Internal-Lower Strength--1) Mineralogy-swell/shrink, 2) Increased hydraulic gradient-decreased particle friction, 3) Weathering-less cohesion, 4) Overconsolidation-lateral and vertical release Fractures, 5) Liquefaction-particle arrangement, 6) Strata dipping to face of slope, 7) Saturation-loss of capillary tension.

The slope failures in the Potomac Group clay can be described as shear slips with some soil flow (Fig. 2 illustrates a typical slide and nomenclature). The slips are characterized by a concave main scarp often with striae on the vertical face. The height of flank scarps decrease towards the foot. The head of the slide may slope inland, sometimes forming small impoundments of water, commonly with fallen trees leaning up hill. The body of the slip has transverse cracks arranged in step fashion. The toe is often an area of earth flow, having the consistency of a thick fluid. Here the fallen trees are generally horizontal. The main scarp and flank scarp often enclose an internal layer of from one to several feet of disturbed clay with very plastic to fluid qualities.

PREVENTION OF SLOPE INSTABILITY

Slope stability in the Potomac Group clay is directly related to angle of slope and groundwater. The solution of these problems lies in terracing the slope or lowering the slope to an angle of between three to one and five to one, and protecting the toe with an erosion-resistant structure. The water problems may be abated at the surface through the use of vegetation and also by the diversion of surface water from local problem areas. Subsurface water problems may also be treated from permeable overburden by wells or even by the use of a siphon type well that can be primed in times of rainy weather.

SUMMARY

Agnes caused a particularly unstable condition in Bay bluffs of outcropping Potomac Group clays of Cretaceous age. Recorded bank failure has been related principally to excess rain rather than to storm-generated waves in Agnes. Future research in coastal erosion on the Chesapeake Bay should include slope stability studies in order to give an accurate picture of the causes of erosion throughout the Bay.

LITERATURE CITED

Barber, E. S. (undated). Engineering properties of Potomac clay. Unpubl. 3 p.

Eckel, E. B., ed. 1958. Landslides and engineering practice. Hwy. Res. Bd. Spec. Rep. #29, NAS-NRC Publ. 544, 232 p.

Engineering News Record. Jan. 10, 1963. Freeze-thaw cycle caused slide. p. 19.

Glaser, J. D. 1969. Petrology and origin of Potomac and Magothy (Cretaceous) sediments, middle Atlantic Coastal Plain. Md. Geol. Surv., R.I. 11, 102 p.

Groot, J. and H. D. Glass. 1958. Some aspects of the mineralogy of the northern Atlantic Coastal Plain. Seventh Natl. Conf. on Clays and Clay Minerals, p. 271-284.

Klingebiel, A. A. 1967. Know the soil you build on. Ag. Inf. Bull. 320, S.C.S. U. S. Department of Agriculture, p. 13.

Knechtel, M. M., H. P. Hamlin, J. W. Hasterman, and D. Carroll. 1961. Physical properties of non-marine Cretaceous clays in Maryland Coastal Plain. Md. Geol. Surv. Bull. 23, 11 p.

Miller, F. P. 1967. Selecting soils for building sites. Univ. Md. Ext. Serv. Fact Sheet 190, 4 p.

Singewald, J. T., Jr. and T. H. Slaughter. 1949. Shore erosion in Tidewater Maryland. Md. Geol. Surv. Bull. 6, 141 p.

Smith, Horace. 1972. Soils investigation-Boykin property. Unpubl. Investigation, SCS, U. S. Department of Agriculture, 2 p.

Smith, Horace. 1973. Soils investigation-General Electric Appliance Park-East. Unpubl. Investigation, SCS, U. S. Department of Agriculture.

Withington, C. F. 1963. Joints in clay and their relation to the slope failure at Greenbelt, Maryland, December 28, 1962. Abstract, U. S. Geological Surv. Open File Report.

Withington, C. F. 1967. Geology: Its role in the development and planning of Metropolitan Washington. J. Washington Acad. Sci. 57:189-199.

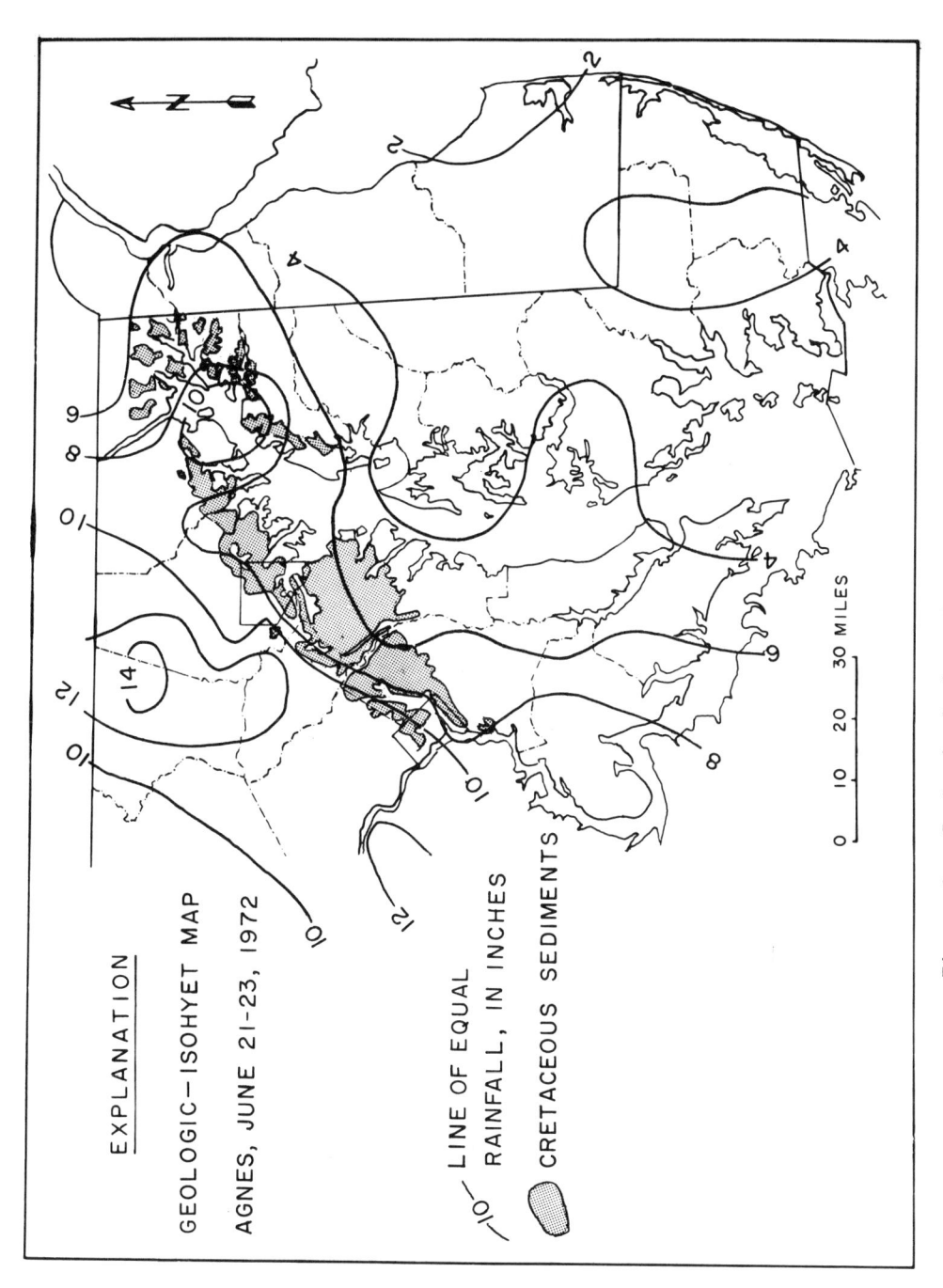

Figure 1. Geologic-isohyetal map, June 21-23, 1972.

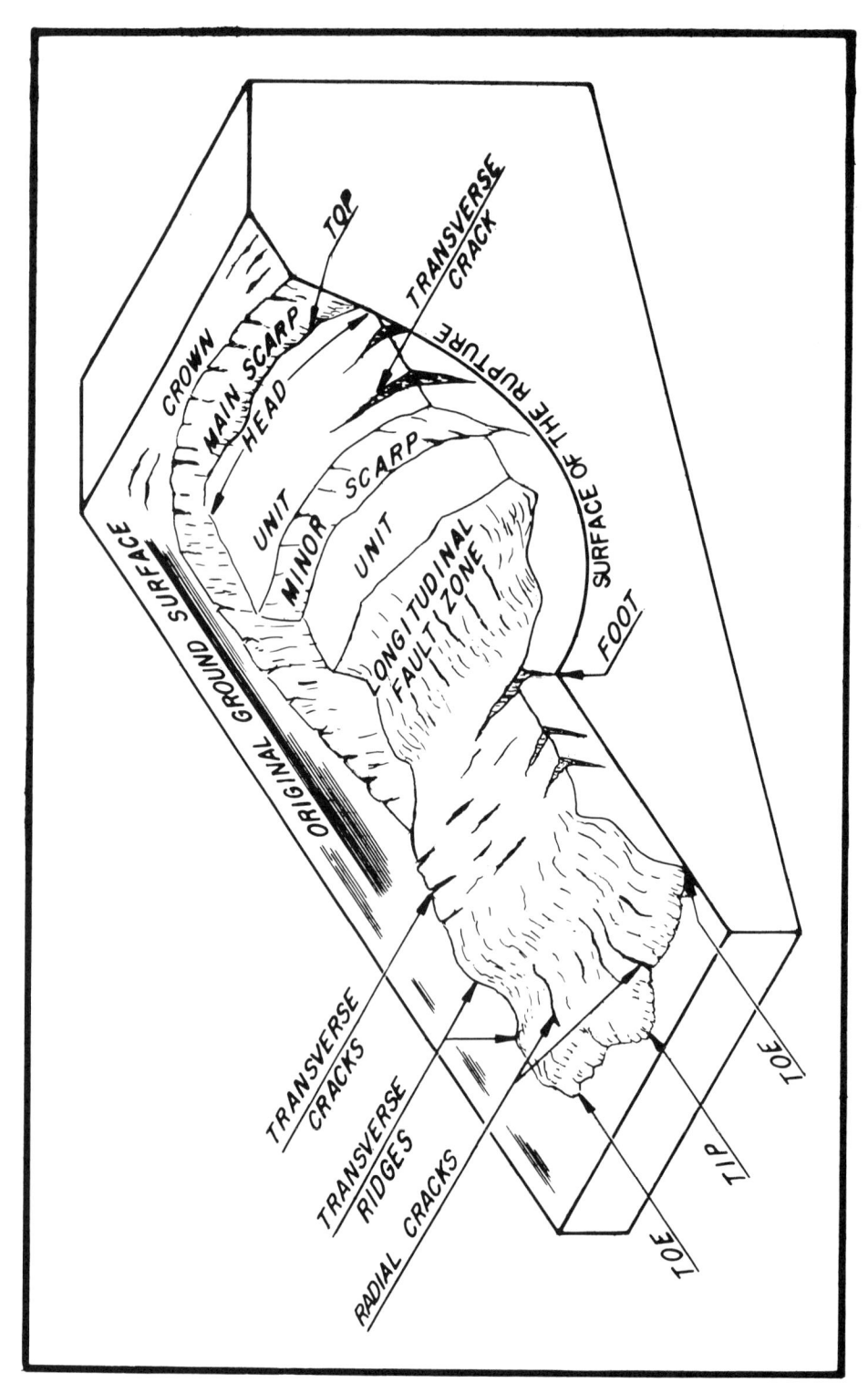

Figure 2. Landslide nomenclature (after Eckel 1958).

CHESTER RIVER SEDIMENTATION AND EROSION: EQUIVOCAL EVIDENCE

H. D. Palmer[1]

ABSTRACT

The Chester River, a major Eastern Shore tributary to Chesapeake Bay, was under scrutiny for hydrologic and geologic data when Tropical Storm Agnes passed through the Upper Bay. Although hydrologic data revealed a pronounced disruption of the normal early summer water structure (reported elsewhere), the bottom sediments and river bank regions throughout the lower portions of this estuary fail to express significant physical evidence related to the passage of this storm. Reconnaissance of over 150 km of river shoreline and comparative analyses of 25 bottom samples obtained on 6 June (before) and 14 July (after) the storm show insignificant changes in sediment texture and shoreline erosion and/or deposition. We conclude that the relatively minor amounts of rainfall, generally less than 16 cm over the period, did not create "rare event" conditions similar to those noted in some estuaries fed by western shore watersheds where 2-3 times this amount fell over a much greater catchment area.

INTRODUCTION

At the time Agnes passed through the Upper Chesapeake Bay, a year-long study was in progress on the Chester River, a major Eastern Shore tributary to Chesapeake Bay (Fig. 1). A portion of this program was devoted to the study of shoreline erosion and sedimentation within the river, and Agnes provided an opportunity to assess the effects of an "extreme" or "rare" event on a local estuary.

As a result of the trajectory of the storm system, Western Shore tributaries to Chesapeake Bay generally received more rainfall in their watersheds than did the tributaries of the Eastern Shore. Consequently, sedimentological effects from flooding, runoff and shoreline erosion were more pronounced in rivers such as the Rappahannock where one million tons of sediment were fed into the middle and upper estuary (Nelson & Nichols 1972).

THE CHESTER RIVER ESTUARY

Bottom Samples

A network of 25 hydrographic and bottom sampling stations was established in the estuary of the Chester River in November 1971. Bimonthly bottom samples were recovered by grab sampler to investigate possible changes in texture which could be related to seasonal events. Fortunately, the period of sampling for late spring occurred three weeks prior to the passage of Agnes. A second set of samples was recovered three weeks after Agnes, after the river had returned to its normal hydrologic regime (water properties remained perturbed for many weeks after the storm).

---

[1]Westinghouse Ocean Research Laboratory, P. O. Box 1488, Annapolis, Md. 21404

Comparison of the "before and after" samples from the axis of the river (Fig. 1) reveals an insignificant change in the mean diameter and in sorting, two parameters which should be sensitive to an influx of a flood-generated sediment load. A slight change in skewness (not shown) values in the middle and deepest portion of the river suggests addition of some silt-sized materials, but this trend is also considered insignificant. There was no detectable change in sediment color, a temporary but effective key used by others (Schubel, personal communication, 1972) to identify sediments of flood origin. There were no definite trends seen in samples to either side of the river axis, and we must conclude that Agnes had a negligible effect on estuarine sedimentation in the Chester River.

The lack of any significant change in these sediments is attributed to the sedimentological regime present in the lower Chester River. As described in detail elsewhere (Clarke et al. 1973; Palmer et al. 1973; Palmer 1974), the net flux of sediment in the lower reaches of the estuary is into the river channel from Chesapeake Bay. In the upper reaches of the estuary, sediments derived from the main channel and several tributaries are being deposited in a wedge at the confluence of these waterways. Flood-generated fine materials may have bypassed this region, and been deposited as normal load downstream. In the lower reaches, Bay muds are encroaching upon the river floor. The lack of any significant change in texture or color at this site may be attributed to the distance of the river mouth (62 km) from the main source of detritus - the Susquehanna River. As noted, others state that much of the sediment was deposited in the uppermost 40 km of the Bay, and that material which reached the Chester River was probably sufficiently fine to remain in suspension in the turbulent flow associated with the flood. Whatever the mechanism, this estuary did not receive a detectable contribution of sediment from the Bay.

Shore Erosion

For purposes of evaluating shoreline erosion along the margins of the Chester River, a supplemental program of shore and beach inspection was maintained during the entire study. The winter conditions were evaluated in February, after the coldest period in which shore ice was noted in some areas. A late spring inspection set for May was postponed, so the opportunity to visit selected sites after Agnes held special importance.

Water levels on 22 and 23 June reached 0.5 meters above predicted high tide, a record height for the period of study. This event, coupled with winds (reaching 21 meters per second), generated both high waves along west-facing shores and a seiche at the mouth of the river which reached 15 cm in amplitude (Tzou & Palmer, this volume). In addition, the heavy rainfall during the 22nd caused some local slumping by percolation through the porous soils forming bluffs bordering the river (Palmer 1973). These combined effects are properly termed a "major event", but again the impact is difficult to discern from a geomorphic viewpoint. It appears that very little damage was done to the beaches and bluffs of the river. A great mass of litter (cans, bottles, plastic objects and driftwood) was left stranded after high tides following the actual storm, but only one site could be described as severely eroded. This area was on the somewhat protected eastern shore of Kent Island, where a property owner claimed he had "lost about seven feet" of his fields adjacent to the beach.

Photographic records of specific shoreline sites were maintained throughout this study, and thus a comparison of pre-and post-Agnes configurations was possible. Inspection of photographs containing reference objects (junked cars, trees, posts, etc) revealed no discernible change in the beaches or in the bluffs, and we must assume that little direct damage was inflicted on the Chester River shoreline as a result of Agnes.

CONCLUSIONS

Evaluation of sediment textural data and shoreline inspection for evidence of erosion attributable to Agnes indicate that the estuary of the Chester River was not significantly modified by this "200-year event" We attribute this to the relatively low precipitation, distance from major sources of sediments, and the hydrologic regime of the river. The evidence for storm-induced effects is equivocal since we have suggestions that slight changes in local shorelines and subtle variations in sediment textures exist, but yet stronger evidence for little or no change throughout most of the estuary. It appears that a survey conducted after the storm, and without knowledge of its passage, would not confirm that a significant regional event had occurred.

ACKNOWLEDGEMENTS

Comparison of Chester River data with those from other sectors of the Upper Chesapeake Bay was facilitated by discussions with Dr. J. R. Schubel. Funds for this study were provided by the Westinghouse Electric Corporation and the State of Maryland Department of Natural Resources.

LITERATURE CITED

Clarke, W. D., H. D. Palmer, and L. C. Murdock. 1972. Chester River Study: Vol. II. State of Maryland Dept. of Nat. Resour. and Westinghouse Ocean Res. Lab. Rep., 251 p.

Nelson, B. W. and M. M. Nichols. 1972. A unique estuarine sedimentological event. Abstracts with Program, Geol. Soc. Am. 4(7):609 (abstract).

Palmer, H. D. 1973. Shoreline erosion in upper Chesapeake Bay: the role of groundwater. Shore and Beach 41(2):18-22.

Palmer, H. D. 1974. Estuarine sedimentation, Chesapeake Bay, Maryland, U.S.A. In: Relations Sedimentaires entre Estuaries et Plateaux Continentaux, Mem. Inst. de Geologie du Bassin d'Aquitaine 7:215-224 (Bordeaux).

Palmer, H. D., R. W. Onstenk, K. T. S. Tzou, and D. M. Dorwart. 1973. Estuarine sedimentation regime, Chester River, Maryland. Abstracts with Program, Geol. Soc. Am. 5(2):204-205 (abstract).

226 Palmer

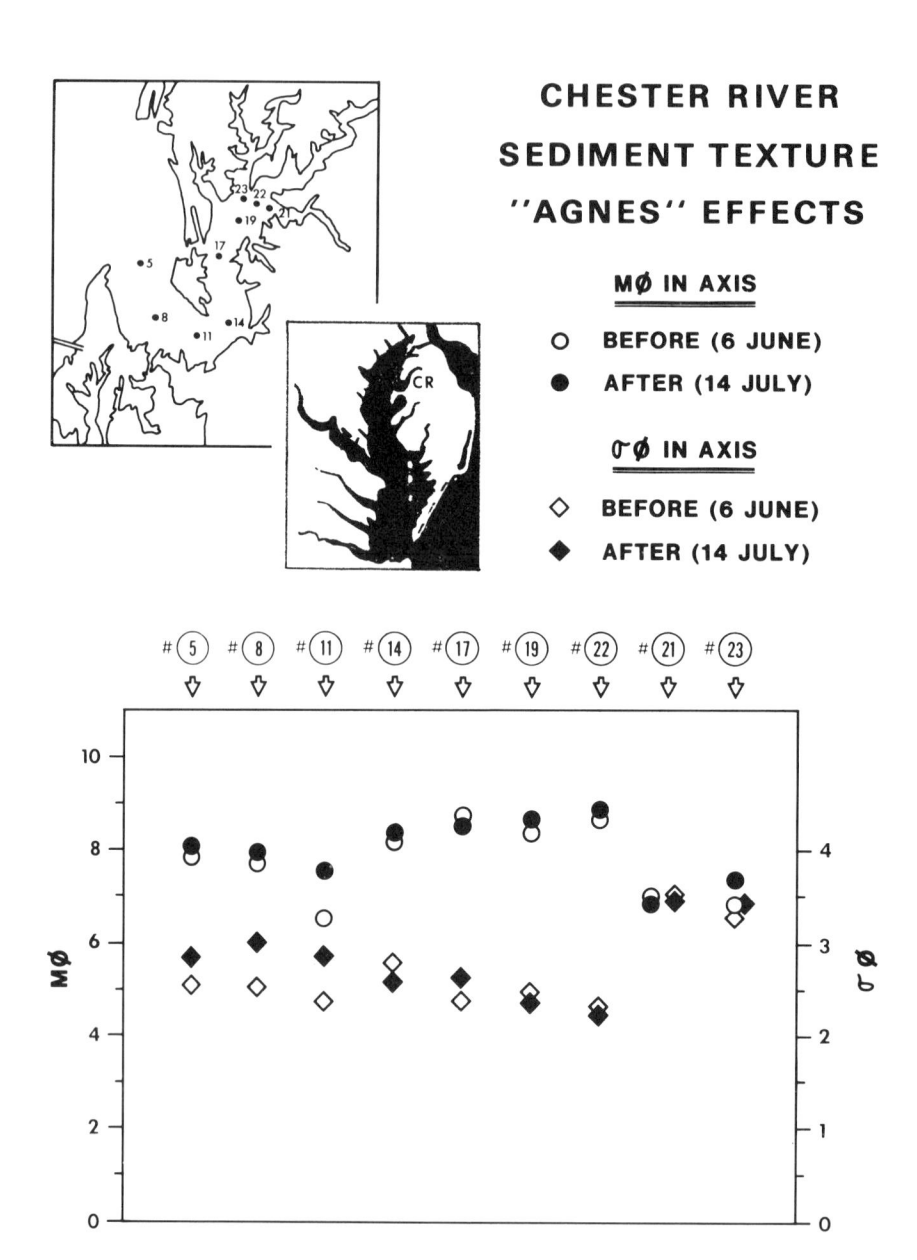

CHESTER RIVER SEDIMENT TEXTURE "AGNES" EFFECTS

M∅ IN AXIS
○ BEFORE (6 JUNE)
● AFTER (14 JULY)

σ∅ IN AXIS
◇ BEFORE (6 JUNE)
◆ AFTER (14 JULY)

Figure 1. Sediment textural data from selected stations, before and after Agnes. Scale for the mean diameter in phi units appears on left ordinate, sorting in phi on right ordinate. Insert map shows station locations, and location of the Chester River (C.R.) within Chesapeake Bay.

EFFECT OF TROPICAL STORM AGNES ON THE BEACH AND NEARSHORE PROFILE

Randall T. Kerhin[1]

ABSTRACT

Beach profiles were measured at two locations in Chesapeake Bay during 1972; a barrier beach complex in Somerset County and a straight, high bluff coastline in Calvert County, Md. Pre- and post-storm profiles were compared to determine the impact of Tropical Storm Agnes on the beach environment. Prior to Agnes, the barrier beach complex was experiencing foreshore deposition of approximately 3000 cubic yards. Profiles collected on June 26 recorded 3500 cubic yards of erosion and 3-5 feet of foreshore recession. The post-storm response was deposition of 2300 cubic yards or 80% of the sediment lost during Agnes. Pre- and post-storm profiles at Chesapeake Beach, Md. show no significant change that can be correlated with Agnes. The quarterly profile schedule did not record the immediate impact of Agnes but long-term comparisons exhibit beach and nearshore stability.

INTRODUCTION

The role of coastal storms is an important modifying agent of the sedimentary processes for the beach and nearshore environment. The degree of modification depends on the size and intensity of the storm, path of the storm with respect to the beach, and the tidal phase and associated storm surge. Beach changes may exhibit spectacular beach retreat which persists with time (Hayes 1967, Goldsmith et al. 1966, Warnke 1969) or temporary changes which show post-storm recovery (Davis & Fox 1971, Hayes et al. 1969). The occurrence of Agnes afforded the opportunity to document storm-related changes in the beach profile. By comparison of pre- and post-storm profile data and the determination of the long-term pattern of beach change, the impact of Agnes was determined for two locations in Chesapeake Bay.

SIGNIFICANT STORM PARAMETERS

The development and history of Agnes has been described by DeAngelis and Hodge (1972). The size of Agnes covered several states but the intensity was a minimum hurricane with respect to the wind parameters. The path of the storm proceeded in a northeasterly direction parallel to the Atlantic coastline and was positioned east of Chesapeake Bay in the Atlantic Ocean by June 22. Cyclonic wind patterns associated with extratropical storms in the northern hemisphere produced a wind shift from a easterly direction to a westerly direction as Agnes proceeded northward. Inspection of wind records at Love Point, Md. recorded this shift of wind direction (Tzou 1972). Fig. 1 is a plot of barometric pressure, wind direction, and wind velocity collected at Crisfield, Maryland. Barometric pressure began falling on June 19 from a high of 30.25 inches to a low of 29.32 inches recorded for June 22. With falling barometric pressure, the wind direction was from the southeast shifting to the northwest with wind velocities of 49 mph. The wind direction is related to the position of Agnes as it proceeded along the

---

[1]Maryland Geological Survey, The Johns Hopkins University, Baltimore, Md. 21218

eastern coast. The significance of wind direction is related to the generation of wind waves and a reversal of wind direction as reported for Chesapeake Bay is not the optimum condition for maximum wave generation. The highest wind velocity reported for Chesapeake Bay was 55 knots from the northwest at Kentmoore Marina, Kent Island, Md. (NOAA 1972).

Associated with coastal storms such as Agnes is the generation of a storm surge. The storm surge is defined as the difference between the observed and the predicted astronomical tide and is considered to be the meteorological effect on sea level (Pore 1965). Tzou (1972) reported a storm surge of 1.1 meters at Love Point, Md. which is comparable to Moyer's (1972) report of 0.5 to 1.0 meters above normal for Chesapeake Bay. This is not considered by Moyer (1972) to be an unusually high storm surge for Chesapeake Bay.

BEACH PROFILE CHANGES

During 1972, a series of beach and nearshore profiles were conducted at two locations in Chesapeake Bay. The two sites were Janes Island State Park in Somerset County and northern Calvert County (Fig. 2). The profiling time period varied from a monthly schedule for Janes Island State Park to a quarterly schedule for northern Calvert County. At the Calvert County site, a survey was not conducted until July 28 or one month after Agnes. It is apparent that the immediate impact of Agnes on this beach environment cannot be readily determined but changes over the long-term and its relationship to the passage of Agnes were analyzed. The Janes Island site was resurveyed four days after Agnes which allowed the immediate response of the beach zone to be evaluated along with the long-term pattern of change.

Janes Island State Park

The Janes Island State Park site is located at the confluence of the Little Annemessex River and Tangier Sound at the southernmost area of Janes Island (Fig. 3). Geomorphically, the area is a barrier beach which developed as a southward migrating recurved spit. Modification by man is present with construction of a bulkhead, three groins and disposal of dredged spoil. The dredged spoil closed a small tidal inlet existing between the nose of the spit and the marshy headland and transformed the area into a barrier beach. A geological cross-section shows a non-barred, shallow nearshore, a narrow barrier beach with a small dune system and tidal flats on the lagoon side.

To evaluate the effects of Agnes, topographic and erosional-depositional maps were constructed and compared. Volumes of sediment eroded and deposited were calculated by planimetering the erosional-depositional maps. From this procedure, the volume computed for erosion and deposition were added to obtain total volume change for each profiling month and the percentage of change associated with either erosion or deposition was calculated and plotted (Fig. 4).

Net deposition was the dominant process for the months of March and April. During March, net deposition of approximately 1300 cubic yards was measured. April profiles recorded a net accumulation of 1700 cubic yards which represents 60% of the total volume change. Deposition occurred as foreshore accumulation with progradation of 3-5 feet. For the May profiles, a month before Agnes, net accumulation of only 190 cubic yards was recorded from a total volume change of 7000 cubic yards. Considering the barrier beach as a single unit, an equilibrium condition of the beach to the coastal processes is suggested by the May profiles. For every cubic yard of sediment eroded, a cubic yard was deposited. The condition of the barrier beach prior to Agnes was one of accumulation and growth (Fig. 4).

The barrier beach was resurveyed on June 26 and compared with the previous monthly profiles. The response of the barrier beach was erosional with an approximate net change of 3500 cubic yards of sediment. This accounted for 69% of the total volume change following Agnes with the remaining 31% being depositional (Fig. 4). The area of extensive erosion was the foreshore and backshore zones with foreshore recession measured at 8-10 feet. Minor erosion occurred on the lagoonal side of the barrier beach in the tidal flats. The only area of deposition was the small dune system paralleling the axis of the barrier beach.

The role of a storm surge in a coastal storm is to expose more of the beach, particularly the backshore, to the high-energy storm waves. This is evident in the foreshore and backshore where storm waves coupled with storm surge cause extensive erosion. Dune deposition indicated that the storm surge did not inundate the barrier beach and deposition along the dune was by aeolian processes. The storm surge level on the lagoon side reached the base of the dune causing minor erosion of the tidal flats, probably through resuspension by small amplitude waves generated in the lagoon. Anderson (1972) reported on this mechanism of resuspension by small amplitude waves in a tidal flat in New Hampshire.

The immediate impact of Agnes at Janes Island was extensive foreshore and backshore erosion but within a month, 80% (2300 cubic yards) of the sediment lost during Agnes returned as foreshore and backshore deposition (Fig. 4). Essentially, the post-storm response of the barrier beach was one of accumulation and rebuilding, mainly the foreshore. No permanent change developed for this area and the impact of Agnes on the barrier beach was no greater than a typical coastal storm in Chesapeake Bay.

North Calvert County

The second area is a reach of shoreline between Chesapeake Beach and Plum Point in northern Calvert County (Fig. 2). Bordering the coastline are the near-vertical cliffs of the Miocene Calvert Formation. The cliffs are composed principally of dense, bluish, clayey fine sand and silt zones with a large percentage of shell and are referred to as the Plum Point Marls. At the base of the cliffs exists a ribbon beach of widths between 20-30 feet which is composed of fine to medium quartz sand. The nearshore is a wide, shallow, gently sloped environment composed of fine to medium sand 1-3 feet thick. Under this thin veneer of sediment lies the basal portion of the Plum Point Marls and the top of the Fairhaven Diatomaceous Earth Member (Slaughter 1967). Slaughter (1967) reported that profiles comparisons between 1944 and 1967 show little change from the 2 foot to the 6 foot contour but the inner nearshore from 0 to 2 feet was wider and shallower in 1944. He reported this may be a result of shore erosion control measures diminishing the volume of sediment available to the littoral drift.

Evident in the nearshore environment are a series of multiple longshore bars (Fig. 5). The longshore bars appear parallel and continuous for the entire reach of shoreline but closer examination shows irregularities in the depth of water over the crests and orientation with respect to each other which is similar to longshore bars in eastern Lake Michigan (Saylor & Hands 1970). Approaching Plum Point, a prominent headland, the longshore bars are diverted offshore and diffused into the Plum Point shoals (Kerhin 1973). The number of bars vary between 6 to 10 with the greatest variation in numbers being in the inner nearshore. There is a general increase in wavelength and amplitude of the longshore bars in the offshore direction. Wavelength averages 80 feet for the innermost bars to 200 feet in the outermost bars with amplitudes of 0.50 feet to 2.0 feet, respectively. The longshore bars show a tendency to be asymmetrical in form with the steep slope in the onshore direction.

To document the impact of Agnes on the longshore bars, determination of the stability of the bars was made by comparing pre- and post-Agnes topography. As shown by Fig. 6, profile comparisons for April and July 1972 show very little significant change in the bar form or relative position with respect to the strandline. Deposition dominated the area during this time period with 0.1 to 0.3 feet of sediment. This depositional pattern may reflect redistribution of offshore sediments since beach and shoreline erosion did not exist during Agnes or an influx of sediment transported by longshore currents. Whether the depositional pattern is related to Agnes cannot be determined from the profile data.

Although the longshore bars show minor fluctuation as indicated by Fig. 6, inspection of summer quarterly topographic maps from 1970 to 1973 indicates persistent characteristics in form and relative position from the strandline (Figs. 7 & 8). The occurrence of Agnes did not significantly change the general topography and the longshore bars remained stable over the long-term. The normal condition appears to be fluctuation within a small zone surrounding the crest of the bars except with the ephemeral bars in the inner nearshore. In this zone are the greatest variation of number of bars which account for the lack of stability. Davis and McGreay (1965) reported fluctuations in the bars in Lake Michigan but within the configuration of the bar and thus concluded a stable bottom configuration. Essentially for this area, a stable bottom configuration exists and the occurrence of Agnes did not significantly disrupt this stability.

CONCLUSION

The only valid conclusion is the lack of significant impact of Agnes on the beach and nearshore profile. Local variations did exist throughout Chesapeake Bay as indicated by the immediate erosional response at Janes Island State Park but the long-term impact with an event such as Agnes did not differ significantly than a typical coastal storm. The minor fluctuations of the longshore bars in Calvert County as shown by pre- and post-Agnes profiles and topography are the "normal" long-term pattern of stability and not a pattern change related to Agnes. Although the impact of Agnes was not apparent, monitoring of pre- and post-storm beach profiles is a valid method of understanding the response of the beach environment to high-energy conditions.

ACKNOWLEDGEMENTS

The author expresses his gratitude to the Park Rangers of Janes Island State Park for their cooperation and assistance in collection of the Janes Island profile data and to the U. S. Army Corps of Engineers, Coastal Engineering Research Center for making available the profile data for Calvert County, Maryland.

LITERATURE CITED

Anderson, F. E. 1972. Resuspension of estuarine sediments by small amplitude waves. J. Sediment. Petrol. 42:602-607.

Davis, R. A., and W. T. Fix. 1971. Beach and nearshore dynamics in eastern Lake Michigan. Office of Naval Res. Tech. Rep. 4, 144 p.

Davis, R. A. and D. F. R. McGreay. 1965. Stability in nearshore bottom topography and sediment distribution, southeastern Lake Michigan. Proc. 8th Conf. Great Lakes Res., pp. 222-231.

DeAngelis, R. M. and W. T. Hodge. 1972. Preliminary climatic data report, Hurricane Agnes, June 14-23, 1972. Natl. Oceanic & Atmos. Admn. Tech. Memo EDS NCC-1.

Goldsmith, V., P. Grose, J. J. Holt, and D. A. Warnke. 1966. Conditions of "spectacular beach retreat" in a low energy environment. Geol. Soc. Am. Spec. Pap. 101, p. 360.

Hayes, M. O. 1967. Hurricanes as geological agents: case studies of Hurricanes Carla, 1961, and Cindy, 1963. Univ. Texas, Bur. Econ. Geol. Rep. Inves. 61:1-38.

Hayes, M. O. and J. C. Boothroyd. 1969. Storms as modifying agents in the coastal environment. SEPM Eastern Section Field Trip, Coastal Environment N. E. Mass. and New Hampshire, pp. 245-253.

Kerhin, R. T. 1973. Recognition of beach and nearshore depositional features of Chesapeake Bay. Symp. Significant Results from ERTS-1, Vol. 1:1269-1274.

Moyer, W. J. 1972. Climatological data, Maryland and Delaware. U. S. Dept. of Commerce, Natl. Oceanic & Atmos. Admn. 26(6).

National Oceanic and Atmospheric Administration. 1972. Storm data. U. S. Dept. of Commerce, Natl. Oceanic & Atmos. Admn. 14(12).

Pore, N. A. 1965. Chesapeake Bay extratropical storms surges. Chesapeake Sci. 6(3):172-182.

Saylor, J. H. and E. B. Hands. 1970. Properties of longshore bars in the Great Lakes. Proc. 12th Conf. on Coastal Engr., Am. Soc. Divil Engr. 2:838-853.

Slaughter, T. H. 1967. Vertical protective structures in Maryland, Chesapeake Bay. Am. Shore and Beach 35:7-17.

Tzou, K. T. S. 1972. Meteorological and hydrographical investigations. Pages 131-211 in Chester River Study, Vol. 2, W. D. Clarke, H. D. Palmer, and L. C. Murdock, eds., State of Maryland and Westinghouse Corp.

Warnke, D. A. 1969. Beach changes at the location of landfall of Hurricane Alma. Southeast. Geol. 10(4):189-200.

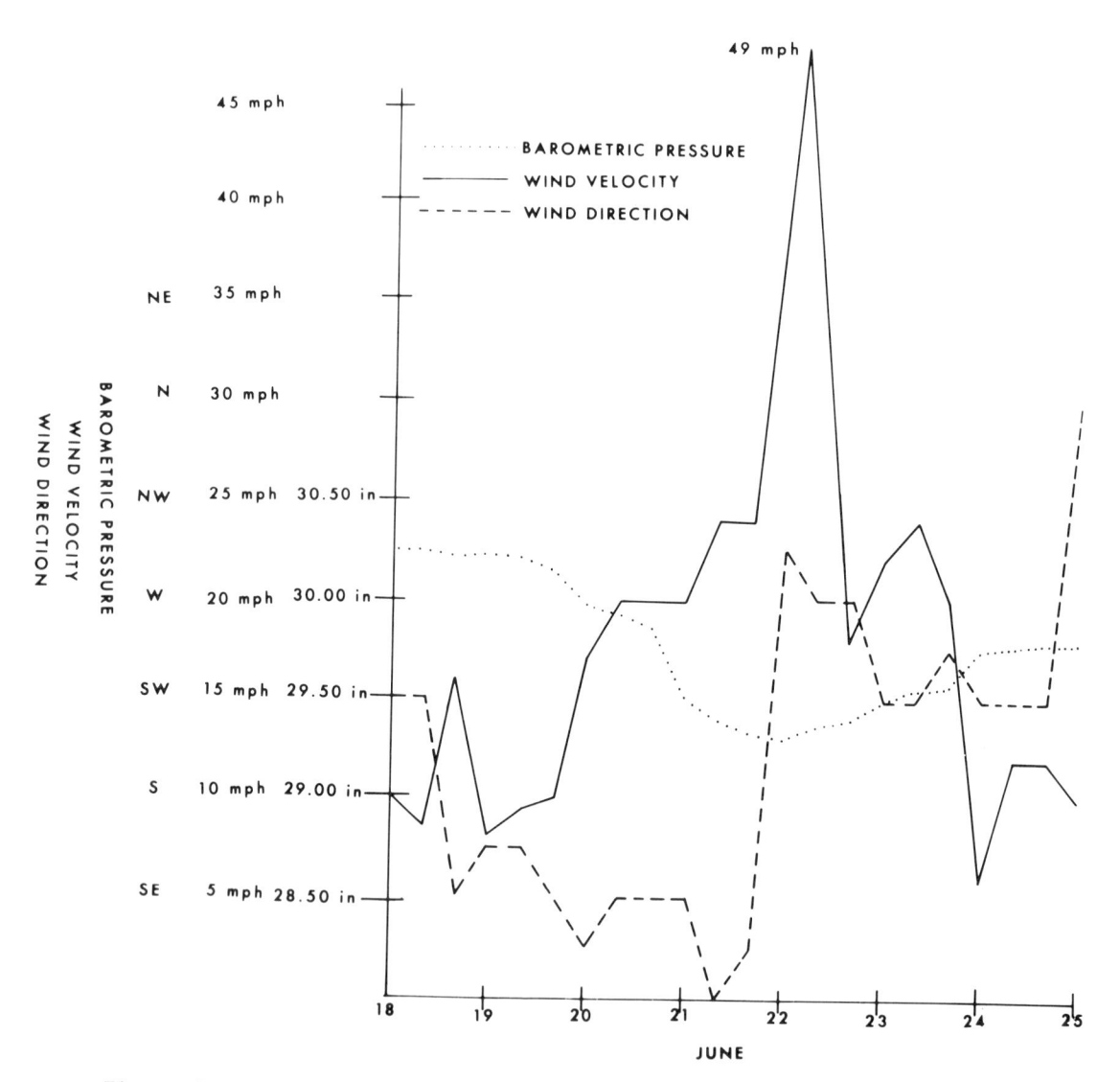

Figure 1. Barometric pressure, wind velocity, and wind direction records for Crisfield, Maryland collected June 18-25, 1972.

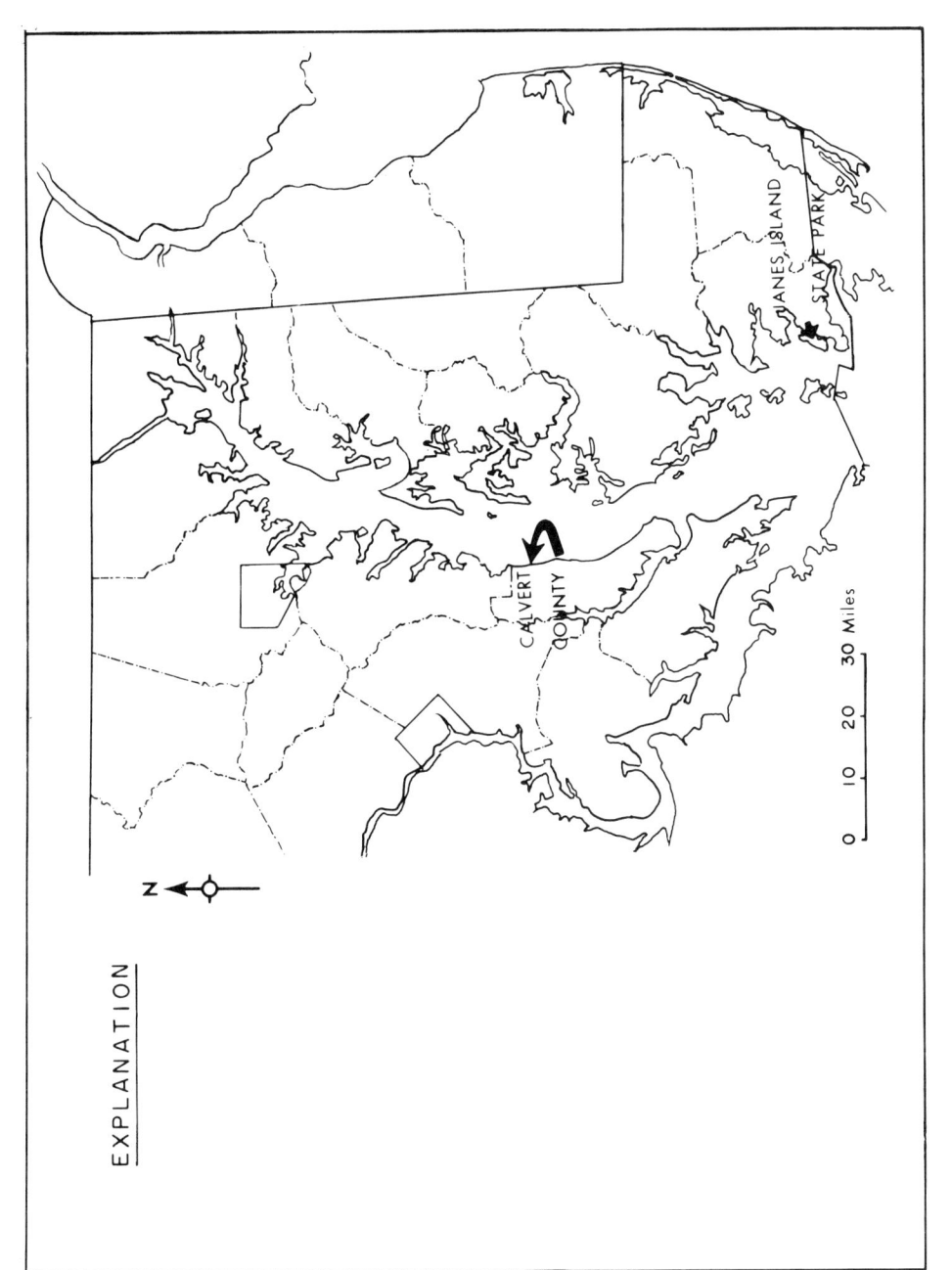

Figure 2. Location map showing the two areas of investigation; Janes Island State Park in Somerset County and northern Calvert County.

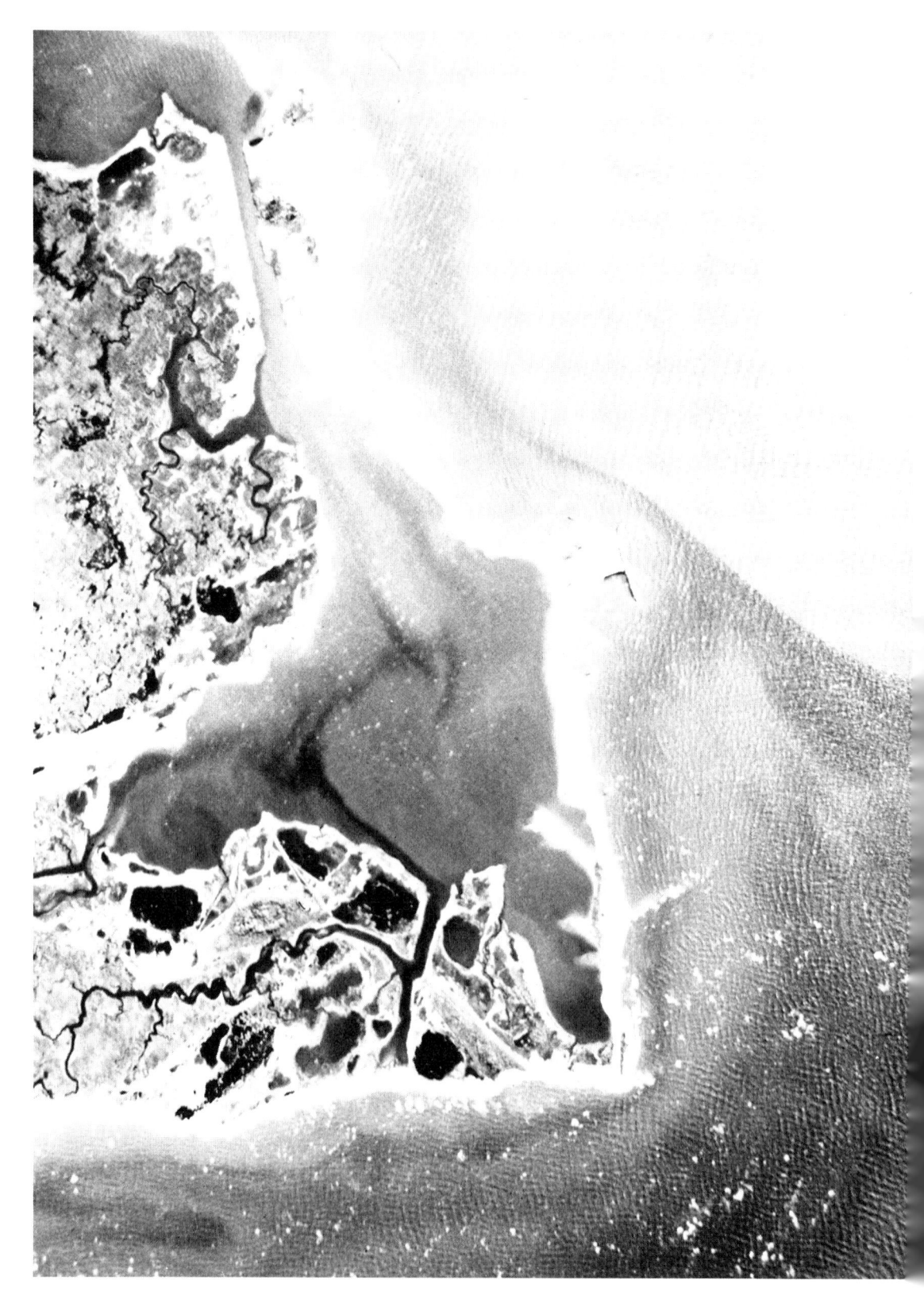

Figure 3. Barrier beach at Janes Island State Park. Twelve locations were selected along the barrier beach for the monthly profiles. North is to the top of the figure and Tangier Sound is to the left.

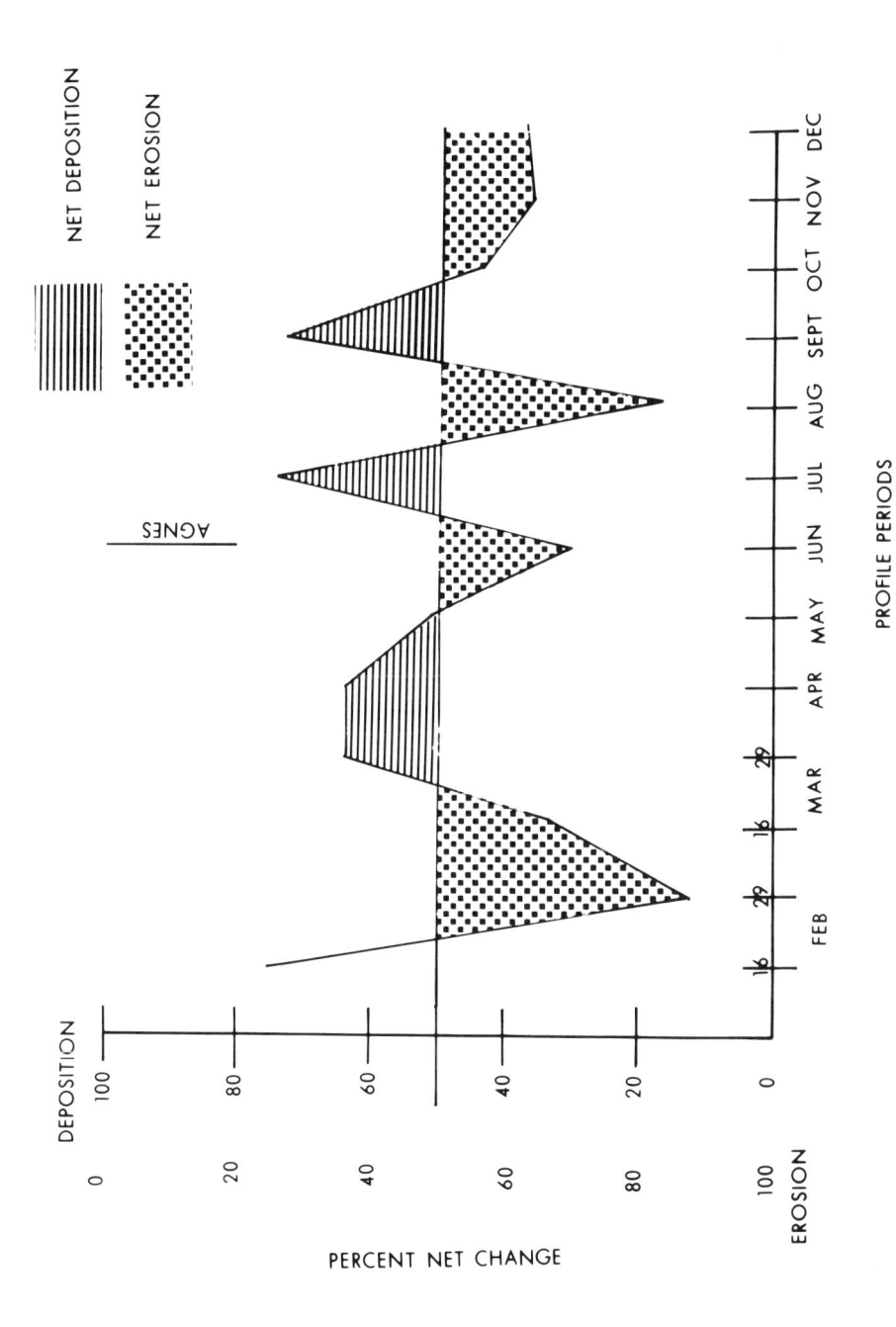

Figure 4. Percent net change for Janes Island State Park during 1972. Net deposition and net erosion are plotted as percentage of total volume change for each profiling period.

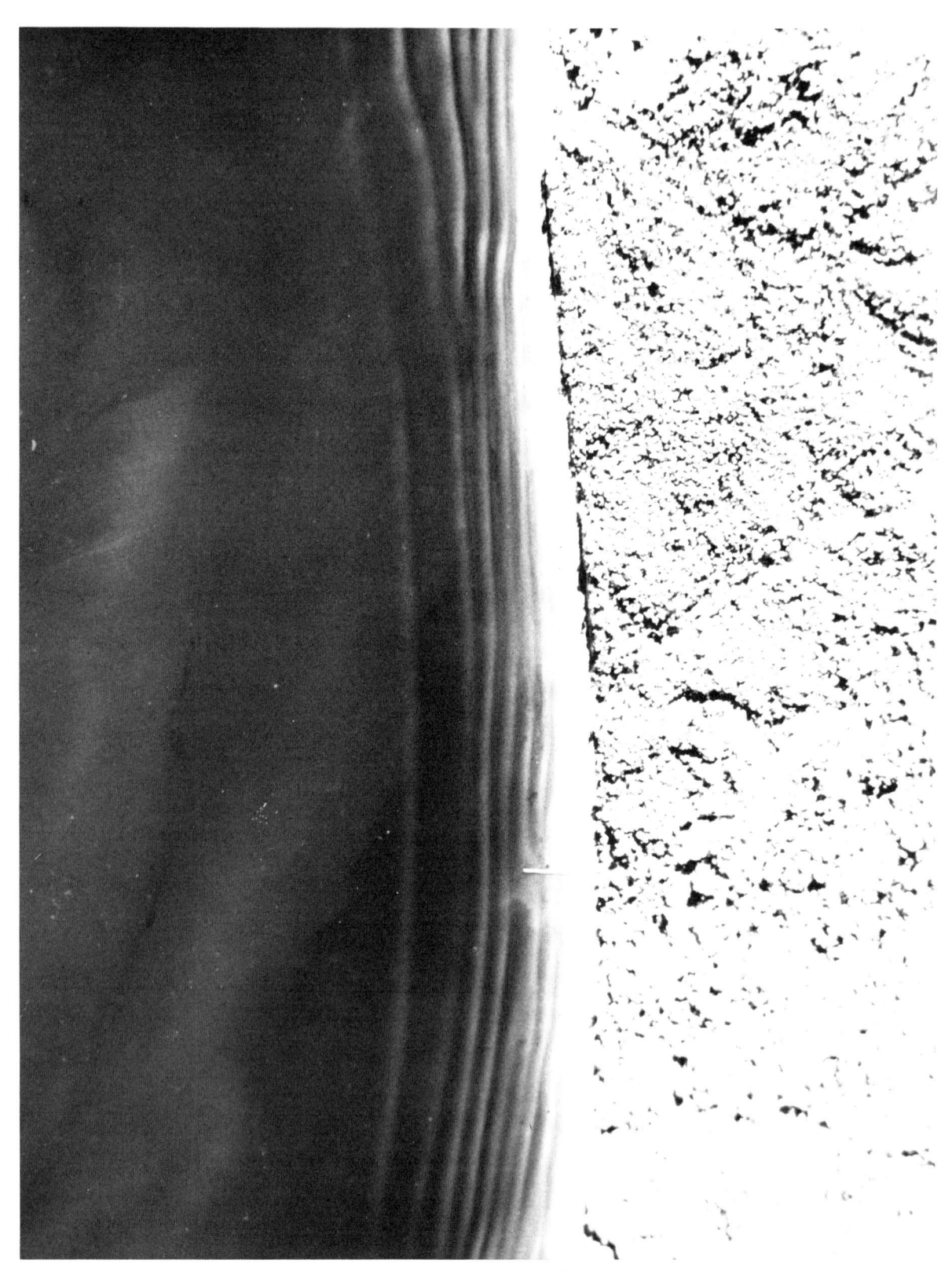

Figure 5. Multiple longshore bars at northern Calvert County. The longshore bars are generally continuous with an increase in wavelength and amplitude in the offshore direction.

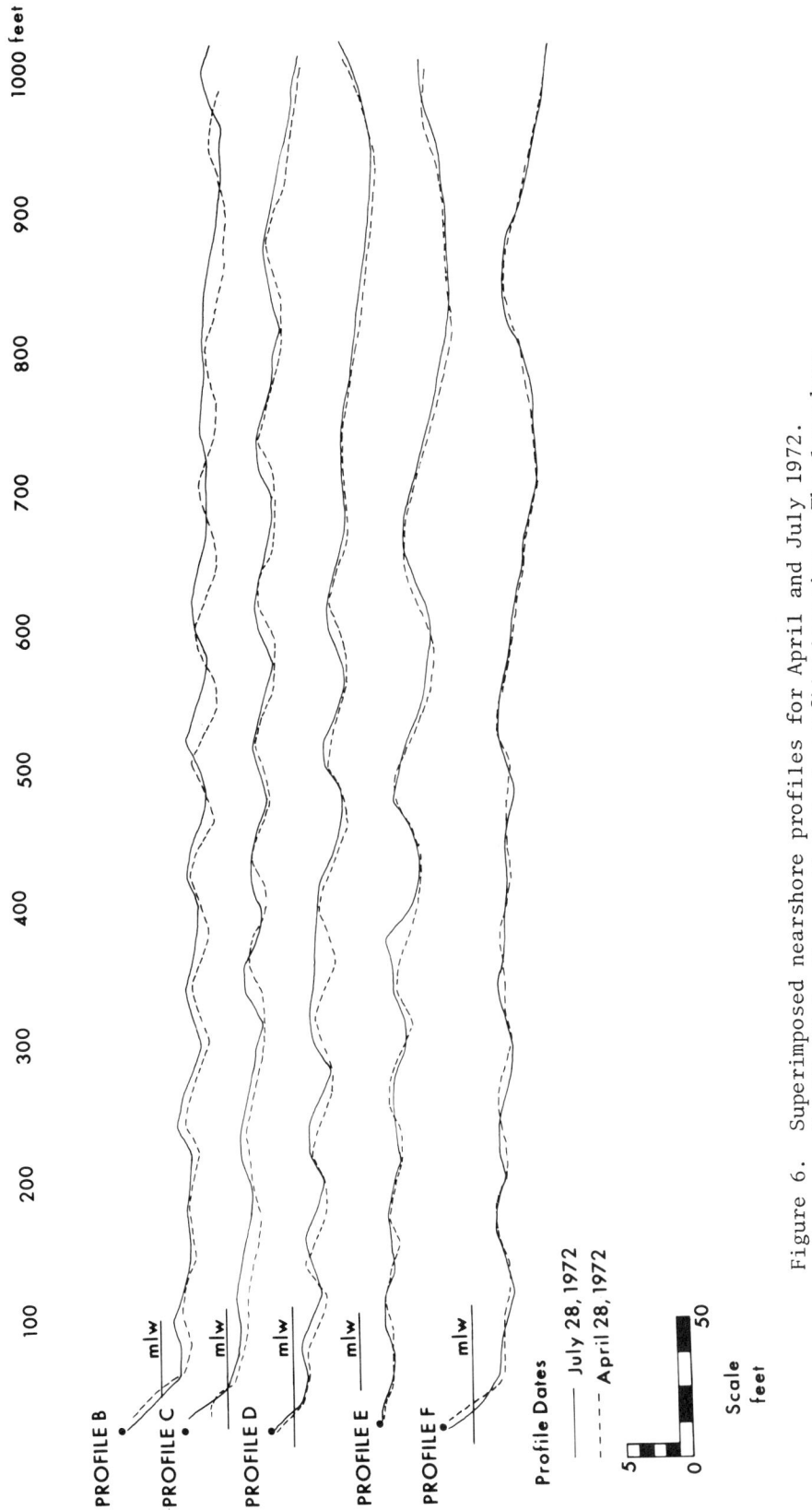

Figure 6. Superimposed nearshore profiles for April and July 1972. Profile B is the northernmost profile location. The longshore bars are persistent in form and position from the strandline.

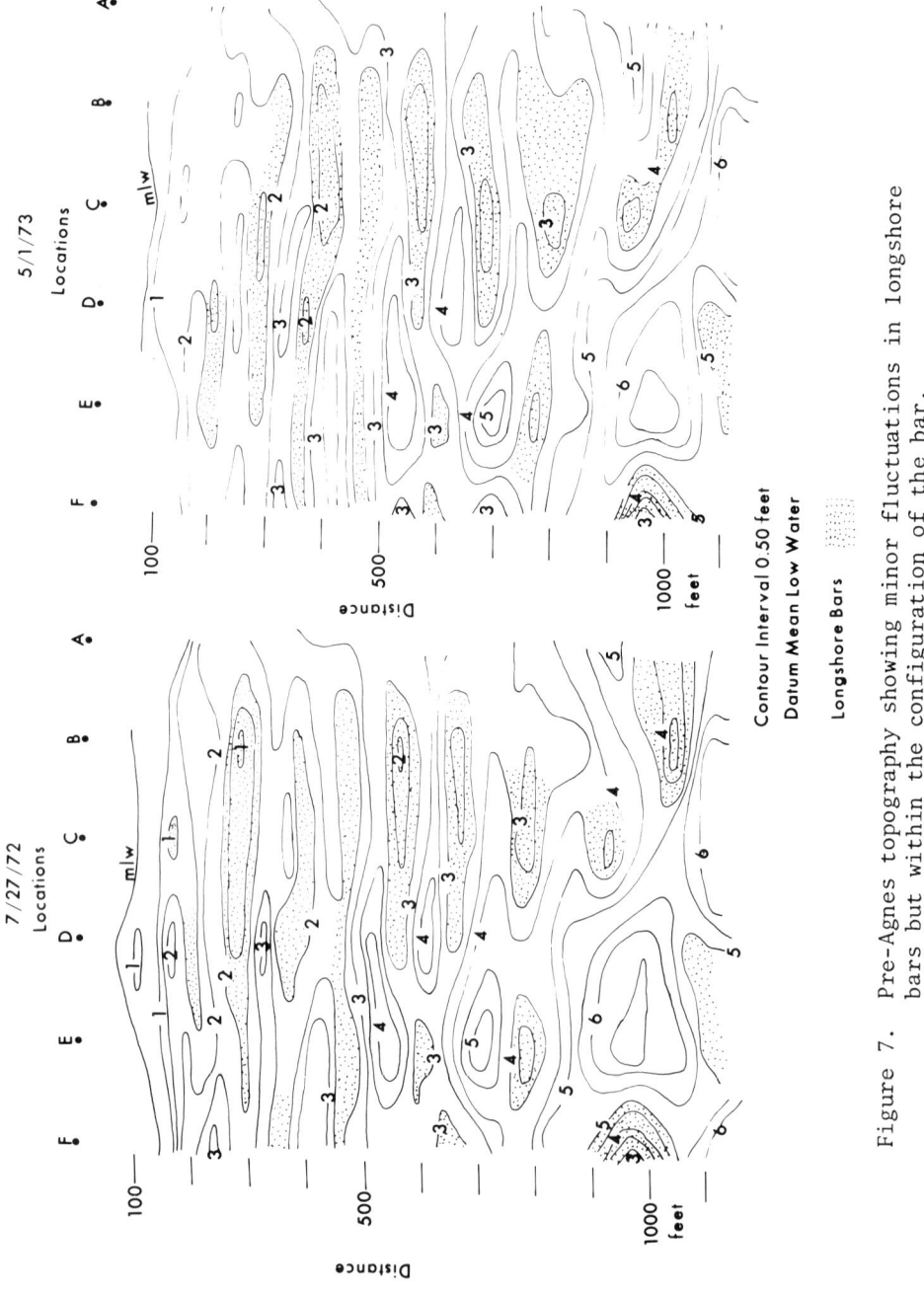

Figure 7. Pre-Agnes topography showing minor fluctuations in longshore bars but within the configuration of the bar.

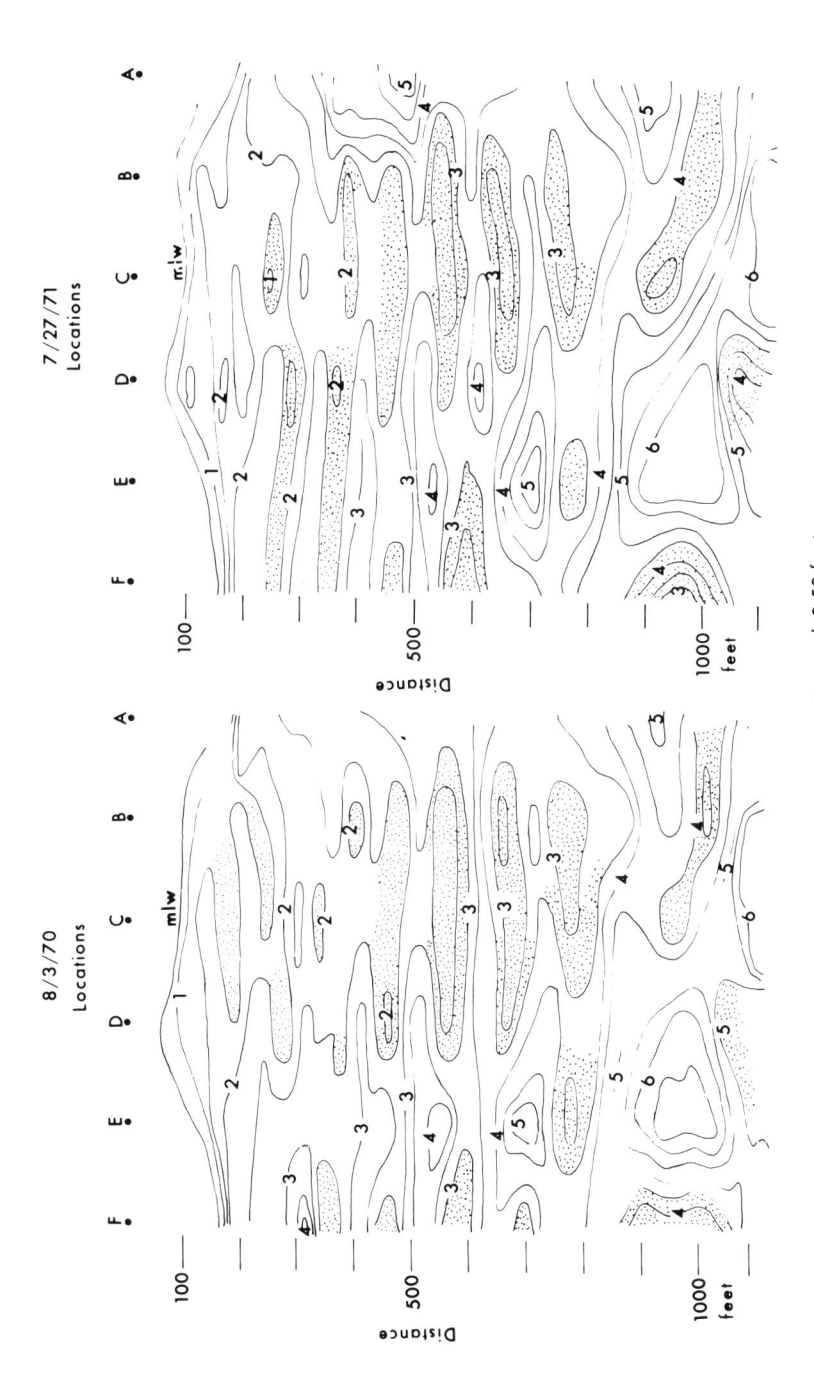

Figure 8. Post-Agnes topography. There is no significant change in topography when compared with pre-Agnes topography. Tropical Storm Agnes did not disrupt the bottom stability of the longshore bars.

AGNES IN THE GEOLOGICAL RECORD OF
THE UPPER CHESAPEAKE BAY[1]

J. R. Schubel[2]
C. F. Zabawa[3]

ABSTRACT

Analyses of cores taken in the upper reaches of Chesapeake Bay in August 1972 showed that the deposit left by Tropical Storm Agnes could be distinguished from underlying deposits by its small-scale sedimentary structures. The Agnes layer was comprised of a series of laminated silts and clays and was separated from older relatively structureless sediments by a sharp contact. Repeated corings at the same stations at 2-4 month intervals over a year revealed that the distinguishing features of the Agnes layer were slowly being obliterated by burrowing organisms.

INTRODUCTION

Tropical Storm Agnes constituted a major episode in the geological history of the Chesapeake Bay, an episode whose impact on the Chesapeake Bay estuary was clearly depositional, and not erosional (see papers in this section). Since floods of this magnitude have an estimated recurrence interval of about 100 years (Kammerer et al. 1972), other such floods have occurred in the Bay's history, and Agnes therefore raises a number of interesting geological questions. Can the layer of sediments deposited by Agnes be distinguished from older, underlying deposits? Will this layer be preserved in the geological record as a recognizable unit, or will it be obliterated by the normal physical, and particularly the biological, mixing processes? Can the Agnes sedimentary layer be used as a guide for identification of other flood deposits in the Bay's geological record?

In an attempt to answer these questions, cores were taken at approximately two to three month intervals between August 1972 and June 1973 at the stations shown in Fig. 1. The stations were selected in the upper reaches of the Bay because this is the segment of the Bay proper where sedimentation rates were highest following Agnes and consequently where the Agnes layer should be most readily identifiable.

METHODS

Cores were taken with a Benthos[4] gravity corer using 2-4 m lengths of 6.3 cm I.D. ^{x} 7.0 cm O. D. clear cellulose acetate butyrate (CAB) tubing. To help minimize physical disturbance of the sedimentary layers during coring, the coring tubes were used without any external core barrel, nose cone, or core catcher.

---

[1]Contribution No. 235, Chesapeake Bay Institute, Johns Hopkins University.
[2]Marine Sciences Research Center, State University of New York, Stony Brook, N. Y., 11794
[3]Dept. of Geology, University of South Carolina, Columbia, S. C. 29208
[4]Benthos, Inc., North Falmouth, Mass. 02556

The cores were kept in a vertical position from the time they were recovered until the time they were analyzed. Immediately after recovery, the cores were examined visually in the coring tubes for any evidence of the interface between the Agnes and the pre-Agnes sediments. More detailed examinations were made in the laboratory of 6 cm wide x 1 cm thick x 25 cm long slices removed from the central portion of each core. Each slice was examined visually with a hand lens and with a low power microscope for color, texture, and organisms. It was then X-rayed with a hospital X-ray unit for evidence of internal sedimentary structures that might characterize the Agnes layer and help identify its lower surface. Using the X-radiograph as a guide, the core slice was carefully dissected under a microscope to attempt to relate the X-ray transmission pattern to variations in the physical character, particularly the texture, of the sediment. Samples were removed from the surface layer and from underlying layers for size analysis and for mineralogical analysis. Size analyses were made with standard pipette techniques, and mineralogical analyses with standard X-ray diffraction techniques.

DISCUSSION

The primary objective of all of these determinations was to identify the lower boundary of the Agnes deposit. Unfortunately, we do not have any cores from this segment of the Bay for the months immediately preceding Agnes. Consequently, identification of the lower boundary of the Agnes sedimentary layer must be based on circumstantial evidence. The criteria we used in attempting to delineate this boundary were changes in color, texture, or in the internal sedimentary structures that would indicate a recent, rapid influx of sediment.

The most useful criteria for delimiting the Agnes layer are the small scale internal structures, both primary and secondary, revealed by X-radiographs. In the first set of cores, taken in August 1972, the Agnes layer could be identified with some assurance at most of the stations. The top layer of sediment, the Agnes deposit, was characteristically a laminated layer of silts and clays, with minor amounts of fine sand. This layer was typically separated from a structureless, bioturbated layer by a sharp boundary. Two of the better examples of this were observed in August 1972 at Stations 2 and 10, Plates 1 and 2. Station 2 is located approximately midway between Pooles Island and Tolchester, at the southern end of the study area, Fig. 1. The layer we interpret as the Agnes deposit is approximately 4-5 cm thick, and is comprised of laminated silts, clays, and some fine sands. The sandy material is primarily quartz and detrital coal with some pellets (fecal?).

Station 10 is located off Howell Point. An X-radiograph of the August 1972 core reveals an upper layer of laminated silts, clays and some fine sands, approximately 15-20 cm thick, underlain by relatively structureless sediments. There is some evidence of erosion at the interface between these two layers. We interpret the upper laminated layer as the Agnes deposit. The lower part of this laminated layer may, however, have been deposited during the normal spring freshet earlier that year.

Size analyses and X-ray diffraction studies failed to reveal any significant differences between the Agnes layer and the underlying sediments. Since the dams along the lower reaches of the Susquehanna trap most of the coarse-grained sediment, it should not be surprising that the Agnes sediments differ little texturally from the sediments discharged into the Bay under the normal range of flow conditions. Prior to construction of the dams, however, flood deposits probably contained considerably more coarse material than the normal estuarine sediments.

While it was generally possible to identify the Agnes sedimentary layer, one of the most striking features revealed by the X-radiographs was the extreme lateral variability of the internal sedimentary structures over relatively small distances in the estuary. Variability was so great at some stations that it was not possible to correlate strata in consecutive cores taken at a given time from an anchored vessel.

All of the cores taken during August were examined in a similar manner to make an estimate of the thickness of the Agnes sedimentary deposit throughout the segment of the Bay from Turkey Point to Pooles Island (Fig. 1). Our best estimate is summarized in Fig. 2. This corresponds very roughly to a volume of about 30×10^6 m^3 of in-place sediment which is equivalent to approximately 13×10^6 metric tons of dry sediment. This estimate is not discordant with the estimated input of sediment from the Susquehanna River during Agnes. Schubel (this volume) estimated that in the 10-day period, 21-30 June 1972, the Susquehanna may have discharged nearly 32×10^6 metric tons of sediment into the upper Chesapeake Bay. This is roughly equivalent to 76×10^6 m^3 of in-place sediment, which if spread uniformly over the segment of the Bay from Turkey Point to Tolchester would form a deposit about 19 cm thick. Not all of the sediment discharged by the Susquehanna was deposited in this segment of the Bay, of course. Significant amounts were deposited farther upstream on the Susquehanna Flats and farther downstream in the Bay (Schubel, this volume).

ACKNOWLEDGEMENTS

We thank M. Glendening, J. Smith, L. Smith, C. Morrow, and T. Kana for their assistance both in the field and in the laboratory. The graphic arts work was done by J. Sullivan and D. Pendleton, and the manuscript was typed by A. Sullivan. This work was supported in part by the U. S. Army Corps of Engineers (Baltimore District); in part by the Oceanography Section, National Science Foundation, NSF Grant GA-36091; and in part by a project jointly funded by the Fish and Wildlife Administration, State of Maryland, and the Bureau of Sport Fisheries and Wildlife, U. S. Department of Interior, using Dingell-Johnson Funds.

LITERATURE CITED

Kammerer, J. C., H. D. Brice, E. W. Coffay, and L. C. Fleshmon. 1972. Water Resources Review for June 1972. U. S. Geological Survey, Washington, D. C., 17 p.

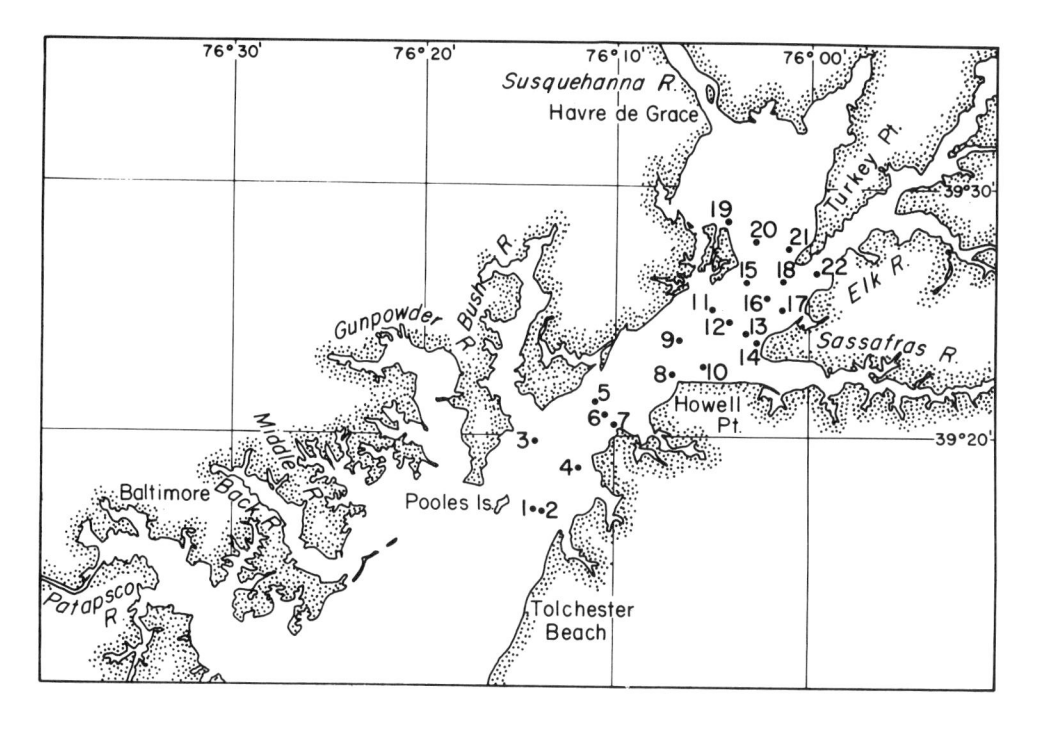

Figure 1. Station location map.

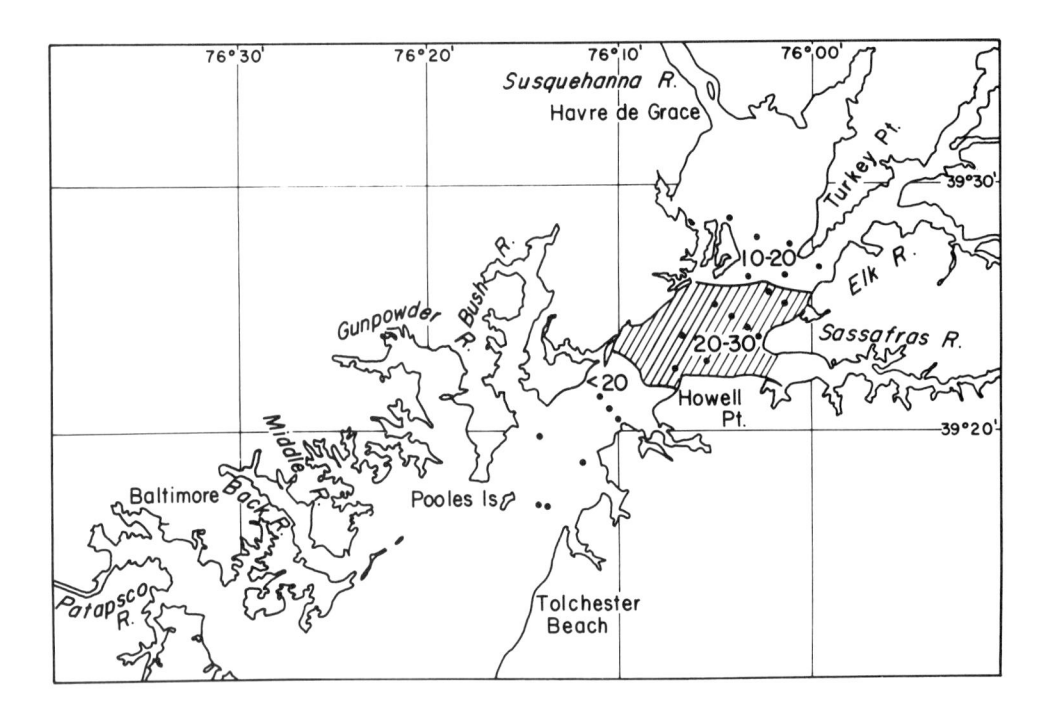

Figure 2. Map of study area showing estimated thickness of Agnes sedimentary layer (cm). Estimates were made from X-radiographs of cores taken in August 1972.

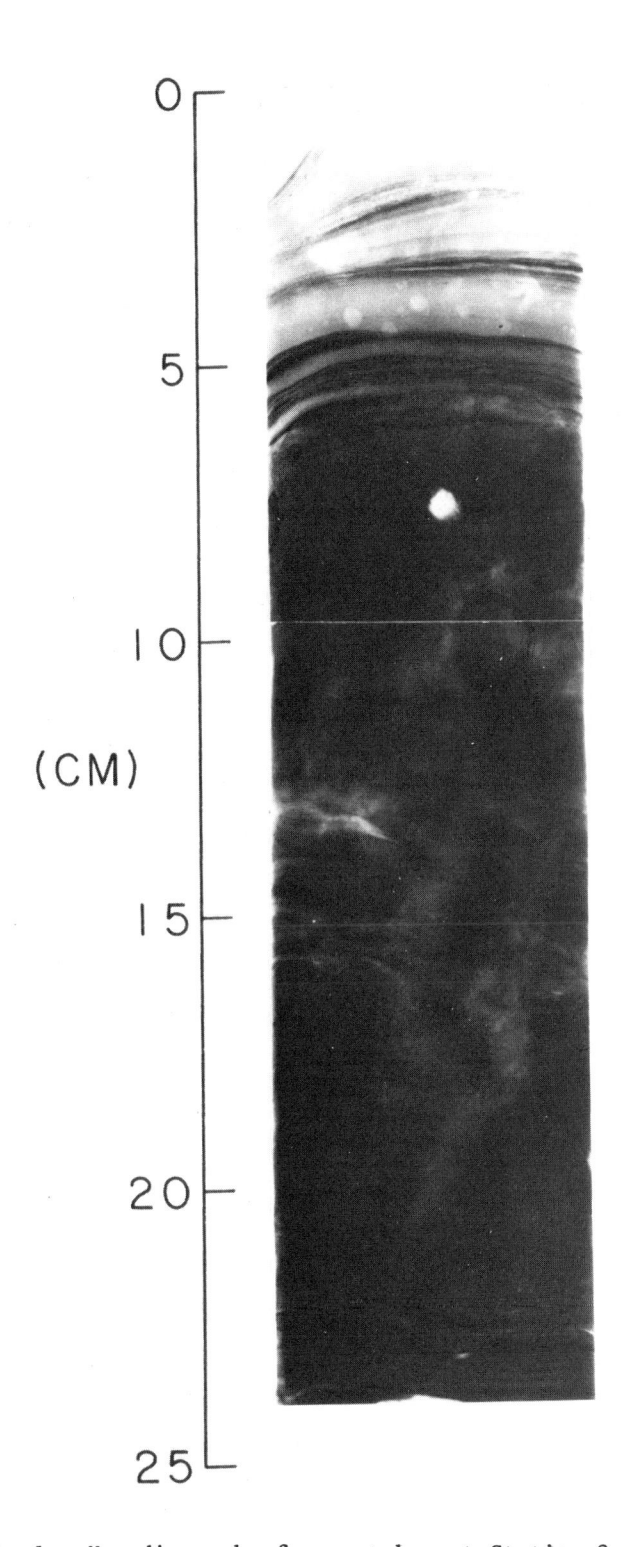

Plate 1. X-radiograph of core taken at Station 2 in August 1972. The Agnes layer is identified as the upper 4-6 cm of the core.

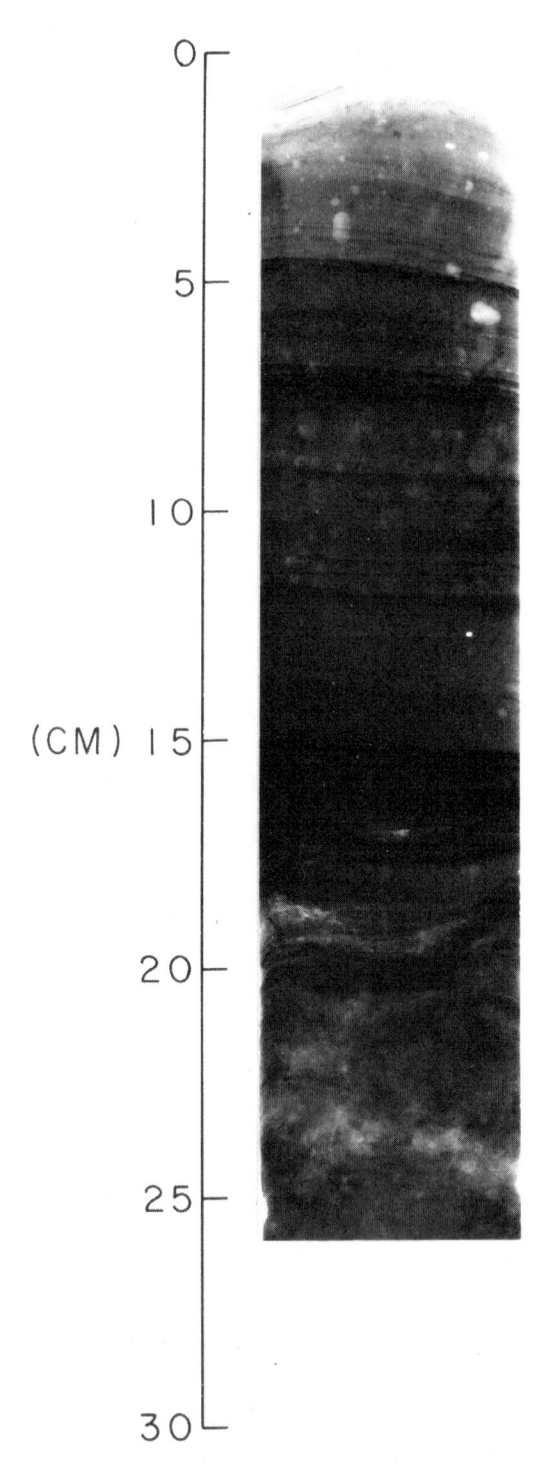

Plate 2. X-radiograph of core taken at Station 10 in August 1972. The Agnes layer is identified as the upper 15-20 cm of the core, although the lower part of the layer may have been deposited during the normal spring freshet earlier in the year.

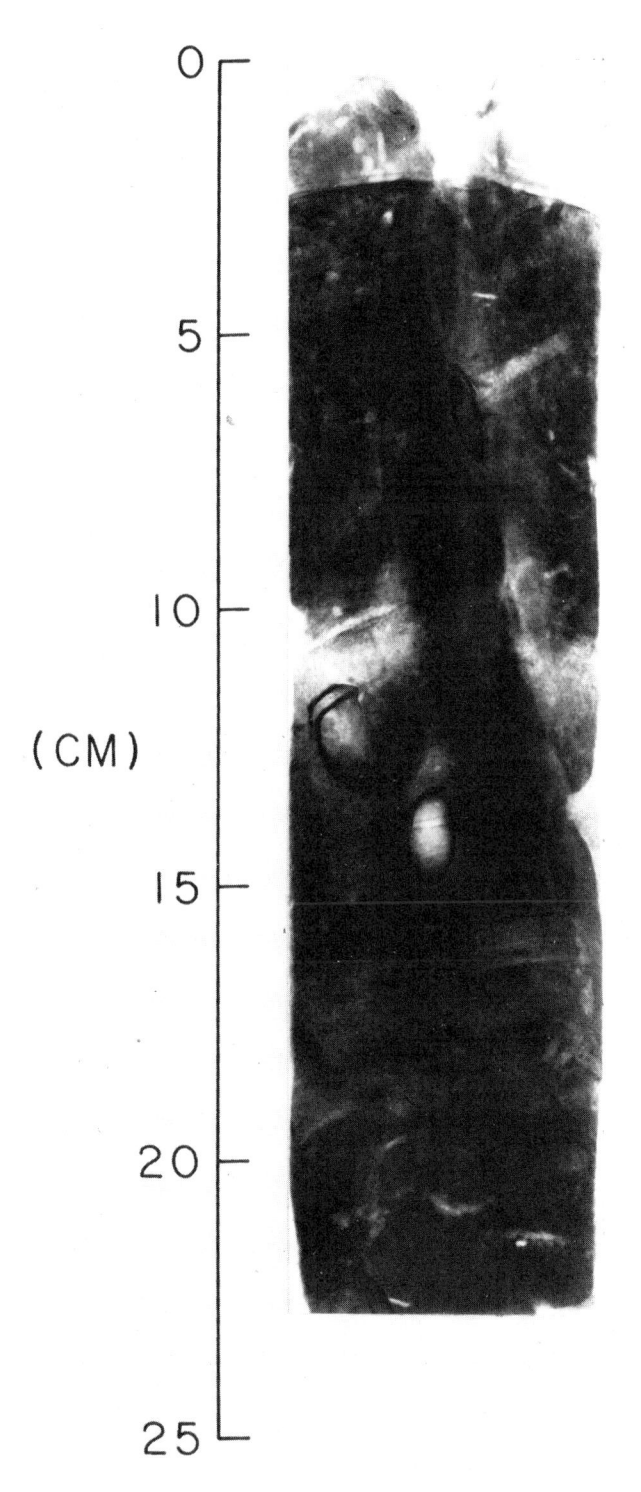

Plate 3. X-radiograph of core taken at Station 2 in January 1972. The features that were originally used to identify the Agnes layer have been nearly obliterated by the action of burrowing organisms.

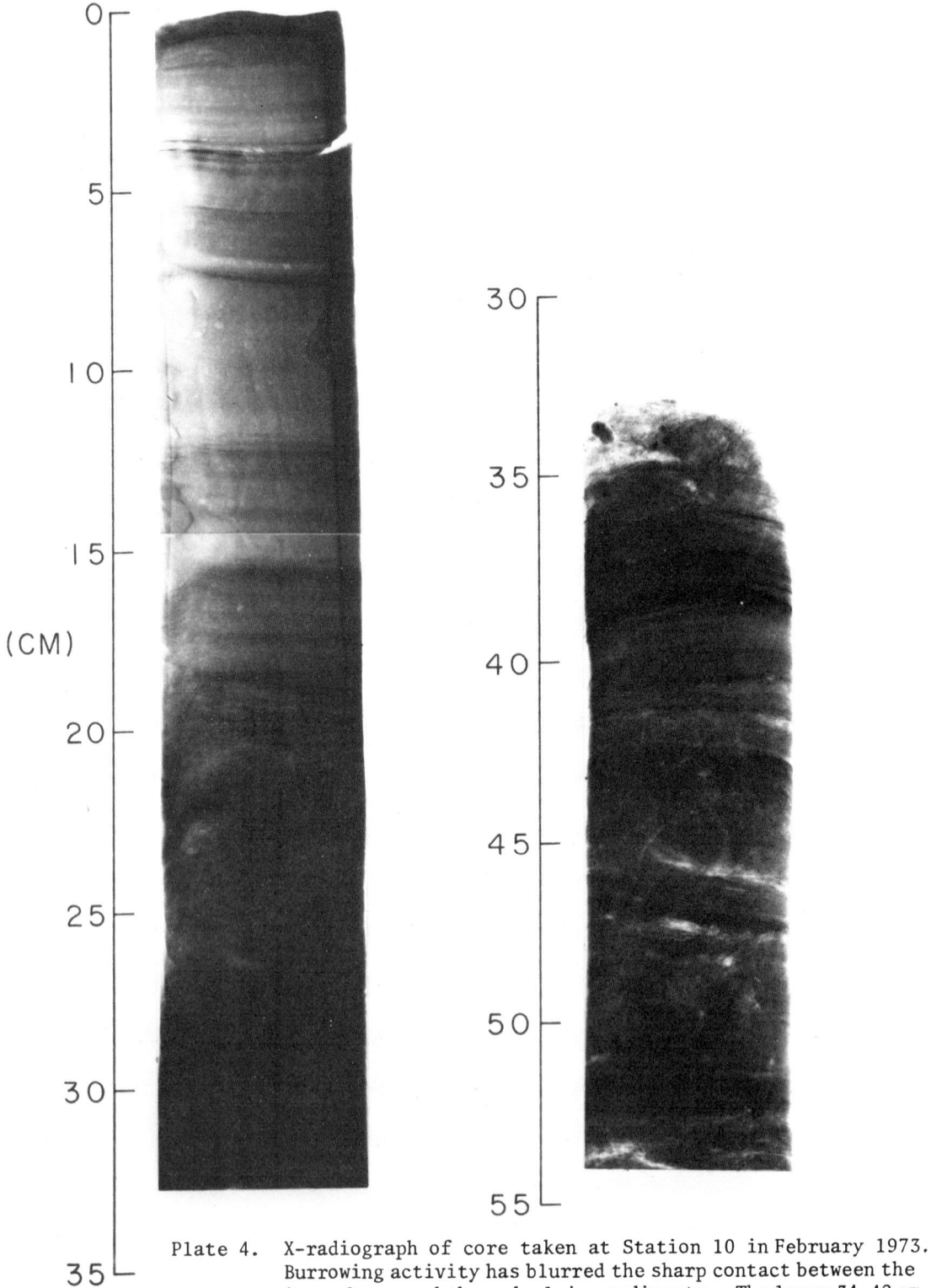

Plate 4. X-radiograph of core taken at Station 10 in February 1973. Burrowing activity has blurred the sharp contact between the Agnes layer and the underlying sediments. The lower 34-42 cm of this core shows the presence of a laminated layer of sediment similar to the Agnes layer above.

WATER QUALITY EFFECTS

Robert J. Huggett, Editor

SOME EFFECTS OF TROPICAL STORM AGNES ON WATER QUALITY IN THE PATUXENT RIVER ESTUARY[1]

David A. Flemer[2]
Robert E. Ulanowicz[2]
Donald L. Taylor, Jr.[3]

ABSTRACT

A post-Agnes study emphasizing environmental factors was carried out on the Patuxent River estuary with weekly sampling at eight stations from 28 June to 30 August 1972. Spatial and temporal changes in the distribution of many factors, e.g., salinity, dissolved oxygen (DO), seston, particulate carbon and nitrogen, inorganic and organic fractions of dissolved nitrogen and phosphorus, and chlorophyll a were studied and compared to extensive earlier records. Patterns shown by the present data were compared especially with a local heavy storm that occurred in the Patuxent drainage basin during July 1969.

Estimates were made of the amounts of material contributed via upland drainage. A first approximation indicated that 14.8×10^3 metric tons of seston were contributed to the head of the estuary between 21 and 24 June. We estimated that 5.6×10^3 metric tons of seston were delivered to the upper estuary between 28 June and 30 August. Particulate carbon was 5% of the seston during the latter period. The particulate carbon:nitrogen ratio (wt/wt) of the material contributed for the 10-week interval, exclusive of the four-day peak flow, was about 4.7:1. From 28 June to 30 August about 135 metric tons of total dissolved nitrogen and 10.8 metric tons of total dissolved phosphorus were added to the estuary. These amounts of nitrogen and phosphorus greatly exceeded the requirement for plant primary production, especially during July.

Some interesting correlations were observed in the data. Particulate nitrogen and active chlorophyll a were usually correlated with a correlation coefficient of ≥ 0.80 (9d.f.) in the lower study area. In the tidal freshwater area (Nottingham), NH_3-N and NO_3-N and total dissolved phosphorus were negatively correlated with chlorophyll a. These correlations were not strong in the lower river. The atomic ratio of dissolved inorganic nitrogen to dissolved inorganic phosphate-phosphorus was usually greater than 15 and often above 30, especially at the most seaward station (near Solomons) and in the tidal fresh waters. During August some of the above ratios were less than 2.0 in the middle portion of the estuary; this suggested that nitrogen would potentially become more limiting to phytoplankton growth than phosphorus. At this time the concentration of active chlorophyll a often approximated 30 to 60 mg m^{-3}.

The minimum surface salinity at Sandy Point, the most seaward station near Solomons was 1 ppt and occurred on 5 July. Strong vertical differences in salinity were noted in the middle portion of the river. In this region, bottom DO values were frequently <1.0 mg liter^{-1}.

---

[1]Contribution No. 573, U. M. Center for Environmental and Estuarine Studies.
[2]Chesapeake Biological Laboratory, Center for Environmental and Estuarine Studies, University of Maryland, Solomons, Md. 20688.
[3]Department of Biology, College of William and Mary, Williamsburg, Va. 23185.

INTRODUCTION

Tropical Storm Agnes entered the Chesapeake Bay area on 21 June 1972 and provided a unique opportunity to study the environmental impact of an event which usually is unavailable to estuarine scientists. Logistically, we were conveniently located on the Patuxent River estuary to engage in a field reconnaissance of several environmental factors, e.g., salinity, DO, nutrients, chlorophyll a, and seston. A field analysis of the above and other factors in the upper estuary had been completed between August 1968 and August 1970 (Flemer et al. 1970, much of this information will be published later), giving our present efforts considerable comparative value.

Though not as unique or as formidable as Agnes, an unusually heavy rainfall in the Patuxent drainage area occurred during late July 1969, and we compared the 1969 data with the present data. Our field studies in this investigation began on 28 June 1972, about one week after Agnes, and were continued at weekly intervals until 30 August 1972.

Extreme conditions imposed by Agnes were believed of general interest in the area of eutrophication. This study was planned to help characterize the relationship between phytoplankton and nitrogen and phosphorus. Questions regarding nutrient removal at wastewater treatment plants is still an unresolved problem in Maryland, and an active research program by the Chesapeake Research Consortium, Inc. in Chesapeake Bay is focusing on the effects of sewage on the estuarine ecosystem.

Detritus is considered an important source of energy for many deposit- and filter-feeding organisms. In the oligohaline area of the Patuxent, we have provided further evidence that the dominant spring copepod, *Eurytemora affinis*, feeds extensively on detritus (Heinle et al. 1974). We have attempted to characterize the input of detritus to the head of the estuary following the large runoff caused by Agnes. The elemental composition of the suspended material in the estuary complements work reported in a recent symposium (Melchiorri-Santolini & Hopton 1972), though relatively little information was presented for estuarine conditions.

DESCRIPTION OF THE STUDY AREA

The drainage basin of about 963 sq. miles (2,494 km^2) lies wholly within the State of Maryland on the western shore of the Bay and forms the next major tributary upstream from the Potomac River (Nash 1947). The basin lies in both the Piedmont Plateau and the Coastal Plain physiographic provinces. Urbanization is occurring, especially in the upper drainage basin near the Fall Line, which runs approximately between Washington, D. C. and Baltimore. Extensive tidal brackish water marshes are located within the upper two-thirds of the study area with the seaward extension of the marsh complex ending just upstream to Trueman Point (Fig. 1). The present study area included eight channel stations from tidal freshwater (Nottingham) to a location adjacent to Solomons (Sandy Point).

Tidal amplitude is small. Cory and Nauman (1967) report a difference of 0.55 m between mean high and mean low tides near Benedict Bridge and Mansueti (1961) gave a tidal range of 0.76 m near Nottingham. The estuary is typically a two-layered system (type B of Pritchard 1955) and occasionally changes to a three-layered system near the mouth.

METHODS AND PROCEDURES

Field Sampling Procedures

Field samples were obtained with a 13-foot Boston Whaler. Sampling began at Sandy Point and proceeded upstream following the node of the same slack current that preceded flood current. Water samples were pumped with a battery powered submersible pump (Teel Manufacturing Co., Model 1P811). DO and temperature were measured with a YSI Model 54 oxygen meter by placing the probe near the end of the sampling hose in a plastic bucket. Salinity was measured with an American Optical Company refractometer prior to 19 July, and a Beckman Model RS-5 salinometer was used after this date. Samples were brought back to the laboratory and processed. Filtrates and unfiltered samples were frozen in polyethylene bottles. Filter pads were desiccated over silica gel and frozen. Chlorophyll a samples collected on filters were directly frozen without desiccation.

Analytical Methods

Samples for chlorophyll a were collected on Whatman GF/C filters and the chlorophyll a was estimated fluorometrically with a Turner fluorometer (Yentsch and Menzel 1963; Holm-Hansen et al. 1965) using our adaptation of these two methods (Flemer et al. 1970).

Seston, or total suspended material, was determined on tared GF/C filters after drying to constant weight over silica gel. Particulate carbon was determined by the method of Menzel and Vaccaro (1964) using a Beckman Model 1R215 Infrared Analyzer and a Coleman CHO Analyzer. Particulate nitrogen was determined with a Coleman Model 29A Nitrogen Analyzer equipped with a Model 29 combustion tube and syringe.

Total phosphorus was determined with the oxidation method of Menzel and Corwin (1965). The same method was used to oxidize dissolved organic phosphorus materials after passing the sample through GF/C filter. Dissolved inorganic reactive phosphorus was determined with the composite reagent method (Strickland & Parsons 1968).

Ammonia nitrogen analysis followed the procedure of Solorzano (1969). Nitrate and nitrite nitrogen were analyzed by the method of Strickland and Parsons (1965). Soluble organic nitrogen analysis employs a modification of the uv light oxidation (Strickland & Parsons 1968). A half-strength seawater solution is used for the solvent, for the blanks, ammonium sulfate, and pyridine standards. The seawater solution is made up according to Strickland's and Parsons' (1968) nitrate method, then diluted by one-half with double distilled, deionized water. This solution is used to dilute river water samples and to add salts to facilitate the uv oxidation (unpublished observations). We dilute, if necessary, 20 ml of river water sample to 100 ml with half-strength sea water. Two drops of 30% hydrogen peroxide are added to the sample in a quartz tube, and the sample is capped and irradiated 7 cm from a 1,200-watt Hanovia-Englehardt 189A lamp for 3 hours. Strickland's and Parsons' (1968) procedure is followed for the remainder of the analysis.

Freshwater Discharge to the Head of the Estuary

Discharge values were calculated for the contribution of fresh water from the drainage basin of the upper Patuxent River above the confluence with Western Branch and for the Western Branch drainage basin. Gaging stations were near the headwaters of the tributaries; thus, the estimates of discharge were based on the ratio of area gaged to the area downstream of the gaging stations. The formulation employed was based on the work of Charles Hall, Maryland Department of Water Resources, where:

$$Q_T = Q_L + 5.9 \ Q_W + 1.42 \ Q_U + 4.2 \ Q_G$$

and

Q_T = total fresh water to head of estuary (main stem of Patuxent, plus Western Branch)

Q_L = discharge at Laurel gaging station

Q_W = discharge at Western Branch gaging station near Largo.

Q_U = discharge near Unity gaging station

Q_G = discharge near Guilford gaging station.

The unpublished gaging station data were provided by the Geological Survey, Water Resources Division, U. S. Department of Interior, College Park, Md.

We determined the total discharge over Δt where the sampling date was near the mid-point. For example, the first sampling date of 28 June with three days before and three days after this date were used to estimate the weekly discharge. The same approach was employed for each succeeding sampling date. The concentration of a factor at Nottingham, e.g., seston, was multiplied by the total weekly discharge to estimate a flux for the interval. The 10 intervals were summed to provide an estimate of total flux.

RESULTS AND DISCUSSION

All data taken in this study are listed in Table 1 and in the appendix.

Temperature, Salinity and DO

Water temperatures measured in this study were characteristic for the summer (Table 1; Flemer et al. 1970). Maximum values occurred upstream in the general vicinity of the Chalk Point Power Plant. Exceptionally high values, some approximating 36°C, were noted in the upper study area on 26 July. Bottom values were sometimes slightly higher than surface values. These data supplement the extensive records of temperature in the upper River (Herman et al. 1968; Cory & Nauman 1967).

Salinities reflected the large rainfall in the drainage basin, and our data record the recovery of the low salinities back to the more normal regime (Table 1). Surface salinities were 0.0 ppt as far seaward as Benedict Bridge during most of July. Strong vertical differences in salinity were noted at Sheridan Point and seaward. The minimum surface salinity at Sandy Point was about 1.0 ppt and occurred on 5 July. Bottom salinities on two occasions at Sandy Point (28 June and 2 August) were about 1.5 ppt less than surface values. We point out these differences as possibly accurate observations since Nash (1947) made similar kinds of observations near Sandy Point.

Some perspective is gained by comparing surface salinities at Sandy Point

Table 1. Temperature (C), salinity (ppt), and dissolved oxygen (mg liter^{-1}), and top (T) and bottom (B), observed during the Patuxent River Post-Agnes Study, summer 1972.

Date			6/28			7/5			7/12			7/19			7/26	
Station	Hours		Temp.	Sal. 14:15	DO	Temp.	Sal. 08:15	DO	Temp.	Sal. 13:35	DO	Temp.	Sal. 07:05	DO	Temp.	Sal. 13:45	DO
Nottingham	T		24.7	0.0	7.7	—	0.0	—	23.6	0.0	6.6	27.0	0.0	4.8	29.9	0.0	7.7
	B		24.0	0.1	7.2	—	0.0	—	23.3	0.0	5.7	26.5	0.0	4.7	34.5*	0.0	6.5
L. Marlboro	T		25.3	0.1	5.2	—	0.0	—	23.9	0.0	7.1	29.0	0.0	6.4	30.0	0.0	9.8
	B		25.0	0.1	5.3	—	0.0	—	23.5	0.0	6.6	30.0	0.0	6.0	36.0*	0.0	6.6
Trueman Point	T		27.2	1.4	5.3	—	0.0	—	25.0	0.0	7.0	29.9	0.0	7.0	30.0	0.5	6.5
	B		27.0	1.9	5.6	—	0.0	—	24.5	0.0	6.6	30.0	0.0	6.3	33.3	1.6	6.1
Benedict Bridge	T		26.0	2.5	5.3	24.0	0.0	5.4	25.4	0.0	7.2	28.0	0.0	7.2	29.7	2.3	6.7
	B		25.8	2.6	5.0	23.9	0.3	4.1	25.2	1.1	6.5	27.0	1.4	6.5	32.0	2.6	6.0
Sheridan Point	T		25.5	3.1	5.3	23.5	0.5	6.1	24.7	2.3	7.6	26.0	1.4	8.2	27.9	3.1	6.6
	B		24.1	4.1	2.8	21.0	2.3	2.9	24.0	2.5	3.3	24.0	4.5	2.0	27.6	4.0	4.0
Queen Tree	T		25.5	3.6	7.0	23.0	1.5	7.8	24.5	2.8	9.1	27.8	2.3	8.5	27.4	3.4	6.9
	B		25.0	4.4	2.6	21.2	4.0	2.5	23.9	3.4	3.5	23.5	5.0	1.5	26.7	4.4	1.9
Broome Island	T		26.0	4.2	7.4	22.5	1.3	7.0	24.2	2.3	9.2	26.0	2.3	8.6	27.0	3.4	6.3
	B		25.5	4.7	2.7	21.0	4.5	3.0	22.9	4.5	4.0	24.5	5.6	2.2	27.2	4.4	1.6
Sandy Point	T		26.0	4.7*	6.8	22.0	1.3	8.6	23.9	4.5	9.1	25.5	4.5	7.9	27.0	5.0	6.3
	B		24.5	3.1	2.3	21.0	4.5	4.2	22.9	5.6	5.4	24.5	5.6	4.6	27.2	5.0	4.3
Hours				10:35			05:30			10:45			04:15			10:10	

*asterisk mark unusual results. Time in hours of initiation of sampling is given at bottom of each column and time of conclusion of sampling at the upstream station is given at the top of each column.

Table 1. Continued. Temperature (C), salinity (ppt), and dissolved oxygen (mg liter^{-1}), and top (T) and bottom (B), observed during the Patuxent River Post-Agnes Study, summer 1972.

Date			8/2			8/9			8/16			8/23			8/30	
Station	Hours	Temp.	Sal.	DO	Temp.	Sal.	DO	Temp.	Sal.	DO	Temp.	Sal.	DO	Temp.	Sal.	DO
			07:10			12:20			07:46			10:43			16:35	
Nottingham	T	24.7	0.0	7.7	26.5	0.1	8.2	24.2	0.2	7.3	–	–	–	26.5	0.1	9.1
	B	24.0	0.1	7.2	25.0	0.2	8.4	23.8	0.2	7.1	–	–	–	26.5	0.1	8.9
L. Marlboro	T	25.3	0.0	5.2	27.0	0.0	6.8	25.0	0.8	6.4	–	–	–	27.5	0.3	7.5
	B	25.0	0.1	5.3	26.0	0.1	6.2	24.3	0.7	6.5	–	–	–	27.5	0.3	7.0
Trueman Point	T	27.2	1.4	5.3	27.5	0.9	7.8	25.1	2.2	6.7	27.2	2.8	6.8	28.5	2.5	6.4
	B	27.5	1.9	5.6	28.0	1.0	6.8	24.5	2.4	6.7	27.3	4.0	7.0	29.5	3.1	6.7
Benedict Bridge	T	26.0	2.5	5.3	27.0	2.7	7.8	24.1	5.4	5.5	27.0	5.3	6.5	29.0	5.7	7.2
	B	25.8	2.6	5.0	27.0	2.9	6.5	24.1	6.2	3.7	26.5	5.6	5.3	28.0	6.3	4.3
Sheridan Point	T	25.5	3.1	5.3	26.5	3.4	9.8	24.5	5.2	5.7	27.9	6.2	7.9	25.5	6.9	7.3
	B	24.1	4.1	2.8	26.0	6.1	1.1	24.0	7.6	0.3	25.3	7.3	1.6	26.0	9.7	0.1
Queen Tree	T	25.5	3.6	7.0	26.0	3.8	9.4	24.8	5.4	6.9	26.9	6.4	7.8	25.5	7.6	7.1
	B	25.0	4.4	2.6	25.0	6.8	0.9	24.0	7.9	0.4	25.0	8.0	0.9	26.5	10.4	0.3
Broome Island	T	26.0	4.2	7.4	25.5	3.9	10.3	24.2	5.9	6.5	27.0	7.2	9.5	27.0	7.8	5.2
	B	24.5	4.7	2.7	20.3	9.3	0.4	24.0	9.3	0.4	24.5	7.3	0.2	26.5	10.4	0.4
Sandy Point	T	26.0	4.7*	6.8	25.1	6.0	8.6	24.0	7.7	6.5	24.5	8.6	7.5	26.0	9.3	2.3
	B	24.5	3.0	2.3	20.3	9.8	1.4	24.0	9.3	5.3	24.5	9.5	3.8	24.5	10.4	2.6
Hours			03:55			09:30			03:35			08:52			13:16	

*asterisk mark unusual results. Time in hours of initiation of sampling is given at bottom of each column and time of conclusion of sampling at the upstream station is given at top of each column.

with those taken nearby at the Chesapeake Biological Laboratory pier. Daily salinities at the pier from 1 June to 20 June 1972 ranged between 7.9 and 10.8 ppt. These data are not continuous as observations are usually not made on weekends; however, on 28 June the salinity was 2.5 ppt and agreed closely with our data at Sandy Point. The monthly mean values at the pier during the period 1938 to 1957 for June, July, and August were 11.0, 12.5, and 13.5 ppt, respectively (Beaven 1960). Agnes resulted in very low salinities at a time when salinities normally proceed toward maximum values in the River.

Salinity data obtained during July 1969 reflect the aftermath of a heavy rainfall that occurred in the drainage basin. In the lower River, 11.85 inches of rain fell at Solomons on 23 July 1969, over a period of a few hours (NOAA, Ashland, N. C., central records for Solomons area). In the upper River, the surface salinity at Nottingham on 8 July 1969 was 2.7 ppt and on 30 July 1969, this salinity level was displaced seaward to a point near Trueman Point (Flemer et al. 1970).

Information is presented on DO that generally characterizes the oxygen resources of the River (Table 1). Only large changes can be evaluated for the summer since the same slack sampling required some cruises to be initiated at night. Very low concentrations of DO, e.g., 1.0 mg liter^{-1}, were measured in bottom waters between Sheridan Point and Broome Island.

Particulate Material

Seston. Highest concentrations of seston occurred upstream of Benedict Bridge (Fig. 2). A maximum value of 170 mg liter^{-1} was measured at Lower Marlboro in the bottom sample on 5 July. Seaward to Benedict Bridge the concentration of seston seldom was greater than 50 mg liter^{-1}. On 2 August, 168 mg liter^{-1} of seston were recorded at Nottingham, and 68 mg liter^{-1} of seston were measured as far seaward as Broome Island. The above pattern of seston in the River suggests that the maximum discharge of suspended solids associated with Agnes occurred before 28 June. The rapid washout of chlorophyll *a*, as discussed later, confirms this conclusion.

High sestonic levels were often encountered during earlier work in the River (Flemer et al. 1970). The upper tidal Patuxent received very high amounts of seston during late July 1969. Several values approached 200 mg liter^{-1}. We have no quantitative data on seston for this event at Solomons, but comparative data for the upper tidal River taken on 8 and 30 July 1969 documented the impact of this unusually large freshwater discharge to the estuary. These data indicated that the Patuxent River has received large concentrations of seston in the past but Agnes was a significant event in the history of the Patuxent system. Normally the so-called "sediment trap," which is well described for the main stem of Chesapeake Bay (Schubel 1968b; Schubel & Biggs 1969), extends seaward in the Patuxent estuary to about Chalk Point. Agnes briefly extended the seaward boundary of the sediment trap downstream to, at least, the mouth of the Patuxent River. The system returned quickly to steady-state conditions, and the seaward extension of the sediment trap moved upstream between Benedict Bridge and Sheridan Point.

Chlorophyll a. The concentration of active chlorophyll *a*, which is an index of the standing crop of phytoplankton (here possibly some mud-dwelling algae), showed a sharp decline between Nottingham and Sheridan Point (Figs. 3a and 3b) during the first two sampling periods. Surface values at this time were 5 to 8 mg m^{-3}. Seaward, between Broome Island and Sandy Point, the surface concentration of chlorophyll *a* ranged between 40 to 50 mg m^{-3}. This pattern probably resulted from the seaward displacement of normally high upstream concentrations.

Earlier work indicated that the high values observed during this study between Nottingham and Trueman Point from 19 July to 30 August are typical for the upper study area (Flemer et al. 1970). Maximum surface concentrations of chlorophyll a observed during the present study were 127 and 105 mg m^{-3}, which occurred at Nottingham and Broome Island on 9 August and 23 August, respectively. Most moderately deep temperate East Coast estuaries show a single sustained maximum concentration of chlorophyll a during the late summer, especially in the sediment trap area (Ryther 1963; Flemer 1970).

Seaward of Lower Marlboro there was a stronger temporal pattern in the distribution of surface chlorophyll a in the present study than that observed for the bottom waters. The low values measured at Queen Tree and Broome Island were associated with the low concentration of DO that was present in the bottom waters. In general, the concentration of chlorophyll a observed in the present study compares well with earlier studies on the River (Flemer & Olmon 1971; Flemer et al. 1970). A feature characteristic of Agnes was for more variations over time to occur in the surface waters at higher salinities.

For perspective, we are able to compare the present distribution of chlorophyll a in the River with that for the summer of 1969. The large amount of rain that occurred in late July 1969 was strongly associated with the low chlorophyll a values measured at that time. For example, values at Lower Marlboro rapidly decreased from 46 to 7 mg m^{-3}. This correlates well with the increased sestonic load received by the River during the July 1969 period and illustrates that the upper Patuxent has experienced an important washout of the phytoplankton at other times.

Particulate Carbon. The concentration of particulate carbon showed a surprising uniformity in the surface waters through most of the River until early August (Fig. 4a). Most values ranged between 100 and 200 μ moles liter^{-1} (1.2-2.4 mg liter^{-1}). Highest concentrations of particulate carbon in the surface and bottom waters (Fig. 4b) occurred during late August when chlorophyll a values were maximal.

We observed that the average values of the percent of carbon relative to seston in the surface waters ranged between 6 and 8% from Nottingham seaward to Queen Tree (Table 2). Exclusive of two unusually high concentrations of particulate carbon noted at Broome Island and Sandy Point, the values averaged 8.6 and 10.7% at these stations. The average values discussed above compare favorably to earlier work on the River (Flemer et al. 1970).

Seaward of Benedict Bridge the percent of carbon relative to seston in bottom samples was about one-half less than the surface average values. Several reasons may help explain these observations. Possibly a differential settling of inorganic material occurred, or the rate of decomposition was greater than that of carbon input from upland drainage, and marsh drainage, and plant production in the River. Also, dilution of the bottom suspended material from the sediments could influence the above pattern (Schubel & Biggs 1969). If we assume that the carbon values represent about 50% of the organic matter on a dry weight basis, then our values agree more closely with the winter and spring data on organic matter in the upper Chesapeake Bay (Schubel 1968a).

Particulate Nitrogen. In contrast to the distribution of particulate carbon, particulate nitrogen in the surface waters varied more with time than it did with position in the river. Most maximum values approximated 40 to 50 μ moles liter^{-1} or 600 to 750 μg liter^{-1} (Fig. 5a). The fairly high correlation between nitrogen

Table 2. The percent particulate carbon of seston in the surface (S) and bottom (B) waters, Post-Agnes Study, Patuxent River Estuary, summer 1972.

Date	Nottingham S	Nottingham B	Lower Marlboro S	Lower Marlboro B	Trueman Pt. S	Trueman Pt. B	Benedict Br. S	Benedict Br. B	Sheridan Pt. S	Sheridan Pt. B	Queen Tree S	Queen Tree B	Broome Is. S	Broome Is. B	Sandy Pt. S	Sandy Pt. B
6/28	4.8	4.0	-	3.3	2.6	2.8	2.6	2.2	2.7	2.3	4.5	1.4	7.4	2.2	8.2	2.2
7/5	3.7	3.5	4.0	3.9	2.1	3.8	3.8	4.3	1.4	5.0	8.1	4.0	6.9	2.9	8.3	3.5
7/13	5.0	4.0	4.4	4.0	3.5	4.2	2.7	2.9	4.3	2.1	2.9	5.0	7.5	2.9	11.5	2.5
7/19	0.9	3.1	5.0	5.0	4.2	5.0	4.7	7.6	3.9	2.8	7.9	2.8	7.9	3.6	12.0	11.2
7/26	4.8	5.6	4.4	3.0	3.0	5.1	4.0	6.0	3.6	5.6	2.8	5.0	4.5	3.6	9.0	7.1
8/2	4.7	3.8	3.6	3.1	3.4	3.0	2.1	2.8	5.6	3.2	5.0	2.7	8.0	3.7	10.8	6.7
8/9	8.7	10.0	7.5	6.7	6.2	2.0	8.1	5.5	17.9	4.6	13.0	5.8	20.0	2.5	13.1	4.2
8/16	-	-	-	-	-	-	-	-	-	-	-	-	-	-	-	-
8/23	-	-	-	-	7.5	6.9	8.9	6.6	10.4	5.7	7.5	5.4	60.4	3.0	12.5	6.8
8/30	17.1	16.7	10.8	6.5	6.9	4.6	21.1	13.5	22.5	2.9	11.4	2.5	6.4	2.9	-	4.4
$\bar{x}$	6.2	6.3	5.7	4.4	4.4	4.2	6.4	5.7	8.0	3.8	7.0	3.8	8.6[1]	3.0	10.7	5.4

[1] average exclusive of high value observed on 8/23.

and chlorophyll a is obvious from the contour diagrams. Conversely, the pattern of particulate nitrogen in the bottom waters was characterized more by position than times in the river; in fact, a relatively sharp decrease occurred throughout the study in the concentration of particulate nitrogen at Sheridan Point (Fig. 5b). A similar trend was noted for chlorophyll a. The concentration of particulate nitrogen measured during this study approximated that measured previously (Flemer et al. 1970).

Nutrients

Phosphorus. Data on several fractions of phosphorus, e.g., total phosphorus, total dissolved phosphorus, and dissolved inorganic reactive phosphate-phosphorus, are illustrated in Fig. 6 for four stations along the River. Total phosphorus approximated 6.0 µg at liter^{-1} at Nottingham and showed a gradual decrease seaward to Sandy Point where the concentration ranged between 2 and 4 µg at liter^{-1}. Inadvertent filtration of samples from 5 July through 19 July prevented analysis for total phosphorus. Total dissolved phosphorus usually ranged between 20 and 50% of the total phosphorus. Thus, by difference, particulate phosphorus constituted a substantial portion of the phosphorus in the river. Dissolved inorganic phosphate usually accounted for most of the total dissolved phosphorus. Therefore, relatively little dissolved organic phosphorus was present. On a few occasions the dissolved inorganic values were analytically greater than values for total dissolved phosphorus. We believe that the total dissolved values should be considered only approximate. Work is under way in an attempt to resolve this problem.

Compared to data obtained during the summer of 1969, we conclude that the general pattern of phosphorus distribution and concentrations in the River did not differ in any important way (see Fig. 6).

We should mention that the inorganic phosphate concentration is largely controlled in turbid estuaries through sorption reactions with suspended sediments (Pomeroy, Smith, & Grant 1965). We expect that such reactions strongly influenced the level of phosphate observed in the present study as the Patuxent qualifies as a turbid estuary. The exchange is apparently near-equilibrium when the water has a phosphorus content about 0.7 to 1.5 µg at liter^{-1} (Butler & Tibbitts 1972). In the Tamar estuary, near the Plymouth Laboratory, England, in contrast to our findings, the total dissolved phosphorus was relatively constant before and after a heavy rain throughout the estuary. The concentration of total dissolved phosphorus in the Patuxent showed a substantial decrease from Nottingham seaward but relatively less change over time during the present study.

Nitrogen. Nitrogen determinations presented in Fig. 7 show the total nitrogen as represented by several fractions. The dissolved organic and inorganic fractions (NH_4^+, NO_2^-, and NO_3^-) are added to the particulate nitrogen. Some comparative data taken during the summer of 1969 are included on the figure (Flemer et al. 1970).

Maximum total nitrogen occurred at Nottingham and during the first six weeks many values ranged between 80 and 100 µg at liter^{-1}. Most values of total nitrogen during the first six weeks at Benedict Bridge and Queen Tree ranged between 60 and 80 µg at liter^{-1}. At Sandy Point during this period there was a slight increase in the total nitrogen. A maximum value for the entire study of 146 µg at liter^{-1} occurred at Sandy Point on 28 June which presumably was related to the displaced upstream material. Total nitrogen decreased about 50% at all stations during the latter half of the study with few exceptions. One notable exception occurred at Nottingham on 30 August where a concentration of 140 µg at liter^{-1} was measured.

The limited data available from the summer of 1969 suggest that total nitrogen was more highly concentrated during the recovery period of Agnes than noted during the summer of 1969. By comparison, the concentration of total phosphorus failed to show a consistent increase over that of the summer of 1969.

We examined the relative proportion of NH_3-N, NO_2-N, and NO_3-N to the total inorganic nitrogen (Appendix). The concentration of NH_3-N decreased more rapidly than the concentration of NO_3-N at Nottingham and Benedict Bridge from 28 June to 19 July. The pattern was not as clearly discernible at Queen Tree as noted at the upstream stations. At Sandy Point, the concentrations of NH_3-N and NO_3-N decreased together proportionally for most of the study with the pronounced exception on 30 August. It is tempting to ascribe the more rapid decrease in NH_3-N relative to NO_3-N to differential uptake by phytoplankton (Harvy 1960). The analysis is complicated by a rapid increase in primary production, unknown changes in the relative rate of supply of the two nutrients and nitrification. Nitrite-nitrogen was quantitatively unimportant compared to the other inorganic nitrogen sources throughout the study.

The concentration of dissolved organic nitrogen often approximated or exceeded the concentration of dissolved inorganic nitrogen (Fig. 7). Most of the dissolved nitrogen was in the form of dissolved organic nitrogen at Benedict Bridge and Queen Tree on 9, 16, and 23 August (Appendix). At this time the significance of dissolved organic nitrogen to the biological system is not apparent but presumably contributes to the available nitrogen pool in the long-term.

Ratios and Correlations

Data on C:N ratios (atomic) for surface and bottom waters are given in Figs. 8a and 8b. Surface ratios showed more variation with time than position along the axis of the estuary. Both surface and bottom ratios usually were within the range of 3-10. Living phytoplankton typically have a C:N (atomic) ratio of about 7 (Strickland 1960). The relative constancy of the ratios is dissimilar to the data reported for the upper Chesapeake Bay (Flemer & Biggs 1971). However, maximum values in the upper Chesapeake Bay usually followed the maximum discharge related to snow melt in the Susquehanna River basin. Possible the small range in values over the summer in the Patuxent resulted from averaging the high C:N ratios associated with higher plant material with the lower ratios associated with phytoplankton (Gucluer & Gross 1964).

Particulate carbon (PC) was highly correlated, $\sqrt{} < 0.80$, with PN at Trueman Point (bottom), Broome Island (surface and bottom), and Sandy Point (surface and bottom) (Table 3). At Sheridan Point and seaward, except Queen Tree, we observed that PC with chlorophyll a and PN with chlorophyll a were significantly correlated (Table 3). These correlations are more likely in the more seaward of the estuary where phytoplanktonic material is relatively more abundant than upstream where considerable material is derived from upland drainage. In earlier work we observed that PN x chlorophyll a were highly correlated at the more seaward stations, (e.g., Trueman Point and seaward to Queen Tree). Brooks (1970) reported for the Brazos River, Texas, that the particulate organic carbon was directly related to river discharge. In the Patuxent autochthonous sources of particulate organic carbon probably mask the relationship between river flow and the concentration of particulate carbon, especially at the more seaward stations.

The measured ratio of PC:chlorophyll a is another way to view the relationship between living algal material and the total suspended particulate carbon

Table 3. Correlation coefficients equal to or greater than +0.80 or -0.70. Post-Agnes study, Patuxent River Estuary, 28 June-30 August 1972.

Station		Factors	Correlation Coefficient	d.f.	P
Nottingham					
	Surface		Positive		
		NO_2 x TDN	0.87	7	0.998
		NO_3 x TDN	0.97	7	1.000
		NO_3 x TDP	0.87	7	0.998
		TDN x DIP	0.95	7	1.990
		TDN x TDP	0.87	7	0.998
		DIP x TDP	0.90	7	0.999
		TCL x ACL	0.996	8	1.000
			Negative		
		NO_3 x PC	-0.86	8	0.999
		NO_3 x TCL	-0.86	8	0.999
		NO_3 x ACL	-0.82	8	0.997
		NH_3 x TCL	-0.85	6	0.992
		NH_3 x ACL	-0.82	6	0.988
		TDN x PN	-0.75	7	0.983
		TDN x PC	-0.83	7	0.994
		TDN x TCL	-0.81	7	0.992
		TDN x ACL	-0.76	7	0.985
		TDN x SAL	-0.77	7	0.987
		TDP x PC	-0.91	7	0.999
		TDP x ACL	-0.74	7	0.982
		TP x SAL	-0.78	5	0.968
	Bottom		Positive		
		PN x TCL	0.92	7	1.000
		PN x ACL	0.92	7	0.994
		TCL x ACL	0.99	8	1.000
			Negative (none)		
Lower Marlboro					
	Surface		Positive		
		TCL x ACL	0.99	8	1.000
			Negative (none)		
	Bottom		Positive		
		PC x SES	0.95	7	1.000
		TCL x ACL	0.99	8	1.000
			Negative (none)		
Trueman Point					
	Surface		Positive		
		TCL x ACL	0.99	9	1.000
			Negative (none)		
	Bottom		Positive		
		PN x PC	0.84	9	0.999
		PN x SES	0.83	8	0.999
		TCL x ACL	1.00	9	1.000
			Negative (none)		
Benedict Bridge					
	Surface		Positive		
		NO_2 x NO_3	0.99	9	1.000
		NO_2 x TDN	0.88	8	0.999
		NO_3 x TDN	0.88	8	0.999

Table 3. (Cont'd)

Station		Factors	Correlation Coefficient	d.f.	P
Benedict Bridge					
	Surface		Positive		
		NO$_3$ x DIP	0.82	9	0.998
		TCL x ACL	0.99	9	1.000
			Negative		
		NO$_2$ x SAL	-0.78	9	0.996
		NO$_3$ x TCL	-0.74	9	0.993
		NO$_3$ x ACL	-0.76	9	0.995
		NO$_3$ x SAL	-0.76	9	0.995
		TDN x PC	-0.71	8	0.984
		TDN x SAL	-0.78	8	0.993
		DIP x SAL	-0.70	9	0.987
	Bottom		Positive		
		TCL x ACL	1.00	9	1.000
			Negative		
		PC x SES	-0.71	8	0.984
Sheridan Point					
	Surface		Positive		
		PN x TCL	0.96	9	1.000
		PN x ACL	0.96	9	1.000
		TCL x ACL	0.98	9	1.000
			Negative (none)		
	Bottom		Positive		
		TCL x ACL	0.96	9	1.000
			Negative (none)		
Queen Tree					
	Surface		Positive		
		NO$_2$ x TDN	0.85	9	0.999
		TDP x SAL	0.84	9	0.999
		TCL x ACL	0.99	9	1.000
			Negative (none)		
	Bottom		Positive		
		TCL x ACL	0.97	9	1.000
			Negative (none)		
Broome Island					
	Surface		Positive		
		PN x PC	0.97	9	1.000
		PN x TCL	0.94	9	1.000
		PN x ACL	0.93	9	1.000
		PN x SES	0.90	8	1.000
		PC x TCL	0.90	9	1.000
		PC x ACL	0.89	9	1.000
		PC x SES	0.89	8	0.994
		TCL x ACL	1.00	9	1.000
			Negative (none)		
	Bottom		Positive		
		PN x PC	0.80	9	0.998
		PN x SES	0.82	8	0.999
		PC x SES	0.95	8	1.000
		TCL x ACL	0.87	9	0.999
			Negative (none)		

Table 3. (Cont'd)

Station	Factors	Correlation Coefficient	d.f.	P
Sandy Point				
Surface		Positive		
	NO_3 x NH_3	0.82	9	0.998
	NO_3 x TDN	0.94	9	1.000
	NH_3 x TDN	0.85	9	0.999
	PN x TP	0.91	6	0.998
	PN x PC	0.84	9	0.999
	PN x TCL	0.87	9	0.999
	PN x ACL	0.86	9	0.999
	TP x TCL	0.93	6	0.999
	TP x ACL	0.94	6	0.999
	TCL x ACL	1.00	9	1.000
		Negative		
	NO_3 x TDP	-0.73	8	0.988
	NO_3 x SAL	-0.70	9	0.980
	TDN x SAL	-0.74	9	0.992
	DON x TP	-0.78	6	0.980
Bottom		Positive		
	PN x PC	0.94	9	1.000
	PN x TCL	0.86	9	0.999
	PN x ACL	0.87	9	0.999
	PC x TCL	0.94	9	1.000
	PC x ACL	0.95	9	1.000
	TCL x ACL	1.00	9	1.000
		Negative (none)		

(Figs. 9a and 9b). High ratios indicate a relatively low contribution of living plant material. Throughout most of the study the ratio was less than 100:1, and frequently values between 35 and 50:1 were observed. These ratios are consistently less than those observed during earlier summer work on the River (Flemer et al. 1970) and these low ratios are believed to be atypical of temperate coastal waters. As discussed in the preceding paragraph, these ratios suggest a high percentage of living algal carbon relative to the total measured particulate carbon. Many laboratory algal cultures under good growth conditions will have C:chl a ratios between 30:1 and 50:1 (Parsons et al. 1961). Ratios greater than 150:1 were associated with the early washout of the phytoplankton, and high ratios followed the sudden decline from relatively high concentrations of chlorophyll a, especially at Benedict Bridge and seaward during the latter part of this study. We attempted to use the ratio of 50:1 of PC:chlorophyll a to partition the measured surface PC into that related to living phytoplankton and a residual (Fig. 10). The residual would presumably contain material from such sources as detritus and small heterotrophs. Some inconsistent results were noted, especially at the most upstream station, Nottingham; however, the use of a reasonable but still lower ratio of 30:1 eliminated most inconsistencies, except at Nottingham. Average values by station, exclusive of Nottingham, of the percent living carbon ranged between 52 and 79%. No clear axial trend along the estuary was noted in this analysis.

An effort was made to assess the impact of Agnes in terms of the ratio of dissolved inorganic nitrogen (DIN) to dissolved inorganic phosphate-phosphorus (DIP) (Table 4). Typically, during the summer in many temperate coastal waters, the ratios of DIN:DIP are less than 15-10:1 (Ryther & Dunstan 1971) which is interpreted that nitrogen would probably become limiting to phytoplanktonic growth before phosphorus. Frequently, the ratio of DIN:DIP was less than 5:1 during the summer studies of 1969 and 1970 (Flemer et al. 1970). The ratios in the River following Agnes were usually very high, often between 15 and 50:1, except at Benedict Bridge and Queen Tree from 9 August to 23 August where several values approximated 1 to 2. These low ratios were generally consistent with high concentrations of chlorophyll a. The rapid decline of this bloom following the very low DIN:DIP ratio is some evidence that nitrogen may have controlled maximum plant biomass. At Sandy Point the ratio of DIN:DIP remained fairly high, a fact which we would not have predicted based on the present and previous work in the River. It is possible that the nutrient regime in the main stem of the Bay influenced the pattern noted at Sandy Point.

Table 4. The ratio of dissolved inorganic nitrogen to dissolved inorganic phosphate-phosphorus, Patuxent River Post-Agnes Study, summer 1972.

Date	Nottingham	Benedict Bridge	Queen Tree	Sandy Point
28 June	31	-	33	179
5 July	28	24	13	95
12 July	25	17	46	74
19 July	22	8	15	86
26 July	17	15	276	66
2 Aug.	28	22	54	60
9 Aug.	-	0.7	1.6	48
16 Aug.	17	1.6	2.0	15
23 Aug.	-	1.1	2.2	37
30 Aug.	73	13	50	34

Many other correlations are given in Table 3. At Nottingham many of the nutrient fractions were correlated, e.g., NO$_3$-N x total dissolved nitrogen (TDN), and TDN x DIP, and nutrients were negatively correlated with several particulate fractions, e.g., NO$_3$-N x ACL (active chlorophyll a). Nutrients would be expected to decrease as particulate material is formed via photosynthesis. The lack of many of these high correlations seaward to Nottingham is evidence that recycling probably dominated the nutrient uptake kinetics, especially from 9 to 23 August.

The C:N:P ratio gives some insight into the relative abundance of elements in the particulate material. Though somewhat variable, living phytoplankton usually are characterized by a C:N:P ratio of 106:10-15:1 (Redfield et al. 1963; Ryther & Dunstan 1971). In this study N:P ratios were 10:1 to 15:1 about half of the time (Table 5) with the remaining comparisons above or below the ratio of 10:1 to 15:1. C to P was more variable than N to P. In 13 out of 25 comparisons, the C:P ratio was less than 84:1, six times the ratio was between 85-126:1 and six times the ratio was greater than 127:1. These ratios indicate that often the particulate material is richer in P relative to C. Below the euphotic zone in the ocean the particulate material usually is phosphorus poor relative to carbon and nitrogen (Menzel & Ryther 1964). The question is still open regarding the relative proportion of dissolved phosphorus that is associated with inorganic material between the open sea and coastal waters.

Table 5. The atomic ratio of C:N:P in particulate material, Patuxent River Post-Agnes Study, summer 1972.

Date	Nottingham C:N:P	Benedict Bridge C:N:P	Queen Tree C:N:P	Sandy Point C:N:P
28 June	39:6:1	-	99:23:1	80:12:1
7 July	-	-	-	-
12 July	-	-	-	-
19 July	-	-	-	-
26 July	42:8:1	48:7:1	49:10:1	78:11:1
2 Aug.	55:7:1	21:6:1	63:11:1	96:17:1
9 Aug.	-	73:10:1	135:16:1	89:10:1
16 Aug.	68:8:1	118:9:1	49:11:1	129:21:1
23 Aug.	-	107:10:1	160:20:1	108:13:1
30 Aug.	69:37:1	218:11:1	132:12:1	434:21:1

Flux of Material to the Head of the Estuary

We estimated that about 5.6 x 10^3 metric tons of seston, 262 metric tons of PC, 46 metric tons of PN, 135 metric tons of TDN, and 11 metric tons of TDP passed Nottingham between 25 June and 2 September 1972 (Table 6). During the peak flow from Agnes between 21 and 24 June, we estimated that 14.8 x 10^3 metric tons of seston were transported to the head of the estuary. This estimate was based on an assumed concentration of 100 mg liter^{-1} since no data on concentrations were available at this time. The estimate is probably conservative when compared to other observations during high flows (Flemer et al. 1970). Further estimates are possible if we assume that the ratio of the various fractions, e.g., PC and TDN, relative to seston contributed between 25 June and 2 September would apply during peak flows of Agnes. This crude approach yielded the following estimates of flux between 21 and 24 June: PC = 696, PN = 121, TDN = 352, and TDP = 28 metric tons, respectively.

Table 6. Flux of materials to the head of the Patuxent River Estuary, Post-Agnes Study, summer 1972.

Date	$Q_T/\Delta t$ m³ x 10⁶	SESTON $\bar{x}$ conc. mgl⁻¹	SESTON Flux g x 10⁹	PC $\bar{x}$ conc. mgl⁻¹	PC Flux g x 10⁶	PN $\bar{x}$ conc. mgl⁻¹	PN Flux g x 10⁶	TDN conc. µg atl⁻¹	TDN Flux g x 10⁶	TDP conc. µg atl⁻¹	TDP Flux g x 10⁶
28 June	42.6	33	1.4	1.4	59.6	0.257	10.9	73.8	44.0	2.87	3.8
5 July	21.1	54	1.1	2.0	42.2	0.257	5.4	84.6	25.0	2.38	1.6
12 July	16.8	32	0.5	1.4	23.5	0.355	6.0	81.2	19.1	2.62	1.4
19 July	18.1	32	0.5	0.6	10.9	0.138	2.5	91.4	23.2	2.99	1.7
26 July	7.9	48	0.4	2.5	19.8	0.548	4.3	56.3	6.2	2.18	0.54
2 Aug	8.1	120	1.0	4.9	39.7	0.600	4.9	64.0	7.3	1.45	0.36
9 Aug	6.9	40	0.3	3.8	26.2	0.400	2.8	40.0[1]	3.9	4.00[2]	0.86
16 Aug	6.0	35[1]	0.2	1.8	10.8	0.433	2.6	38.5	3.2	1.09	0.20
23 Aug	3.8	20[1]	0.08	3.0[1]	11.4	0.6[1]	2.3	20.0[1]	1.1	1.17[1]	0.14
30 Aug	4.4	24	0.1	4.0	17.6	1.030	4.5	33.1	2.0	1.21	0.17
Total			5.6		261.7		46.2		135		10.8

[1] estimates based on flow and values preceding and following missing data.
[2] based on DIP
Note -Q_T is an estimate of total freshwater discharge to the head of the estuary.

It is only possible to give a semiquantitative comparison of the flux of material between Agnes and July 1969 since only three samples were taken between the end of June and the end of August 1969. However, for simplicity, the relative flows are instructive. We estimated that for 28 July 1969 and 22 June 1972, times of peak flows, that 3,104 and 42,554 cfs of water, respectively, were delivered to the head of the estuary. As a minimum, the impact of Agnes was 14 times that of the rainfall of July 1969 in terms of water transported to the upper estuarine area.

In the Patuxent, upland drainage is a significant source of nutrients. In the Ythan estuary, just north of Aberdeen, Scotland, marine water contributed about 70% of the phosphate, and fresh water supplied about 70% of the nitrate (Leach 1971). This author presents interesting comparative data on the contribution of nutrients via fresh water. Based on a semidiurnal tidal cycle (like that of the Chesapeake Bay), the Ythan estuary received about 16.8 kg of inorganic phosphate during the summer. By comparison, we estimated that about 360 kg of total dissolved phosphorus per tidal cycle on the average entered the Patuxent estuary from upstream between 25 June and 2 September. We should emphasize that most of the dissolved phosphorus in the Patuxent is apparently inorganic phosphate, thus the comparison has validity.

GENERAL DISCUSSION

Earlier work in the Patuxent River estuary showed that the upper tidal system probably borders on hypertrophication (Stross & Stottlemyer 1965; Herman et al. 1968; Flemer et al. 1970). Important increases in recent years have occurred in primary productivity, the concentration of chlorophyll a, nitrogen, and phosphorus. The increase in available nitrogen seems especially striking. In view of this information, we were especially interested in the impact of Agnes as a possible stimulant to further overenrichment. It is surprising that the Patuxent is still free of the massive bluegreen algal growths so characteristic of the upper Potomac estuary (Jaworski et al. 1972). The dilutions of the wastewaters received by the Patuxent and Potomac Rivers are quite similar (Brush 1972). Other factors surely play important roles. For example, the extensive tidal marshes characteristic of the Patuxent, but not abundant in the upper Potomac, may play the role of a tertiary treatment system. Salt marsh plots near Woods Hole, Mass., have been shown to retain a large fraction of the nitrogen and phosphorus that were experimentally added as sewage sludge (Valiela et al. 1973). We are currently studying the flux of nutrients between a marsh and its dominant tidal creek in the upper Patuxent; however, results are unavailable for this paper (Heinle et al. 1974). Partial information is available for the Patuxent River on the rate of grazing as an important controlling mechanism to excessive biomass (Heinle 1974). If an additional trophic level occurs between phytoplankton and copepods, then it is suspected that grazing will be an important regulator to the standing crop of phytoplankton. Unfortunately, we were unable to obtain data on zooplankton during this study to aid in interpretation of their rate of grazing during the extreme hydrographic conditions imposed by Agnes. It must be emphasized that conditions that lead to massive bluegreen algal growths are a problem in ecological succession and not simply one of uptake kinetics or development of algal biomass.

The fact that moderately high concentrations of phytoplanktonic biomass occurred during this study compared to previous studies in the Patuxent probably is strongly linked to washout. The apparent displacement of high chlorophyll levels typical of the sediment trap area to the mouth of the River substantiates this conclusion for the initial phase following Agnes. The sustained high levels of chlorophyll a at the two upper stations after 26 July suggest that washout was of much less significance at this time. The periods of low and high chlorophyll a concentrations seaward to Lower Marlboro after about 26 July reflect the dominance of other factors that are known to control phytoplankton biomass. Though other nutrients were not measured, we would suspect from the N:P ratios that nitrogen played an important role. Welch et al. (1972) have shown that phytoplankton blooms in the Duwamish estuary (Seattle, Washington) are strongly influenced by hydrographic conditions.

Comparative information obtained in our post-Agnes study shows that factors measured in the Patuxent are high relative to most temperate estuaries; Thayer (1971) has summarized much of the pertinent data for these systems. Little information is published on the effects of large floods on water quality in temperate estuarine systems. A phenomenon of comparative interest is the monsoon rain that occurs in some tropical areas. These disturbances are a partial natural analog to Agnes in the Patuxent. In the Cochin Backwater, S. W. India, the depth profile of nutrients, e.g., nitrogen and phosphorus, showed a marked seasonal change induced by land runoff (Sankaranarayanan & Qasim 1969). The system changes annually from a marine estuary during the premonsoon period to a freshwater system during the monsoon period. At times of maximum discharge and turbidity, the quantity of settled detritus was comparatively low. This resulted from the strong stratification or halocline that developed in the estuary (Qasim and Sankaranarayanan 1972). In S. E. India the nutrients were increased with monsoon season in the Vellar estuary (Krishnamurthy 1967). For example, total phosphorus ranged between 1.01 and 5.05 µg at liter^{-1} near the mouth of the Vellar estuary. We would not like to overdraw the above comparison, but the partial environmental parallel seemed worthy of mention.

Ratios of C:N:P and C:chl $\underline{a}$ and N:chl $\underline{a}$ used to characterize the particulate material suggest that some important effects resulted from Agnes. Generally, the emergent picture shows that much of the suspended material found in coastal waters, exclusive of temporary algal blooms, is in the form of organic detritus. Compared to previous data for summer conditions in the Patuxent, the relative amount of living algal carbon to total particulate carbon seemed quite high. Some speculations might prove useful, and they may be tested as hypotheses under experimental conditions. It is surmised that the large-scale flushing of the estuary following Agnes reduced the numbers of many grazers and, consequently, the abundant fecal pellets usually observed in water samples from the Patuxent (Heinle, pers. com.). Also, the detrital carbon derived from upland drainage apparently was diluted. It should be noted that we did not observe exceptionally high levels of particulate carbon. Consequences of a high percent of living carbon in a system usually dominated by detrital carbon would be interesting to examine in terms of food web dynamics.

In the wake of an event such as Agnes we were interested to learn if the flux of particulate carbon that entered the tidal fresh waters above Nottingham was a significant fraction compared to primary production in the estuary. Cronin (1971) gives the mean low water area of the Patuxent from the mouth to approximately Nottingham as $137 \times 10^6 m^2$. As a first approximation, we can assume that net primary production averaged about 1 gCm^{-2} day^{-1} (Stross & Sotttlemyer 1965). Turbidity caused the rate to become somewhat higher downriver and less in the upper portion of the estuary. Thus 10.1×10^9 gC were estimated to be fixed photosythetically in the estuary from 21 June to 2 September 1972. Thus, 10.1×10^3 metric tons of fixed carbon from phytoplankton activity far exceed the estimated 958 metric tons of PC derived from upland drainage above Nottingham. It is reasonable to assume that Agnes contributed much organic material to the head of the estuary where photosynthesis was minimal on an areal basis, but over the 10-week period following the storm, the allochthonous sources of PC were quantitatively minor for the whole estuary.

Compare this result with that obtained from the upper Chesapeake Bay, where on an annual basis the PC derived from upland drainage constituted about 90% of the PC pool (Biggs & Flemer 1972). In the Strait of Georgia, British Columbia, the allochthonous organic material contributed per year by upland drainage approximated the natural primary production of the area (Seki, Stephens, & Parsons 1969). Thus, for coastal bodies of water, there is a broad range in the relative amount of PC derived from land sources compared to natural primary production.

ACKNOWLEDGEMENTS

We thank Drew Brown and Bruce Lindstrom for their able field and laboratory assistance. Shelley Sulkin and Linton Beaven helped with several of the chemical analyses. Frances Younger provided the illustrations. Partial financial support was provided by the U. S. Army Corps of Engineers under Contract No. DACW31-73-C-0189.

LITERATURE CITED

Beaven, G. F. 1960. Temperature and salinity of surface water at Solomons, Maryland. Chesapeake Sci. 1(1):2-11.

Biggs, R. B. and D. A. Flemer. 1972. The flux of particulate carbon in an estuary. Mar. Biol. 12:11-17.

Brooks, J. M. 1970. The distribution of organic carbon in the Brazos River basin. M.S. Thesis, Texas A&M Univ., 90 p.

Brush, L. M., Jr. 1972. Domestic and municipal waste loading to Chesapeake Bay. Pages 673-685 in Chesapeake Research Consortium, Inc., Ann. Rep. June 1971-May 1972, Johns Hopkins Univ., Baltimore, Md.

Butler, E. I. and Susan Tibbitts. 1972. Chemical survey of the Tamar estuary. 1. Properties of the waters. J. Mar. Biol. Assoc. U. K. 52:681-699.

Cory, R. L. and J. W. Nauman. 1967. Temperature and water quality conditions for the period July 1963 to December 1965 in the Patuxent River estuary, Maryland. U. S. Geol. Surv. Open File Rep.

Cronin, W. B. 1971. Volumetric, areal, and tidal statistics of the Chesapeake Bay estuary and its tributaries. Johns Hopkins Univ. Chesapeake Bay Inst., Spec. Rep. 20.

Flemer, D. A. 1970. Primary production in the Chesapeake Bay. Chesapeake Sci. 11:117-129.

Flemer, D. A. and R. B. Biggs. 1971. Particulate carbon:nitrogen relationships in northern Chesapeake Bay. J. Fish. Res. Bd. Canada 28:911-918.

Flemer, D. A. and Janet Olmon. 1971. Daylight incubator estimates of primary production in the mouth of the Patuxent River, Maryland. Chesapeake Sci. 12(2):105-110.

Flemer, D. A., D. H. Hamilton, Carolyn W. Keefe, and J. A. Mihursky. 1970. The effects of thermal loading and water quality on estuarine primary production. Final Tech. Rep., August 1968-August 1970. Submitted to Office of Water Resources Research, U. S. Dept. of Interior; Univ. Md. Nat. Res. Inst. Ref.

71-6. (Can be obtained from the National Technical Information Center, Springfield, Virginia, PB 209-811, $3.00).

Gucluer, S. M. and M. G. Gross. 1964. Recent marine sediments in Saanich Inlet, a stagnant marine basin. Limnol. Oceanogr. 9:359-376.

Harvey, H. W. 1960. The Chemistry and Fertility of Sea Water. Cambridge Univ. Press, New York, 240 p.

Heinle, D. R. 1974. An alternate grazing hypothesis for the Patuxent Estuary. Chesapeake Sci. 15(3):146-150.

Heinle, D. R., D. A. Flemer, J. F. Ustach, R. A. Murtagh, and R. P. Harris. 1974. The role of organic debris and associated micro-organisms in pelagic estuarine food chains. Final Report to Office of Water Resources Research, U. S. Dept. of Interior; Univ. Md. Nat. Res. Inst. Ref. 74-29.

Herman, S. S., J. A. Mihursky, and A. J. McErlean. 1968. Zooplankton and environmental characteristics of the Patuxent River estuary 1963-1965. Chesapeake Sci. 9:67-82.

Holm-Hansen, O., C. J. Lorenzen, R. W. Holmes, and J. D. H. Strickland. 1965. Fluorometric determination of chlorophyll. J. Cons. Perm. Intern. Explor. Mer 30:3-15.

Jaworski, N. A., D. W. Lear, Jr., and O. Villa, Jr. 1972. Nutrient management in the Potomac estuary. Pages 246-273 in Nutrients and Eutrophication: the Limiting Nutrient Controversy (G. E. Likens, ed.), Limnol. Oceanogr. Spec. Symp. 1.

Krishnamurthy, K. 1967. The cycle of nutrient salts in Porto Nova (India) water. Int. Rev. ges. Hydrobiol. 52:427-436.

Leach, J. H. 1971. Hydrology of the Ythan estuary with reference to distribution of major nutrients and detritus. J. Mar. Biol. Assoc. U. K. 51:137-157.

Mansueti, R. J. 1961. Movements, reproduction, and mortality of the white perch, *Roccus americanus*, in the Patuxent estuary, Maryland. Chesapeake Sci. 2:142-205.

Melchiorri-Santolini, U. and J. W. Hopton, eds. 1972. Detritus and its role in aquatic ecosystems. Mem. Dell'Ist. Ital. Idrobiol. 29(Suppl.), 1, 540 p.

Menzel, D. W. and R. F. Vaccaro. 1964. The measurement of dissolved organic and particulate carbon in seawater. Limnol. Oceanogr. 9:138-142.

Menzel, D. W. and J. H. Ryther. 1964. The composition of particulate organic matter in the western north Atlantic. Limnol. Oceanogr. 9:179-186.

Menzel, D. W. and N. Corwin. 1965. The measurement of total phosphorus in seawater based on the liberation of organically bound fractions by persulfate oxidations. Limnol. Oceanogr. 10:280-282.

Nash, C. B. 1947. Environmental characteristics of a river estuary. J. Mar. Res. 6:147-174.

Parsons, T. R., K. Stephens, and J. D. H. Strickland. 1961. On the chemical composition of eleven species of marine phytoplankters. J. Fish. Res. Bd. Canada 18:1001-1016.

Pomeroy, L. R., E. E. Smith, and C. M. Grant. 1965. The exchange of phosphate between estuarine water and sediments. Limnol. Oceanogr. 10:167-172.

Pritchard, D. W. 1955. Estuarine circulation patterns. Proc. Am. Soc. Civil Engr. 81:1-11.

Qasim, S. Z. and V. N. Sankaranarayanan. 1972. Organic detritus of a tropical estuary. Mar. Biol. 15:193-199.

Redfield, A. C., B. H. Ketchum, and F. A. Richards. 1963. The influence of organisms on the composition of sea-water. Pages 26-77 in The Sea (M. N. Hill, ed.), Vol. 2, Interscience Publ., New York.

Ryther, J. H. 1963. Geographic variations in productivity. Pages 347-380 in The Sea (M. N. Hill, ed.), Vol. 2, Interscience Publ., New York.

Ryther, J. H. and W. M. Dunstan. 1971. Nitrogen and phosphorus, and eutrophication in the coastal marine environment. Science 171(3975):1008-1013.

Sankaranrayanan, V. N. and S. Z. Qasim. 1969. Nutrients of the Cochin Backwater in relation to environmental characteristics. Mar. Biol. 2:236-247.

Seki, H., K. V. Stephens, and T. R. Parsons. 1969. The contribution of allochthonous bacteria and organic materials from a small river into a semi-enclosed area. Arch. Hydrobiol. 66:37-47.

Schubel, J. R. 1968a. Suspended sediment of northern Chesapeake Bay. Johns Hopkins Univ., Chesapeake Bay Inst. Tech. Rep. 35, 264 p.

Schubel, J. R. 1968b. Turbidity maximum of the northern Chesapeake Bay. Science 161(3845):1013-1015.

Schubel, J. R. and R. B. Biggs. 1969. Distribution of seston in upper Chesapeake Bay. Chesapeake Sci. 10:18-23.

Solorzano, L. 1969. Determination of ammonia in natural waters by the phenol-hypochlorite method. Limnol. Oceanogr. 14:799-801.

Strickland, J. D. H. 1960. Measuring the production of marine phytoplankton. Fish. Res. Bd. Canada Bull. 122, 172 p.

Strickland, J. D. H. and T. R. Parsons. 1965. A practical handbook of sea-water analysis. Fish. Res. Bd. Canada Bull. 125, 203 p.

Strickland, J. D. H. and T. R. Parsons. 1968. A practical handbook of sea-water analysis. Fish. Res. Bd. Canada Bull. 167, 311 p.

Stross, R. G. and J. R. Stottlemyer. 1965. Primary production in the Patuxent River. Chesapeake Sci. 6:125-140.

Thayer, G. W. 1971. Phytoplankton production and the distribution of nutrients in a shallow unstratified estuarine system near Beaufort, N. C. Chesapeake Sci. 12:240-253.

Valiela, I., J. M. Teal, and W. Sass. 1973. Nutrient retention in salt marsh plots experimentally fertilized with sewage sludge. Estuar. Mar. Sci. 1:261-269.

Welch, E. B., J. A. Buckley, and R. M. Bush. 1972. Dilution as an algal bloom control. J. Water Pollut. Control Fed. 44(12):2245-2265.

Yentsch, C. S. and D. W. Menzel. 1963. A method for the determination of phytoplankton chlorophyll and phaeophytin by fluorescence. Deep Sea Res. 10:221-231.

APPENDIX
PATUXENT RIVER POST-AGNES DATA SUMMARY, SUMMER OF 1972

Seston (mg liter^{-1})

Station		6/28	7/5	7/12	7/19	7/26	8/2	8/9	8/16	8/23	8/30
Sandy Pt.	T	33	12	20	20	20	12	16	--	24	12
	B	45	20	28	32	28	24	24	--	25	16
Broome Is.	T	27	16	16	24	20	20	16	--	48	28
	B	45	24	28	28	25	68	24	--	70	24
Queen Tree	T	40	16	28	24	25	20	20	--	20	28
	B	35	20	36	32	20	56	24	--	24	24
Sheridan Pt.	T	40	64	35	28	25	16	24	--	24	32
	B	35	36	48	36	25	25	28	--	28	28
Benedict Br.	T	50	24	56	36	45	24	36	--	35	36
	B	55	44	48	25	35	64	40	--	35	20
Trueman Pt.	T	87	67	48	50	64	35	32	--	48	32
	B	80	120	48	40	35	64	64	--	48	48
L. Marlboro	T	40	47	32	32	48	56	24	--	--	24
	B	60	170	40	40	40	136	24	--	--	40
Nottingham	T	27	54	24	32	48	72	32	--	--	24
	B	40	54	40	32	48	168	48	--	--	24

T=Top; B=Bottom

Active chlorophyll a (mg m^{-3})

Station		6/28	7/5	7/12	7/19	7/26	8/2	8/9	8/16	8/23	8/30
Sandy Pt.	T	46.0	13.2	33.5	35.9	31.2	19.2	38.3	9.3	44.9	54.5
	B	2.0	5.7	3.6	38.3	18.0	26.3	10.2	55.1	9.9	6.3
Broome Is.	T	44.0	18.0	18.0	22.4	9.8	31.2	74.2	16.2	127.2	18.0
	B	1.5	2.2	5.6	7.3	6.9	7.2	5.7	4.6	4.3	4.5
Queen Tree	T	24.0	21.6	25.2	34.7	10.7	14.4	55.1	19.2	15.0	36.5
	B	1.5	6.9	3.2	6.8	14.4	6.5	18.0	3.6	7.3	1.1
Sheridan Pt.	T	9.5	11.0	14.7	29.9	10.8	19.2	53.8	26.3	31.5	43.1
	B	3.5	4.2	5.2	7.2	20.3	7.3	12.0	5.2	11.6	1.6
Benedict Br.	T	3.0	8.7	15.6	29.9	35.3	13.2	49.4	19.2	29.2	38.3
	B	2.0	6.8	9.6	39.0	49.1	10.8	40.4	27.5	15.5	18.0
Trueman Pt.	T	6.0	8.6	16.3	29.9	28.7	10.8	46.4	31.1	23.9	18.9
	B	4.0	8.7	14.9	28.7	38.3	8.2	34.4	28.7	27.5	19.2
L. Marlboro	T	6.0	5.0	17.8	43.1	69.4	20.3	33.4	35.9	--	33.5
	B	5.0	6.6	19.7	38.3	20.3	26.4	23.9	31.1	--	35.5
Nottingham	T	4.0	5.5	15.5	7.0	81.4	71.8	105.3	45.4	--	74.2
	B	4.0	6.0	10.2	6.6	103.0	105.3	35.9	52.6	--	75.3

APPENDIX (Cont'd)

Total chlorophyll a (mg m^{-3})											
						Date					
Station		6/28	7/5	7/12	7/19	7/26	8/2	8/9	8/16	8/23	8/30
Sandy Pt.	T	52.7	17.6	38.5	40.3	36.6	23.4	44.6	11.9	54.4	66.6
	B	3.5	8.7	5.7	43.9	22.0	31.5	13.0	64.4	13.2	8.0
Broome Is.	T	51.3	21.2	21.2	28.2	14.2	38.8	83.4	19.0	149.7	22.0
	B	2.5	4.9	8.9	11.9	11.0	7.8	11.4	6.0	5.6	6.6
Queen	T	29.5	25.6	30.8	43.2	14.7	19.8	64.4	23.4	18.7	39.9
Tree	B	2.5	11.7	7.0	11.4	19.0	13.2	23.4	5.2	9.1	1.1
Sheridan	T	11.0	14.4	20.5	36.6	16.8	24.2	71.7	34.4	36.1	46.4
Pt.	B	4.5	7.9	10.2	14.3	26.4	11.7	16.1	8.0	12.6	2.6
Benedict	T	4.0	11.6	20.5	38.5	44.6	17.6	58.6	23.4	35.2	41.2
Br.	B	3.5	10.4	14.6	48.5	60.0	17.6	51.2	32.9	19.5	19.3
Trueman	T	9.0	11.2	22.5	38.5	37.3	14.6	57.6	42.5	30.7	24.4
Pt.	B	6.0	12.8	20.7	36.6	46.8	13.5	45.8	36.6	35.1	24.9
L. Marl-	T	7.0	7.5	21.6	51.3	82.0	27.8	42.5	51.2	--	46.8
boro	B	6.0	9.5	24.9	46.9	27.1	39.5	32.0	41.0	--	46.8
Notting-	T	5.0	8.3	19.5	9.9	92.2	84.9	120.1	61.5	--	95.2
ham	B	5.0	8.2	13.7	9.5	115.6	118.6	45.4	70.3	--	96.6

Particulate Carbon (mg liter^{-1})											
						Date					
Station		6/28	7/5	7/12	7/19	7/26	8/2	8/9	8/16	8/23	8/30
Sandy Pt.	T	2.7	1.0	2.3	2.4	1.8	1.3	2.1	1.1	3.0	14.6
	B	1.0	0.7	0.7	3.6	2.0	1.6	1.0	4.0	1.7	0.7
Broome Is.	T	2.0	1.1	1.2	1.9	0.9	1.6	3.2	1.3	29.0	1.8
	B	1.0	0.7	0.8	1.0	0.9	2.5	0.6	0.8	2.1	0.7
Queen	T	1.8	1.3	0.8	1.9	0.7	1.0	2.6	1.1	1.5	3.2
Tree	B	0.5	0.8	1.8	0.9	1.0	1.5	1.4	1.1	1.3	0.6
Sheridan	T	1.1	0.9	1.5	1.1	0.9	0.9	4.3	3.1	2.5	7.2
Pt.	B	0.8	1.8	1.0	1.0	1.4	0.8	1.3	1.5	1.6	0.8
Benedict	T	1.3	0.9	1.5	1.7	1.8	0.5	2.9	2.3	3.1	7.6
Br.	B	1.2	1.9	1.4	1.9	2.1	1.8	2.2	2.5	2.3	2.7
Trueman	T	2.3	1.4	1.7	2.1	1.9	1.2	2.0	3.0	3.6	2.2
Pt.	B	2.2	4.6	2.0	2.0	1.8	1.9	1.3	3.8	3.3	2.2
L. Marl-	T	--	1.9	1.4	1.6	2.1	2.0	1.8	1.1	--	2.6
boro	B	2.0	6.6	1.6	2.0	1.2	4.2	1.6	2.0	--	2.6
Notting-	T	1.3	2.0	1.2	0.3	2.3	3.4	2.8	2.7	--	4.1
ham	B	1.6	1.9	1.6	1.0	2.7	6.4	4.8	0.9	--	4.0

APPENDIX (Cont'd)

Station		\multicolumn{10}{c}{Particulate Nitrogen (mg liter^{-1})}
		6/28	7/5	7/12	7/19	7/26	8/2	8/9	8/16	8/23	8/30
Sandy Pt.	T	0.567	0.157	0.436	0.497	0.281	0.265	0.285	0.209	0.441	0.823
	B	0.131	0.193	0.181	0.628	0.369	0.287	0.156	0.484	0.250	0.162
Broome Is.	T	0.521	0.251	0.275	0.487	0.135	0.328	0.486	0.200	1.584	0.316
	B	0.306	0.215	0.140	0.323	0.157	0.387	0.148	0.121	0.346	0.166
Queen Tree	T	0.484	0.215	0.328	0.437	0.157	0.202	0.368	0.284	0.231	0.356
	B	0.083	0.189	0.216	0.225	0.207	0.271	0.230	0.135	0.144	0.100
Sheridan Pt.	T	0.111	0.179	0.282	0.407	0.163	0.189	0.595	0.284	0.352	0.465
	B	0.354	0.428	0.299	0.183	0.224	0.047	0.203	0.138	0.202	0.164
Benedict Br.	T	0.196	0.193	0.501	0.470	0.327	0.174	0.474	0.210	0.342	0.458
	B	0.128	0.280	0.483	0.467	0.409	0.448	0.342	0.378	0.258	0.295
Trueman Pt.	T	0.345	0.362	0.520	0.329	0.326	0.285	0.373	0.382	0.170	0.511
	B	0.367	0.778	0.375	0.363	0.327	0.278	0.408	0.535	0.385	0.317
L. Marlboro	T	0.422	0.448	0.620	0.608	0.638	0.189	0.319	0.296	--	0.370
	B	0.455	0.642	0.378	0.454	0.090	0.546	0.195	0.445	--	0.396
Nottingham	T	0.259	0.257	0.387	0.098	0.503	0.502	0.266	0.394	--	1.504
	B	0.254	--	0.323	0.179	0.593	0.698	0.534	0.472	--	0.555

	\multicolumn{10}{c}{Dissolved inorganic phosphate-phosphorus (DIP), total dissolved phosphorus (TDP) and total phosphorus (TP) at the surface (µg at liter^{-1})}
	6/28	7/5	7/12	7/19	7/26	8/2	8/9	8/16	8/23	8/30
\multicolumn{11}{c}{Sandy Pt.}
DIP	0.56	0.66	0.56	0.41	--	0.56	0.41	0.76	0.20	0.71
TDP	--	0.26	0.58	0.34	--	0.57	0.55	0.82	0.58	1.25
TP	3.79	--	--	--	--	1.64	2.51	1.54	2.91	4.04
\multicolumn{11}{c}{Queen Tree}
DIP	0.81	1.83	0.36	0.41	--	0.25	0.56	0.61	0.40	0.60
TDP	0.63	0.35	0.46	0.63	--	0.58	0.40	0.59	0.87	1.22
TP	2.15	--	--	--	--	1.88	2.03	2.39	1.69	3.20
\multicolumn{11}{c}{Benedict Br.}
DIP	1.42	1.83	1.98	1.68	--	1.02	0.76	0.81	1.21	1.01
TDP	--	1.70	1.68	1.11	--	1.44	0.88	0.93	1.84	1.53
TP	4.31	--	--	--	--	3.44	4.22	2.49	4.21	4.41
\multicolumn{11}{c}{Nottingham}
DIP	1.88	2.19	2.34	2.85	--	1.17	3.97	0.92	--	0.71
TDP	2.87	2.38	2.62	2.99	--	1.45	--	1.09	--	1.21
TP	5.68	--	--	--	--	6.55	4.76	4.36	--	6.09

APPENDIX (Cont'd)

	Ammonia (NH$_3$), nitrite (NO$_2$), nitrate (NO$_3$), dissolved organic nitrogen (DON) and total dissolved nitrogen (TDN) at the surface (µg at liter^{-1})									
	\multicolumn{10}{c}{Date}
	6/28	7/5	7/12	7/19	7/26	8/2	8/9	8/16	8/23	8/30
	\multicolumn{10}{c}{Sandy Pt.}
NH$_3$	48.1	29.6	16.7	16.0	--	9.8	7.7	3.1	2.7	21.5
NO$_2$	1.1	0.9	1.4	1.3	--	1.8	1.0	0.7	0.2	0.4
NO$_3$	51.1	32.1	23.4	18.1	--	22.2	11.0	7.6	4.6	2.6
DON	5.0	13.4	20.3	12.6	--	25.8	25.6	22.4	12.1	6.3
TDN	105.3	76.0	61.8	48.0	--	59.6	45.2	33.8	19.6	30.8
	\multicolumn{10}{c}{Queen Tree}
NH$_3$	7.7	3.0	6.6	4.2	--	5.5	0.4	0.7	0.6	--
NO$_2$	0.7	1.0	0.7	0.1	--	0.8	0.0	0.0	0.1	0.0
NO$_3$	18.2	20.7	9.2	1.8	--	7.3	0.5	0.5	0.2	0.1
DON	15.6	11.9	23.7	29.2	--	25.4	11.0	19.6	22.2	--
TDN	42.2	36.6	40.2	35.4	--	39.0	11.9	20.8	23.1	22.8
	\multicolumn{10}{c}{Benedict Br.}
NH$_3$	--	12.5	2.9	1.5	--	10.8	0.4	1.1	0.0	--
NO$_2$	0.8	1.5	1.0	0.4	--	0.6	0.0	0.0	0.0	0.1
NO$_3$	23.1	30.3	29.4	12.8	--	10.9	0.1	0.2	0.0	0.0
DON	--	11.6	27.4	11.5	--	28.1	26.4	14.9	20.6	--
TDN	--	55.9	60.7	26.2	--	50.4	26.9	16.2	20.6	11.1
	\multicolumn{10}{c}{Nottingham}
NH$_3$	16.0	13.9	6.9	9.3	--	3.1	--	0.6	--	--
NO$_2$	1.4	2.0	2.3	2.3	--	2.1	1.9	1.0	--	0.9
NO$_3$	41.7	45.0	49.7	50.8	--	27.6	19.2	14.4	--	7.2
DON	13.9	23.8	22.3	29.0	--	31.3	--	22.6	--	--
TDN	73.0	84.9	81.2	91.4	--	64.1	--	38.6	--	33.1

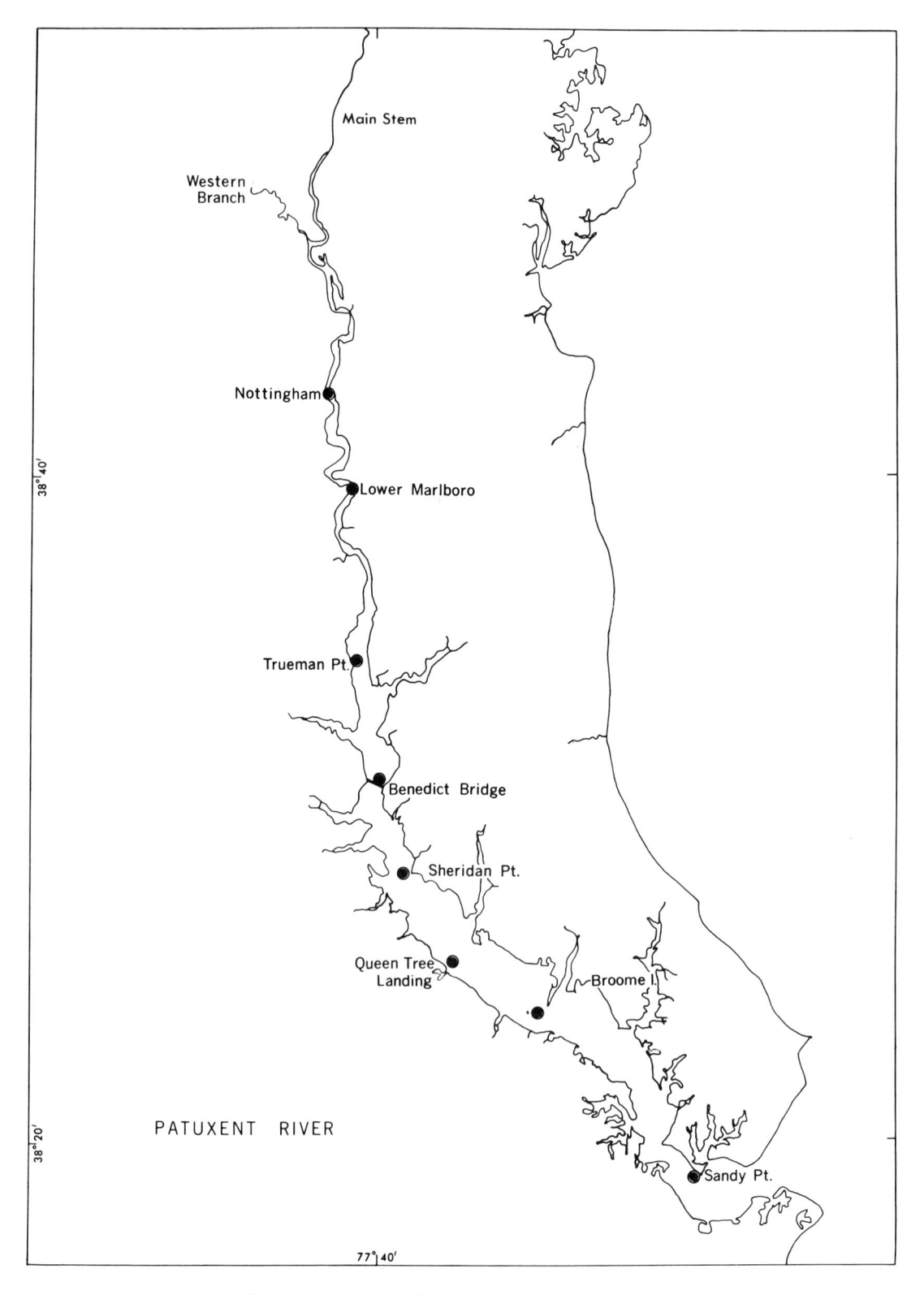

Figure 1. Map of the Patuxent River estuary showing station locations.

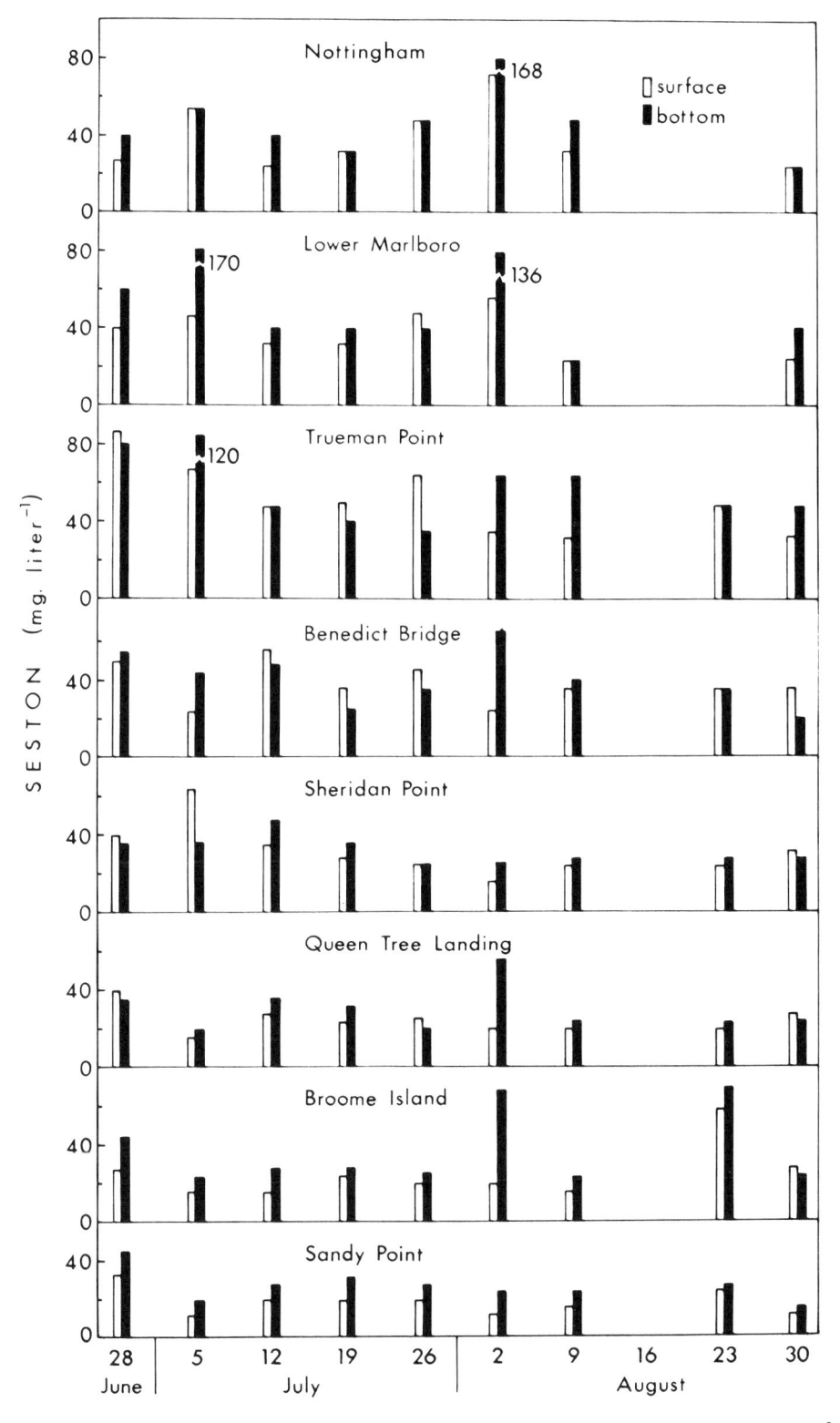

Figure 2. Surface and bottom seston concentrations (mg liter^{-1}) observed in the Patuxent River estuary during the Post-Agnes study, Summer 1972.

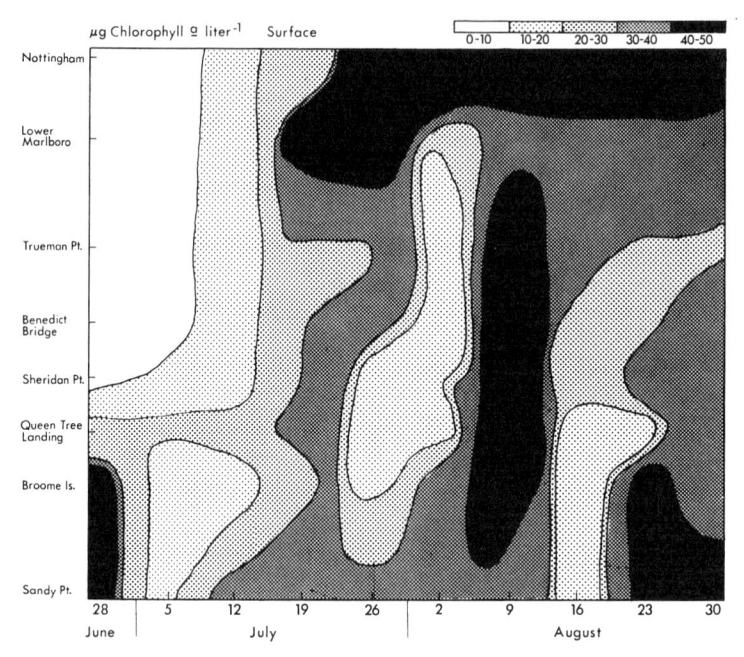

Figure 3a. Surface concentration of active chlorophyll a (mg m^{-3}) observed in the Patuxent River estuary during the Post-Agnes study, Summer 1972.

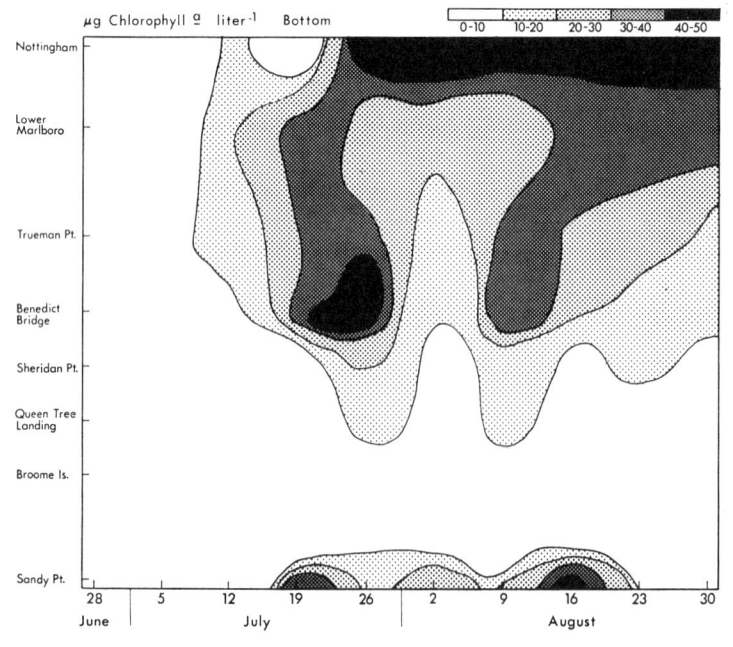

Figure 3b. Bottom concentration of active chlorophyll a (mg m^{-3}) observed in the Patuxent River estuary during the Post-Agnes study, Summer 1972.

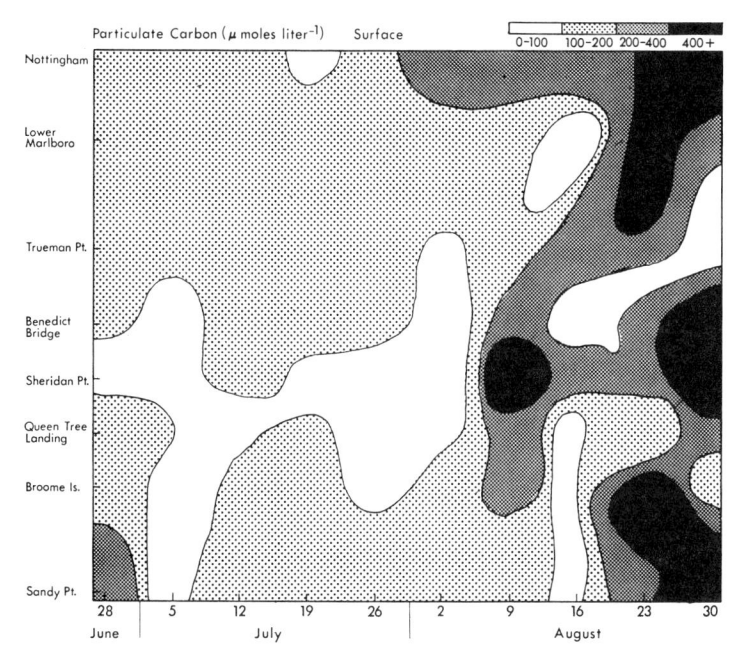

Figure 4a. Surface concentration of particulate carbon
(μ moles liter^{-1}) observed in the Patuxent River
estuary during the Post-Agnes study, Summer 1972.

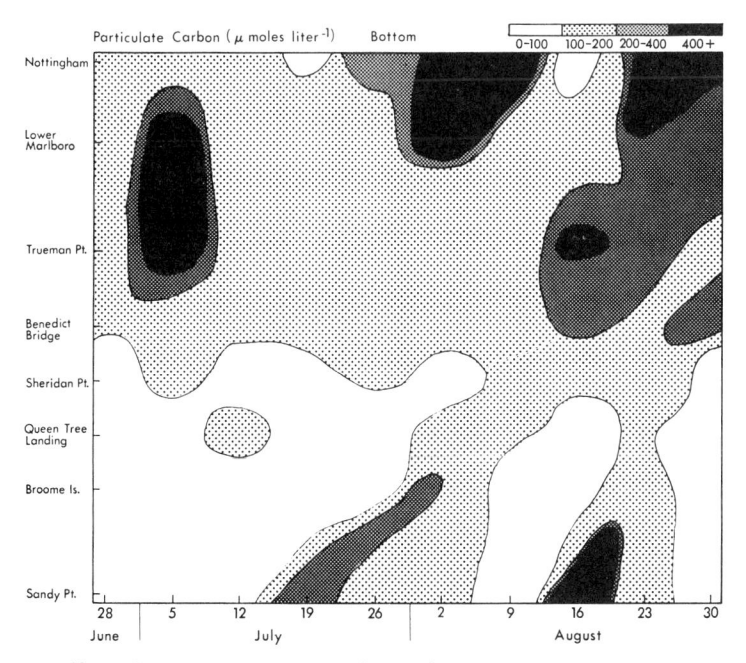

Figure 4b. Bottom concentration of particulate carbon
(μ moles liter^{-1}) observed in the Patuxent River
estuary during the Post-Agnes study, Summer 1972.

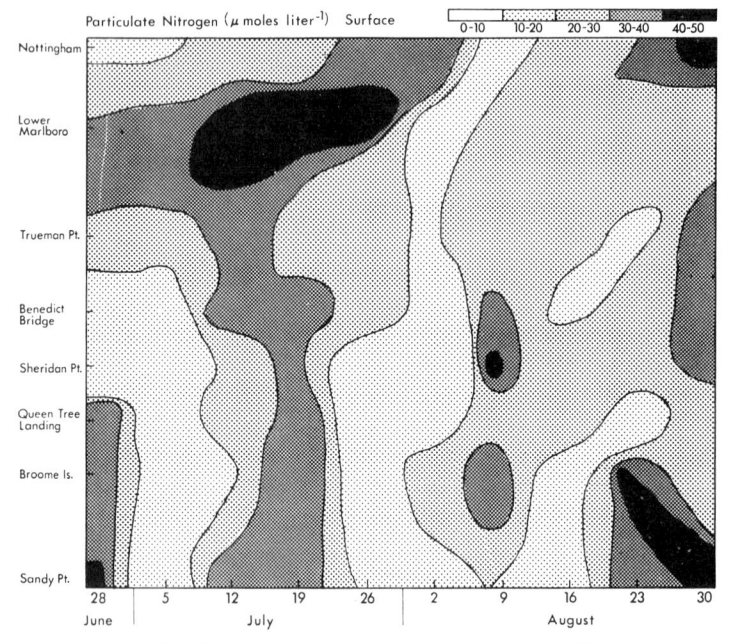

Figure 5a. Surface concentration of particulate nitrogen
(μ moles liter^{-1}) observed in the Patuxent River
estuary during the Post-Agnes study, Summer 1972.

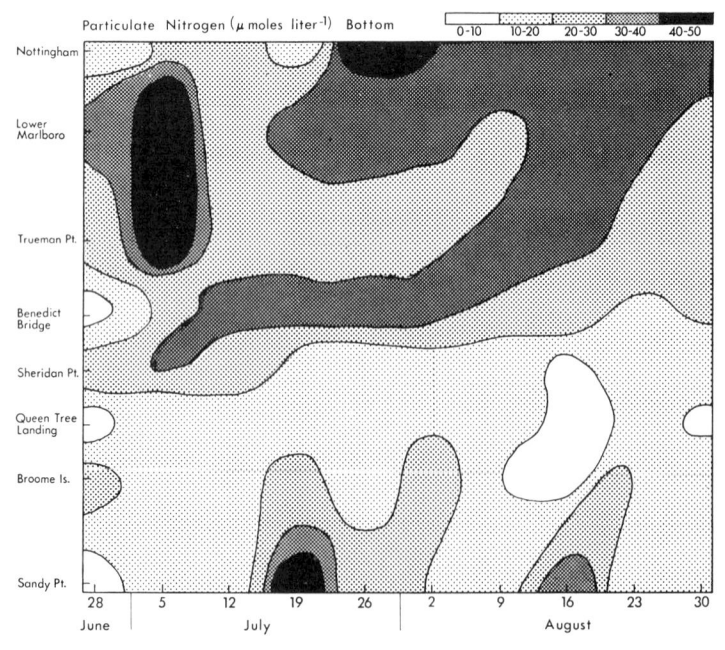

Figure 5b. Bottom concentration of particulate nitrogen
(μ moles liter^{-1}) observed in the Patuxent River
estuary during the Post-Agnes study, Summer 1972.

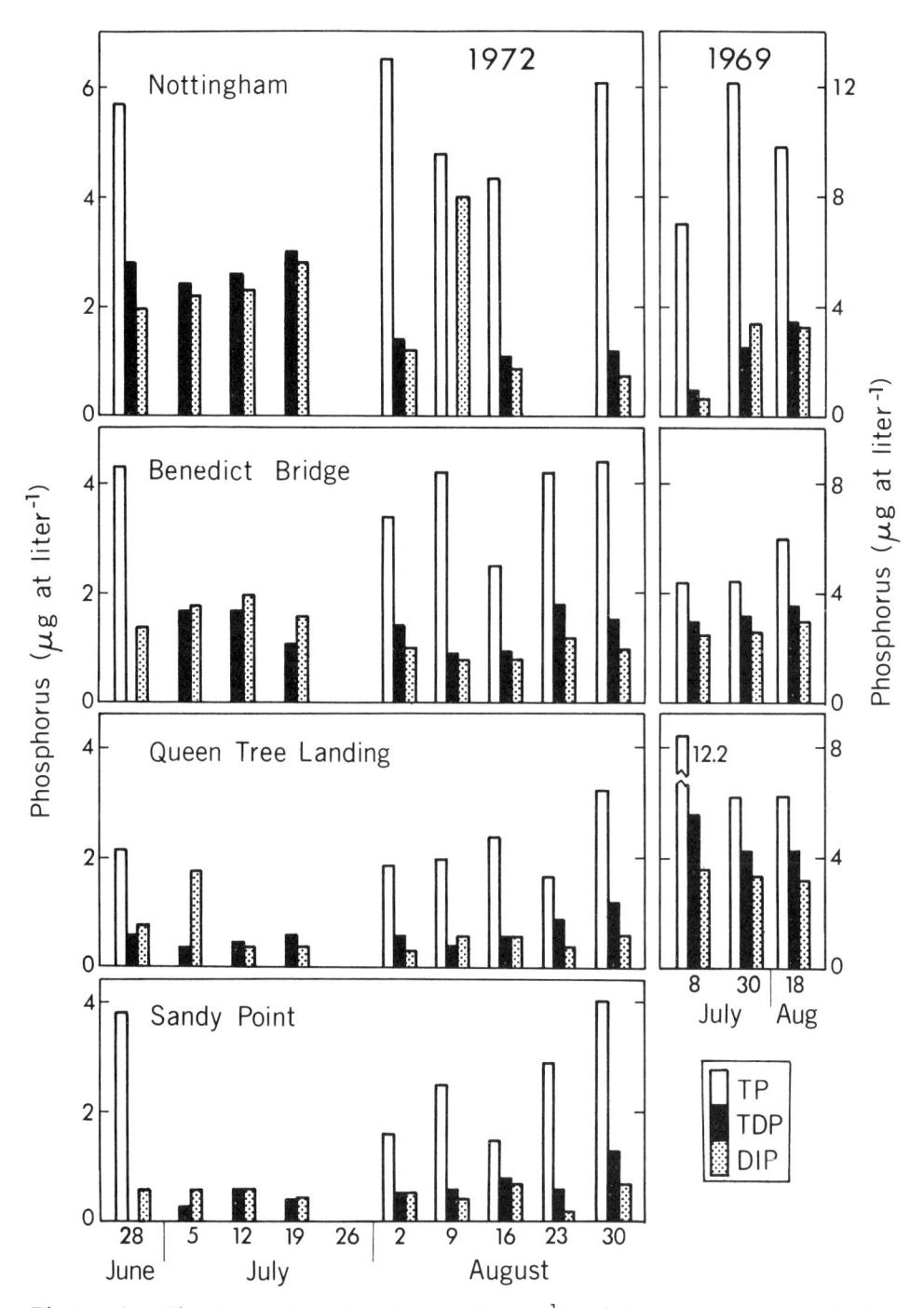

Figure 6. The concentration (µg at liter^{-1}) of total phosphorus (TP), total dissolved phosphorus (TDP), and dissolved inorganic phosphorus (DIP) at selected stations, Patuxent River Post-Agnes study, Summer 1972. Shown on the figures are comparative data for July and August 1969 (from Flemer et al. 1970).

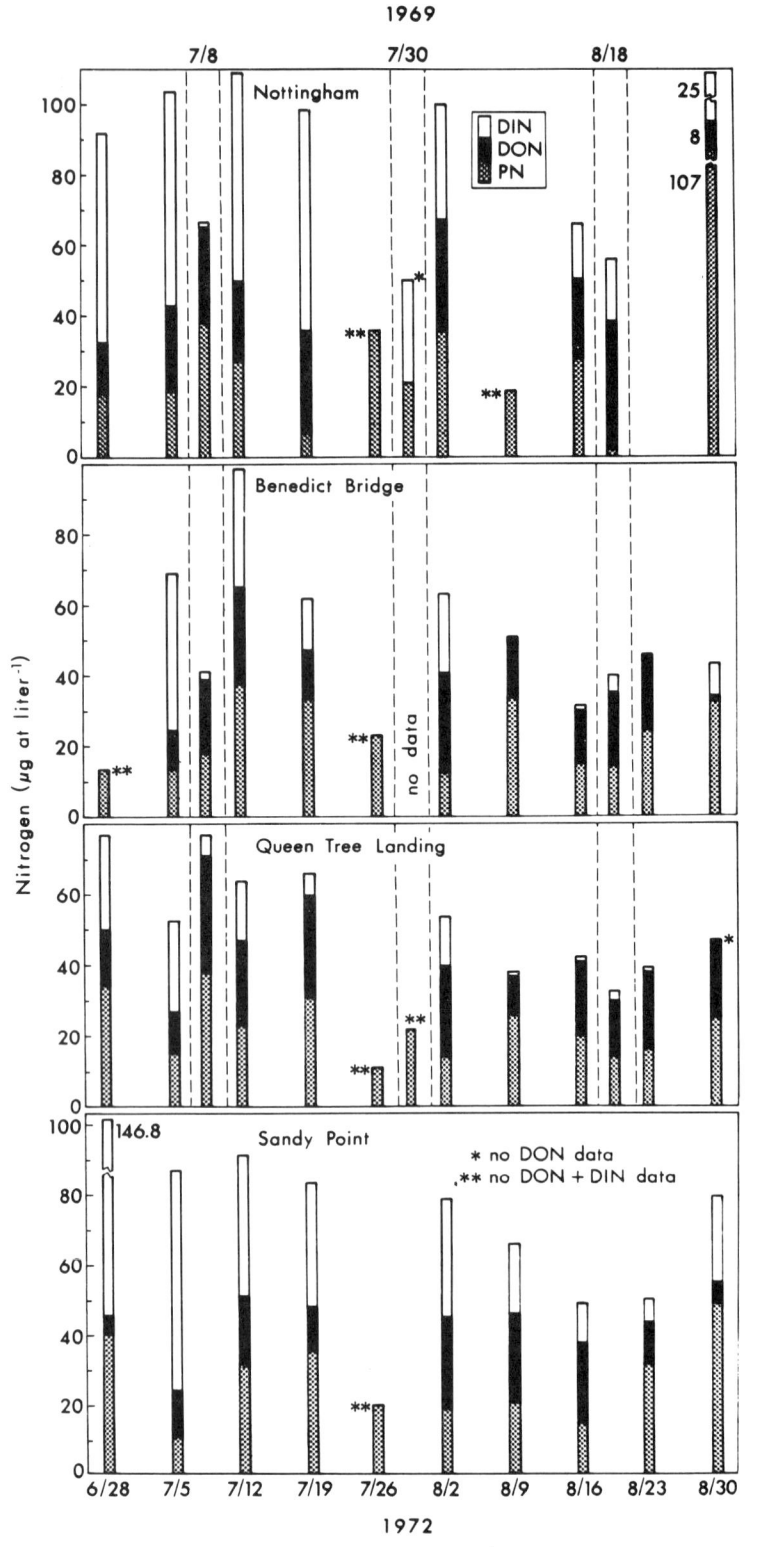

Figure 7. The concentration (µg at liter^{-1}) of particulate nitrogen (PN), dissolved inorganic nitrogen (DIN), and dissolved organic nitrogen (DON) at selected stations, Patuxent River Post-Agnes study, Summer 1972. Shown in the figures are comparative data for July and August 1969 (from Flemer et al. 1970).

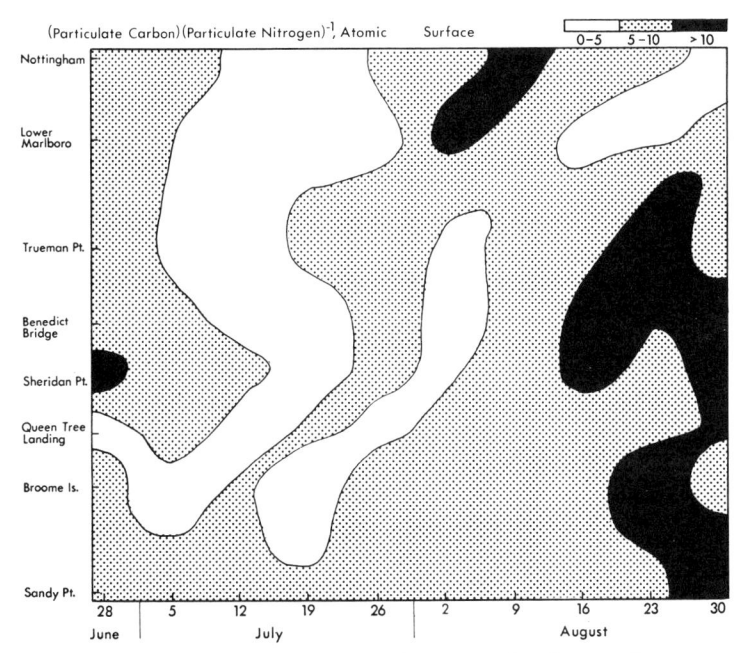

Figure 8a. The particulate carbon:particulate nitrogen ratio (atomic) for the surface waters, Patuxent River Post-Agnes study, Summer 1972.

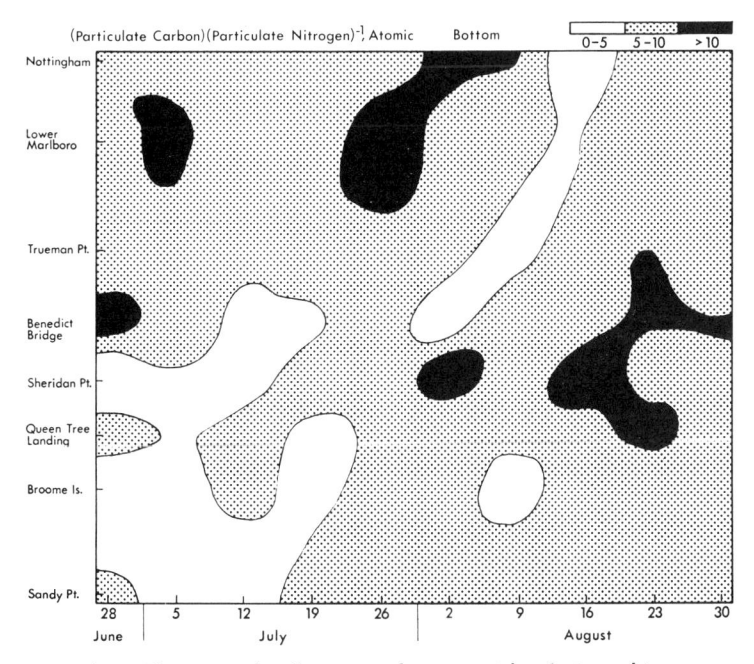

Figure 8b. The particulate carbon:particulate nitrogen ratio (atomic) for the bottom waters, Patuxent River Post-Agnes study, Summer 1972.

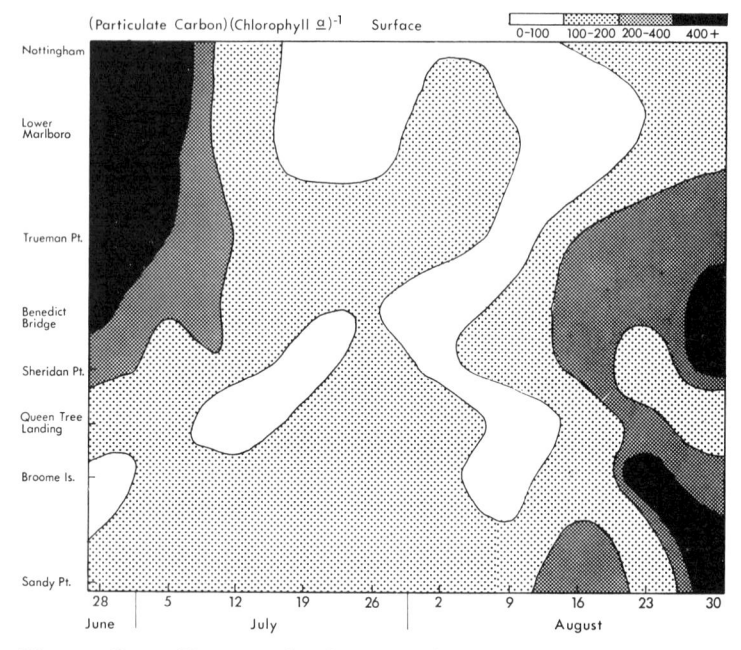

Figure 9a. The particulate carbon: chlorophyll a ratio for the surface waters, Patuxent River Post-Agnes study, Summer 1972.

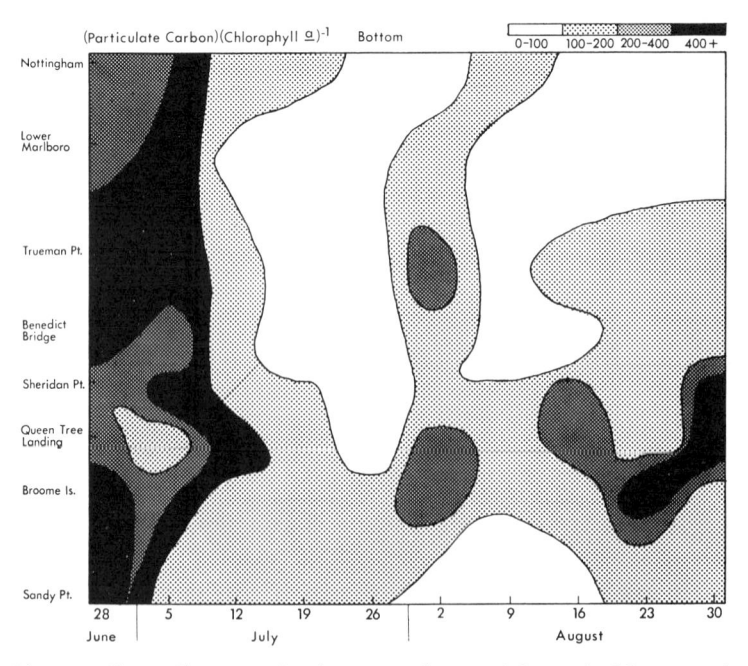

Figure 9b. The particulate carbon: chlorophyll a ratio for the bottom waters, Patuxent River Post-Agnes study, Summer 1972.

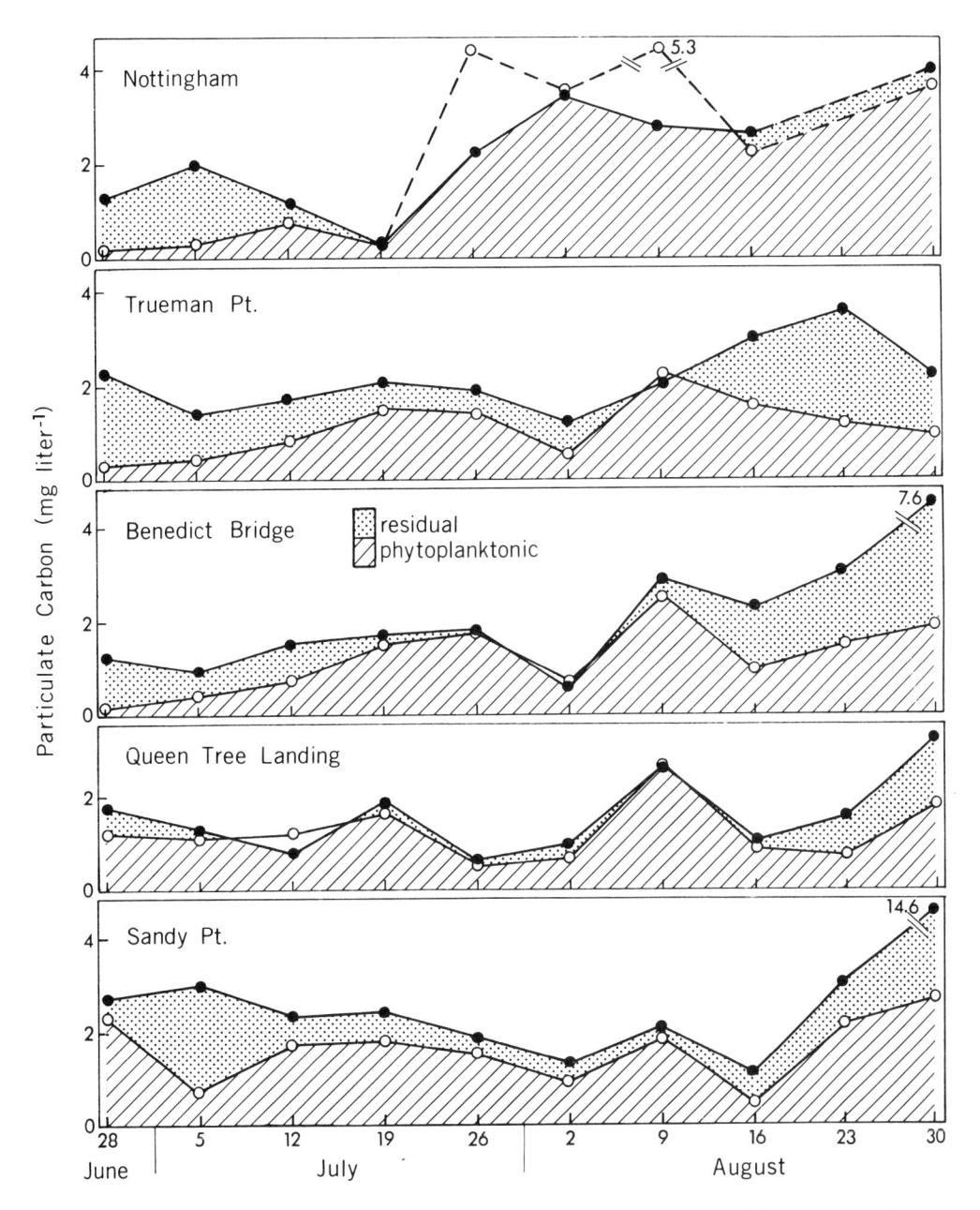

Figure 10. The distribution of particulate carbon (PC)(top line) and the estimated living phytoplanktonic carbon (bottom line) for selected stations, Patuxent River Post-Agnes study, Summer 1972. The residual PC is indicated.

INDIRECT EFFECTS OF TROPICAL STORM AGNES UPON THE RHODE RIVER

David L. Correll[1]

ABSTRACT

The storm's effects upon much of the Chesapeake Bay were indirect, as illustrated by a case study of the Rhode River. Only 9.7 cm of rain fell upon the Rhode River watershed at the time (20 to 24 June 1972). However, the salinity of the Rhode River reached a minimum about 12 July as a result of the Susquehanna River flood. This was coincident with the year's highest water temperatures (30-31°C) and resulted in severe mortalities in the biota. Periphyton (attached microbial communities) experienced a nearly complete die-off. High levels of sediments and of nutrients, especially nitrate and total phosphorus, were delivered to Rhode River via the bay proper and from local runoff. These nutrients were deposited in Rhode River bottom sediments. This reservoir released nutrients during 1973, especially in May at a time of low dissolved oxygen in the bottom water and of intensive dinoflagellate blooms. In a 13-day period it is estimated that over 900 Kg phosphorus was released from the bottom sediments.

INTRODUCTION

The Rhode River and its watershed were only on the fringes of the rain pattern associated with Agnes. However, 25.5 cm of rainfall was recorded on the Rhode River from 18 June to 16 July 1972. This was above average for that time of year (9.9 cm was recorded for this period in 1973) and must be taken into consideration in an analysis of events at Rhode River. It is estimated, for example, that during the months of June and July local watershed runoff carried 870 Kg P into the Rhode River.

The following investigation was undertaken to determine if the effects of the storm on such locations as the Rhode River were due to long distance effects caused by the flooding of large rivers such as the Susquehanna River and mediated via the open bay (low salinity, high nitrate, turbidity levels) combined with local events (above average runoff with high phosphorus, turbidity levels, high water temperatures).

METHODS

Water Chemistry

Samples were analyzed for total phosphorus by potassium persulfate digestion followed by reaction with ammonium molybdate and reduction with stannous chloride (American Public Health Association 1971). Nitrate was reduced to nitrite on a column of mercury amalgamated cadmium filings (Strickland & Parsons 1965). Nitrite concentrations were determined colorimetrically after coupling to sulfanilamide (American Public Health Association 1971). In the subsequent text nitrate plus nitrite will be referred to as nitrate. Turbidity was measured in the

---

[1] Radiation Biology Laboratory, Smithsonian Institution, 12331 Parklawn Drive, Rockville, Md. 20852

field to avoid rapid changes when samples are bottled. A Hach, Model 2100A, turbidimeter calibrated in Jackson units was used. Sodium chloride concentrations were determined with an Orion, Model 407, specific ion meter using Corning reference and sodium ion electrodes. Molarity values for sodium chloride may be converted to parts per thousand salinity by multiplication by 64.15. Samples for dissolved oxygen determination were fixed in the field and titrated in the laboratory using the azide modification of the Winkler method (American Public Health Association 1971).

Periphyton

Clear Plexiglas plates (artifical substrates) were used essentially as described by Grzenda and Brehmer (1960). The plates were 6 cm x 12 cm x 0.3 cm and were mounted on wooden bars with the 12 cm edge directed upward. The bars were positioned so that the plates were just above the surface of the bottom sediments at stations 5 and 8 (Fig. 1). Both locations were shallow water tidal mud flats. Normally, a set of 70 plates was placed at each station and three were removed from each station at each time interval for dry weight and ash-free dry weight determinations. Samples were scraped from the surface with a spatula and rinsed into a tared gooch crucible. The crucible was brought to constant weight at $60°$ for dry weight and was then ashed in an electric furnace at $600°$ for two hours, cooled in a desiccator and brought to constant weight again at $60°$ for ash-weight.

RESULTS AND DISCUSSION

The effects of Agnes on water quality in Rhode River are shown in two types of graphs. The changes in five parameters at station 12 during the 1972 calendar year are given in Figs. 2 and 3. Station 12 was selected to be representative of the main basin of Rhode River. The gradients of three of these parameters along the channel axis of Rhode River on five days are shown in Fig. 4. These days were chosen to represent periods prior to the storm (Day 166; 14 June), at the peak of the recovery from the first effects of the storm (Day 188; 6 July), at a time when surface water salinities were at a second minimum (Day 201; 19 July), and at two times of increasing encroachment of more saline bottom water (Day 222; 9 August and Day 234; 21 August). Salinity at stations 10 and 13 for the critical period of June and July is compared with local rainfall and watershed runoff in Fig. 5. It is clear from Fig. 2 that a peak in phosphorus and nitrate concentrations and turbidity occurred about 27 June (Day 179). This peak was not as large for nitrate as the annual spring peak, but it was unusual for that time of year. In the case of phosphorus and turbidity it was unusual for any time of year. Turbidity and total phosphorus normally increase up the axis of the Rhode River but on 27 June they were highest at each end of the axis in addition to being present in unusually high concentrations. Surface water temperatures (Fig. 3) reached a maximum of $30°C$ on 19 July, the same time period as sodium chloride values had reached their minimum (Figs. 3 & 5). In general sodium chloride concentrations decreased from 21 June to a first minimum on 27 June, recovered somewhat by 6 July, then declined to a second minimum by 12 July, then slowly recovered, due to encroachment of more saline bottom waters from the bay and vertical mixing (Fig. 4, 9 August and 21 August). It is apparent from Fig. 5 that local runoff could have been influential in the establishment of a first surface water salt concentration minimum. Local runoff from 21 June to 24 June amounted to $8.8 \times 10^5 m^3$. This is about equal to the volume of the Rhode River upstream from the island near station 10 in Fig. 1. However, it was not sufficient to bring about the salinity effects observed downstream from this island in the Rhode River (about 90% of the volume of the Rhode River is below this point). Changes in the bay proper, brought about by Agnes, must be the cause of these changes. The axial gradient plots in Fig. 4 show a normal pattern (14 June) for sodium chloride, nitrate and total phosphorus. Nitrate in the spring is normally bimodal and on 14 June still showed this pattern. Phosphorus concentrations almost always increase up the axis and most phosphorus in the Rhode River originates from the local watersheds. By 27 June surface sodium chloride concentrations were very low; nitrate and total phosphorus were much higher in general and were both bimodal. By 6 July sodium chloride concentrations had increased in surface waters; total phosphorus was no longer bimodal, but was higher than usual in the deeper waters. Nitrate was still bimodal but the minimum had increased from 3 µg N/l on 27 June to 92 µg N/l on 6 July. By 19 July sodium chloride had reached essentially zero concentrations (less than 0.003 molar throughout the axis), nitrate values were very low except in the upper part of the axis, and phosphorus values were coming down. By 9 August sodium chloride concentrations were increasing some in surface waters and a lot in bottom waters, nitrate was still low in surface waters but was higher in bottom water and phosphorus was about the same as on 19 July. Bottom waters were not higher in total phosphorus. By 21 August surface water sodium chloride concentrations had reached those of the bottom waters. Nitrate was low in both surface and bottom water and phosphorus was still above normal and had even increased somewhat. Much of the phosphorus and nitrate brought into the Rhode River during this time was probably deposited in the bottom sediments. This is especially true of phosphorus in the area below Rhode River axial Km 4 which normally is not exposed to such high levels.

Growth curves (biomass) for periphyton on artificial substrates are shown in Fig. 6. The slopes of these growth curves are plotted in Fig. 7. It is apparent that between 23 June and 6 July a severe mortality occurred in the periphyton of both stations, particularly at station 8. Biomass growth rates were going into an annual late spring peak at station 8 (over 650 mg ash-free dry wt/m^2-day in the period ending on 23 June). By 6 July the population at station 8 had declined to 24% of the value on 23 June. At station 5 the population declined by 6 July to 57% of the value on 23 June. In Fig. 8 the sodium chloride concentrations at stations 5 and 8 are shown. Between 23 June (Day 175) and 6 July (Day 188) sodium chloride concentrations dropped 2.5 orders of magnitude at station 8 and 1.7 orders of magnitude at station 5. This obvious physiological stress was presumably at least an important cause of this die-off. In over two years of measuring this parameter no other significant, rapid die-offs have been observed. These effects of Agnes upon the Rhode River were short term, only affecting the system for one season. However, they certainly reduced the primary production of the periphyton component of the system drastically for a considerable part of one season.

One of the longer term effects of the storm upon the bay as a whole and Rhode River specifically has been a widely observed increase in surface water phosphorus levels in 1973. Mean phosphorus concentrations were tabulated for Rhode River surface water at station 11 between 8 August (Day 220) and 27 October (Day 300) for 1971, 1972, and 1973. The corresponding means for total phosphorus were 90, 84, and 132 µg P/l. For dissolved orthophosphate these means were 12, 25, and 29 µg P/l, respectively. These data indicate that the surface waters in the main basin of Rhode River have indeed maintained higher phosphorus levels since Agnes. One possible mechanism for this fact is illustrated in Fig. 9. If total phosphorus concentrations are plotted against salt concentrations, any uptake or release of phosphorus into the water mass within the Rhode River should appear as a bowing of the axial profiles down or up respectively. A series of profiles in time allow the conclusion to be reached, in this case, that the profiles on Days 108 and 123 (1973) were close to those expected for no net uptake or release within the system. Only dilution of freshwater runoff, which was higher in total phosphorus, with bay water was evident. On Day 131 some significant deviation was present and on Day 136 a large bulge was evident. This was again replaced with an almost normal profile by Day 149. The profile of Day 136 has consecutive axial data points connected as they should be by virtue of their geographical order. It is evident that the salt concentrations near the mouth of the Rhode River are lower than in the middle of the river axis. At this time fresher bay water was beginning to exchange with the Rhode River and the subsequent turnover flushed this large mass of high phosphorus water into the bay. This release was calculated to be 916 Kg of phosphorus on a volume-average basis for both surface and bottom water phosphorus concentrations in the period from Day 123 to Day 136. It corresponded to the time of an intensive bloom of the dinoflagellate *Prorocentrum*. Its release was associated with the season's first observed low values for dissolved oxygen in Rhode River bottom waters. Dissolved oxygen values measured at about 11 a.m. on Day 131 varied from 3.9 to 5.8 parts per million for stations 10 to 12. Values of 5 to 6.5 parts per million were observed at the same time and placed on Day 136. It can only be suggested that the dissolved oxygen values at dawn and on other days would have been lower and may have reached or nearly reached zero, allowing a major pulse of phosphorus release from the bottom sediments to occur.

ACKNOWLEDGEMENTS

Research was supported in part by a grant from the Program for Research Applied to National Needs of the National Science Foundation to the Chesapeake Research Consortium and by the Smithsonian Institution's Environmental Sciences Program. Preparation of this report was supported by a grant from the U. S. Army Corps of Engineers to the Chesapeake Research Consortium. Published with the approval of the Secretary of the Smithsonian Institution.

Rainfall data were provided by Mr. Daniel Higman, Chesapeake Bay Center for Environmental Studies. Streamflow data were provided by Dr. Edward Pluhowski, U. S. Geological Survey. Rhode River hydrodynamic data were provided by Mr. Gregory Han, Chesapeake Bay Institute, Johns Hopkins University.

LITERATURE CITED

American Public Health Association. 1971. Standard methods for the examination of water and waste water. 13th ed., APAA, New York.

Grzenda, A. R. and M. L. Brehmer. 1960. A quantitative method for the collection and measurement of stream periphyton. Limnol. Oceanogr. 15:190-194.

Strickland, J. D. H. and T. R. Parsons. 1965. A manual of sea water analysis. Bull. Fish. Res. Bd. Canada 125, 2nd ed.

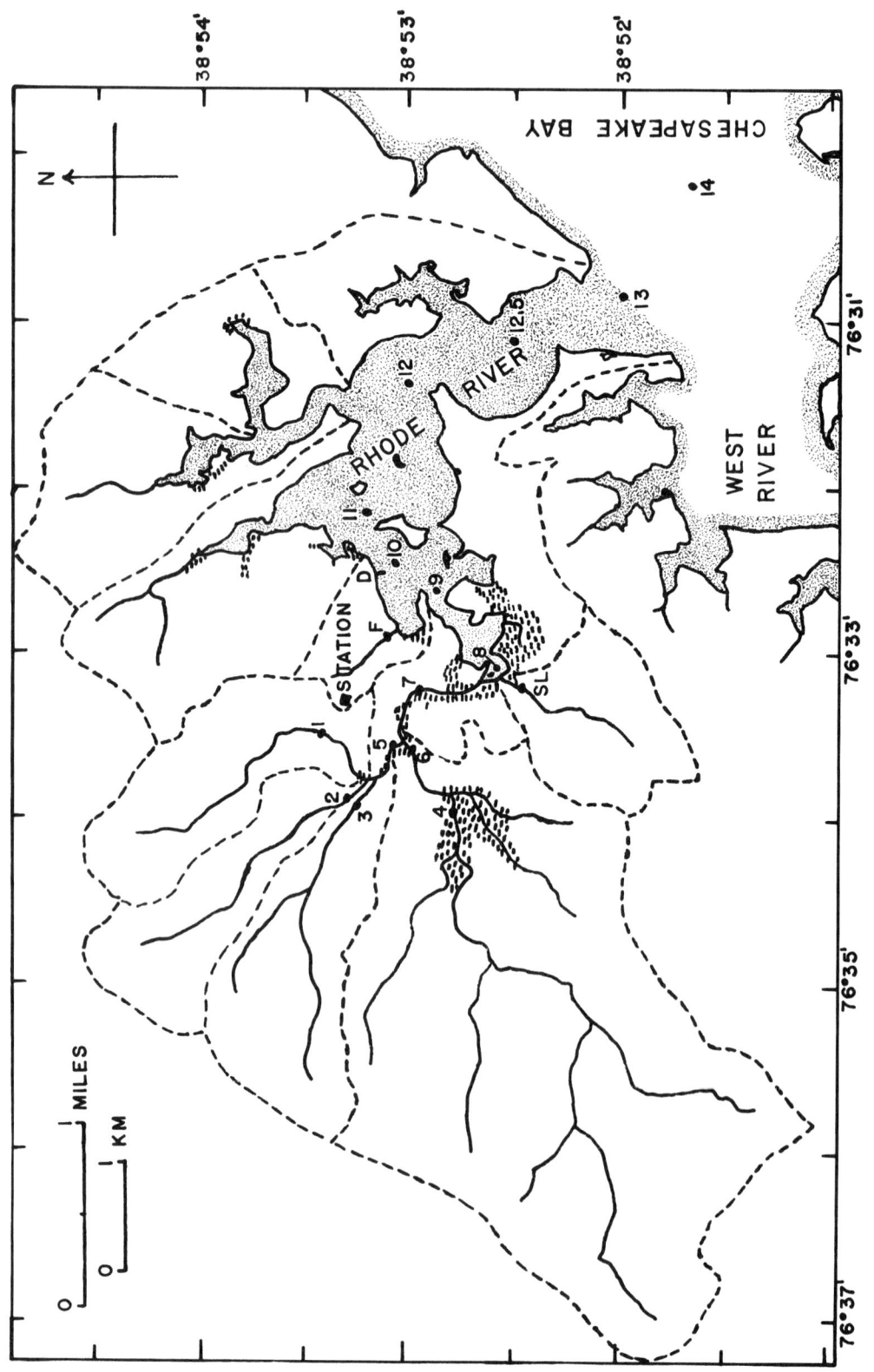

Figure 1. Map of the Rhode River and its watershed. Tributary streams and their individual watersheds are indicated. Sampling stations for water chemistry are designated with numbers. Periphyton experiments were done at stations 5 and 8. The channel of Muddy Creek is along stations 4, 6, 7, 8, and 9.

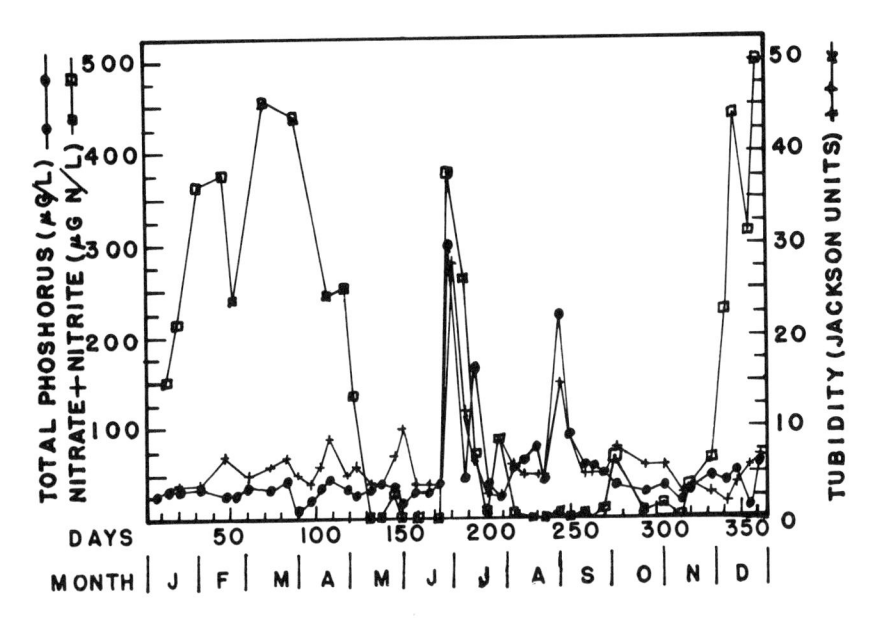

Figure 2. Levels of total phosphorus, nitrate plus nitrite and turbidity in the surface waters of Rhode River, station 12 for 1972.

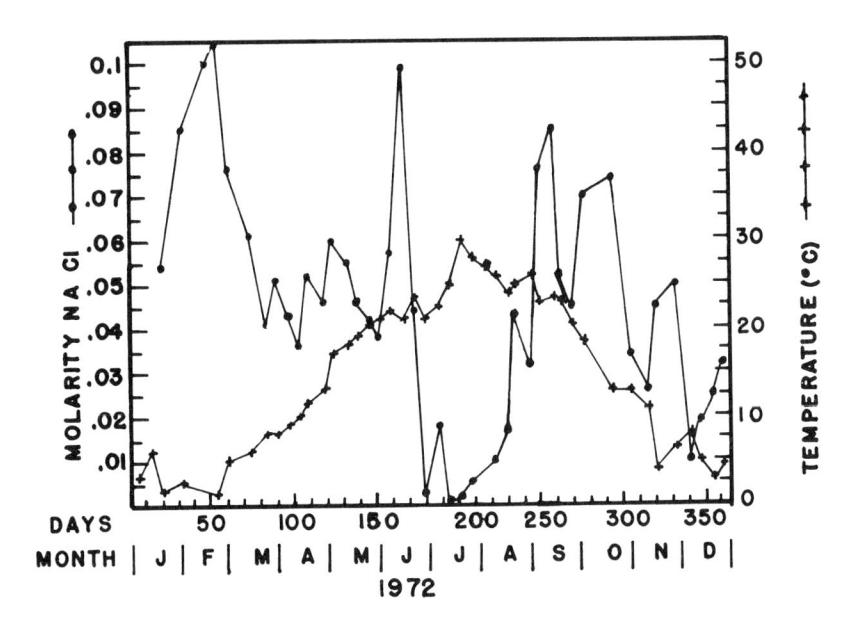

Figure 3. Water temperatures and sodium chloride concentrations in the surface waters of Rhode River, station 12 for 1972.

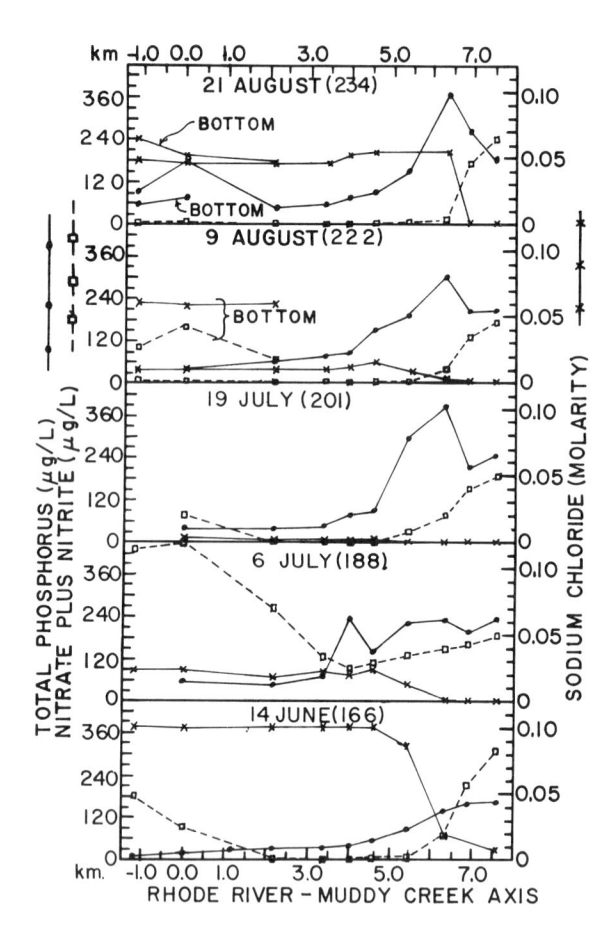

Figure 4. Axial gradients of sodium chloride, nitrate plus nitrite, and total phosphorus for the Rhode River on five dates of 1972. The station sequence (Fig. 1) is 14 (-1.0 Km), 13, 12, 11, 10, 9, 8, 7, 6, 4 (7.5 Km).

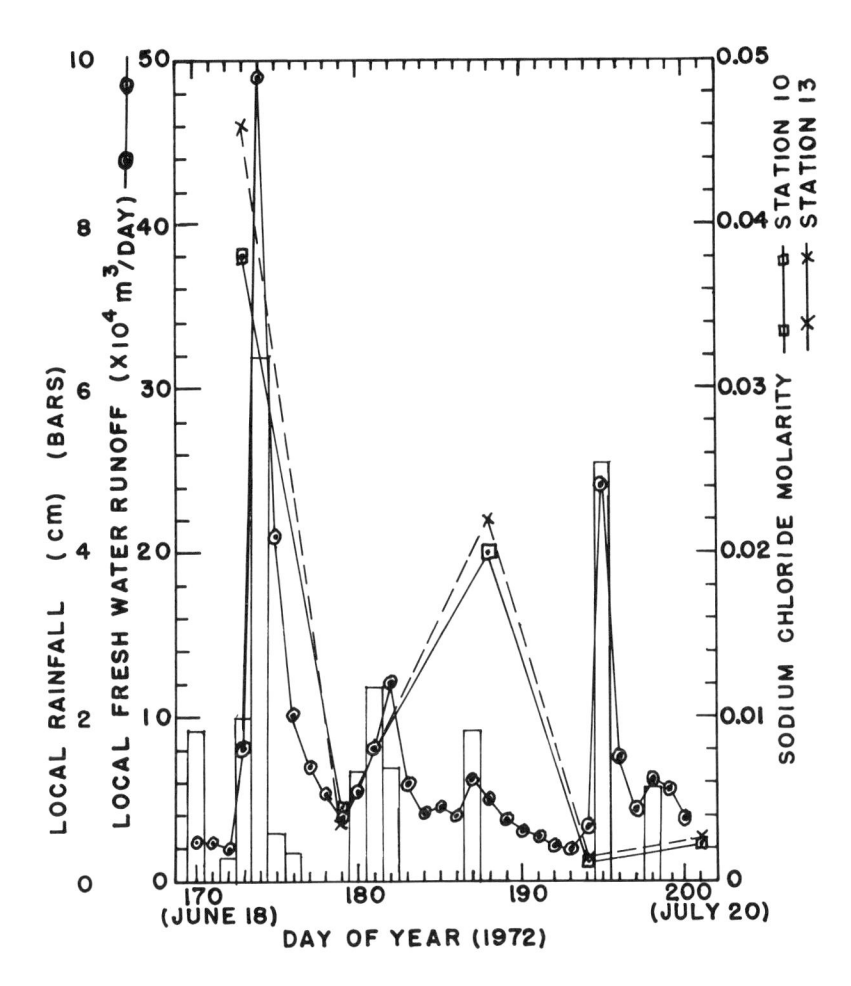

Figure 5. Changes in surface water sodium chloride concentration at stations 10 and 13 of Rhode River compared with local watershed rainfall and runoff for the period from 18 June to 20 July, 1972.

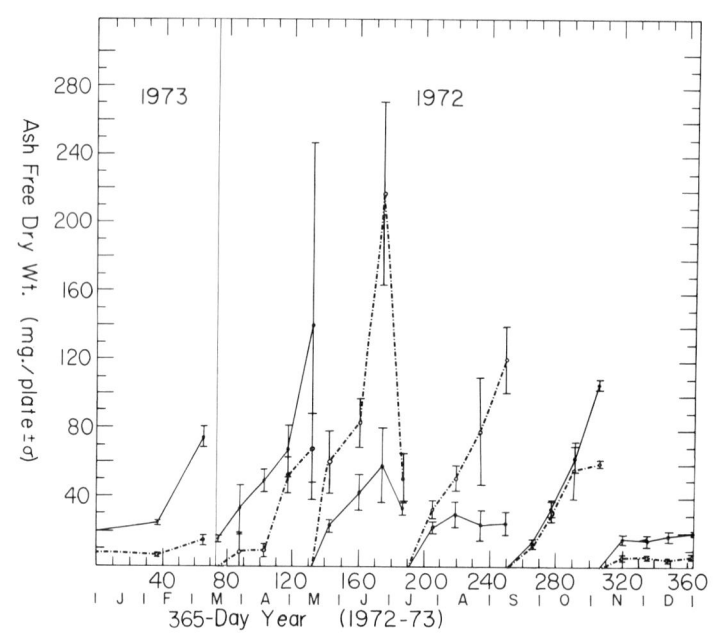

Figure 6. Growth curves for periphyton on artificial substrates in Rhode River. Each point is a mean and is shown with a bar to represent one standard deviation about the mean. Broken lines are for data from station 8, solid lines are for data from station 5.

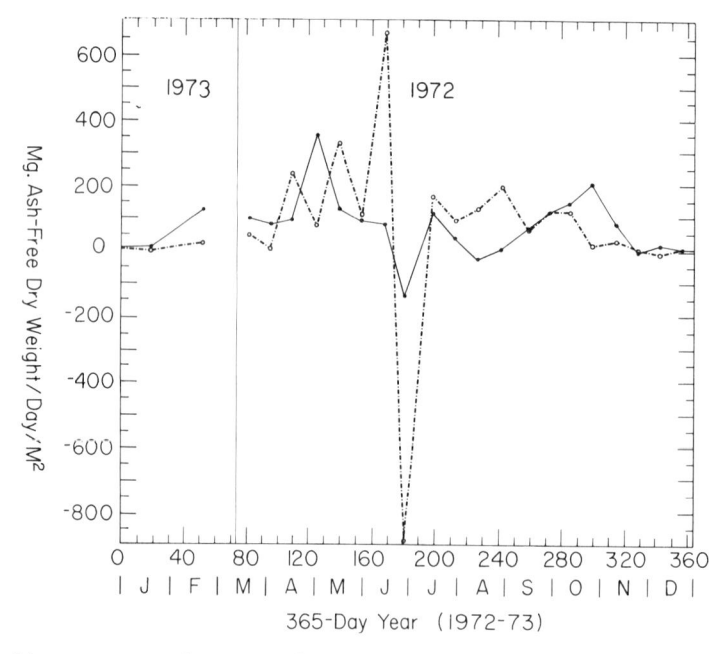

Figure 7. Biomass growth rates for periphyton on artificial substrates per total exposed surface area of substrate in Rhode River. Broken line is for data from station 8, solid line is for data from station 5.

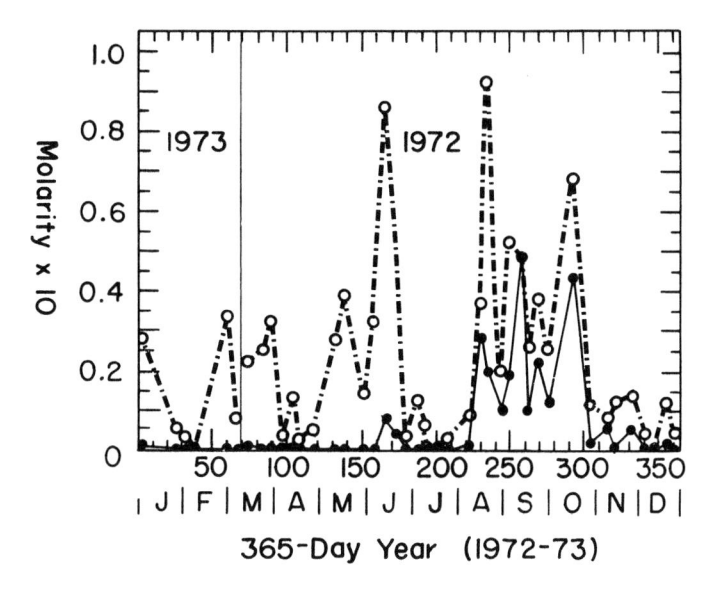

Figure 8. Sodium chloride concentrations in the Rhode River. Broken line for data from station 8, solid line is for data from station 5.

298   Correll

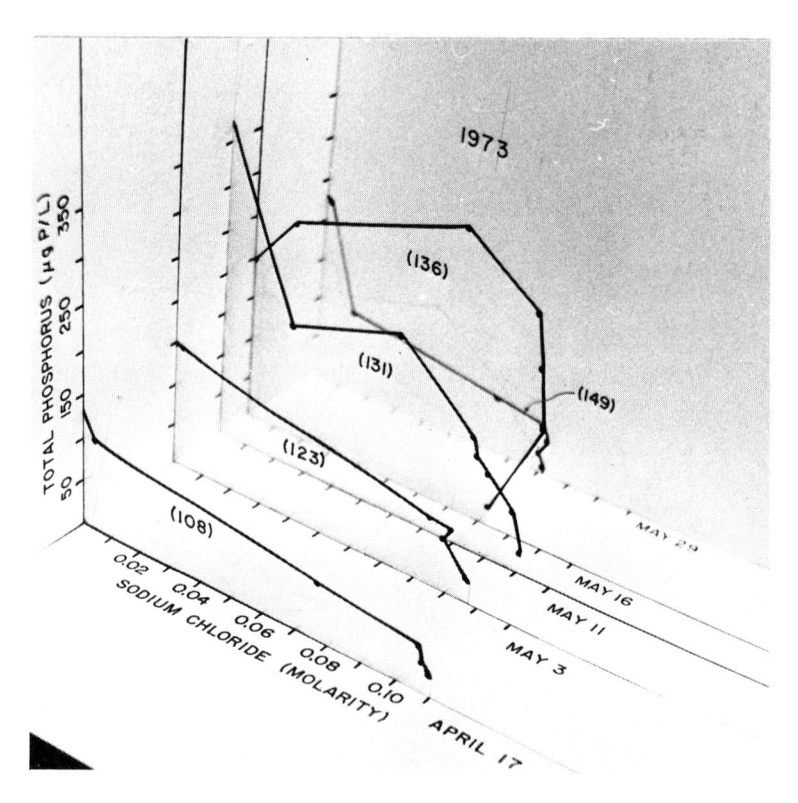

Figure 9. A series of profiles of total phosphorus concentrations versus sodium chloride concentrations in the Rhode River at consecutively later dates in 1973 showing the release of over 900 Kg phosphorus between 3 May and 16 May, followed by its disappearance from the Rhode River. The numbers in parentheses on each profile are the day of the year.

EFFECTS OF TROPICAL STORM AGNES ON NUTRIENT FLUX AND DISTRIBUTION IN LOWER CHESAPEAKE BAY[1]

C. L. Smith[2]
W. G. MacIntyre[2]
C. A. Lake[2]
J. G. Windsor, Jr.[2]

ABSTRACT

Nutrient concentrations measured in lower Chesapeake Bay in the summer of 1972 immediately following the flooding associated with Tropical Storm Agnes are compared with those in the summer of 1973, a season of more normal rainfall. The large amount of land runoff produced unseasonably high concentrations of dissolved inorganic nitrogen in the Bay near the mouth of the Potomac River. Phosphate concentrations were essentially unaffected by the flooding. Fluxes of total nitrogen and total phosphorus nutrients through the mouth of Chesapeake Bay were calculated for both summers. The calculated net export of nutrients from the Bay in both August 1972, and June 1973 was found to be small in comparison to nutrient inputs.

INTRODUCTION

The passage of Agnes in late June 1972, produced unusually heavy rainfall on the drainage basins of tributaries of Chesapeake Bay, providing a unique opportunity to investigate the effects of a major flood on nutrients in the Bay. The volume of water entering the Bay during the month of June 1972 was estimated to be nearly 23.9 billion cubic meters, or about six times the normal streamflow for June (Kammerer et al. 1972). This volume of water, most of which was discharged into the Bay during the last week of June 1972, is nearly half the mean low water volume of the Bay--50 billion cubic meters (Cronin 1971).

One can envision two possible consequences of the rapid addition of such a large amount of water on Chesapeake Bay nutrient concentrations. If the added water was very low in nutrients, then the Bay would have experienced dilution of the nutrients already present. If, on the other hand, the added water had high nutrient loadings, nutrient concentrations in the Bay would have increased. The latter case is the more probable, as the flood waters should have contained nutrients from sewage system overflows, scour from nutrient rich sediments, and nutrients leached from the land, particularly agricultural land.

The purpose of this study was twofold. First, to document the effects of Agnes on nutrients in lower Chesapeake Bay. Measurements of nutrient concentrations along two transects across the Bay were conducted during the two months following Agnes for the documentation. Second, an attempt was made to measure the flux of nutrients out of the Bay through the Bay mouth. Because the Chesapeake Bay mouth is, with the minor exception of the Chesapeake and Delaware Canal, the only connection between the Chesapeake estuarine system and the Atlantic Ocean, it is the strategic location for such measurement. Knowledge of the magnitude of nutrient flushing at the Bay mouth enables better understanding of the distribution of nutrients observed in the Bay, and on its ability to accept nutrient loadings from wastewater treatment plants and from land runoff.

---

[1]Contribution No.773, Virginia Institute of Marine Science
[2]Virginia Institute of Marine Science, Gloucester Point, Va. 23062

METHODS

Sampling

Most work was conducted in lower Chesapeake Bay, south of the Potomac River mouth. Stations were established along two transects; one between Smith Point and Tangier Island, and one between Cape Henry and Fisherman Island (Fig. 1). Stations on both transects were occupied periodically during the two months following Agnes, and the stations on the Bay mouth transect were re-occupied in June 1973. Two slack water runs were made between the two transects during the summer of 1973, and one sampling cruise was conducted on the continental shelf offshore from Chesapeake Bay.

Current meters were deployed at each station of both transects at approximately 3 meter depth intervals, and current velocities were recorded at 20 minute intervals. Water samples collected with plastic Frautschy bottles were stored in Nalgene containers at about 4°C before processing the same in the laboratory. One aliquot from each sample was filtered through a .8µ membrane filter and preserved by the addition of 40 mg/l $HgCl_2$ for later analysis of dissolved nutrients. A second aliquot was removed for determination of salinity by induction salinometer, and a third unfiltered aliquot was frozen and stored at -20°C in plastic bags for later analysis for total nitrogen and total phosphorus. Slack water run stations were sampled on the same low slack tide, and samples treated as above. Continental shelf stations were sampled from the R/V *Ridgely Warfield* in July 1972. Samples were treated as above.

Chemical Analyses

Nutrient analyses were conducted using standard methods of seawater analysis (Strickland & Parsons 1968). Total phosphorus was determined by spectrophotometry of the reduced phosphomolybdic heteropoly acid after digestion of the unfiltered water sample with perchloric acid. Total Kjeldahl nitrogen was determined by the Griess reaction following H_2SO_4-SeO_2 digestion of the unfiltered water sample and oxidation of the resulting ammonium ion by HClO. Dissolved nutrients in the filtered water samples were analyzed by an automated reagent mixing and spectrophotometry system (Technicon Auto-Analyzer): dissolved orthophosphate by spectrophotometry of the reduced phosphomolybdic heteropoly acid; dissolved nitrite by the Griess reaction; and dissolved inorganic nitrogen by the Griess reaction following reduction of nitrate to nitrite on a Cd column. Total nitrogen was calculated as:

$$(\text{Total N}) = (\text{Kjeldahl N}) + (NO_2^- + NO_3^-) - (NO_2^-)$$

Analytical methods were tested with standards and shown to be reliable within ± 10% of the reported value.

Flux Calculations

Two methods were used for the calculation of nutrient flux at the Bay mouth. In the first, the cross-sectional area along the Bay mouth transect was divided into 20 subsections, each associated with a current meter. The product of the nutrient concentration, the current velocity, and the cross-sectional area of the subsection provided the instantaneous nutrient flux through the subsection. Total nutrient flux through the Bay mouth transect per tidal cycle was obtained by summing the instantaneous nutrient fluxes for all subsections throughout one tidal cycle. Because adequate current data for this type of calculation were available only for the June 1973 section, a second method for estimation of nutrient flux was employed.

In the second method, the non-tidal transport of water through the Bay mouth is used to estimate the nutrient flux. This non-tidal transport is due to 1) the net discharge of water from the Bay, and 2) the gravitational circulation at the Bay mouth. The net discharge of water from the Bay was assumed to be equal the mean streamflow entering the Bay, which is published by the U. S. Geological Survey (Kammerer, et al. 1972 & U. S. Geological Survey 1972-1973). Mean streamflow for August 1972 was used for the August 1972 calculation, and mean streamflow for May and June 1973 was used for the June 1973 calculation. The nutrient flux was calculated by summing the product of mean streamflow and the average nutrient concentrations at the Bay mouth over a tidal cycle.

The transport of water due to gravitational circulation was calculated from a model based on Rattray and Hansen's theory of estuarine circulation (1965) which was verified and calibrated from salinity gradient data (Kuo, personal communication). According to this model, the water mass above a certain critical depth of no net motion is transported out of the Bay, and that below the critical depth is transported into the Bay. The available water--that net volume of water in either layer which passes through the transect during one tidal cycle--was multiplied by the tidal cycle averaged total nitrogen and total phosphorus concentrations in the upper and lower water layers to give nutrient fluxes. The difference in flux between the upper and lower layers is the net nutrient flux produced by the gravitational circulation. The sum of nutrient flux produced by net discharge of water and by gravitational circulation was taken to be the total flux for the Bay mouth.

RESULTS

Nutrient concentrations measured at both transects in both 1972 and 1973 showed little or no systematic variation with depth, tidal stage, or distance along the transect. Therefore, nutrient concentrations presented in Table 1 are cross-sectional and tidal cycle averages.

The total nitrogen concentrations measured at the Smith Point transect in 1972 were generally higher than those at the Bay mouth transect for the same period (Table 1), but were well within the range of values measured on slack water runs in 1973 (Tables 2 and 3). Dissolved inorganic nitrogen concentrations (primarily NO_3^-) measured along the Smith Point transect in 1972 were considerably higher than those measured at the Bay mouth. These high concentrations, ranging from 23-27 µg-at N/l, were not observed in 1973, even at slack water run stations near the Smith Point transect, where dissolved inorganic nitrogen did not exceed 5 µg-at N/l.

Total phosphorus concentrations were similar at both transects in 1972, and were comparable to those measured in 1973. Orthophosphate concentrations at the Smith Point transect in 1972 were somewhat lower than those measured at the Bay mouth during the same period. Orthophosphate concentrations measured on slack water runs in 1973 exhibited considerable patchiness; no consistent trend up the Bay could be discerned.

Fluxes of total nitrogen and total phosphorus at the Bay mouth estimated for June 1973 using current meter measurements (Table 5, A) were not only quite small in magnitude, but indicated transport of nutrients into the Bay. Fluxes calculated by the second method using mean streamflow and the gravitational circulation model were considerably larger, and nutrient transport for both August 1972

Table 1. Tidal cycle averaged nutrient concentrations for transects.

Station	Total Nitrogen (µg-at N/l)	Total Phosphorus (µg-at P/l)	Dissolved Inorganic Nitrogen (µg-at N/l)	Dissolved Orthophosphate (µg-at P/l)
Smith Point Transect				
10-11 Jul 72	45.0 ± 4.1	1.19 ± 0.28	23.0 ± 1.5	0.20 ± 0.03
17 Jul 1972	45.3 ± 3.9	0.89 ± 0.37	24.2 ± 3.3	0.24 ± 0.03
24 Jul 1972	50.2 ± 4.2	1.03 ± 0.30	26.6 ± 2.1	0.34 ± 0.08
Cape Henry Transect				
5 Jul 1972	49.4 ± 14.1	3.48 ± 1.51	1.9 ± 2.1	0.41 ± 0.24
12 Jul 1972	40.3 ± 5.5	1.23 ± 0.23	2.8 ± 1.4	0.44 ± 0.14
14 Jul 1972	34.2 ± 5.2	1.26 ± 0.15	3.5 ± 1.5	0.46 ± 0.13
20 Jul 1972	38.7 ± 9.2	1.00 ± 0.13	1.5 ± 0.92	0.42 ± 0.12
27 Jul 1972	35.8 ± 5.2	1.23 ± 0.15	1.4 ± 0.56	0.52 ± 0.10
17-18 Aug 1972	30.3 ± 5.1	1.61 ± 0.14	0.9 ± 0.09	0.64 ± 0.16
5 Jun 1973	29.1 ± 5.2	1.29 ± 0.65	0.32 ± 0.14	0.48 ± 0.14

Table 2. Nutrients from slack water run, 21 June 1973.

Station	Total Nitrogen (µg-at N/l)	Total Phosphorus (µg-at P/l)	Dissolved Inorganic Nitrogen (µg-at N/l)	Dissolved Orthophosphate (µg-at P/l)
1 Surface	28.6	0.58	1.1	0.39
2 Surface	61.8	1.36	1.0	0.37
3 Surface	39.8	0.67	0.5	0.35
4 Surface	43.4	0.82	0.2	0.15
5 Surface	48.9	0.81	0.1	0.24
6 Surface	42.7	0.92	-	0.16
7 Surface	48.2	0.82	0.5	0.40
8 Surface	39.8	0.88	0.5	0.36
9 Surface	40.1	1.33	0.3	0.16
10 Surface	-	0.94	1.4	0.16
11 Surface	59.8	1.02	0.3	0.38
12 Surface	54.5	2.79	0.5	0.44
13 Surface	19.2	1.36	2.5	0.44
14 Surface	50.8	1.33	4.0	0.48
15 Surface	55.2	1.77	3.5	0.22

Table 3. Nutrients from slack water run, 25 July 1973.

Station	Total Nitrogen (μg-at N/l)	Total Phosphorus (μg-at P/l)	Dissolved Inorganic Nitrogen (μg-at N/l)	Dissolved Orthophosphate (μg-at P/l)
1 Surface	20.0	1.52	-	0.17
Bottom	41.8	1.47	0.6	0.19
2 Surface	43.4	0.89	0.6	0.10
Bottom	-	1.83	1.0	0.39
3 Surface	23.3	1.53	-	0.15
Bottom	69.4	1.31	0.8	0.63
4 Surface	52.6	1.69	2.0	0.29
Bottom	-	1.91	1.4	0.49
5 Surface	26.2	2.03	2.4	0.29
Bottom	43.0	1.85	1.9	0.50
6 Surface	39.6	1.95	1.7	0.20
Bottom	45.9	2.25	1.7	0.90
7 Surface	55.4	1.95	1.6	0.29
Bottom	42.0	1.52	1.8	0.86
8 Surface	35.7	1.51	1.4	0.26
Bottom	55.5	1.51	0.9	0.33
9 Surface	-	2.32	2.0	0.29
Bottom	64.2	2.66	1.1	0.21
10 Surface	62.9	2.17	2.7	0.36
Bottom	49.2	1.56	1.4	0.24
11 Surface	52.5	1.86	1.5	0.29
Bottom	32.6	1.86	1.5	0.29
12 Surface	40.1	1.53	1.7	0.31
Bottom	43.5	1.95	-	-
13 Surface	49.8	2.69	0.8	0.22
Bottom	57.8	2.13	1.6	0.66
14 Surface	-	2.05	1.4	0.23
Bottom	70.8	1.85	0.8	0.14
15 Surface	34.3	1.13	0.8	0.08
Bottom	40.9	1.93	0.6	0.68

and June 1973 was out of the Bay (Tables 4 and 5), as would be expected.

DISCUSSION

Distribution of Nutrients

The major effect of Agnes on nutrient distribution in lower Chesapeake Bay was an elevated concentration of dissolved inorganic nitrogen, not common to that area in that season. Levels were highest near the mouth of the Potomac River, where dissolved inorganic nitrogen comprised nearly half the total nitrogen, and considerably lower near the Bay mouth. Levels of dissolved orthophosphate near the mouth of the Potomac River were somewhat depressed relative to those at the Bay mouth. These effects persisted for at least a month following the passage of Agnes near the Potomac, but returned to normal during that time at the Bay mouth. The large input of nutrients from the Agnes flood-flows, mostly in the form of dissolved inorganic nitrogen, produced a situation in lower Chesapeake Bay similar to that normally observed much earlier in the year. For example, in upper Chesapeake Bay, Carpenter, et al. (1969) found strong seasonal variation of both total and dissolved inorganic nitrogen concentrations. Total nitrogen concentrations in the spring ranged from 80-105 µg-at N/l, but dropped to around 50 µg-at N/l in other seasons. Likewise, dissolved inorganic nitrogen concentrations averaging near 45 µg-at N/l in mid-April had dropped to less than 1µg-at N/l by September. Scattered data from the lower portion of Chesapeake Bay show that this basic seasonal variation occurs, but to a lesser degree. Dissolved inorganic nitrogen concentrations rarely exceed 20 µg-at N/l in any season (Grant, unpublished data). The regular seasonal increase of nitrate concentrations presumably is due to the higher average rainfall, with associated higher streamflows, and the decreased level of primary production in late winter and early spring.

The small depression in orthophosphate concentration observed at the Smith Point transect relative to that at the Bay mouth might be due to the adsorption of orthophosphate on sediment particles associated with the high river discharge. These particles mostly settle out of the water column before reaching the lower Bay, thus removing orthophosphate from the water. Such a phenomenon has been documented for the upper portion of the Potomac River (Jaworski, Lear, & Villa 1971).

Flux of Nutrients

Fluxes of total nitrogen and total phosphorus calculated by the first method employing current meter velocity approach to flux measurement must be regarded as invalid. For such a large cross-sectional area as that of the Bay mouth, it is impractical to deploy a sufficient number of current meters to accurately monitor the total flow through the section. Furthermore, strong oscillatory tidal currents at the Bay mouth tend to obscure measurement of any net flow of water through the section. Consequently, a better estimate of nutrient flux must be obtained by the second method, employing published streamflow data and the gravitational circulation model. Nutrient fluxes calculated by the second method for August 1972 (Table 4) are comparable in magnitude to those for June 1973 (Table 5). By August 1972, streamflow into the Bay had not only decreased from the June 1972 record discharges, but were even smaller than the June 1973 flows. Sufficient data were not collected to enable complete calculation of nutrient fluxes for June 1972. However, it is expected that nutrient flux was considerably enhanced by the massive input of water to the Chesapeake Bay system. If one uses the nutrient concentrations measured at the Bay mouth in early July 1972, and the peak streamflow of June 24, 1972, the fluxes of total nitrogen and

and total phosphorus due to the net discharge of water are two orders of magnitude larger than those calculated for August 1972. Unfortunately, the contribution due to gravitational circulation cannot be estimated for that time.

Table 4. Nutrient flux for Chesapeake Bay mouth - August 17-18, 1972.

Station	Flux of total P* (gP/tidal cycle)	Flux of total N* (gN/tidal cycle)
A. Due to net discharge of water.		
Bay Mouth	$+2.70 \times 10^6$	$+2.30 \times 10^7$
B. Due to gravitation circulation		
A & B	$+9.30 \times 10^5$	$+8.16 \times 10^5$
C	$+3.44 \times 10^6$	$+5.36 \times 10^7$
D	$+1.55 \times 10^5$	$+3.22 \times 10^6$
E	$+0.0 \times 10^5$	-2.39×10^7
Bay Mouth	$+4.53 \times 10^6$	$+3.37 \times 10^7$
C. Total Flux		
Bay Mouth	$+7.23 \times 10^6$	$+5.67 \times 10^7$

*Positive sign indicates net flux out of Bay, negative into Bay.

Table 5. Nutrient flux for Chesapeake Bay mouth - June 5-6, 1973.

Station	Flux of total P* (gP/tidal cycle)	Flux of total N* (gN/tidal cycle)
A. Total flux using current meter velocities		
A	+3.08	$+1.75 \times 10^1$
B	-5.58	-7.98×10^1
C	+1.77	+6.49
D	-5.20	$+7.25 \times 10^1$
E	+3.50	-2.22×10^1
Bay Mouth	-2.44	-1.50×10^2
B. Due to net discharge of water		
Bay Mouth	$+5.22 \times 10^6$	$+5.28 \times 10^7$
C. Due to gravitational circulation		
A	-8.49×10^5	-6.17×10^5
B	-1.26×10^5	$+1.02 \times 10^7$
C	-1.22×10^5	-7.31×10^6
D	-3.22×10^5	-5.80×10^5
E	$+4.31 \times 10^4$	-5.82×10^5
Bay Mouth	-1.38×10^6	$+2.28 \times 10^6$
D. Total flux		
Bay Mouth	$+3.84 \times 10^6$	$+5.50 \times 10^7$

*Positive sign indicates net flux out of Bay, negative into Bay.

The most striking result of the nutrient fluxes calculated for the Bay mouth is the relatively small extent of nutrient flushing. In part, this must be due to the gentle gradient of nutrient concentrations on the Continental shelf offshore from the Bay (Figs. 3 and 4). Without a water mass deficient in nutrients for dilution, tidal flushing is less effective. The net total fluxes of nitrogen and phosphorus out of the Bay calculated for June 1973 are respectively two and ten times smaller than the normal rates of addition of those nutrient species to the Bay via the Potomac River from the wastewater treatment facilities in Washington, D. C. alone (Jaworski, Lear, & Villa 1971). When the additional loading from land runoff and wastewater discharges from other metropolitan areas (Baltimore, Hampton Roads, etc.) are considered, it is apparent that in times of normal streamflow, vastly greater amounts of nutrients are being added to the Bay than are being removed at the Bay mouth. Since the Bay waters are not drastically increasing in concentration of nitrogen and phosphorus, there must be other mechanisms operating to remove these nutrients. Excess phosphate is probably removed by adsorption on suspended sediment, and subsequent deposition on the bottom. Nitrogen may behave similarly, and be deposited on the bottom with organic detritus, or may be converted to volatile compounds (i.e. NH_3, N_2) and lost to the atmosphere.

ACKNOWLEDGEMENTS

This work was supported in part by the Oceanographic Section of the National Science Foundation (GA 35446).

LITERATURE CITED

Carpenter, J. H., D. W. Pritchard, and R. C. Whaley. 1969. Observations of eutrophication and nutrient cycles in some coastal plain estuaries. Pages 210-221 in Eutrophication: Causes, Consequences, Correctives. National Academy of Sciences, Washington, D. C.

Cronin, W. B. 1971. Volumetric, areal, and tidal statistics of Chesapeake Bay and its tributaries. Spec. Rep. 20, Chesapeake Bay Inst., Johns Hopkins Univ., Baltimore, Md.

Jaworski, N. A., D. W. Lear, Jr., and O. Villa, Jr. 1971. Nutrient management in the Potomac estuary. Tech. Rep. 43, U. S. Environmental Protection Agency, Middle Atlantic Region, Water Quality Office.

Kammerer, J. C., H. D. Brise, E. W. Coffay, and L. C. Fleshmon. 1972. Water Sources Review, June. U. S. Geological Survey, Washington, D. C.

Rattray, M., Jr. and D. V. Hansen. 1965. The physical classification of estuaries. Page 543 in Ocean Science and Ocean Engineering, Trans. Joint Conf. and Exhibition, June 1965, Washington, D. C., sponsored by Marine Tech. Soc. and Am. Soc. Limnol. & Oceanogr.

Strickland, J. D. H. and T. R. Parsons. 1968. A practical handbook of seawater analysis. Bull. 167, Fish. Res. Bd. Canada.

U. S. Geological Survey. 1972-73. Estimated stream discharge entering Chesapeake Bay. Washington, D. C.

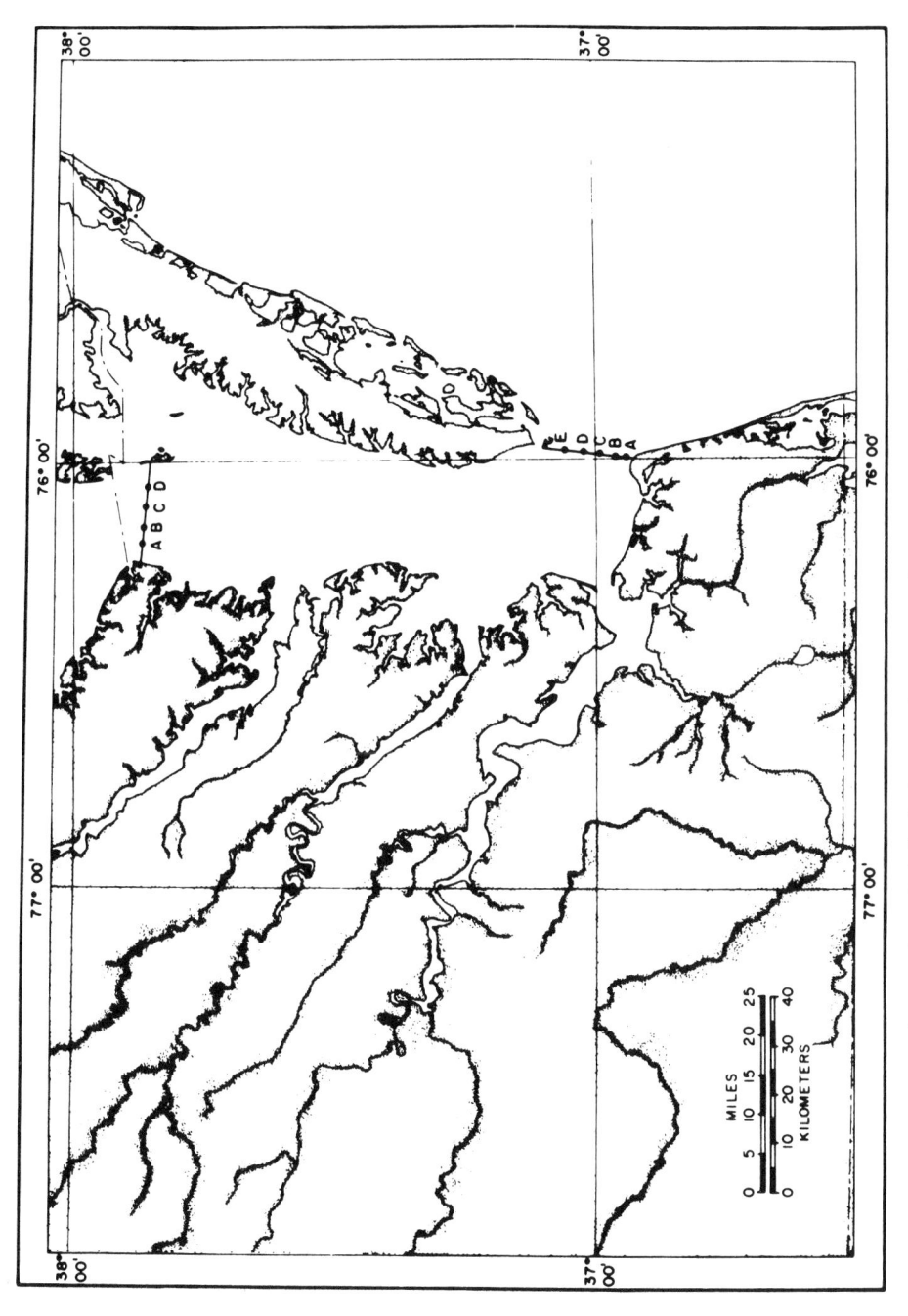

Figure 1. Smith Point and Bay mouth transect stations.

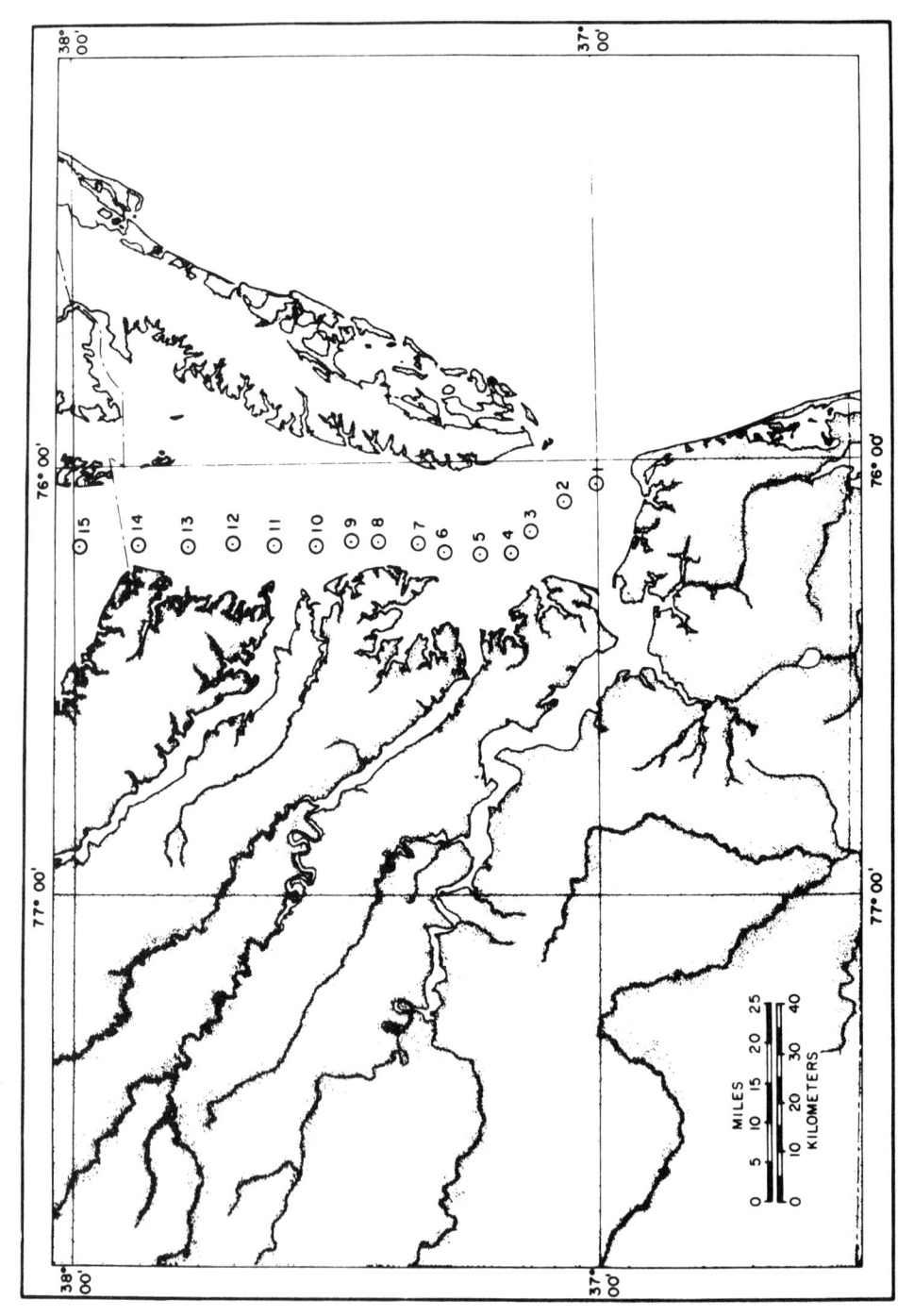

Figure 2. Slack water run stations.

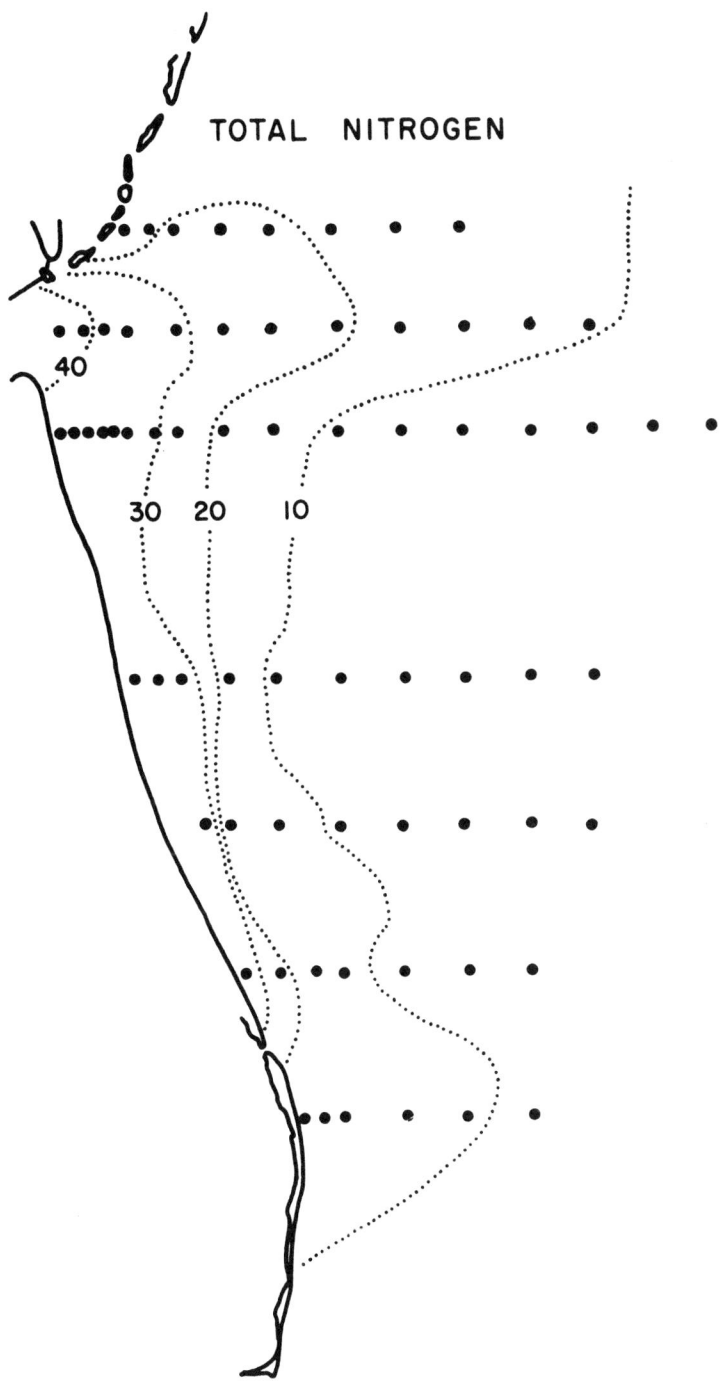

Figure 3. Distribution of total nitrogen concentration on continental shelf, July 1972, 2 m depth, in µg-at N/l.

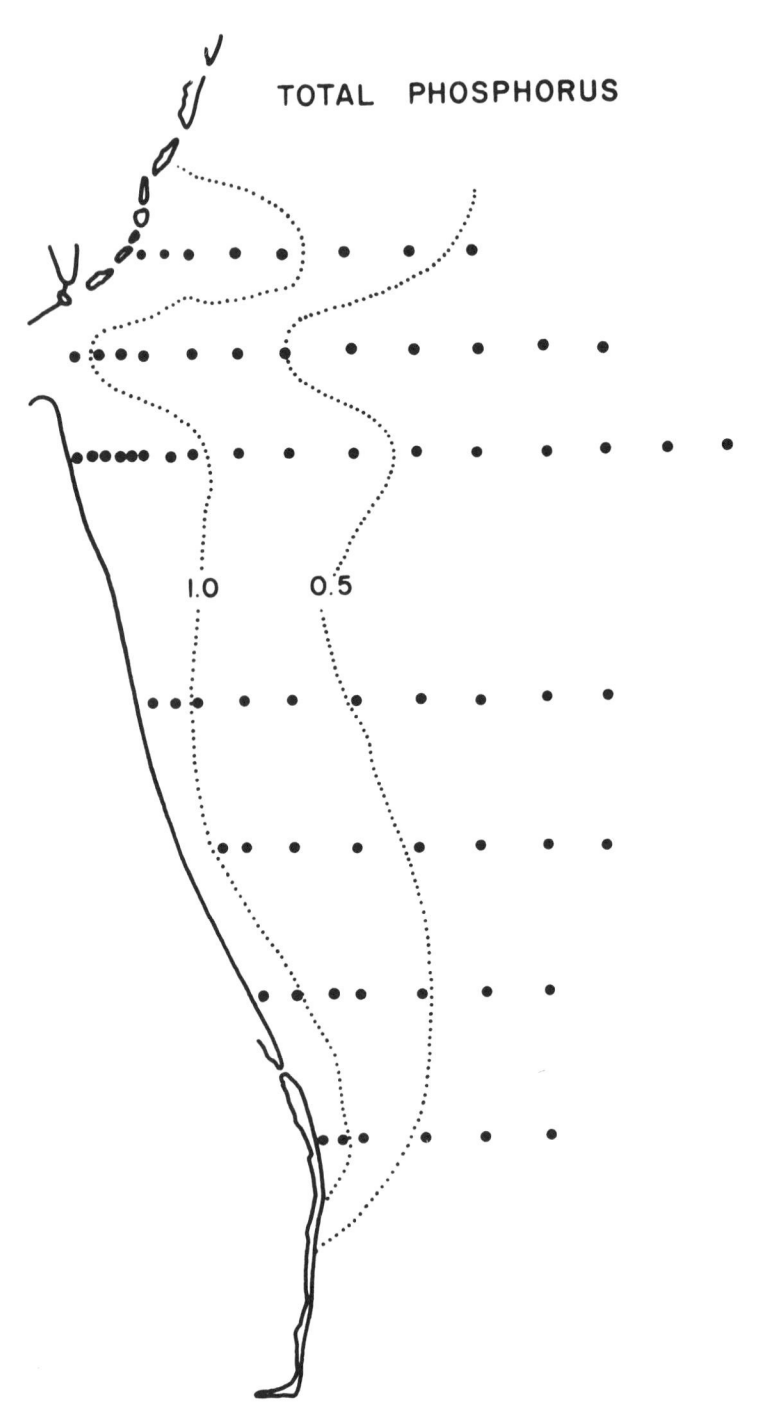

Figure 4. Distribution of total phosphorus concentration on continental shelf, July 1972, 2 m depth, in µg-at P/l.

EFFECTS OF AGNES ON THE DISTRIBUTION OF NUTRIENTS IN UPPER CHESAPEAKE BAY [1]

J. R. Schubel[2]
W. R. Taylor[2]
V. E. Grant[2]
W. B. Cronin[2]
M. Glendening[2]

ABSTRACT

Water samples were collected before and after Agnes from thirteen stations in the Maryland portion of the Chesapeake Bay. Comparisons of the nutrient concentrations in these samples before and after the flooding show the effect of the storm on the Chesapeake Bay's nutrient budget. Nitrate and nitrite were increased 2-3 times in the upper Bay and appeared to return to near normal approximately two months later. Ammonia concentrations were also elevated due to the storm and were shown to increase with increasing depth in the water column. Comparison of the distribution of phosphate after Agnes with those observed during more normal years, shows that the concentrations of phosphate in the Maryland portion of the Bay were not abnormal for that time of the year.

INTRODUCTION

There have been a number of investigations of the distribution of nutrients in the upper Chesapeake Bay and tributary estuaries (see e.g. Whaley, et al., 1966; Carpenter, et al., 1969; and the numerous reports of the U. S. Environmental Protection Agency, Annapolis Field Office Region III). These observations have characterized the "normal" range of the distributions of nitrogen and phosphorus in both time and space. But, in addition to these "normal" variations, marked fluctuations can result from catastrophic events such as floods. There was, until Agnes, a dearth of direct observations of the effects of such rare events on the distribution of nutrients.

Agnes presented scientists with an unusual opportunity to document the effects of a major flood on the distribution of nutrients in the Chesapeake Bay estuarine system. Since there was little wind associated with Agnes when she reached this area, seas were not abnormally rough and sampling difficulties were minimal. But the torrential rains associated with Agnes sent flows of the major tributaries to record or near-record levels and it was anticipated that the concentrations of nutrients would be anomalously high.

The Chesapeake Bay Institute made a series of observations to document the impact of Agnes on the distributions of a variety of nutrients both in the Maryland portion of the Bay proper, and in the Potomac estuary. Sampling was initiated on 13 July 1972 and extended over about six months to chronicle the impact of Agnes, and the subsequent recovery of the Bay to "normal" conditions. The primary purpose of this paper is to summarize some of the nutrient data from the

---

[1]Contribution No. 236, Chesapeake Bay Institute, Johns Hopkins University
[2]Chesapeake Bay Institute, The Johns Hopkins University, Baltimore, Md. 21318

Maryland portion of the main body of the Bay. These observations were compared with data in the CBI data bank from more "normal" years to determine whether they fell outside of the range of "normal" year to year variations. In addition, the post-Agnes observations are compared to the pre-Agnes sampling (12-16 June 1972) period.

METHODS

Table 1 is a summary of the sampling frequencies in both time and space; the stations are plotted in Fig. 1. Water samples were collected with a submersible pump and immediately frozen in acid-cleaned milk dilution bottles for later analysis. Two water samples were preserved from each sampling depth; one a raw water sample, the other a sample filtered through a Whatman GF/C glass filter. All analyses reported here were made on the filtered samples. The concentrations of suspended solids were so great in the raw water samples that they interferred with the analytical procedures producing spurious results. Analyses of dissolved ammonia, nitrite, nitrate, and soluble reactive phosphate (SRP) were done on a Technicon Autoanalyzer using standard procedures [1]. Determinations of dissolved organic carbon were made with a Beckman Carbon Analyzer. All of these data are stored in the CBI data bank, and are available upon request. Only a portion of the data are presented here to illustrate the general trends in nutrient distributions as a function of time.

Table 1. Summary of nutrient sampling locations and dates, 12 June 1972 - 9 October 1972.* See Fig. 1 for station locations.

	927SS	921W	917S	913R	909	904N	903A	858C	848E	834F	818P	804C	745
12-16 June 72		X						X			X		X
13 July 72								X		X		X	X
20-22 July 72				X				X		X	X	X	X
3-5 Aug 72		X	X	X		X		X	X	X	X	X	X
9 Aug 72		X		X			X	X					
15-17 Aug 72		X		X			X	X	X	X	X	X	
28-31 Aug 72	X	X		X			X	X	X		X	X	X
16-18 Sept 72		X		X			X	X	X		X	X	X
25-26 Sept 72	X	X		X			X	X	X				
9 Oct 72	X	X		X	X		X						

*Samples collected from 9 Oct-31 Dec 1972 have not been analyzed.

OBSERVATIONS AND DISCUSSION

The distributions of $NO_2 + NO_3$ in the surface waters along the axis of the the upper Bay approximately two weeks before Agnes (12-16 June 1972), approximately three weeks after Agnes (13 July 1972), and on later representative dates following

---

[1] Ammonia was analyzed by the method described in A Practical Handbook of Seawater Analysis, J. D. H. Strickland and T. R. Parsons (Eds.), Bull. 167 (2nd edition) of the Fisheries Research Board of Canada, Ottawa, 1972, 310 p. Nitrite, nitrate, and soluble reactive phosphate were determined according to procedures outlined in the Technicon Industrial Methods for AAII, Jan. 1971.

Agnes are plotted in Figs. 2 and 3. Comparison of the data from 12-16 June with those from 13 July 1972 clearly shows the effects of the high discharges of nutrient-rich waters on the distribution of $NO_2 + NO_3$. Following Agnes, concentrations increased 2 - 3 times throughout the upper Bay. Comparison of the Agnes data with data from other more "normal" years, indicates that the levels of $NO_2 + NO_3$ observed in July-August 1972 were clearly anomalous for that time of year. They were typical of concentrations that might be observed during a "normal" spring freshet. Most of the $NO_2 + NO_3$, more than 90 percent of it, was in the form of nitrate. On any given cruise the concentrations of nitrate decreased with depth at a station, particularly in the middle reaches of the Bay. On the 20-22 July 1972 cruise, for example. the concentration of NO_3 at Station 913R decreased from 72 µg.at. NO_3-N/L at the surface to 66 µg.at.NO_3-N/L at 10 m. On that same cruise the concentration of NO_3 at Station 804C farther downstream decreased from 29 µg.at.NO_3-N/L at the surface to only 4 µg.at.NO_3-N/L at a depth of 24 m.

The distributions of ammonia in the surface waters of the upper Bay before, and on selected dates following Agnes, are plotted in Figs. 4 and 5. The effects of Agnes are once again apparent. Comparison with data from other years indicates that concentrations of ammonia decreased to "normal" levels within about a month after Agnes.

The concentrations of ammonia increased with depth at most stations from July-September. On 13 July 1972, for example, the concentration of ammonia at Station 858C increased from about 22 µg.at.NH_4-N/L at the surface to nearly 37 µg.at.NH_4-N/L at a depth of 20 m. On 20 July 1972 at the same station, the concentration of ammonia increased from about 2 µg.at.NH_4-N/L at the surface to more than 44 µg.at.NH_4-N/L at 25 m. Appreciable vertical gradients persisted throughout most of September.

Comparison of the distributions of phosphate after Agnes with those observed during more normal years shows that the concentrations of phosphate in the Maryland portion of the Bay proper in July and August of 1972 were not abnormal for that time of year. The uptake of phosphate was apparently so rapid that no SRP pulse was detected. During most of the year phosphorus enters the Bay predominately in the form of particulate material. The small quantities of dissolved orthophosphate introduced by the Susquehanna River are quickly taken up by the estuarine phytoplankton. Therefore, the magnitude of the phosphorus input resulting from Agnes cannot be determined without total and particulate phosphorus data.

ACKNOWLEDGEMENTS

We are indebted to I. Hopkins, C. F. Zabawa, Y. Mazurek, T. W. Kana, C. H. Morrow, L. Smith and R. Mervine for their assistance in the field and in the laboratory. The figures were drafted by D. Pendleton and photographed by J. Sullivan. The manuscript was typed by A. Sullivan.

This research was supported in part by the Oceanography Section, National Science Foundation, NSF Grant GA-36091; in part by a project jointly funded by the Fish and Wildlife Administration, State of Maryland, and the Bureau of Sport Fisheries and Wildlife, U. S. Department of the Interior, using Dingell-Johnson Funds; and in part by the U. S. Army Corps of Engineers, Baltimore and Philadelphia Districts.

LITERATURE CITED

Carpenter, J. H., D. W. Pritchard, and R. C. Whaley. 1969. Observations of eutrophication and nutrient cycles in some coastal plain estuaries. Pages 210-221 in Eutrophication: Causes, Consequences, Correctives. National Academy of Sciences, Washington, D. C.

Whaley, R. C., J. H. Carpenter, and R. L. Baker. 1966. Nutrient data summary 1964, 1965, 1966: Upper Chesapeake Bay (Smith Point to Turkey Point) Potomac, South, Severn, Magothy, Back, Chester, and Miles Rivers; and Eastern Bay. The Johns Hopkins Univ., Chesapeake Bay Inst., Spec. Rep. 12, Ref. 66-4.

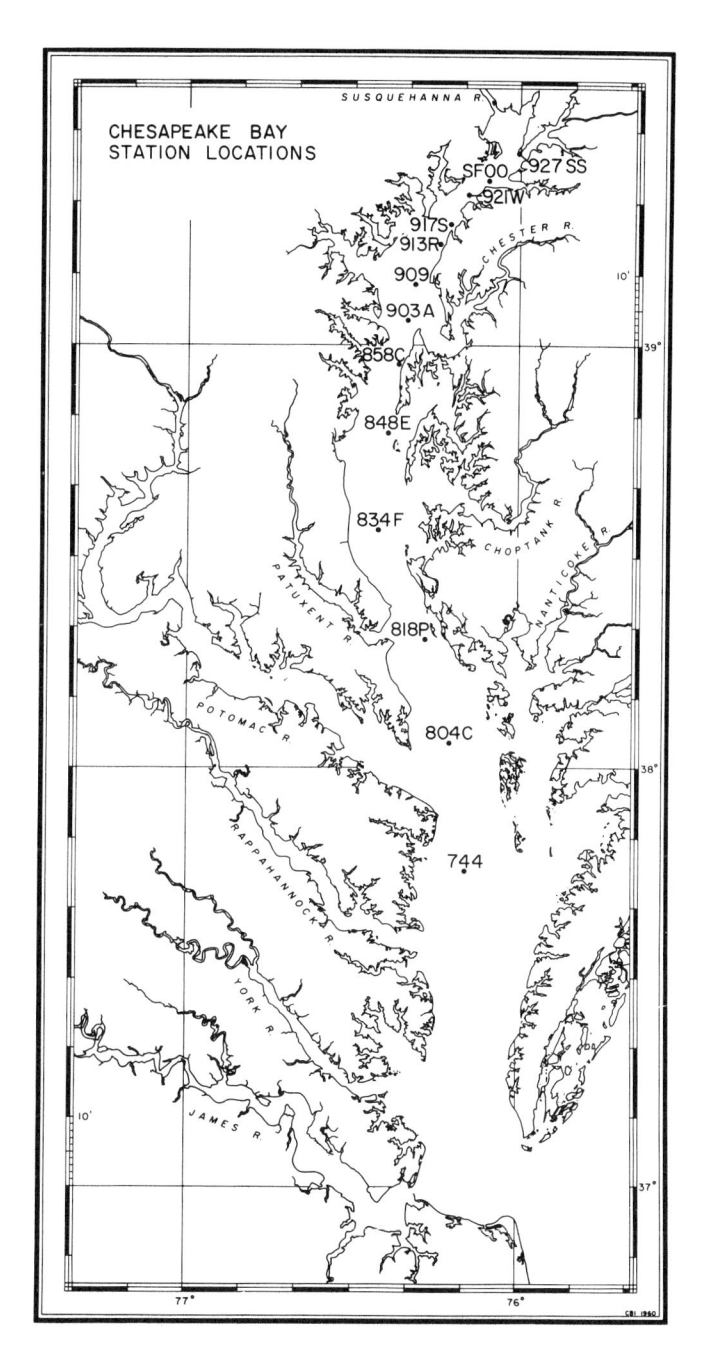

Figure 1. Map showing locations of stations made most frequently. See Table 1 for a summary of sampling frequencies.

316 Schubel et al.

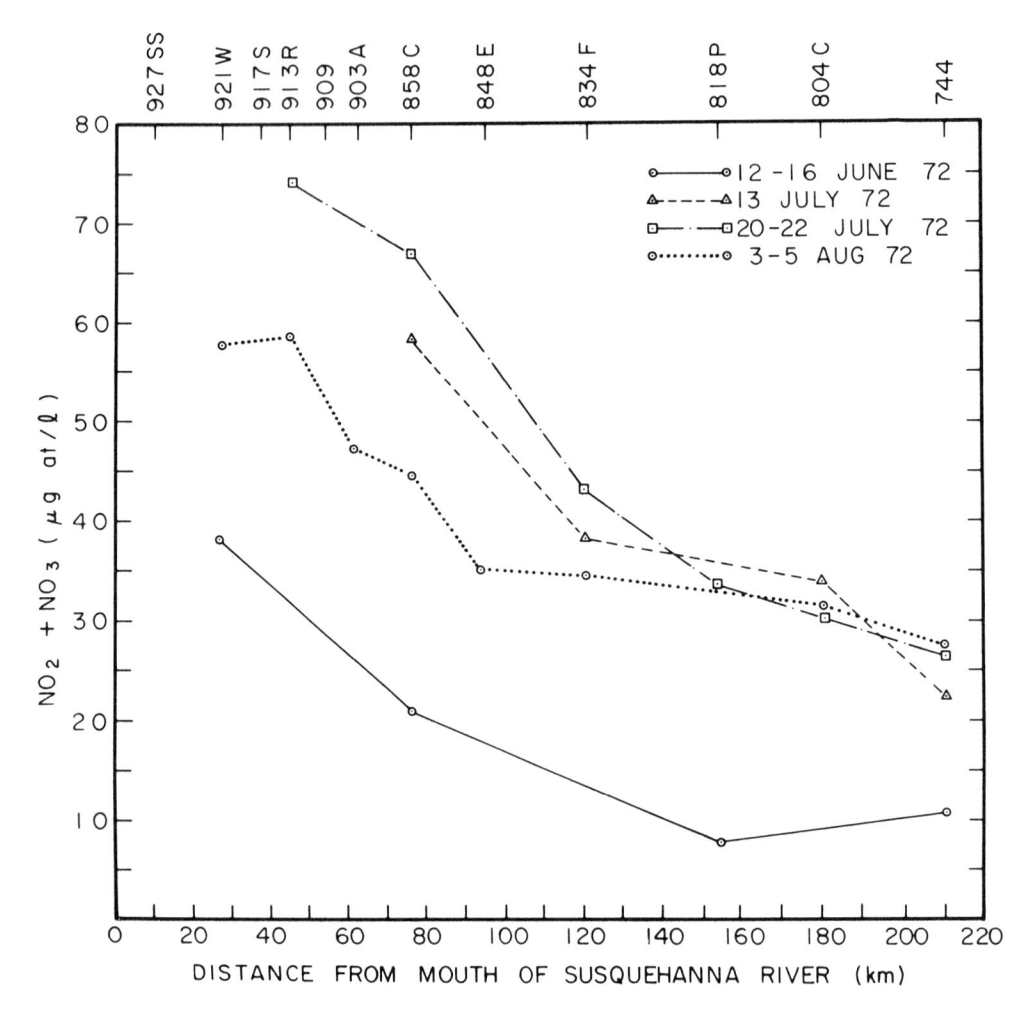

Figure 2. Longitudinal distributions of $NO_2 + NO_3$ in the surface waters of the upper Bay on selected dates before and after Agnes. Data expressed in µg.at. (NO_2+NO_3)-N per liter.

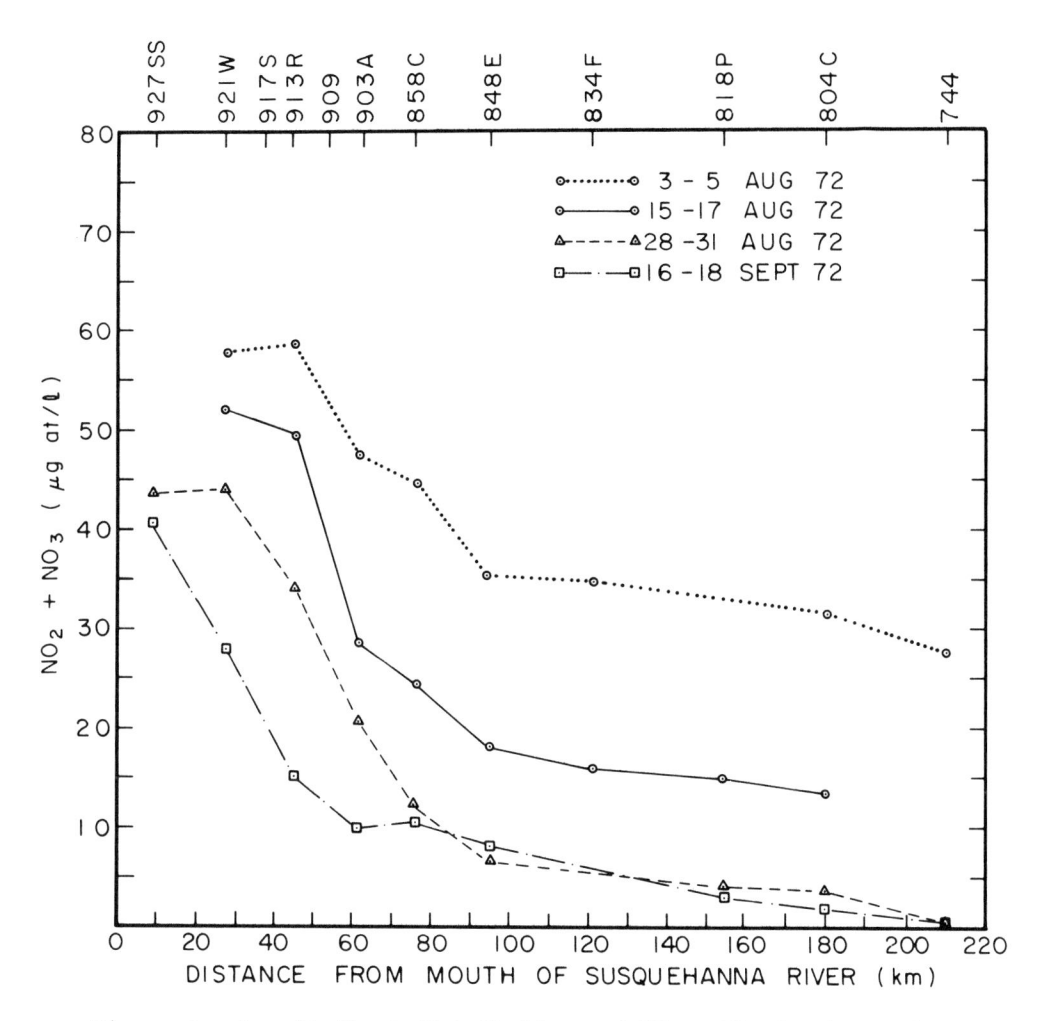

Figure 3. Longitudinal distributions of NO$_2$ + NO$_3$ in the surface waters of the upper Bay on selected dates after Agnes. Data expressed in μg.at.(NO$_2$+NO$_3$)-N per liter.

318   Schubel et al.

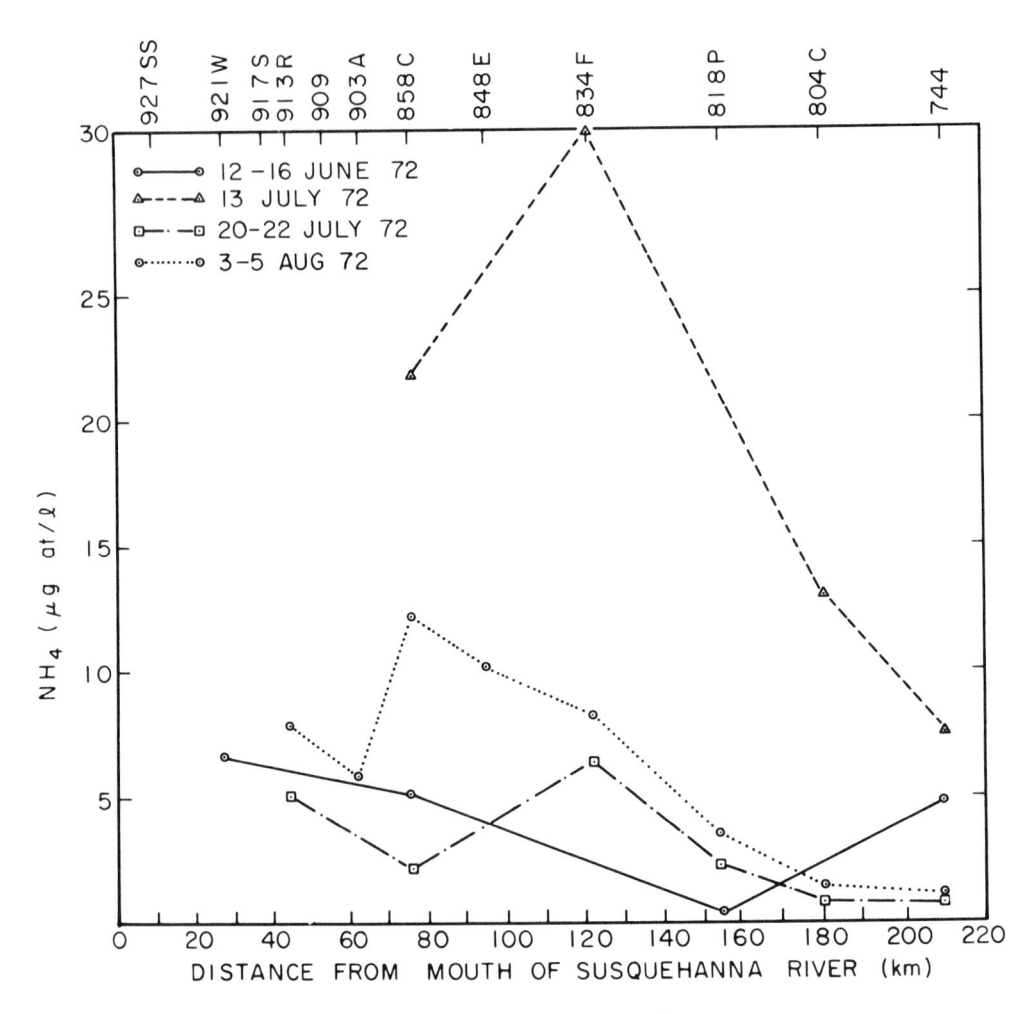

Figure 4. Longitudinal distributions of NH_4^+ in the surface waters of the upper Bay on selected dates before and after Agnes. Data expressed in µg.at.NH_4-N per liter.

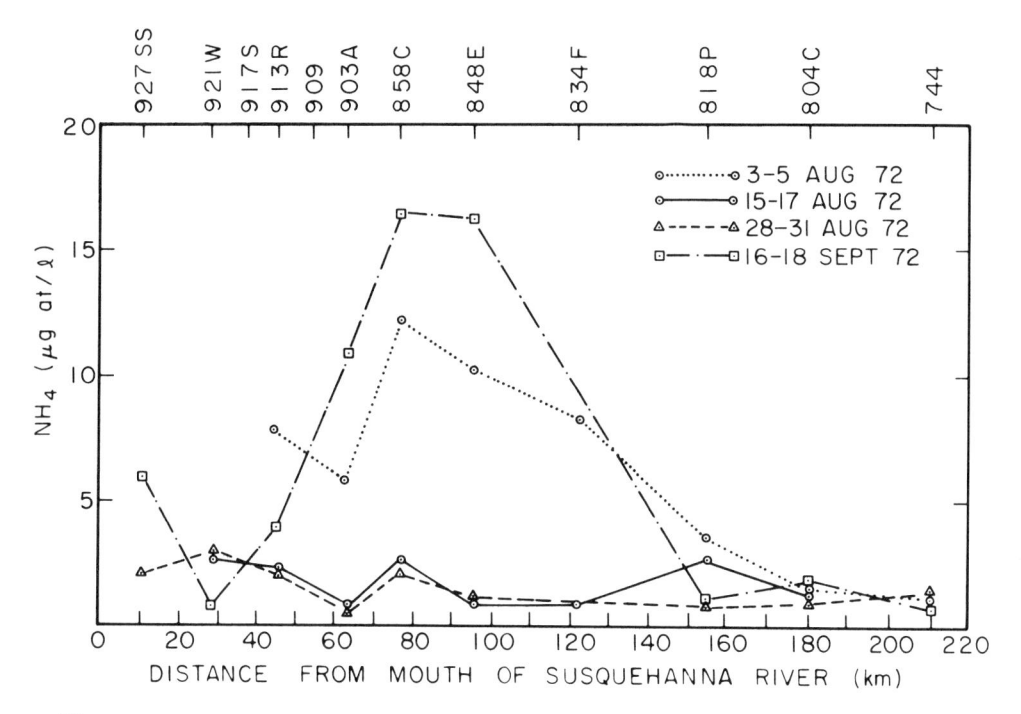

Figure 5. Longitudinal distributions of NH_4^+ in the surface waters of the upper Bay on selected dates after Agnes. Data expressed in µg.at.NH_4-N per liter.

THE EFFECT OF TROPICAL STORM AGNES ON HEAVY
METAL AND PESTICIDE RESIDUES IN THE EASTERN
OYSTER FROM SOUTHERN CHESAPEAKE BAY [1]

M. E. Bender [2]
R. J. Huggett [2]

ABSTRACT

The concentrations of cadmium, copper, and zinc in the eastern oyster, *Crassostrea virginica*, are compared for samples collected before and after Tropical Storm Agnes. The "before" samples consisted of 475 animals from 95 stations collected in January 1971 and the "after" samples of 285 animals from 57 stations collected in January 1973. Shifts in the areal concentrations distributions were observed, apparently due to Agnes.
Analyses of hard clams, blue crabs and oysters for chlorinated hydrocarbon pesticides showed influx of these compounds to be minimal as reflected by the residue levels observed. Comparison of residue levels in oysters to pre-Agnes conditions revealed a decrease in pesticide body burden.

INTRODUCTION

The rainfall and subsequent runoff from Agnes modified the chemical environment of organisms capable of concentrating pollutants by altering ambient exposure levels and by changing their "normal" physiological regime. Metals and pesticides contained in bottom sediments were both resuspended and in the case of some metals possibly desorbed. In addition to mobilization of these once "stored" substances, their levels were augmented by massive amounts of erosional products.

In order to assess the actual impact of the storm on these pollutants, residue analyses for the metals Cu, Cd and Zn and for chlorinated hydrocarbon pesticides were performed on the eastern oyster (*Crassostrea virginica*). Comparisons were made with the previously published studies of Huggett, Bender and Slone (1973) for metals and with those of Butler (1973) for pesticides.

METHODS

Sampling locations were chosen to correspond where possible to those of previous studies (Huggett, Bender & Slone 1973; Butler 1973). Five oysters were sampled from each of the stations shown in Figs. 1, 2 and 3, during January 1973. In some cases samples were difficult to obtain due to high mortalities at the upstream stations and in the Hampton Roads segment of the James River.

The analytical procedures utilized to determine residue levels have been described previously (Huggett, Bender & Slone 1973; Butler 1973)

---

[1] Contribution No.774, Virginia Institute of Marine Science
[2] Virginia Institute of Marine Science, Gloucester Point, Va. 23062

RESULTS

Average levels of each metal at a station were determined to ascertain the areal distribution of metals in the various river systems. The means showed that a concentration gradient similar to that observed previously existed in all systems and that each metal generally increased in concentration as fresh water was approached (Figs. 4, 5 & 6).

To test whether Agnes had measurably altered these concentration gradients, those stations for which data from 1973 were available were statistically compared with the pre-Agnes period. Comparisons could be made on 9 stations in the James, 7 in the Rappahannock and 5 in the York (Table 1). Significantly different concentrations (t test, $\alpha = 0.05$) for copper were observed in 4 of 21 comparisons. Of these, one station decreased while the other 3 showed increased levels. Those showing increased concentrations had levels less than 5 ppm prior to Agnes with increases to about 12 ppm after Agnes.

Zinc levels showed a significant increase at only one location in the upper York River.

The following is based upon a comparison of the areal concentration pattern as shown in Figs. 4, 5 and 6 with those previously published by Huggett, Bender and Slone (1973). Although it is not possible to statistically evaluate these data, we believe some insights may be gained from this descriptive approach.

James River

Cadmium. The upper river appears to be "cleaner" with respect to cadmium since Agnes. In 1971 concentrations in the Deep Water Shoals and Burwell Bay areas were greater than 1.6 ppm but after Agnes fell to between 1.1 and 1.5 ppm while the lower part of Burwell Bay decreased to less than 1 ppm. Between Burwell Bay to about five miles below the Warwick River concentrations ranged from 1.1 to 1.5 ppm in 1971, but fell to between 0.6 to 1.0 in 1973.

Prior to Agnes the Nansemond River showed levels between 1.1 and 1.5 ppm, while after the storm the levels decreased to between 0.6 and 1.0 ppm. Concentrations in the Nansemond Ridge area increased from 0.6-1.0 to 1.1-1.5 ppm.

Levels in the lower portion of the Elizabeth River decreased from 1.1-1.5 ppm in 1971 to 0.6-1.0 ppm after Agnes. In the upper portion the levels were unchanged.

Residues in Willoughby Bay increased considerably with levels averaging 4.6 ppm in 1973 and ~2.5 in 1971.

Copper. As was seen for cadmium, oysters in the upper James River appear to have decreased in copper concentrations since 1971. Except for a small area in the upper section of Burwell Bay, the levels have dropped from 101-150 ppm to 51-100 ppm. The area of high residues, from 151-200 ppm, found in the middle of Burwell Bay in 1971 is now absent. The concentrations from Burwell Bay to the James River Bridge were unchanged since the 1971 survey.

The Nansemond Ridge oyster beds show that this region has worsened since the previous survey with concentrations increasing from 26-50 ppm to 101-150 ppm.

The Nansemond River showed an increase with levels after Agnes between 51-100 ppm as compared to 26-50 ppm in 1971.

Table 1. Comparison of Zn & Cu Levels in Oysters 1971 vs 1973.

River	Station	Mean Copper (ppm) 1971	1973	DF	t	Mean Zinc (ppm) 1971	1973	DF	t
James	1	141.0	97.8	8	1.42	1465.0	1097.2	8	1.22
	2	172.6	124.1	8	1.66	1253.0	1090.4	8	0.55
	3	95.7	79.2	7	0.51	643.2	902.0	8	1.15
	4	142.2	91.1	8	2.09	1030.8	987.6	8	0.19
	7	82.0	82.7	8	0.03	858.4	903.2	8	0.21
	8	91.8	92.8	8	0.07	887.8	1184.4	8	1.29
	9	84.1	84.7	8	0.04	1350.6	1182.0	8	0.78
	10	51.9	71.3	7	0.96	905.8	1159.0	7	0.61
	11	73.1	112.1	7	1.84	881.3	1362.2	7	1.60
Rappahannock	5	16.2	12.5	6	1.15	324.0	376.6	7	0.74
	6	6.4	12.2	8	1.96	253.6	311.8	8	0.94
	7	4.2	12.8	7	*9.77	213.8	298.8	8	1.56
	8	39.8	26.6	9	0.71	454.8	487.0	9	0.22
	9	28.8	31.3	8	0.39	440.0	410.8	8	0.76
	10	34.9	21.7	7	1.21	456.5	232.6	7	1.58
	14	4.6	20.3	5	*7.15	376.7	502.4	7	2.51
York	1	21.1	33.6	8	0.49	266.6	448.2	8	1.23
	2&3	19.5	17.5	18	*3.02	604.9	875.4	18	1.81
	4,5&6	21.7	15.6	25	1.70	486.9	892.5	25	*4.51
	7	3.8	9.5	7	*4.65	315.3	335.6	7	0.34
	8	6.4	9.2	5	1.75	424.5	330.0	5	1.81

*Significant difference at 95% level.

The Elizabeth River samples indicate an increase of copper. In 1971 the levels ranged between 26-50 ppm throughout most of the river, but in 1973 the concentrations were all greater than 50 ppm.

Zinc. The upper James River decreased with respect to its oyster-zinc levels since the previous sampling in 1971. The upstream part of Burwell Bay and Deep Water Shoals dropped from 1201-1600 ppm to 801-1200 ppm.

The remainder of the river appears unchanged with the exception of Nansemond Ridge and Willoughby Bay. The samples from Nansemond Ridge showed an increase in zinc since Agnes with levels increasing from 401-800 ppm in 1971 to 1201-1600 ppm in 1973. Willoughby Bay samples indicated a decrease since Agnes of from 801-1200 ppm to 401-800 ppm.

The Elizabeth River remained the most contaminated area sampled. Although not obvious from the figures, one station yielded an animal with 20,000 ppm zinc. This is the highest level ever recorded by this laboratory and as far as we can ascertain, the highest ever recorded anywhere.

York River

Cadmium. The cadmium levels in oysters from the York River showed no apparent change from the 1971 study. The concentration range and distributions were approximately the same.

Copper. The copper levels in oysters from this river are generally higher in the middle and lower segments of the stream relative to the 1971 samples. Levels have increased from 26-50 ppm in 1971 to a range of between 51-100 ppm in 1973.

Zinc. As was the case with copper, the zinc concentrations have increased from 401-800 ppm in 1971 to 801-1200 ppm in 1973.

Rappahannock River

Cadmium. The cadmium distribution appears to have changed since the 1971 study. The concentrations were lower in the upper river decreasing to between 0.6-1.0 ppm from 1.1-1.5 ppm in 1971. In addition, an apparent anomaly exists at the mouth of the estuary. This "high" cadmium level may be due to the influence of the storm waters that came down the Bay from the Potomac and Susquehanna Rivers.

Copper. The copper data indicate that the concentration range was not different from that found in 1971; however, the distribution changed in the upper estuary. From previous work, we expected the highest concentrations to appear in the low salinity waters. After Agnes, residues in the upper portion were lower followed by an area of higher concentrations with a return to lower levels in the downstream segment.

Zinc. The zinc distribution followed a pattern very similar to that of copper. The concentration range was unchanged since 1971, but the distribution was altered.

Mobjack Bay, Poquoson River, Back River

Samples from these areas did not indicate any significant changes in either concentration ranges or distributions since the 1971 study. This was most likely due to the immediate proximity of these areas to the Chesapeake Bay proper and therefore the lesser effects of the storm waters from Agnes.

It has been shown (Huggett, Bender & Slone 1973) that within similar drainage basins a relationship exists between the amount of zinc and copper taken up by oysters. Regression analyses were performed on the 1971 and 1973 data (Cu vs. Zn) for each river system in an attempt to ascertain if Agnes had modified the relationships shown previously. The results of these analyses are presented in Table 2 and depicted graphically in Fig. 7. As can be seen from the figure, generally good agreement exists between the Cu and Zn relationships in all three river systems. Statistical tests cannot be applied to the data because both variables are independent; however, the data suggest that copper levels relative to zinc are slightly lower in the post-Agnes period.

Table 2. Regression Equations, Metals in Oysters.

	James 1971	James 1973	York 1971	York 1973	Rappahannock 1971	Rappahannock 1973
Cu(y) vs Zn(x)						
Slope	0.033	0.075	0.088	0.075	0.090	0.035
Intercept	57.47	4.36	3.18	-12.21	-12.86	3.76
N	95	75	53	124	45	70
Correlation(r)	0.53	0.78	0.71	0.92	0.79	0.66
Cd(y) vs Zn(x)						
Slope		0.0004		0.0008		0.0007
Intercept		0.59		0.17		0.33
N		70		49		71
Correlation		0.43		0.75		0.45
Cd(y) vs Cu(x)						
Slope		0.005		0.010		0.018
Intercept		0.51		0.26		0.28
N		70		49		71
Correlation		0.59		0.83		0.64

Regressions relating the concentrations of Cu to Cd and Cd to Zn calculated on the 1973 data are tabulated in Table 2 and are graphically shown in Fig. 8.

A comparison of pesticide residues found in oysters at four stations is presented in Table 3. Although the data are limited, the conclusion suggested, i.e. that Agnes did not elevate pesticide residues, is supported by the fact that in 9 samples of 12 hard clams each, no residue higher than 0.001 ppm was detected.

In addition, in 5 samples, each containing the meats of 12 blue crabs, the highest residue obtained was 0.04 ppm with an average of 0.02 ppm.

Table 3. Comparison of Pesticide Residues in Oysters 1971 vs. 1972.

Location	DDT (family) ppm 1971	DDT (family) ppm 1972	PCB's ppm 1971	PCB's ppm 1972
Lynnhaven Bay	0.047	0.002	ND	ND
Cherrystone Inlet	0.052	0.024	0.350	ND
Old Plantation Creek		0.033		ND
Poquoson River		0.001		ND
Pages Rock	0.010	0.001	ND	ND
Bells Rock	0.010	0.015	0.450	ND

DISCUSSION

The data indicate that the residues of cadmium, copper, and zinc in oysters from the upper segments of the James and Rappahannock Rivers have decreased since Agnes. The residues in the middle segments of these rivers have remained nearly unchanged since the previous sampling in 1971. A plausible hypothesis to account for the observed decreases in the upper reaches is that the oysters metabolized stored food reserves, i.e. proteins, containing nonspecifically bound metals at a faster rate than they picked up residues from the water. This occurred because their pumping rates were slowed due to depressed salinities. Support for this hypothesis can be obtained from the results of Huggett and Lunz (1974) who showed that metals were depurated from oysters when subjected to unknown stresses caused by high concentrations of sewage effluent.

In the Rappahannock, oysters were subjected to lowered salinities not only in the upper reaches but also in the lower river due to large influxes of fresh water from the Susquehanna and Potomac drainages. Oysters from this region showed higher than expected levels of cadmium and zinc. These elevations may be attributed to metals either being transported to the system with the fresh waters from up-Bay or by mobilization of sediment-stored metals by the low salinities and accompanying low pH's.

Although salinity changes were less dramatic in the York than in the other river systems, zinc levels in the upper portion of the system were higher than previously recorded. Agnes may have been partly responsible for the observed elevation in residues by moving polluted sediments with their associated metals further downstream (Huggett, Cross & Bender 1974).

ACKNOWLEDGEMENTS

The project was funded, in part by contract funds from the Food and Drug Administration (Contract No. 73-10).

LITERATURE CITED

Butler, P. A. 1973. Organochlorine residues in estuarine mollusks, 1965-1972 national pesticide monitoring program. Pest. Monit. J. 6:238-362.

Huggett, R. J., M. E. Bender, and H. D. Slone. 1973. Utilizing metal concentration relationship in the eastern oyster (*Crassostrea virginica*) to detect heavy metal pollution. Water Res. 7:451-460.

Huggett, R. J., F. A. Cross, and M. E. Bender. 1974. Trace metals in tidal rivers and biota. In: Mineral Cycling in Southeastern Ecosystems, Augusta, Georgia.

Huggett, R. J. and J. L. Lunz. 1974. Heavy metal uptake by oysters from sewage. Interim Progress Rep. to National Science Foundation (Research Applied to National Needs Program).

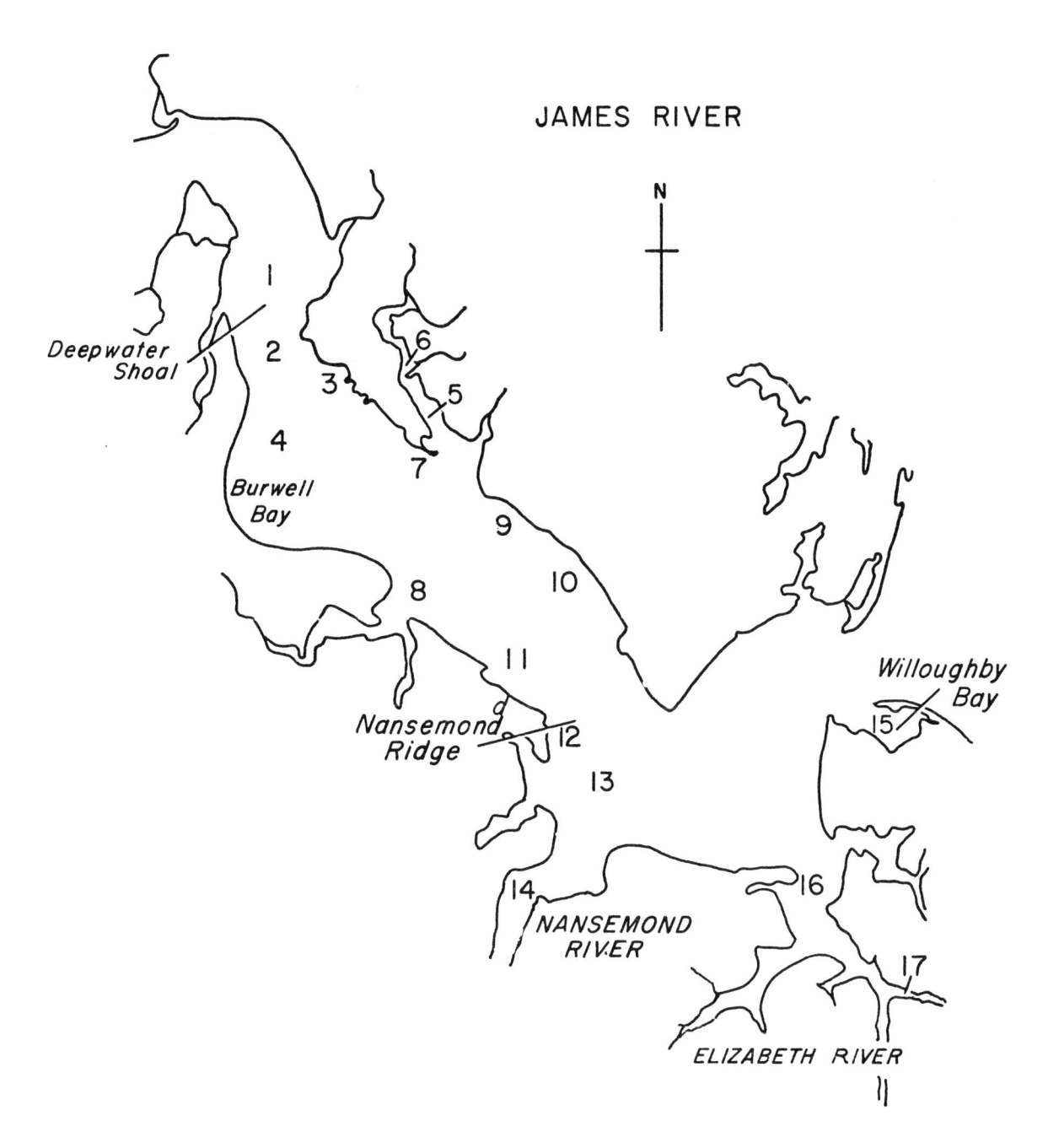

Figure 1. Sampling locations in the James, Elizabeth, and Nansemond Rivers.

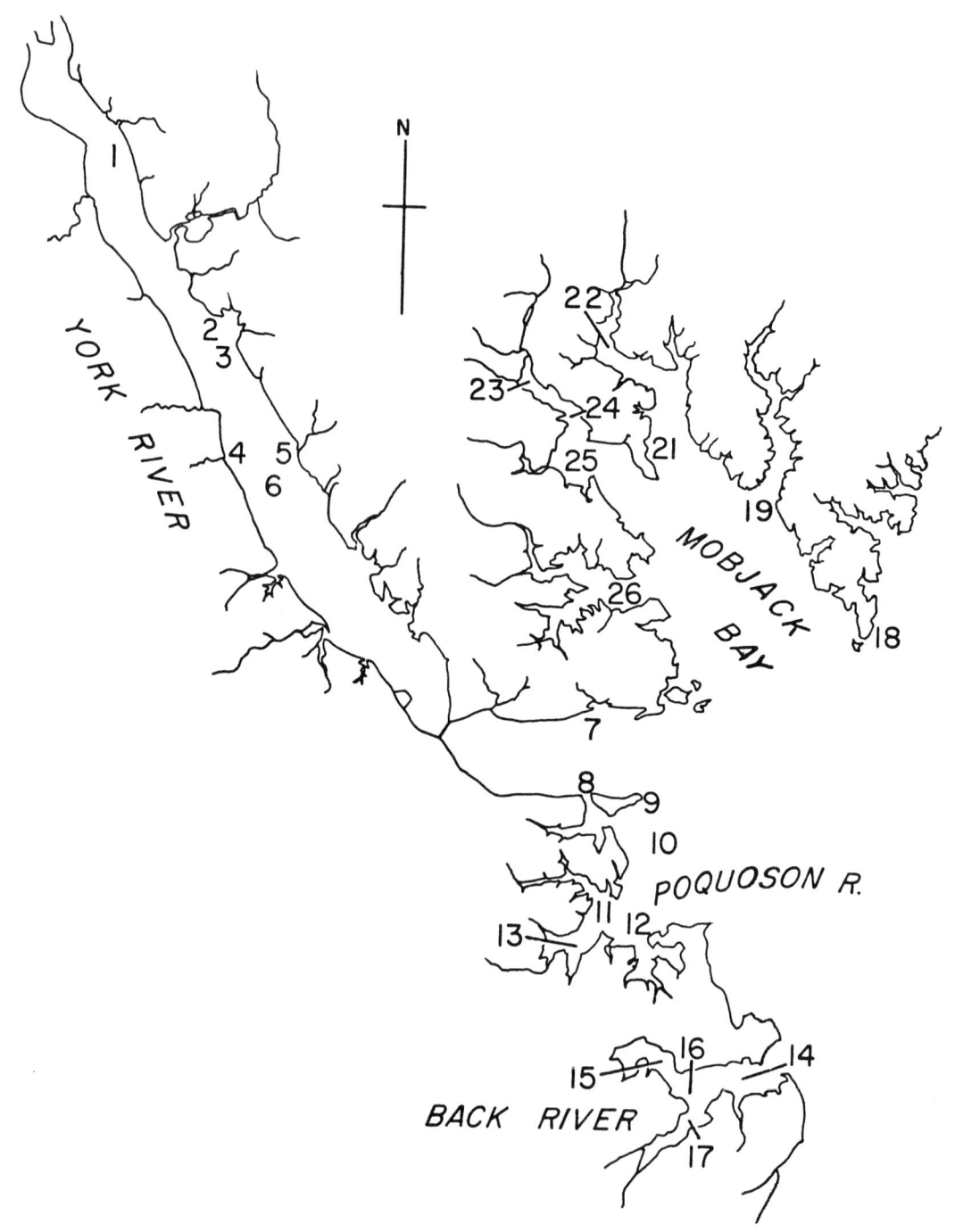

Figure 2. Sampling locations in the York, Back, and Poquoson Rivers and Mobjack Bay.

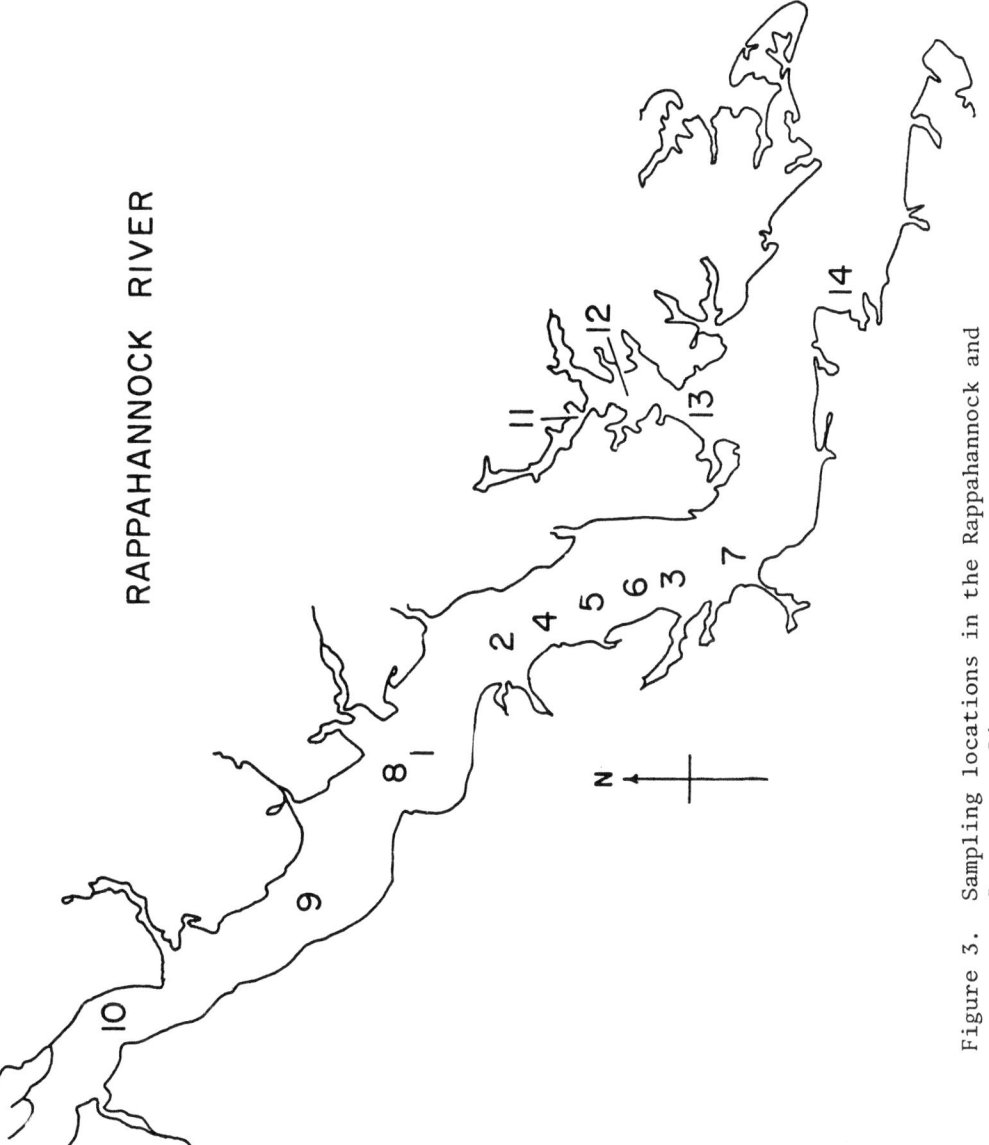

Figure 3. Sampling locations in the Rappahannock and Corrotoman Rivers.

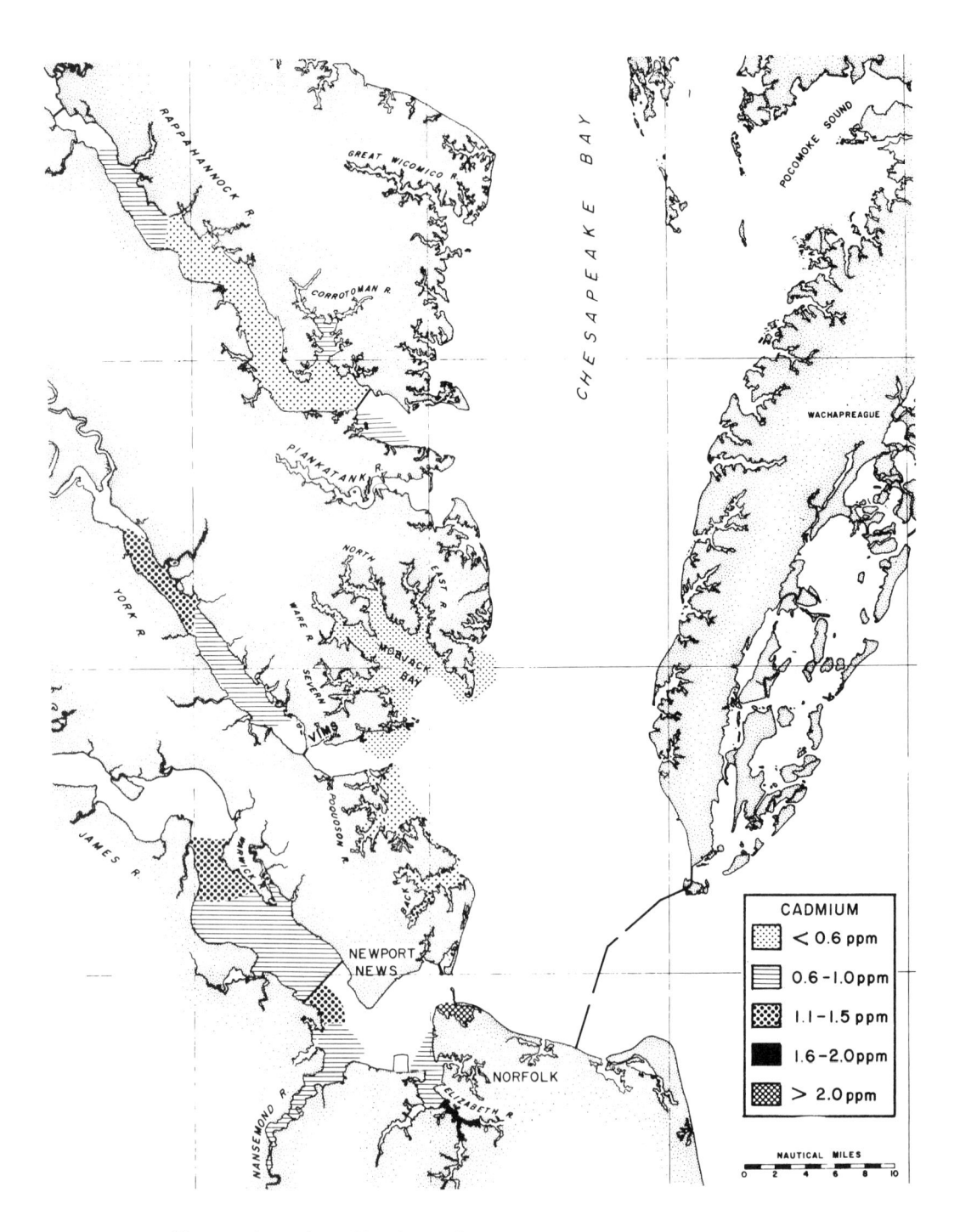

Figure 4. Distribution of cadmium in oysters sampled in January 1973.

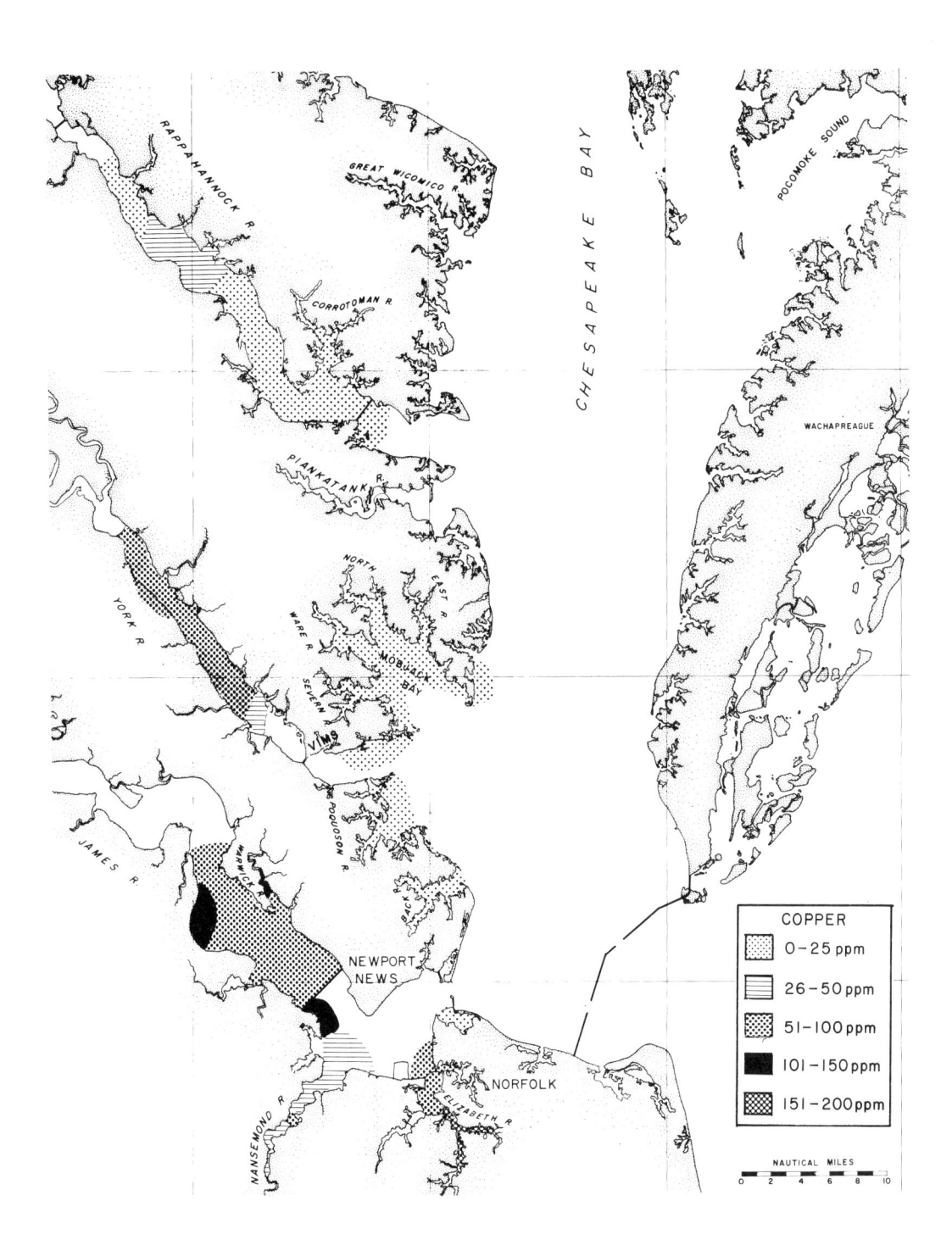

Figure 5. Distribution of copper in oysters sampled in January 1973.

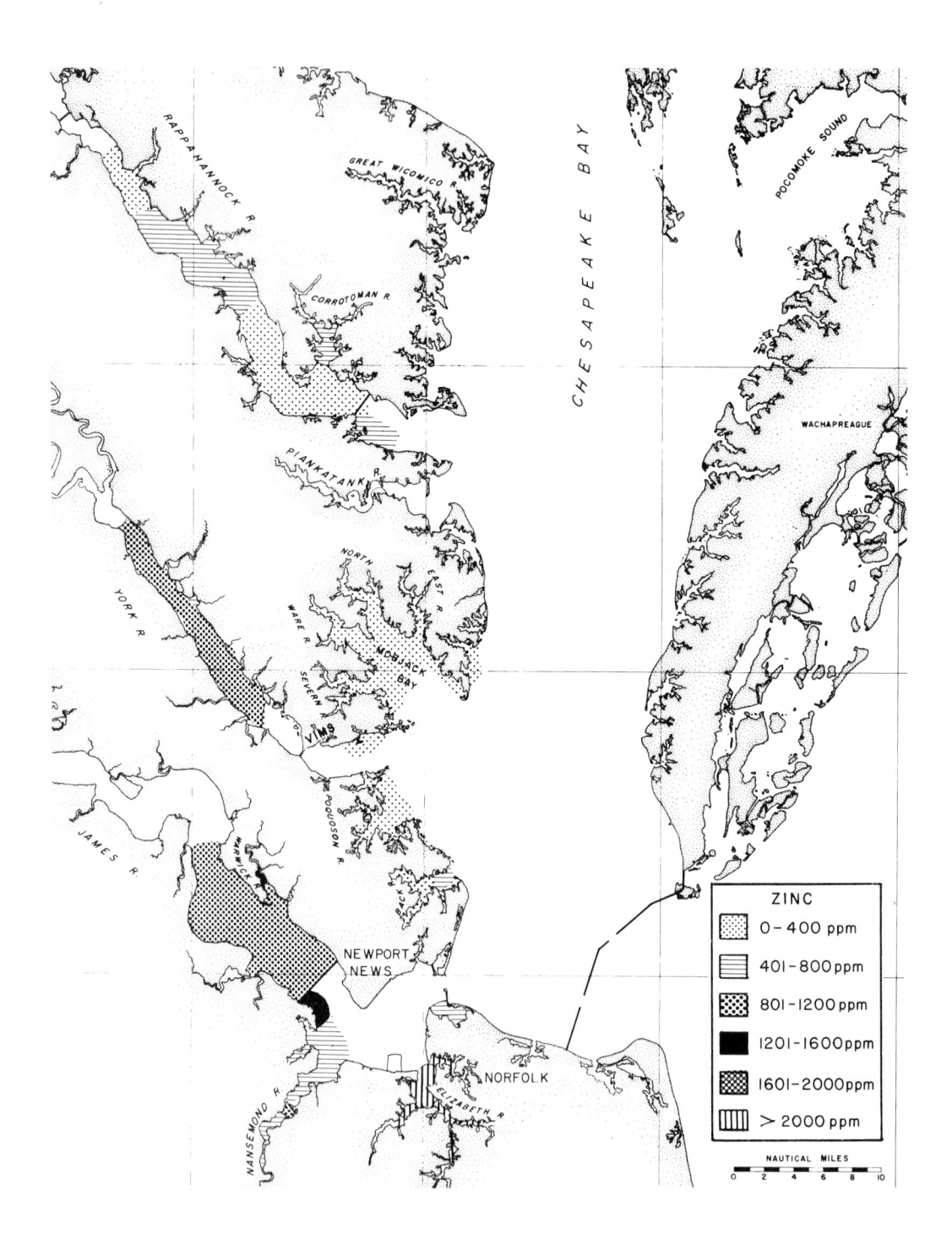

Figure 6. Distribution of zinc in oysters sampled in January 1973.

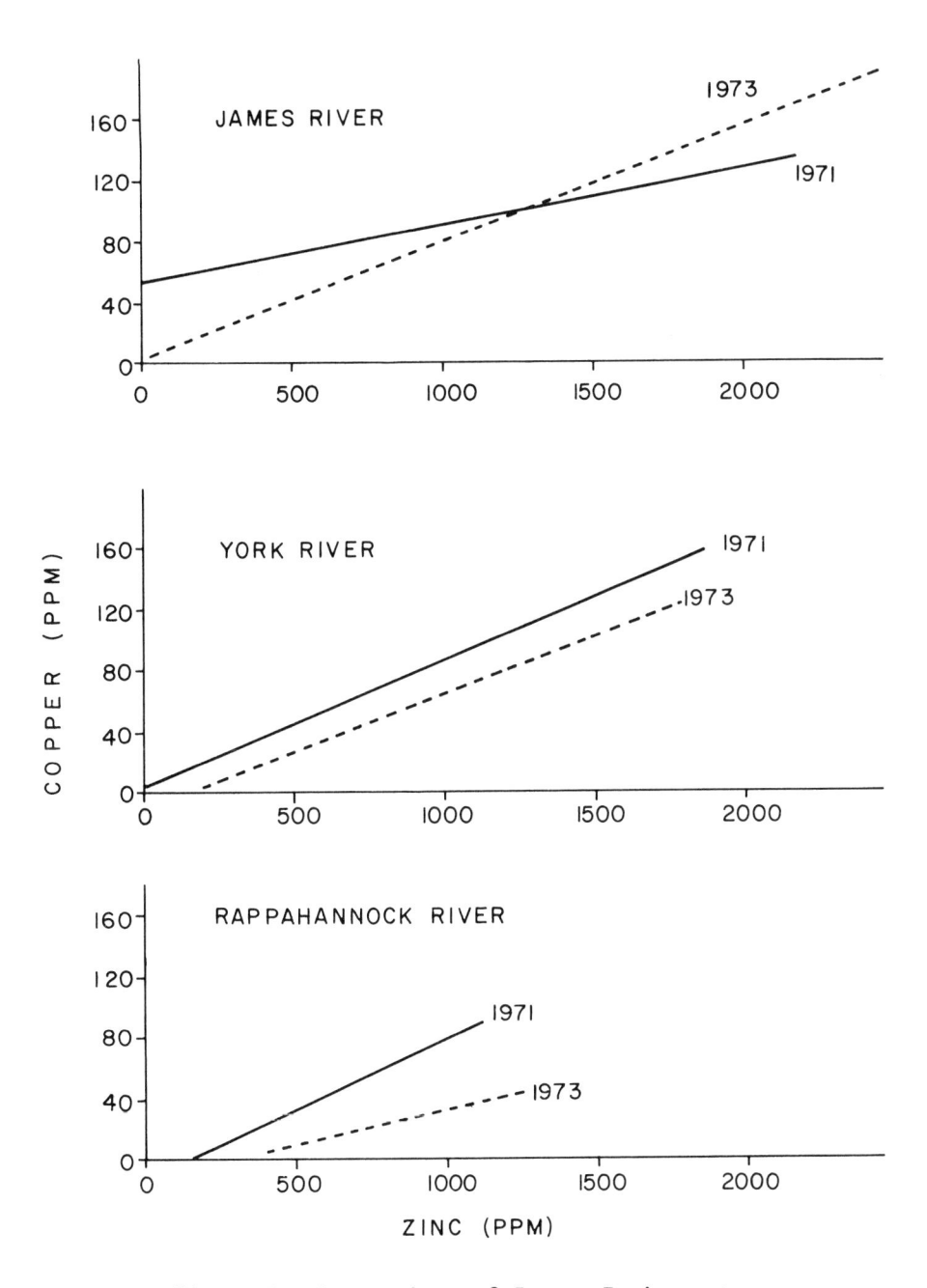

Figure 7. Regressions of Cu vs. Zn in oysters.

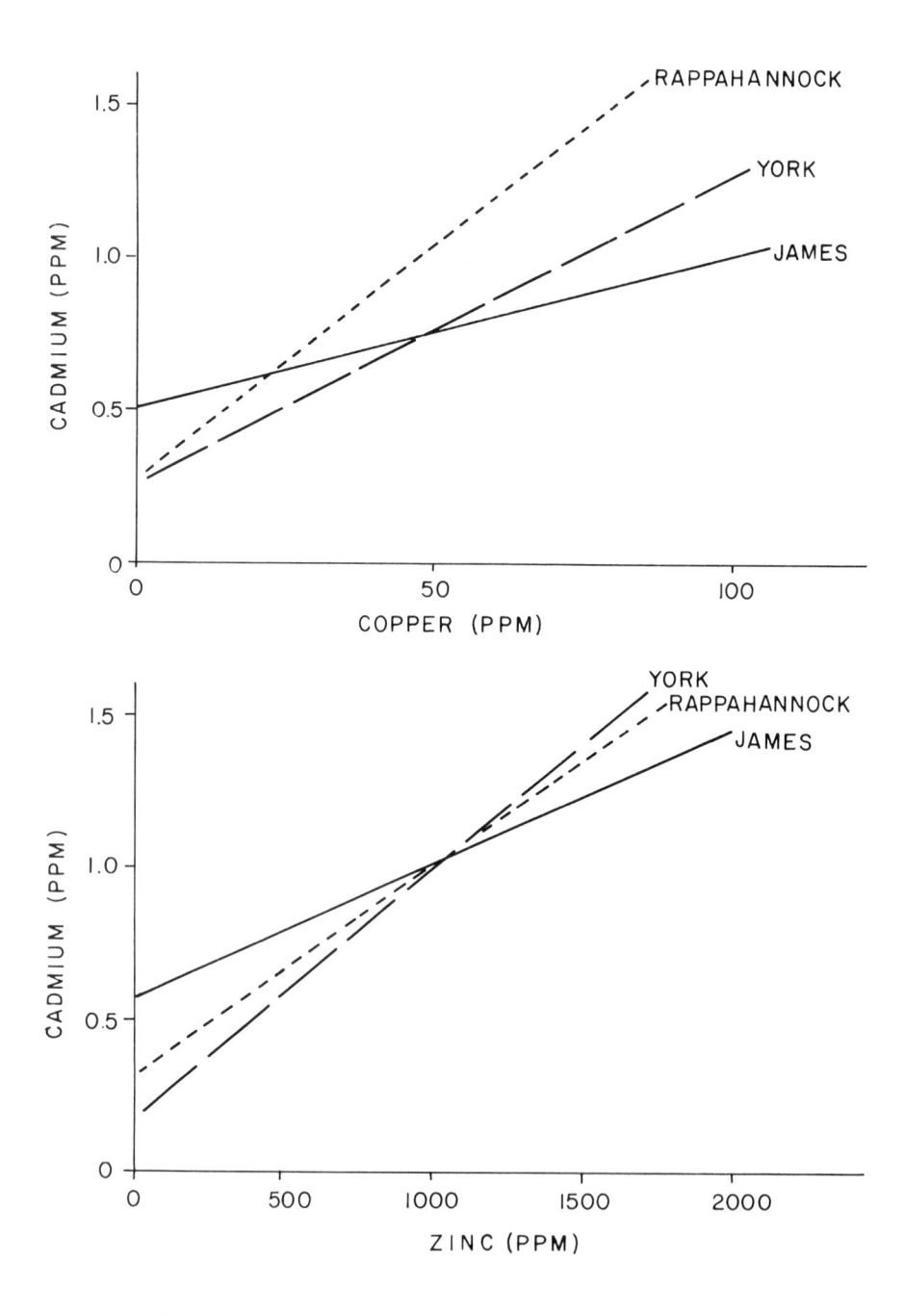

Figure 8. Regression of Cu vs. Cd and Cd vs. Zn in oysters.

EFFECTS OF AGNES ON THE DISTRIBUTION OF DISSOLVED OXYGEN ALONG THE MAIN AXIS OF THE BAY[1]

J. R. Schubel[2]
W. B. Cronin[2]

ABSTRACT

Determinations were made of the distribution of dissolved oxygen (DO) along the axis of the Bay at approximately two to four week intervals from 20 July 1972 through 29 June 1973 to chronicle the impact of Agnes on the distribution of DO. The results are equivocal. On only one cruise, 28-31 August 1972, did the results appear anomalous. At that time there was convincing evidence of an extensive anaerobic layer in the deeper waters between the mouth of the Potomac and the northern end of Kent Island. The layer did not persist. By the next cruise, 19-21 September 1972, the lower layer had been replenished with oxygen and the distribution of DO remained near "normal" throughout the remainder of the sampling period.

INTRODUCTION

It was anticipated that in 1972 the zone of low oxygen that characterizes the deeper portions of the Bay in mid and late summer would, as a result of Agnes, appear earlier, be more extensive, and persist longer than in most years. To test this hypothesis, analyses were made of the distribution of DO along the axis of the Bay at approximately two to four week intervals for nearly a year following Agnes -- 20 July 1972 through 29 June 1973. To assess the probable impact of Agnes on the distribution of DO these data were compared with data from "normal" years to determine whether the distributions of DO following Agnes fell outside of the "normal" range of observations.

The spatial and temporal variations of the concentration of DO in the Bay proper during "normal" years have been graphically described by Hires, Stroup and Seitz (1963). Other observations of the distribution of DO made by the Chesapeake Bay Institute (CBI) have been summarized in tabular form in special summary data reports (Whaley, Carpenter & Baker 1966), and in the regular CBI data report series. In addition, CBI has numerous unpublished observations of the concentration of DO made over the past 25 years in its data bank.

RESULTS AND DISCUSSION

Water samples were collected with a submersible pump, and determinations of the concentration of DO were made using a modified Winkler technique (Carpenter 1965b). The method has an accuracy of about 0.1% (Carpenter 1965a).

Data from selected cruises are presented as a series of longitudinal distributions of DO along the axis of the Bay (Figs. 2-13). The station locations are shown in Fig. 1, and data on sampling frequency are summarized in Table 1.

---

[1] Contribution No. 237, Chesapeake Bay Institute, Johns Hopkins University
[2] Chesapeake Bay Institute, Johns Hopkins University, Baltimore, Md. 21218

Table 1. Summary of sampling locations and dates. Data from selected cruises are plotted in Figs. 2-13.

	927SS	SF00	921W	917S	913R	909	903A	858C	848E	834F	818P	804C	745	724R	707Ø	659W
1972																
20-24 July	X	X	X	X	X	X	X	X		X	X	X	X			
3-4 Aug.	X	X	X	X	X	X	X	X								
9 Aug.	X		X		X		X	X								
16-17 Aug.	X	X	X	X	X	X	X	X	X							
28-31 Aug.	X	X	X	X	X	X	X	X	X		X	X	X	X	X	X
16-19 Sept.	X	X	X	X	X	X	X	X	X		X	X	X	X	X	X
25-26 Sept.		X	X	X	X	X	X	X								
9-12 Oct.	X	X	X	X	X	X	X	X	X	X	X	X	X	X	X	X
7 Nov.	X	X	X	X	X	X	X	X	X							
16-20 Nov.				X			X			X	X	X	X	X	X	
27-28 Nov.	X	X	X	X	X	X	X									
11-15 Dec.	X	X	X	X	X	X	X	X	X		X	X	X	X	X	X
1973																
8 Jan.		X	X	X	X	X	X	X	X							
22-24 Jan.	X	X	X	X	X	X	X	X	X	X	X	X	X	X		
5 Feb.	X	X	X	X	X	X	X	X	X							
27 Feb-3 Mar	X	X	X	X	X	X	X	X	X	X	X	X	X	X	X	X
26-30 Mar	X	X	X	X	X	X	X	X	X	X	X	X	X	X	X	X
23-26 April	X	X	X	X	X	X	X	X	X	X	X	X	X	X	X	X
28-30 May	X	X	X	X	X	X	X	X	X	X	X	X	X	X	X	X
5 June	X	X	X	X	X	X	X	X	X							
25-29 June	X	X	X	X	X	X	X	X	X	X	X	X	X	X	X	X

The first post-Agnes oxygen determinations were made between 20-24 July 1972, approximately one month after the rivers crested (Fig. 2). Comparison of these data from other years indicates that the distribution of DO on 20-24 July 1972 was within the normal year to year variations observed over the past 25 years. Fig. 14, for example, depicts the distribution of DO in July 1969.

On only one cruise did the distribution of DO appear to be anomalous, and even then the evidence is equivocal. Our data for 28-31 August 1972 (Fig. 3) show that during that period there was a fairly extensive layer in which no oxygen was detected. This anaerobic zone extended from Station 804C off the mouth of the Potomac north to Station 903A off the northern end of Kent Island, and came to within 12 m of the surface (Fig. 3). A query of the CBI data bank showed that this was the first time we had had such convincing evidence of an extensive anaerobic layer in the Bay proper. Previously there had been some indication of such a zone (see, for example, Fig. 15), but the evidence was always equivocal because of the small number of determinations made within the layer in question.

The anaerobic layer reported on 28-31 August 1972 did not persist. The next set of observations taken about three weeks later (Fig. 4), showed that the oxygen levels in this zone had increased, and that the longitudinal distribution of DO on 19-21 September 1972 was not significantly different from that observed in the month of September during other more "normal" years (Fig. 16).

The remainder of the post-Agnes distributions of DO (Figs. 5-12) fall within the normal range of year to year variations that have been observed for comparable seasons over the past 25 years.

The relatively rapid increase in the levels of DO in the lower layer that were observed between the 28-31 August 1972 cruise and the 19-21 September 1972 cruise can easily be accounted for by the turbulent exchange of DO from the upper to the lower layers. Consider the following: Define $F_{D.O.}$ as the vertical flux of dissolved oxygen through the imaginary horizontal plane in the pycnocline. Then,

$$F_{D.O.} = -K_z \frac{\partial \xi_{D.O.}}{\partial z}$$

where K_z is the vertical eddy diffusivity (cm^2/sec), $\xi_{D.O.}$ is the concentration of dissolved oxygen (mg/kg), and z is depth (m) taken in a positive direction downward. K_z is a minimum in the pycnocline, and generally falls between 1 and 10 cm^2/sec in this zone of maximum stability (Pritchard 1960). For a conservative estimate of the downward flux of dissolved oxygen take $K_z = 1$ cm^2/sec, which is equal to 8.64 m^2/day. From Fig. 3 a reasonable estimate of the vertical gradient of the concentration of dissolved oxygen in the pycnocline in the vicinity of Station 834F during the period 28-31 August 1972 was about 4 mℓ/ℓ in 4 m, or 1 mℓ/ℓ/m. This is approximately equivalent to 1.4 mg/kg/m. Substituting in (1) we have

$$F_{D.O.} = -K_z \frac{\partial \xi_{D.O.}}{\partial z} \approx -\frac{(8.64 \text{ m}^2)}{\text{day}} \left(\frac{-1.4 \text{ mg/kg}}{\text{m}}\right)$$
$$\approx 12 \text{ m/day} \times \text{mg/kg}$$

Since 1 mg/kg is approximately equal to 1 gm/m^3, the downward flux of oxygen through an imaginary horizontal plane in the pycnocline is approximately 12 gm/m^2/day. If we take the average thickness of the oxygen depleted layer as 10 m, then the volume of the anaerobic layer under each m^2 of this imaginary plane is 10 m^3. Therefore, the rate of oxygen replenishment per unit volume of water in this lower layer is about 1.2 gm/m^3/day, which is nearly 1 mℓ/ℓ/day. Since the "average" rate of oxygen utilization is also in the range of 0.5-1 mℓ/ℓ/day, relatively small changes in either the rate of oxygen utilization or in the rate of replenishment of oxygen by turbulent diffusion across the pycnocline could easily account for the increases in the levels of DO observed between 28-31 August and 19-21 September 1972.

In summary, Agnes certainly imposed an increased oxygen demand on the Bay, and this demand certainly affected the distribution of DO. These effects however were not persistent, nor were they even unequivocally documented. All of the observed distributions, except perhaps that of 28-31 August 1972, fell within the "normal" range--the range of distributions delimited by the normal year to year variations for that same season. On the 28-31 August 1972 cruise the distribution of DO did appear to be anomalous; a fairly extensive anaerobic layer characterized the deeper portions of the middle Bay. Within the three weeks however, the lower layer was replenished with oxygen and the distribution of DO remained near "normal" throughout the remainder of the sampling period.

There is a feeling among some Bay investigators that, as a result of man's activities there is a general trend for the zones of low oxygen that characterize the deeper waters of the Bay and its tributaries in summer to appear earlier, be more extensive, and persist longer. While such trends would be difficult to either prove or disprove because of the large year to year variations in the distribution of DO there is a fairly extensive set of historical data that should be critically examined.

ACKNOWLEDGEMENTS

We thank M. Glendening, J. Oland, C. Zabawa, and T. Kana for the chemical analyses, and D. W. Pritchard and H. H. Carter for their suggestions. The figures were drawn by Dean Pendleton and photographed by J. Sullivan. The manuscript was typed by A. Sullivan. This research was supported in part by the Oceanography Section, National Science Foundation, NSF Grant GA-36091; in part by a project jointly funded by the Fish and Wildlife Administration, State of Maryland, and the Bureau of Sport Fisheries and Wildlife, U. S. Department of Interior, using Dingell-Johnson Funds; and in part by the U. S. Army Corps of Engineers (Baltimore and Philadelphia Districts).

LITERATURE CITED

Carpenter, J. H. 1965a. The accuracy of the Winkler method for dissolved oxygen analysis. Limnol. Oceanogr. 10(1):135-140.

Carpenter, J. H. 1965b. The Chesapeake Bay Institute technique for the Winkler dissolved oxygen method. Limnol. Oceanogr. 10(1):141-143.

Hires, R. I., E. D. Stroup, and R. C. Seitz. 1963. Atlas of the distribution of dissolved oxygen and pH in Chesapeake Bay 1949-1961. Johns Hopkins Univ., Chesapeake Bay Inst. Graph. Summ. Rep. 3, Ref. 63-4, 411 p.

Pritchard, D. W. 1960. The movement and mixing of contaminants in estuaries. Pages 512-525 in Proc. 1st Int. Conf. Waste Disposal in the Marine Environment (E. A. Pearson, ed.), Univ. of California, Berkeley, July 22-25, 1959. Pergamon Press, Inc., New York, N. Y., 569 p.

Whaley, R. C., J. H. Carpenter, and R. L. Baker. 1966. Nutrient Data Summary 1964, 1965, 1966: Upper Chesapeake Bay (Smith Point to Turkey Point) Potomac, South, Severn, Magothy, Back, Chester, and Miles Rivers; and Eastern Bay. Johns Hopkins Univ., Chesapeake Bay Inst. Spec. Rep. 12, Ref. 66-4, 88 p.

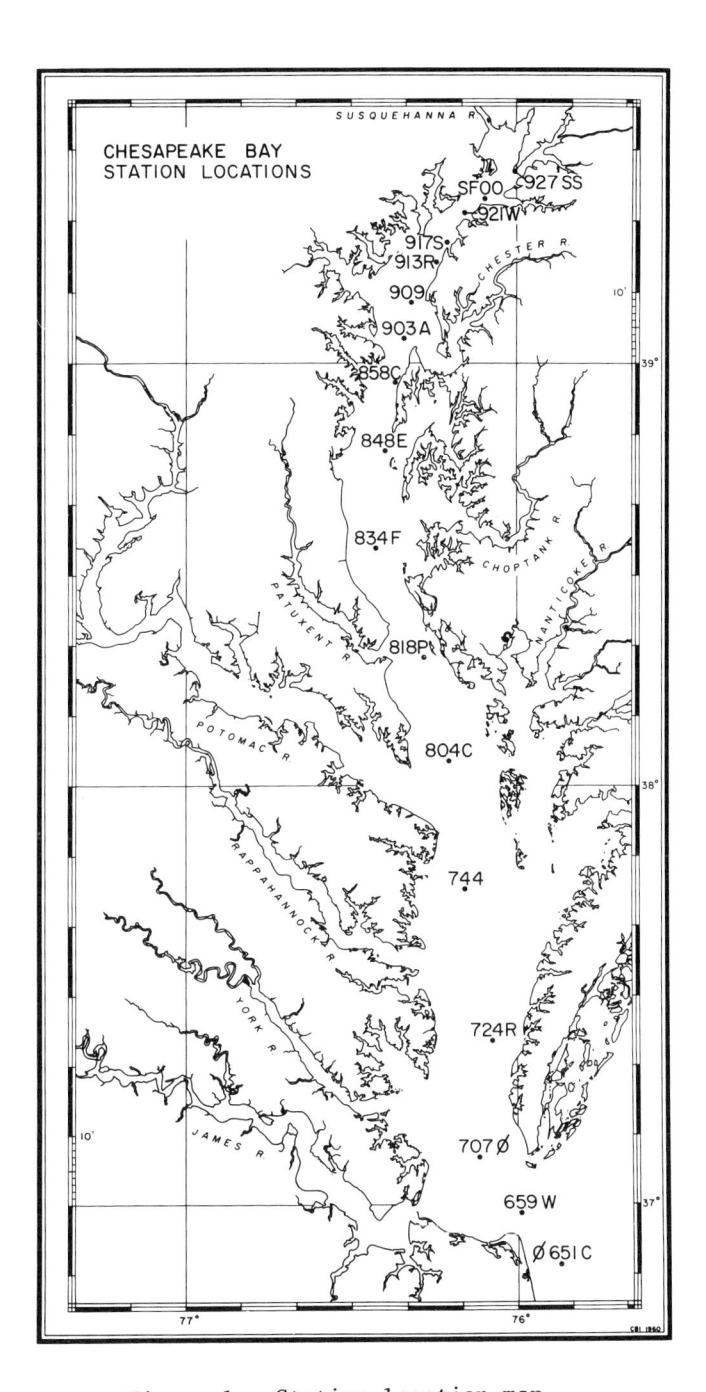

Figure 1. Station location map.

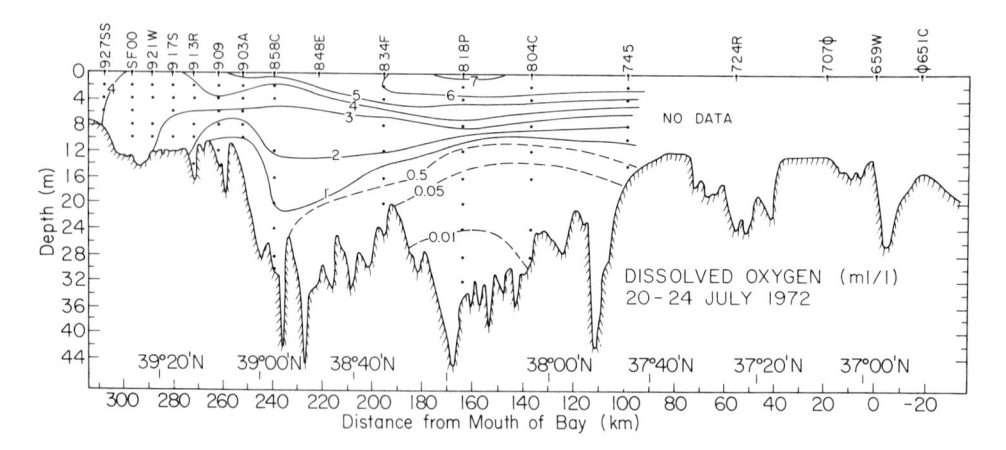

Figure 2. Distribution of dissolved oxygen along the axis of the Bay, 20-24 July 1972.

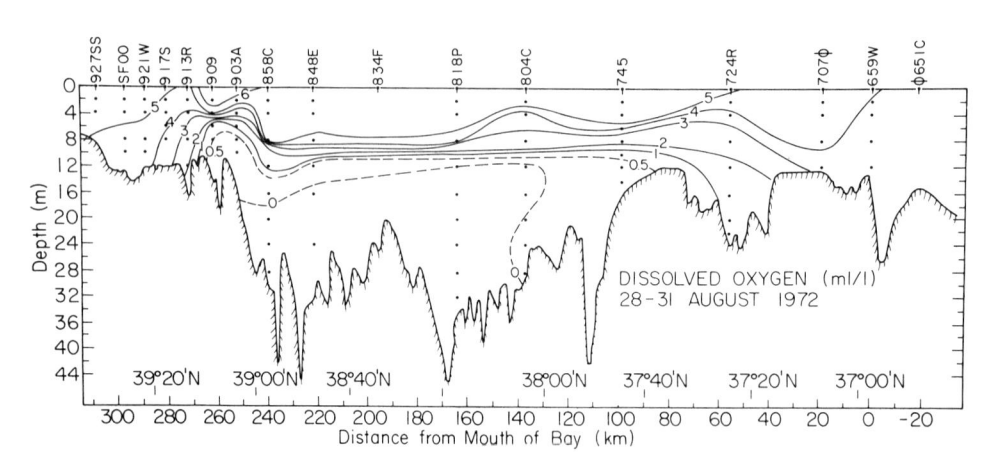

Figure 3. Distribution of dissolved oxygen along the axis of the Bay, 28-31 August 1972.

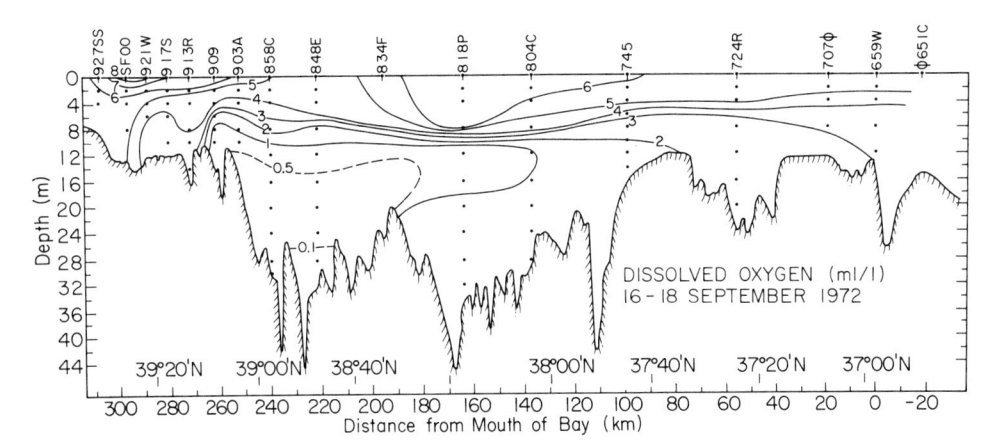

Figure 4. Distribution of dissolved oxygen along the axis of the Bay, 16-18 September 1972.

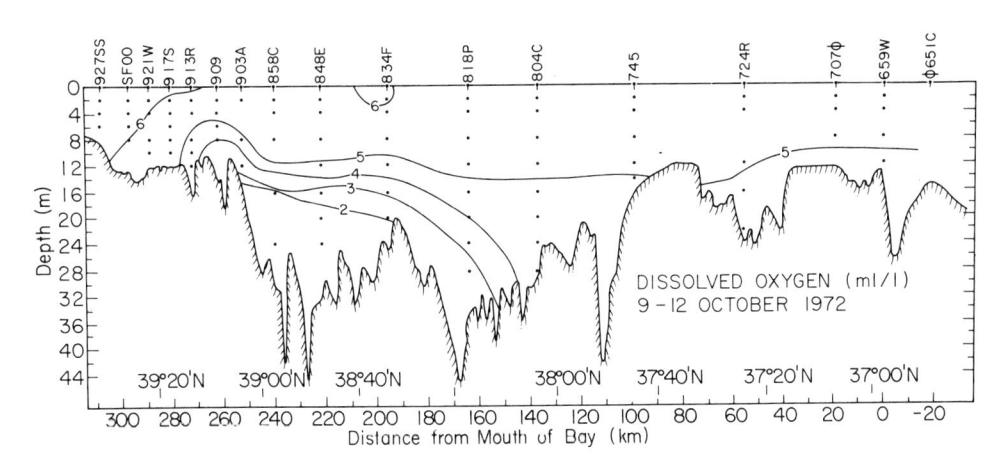

Figure 5. Distribution of dissolved oxygen along the axis of the Bay, 9-12 October 1972.

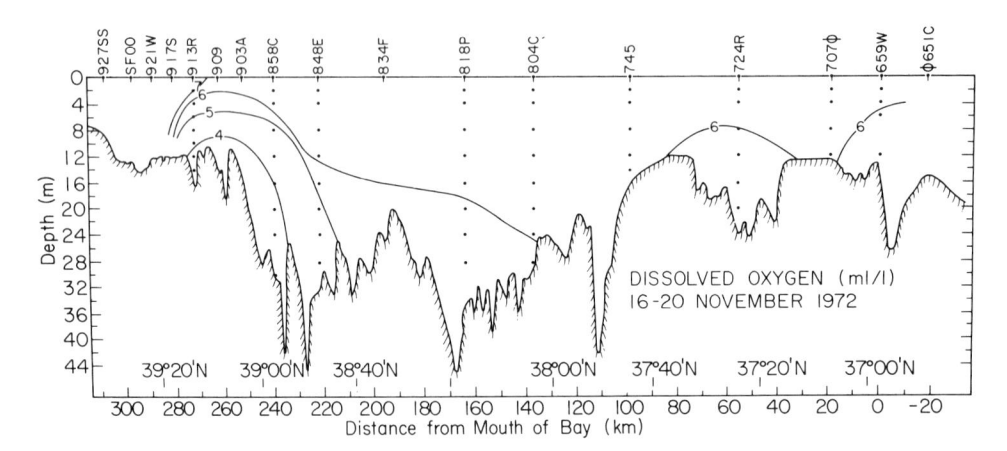

Figure 6. Distribution of dissolved oxygen along the axis of the Bay, 16-20 November 1972.

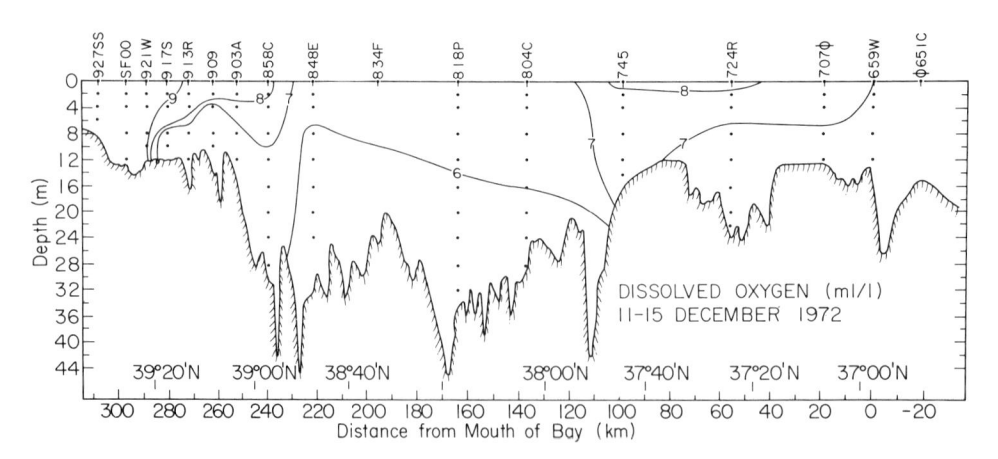

Figure 7. Distribution of dissolved oxygen along the axis of the Bay, 11-15 December 1972.

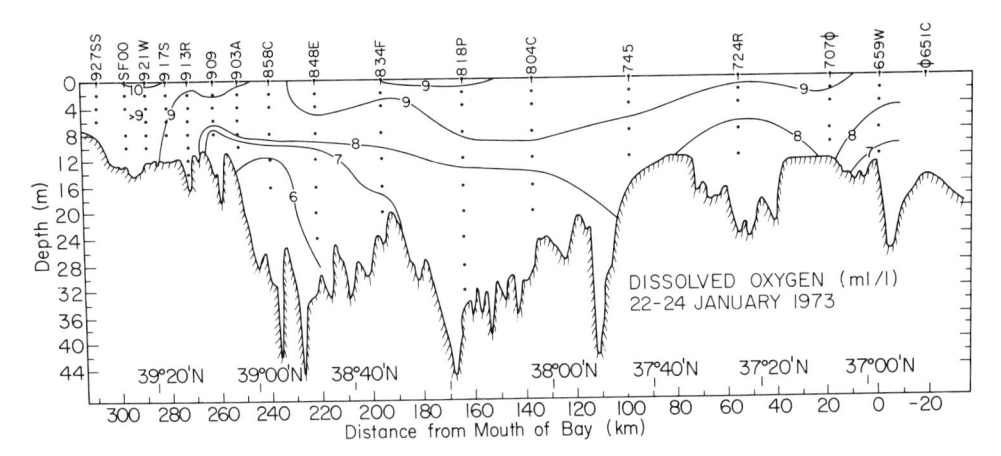

Figure 8. Distribution of dissolved oxygen along the axis of the Bay, 22-24 January 1973.

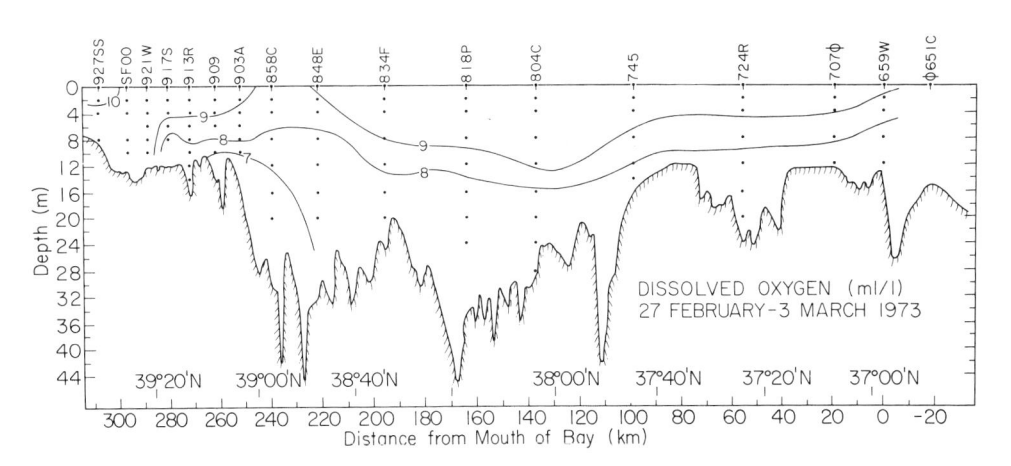

Figure 9. Distribution of dissolved oxygen along the axis of the Bay, 27 February-3 March 1973.

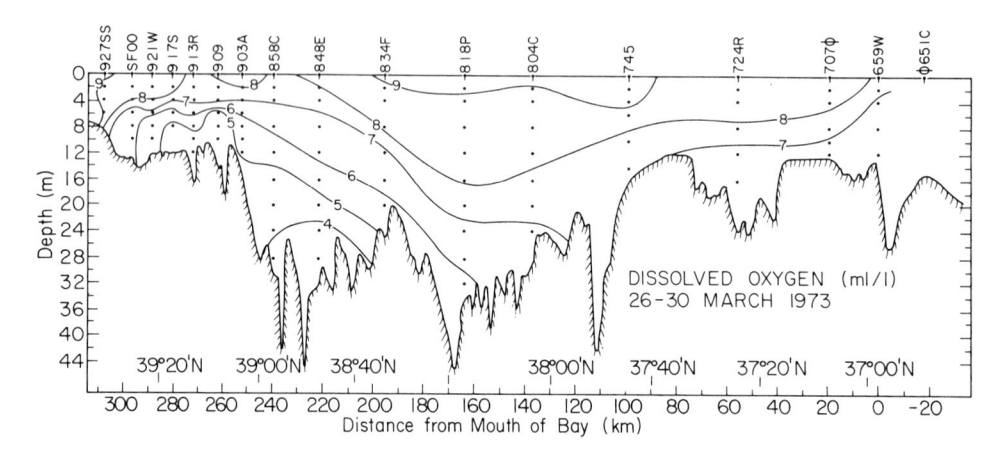

Figure 10. Distribution of dissolved oxygen along the axis of the Bay, 26-30 March 1973.

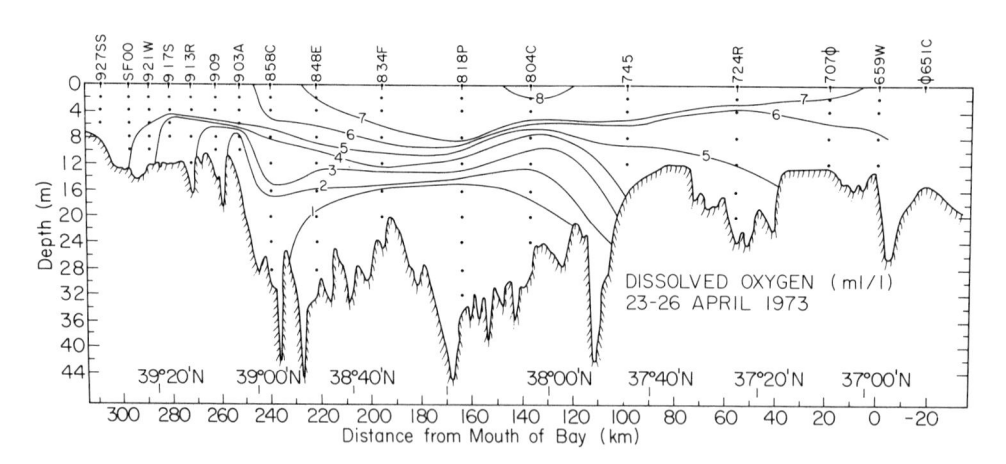

Figure 11. Distribution of dissolved oxygen along the axis of the Bay, 23-26 April 1973.

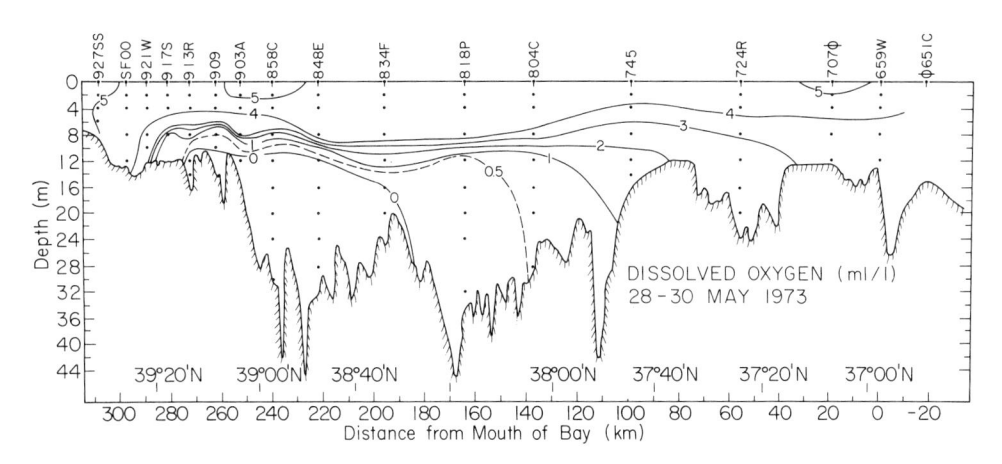

Figure 12. Distribution of dissolved oxygen along the axis of the Bay, 28-30 May 1973.

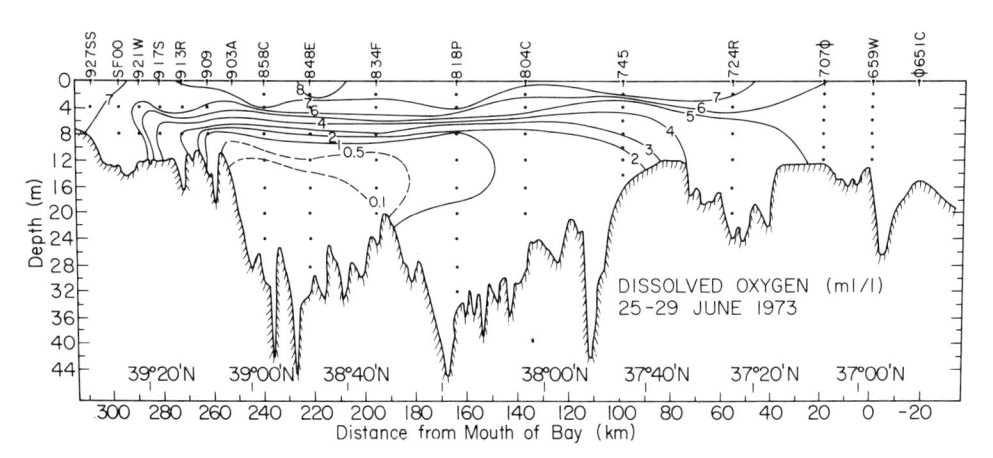

Figure 13. Distribution of dissolved oxygen along the axis of the Bay, 25-29 June 1973.

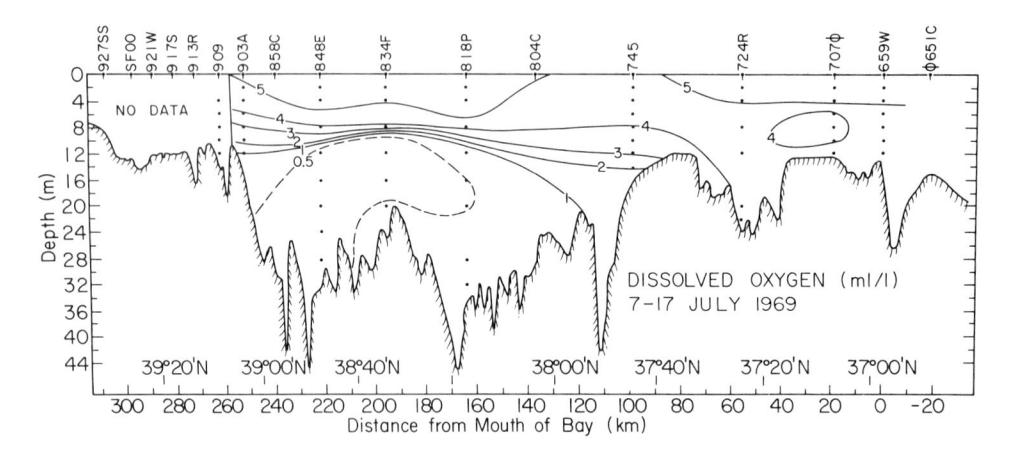

Figure 14. Distribution of dissolved oxygen along the axis of the Bay, 7-17 July 1969.

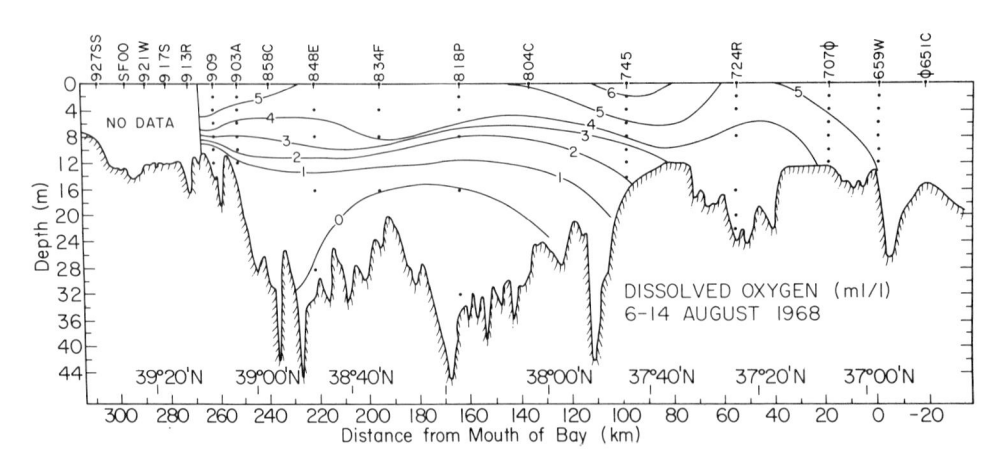

Figure 15. Distribution of dissolved oxygen along the axis of the Bay, 6-14 August 1968.

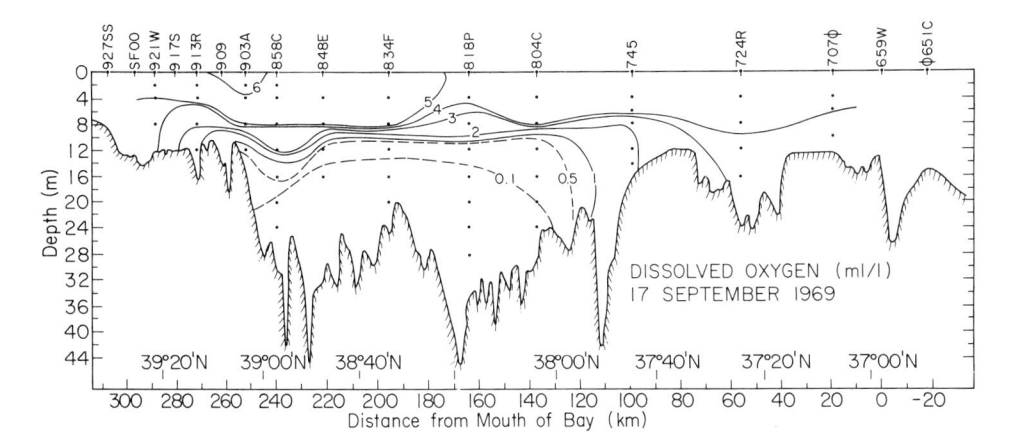

Figure 16. Distribution of dissolved oxygen along the axis of the Bay, 17 September 1969.

OBSERVATIONS ON DISSOLVED OXYGEN CONDITIONS IN THREE
VIRGINIA ESTUARIES AFTER TROPICAL STORM AGNES (SUMMER 1972)[1]

Robert A. Jordan [2]

ABSTRACT

Dissolved oxygen (DO) and salinity levels in the James, York, and Rappahannock estuaries were monitored for approximately two months (June 24-August 31, 1972) following Tropical Storm Agnes. DO depressions developed more rapidly and were more severe in the deep waters of the York and Rappahannock than in the James. Depressions that developed immediately after the storm were followed by recoveries and subsequent, more severe depressions. In late July, bottom water DO concentrations below 1 mg/l were found at stations covering 15 miles of the York and 25 miles of the Rappahannock. Comparison of river data with Chesapeake Bay data suggests that the rivers contributed oxygen poor water to the Bay during the post-Agnes period. Comparison of 1972 river data with data from other years suggests that the post-Agnes oxygen depressions were more severe than those that occur in normal years.

INTRODUCTION

Among the environmental disturbances that occurred in Virginia waters in the aftermath of Agnes were episodes of depressed DO concentrations. This account consists of descriptions of DO fluctuations that were observed in the estuarine reaches of the James, York, and Rappahannock Rivers during the two months (June 24-August 31, 1972) following the storm.

METHODS

In the immediate post-Agnes period, VIMS personnel engaged in an intensive effort to monitor the major hydrographic parameters in these rivers as well as in portions of lower Chesapeake Bay and at certain stations off the Bay mouth. Anchor stations were manned continually from June 24 through the first week of July for measurement of DO and salinity levels at one or two hour intervals. Slack water sampling runs, covering the same parameters at a longitudinal series of stations in each river, were conducted frequently beginning June 24 and extending into August.

Salinities were measured in terms of electrolytic conductivity on a Beckman model RS-7B Induction Salinometer. DO concentrations were measured by the standard Winkler iodometric method (American Public Health Association 1971).

RESULTS

James River Anchor Station

One of the anchor stations occupied in the James River was located at $36°59.5'$N latitude and $76°27.1'$W longitude, approximately 10 nautical miles upstream

---
[1] Contribution No. 775, Virginia Institute of Marine Science
[2] Virginia Institute of Marine Science, Gloucester Point, Va. 23062

from the river mouth. Maximum water depth was 12 m.

Fig. 1 shows the temporal distributions of surface and bottom salinity and DO at this station from June 24 to July 6, 1972. Salinity stratification was initially strong, with surface levels of approximately 1 ppt and bottom levels of 14-16 ppt. Bottom salinity declined and surface salinity remained relatively stable until June 30, when salinity at both depths began to rise. By July 6 salinity stratification had been reduced to a difference of 5 ppt between surface and bottom.

DO concentrations at both the surface and the bottom varied diurnally in relation to the solar cycle, but showed little day to day change during the 13 day period. Bottom concentrations rarely dropped below 4.5 mg/l, and were usually within 3 mg/l of the surface concentrations.

James River Slack Water Stations

Slack water stations were located at approximate three mile intervals, starting at the river mouth (Mile 0). Fig. 2 shows the temporal patterns of surface and bottom salinity and DO measured at Mile 0 and Mile 12. Maximum depth was 14 m at both stations.

The salinity distribution at Mile 12 closely follows the pattern observed at the anchor station through July 6. At Mile 0 salinities at both depths were higher and varied more widely than at Mile 12 during this initial period. At both stations there was a period in mid-July (11-14) of reduced salinity stratification, followed by an abrupt rise in bottom salinity, which peaked on July 21. Salinity stratification remained strong at Mile 0 into August, while it weakened at Mile 12 as surface salinity rose.

The DO patterns observed at both stations were similar. Concentrations ranged for the most part between 4 and 9 mg/l. A major decline in bottom DO concentrations, to levels below 4 mg/l, occurred in conjunction with the salinity rise in the latter part of July. Subsequent recovery was rapid.

Figs. 3 and 4 show the spatial distributions of DO for selected slack water runs. The June 1971 run, included for reference in Fig. 3, shows essentially no variation with depth, but a tendency for DO to decline with distance upstream from the mouth. The June 24 and July 5, 1972 runs show the latter trend, but also some decline with depth at deeper stations. The July 24 run (Fig. 4) shows that the DO depression during the July peak in bottom salinity affected most of the river below Mile 24, while the August 25 run shows that the August recovery was equally widespread.

York River Anchor Station

The temporal salinity and DO distributions presented in Fig. 5 were observed at an anchor station located at $37°14.7'N$ latitude and $76°31.1'W$ longitude, approximately 6 nautical miles upstream from the York River mouth. Maximum water depth was 18 m.

Surface salinity dropped sharply between June 24 and June 27, while bottom salinity declined gradually until July 2. Salinity at both depths then began increasing. Surface and bottom DO concentrations fluctuated diurnally more strongly than they did day to day. Bottom concentrations frequently fell below 3 mg/l.

York River Slack Water Stations

The salinity patterns at two slack water stations (Fig. 6) are similar to the anchor station patterns for the June 24-July 6 period. Maximum depth at both stations was 18 m. Both stations show subsequent sharp declines in salinity stratification (July 7-15) followed by sharp rises in bottom salinity similar to those observed in the James River slack water runs.

The surface DO distributions show large short-term fluctuations, but no long-term trends. Bottom DO declined initially to minima at both stations in early July. Concentrations subsequently increased, peaking July 13, before plunging rapidly to levels below 1 mg/l in late July, in conjunction with the maximum in salinity stratification. Bottom DO concentrations fluctuated between 1 and 3 mg/l at both stations for the remainder of July and August.

Fig. 7, 8, and 9 present slack water spatial distributions of DO for selected dates. The three plots in Fig. 7 cover the initial period of bottom DO decline and recovery. Concentrations were between 2 and 3 mg/l in the bottom water at all stations below mile 20 on June 29. Fig. 8 shows the distribution of DO during the second period of declining bottom concentrations. Levels below 1 mg/l occurred over much of the bottom between Mile 20 and Mile 5 on July 24. The August 29 distribution (Fig. 9) was similar to the one observed two months earlier (Fig. 7), and bottom water concentrations were similar to those observed in August 1973 (Fig. 9). The July 1973 plot shows bottom concentrations much above those found in August 1973, and similar to those observed on July 13, 1972 at the end of the initial post-Agnes recovery.

Rappahannock River Anchor Station

An anchor station located at 37°40.0'N latitude and 76°33.2'W longitude, approximately 15 nautical miles upstream from the Rappahannock River mouth, was sampled hourly on a 24 hour basis from June 24 to June 30, and hourly during the day from July 1 through July 7, 1972. Maximum water depth was 15 m.

The salinity time distribution (Fig. 10) shows initially strong vertical stratification which weakened gradually during the observation period. Surface and bottom DO fluctuated diurnally, while bottom concentrations declined, on the average, from June 27 through July 4, to a minimum of 2 mg/l.

Rappahannock River Slack Water Stations

Fig. 11 shows the temporal distributions of salinity and DO at two slack water stations, Mile 0 (maximum depth 12 m) and Mile 10 (maximum depth 20 m). Salinity stratification was strong at the outset, reduced in early July, then became stronger and was relatively stable from mid-July to early August. Bottom DO at Mile 10 recovered briefly after its initial decline in June, then declined below 3 mg/l in early July and below 1 mg/l in late July. At the shallower Mile 0 station bottom DO remained above 4 mg/l until mid-July, when it dropped below 3 mg/l. Concentrations at this station did not fall below 1 mg/l during the period of observations.

The spatial distributions of DO for six selected slack water runs are shown in Figs. 12 and 13. The June 24 and July 3, 1972 plots illustrate the development of vertical DO stratification immediately following the storm. The July 11, 1973 distribution indicates that stratified DO conditions develop in more normal summers as well. The July 24 plot shows the intense vertical concentration gradient observed during late July, with levels below 1 mg/l in deeper layers at all stations below Mile 25. Slight recovery had occurred by August 14. The August 1973 plot also shows strongly stratified conditions, with concentrations below 3 mg/l in the deeper layers.

Chesapeake Bay Slack Water Stations

Selected Chesapeake Bay slack water data are included for comparison with river data. Fig. 14 presents time series plots of salinity and DO measurements made at a station located at 37°15.8'N latitude and 76°8.6'W longitude, approximately 19 miles above the Bay mouth and 12 miles east of the York River mouth. Maximum water depth was 12 m.

Surface salinity varied between 10 and 14 ppt. Bottom salinity fluctuated in a manner similar to the patterns observed for the James and York Rivers, with an initial decline followed by a sharp increase to a peak in late July. Bottom DO dropped from 5 mg/l to 2 mg/l during the period of rising bottom salinity.

The plots in Fig. 15 show the spatial distributions of salinity and DO in the Bay during three slack water runs conducted during the period of rising bottom water salinity. Pronounced vertical stratification of both parameters was evident on both July 20 and July 25, with bottom water DO concentrations below 2 mg/l at four of the upper stations.

DISCUSSION

Following Agnes, bottom water DO fluctuations in the three Virginia estuaries, as well as in lower Chesapeake Bay, were closely related to fluctuations in vertical salinity gradients. The James River experienced less severe DO depressions than did the other two rivers. There was no marked depression immediately after the storm in this river, but one occurred in late July in conjunction with an influx of highly saline water from the Bay. Bottom concentrations during the depression remained above 3 mg/l, and recovery was rapid.

In both the York and Rappahannock Rivers, bottom DO declined immediately after Agnes, recovered briefly, then dropped again to levels lower than those reached in the initial decline. In the York River the major depression occurred in conjunction with the influx of highly saline Bay water in late July, while in the Rappahannock no massive salinity peak occurred. In late July, DO concentrations below 1 mg/l were observed in bottom waters at stations covering 15 miles of the York River and 25 miles of the Rappahannock. Recovery was much slower in these rivers than in the James.

Bottom water DO concentrations at Chesapeake Bay slack water stations were similar to concentrations found at the mouths of adjacent rivers. On July 20, for example, at Mile 37 in the Bay the bottom DO was 1.3 mg/l, while on July 24 at the mouth of the Rappahannock River, eight miles to the west, the concentration was 1.1 mg/l. At Mile 19 in the Bay on July 20 the bottom DO was 2.5 mg/l, while at the York River mouth on July 24 the bottom level was 2.8 mg/l. These similarities suggest that the DO decline that was observed in the lower Bay during July was due in part to contributions of oxygen-poor water from the tributaries.

Low bottom water DO concentrations are known to occur in the deeper sections of Virginia's estuaries during normal summers. Comparisons of post-Agnes slack water data with data from other years suggest that for the York and Rappahannock Rivers, Agnes was followed by abnormally severe DO depressions.

LITERATURE CITED

American Public Health Association, American Water Works Association, Water Pollution Control Federation. 1971. Standard Methods for the Examination of Water and Wastewater. 13th ed., American Public Health Association, Washington, D. C., 874 p.

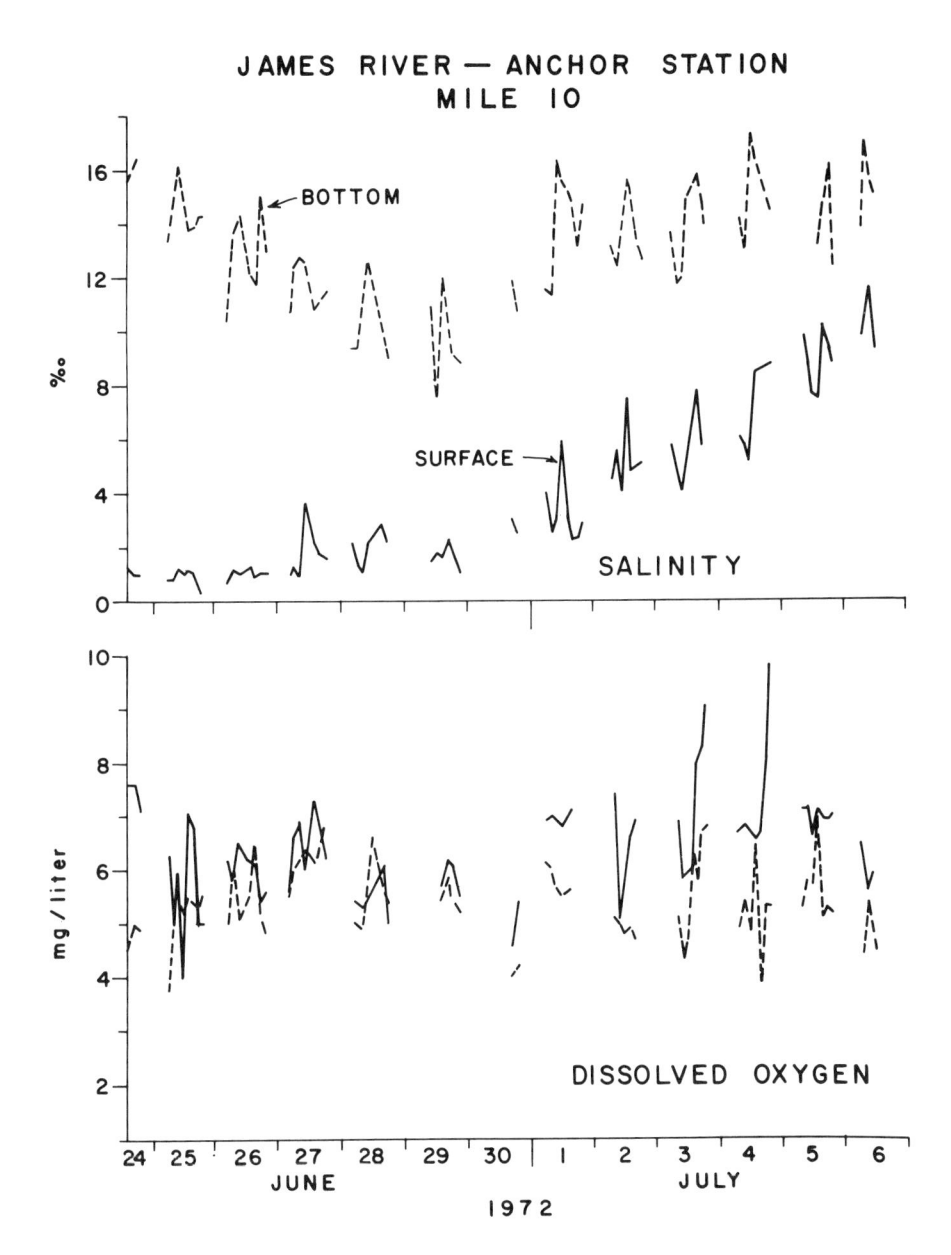

Figure 1. James River anchor station salinity and dissolved oxygen temporal distributions.

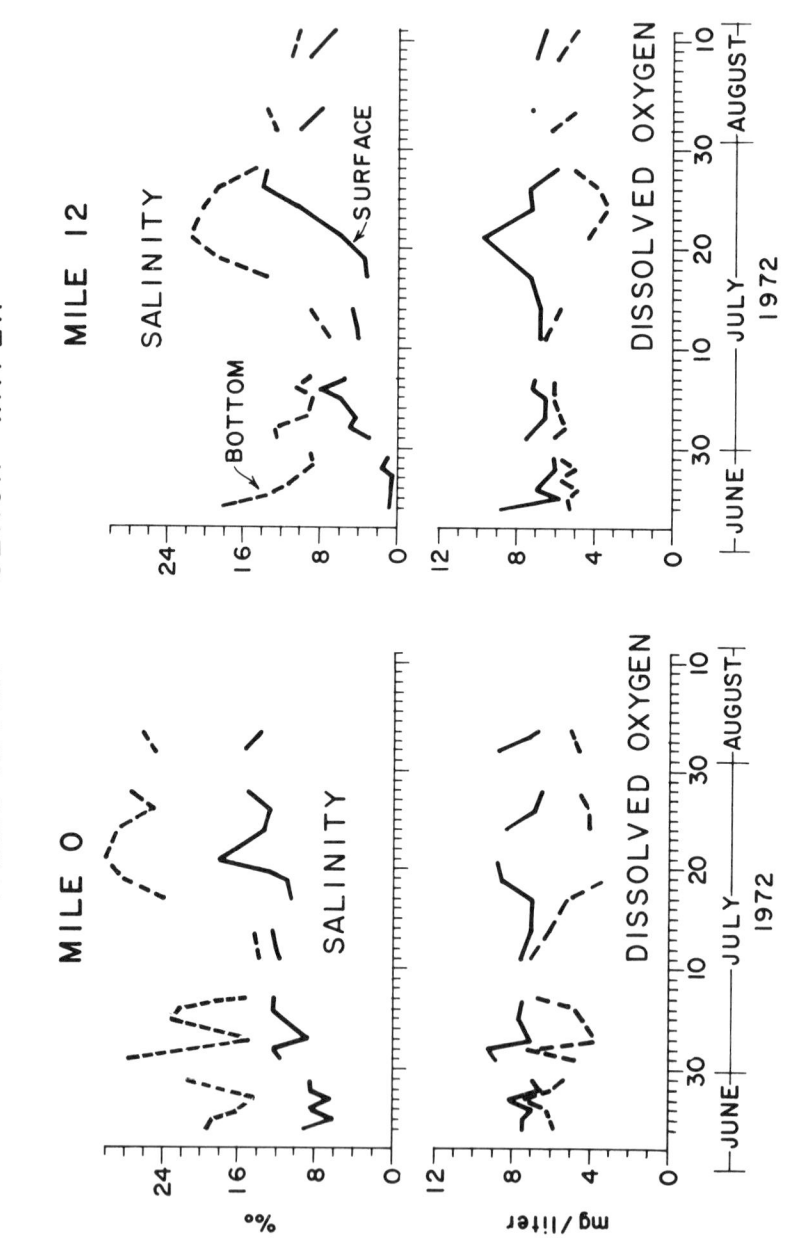

Figure 2. James River slack water salinity and dissolved oxygen temporal distributions.

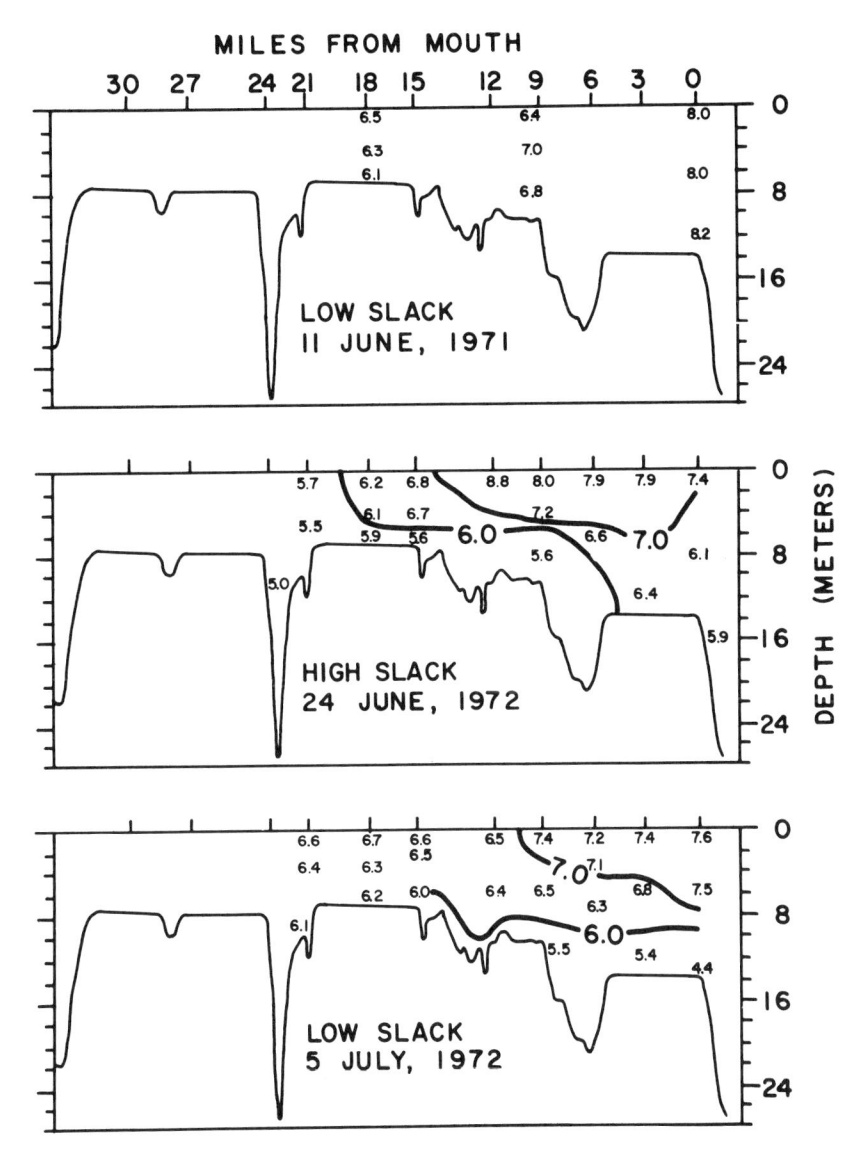

Figure 3. James River slack water dissolved oxygen spatial distributions I.

**JAMES RIVER
DISSOLVED OXYGEN (mg/liter)**

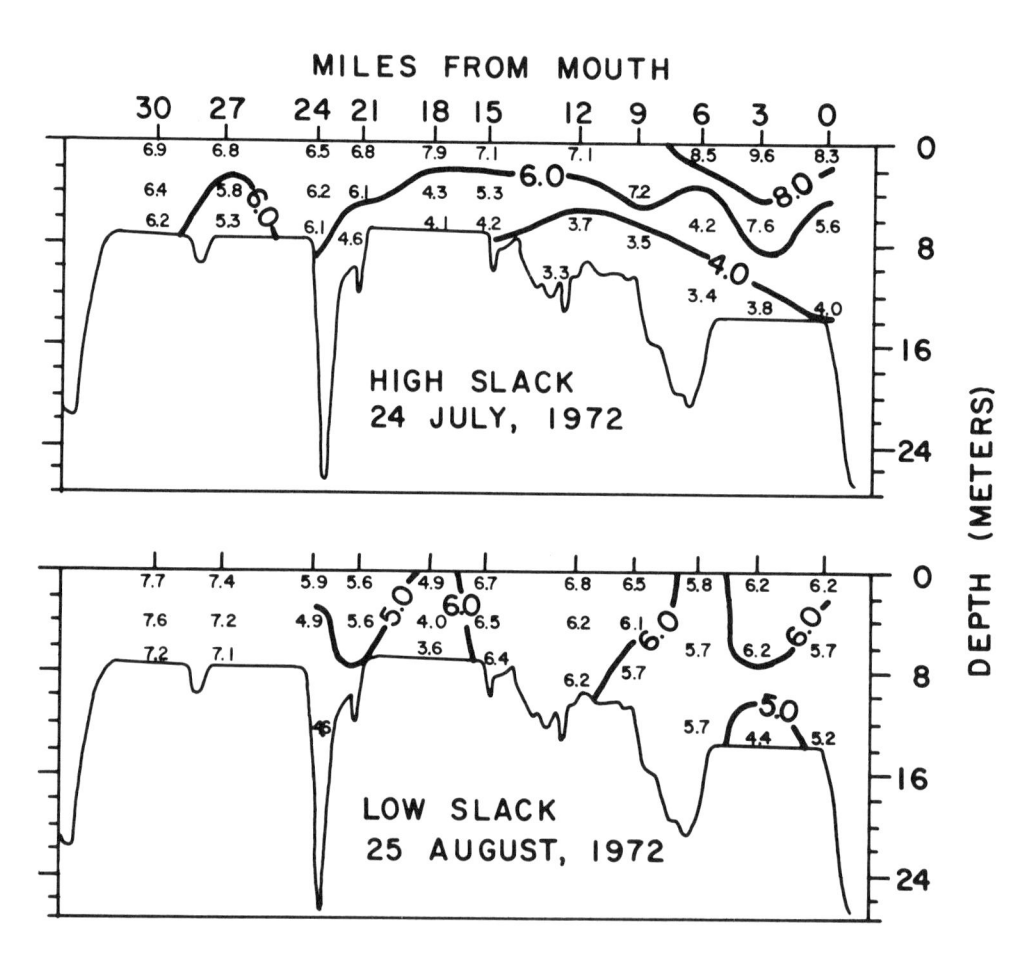

Figure 4. James River slack water dissolved oxygen spatial distributions II.

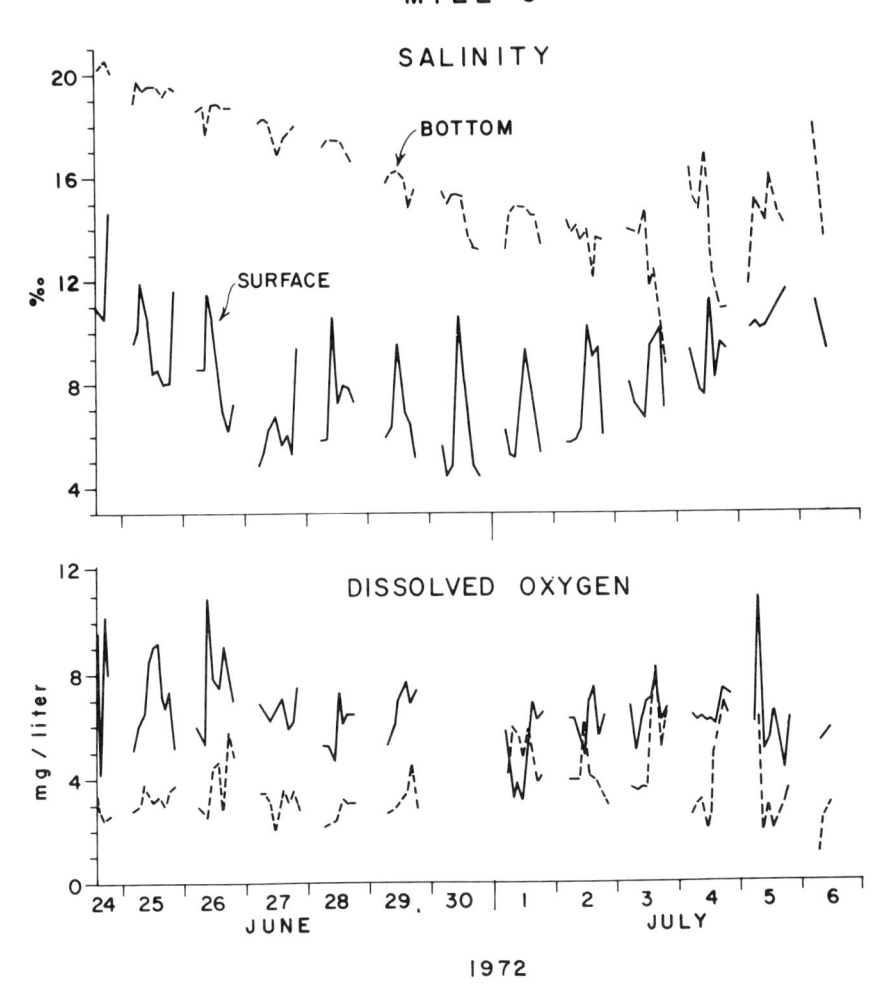

Figure 5. York River anchor station salinity and dissolved oxygen temporal distributions.

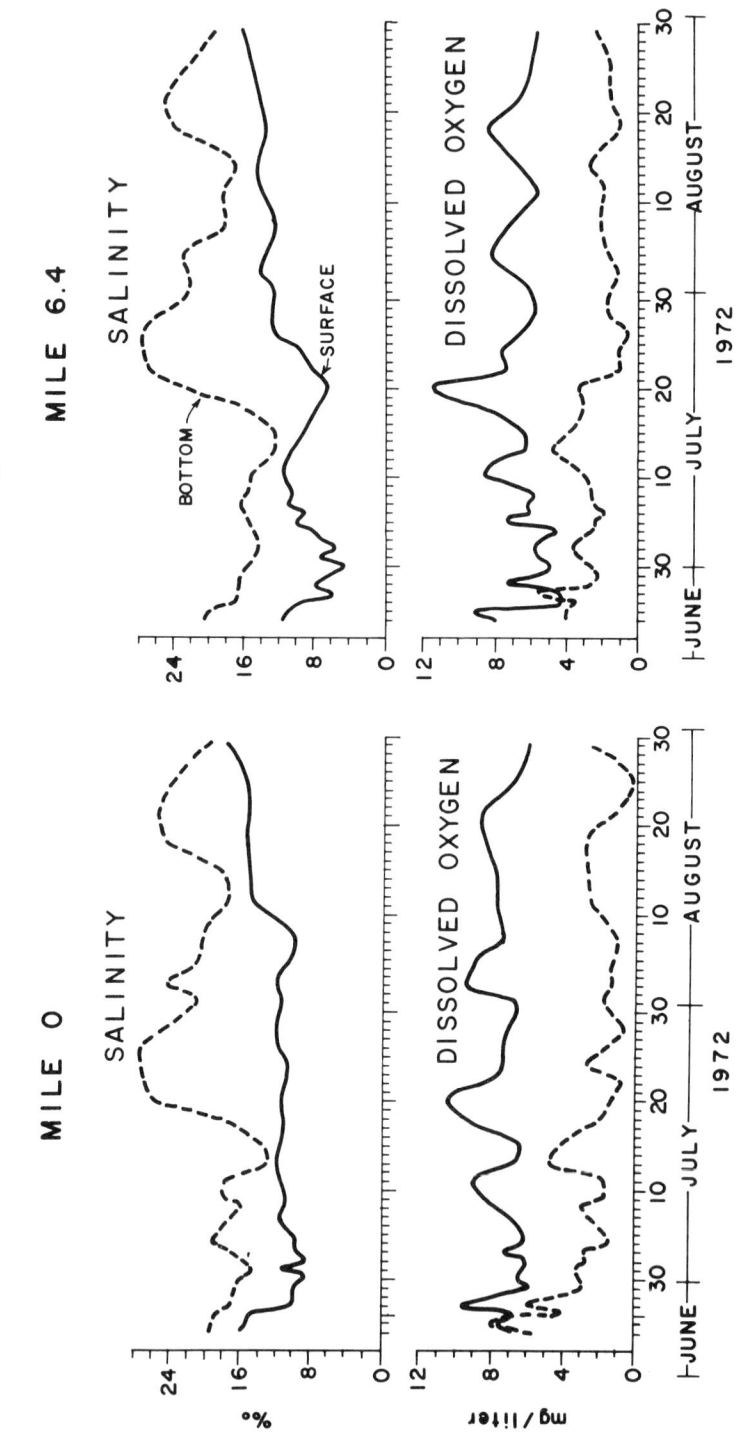

Figure 6. York River slack water salinity and dissolved oxygen temporal distributions.

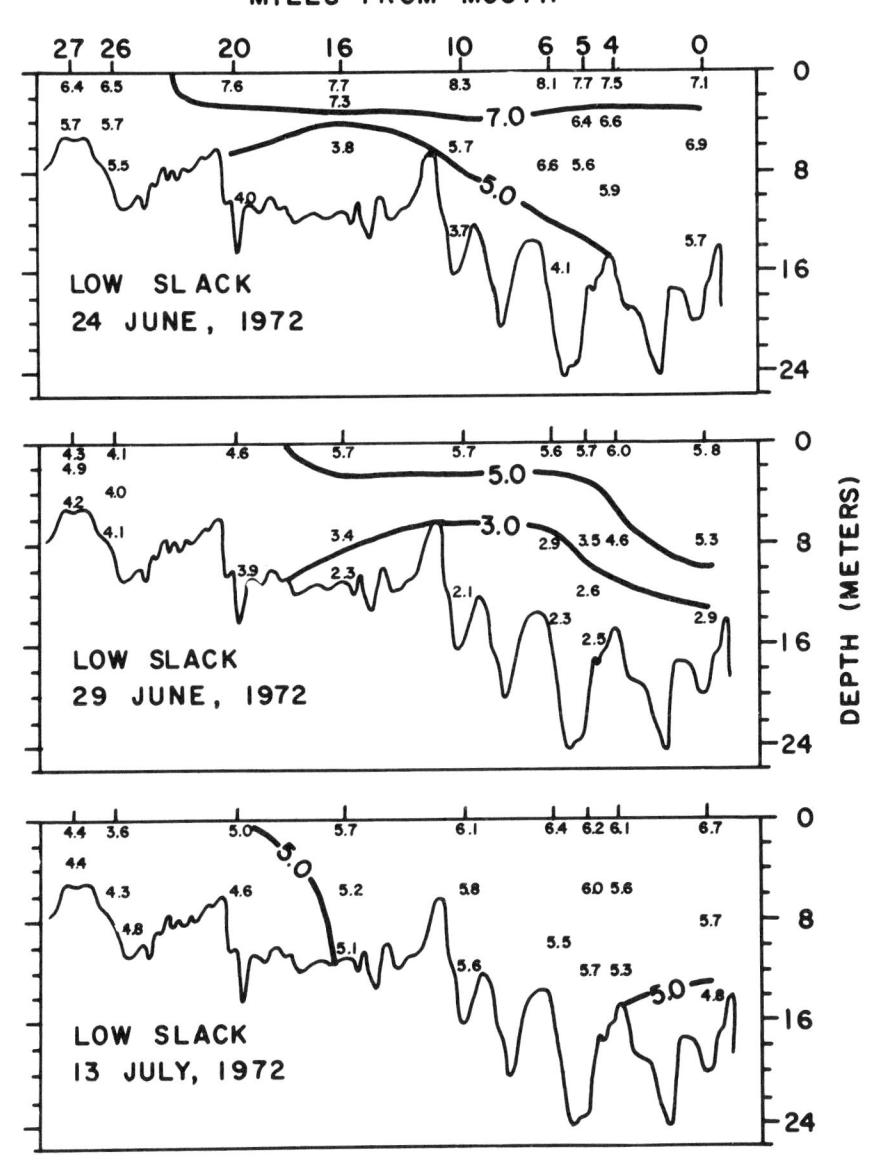

Figure 7. York River slack water dissolved oxygen spatial distributions I.

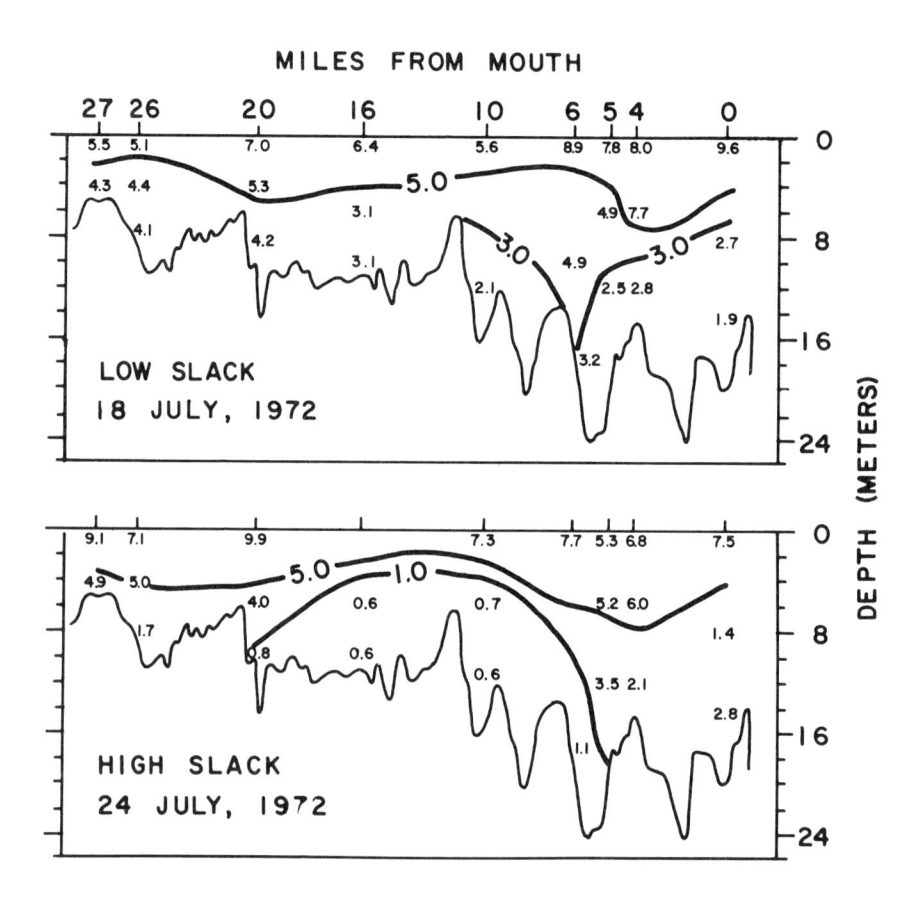

Figure 8. York River slack water dissolved oxygen spatial distributions II.

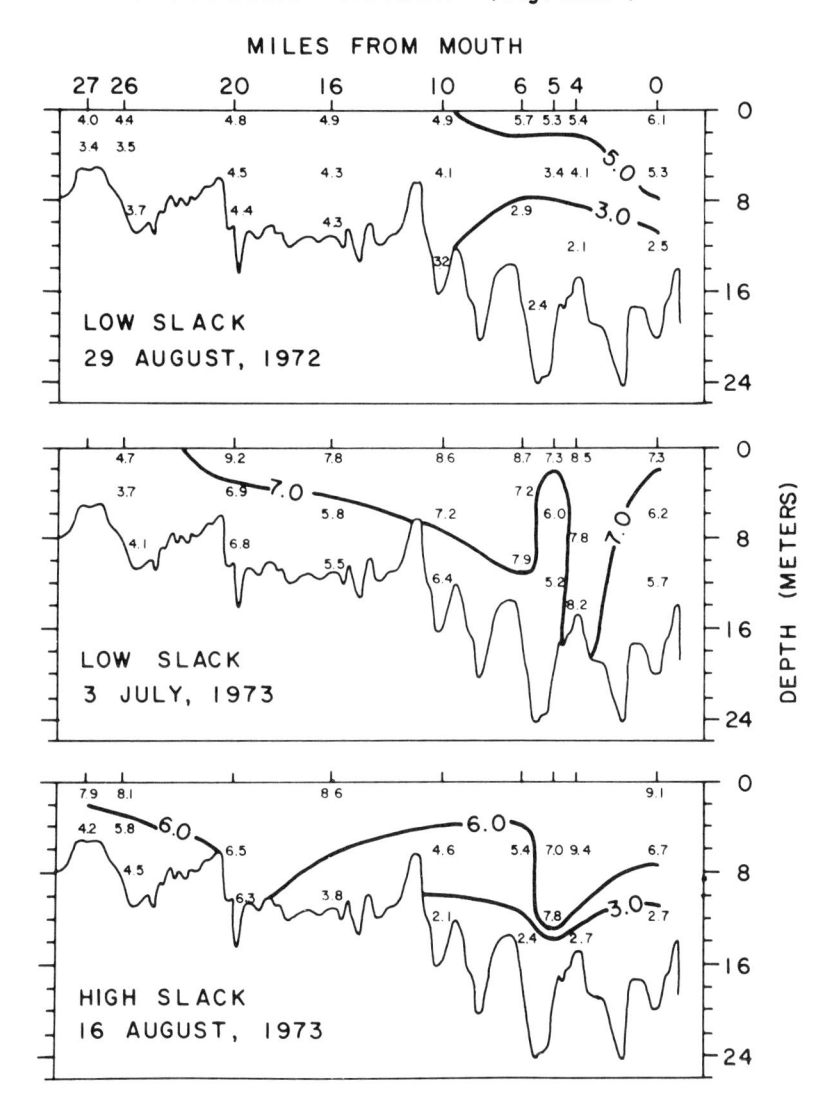

Figure 9. York River slack water dissolved oxygen spatial distributions III.

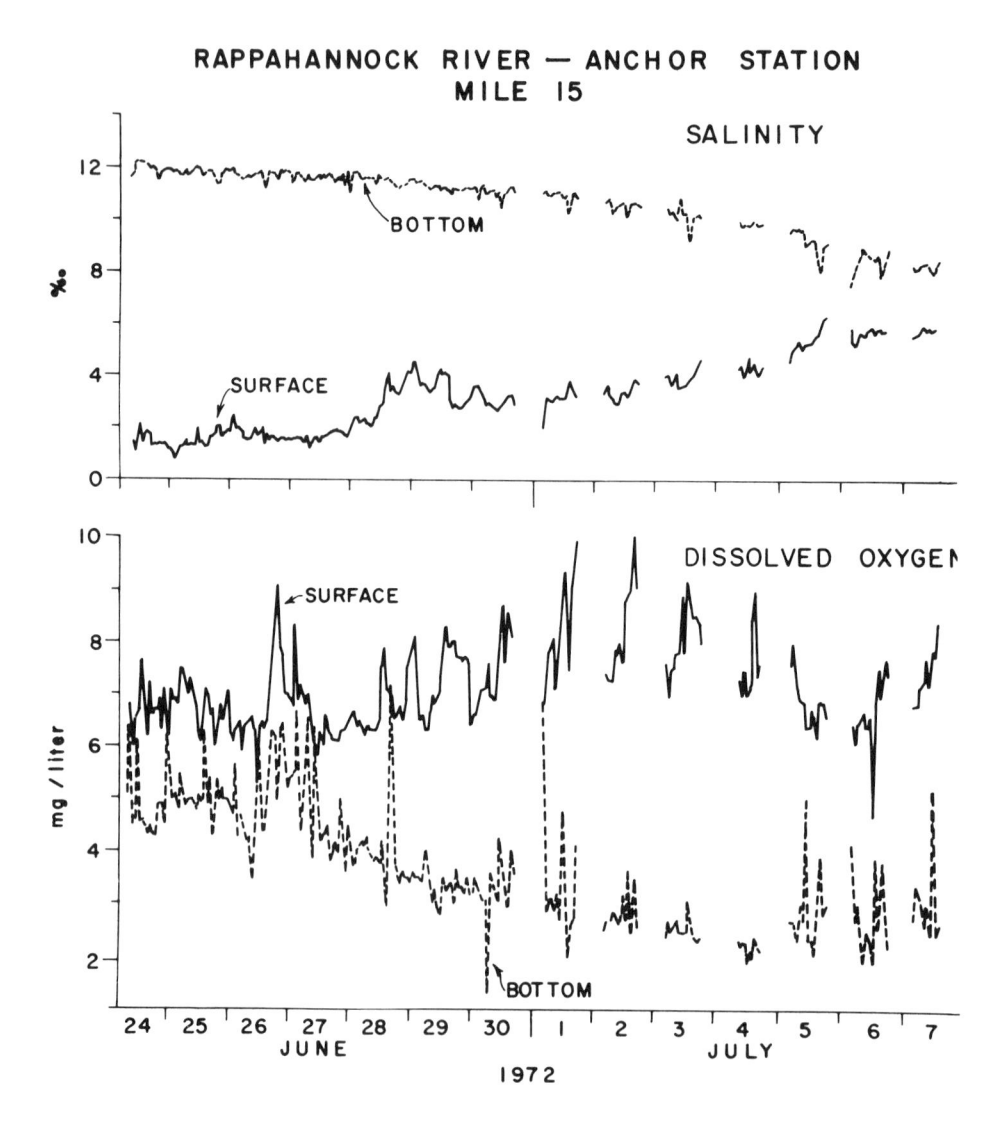

Figure 10. Rappahannock River anchor station salinity and dissolved oxygen temporal distributions.

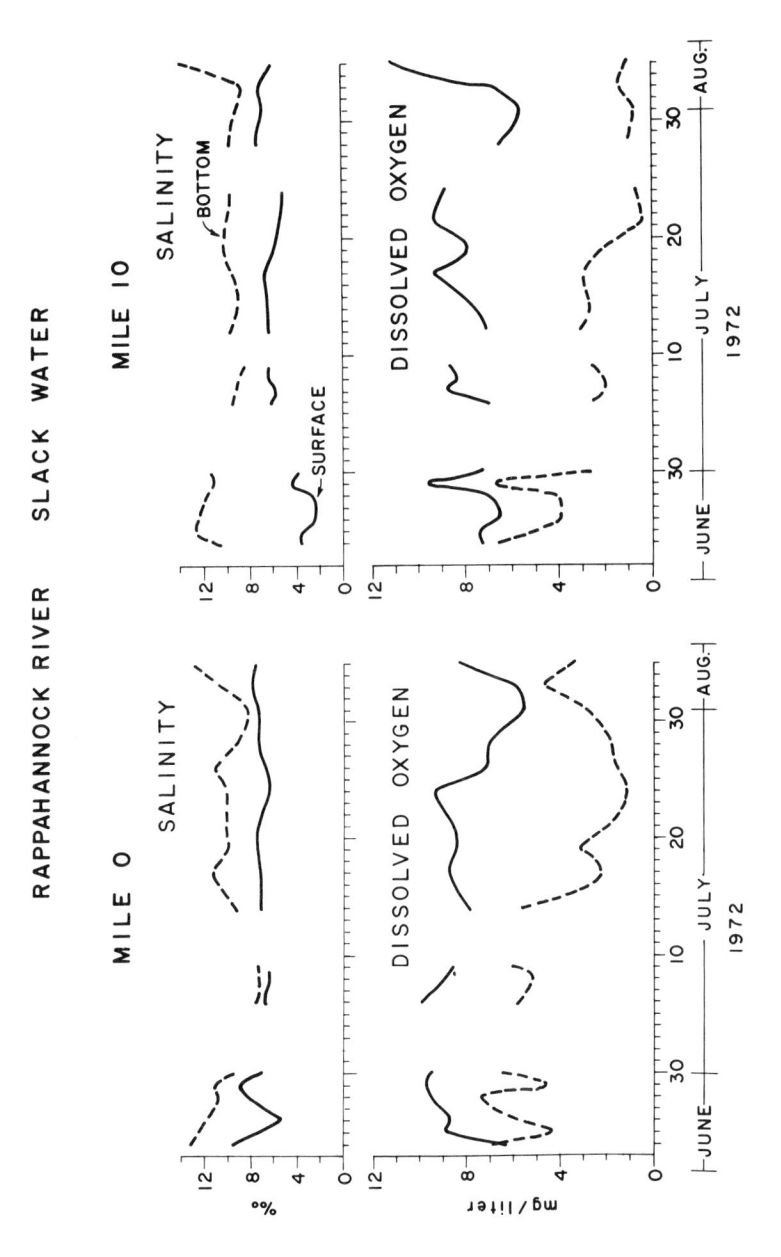

Figure 11. Rappahannock River slack water salinity and dissolved oxygen temporal distributions.

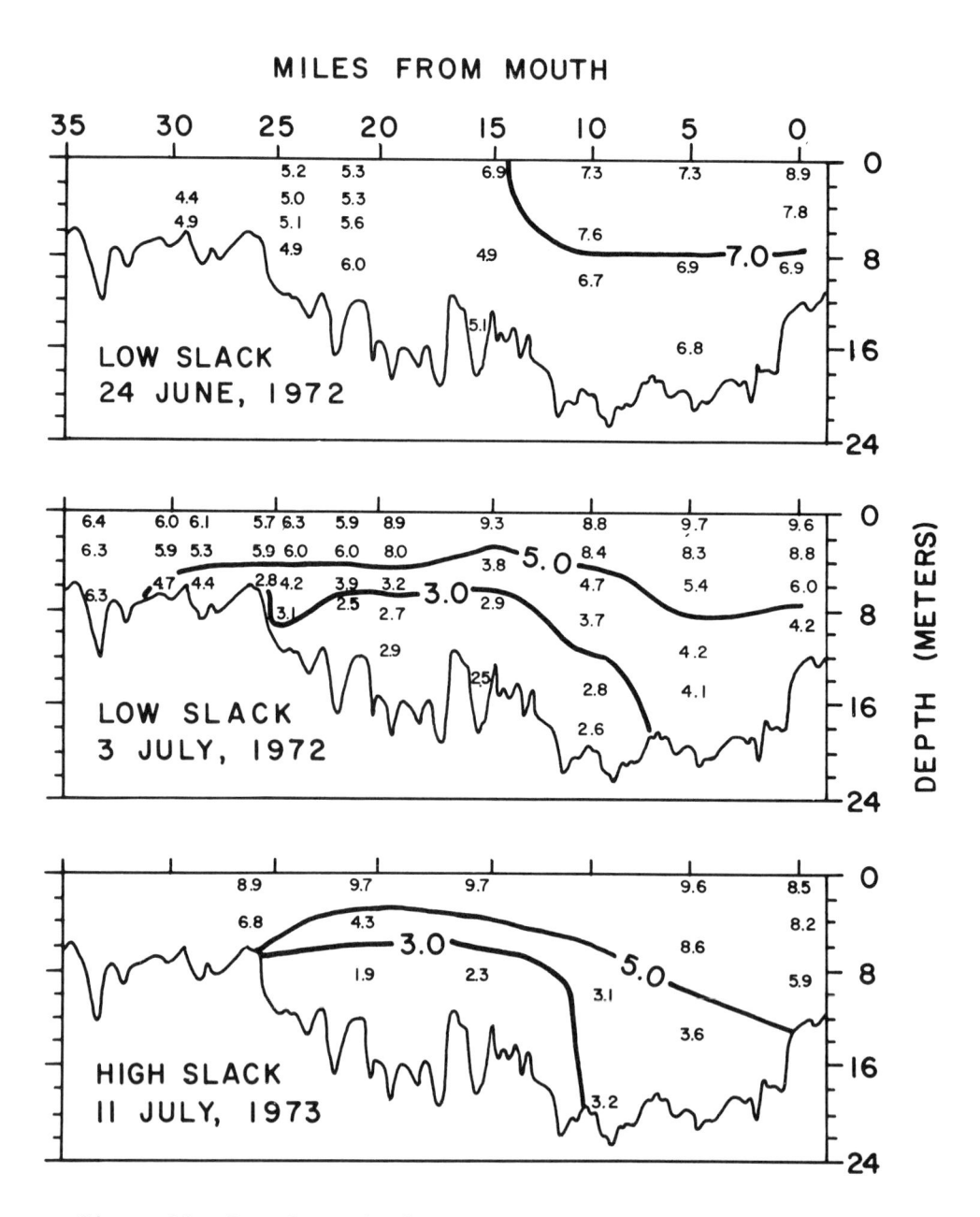

Figure 12. Rappahannock River slack water dissolved oxygen spatial distributions I.

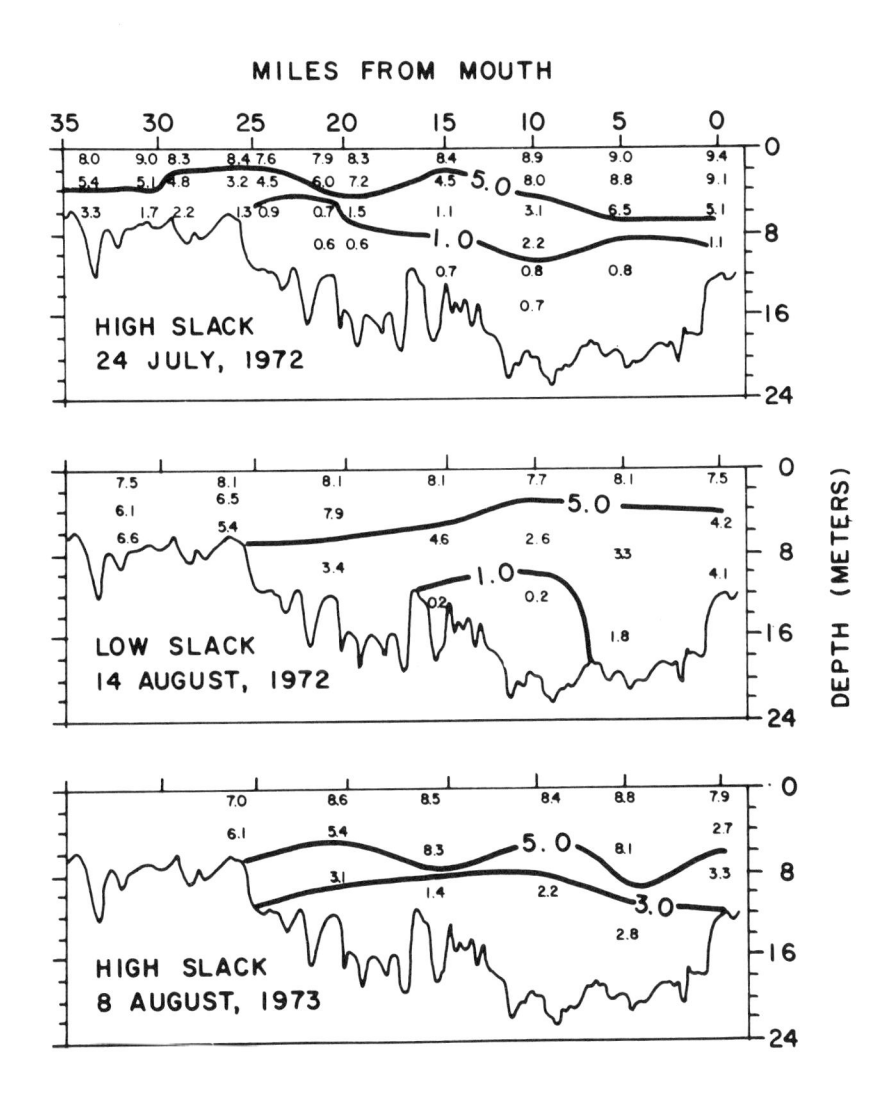

Figure 13. Rappahannock River slack water dissolved oxygen spatial distributions II.

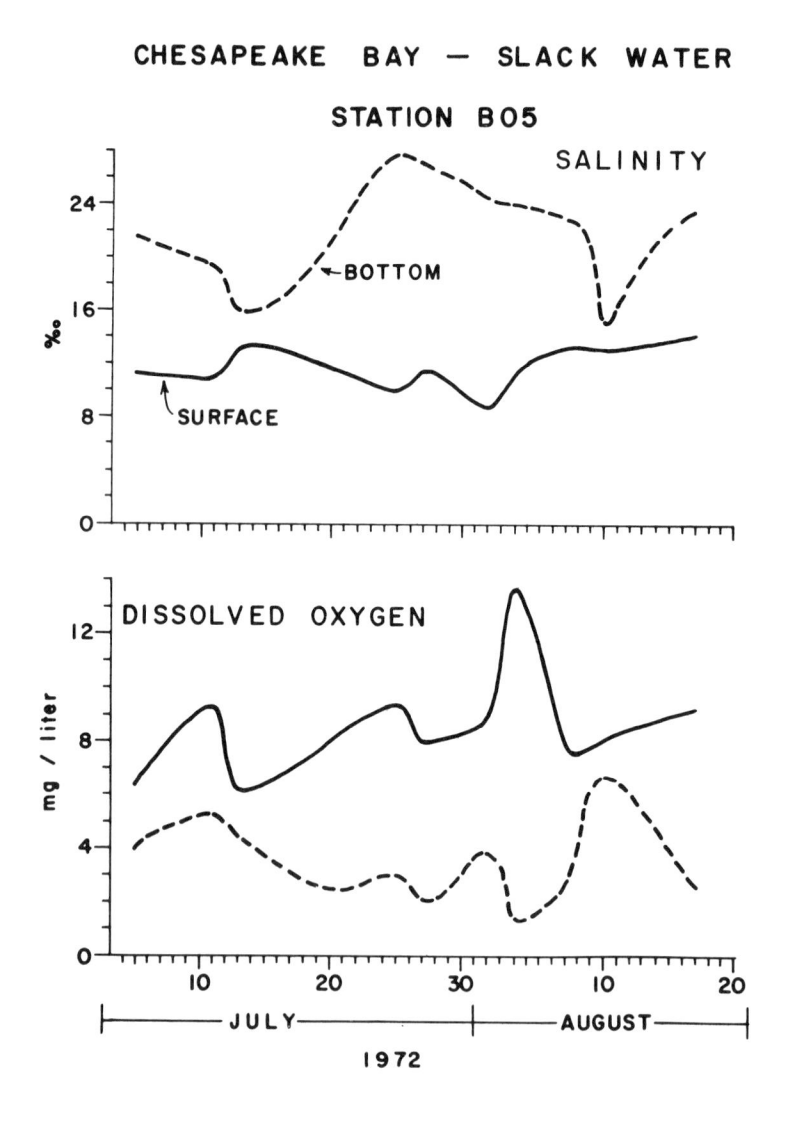

Figure 14. Chesapeake Bay slack water salinity and dissolved oxygen temporal distributions.

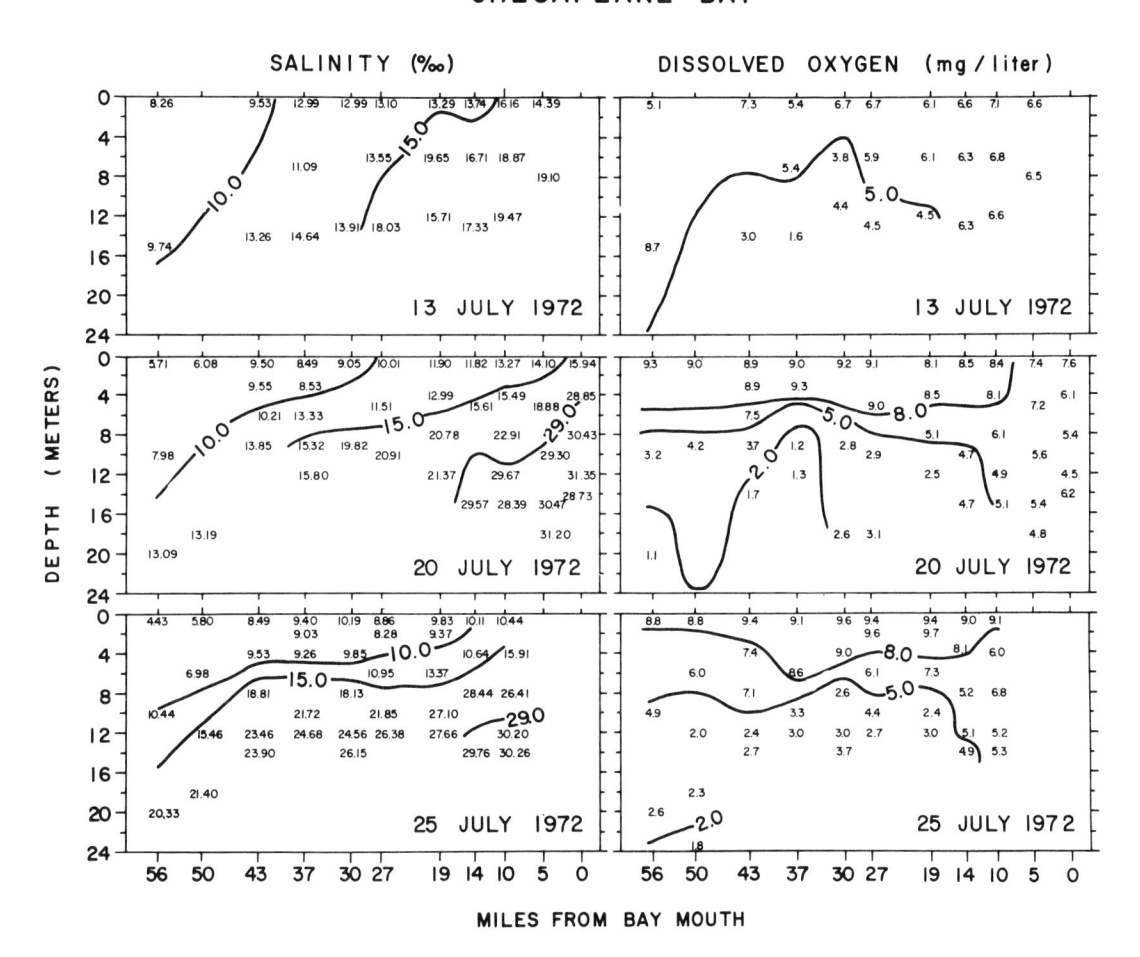

Figure 15. Chesapeake Bay slack water salinity and dissolved oxygen spatial distributions.

THE EFFECT OF TROPICAL STORM AGNES AS REFLECTED IN CHLOROPHYLL A
AND HETEROTROPHIC POTENTIAL OF THE LOWER CHESAPEAKE BAY[1]

Paul L. Zubkoff[2]
J. Ernest Warinner, III[2]

ABSTRACT

A hydrographic station (Station Y) at the mouth of the York River ($37°14.6'N$, $76°23.4'W$) was under biological surveillance for one year prior to the arrival of Tropical Storm Agnes. For one full year following this storm, these measurements were continued. In addition, the chlorophyll a and heterotrophic potential measurements were incorporated into an ongoing zooplankton sampling program of the lower Chesapeake Bay below $37°40'N$ latitude.

In the sub-surface waters (0.5-1.0 meter below the surface) at Station Y, chlorophyll a distributions for the year (June 1971 to June 1972) varied seasonally between 4 and 20 µg chlorophyll a l^{-1}. Within 4 weeks following Agnes, a maximum of 22 µg chlorophyll a l^{-1} was reached, which dropped to a minimum of 6 µg chlorophyll a l^{-1} in November 1972. In lower Chesapeake Bay, post-Agnes chlorophyll a distributions were greater than 6-8 µg^{-1} at the lower limit; the upper limit for the post-Agnes summers (1972 and 1973) was considerably greater than 20 µg l^{-1} pre-Agnes maximum at Station Y.

The heterotrophic potential (µg glucose $l^{-1}h^{-1}$) reflects the activity of microorganisms capable of growth and reproduction when dissolved organic matter is available. In the two-year Station Y study, the heterotrophic potentials fell into 4 ranges: low 0.04-0.25; moderate 0.26-0.82; high 0.83-1.70; and very high 1.70-3.00. In the immediate post-Agnes period (July-August 1972) the heterotrophic potential for approximately one-half of the Lower Bay stations was moderate (0.26-0.82), whereas the other half was very high (1.7-3.0).

INTRODUCTION

The York River, a mesohaline estuary in southeastern Virginia with a sizeable fishing industry, has, until recent times, been little influenced by industrial and municipal development. Because populations in the immediate vicinity are increasing rapidly, the York River will undoubtedly become impacted by human endeavors. In order to provide a stronger reference base for helping management agencies to make informed judgements with respect to the lower York River ecosystem, an investigation into the producer trophic levels was initiated in the summer of 1971. Specific objectives of this project included the development of techniques for the measurement of the autotrophic (photosynthesizing) and heterotrophic (microbial) communities and the testing of these techniques at a hydrographic station at the mouth of the York River (Station Y, $37°14.6'N$, $76°23.4'W$; Fig. 1). In addition, the selected techniques were to be amenable to the processing of large numbers of samples with a time for analysis less than that more commonly employed in ecological studies (Patten, Mulford, & Warinner 1963; Mackiernan 1968; Manzi 1973; Stofan 1973).

---
[1]Contribution No. 772, Virginia Institute of Marine Science.
[2]Virginia Institute of Marine Science, Gloucester Point, Virginia 23062

When Agnes passed through the Chesapeake Bay drainage basin, a zooplankton investigation using a stratified sampling regime for the entire lower Chesapeake Bay was underway (Grant, et al. this volume). Subsequently, these projects were coordinated in order to better assess the effects of the massive freshwater input on the distributions of plankton in the lower Chesapeake Bay and more effectively utilize limited personnel and vessel availability. There were no published synoptic measurements for chlorophyll a and heterotrophic potential for the lower Chesapeake Bay; therefore these measurements were continued for one full year after the occurrence of the tropical storm in order to assess the aftermath of the perturbation.

Nineteen Station Y and 276 lower Chesapeake Bay (Lower Bay) hydrographic stations were sampled in the year following the passage of Agnes. Although several hydrographic and biological observations were made, this report summarizes only the chlorophyll a and heterotrophic potential in the surface waters.

SAMPLING AND METHODS

Six sub-strata, A-F, in lower Chesapeake Bay were selected according to morphometry of the region, predominating salinity regimes, and major circulation patterns (Grant 1972). Sub-strata A and D are predominantly shallow areas of lower salinity waters; sub-strata C and E are deep channel areas of intermediate salinities; and sub-strata B and F are higher salinity waters (Stroup & Lynn 1963). Station Y, located in the channel of the York River, is approximately 20 meters deep. Stations for the Lower Bay were sampled at monthly intervals before and after Agnes, with three additional sampling cruises during the weeks following the storm (Grant, et al. this vol.). Although samples were routinely obtained by either a Van Dorn bottle or submersible pump for sub-surface (0.5-1.0 meter), intermediate, and bottom depths, only the measurements obtained for the sub-surface waters are reported herein.

Chlorophyll a, determined either spectrophotometrically or fluorimetrically (Biological Methods Panel on Oceanography 1969) serves as a first order approximation of the standing crop of phytoplankton. As with any single chemical measurement of biomass, chlorophyll a does not reveal species identification, enumeration, nor does it reflect the state of viability of the organisms. However, chlorophyll a is a useful approximation, within limits, of algal biomass.

The heterotrophic potential (V_{max}) is a kinetic parameter which estimates the potential activity of the microbial populations (heterotrophic microorganisms) and is useful for spatial and temporal comparisons of the relative microbial populations of an aquatic ecosystem (Wright & Hobbie 1966; Williams 1973).

The heterotrophic potential of the estuarine plankton community is measured by the uptake of simple ^{14}C-labeled dissolved organic substrates added to natural water samples. When varying concentrations of substrate are employed, the response to an increase in substrate concentration resembles that of an enzyme-catalyzed reaction. As the concentration of a substrate such as glucose is increased, there is initially a linear increase in the rate of uptake by the heterotrophic organisms at low substrate concentration; at higher substrate concentrations, the rate of uptake reaches a maximum.

The rate of uptake of the substrate at any given concentration is calculated using the pseudo-first order equation of enzyme kinetics (Parsons & Strickland 1962):

$$v = \frac{c(Sn + Sa)}{C\mu t}$$

where

    v = velocity of uptake of the substrate ($\mu g\ l^{-1}h^{-1}$)
    c = counts taken up by heterotrophic population
   Sn = natural substrate concentration
   Sa = added substrate concentration
    C = number of counts per μCi of substrate
    μ = number of μCi added to incubation medium
    t = time of incubation (hours)

When this equation is combined with a modified form of the Michaelis-Menten equation:

$$\frac{(Sn + Sa)}{v} = \frac{K_t}{V_{max}} + \frac{(Sn + Sa)}{V_{max}}, \text{ where}$$

   K_t = transport constant
  V_{max} = maximum velocity of uptake at substrate saturation

the resulting equation is:

$$\frac{C\mu t}{c} \text{ or } \frac{(Sn + Sa)}{v} = \frac{(K_t + Sn)}{V_{max}} + \frac{1}{V_{max}}(Sa)$$

When $C\mu t/c$ is plotted against Sa, the result is a straight line with slope equal to $1/V_{max}$. The value of V_{max} has the units of μg glucose $l^{-1}h^{-1}$. Although several kinetic parameters may be calculated, V_{max} is the most useful one for describing the relative functional microbial activity in this study.

The kinetic data obtained are undoubtedly the result of heterogeneous assemblages of organisms which have active transport systems (Williams 1973). V_{max} is the rate of uptake observed at a substrate concentration high enough to completely saturate the transport mechanisms of the natural microbial populations under the experimental conditions (Vaccaro & Jannasch 1966). In these studies, V_{max}, glucose is interpreted as an indicator which reflects the activity of the viable natural population of microbial organisms at the time of sampling; it is an experimentally measured number which is a resultant of the endemic community's cell size, number, and state of viability as a function of temperature.

The V_{max}, glucose was determined by incubating 10 ml aliquots of an estuarine water sample with labeled ^{14}C-glucose and carrier at final concentrations of 37.5, 75.0, 187.5 and 375 $\mu g\ l^{-1}$ in the dark at ambient temperature for two hours. Approximately 0.1 ml of 2% neutralized formalin was added for the inactivated control sample and to terminate the reaction. The ^{14}C-labeled particulate fraction was then collected on cellulose-acetate filters (Millipore[R] EH, 0.5μ), treated with NCS[R] tissue solubilizer, and dissolved in a toluene-based scintillation fluid containing 2,5-diphenyloxazole. Counting was completed at 87-95% efficiency using a liquid scintillation counter with external standardization. The calculation of V_{max}, glucose using linear regression analysis has an r value of 0.85 or greater for at least three of the four concentrations of substrate used.

Since no provision was made to trap and measure the respired CO_2 from the assimilated ^{14}C-glucose, the calculated V_{max}, glucose values represent only that portion of labeled substrate transformed into particulate form, and is therefore, a minimum estimate of the functional microbial community (Wright 1973).

RESULTS

Hydrography

After the passage of Agnes, the measured sub-surface salinity at the mouth of the York River ranged from 10.28-21.82 ppt (Table 1, Fig. 2). The seasonally low salinity of the sub-surface waters of May-June 1972 were below those of 1971, a result of the unusually high freshwater discharge into the York River from its drainage basin during the spring of 1972 (Fig. 3). Discharge was estimated by summing the discharge data obtained from the U. S. Geological Survey for the Mattaponi and Pamunkey Rivers and adjusting for drainage area below the gauging stations.

Chlorophyll a

In the year prior to the passage of Agnes, chlorophyll a at the mouth of the York River ranged from 4.3-20.0 µg l^{-1} (mean = 8.80, s.d. = 3.71, n= 27, Table 1), with isolated low values in the winter. Highest values occurred in late spring and summer, periods of high insolation and temperature. In 1972, prior to Agnes, all values measured were generally low, between 5.0 and 13.0 µg $^{-1}$, including the commonly recognized "spring pulse" (maximum) in May. However, three to four weeks following Agnes, a maximum at 24.5 µg $^{-1}$ chlorophyll a was observed, with a peak of 27.9 µg l^{-1} in August (Fig. 4A).

For the year following Agnes, chlorophyll a at the mouth of the York River ranged from 5.7-24.6 µg l^{-1} (mean = 14.26, s.d. = 5.60, n = 18). Chlorophyll a reached a low observed value of 6 µg l^{-1} during the late fall (November), with generally higher values throughout winter and spring of 1973 in comparison to 1972. The chlorophyll a concentrations for the summer of 1973 were approximately the same as those for the post-Agnes period of 1972. It appears that for a period after Agnes, the values were generally higher than for the year before. It should be noted, however, that climatological parameters of 1971, 1972, and 1973 were quite different, springs of 1971 and 1972 being relatively "wet" and the summers of 1971 and 1973 being relatively "dry".

The chlorophyll a measurements for the lower Chesapeake Bay sub-strata A-F (Tables 2A, 2B, and 2C) are plotted according to monthly sampling periods and are presented in Fig. 4B. Most values are greater than 6-8 µg l^{-1} at the lower limit (in comparison to 4-6 µg l^{-1} for the York River Mouth prior to Agnes). The upper limits for the post-Agnes summers of 1972 (A = 19.67, B = 30.13, C = 15.43, D = 38.23, E = 24.53 and F = 23.80) and 1973 (A = 16.87, B = 11.23, C - 7.97, D = 28.63, E = 18.73, and F = 14.20) are considerably greater than the 20.0 µg l^{-1} 1971 maximum for the mouth of the York River.

In Fig. 4C, the approximate range of values (within 2 standard deviations of the mean), obtained by grouping the entire Lower Bay values for a sampling period, is superimposed on those values for the mouth of the York River reported for the same sampling period. It should be noted that most of the values, both for the Lower Bay and Station Y, are in the same range.

When the post-Agnes chlorophyll a measurements are compared with those of the previous year (Fig. 4D), it is noted that the 1972 values far exceed those for the pre-Agnes period (cross-hatched area) whereas the summer values for 1972 and 1973 (both post-Agnes) are quite similar.

Table 1. Selected parameters for the surface waters at the York River Mouth (August 1971 - August 1973).

Date	Salinity (ppt)	Temperature (°C)	Chlorophyll a µg l^{-1}	V_{max} µg G l^{-1}h^{-1}
5 May 1971	17.34	16.0	7.1	-
19	17.67	20.3	7.8	-
4 June	14.68	23.5	20.0	-
12	15.27	24.5	14.8	-
18	16.77	23.5	8.6	-
23	15.74	25.5	9.5	-
2 July	16.47	26.5	11.9	-
9	17.27	27.0	10.5	-
20	19.21	27.0	-	-
4 Aug	-	28.0	5.1	2.90
10	21.30	26.2	11.0	1.39
17	-	26.0	7.1	-
25	20.79	25.5	10.0	0.54
1 Sept	19.70	25.0	9.2	0.48
8	19.72	27.0	5.1	0.97
17	20.29	26.0	11.7	1.00
23	20.82	24.0	13.2	0.21
8 Oct	19.24	20.7	7.0	0.30
13	18.84	-	7.3	0.35
22	19.13	20.5	3.9	0.38
13 Nov	17.62	15.0	4.4	0.28
7 Dec	18.31	10.0	5.2	0.10
24 Jan 1972	18.84	11.5	7.5	0.08
16 Feb	19.64	6.0	10.0	0.21
10 Mar	15.57	7.0	6.7	0.41
4 Apr	16.07	11.0	12.9	0.46
2 May	15.14	18.0	5.9	1.03
6 June	13.83	18.5	4.3	1.69
15	15.81	20.4	-	-
21-22		TROPICAL STORM AGNES		
26	-	23.0	7.0	0.67
27	10.28	22.9	-	-
28	-	24.5	15.8	1.38
6 July	11.67	23.0	7.9	0.68
13	11.93	24.5	11.8	0.81
27	-	26.7	21.6	2.44
8 Aug	11.83	27.9	21.6	2.30
21	15.17	26.5	16.8	2.46
14 Sept	17.20	24.2	12.9	0.75
13 Oct	17.43	18.4	9.3	0.38
16 Nov	-	12.5	5.7	0.15
11 Dec	17.75	9.4	-	-
12 Jan 1973	-	-	9.5	0.04
13 Feb	12.62	-	13.8	-
19 Mar	17.69	8.3	13.8	0.63
16 Apr	14.77	11.0	17.0	0.69
14 May	15.90	18.9	14.3	0.89
25 June	16.69	24.9	10.8	0.63
23 July	19.66	26.1	24.6	1.87
13 Aug	21.82	26.7	22.5	-

Table 2A. Selected hydrographic parameters for the western sub-surface waters (Lower Chesapeake Bay). Cruise mean values and standard deviation.

1972	Sub-Stratum A					Sub-Stratum D				
	S ppt	T °C	Chl. a µg G l^{-1}	V_{max} µg G l^{-1}h^{-1}	n	S ppt	T °C	Chl. a µg l^{-1}	V_{max} µg G l^{-1}h^{-1}	n
June 29-30	13.50 1.35	22.37 0.21	15.60[3] 1.97	0.75 0.06	3	12.87 0.26	22.70 0.14	12.60[1] 0.85	1.01 0.06	2
July 6-7	16.11 0.89	21.40 0.44	10.37 0.40	0.52 0.10	3	10.59 0.50	23.05 0.35	14.25[2] 3.61	0.86 0.03	2
July 13-14	14.24 0.81	23.67 0.35	9.30[2] 0.14	0.73 0.16	3	12.11 0.01	23.25 0.07	11.05[1] 0.07	1.13 0.13	2
July 24-27	16.77 1.51	27.07 0.46	14.60[2] 10.83	1.37 0.27	3	9.28 1.33	28.73 0.06	17.00[3] 2.19	2.02 0.19	3
Aug. 15-21	18.43 1.75	24.60 1.40	9.70 *	0.27 *	2	14.82 0.74	25.87 0.06	38.23[3] 13.61	1.27 0.31	3
Sept. 12-14	17.34 0.29	23.10 0.10	19.67[2] 2.19	0.71 0.51	3	17.15 0.84	23.37 0.81	14.03[1] 2.11	0.61 0.10	3
Oct. 16-24	21.92 1.00	17.23 0.21	9.13 0.90	0.69 0.22	3	19.47 0.15	17.20 0.30	8.63 1.37	0.28 0.23	3
Nov. 13-16	21.75 0.50	14.43 0.25	9.27 4.97	0.20 0.05	3	19.16 0.36	12.60 1.18	6.73 1.01	0.65 0.41	3
Dec. 11-13	19.35 0.25	9.53 0.35	8.67 2.48	0.19 0.09	3	17.98 0.06	8.65 0.07	10.20 2.12	0.43 0.13	2
1973										
Jan. 3-4	19.42 4.59	7.27 0.31	5.50 0.36	0.63 0.05	3	15.25 0.28	6.53 0.23	7.20 1.20	-	3
Feb. 13-14	14.80 0.93	3.20 0.40	9.27 2.11	0.33 *	3	13.97 0.66	2.97 0.23	10.90 1.39	-	3
March 19-20	20.58 1.36	8.13 0.21	10.80 2.31	0.67 *	3	17.65 0.89	7.77 0.25	10.83 5.47	-	3
April 16-17	17.48 0.91	11.20 0.56	9.50 3.21	0.80 *	3	16.49 0.15	11.40 0.44	13.53 7.09	-	3
May 14-16	19.58 2.14	17.40 0.36	10.00 1.56	0.75 *	3	15.85 0.31	18.40 0.36	12.93 0.85	1.12 *	3
June 25-27	17.01 1.13	25.07 0.25	8.33 0.64	0.66 *	3	15.54 0.25	25.17 0.38	10.93 3.54	0.80 *	3
July 23-24	23.58 0.96	23.83 0.49	7.75 0.49	0.32 *	3	18.50 2.26	25.80 0.20	28.63[2] 8.29	1.59 *	3
Aug. 13-14	23.60 1.25	26.97 1.93	16.87[1] 10.21	1.14 *	3	21.01 1.22	26.60 0.14	25.53[3] 2.17	1.56 *	3

Lower number in each pair is standard deviation
- = no data taken
* = single value used
[1] [2] [3] = number of plankton blooms observed at stations.

Table 2B. Selected hydrographic parameters for the mid-channel sub-surface waters (Lower Chesapeake Bay). Cruise mean values and standard deviation.

1972	Sub-Stratum C					Sub-Stratum E				
	S ppt	T °C	Chl. a µg l^{-1}	V_{max} µg G l^{-1}h^{-1}	n	S ppt	T °C	Chl. a µg l^{-1}	V_{max} µg G l^{-1}h^{-1}	n
June 29-30	15.51 1.89	22.45 1.34	12.15 2.19	0.58 0.11	2	11.92 *	22.7 *	6.4 *	0.68 *	1
July 6-7	13.85 *	21.10 *	9.30 *	1.00 *	1	11.10 *	22.1 *	16.3[1] *	0.63 *	1
July 13-14	16.84 4.20	23.40 0.85	9.30 0.42	0.80 0.21	2	11.88 0.07	24.40 1.27	13.65 0.78	0.99 0.06	2
July 24-27	16.07 1.14	26.53 1.31	12.43 2.15	1.64 0.80	3	8.46 1.95	29.23 2.01	15.97[2] 4.27	3.30 2.84	3
Aug. 15-21	19.34 2.43	24.23 1.55	15.43[1] 15.12	0.98 0.81	3	14.87 0.66	25.53 0.40	24.53[3] 4.51	2.50 *	3
Sept. 12-14	17.79 0.94	22.90 0.00	14.90 3.24	2.26 2.52	3	16.17 0.51	23.87 1.01	14.20[3] 1.22	1.09 1.00	3
Oct. 16-24	21.10 0.12	17.15 0.07	7.15 2.19	0.49 0.01	2	18.42 1.92	16.60 0.95	9.80 0.69	0.50 0.12	3
Nov. 13-16	23.82 2.28	14.47 0.15	7.50 0.95	0.20 0.03	3	20.09 0.70	12.73 0.42	7.33 0.42	0.26 0.12	3
Dec. 11-13	19.48 0.81	8.70 0.35	9.20 1.87	0.17 0.04	3	17.95 0.76	9.07 0.29	8.30 1.08	0.21 0.00	3
1973										
Jan. 3-4	23.97 0.87	7.65 0.07	4.40 0.42	-	2	17.12 2.21	6.57 0.25	5.60 0.53	-	3
Feb. 13-14	15.98 1.35	2.97 0.15	9.37 1.55	-	3	13.97 1.00	3.60 0.56	9.00 4.46	-	3
March 19-20	24.09 0.80	7.67 0.06	14.80 6.85	0.18 *	3	16.35 0.78	7.80 0.17	8.43 1.30	0.36 *	3
April 16-17	21.15 2.98	11.40 0.72	3.33 0.45	-	3	16.36 1.27	11.00 0.46	13.13 6.01	0.63 *	3
May 14-16	20.77 2.12	17.33 0.25	7.90 1.01	0.68 *	3	16.67 0.62	17.90 0.17	12.30 2.19	0.91 *	3
June 25-27	18.98 1.85	24.10 0.98	4.90 0.60	-	3	15.42 0.47	25.37 0.31	9.03 2.30	0.61 *	3
July 23-24	25.97 1.19	23.47 0.83	6.87 3.17	0.16 *	3	19.09 1.55	25.80 0.36	18.73 0.42	1.56 *	3
Aug. 13-14	27.92 5.20	24.03 1.19	7.97 3.41	0.66 *	3	21.54 2.57	26.70 0.28	17.40 0.52	1.43 *	3

Lower number in each pair is standard deviation
- = no data taken
* = single value used
[1] [2] [3] = number of plankton blooms observed at stations

Table 2C. Selected hydrographic parameters for the eastern sub-surface waters (Lower Chesapeake Bay). Cruise mean values and standard deviation.

	Sub-Stratum B					Sub-Stratum F				
1972	S ppt	T °C	Chl. a µg l^{-1}	V_{max} µg G l^{-1}h^{-1}	n	S ppt	T °C	chl. a µg l^{-1}	V_{max} µg G l^{-1}h^{-1}	n
June 29-30	15.90 1.54	22.30 0.28	10.60 2.40	1.08 0.21	2	13.17 1.87	22.85 0.92	8.70 1.13	1.05 0.78	2
July 6-7	15.43 4.42	21.35 0.35	8.35[1] 3.61	2.91 2.46	2	12.92 1.17	21.85 0.21	23.80[1] 4.10	1.03 0.10	2
July 13-14	17.58 5.11	24.03 1.86	12.73[3] 5.78	0.87 0.22	3	14.92 2.54	23.20 0.14	8.75 2.05	1.00 0.04	2
July 24-27	14.80 *	25.60 *	11.90 *	- -	1	9.31 1.73	28.80 1.06	12.00 1.56	2.16 1.05	3
Aug. 15-21	19.40 3.00	23.77 0.21	30.13[2] 31.09	0.90 0.15	3	14.25 0.39	24.17 0.98	16.60 2.01	1.45 0.33	3
Sept. 12-14	19.74 1.28	23.20 0.10	13.43[3] 0.47	0.67 0.25	3	16.60 0.49	24.03 0.21	12.17 1.76	1.44 *	3
Oct. 16-24	- -	- -	- -	- -	0	18.33 *	15.90 *	9.5 *	0.67 *	1
Nov. 13-16	22.83 2.78	13.53 0.95	8.07[1] 1.44	0.22 0.04	3	21.20 0.64	12.40 0.20	6.50 0.00	0.20 0.13	3
Dec. 11-13	20.05 1.18	9.40 0.46	8.40 0.95	0.18 0.05	3	17.71 0.33	9.60 0.26	9.33 0.81	0.24 0.04	3
1973										
Jan. 3-4	21.60 2.05	7.33 0.38	7.10 1.04	-	3	16.69 0.18	7.27 0.12	10.93 3.07	-	3
Feb. 13-14	18.30 3.16	3.60 0.78	8.30 2.69	-	3	17.18 2.24	3.27 0.12	12.17 1.42	-	3
March 19-20	20.37 0.86	7.87 0.21	11.20 0.89	-	3	17.19 1.33	7.67 0.12	14.97 2.75	-	3
April 16-17	21.92 1.68	11.37 0.25	5.03 2.41	-	3	17.27 0.54	11.30 0.26	7.47 2.18	0.67 *	3
May 14-16	21.71 6.49	17.33 0.60	11.23 7.44	0.98 *	3	18.41 1.38	17.47 0.32	11.50 1.25	0.46 *	3
June 25-27	18.34 0.90	24.83 0.21	6.50 4.59	1.01 *	3	16.29 4.58	24.63 0.51	11.07 3.42	0.98 *	3
July 23-24	26.67 0.90	23.33 0.57	8.80 1.15	0.50 *	3	21.02 0.04	25.25 0.07	14.10 *	0.23 *	2
Aug. 13-14	23.43 1.63	23.10 5.09	8.77 2.56	0.48 *	3	24.06 2.18	- -	14.20 1.41	- -	2

Lower number in each pair is standard deviation
- = no data taken
* = single value used
[1] [2] [3] = number of plankton blooms observed at stations.

It is apparent that approximately 68% of the Lower Bay stations sampled had very high chlorophyll a values in both the period immediately after Agnes and the year following. It should be emphasized that although there is only a single reference location for comparison (Station Y), this conclusion is a conservative interpretation.

Heterotrophic Potential (V_{max}, glucose)

The heterotrophic potential is a kinetic parameter which reflects the activity of a population of microorganisms. For the two-year study at the mouth of the York River (Table 1), the heterotrophic potentials may be divided into 4 ranges: low, moderate, high, and very high (Table 3). The highest values were observed in summer months and the lowest values in the winter.

Table 3. Frequency of Heterotrophic Potential ranges for the sub-surface waters (mouth of the York River and Lower Chesapeake Bay).

	Low 0.04-0.25 µg G $l^{-1}h^{-1}$	Moderate 0.26-0.82 µg G $l^{-1}h^{-1}$	High 0.83-1.50 µg G $l^{-1}h^{-1}$	Very High >1.50 µg G $l^{-1}h^{-1}$	n
York River Mouth		All Months			
Pre-Agnes 1971-1972	4	8	4	2	18
Post-Agnes 1972-1973	2	8	2	4	14
		June- August			
Pre-Agnes 1971	0	1	1	1	3
1972	0	0	0	1	1
Post-Agnes 1972	0	2	1	3	6
1973	0	0	1	1	2
Lower Chesapeake Bay					
Sub-Stratum		June-August (1972)			
A	0	9	3	0	12
B	0	6	2	0	8
C	0	2	6	3	11
D	0	0	9	3	12
E	0	2	2	4	8
F	0	7	3	0	10

In the immediate post-Agnes period (July-August 1972) the heterotrophic potential for approximately one-half of the Lower Bay stations was moderate (0.26-0.82) and the other half was very high (1.7-3.0) (Fig. 5B). In other words, the activity of the microbial flora of the Lower Bay waters was abundant. These abundant populations continued throughout the autumn and low populations occurred in the winter (Figs. 5A and 5B).

Although there is a paucity of data, it can be seen that, for the most part, the microbial populations of the York River Mouth and approximately 68% of those of the Lower Bay stations were quite similar during the post-Agnes period (1972-1973) (Fig. 5C).

When the 68% range of Lower Bay heterotrophic potentials of 1972-1973 is superimposed on that of the Station Y for 1971-1972, it appears that these population densities are quite similar (Fig. 5D). It may be inferred that the 1972 York River microbial population was near its maximum just prior to Agnes, that Agnes disrupted it, and that the population recovered rapidly and approached its greatest value. Furthermore, this population underwent its normal decline to the minimum in the winter. The exception is the response of the York River Mouth station following Agnes which more nearly follows the higher values shown in Fig. 5B for June and July 1972. Examination of these data reveals that the high V_{max}, $glucose$ values obtained in the Lower Bay on the July 24-27 cruise and the August 15-21 cruise were from stations in sub-strata D and E, adjacent to the York River Mouth.

With respect to spatial distribution, the major observation on heterotrophic populations present in the Lower Bay was the extreme abundance that occurred in sub-strata D, E, and F when sub-surface waters reached their maximum temperature four to six weeks following the passage of Agnes. Some increased abundance was also observed in the deep channel sub-strata of the Lower Bay and in sub-stratum A which may represent discharge from the James River.

An illustration of the variability in the processed data is presented in Table 4. Although variations in the salinity and temperature are evident in July and August, the great disparity between the mean and standard deviation for chlorophyll a is evident because the concentrations measured do not follow a statistically normal distribution due to the occurrence of phytoplankton blooms which were noted at the time of sampling (Table 4). More consistent data are reported for the 4 parameters in October, November, and December, whereas greater variation in the heterotrophic potential is seen for September. This detailed presentation of the original data is provided for making a judgement on the similarity or dissimilarity of the producer communities. Identifying a finer gradient of planktonic communities on the basis of these few parameters is still premature.

DISCUSSION

Several earlier studies have reported a seasonal trimodal distribution of phytoplankton populations in the relatively higher salinity waters of the lower Chesapeake Bay (Patten, Mulford, & Wariner 1963; Mackiernan 1968; Taylor 1972; Manzi 1973). There is a prominent abundance of phytoplankton seasonally in the spring, early summer and fall. This may in part be related to the grazing patterns of the zooplankton (Burrell 1968; Grant, et al. this volume).

This trimodal abundance of phytoplankton is also indicated by chlorophyll a distributions for the York River Mouth (Fig. 4A). It should be noted that the range of chlorophyll a values for the York River was rather small (4-20 µg l^{-1}) for the year prior to Agnes. However, there was a prolonged peak in the late

Table 4. Selected hydrographic parameters for Lower Chesapeake Bay sub-surface waters.

Date	Sub-Strata	S ppt	T °C	Chl. a µg l^{-1}	V_{max} µg G $l^{-1} h^{-1}$	n
26 July 1972	A90	15.03	27.6	2.1	1.61	
	A46	17.55	26.8	20.6*	1.42	
	A09	17.73	26.8	21.1*	1.07	3
	Mean	16.77	27.07	14.60[2]	1.37	
	S. D.	1.51	0.46	10.83	0.27	
17 Aug. 1972	B13	16.13	23.6	65.6*	0.84	
	B73	22.01	23.7	7.6	0.78	
	B44	20.06	24.0	17.2*	1.07	3
	Mean	19.40	23.77	30.13[2]	0.90	
	S. D.	3.00	0.21	31.09	0.15	
12 Sept. 1972	A05	17.38	23.0	17.2*	0.62	
	A22	17.03	23.1	21.4*	0.25	
	A39	17.61	23.2	20.4	1.26	3
	Mean	17.34	23.10	19.67[2]	0.71	
	S. D.	0.29	0.10	2.19	0.51	
18 Oct.	C08	21.18	17.2	8.7	0.50	
	C12	21.01	17.1	5.6	0.48	2
	Mean	21.10	17.15	7.15	0.49	
	S. D.	0.12	0.07	2.19	0.01	
13 Nov. 1972	A13	21.26	14.2	6.3	0.25	
	A74	21.75	14.4	15.0	0.17	
	A83	22.28	14.7	6.5	0.17	3
	Mean	21.75	14.43	9.27	0.20	
	S. D.	0.50	0.25	4.97	0.05	
12 Dec. 1972	07	19.24	8.9	7.2	0.18	
	08	18.81	8.9	10.9	0.13	
	16	20.38	8.3	9.5	0.20	3
	Mean	19.48	8.70	9.20	0.17	
	S. D.	0.81	0.35	1.87	0.04	

*phytoplankton bloom observed at station
[2]number of plankton blooms observed at stations.

summer after Agnes. The range of chlorophyll a values we report for the Lower Chesapeake Bay is quite similar to that reported by Taylor (1972). For comparison, chlorophyll a values for other mid-Atlantic estuaries are listed in Table 5.

Table 5. Annual ranges of Chlorophyll a values for the mid-Atlantic Region (values in µg liter^{-1}).

Location	Annual Range	Blooms	Citation
Long Island Sound (N. Y.)	3-25	-	Horne 1969
Patuxent River (Md.)	1-120	-	Flemer et al. 1970
Severn River (Md.)	13-30	386	Loftus et al. 1972
Pamlico River (N.C.)	1-40	250	Hobbie et al. 1972
Potomac River (Md.)	20-30	-	Jaworski et al. 1972
Chesapeake Bay (Md.)	6-59	-	Taylor 1972

The V_{max}, $glucose$ for the York River Mouth varied between 0.04 µg $l^{-1}h^{-1}$ in the winter to as high as 2.9 µg $l^{-1}h^{-1}$ in the late summer. These values are lower than those reported for the Pamlico Sound estuary, one of the most microbially active aquatic environments where, V_{max}, $glucose$ was reported to be 0.15-24.10 µg glucose $^{-1}h^{-1}$ during 1966, 1967 and 1968 (Crawford, Hobbie, & Webb 1973).

It should be noted that for most of the sub-surface waters of the Lower Bay (204 stations) that the V_{max}, $glucose$ values fell within the range for that of the York River (0.04-3.0), whereas 72 stations exceeded the upper limit, particularly during the summer months of 1972. These values reached a maximum in late July, which coincided with maximum water temperatures for the Lower Bay. However, this peak of heterotrophic potential preceded that for primary productivity potential which occurred in mid-August (Zubkoff & Warinner, unpublished).

The reasons for the increases in heterotrophic potential may be several:
1) Responses of the phytoplankton and heterotrophic communities to freshwater runoff and associated enrichment by inorganic nutrients (Loftus, SubbaRao, & Seliger 1972).
2) Increase in suspended sediment with its associated microbial flora.
3) Infestation of the regions of normally higher salinity water by a microbial flora which is adapted to lower salinity waters.
4) Shift from the usual heterotrophic population to another population which is resistant to grazers.
5) Relatively high water temperatures (24-27°C) with prolonged sunlight which would favor both microbial and phytoplankton growth and reproduction.

CONCLUSION

From the chlorophyll a and heterotrophic potential data of the lower Chesapeake Bay, and with respect to a single hydrographic station (Station Y) followed on a seasonal basis, it may be concluded that Agnes produced:
1) An enriching effect on the phytoplankton populations of the Lower Bay waters during the period immediately following the storm.

2) The enrichment persisted throughout the year (in the fall and winter) and continued into the summer of 1973.

3) An enriching and stimulatory effect on heterotrophic activity (York River Mouth and Lower Bay) which subsequently declined to their probable winter levels.

4) Activities of the Lower Bay heterotrophic populations of 1973 spring and summer were probably similar to those populations which occurred prior to Agnes.

ACKNOWLEDGEMENTS

We gratefully acknowledge the technical assistance of P. Crewe, D. Francis, G. Fulton, M. Gorey, and L. Jenkins. This research was supported, in part, by the Chesapeake Research Consortium, Inc. through funds supplied by the National Science Foundation RANN (Research Applied to National Needs) Grant, GI 34869 and by Public Law 89-720, The Jellyfish Act, Contract No. N-043-226-72(G) from the National Marine Fisheries Service of the National Oceanic and Atmospheric Administration, U. S. Department of Commerce and the Commonwealth of Virginia.

LITERATURE CITED

Biological Methods Panel Committee on Oceanography. 1969. Recommended procedures for measuring the productivity of plankton standing stock and related oceanic properties. Natl. Acad. Sci., Washington, D. C., 59 p.

Burrell, V. G., Jr. 1968. The ecological significance of the ctenophore, *Mnemiopsis leidyi* (A. Agassiz) in a fish nursery ground. M. S. Thesis, College of William and Mary, Williamsburg, Va., 61 p.

Crawford, C. C., J. E. Hobbie, and K. L. Webb. 1973. Utilization of dissolved organic compounds by microorganisms in an estuary. Pages 169-180 in Estuarine Microbial Ecology, L. H. Stevenson and R. R. Colwell, eds., Univ. South Carolina Press, Columbia, S. C.

Flemer, D. A., D. H. Hamilton, C. W. Keefe, and J. A. Mihursky. 1970. The effects of thermal loading and water quality on estuarine primary production. OWRR Project No. 14-01-0001-1979, 217 p.

Grant, G. C. 1972. Plankton program. I. Zooplankton. Ann. Rep. Chesapeake Research Consortium to National Science Foundation (RANN Program), pp. 28-76.

Hobbie, J. E., B. J. Copeland, and W. G. Harrison. 1972. Nutrients in the Pamlico River Estuary, N. C., 1969-1971. OWRR Project No. B-020-NC, 242 p.

Horne, R. A. 1969. Marine Chemistry, the structure of water and the chemistry of the hydrosphere. Wiley-Interscience, N. Y., 568 p.

Jaworski, N. A., D. W. Lear, and O. Villa, Jr. 1972. Nutrient management in the Potomac Estuary. Pages 246-273 in Nutrients and Eutrophication: The Limiting Nutrient Controversy, Spec. Symp. Vol. 1, G. E. Likens, ed., 327 p.

Loftus, M. E., D. V. SubbaRao, and H. H. Seliger. 1972. Growth and dissipation of phytoplankton in Chesapeake Bay. I. Response to a large pulse of rainfall. Chesapeake Sci. 13:282-299.

Mackiernan, G. 1968. Seasonal distribution of dinoflagellates in the lower York River, Virginia. M. S. Thesis, College of William and Mary, Williamsburg, Va., 91 p.

Manzi, J. J. 1973. Temporal and spatial heterogeneity in diatom populations at the lower York River, Virginia. Ph.D. Dissertation, College of William and Mary, Williamsburg, Va., 163 p.

Parsons, T. R. and J. D. H. Strickland. 1962. On the production of particulate organic carbon by heterotrophic processes in sea water. Deep-Sea Res. 8: 211-222.

Patten, B. C., R. A. Mulford, and J. E. Warinner. 1963. An annual phytoplankton cycle in the lower Chesapeake Bay. Chesapeake Sci. 4:1-20.

Stofan, P. E. 1973. Surface phytoplankton community structure of Mobjack Bay and York River, Virginia. Ph.D. Dissertation, College of William and Mary, Williamsburg, Va., 116 p.

Stroup, E. P. and R. J. Lynn. 1963. Atlas of salinity and temperature distributions in Chesapeake Bay, 1949-1961. Graphical Summary Rep. No. 2, Johns Hopkins Univ., Baltimore, Md., 409 p.

Taylor, W. R. 1972. The ecology of the plankton of the Chesapeake Bay estuary. Progress Rep. (1 December 1970-31 August 1972), Contract No. AT(11-1)-3279, Johns Hopkins Univ., Baltimore, Md., 86 p.

U. S. Geological Survey. 1968-72. Water resources data for Virginia. U. S. Dept. of Interior, Geological Survey, Washington, D. C.

Vaccaro, R. F. and H. W. Jannasch. 1966. Studies on heterotrophic activity in seawater based on glucose assimilation. Limnol. Oceanogr. 11:596-607.

Williams, P. J. LeB. 1973. The validity of the application of simple kinetic analysis to heterogeneous microbial populations. Limnol. Oceanogr. 18: 159-165.

Wright, R. T. 1973. Some difficulties in using ^{14}C-organic solutes to measure heterotrophic bacterial activity. Pages 199-217 in Estuarine Microbial Ecology, L. H. Stevenson and R. R. Colwell, eds., Univ. South Carolina Press, Columbia, S. C.

Wright, R. T. and J. E. Hobbie. 1966. Use of glucose and acetate by bacteria and algae in aquatic ecosystems. Ecology 47:447-464.

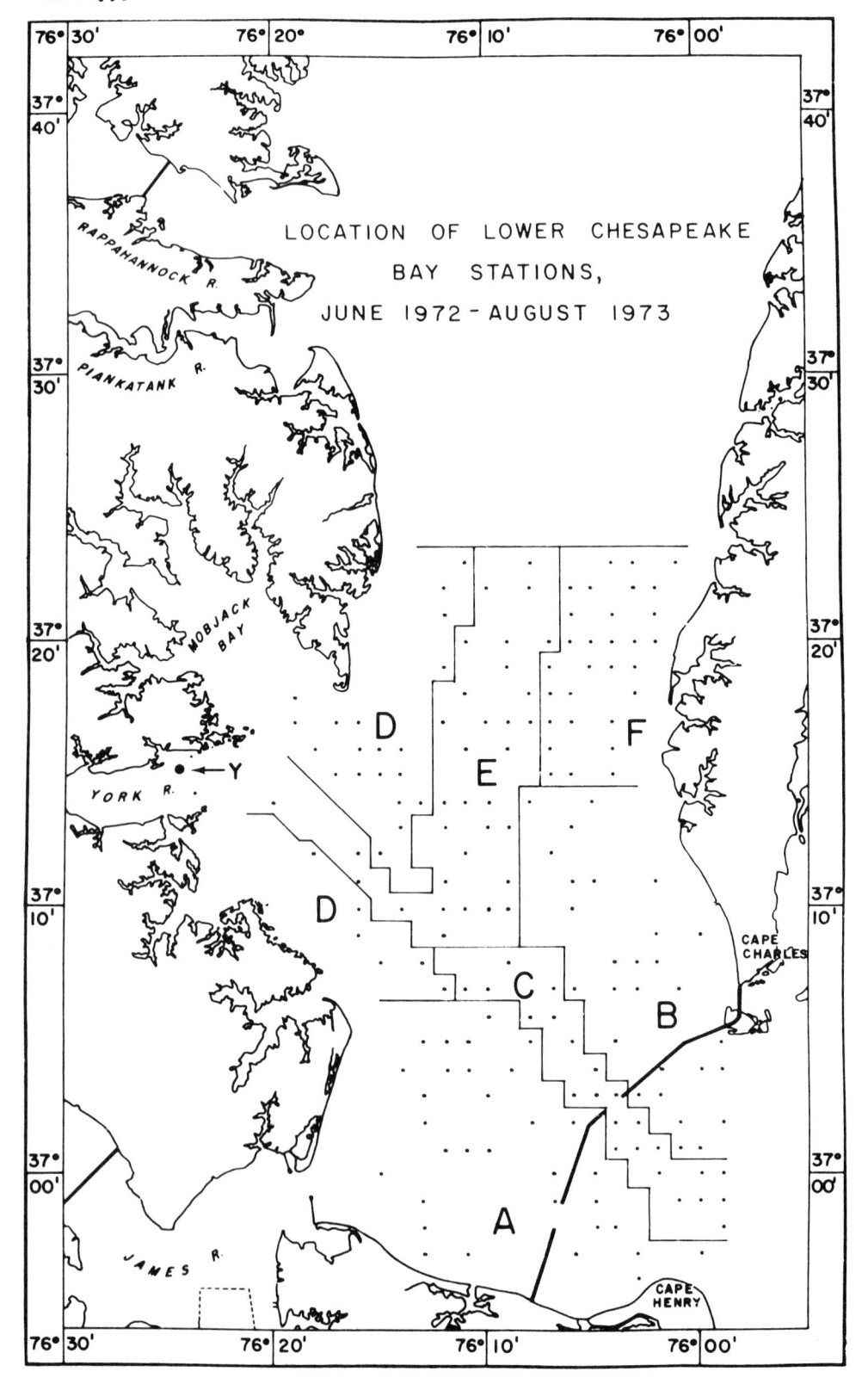

Figure 1. Area of study during one year study of lower Chesapeake Bay following Agnes. Y designates station at the mouth of the York River.

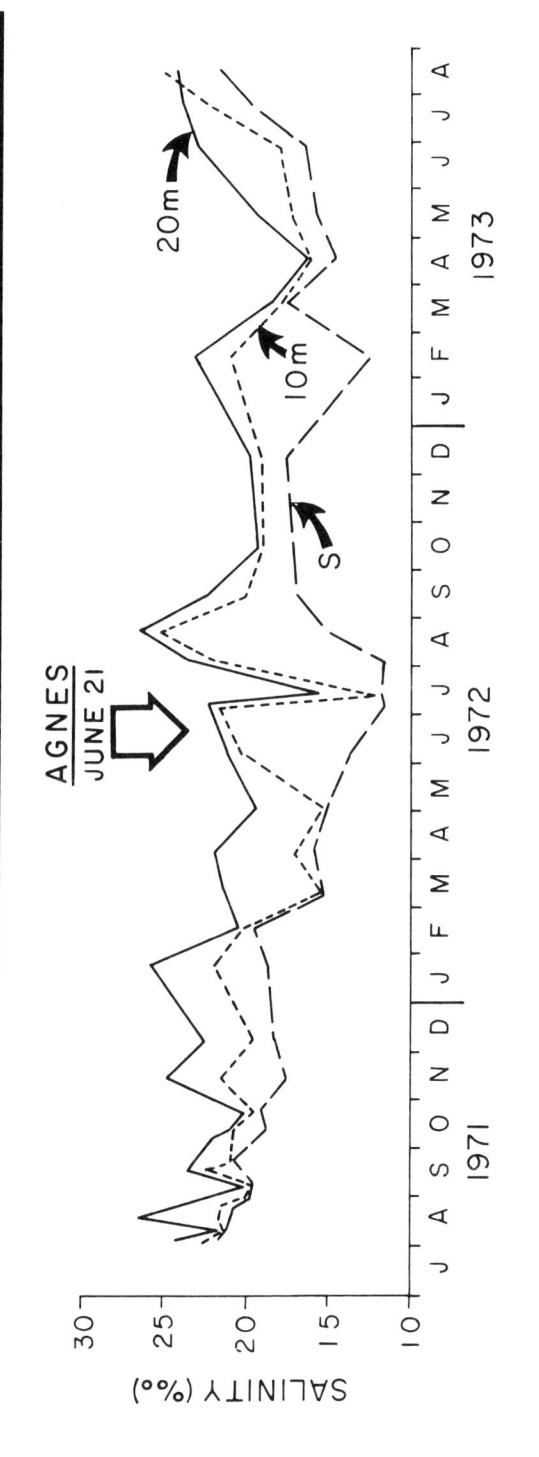

Figure 2. Salinity at the mouth of the York River (June 1971-August 1973) S = waters within 1 meter of the surface; 10 m = water at 10 meters below surface; 20 m = bottom waters.

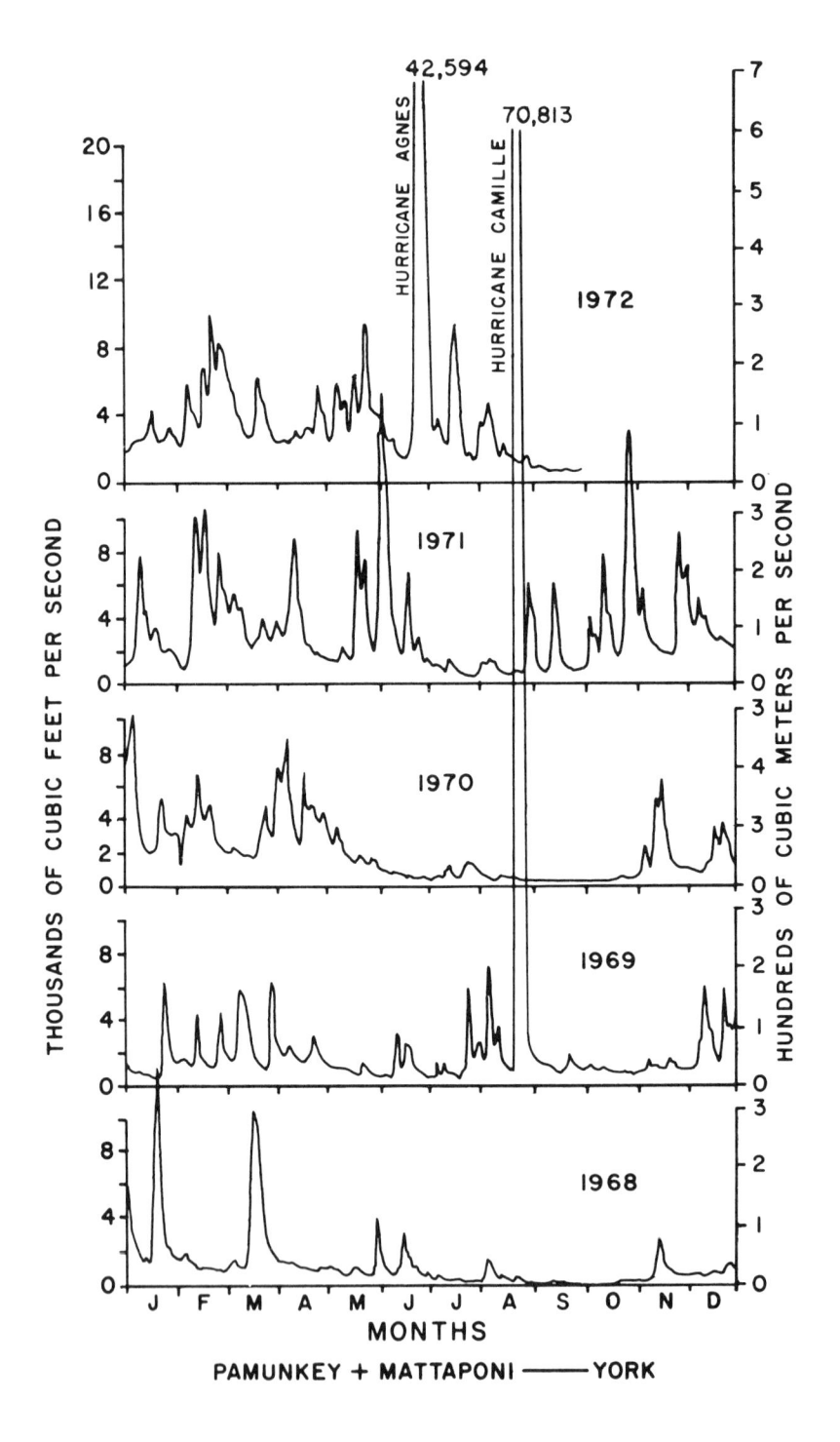

Figure 3. Freshwater discharge into the York River (1968-1972).

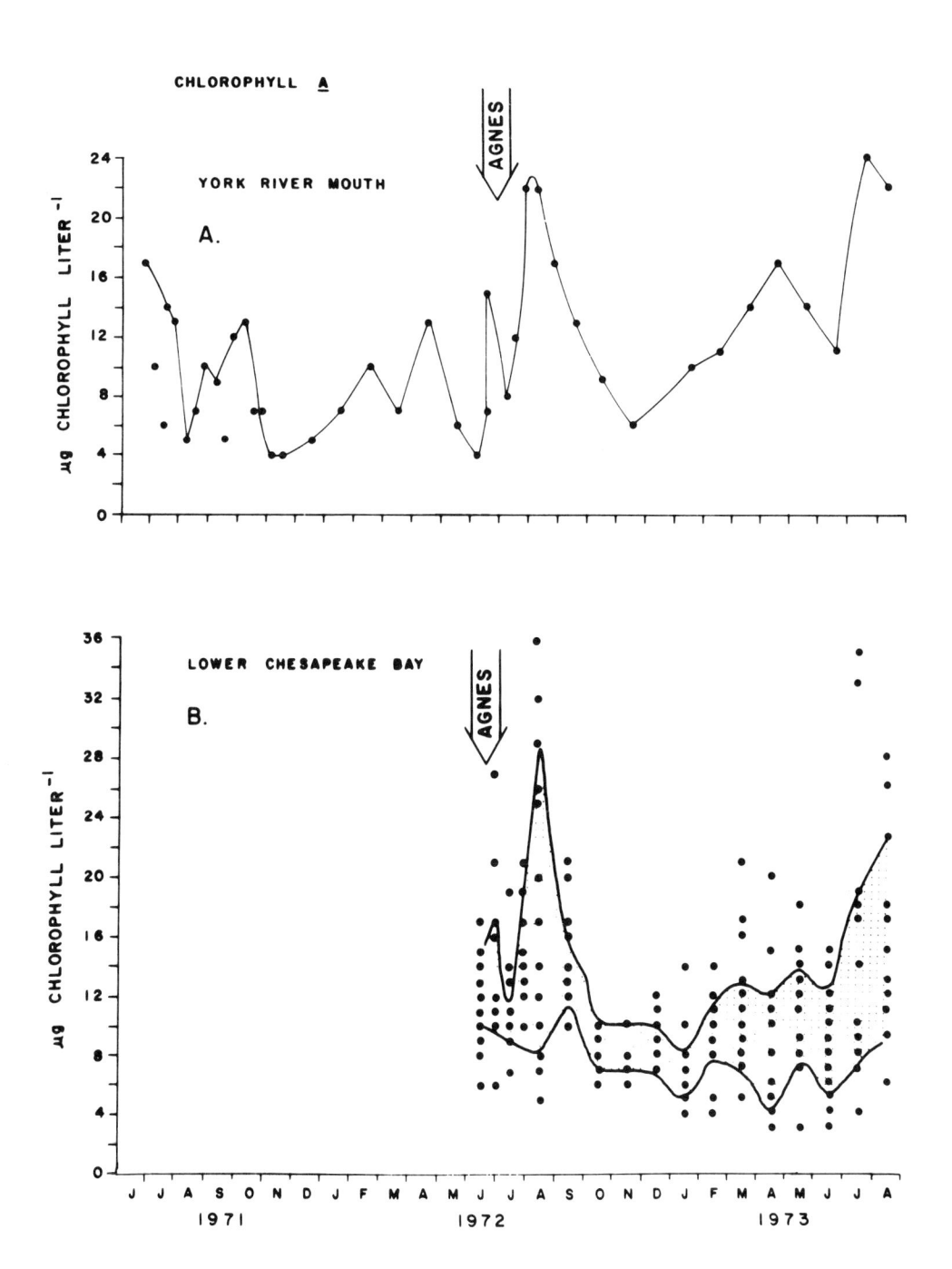

Figure 4A&B. Concentrations of chlorophyll a in the sub-surface waters. The shaded area bounded by heavy lines indicates the range in which 68% (2 standard deviations) of the observed values occur.
A. York River Mouth (June 1971-August 1973)
B. Lower Chesapeake Bay (June 1972-August 1973).

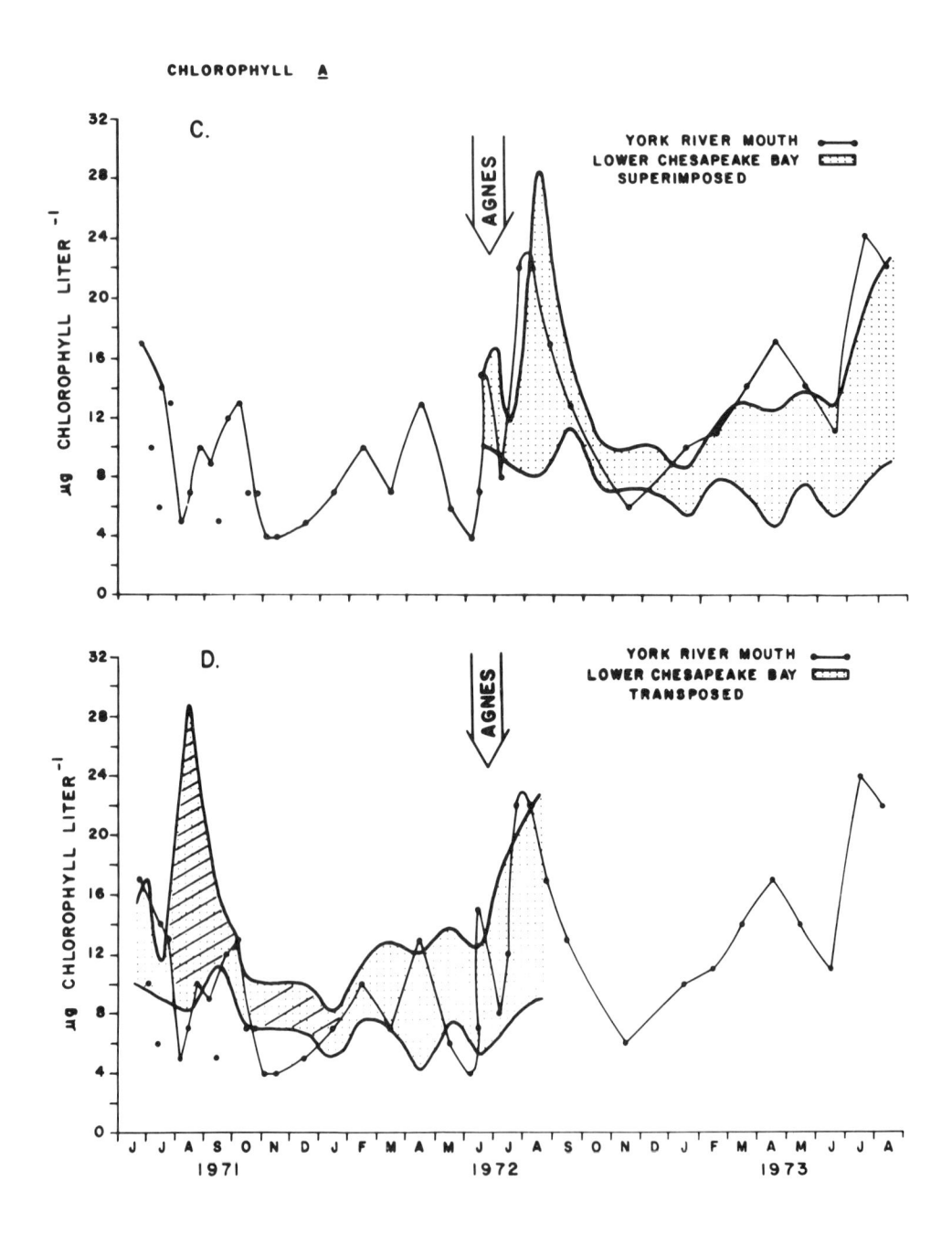

Figure 4C&D. Concentrations of chlorophyll a in the sub-surface waters. The shaded area bounded by heavy lines indicates the range in which 68% (2 standard deviations) of the observed values occur.
    C. Lower Chesapeake Bay (1972-1973) superimposed on York River Mouth (1972-1973).
    D. Lower Chesapeake Bay (1972-1973) transposed on York River Mouth (1971-1972).

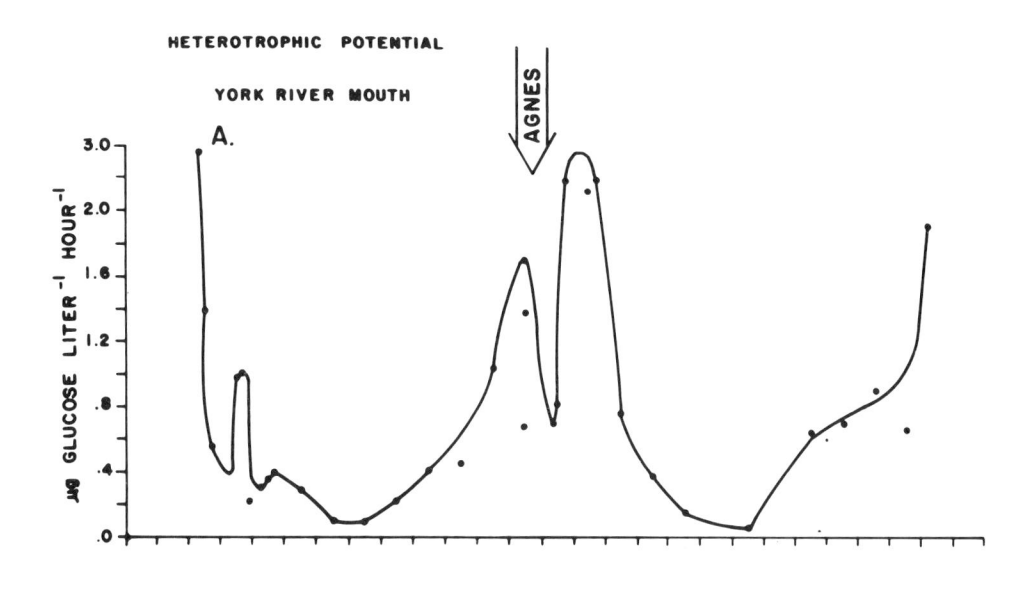

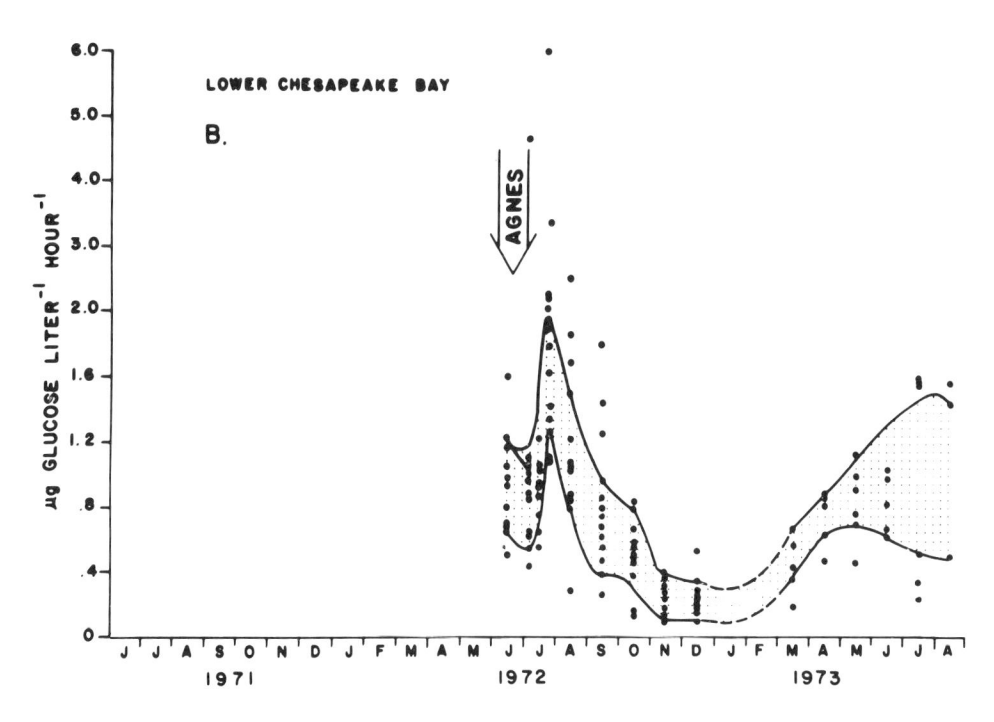

Figure 5A&B. Heterotrophic potential (V_{max}, $glucose$) in the sub-surface waters. The shaded area bounded by heavy lines indicates the range in which 68% (2 standard deviations) of the observed values occur.
A. York River Mouth (June 1971-August 1973)
B. Lower Chesapeake Bay (June 1972-August 1973).

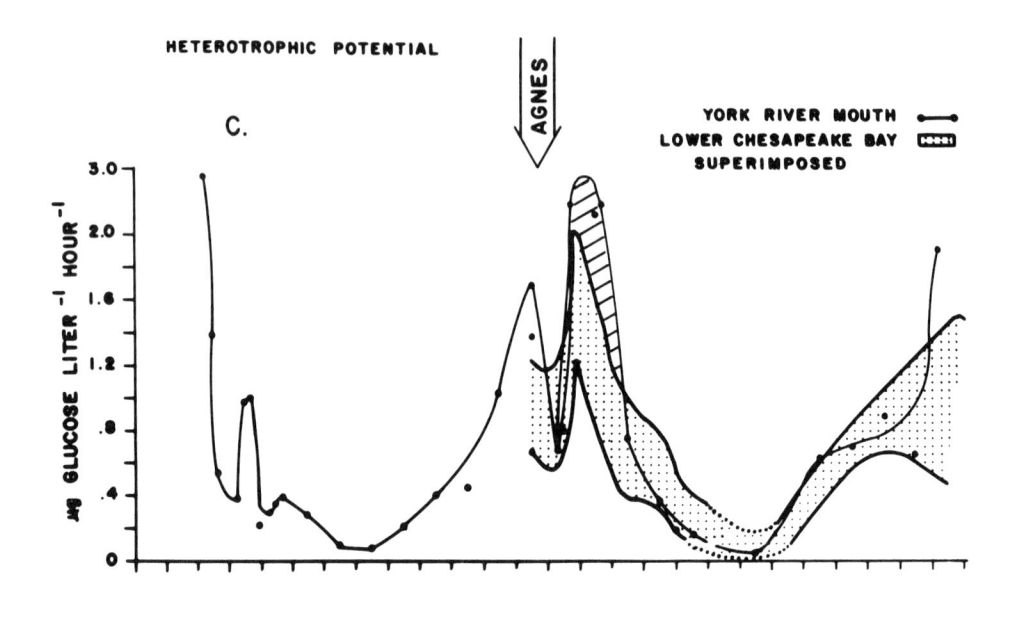

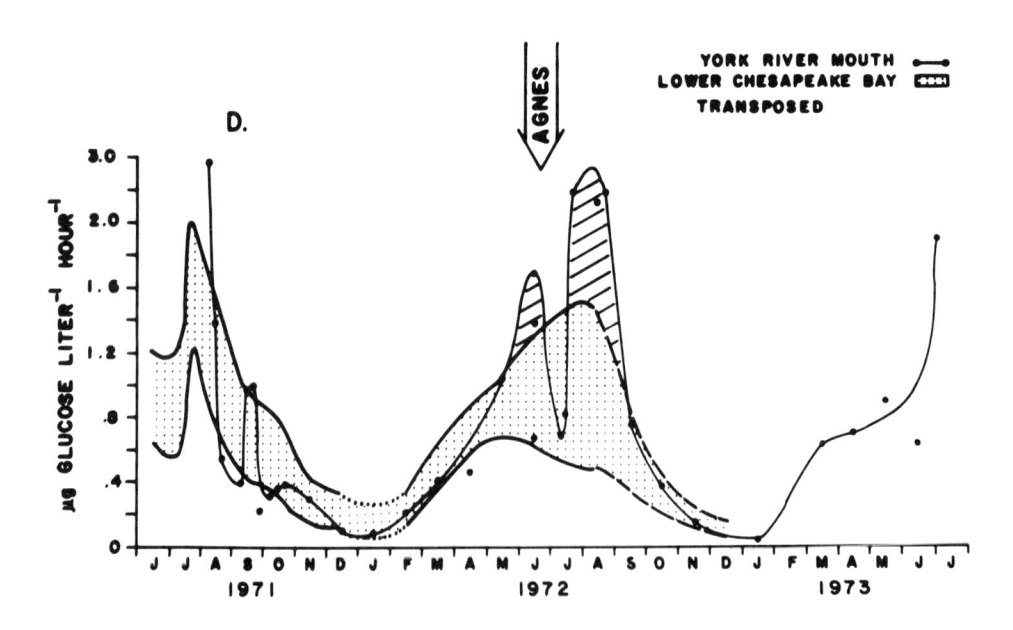

Figure 5C&D. Heterotrophic potential (V_{max}, glucose) in the sub-surface waters. The shaded area bounded by heavy lines indicates the range in which 68% (2 standard deviations) of the observed values occur.
C. Lower Chesapeake Bay (1972-1973) superimposed on York River Mouth (1972-1973).
D. Lower Chesapeake Bay (1972-1973) transposed on York River Mouth (1971-1972).

CALVERT CLIFFS SEDIMENT RADIOACTIVITIES BEFORE AND AFTER
TROPICAL STORM AGNES

Philip J. Cressy, Jr.[1]

ABSTRACT

Natural U, Th, and K, and man-made ^{60}Co, ^{106}Ru and ^{137}Cs, have been measured in surface sediments from 19 sites near the Calvert Cliffs Nuclear Power Plant on the Chesapeake Bay. Use of a low background gamma-gamma coincidence counting system permitted positive identification and quantitative measurement of some contaminants down to a level of 0.005-0.01 pCi/g. Samples taken in August, 1972 show some pronounced differences in radionuclide concentrations when compared with samples taken in May or earlier. The radionuclides act as tracers, permitting detection of sediment transport caused by Tropical Storm Agnes.

---

[1] Earth Resources Branch, Goddard Space Flight Center, Greenbelt, Md. 20771

BIOLOGICAL EFFECTS

Commercial
Aven M. Andersen, Editor

Non-Commercial
Marvin L. Wass, Editor

DISTRIBUTION AND ABUNDANCE OF AQUATIC VEGETATION IN
THE UPPER CHESAPEAKE BAY, 1971-1974

James A. Kerwin[1]
Robert E. Munro[1]
W. W. Allen Peterson[1]

ABSTRACT

Submersed aquatic plants were sampled in shoal waters of Chesapeake Bay during August-September 1971 through 1974. In 1972, sampling indicated a decrease in distribution of submersed aquatic plants most likely due to the effects of Tropical Storm Agnes. In 1973 and 1974, the frequency occurrence of plants was still considerably less than in 1971.

Dominant plants in shoal waters less than 8 feet deep included widgeongrass *(Ruppia maritima)*, redhead-grass *(Potamogeton perfoliatus)*, and eelgrass *(Zostera marina)*.

INTRODUCTION

Estuaries are extremely important natural areas because they provide habitat for higher plant and animal life, serve as nursery grounds for marine fish, and abound in lower plant and animal life. They are among the most productive habitat types on earth. Moreover, they are aesthetically and economically important as recreation areas.

Important submersed aquatic plants in the marine and estuarine ecosystem represent 2 families (Hydrocharitaceae and Potamogetonaceae), 7 genera and 13 species (Dawson 1966). Other taxa significant in the estuarine environment but not necessarily confined to brackish waters have been discussed by Fernald (1950) and Sincock et al. (1966). Few of these species have been studied intensively and most of the environmental requirements are unknown.

Partially responsible for the paucity of knowledge regarding estuarine vascular plants have been a lack of interest and an inadequacy of sampling methods. The lack of interest is, in some measure, due to a failure to recognize aquatic vascular plants as an important estuarine component. The inadequacy of sampling methods may be attributed to the relatively small numbers of studies conducted, although methods were extensively developed during an ecological study of the Back Bay-Currituck Sound estuary in 1958-1964 (Sincock et al. 1966).

Objectives of the present study were the following: 1) to measure annual trends in abundance, composition, and distribution of submersed aquatic plants, 2) to determine the major environmental factors (e.g., salinity and turbidity) effecting changes in the abundance and composition of aquatic vegetation.

Because of the familiarity of the Chesapeake Bay to the scientific community and the public, a description of the area will be omitted. The upper Chesapeake refers to that portion of the Bay north of Point Lookout on the western and Cedar Island on the eastern shores.

---

[1]Migratory Bird and Habitat Research Laboratory, U. S. Fish and Wildlife Service, Laurel, Md. 20811

PROCEDURES

Results of preliminary sampling in 1970 were used to estimate variances associated with plant species composition and sample volume over localized areas. Sampling intensity was thereby derived so that a 5% change in frequency occurrence of vegetation over the upper Bay would be detectable. The acreage of shoal water (8 feet or less in depth at mean low water) over the upper Bay and its tributaries was determined from nautical charts of 1:40,000 scale. The acreages of shoal water represented on each chart were summed and proportions to the total were calculated. More than 600 sample stations were distributed on the charts according to these proportions (i.e., a chart with 5% of the total shoal water acreage received 5% of the sampling intensity). The charts were lined off and numbered at one-half minute intervals of latitude and longitude. Sets of coordinates were randomly chosen and plotted on the charts as long as the 8 feet depth limitation was satisfied. Sets of sample stations were logically grouped by river system where possible to facilitate discussion.

The same sample stations were visited each summer. It was realized that precise relocation of sample stations was impossible, although all landmarks and chart features were used. Because of the inherent patchiness of vegetation beds, it is believed that relatively small location errors, particularly when parallel to the contour, do not materially affect species composition. However, sample volume measurements are much more variable and these have not been stressed.

The boat crews consisted of two men each in two 17-foot Boston Whalers. Salinities were determined in 1971 with a salinometer and in following years with a hydrometer. A Secchi disc was used to measure turbidity. Samples were obtained with modified oyster tongs, designed after those used in the Bay Bay-Currituck Sound study (Sincock et al. 1966), unpublished data). The oyster tongs had a metal plate welded to the "teeth" and the lower "biting" edge was sharpened so they could effectively remove entire plants. Each sample consisted of the plants growing on 1-square foot of bottom. Three samples were collected, recorded separately and later averaged at each sample station. About 25-30 sample stations were covered by each boat crew in a day's work during the peak development of aquatic plants (August-September).

Each vegetation sample was placed in a ¼-inch mesh basket and washed to remove soil. The plant material was removed from the basket and spread loosely on the deck of the boat, and a visual estimate was made of the percent composition of each species. Any species comprising less than 5% of the total was listed as a trace. The vegetation was gently squeezed into a ball, shaken to remove excess water, then submersed in a cylindrical volumetric displacement device. A 10 ml pipette attached to the measuring device provided the sample volume; species volumes, because of the normally small sample volume, were not routinely measurable.

RESULTS

A few comments are in order in relation to environmental measurements discussed below. Time of day and tidal stage were totally disregarded during all field collections. Water depth, surface temperature, and salinity therefore behave as random variables. When results from one season to the next are compared over the upper Bay, normal daily variance in these measurements from station to station are in essence cancelled out and seasonal differences are therefore compared.

With the exception of salinity, environmental parameters changed little from 1971 to 1974 (Table 1). Salinity decreased uniformly and significantly over the upper Bay by an average of 5.78 ppt from 1971 to 1972. By later summer of 1973 salinities had increased somewhat but were still significantly lower than in 1971. The minimum detectable salinity was in the range of 1-2 ppt, and samples falling in this range were arbitrarily assigned a value of 1.0 ppt.

Table 1. Mean water depth, salinity, temperature, and turbidity in the upper Chesapeake Bay, 1971-1974. (Number of samples in parentheses).

Parameter[1]	1971	1972	1973	1974	1971-74	1972-74	1973-74
Water depth (Ft.)	5.05 (624)	5.06 (615)	5.28 (629)	4.98 (613)			2.18*
Salinity (ppt)	15.44 (556)	9.66 (617)	10.37 (634)	13.49 (566)	5.07**	-10.02**	-8.80**
Temperature (°C)	24.17 (480)	24.32 (549)	26.82 (634)	25.34 (612)	-7.51**	-6.47**	12.80**
Light Penetration (inches)		27.6 (585)	28.0 (607)	31.3 (547)		-5.22**	-4.97**

[1]Analyzed for upper Bay as a whole, regardless of river system boundaries.
* .05 significance level.
** .01 significance level.

Water depths (measured to the nearest 0.5 foot) varied little over the 4 years and averaged just over 5 feet at the sampling stations (Table 1). Mean surface water temperatures ranged from 24 to 26°C during the 4-year period. Secchi disc readings (not recorded in 1971) averaged 28 inches for both 1972 and 1973, with a small increase noted in 1974.

Data on the relationship between sample water depth and occurrence of vegetation (all species and 4 years combined), the distribution of vegetated samples by water depth, and the sample distribution by water depth for each year are shown in Fig. 1. Almost 50% of the sampling occurred in the 3- to 5-foot depth range each year or all years combined. More than 30% of all stations revealing vegetation occurred in 3 feet of water each year, and of the stations at 3 feet, more than 40% yielded vegetation in 1971 or 1972. Only 25% of samples taken in 3 feet of water produced vegetation in 1973 or 1974.

A qualitative decrease in aquatic vegetation was noted in certain species. Widgeongrass *(Ruppia maritima)*, Eurasian watermilfoil *(Myriophyllum spicatum)*, wildcelery *(Vallisneria americana)*, naiad *(Najas guadalupensis)*, and algae (various genera grouped) all decreased significantly from 1971 to 1972 in frequency of occurrence, while sago pondweed *(P. pectinatus)* and horned pondweed *(Zannichellia palustris)* increased significantly (Table 2). For instance, widgeongrass occurred at 14.7% of all sample stations in 1971 and 9.9% in 1972. Frequencies of occurrence of widgeongrass, watermilfoil, wildcelery, and algae in 1973 and 1974 were still significantly lower than 1971 levels. Highly significant decreases were noted over the 4 years, when presence or absence of any amount of vegetation is considered.

Table 2. Frequency of occurrence and indicated change in species of submersed aquatic plants from 1971 to 1974 in upper Chesapeake Bay (regardless of river system boundaries).

Species[1]	Percentage occurrence 1971	1972	1973	1974	Chi-square 1971-74	1972-74	1973-74
Redhead-grass (*Potamogeton perfoliatus*)	5.29	3.41	4.13	3.43			
Sago pondweed (*P. pectinatus*)	1.28	5.69	2.86	1.31		16.20**	
Slender pondweed (*P. pusillus*)	0.32	0.00	0.16	0.00			
Horned-pondweed (*Zannichellia palustris*)	0.00	2.76	0.16	0.16		12.64**	
Widgeongrass (*Ruppia maritima*)	14.74	9.92	6.04	9.79	6.59*		5.49*
Eelgrass (*Zostera marina*)	10.26	7.32	1.11	2.45	30.25**	14.64**	
Wildcelery (*Vallisneria americana*)	2.72	0.16	0.47	0.82	5.40*		
Eurasian watermilfoil (*Myriophyllum spicatum*)	3.85	1.14	0.48	1.14	8.18**		
Common elodea (*Elodea canadensis*)	3.04	1.46	0.32	0.49	10.14**		
Naiad (*Najas guadalupensis*)	1.60	0.33	0.95	1.47			
Coontail (*Ceratophyllum demersum*)	0.64	0.00	0.00	0.16			
Muskgrass (*Chara* spp.)	0.32	0.33	0.64	0.49			
Algae	8.97	3.90	1.43	2.45	23.16**		
Species disregarded	28.53	20.98	10.49	14.85	33.21**	7.34**	4.94*

[1]Nomenclature follows that of Hotchkiss (1967).
* = .05 significance level.
** = .01 significance level.
[4]Included genera *Ulva*, *Enteromorpha*, *Ceramium*, *Polysiphonia*, etc.

The lowest unconverted volumetric displacement measurement was .05 ml. Sample volumes less than "measurable" were recoreded as "trace", and because of this another comparison can be made. A comparison of the number of sample stations with measurable volumes (species ignored) for 1971 and 1972 revealed a high significant decrease (Table 3). A 1972-1973 comparison also yielded a significant decrease. Results of 1974 sampling indicated that five species were still below 1971 levels.

Over the upper Bay, widgeongrass was the most common vascular plant encountered each year, followed by eelgrass (*Zostera marina*) in 1971-1972 and by redheadgrass (*Potamogeton perfoliatus*) in 1973-1974. In 1971 widgeongrass was present in 15 of 26 river systems, followed by eelgrass in 12 of 26 systems. Widgeongrass was found in 12 river systems in 1972 and 10 in 1973, while eelgrass occurred in 10 in 1972 and 4 in 1973.

Table 3. Occurrence of plant species in measurable quantities, and indicated change from 1971 to 1974 in the upper Chesapeake Bay (regardless of river system boundaries).

Species[1]	Number with measurable volume				Chi-square		
	1971	1972	1973	1974	1971-74	1972-74	1973-74
Redhead-grass	22	8	16	12			
Sago pondweed	4	20	10	1		15.64**	5.66*
Slender pondweed	1	0	0	0			
Horned-pondweed	0	10	0	0		8.37**	
Widgeongrass	57	33	20	26	11.10**		
Eelgrass	45	13	2	7	26.80**		
Wildcelery	13	0	3	3	5.20*		
Eurasian water-milfoil	10	3	0	2	4.00*		
Common elodea	11	5	2	2	4.83*		
Naiad	5	1	2	3			
Coontail	3	0	0	0			
Muskgrass	0	2	3	2			
Algae	10	3	1	3			
Species disregarded	119	59	39	44	37.20**		

[1]Scientific names are given in Table 2.
* = .05 significance level.
** = .01 significance level.

Four rivers revealed significant declines in numbers of samples with vegetation, when comparing 1971 and 1972 (Table 4). Where possible, rivers were analyzed by comparing numbers of samples with measurable, trace, and zero vegetation. Significant declines were noted in seven rivers by late summer 1973.

In terms of average sample volume of vegetation as measured by volumetric displacement, no differences were detected in the overall means for the 4 years. As indicated earlier, little can be said in relation to the sample volumes recorded. Average measurable volumes of plant species (river systems disregarded) and average measurable sample station volumes by river system (species disregarded) are available from the authors.

DISCUSSION

Major factors limiting the growth, distribution, and abundance of pondweeds in the Chesapeake Bay estuary appear to be salinity and turbidity. Salinity limits the penetration of slightly brackish water species (e.g., redhead-grass) into the more saline reaches of the Bay (above 15 ppt salinity). Species adapted to living in waters above 15 ppt total chlorides are widgeongrass and eelgrass. Turbidity is particularly limiting in the slightly brackish tributaries of the estuary. It is not unusual to find Secchi disc readings of 12 inches or less. The lowest Secchi disc recording during the 1971-1973 surveys was 8 inches. Pollution may be limiting in some tributaries of the Bay; however, factors other than salinity and turbidity were not measured during this study.

Table 4. Frequency of occurrence of vegetated samples and indicated change by river systems from 1971 to 1974 in upper Chesapeake Bay (species disregarded).

River System	1971 % Veg.	1971 N[1]	1972 % Veg.	1972 N	1973 % Veg.	1973 N	1974 % Veg.	1974 N	Chi-square[2] 1971-74	1972-74	1973-74
Elk & Bohemia Rivers	6.67	15	0.0	16	0.0	16	0.0	16			
Sassafras River	30.0	10	0.0	10	0.0	10	0.0	10			
Howell Pt. - Swan Pt.	16.67	12	0.0	6	0.0	12	0.0	12			
Eastern Bay	34.04	47	46.51	43	34.04	47	36.17	47			
Choptank River	35.0	60	39.66	58	19.3	57	27.59	58			
Little Choptank River	21.05	19	21.05	19	0.0	19	0.0	19			
James Is. - Honga River	44.12	34	35.29	34	2.94	34	5.88	34	13.77**	7.29**	
Honga River	50.0	30	40.0	30	13.33	30	16.66	30	9.30**		
Bloodsworth Island	37.5	40	22.73	44	10.87	46	11.63	43	7.64*		
Susquehanna Flats	44.44	27	2.70	37	0.0	37	12.82	39	9.23**		
Fishing Bay	8.0	25	4.0	25	0.0	25	0.0	25			
Nanticoke & Wicomico Rivers	0.0	30	0.0	30	0.0	30	0.0	31			
Manokin River	40.0	15	46.67	15	13.33	15	20.0	15			
Patapsco River	0.0	21	5.0	20	4.76	21	9.52	21			
Big & Little Annemessex Rivers	70.0	20	60.0	20	30.0	20	57.89	19	9.53**		6.22*
Gunpowder & Bush R. Hdwts.	11.11	9	0.0	8	0.0	7	0.0	9			
Pocomoke Sound, Md.	18.18	22	10.0	20	4.76	21	-	0			
Magothy River	33.33	12	0.0	12	16.67	12	16.66	12			
Severn River	40.0	15	20.0	15	26.67	15	26.67	15			
Patuxent River	2.0	50	4.26	47	0.0	50	4.00	50			
Back, Middle & Gunpowder Rivers	13.64	22	4.55	22	4.55	22	4.55	22			
Curtis Pt. - Cove Pt.	0.0	20	0.0	19	0.0	19	0.0	19			
South, West & Rhode Rivers	0.0	8	0.0	10	0.0	10	0.0	8			
Chester River	61.11	36	36.11	36	26.47	34	23.52	34	10.74**		
Love Pt. - Kent Pt.	0.0	8	0.0	8	0.0	8	12.50	8			
Smith Island, Md.	64.71	17	45.46	11	25.0	12	35.29	17			
River systems disregarded	28.53	624	20.98	615	10.49	629	14.85	613	41.74**	7.43**	7.05*

[1]Number of sample stations visited in river systems.
[2]Where possible, differences in "trace" or "measurable volume" samples were considered which resulted in more powerful tests.
* = .05 significance level.
** = .01 significance level.

In 1971, pondweeds exhibited vigorous growth and diversity in many of the tributaries of the Bay. Following Agnes the growth of pondweeds decreased particularly in northern Bay areas, due in some measure to the flushing action of the storm. In 1972, only two species of plants (sago pondweed and horned-pondweed) increased in frequency. It is possible that freshwater flushing caused these fresh and slightly brackish water species to extend their range into the normally higher salinity reaches of the Bay. After the storm all other species decreased in frequency. It is highly unlikely that the continued decrease noted in 1973 and 1974 may be attributed to the adverse effects of the extratropical storm.

CONCLUSIONS

Salinity and turbidity were assumed to be the major limiting factors affecting the distribution and abundance of aquatic vascular plants in the Chesapeake Bay estuary. Salinities were less in 1972 through 1974 than in 1971.

The three dominant species of plants during the 1971-1974 surveys were widgeongrass, redhead-grass, and eelgrass.

The frequency occurrence of aquatic plants decreased from 1971 to 1973. The plants occurred at 29% of the sample stations in 1971, 21% in 1972, and 10% in 1973, with an increase to 15% noted in 1974. Reductions in frequency occurrence of vegetated stations were most apparent in northern Bay areas, although the status of submersed vegetation in 1974 was still far below 1971 levels. Decreases in the frequency occurrence of vegetation in more southern areas of the upper Bay, such as the Choptank, and Big and Little Annemessex Rivers, were more obvious from 1972-1973. It seems, therefore, that Agnes caused much damage to submersed aquatic vegetation during 1972 in northern areas, but subsequent decreases further south should not be attributed to Agnes.

ACKNOWLEDGEMENTS

The authors are indebted to V. D. Stotts, Waterfowl Section Leader with the Maryland Wildlife Administration, for cooperative assistance during the 1973 and 1974 surveys. Mr. Stotts was assisted by J. E. Schauber, S. Willing, and J. B. Bright. We are further indebted to R. Andrews, R. Anglin, and B. Nichols of the Migratory Bird and Habitat Research Laboratory.

LITERATURE CITED

Dawson, E. Y. 1966. Marine Botany: An Introduction. Holt, Rinehart and Winston, New York, 371 p.

Fernald, M. L. 1950. Gray's Manual of Botany. 8th ed., American Book Co., N. Y., 1632 p.

Hotchkiss, N. 1967. Underwater and floating-leaved plants of the United States and Canada. Resour. Publ. 44, Bur. Sport Fish. & Wildl., Washington, D. C. 124 p.

Sincock, J. L., et al. 1966. Data Report, Proc. Back Bay, Virginia-Currituck Sound, North Carolina Cooperative Investigations. Patuxent Wildl. Res. Center, Laurel, Md., 1600 p.

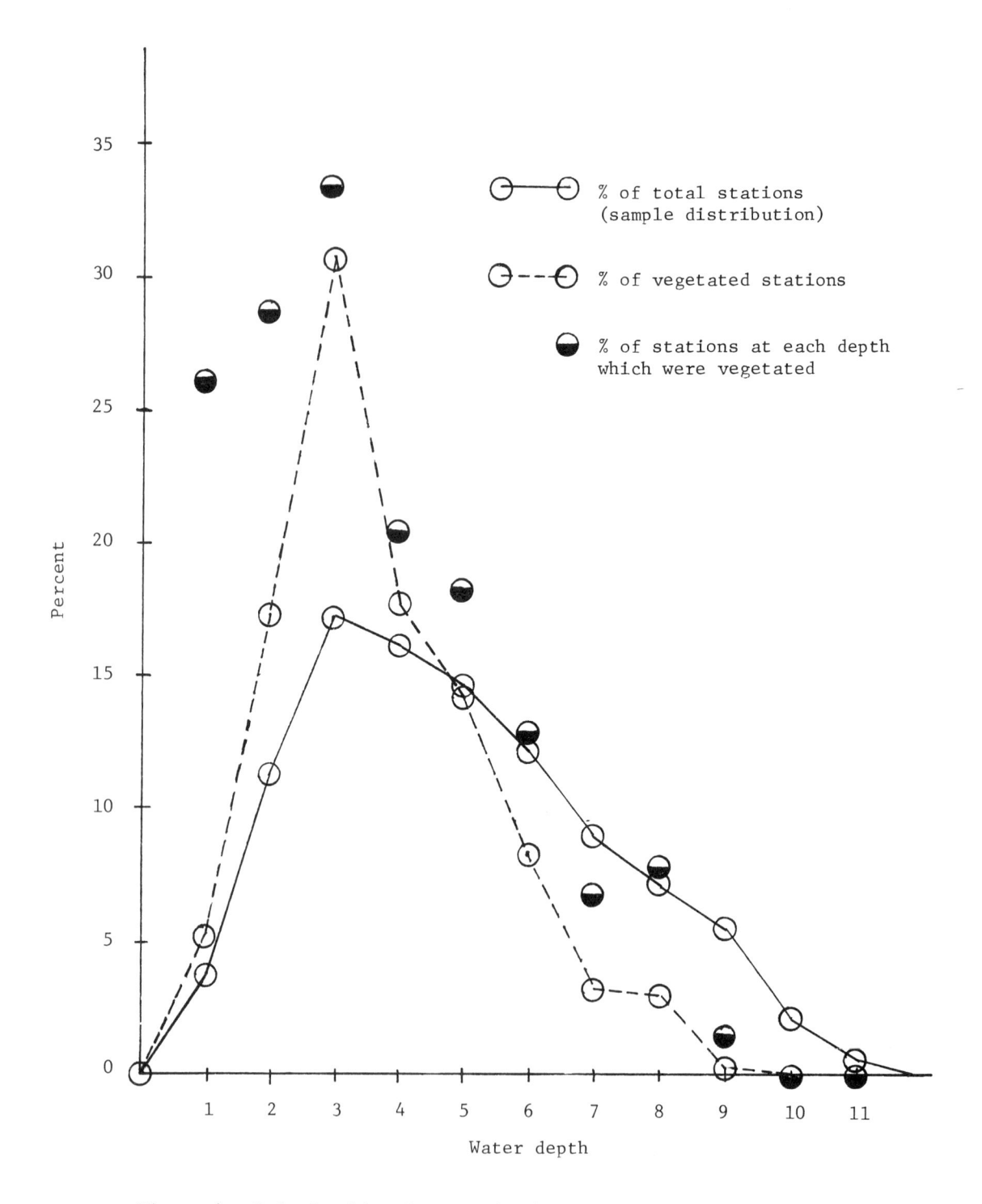

Figure 1. Relationship of water depth, sample distribution, and vegetation in the upper Bay (1971-74 seasons combined).

EFFECTS OF TROPICAL STORM AGNES AND
DREDGE SPOILS ON BENTHIC MACROINVERTEBRATES
AT A SITE IN THE UPPER CHESAPEAKE BAY

J. Gareth Pearson[1]
Edward S. Bender[1]

ABSTRACT

Populations of benthic macroinvertebrates were sampled 14 times from April to August 1972 to determine the effects of discharges from a dredge spoils site on the western shore of the upper Chesapeake Bay.

During the study, 8.78 inches of rain fell in the region from 21 to 23 June 1972 as a result of Tropical Storm Agnes. Nine samples were collected prior to Agnes and five after the storm. Twenty-two species (20 invertebrates and 2 vertebrates) were collected, and 81.5% of the individuals collected comprised three species: the brackish-water clam, *Rangia cuneata*; the amphipod, *Leptocheirus plumulosus*; and the polychaete, *Scolecolepides viridis*. The number of small *R. cuneata* (1.3-4.3 mm) increased significantly after the storm, indicating that an influx of juvenile clams occurred. Despite the extreme hydrologic perturbations caused by the tropical storm, most changes in the numbers of individuals and species in the study area could be attributed to the effects of the discharge of dredge spoils and to seasonal changes. Possible exceptions were the freshwater snail, *Hydrobia* sp., which was found commonly after the storm but was collected only once before the storm, and the amphipod, *Corophium lacustre*, and an isopod, *Edotea triloba*, which decreased in abundance after the storm.

INTRODUCTION

Much research has been done on the effects of dredging on various aquatic ecosystems. Most work, however, has been limited to the actual dredge site or areas of overboard spoils deposition (Pfitzenmeyer 1970). Little is known of the effects of discharges from diked spoils deposition areas. This study was designed to evaluate the biological effects of discharges from a dredge site on Carroll Island, Maryland in the upper Chesapeake Bay.

During the course of the study, 8.78 inches of rain fell within 48 hours as a result of Agnes (Atmospheric Science Laboratory 1972) causing the most serious flooding in this area in recent history. Therefore, our results reflected both the effects of discharge from the spoils area and the effects of the storm.

The dredging site was in the tidal portion of Seneca Creek (Fig. 1) adjacent to a power plant owned by the Baltimore Gas and Electric Company. The dredge spoils (88,030 cubic yards) were pumped into a diked retention area on Carroll Island (Fig. 1) which had also been used to deposit spoils from earlier dredging in the same general area. Dredging began on 5 May 1972 and continued through 20 June 1972. The discharge from the weir at the spoils site began on 5 June 1972 and ended on 20 June 1972.

---
[1]Ecological Research Office, Biomedical Laboratory, Edgewood Arsenal, Aberdeen Proving Ground, Md. 21010

METHODS AND MATERIALS

Two samples were collected with a 0.02m² Ekman grab at each of six sites 14 times between 21 April and 3 August 1972 in the tidal portion of Seneca Creek south of Carroll Island (Fig. 1). Five stations (1-5) were in an area directly south of the discharge site, and one station in an adjacent cove was selected as a control (Fig. 1). Each sample from the grab was placed on a standard #30 sieve (.0232", .0584 mm mesh opening) and was washed with bay water. The material remaining on the sieve was placed in a jar and fixed with 10% formalin. After about three days, the samples were rinsed and the jars filled with 70% ethanol and congo red dye. The samples were sorted in a white enamel pan with the aid of a stereoscopic microscope. All specimens were removed and placed in small jars with 70% ethanol for later identification and enumeration.

Salinity was measured with a conductivity meter and dissolved oxygen (DO) using a modified Winkler method. A method of cluster analysis was used in this study to measure the similarity in macroinvertebrate community structure between stations and species distributions. The first step in this method is to calculate a matrix of similarity values. Sneath and Sokal (1973) provided an excellent review of many of the similarity values proposed in the literature.

For our analyses we chose the Pinkham-Pearson index of biotic similarity (Pinkham & Pearson 1974) which allows for comparison of station similarity based on species abundance and occurrence. The equation for this index is:

$$B = \frac{1}{k} \sum_{i=1}^{k} \frac{\min(X_{ia}, X_{ib})}{\max(X_{ia}, X_{ib})}$$

where k is the number of different species in the samples and X_{ia} and X_{ib} are the number of individuals of the ith species in samples a and b, respectively.

An important feature of this index is that it allows the user to omit matches in which a particular species is absent from both stations (zero-zero matches ignored) or to assign them the maximum values (zero-zero match equal one) for a taxon which is abundant at some locations and absent at others. This allows the user to incorporate or exclude mutual-absence data when generating the similarity index, depending upon the thoroughness of the sampling, the reliability of the data, and the user's judgment. For example, in a set of data in which there are many rare species there would also be numerous mutual-absence matches because of rarity rather than a causal factor, therefore the option to ignore zero-zero matches would be preferred. However, if the rare occurrences were removed from the data for this type of analysis, then remaining mutual-absence matches would reflect actual similarity between stations based on abundance and distribution, and therefore, the option to make zero-zero matches equal to one would be appropriate.

The second step in cluster analysis is the actual clustering of the resultant similarity values. We have used an unweighted pair-group method (Pinkham et al. 1975) which generally introduces less distortion into the clusters. A detailed discussion of this method and cluster analysis in general can be found in Sneath and Sokal (1973).

The next step is to display graphically the results of the cluster analysis, and dendrograms are constructed for this purpose. The scale of similarity is shown across the top of the dendrogram, and the similarity level of any cluster can be determined by drawing a vertical line from the level of branching to the scale.

Finally the amount of distortion in the dendrogram caused by the clustering method is measured. This is necessary because the clustering method involves averaging of similarity values in order to express the multidimensional similarity matrix in a two-dimensional, heirarchical relationship. This is accomplished by computing a correlation coefficient between the original similarity values and the ones from the dendrogram (Sneath & Sokal 1973).

In cluster analysis of our data, rare species (those which accounted for less than 1% of the total number of individuals) were dropped, and an index of biotic similarity was computed using the option of zero-zero matches equal to one.

Chi-square analyses were performed to test the statistical significance of the difference between the number of individuals observed and expected before and after Agnes (Dixon & Massey 1969). The test was performed for each taxon which represented greater than 1% of the total number of individuals collected at all stations and at each station separately. The hypothesis was that the observed catch was not different than the expected catch, i.e., 64.3% of the total before Agnes and 35.7% of the total after Agnes. The catches during the period of discharge of dredge spoils before Agnes was not tested separately because it included only two sampling dates.

Shannon diversity indices ($\bar{d}$) were calculated according to Wilhm and Dorris (1968).

RESULTS

During the study period, salinity remained relatively constant, ranging from 1 ppt to 4 ppt with a mean of 2 ppt. This is shown in Fig. 2 along with water temperature and weekly rainfall. Water temperature ranged from 13 to 31°C, and increased during the study period, fluctuating directly with the ambient air temperature ($r = .891$, $p \leq .01$). DO concentrations ranged from 7 to 12 mg/l with an average of 8.6 mg/l. Weekly rainfall ranged from 0 to 8.51 inches (average 1.52 inches/week) with the high occurring during the week of Agnes (Atmospheric Science Laboratory 1972) (Fig. 2).

Solids from the spoils discharge were visually distinguishable from the natural sediments by their higher water content and uniform gray color. These solids were observed at station 3 on 8 June 1972, three days after the beginning of discharges from the spoils site, and were observed for the duration of the study. Eurasian milfoil, *Myriophyllum spicatum*, was collected in samples at stations 2, 5, and the control.

Twenty two species of organisms, 20 macroinvertebrates, and two vertebrates were collected during the study. Over 85% of the individuals by number were amphipods, polychaetes, and mollusks. Three species were abundant throughout the study period, *Leptocheirus plumulosus*, *Scolecolepides viridis*, and *Rangia cuneata*.

Tubificidae and other small specimens were generally lost or damaged during washing so that their numbers are not representative of their relative abundance. Unidentifiable tubificid fragments were found in most samples except those from station 4.

The mean number of individuals collected on each sampling date immediately after Agnes (287.2) was greater than the mean number of individuals collected

Table 1. Data matrix of species abundance on each sampling date for all stations.

Species Names	Apr 21	Apr 27	May 4	May 10	May 18	May 25	Jun 1	Jun 8	Jun 15	Jun 28	Jul 6	Jul 13	Jul 20	Aug 3
Cumacea sp.	0	0	0	2	0	0	0	2	0	0	0	0	0	0
Menidia sp.	0	0	0	0	0	0	0	1	0	0	0	0	0	0
Agrionidae	0	0	0	1	0	0	0	0	0	0	0	0	0	0
Lirceus fontinalis	1	0	0	0	0	0	0	0	0	0	0	0	0	0
Rhithropanopeus harrisi	0	0	1	0	0	0	0	0	0	0	0	0	0	0
Diptera immago	0	0	0	0	1	1	0	0	0	0	0	0	0	0
Anguila rostrata	0	0	0	0	0	1	0	0	0	0	0	0	0	0
Chrysops sp.	0	0	0	0	0	0	0	0	0	0	0	0	1	1
Hydrobia sp.	0	0	0	0	0	1	0	0	0	25	13	0	11	19
Nematode	0	6	1	3	2	0	0	2	0	1	3	0	0	0
Macoma balthica	0	6	0	1	0	0	0	1	0	0	3	0	0	0
Congeria leucophaeta	0	2	0	0	1	0	0	1	0	2	3	2	2	0
Hypaniola grayi	6	8	4	2	13	8	5	12	1	0	2	1	3	0
Chironomidae	6	5	4	6	9	13	5	9	12	0	9	2	13	0
Nereis succinea	3	9	5	3	12	4	1	6	5	5	3	2	0	0
Cyathura polita	4	2	2	5	0	3	5	6	5	5	4	2	8	0
Gammarus sp.	11	38	3	22	10	8	10	3	2	3	1	7	2	0
Corophium lacustre	3	11	1	22	12	7	19	1	4	0	1	0	2	0
Edotea triloba	7	5	1	8	4	1	2	0	0	2	0	1	5	2
Scolecolepides viridis	63	25	18	36	43	48	10	27	10	18	27	20	47	3
Leptocheirus plumulosus	77	60	52	56	65	134	105	90	83	25	20	16	12	13
Rangia cuneata	27	68	79	40	40	42	43	50	70	325	248	214	136	148
Total Number Individuals	208	245	171	207	212	271	205	211	188	406	334	266	244	186
Total Number Species	11	13	12	14	12	13	10	14	9	9	12	10	13	6
Average Diversity	1.82	2.21	1.65	2.56	1.97	1.79	1.50	1.28	1.58	0.75	1.27	0.92	1.50	0.77

on each date before the storm (213.0) (Fig. 3, Table 1).

Excluding *R. cuneata*, significantly fewer organisms ($\alpha \leq .05$) were found after Agnes than before (Fig. 3, Table 1). Therefore, the increase in total number of individuals was due to an increase in the number of *R. cuneata*, especially those less than 4.0 mm.

The changes in total number of individuals for all stations are presented on Fig. 4. Station 3 showed the greatest decrease in number of individuals. Low numbers of individuals were found throughout the study at station 4. For stations 1, 2, and 5, all within the discharge cove, the average number of individuals was relatively stable throughout the study. Trends at the control station (Fig. 4) were similar to those found for the pooled data (Fig. 3).

The results of chi-square tests on the number of total individuals and individuals of the most abundant taxa at all stations before and after Agnes are shown in Table 2. The total number of individuals and the number of individuals of the mollusks *Hydrobia* sp. and *R. cuneata* increased significantly after Agnes. The number of individuals of Chironomidae, *S. viridis*, *H. grayi*, *Gammarus* sp., *L. plumosus*, and *C. lacustre* each had highly significant declines ($\alpha = \leq .01$) after Agnes. Additional statistical tests conducted on taxa at each station before and after Agnes showed that most taxa had statistically significant changes similar to those found at all stations combined. The number of *S. viridis* collected after Agnes decreased significantly at all stations, except at the control, where it was more abundant after Agnes.

A Shannon diversity index was calculated for all samples, dates, and stations (Tables 1 & 3). Average diversity for all stations was higher for all dates before Agnes ($\bar{d} = 1.82$) than after the storm ($\bar{d} = 1.04$). Diversity decreased slightly at all stations, except at the control and station 1 after the spoils discharge began. The average diversity throughout the study was greatest for stations 1 and 2 (Table 3). Stations 3 and 4 had the lowest average diversity and density of all stations after the discharge occurred.

A dendrogram (Fig. 5) from cluster analysis of stations based on species occurrence and abundance showed two clusters: stations 3 and 4; and 1, 2, 5, and control. Stations 3 and 4 had the highest degree of similarity with one another (B = 0.50). The remaining cluster had an average similarity of 0.44. The first cluster represents stations (3 and 4) which had relatively low numbers of individuals, whereas the remaining stations had approximately the same proportions of individuals of each major taxon.

Hydrobia sp., a common hydrobiid snail in freshwater creeks in the area, was found once prior to Agnes. Seventy specimens were collected after Agnes, some on each date, except 13 July 1972 (Table 1). *Hydrobia* sp. occurred at all stations except station 4 (Table 3).

After Agnes, *R. cuneata* population sizes increased (Fig. 6), but most of the increase was caused by individuals in the smaller size classes (peak 2.7 and 3.0 mm). In following samples this peak gradually shifted to the right (4.0 mm) and total population sizes decreased.

DISCUSSION

Any physical disruption to an ecological community places stress on the members of that system. Depending on their tolerances to this stress, organisms

Table 2. Results of Chi Square test on the number of individuals for major taxa collected before and after Agnes for individual and all stations combined. [(+) indicates an increase, (-) indicates a decline].

Species	Control	Station 1	Station 2	Station 3	Station 4	Station 5	All Stations
Scolecolepides viridis	8.05(+)[2]	NS(-)	37.53(-)[2]	20.17(-)[2]	NS	NS(+)	7.40(-)[2]
Hypaniola grayi	NS	NS	12.82(-)[2]	NS	NS	NS(-)	8.70(-)[2]
Rangia cuneata	1081.95(+)[2]	5.79(+)[1]	148.28(+)[2]	NS(+)	NS(+)	224.40(+)[2]	783.8(+)[2]
Gammarus spp.	5.58(-)[1]	6.63(-)[2]	NS(-)	20.05(-)[2]	NS(-)	13.30(-)[2]	38.7(-)[2]
Leptocheirus plumulosus	18.45(-)[2]	55.30(-)[2]	40.12(-)[2]	105.12(-)[2]	14.34(-)[2]	17.86(-)[2]	221.07(-)[2]
Corophium lacustre	NS	8.18(-)[2]	7.73(-)[2]	NS	NS	16.67(-)[2]	35.1(-)[2]
Chironomidae	NS	NS(-)	NS	7.85(-)[2]	NS(-)	NS(-)	8.6(-)[2]
Hydrobia sp.	51.92(+)[2]	NS(+)	34.89(+)[2]	NS	NC	NS(-)	119.00(+)[2]

[1]Significant (df = 1, P_{95} = 3.84)
[2]Highly significant (df = 1, P_{99} = 6.63)
NS: not significant
NC: not collected

Table 3. Species abundance by stations for the entire study.

| Species Names | \multicolumn{5}{c}{Stations} | Total # of Individuals | Relative Abundance |
	5	C	1	2	3	4		
Cumacea sp.	0	0	2	2	0	0	4	.001
Menidia sp.	0	0	1	0	0	0	1	<.001
Agrionidae	0	0	0	1	0	0	1	<.001
Lirceus fontinalis	0	0	1	0	0	0	1	<.001
Rhithropanopeus harrisi	0	0	0	1	0	0	1	<.001
Diptera immago	0	0	0	2	0	0	2	<.001
Anguila rostrata	0	0	0	0	1	0	1	<.001
Chrysops sp.	0	2	0	0	0	0	2	<.001
Hydrobia sp.	5	38	3	21	2	0	69	.021
Nematode	3	3	8	4	0	0	18	.005
Macoma balthica	3	7	0	0	0	1	11	.003
Congeria leucophaeta	3	4	1	3	0	0	11	.003
Hypaniola grayi	10	11	27	23	1	1	73	.022
Chironomidae	5	17	5	39	16	5	87	.026
Nereis succinea	12	14	24	9	4	2	65	.019
Cyathura polita	4	14	7	9	1	2	37	.011
Gammarus spp.	30	23	10	34	18	4	119	.035
Corophium lacustre	25	5	21	17	6	5	79	.024
Edotea triloba	2	25	6	4	1	1	39	.012
Scolecolepides viridis	32	50	128	139	37	9	395	.118
Leptocheirus plumulosus	58	157	149	218	195	31	808	.241
Rangia cuneata	300	855	121	144	47	63	1530	.456

Total Number								
Individuals	482	1226	514	670	328	124	3354	
Total Number Species	14	15	16	17	12	11		
Average Diversity ($\bar{d}$)	1.57	1.73	1.95	1.86	0.97	1.13		

can respond through a decline or increase in their population sizes, hence causing changes in relative abundances and community structure. Changes in species abundance are most easily detected by calculating relative abundance and community structure. However, these changes must be separated from natural seasonal peaks in migration, recruitment, and mortality to determine the effects of dredge spoils discharge and Agnes. Low salinity probably had little impact in this study due to its slight variation; therefore, the effects of temperature changes, flow rates, and sedimentation caused by Agnes and dredging are the most important stress factors to be considered.

The species collected here were also found in studies by Pfitzenmeyer (1970, 1973) in the upper Chesapeake Bay, and Jensen et al. (1973) in the Bush River. Tubificids comprised 5 to 24% of total animals collected by Pfitzenmeyer and Jensen. This group was not counted in our study.

Contrary to the results of other studies on dredge spoil disposal (Pfitzenmeyer 1970), faunal density in our investigation increased after dredge spoil discharge and Agnes. The local increase here was due to *R. cuneata*. *Scolecolepides viridis* was collected in the same densities before and after the storm.

Pfitzenmeyer (1970) considered *Leptocheirus plumulosus*, *Scolecolepides viridis*, and *Cyathura polita* to be the dominant fauna of the upper bay. Although present, *C. polita* was not one of the most abundant species found and a close examination of his data shows that *C. polita* was uncommon or absent in areas of dredge spoil disposal until 6 to 10 months after dredge spoil disposal ended. *S. viridis* recovered quickly and reached high densities in the disposal area. *L. plumulosus* was absent in some areas after overboard spoils disposal and very dense in other areas immediately after disposal. *L. plumulosus* reaches its peak density in January (Pfitzenmeyer 1970), so that the decrease found there may be part of a natural seasonal decline. All three species are considered to be tolerant of salinity changes to even fresh water (Pfitzenmeyer 1970). Jensen et al. (1973) found *L. plumulosus* and *C. lacustre* in all but the freshest portions of the Bush River. Pfitzenmeyer's data (1973) show that isopods and chironomids were concentrated at shallow-water stations (2-4 feet) in June and most abundant in intermediate and deep-water stations (7-13 feet and 27-40 feet, respectively) during the fall and winter months. However, after Agnes, Boesch (personal communication) found *C. lacustre*, *E. triloba*, and chironomids at greater depth and in mesohaline waters of the James and York Rivers where they were previously absent. He concludes high discharges caused by Agnes transported them into these areas. Our results show that the epibenthic crustaceans were less abundant in Seneca Creek immediately after Agnes, and this supports Boesch's conclusion.

Burbanck (1961) found similar faunal declines in the Ashepoo River after Hurricane Gracie; *C. polita* was still absent 10 months later. Burbanck postulated that hydrogen sulfide released from the sediments during the storm caused the loss of crustaceans and polychaetes. Robins (1957) found that silt suspended by storms may erode the filaments of fish gills and cause death by suffocation. He suggested that similar morphological damage may occur among the epibenthic crustaceans and annelids.

The collections of *Hydrobia* sp. suggest that many snails were transported from the tributary creeks and marshes during the rain from Agnes. Pfitzenmeyer (1973) found *Hydrobia* spp. abundant in Elk River during June, but absent during March, September, and December of the same year. From his quarterly sampling he assumed that recruitment of a new year-class occurred in the early summer when the individuals are most active. *Hydrobia* sp. could probably survive in this new environment if competition for food and space were reduced. *Hydrobia* sp., a deposit feeder, would encounter little interspecific competition with the abundant *R. cuneata* which is a suspension feeder.

Siltation might have covered food sources for *L. plumulosus* at station 3, whereas, *S. viridis* is adapted to chew through the sediment to feed. Station 4 showed little change in density despite its close proximity to the discharge site.

Reproduction is probably restricted in this area for both *S. viridis* and *Rangia* by low salinity and heavy siltation. George (1966) stated that fertilization for *S. viridis* will not take place in salinities below 5 ppt. Such salinities occur at Carroll Island during low flow from August through October. Since *S. viridis* spawns in the spring in the upper bay when salinity is low in Seneca Creek, reproduction probably does not occur there for *S. viridis*. Salinity of 3 to 5 ppt is considered to be critical to initiate spawning in *R. cuneata* (Hopkins et al. 1973) although gametes are ripe from April to September in the bay (Pfitzenmeyer, personal communication).

In response to catastrophic changes in salinity, silt load, or temperature, animals may survive stress by burrowing or closing valves, avoid stresses by leaving the disturbed area, or succumb to stresses (Nicol 1969). The changes in species abundance found here probably represent organism responses to local changes and sample variability. Most upper bay fauna are relatively tolerant to chronic siltation (Pfitzenmeyer 1970) and some such as *S. viridis* and *R. cuneata* are probably nonreproductive populations due to the dilute environment (Hopkins et al. 1973). After stress from Agnes, station 5 and the control showed an increase of first-year *Rangia*. It seems likely that recruitment must be significant to sustain populations in this area. By extrapolating from data in Hopkins et al. (1973), *R. cuneata* larvae reach 2.0 mm size within three months after they are spawned. Therefore spawning probably occurred in early spring and Agnes might have affected their recruitment.

Pfitzenmeyer found that *S. viridis* cannot survive siltation or storm-caused disturbances; thus recruitment must occur from areas where salinity does not limit reproduction.

The effect of discharge from the dredge spoil site was localized primarily at station 3. Increases in the relative abundance of juvenile *R. cuneata* and *Hydrobia* sp. may be related to transport and disturbance caused by Agnes. In high siltation *L. plumulosus* populations declined while *S. viridis* remained relatively constant.

ACKNOWLEDGEMENTS

We thank Drs. C. F. A. Pinkham, J. V. Martin, and F. P. Ward for their assistance in field collections, and Mrs. J. Dierdorf for the illustrations; special thanks are due to Mrs. P. K. Schultz for editing and typing the manuscript.

LITERATURE CITED

Atmospheric Science Laboratory. 1972. Meteorological Team Data, Edgewood Arsenal, Maryland, April, May, June, July, August 1972. Atmospheric Sciences Lab., White Sands Missile Range, New Mexico.

Burbanck, W. D. 1961. The disappearance of *Cyathura polita* from the Ashepoo River, South Carolina after Hurricane "Gracie" 1959. Ecology 42(3):606-607.

Dixon, W. J. and F. J. Massey, Jr. 1969. Introduction to Statistical Analysis. McGraw Hill, Inc., New York, pp. 240-243, 465.

George, J. D. 1966. Reproduction and early development of the spionid polychaete, *Scolecolepides viridis* (Verrill). Biol. Bull 130(1):76-93.

Hopkins, S. H., J. W. Anderson, and K. Horvath. 1973. The brackish water clam, *Rangia cuneata*, as indicator of ecological salinity changes in coastal waters. Report to Office of Chief Engineers, U. S. Army. Dept. of Biological Research Foundation, Texas A & M Univ., Contract Rep. H-73-1, 251 p.

Jensen, L. D., et al. 1973. Biota of the Bush River. Power Plant Site Evaluation Interim Report for the Maryland Power Plant Siting Program. Dept. of Geography & Environmental Engineering, The Johns Hopkins Univ. (PPSE-2-M-1), 162 p.

Pfitzenmeyer, H. T. 1970. Benthos. Project C. In: Gross physical and biological effects of overboard spoils disposal in the upper Chesapeake Bay. Final Rep. to Bureau of Sport Fish. & Wildl. Natural Resources Inst., Univ. of Maryland, NRI Spec. Rep. 3:26-38.

Pfitzenmeyer, H. T. 1973. Appendix III. Benthos of Maryland waters in and near C and D Canal. In: Hydrographic and Ecological Effects of Enlargement of The Chesapeake and Delaware Canal. Natural Resources Inst., Univ. of Maryland, NRI Ref. 73-113, 40 p.

Pinkham, C. F. A. and J. G. Pearson. 1974. A new measure of biotic similarity between samples and its applications with a cluster analysis program. Edgewood Arsenal Techn. Rep. EB-TR-74062.

Pinkham, C. F. A., Pearson, J. G., Clontz, W. L., and A. E. Asaki. 1975. A computer program for the calculations of measures of biotic similarity between samples and the plotting of the relationship between these measures. Edgewood Arsenal Techn. Rep. EB-TR-75013.

Robins, C. R. 1957. Effects of storms on the shallow water fish fauna of south Florida with new records of fishes from Florida. Bull. Mar. Sci. Gulf Caribb. 7(3):266-276.

Sneath, P. H. A. and R. R. Sokal. 1973. Numerical Taxonomy. W. H. Freeman & Co., San Francisco, 573 p.

Wilhm, J. L. and T. C. Dorris. 1968. Biological parameters for water quality criteria. Bioscience 18:477-481.

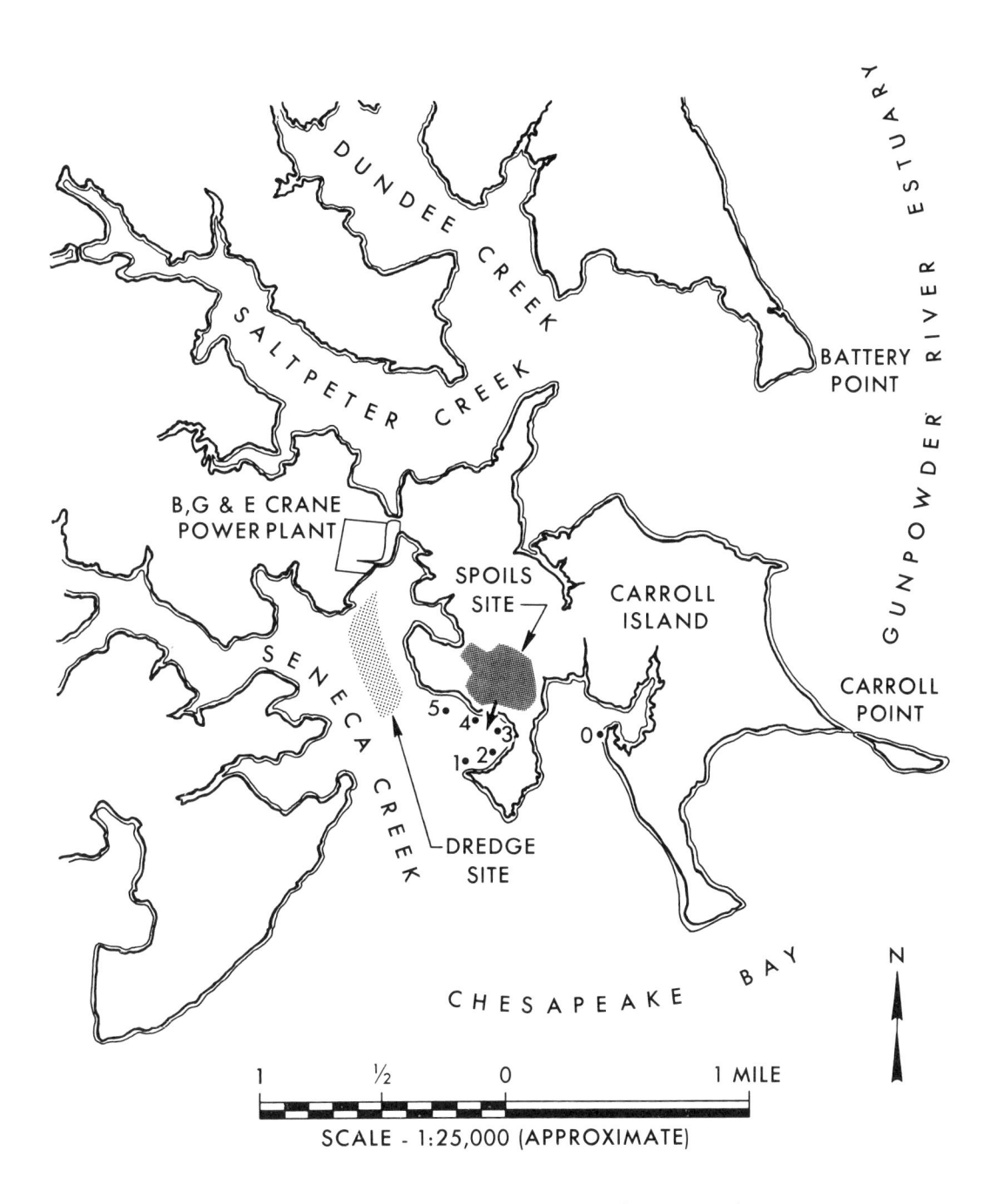

Figure 1. Location of study area and sample sites.

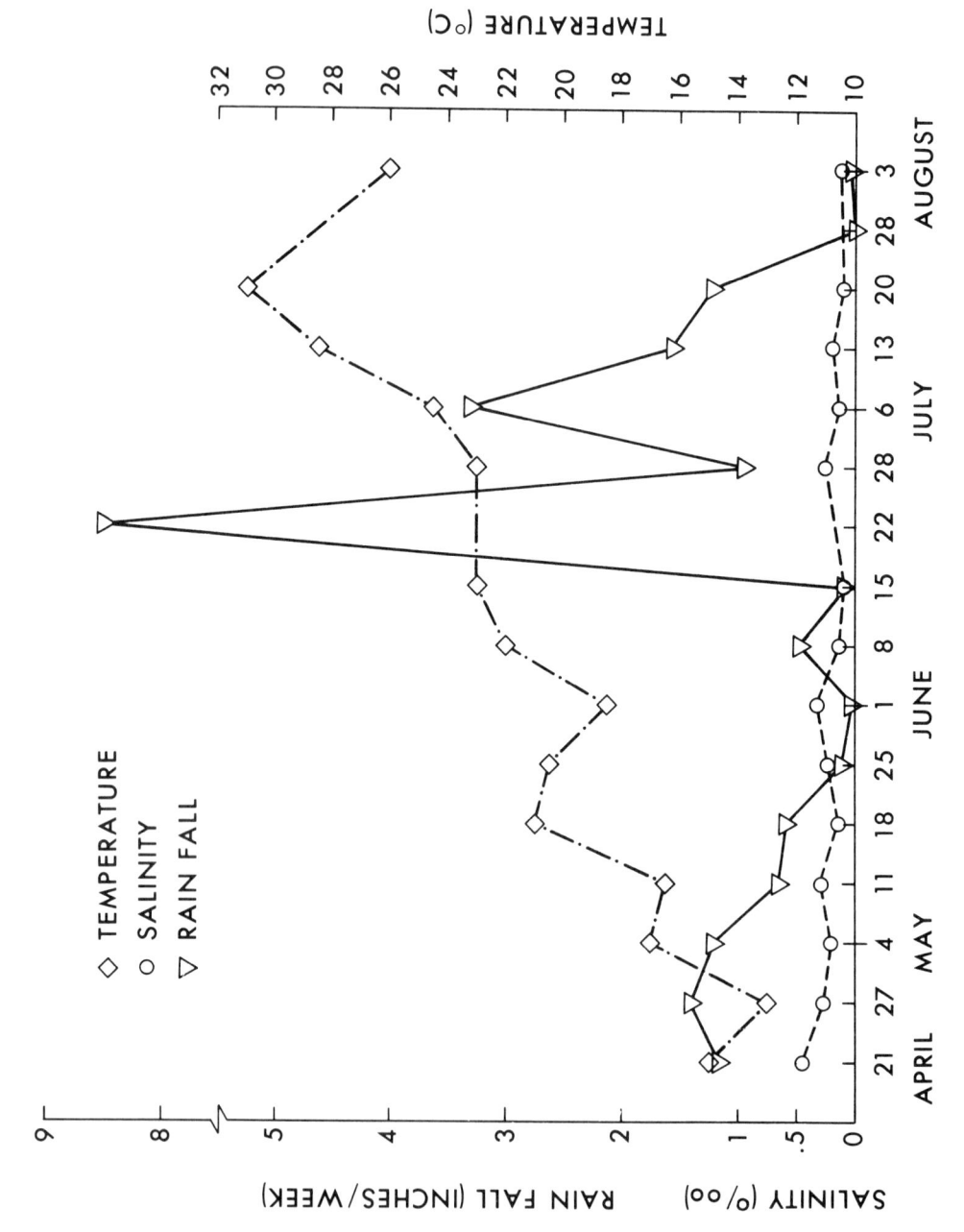

Figure 2. Water temperature, salinity, and rainfall per week from 21 April through 3 August 1972.

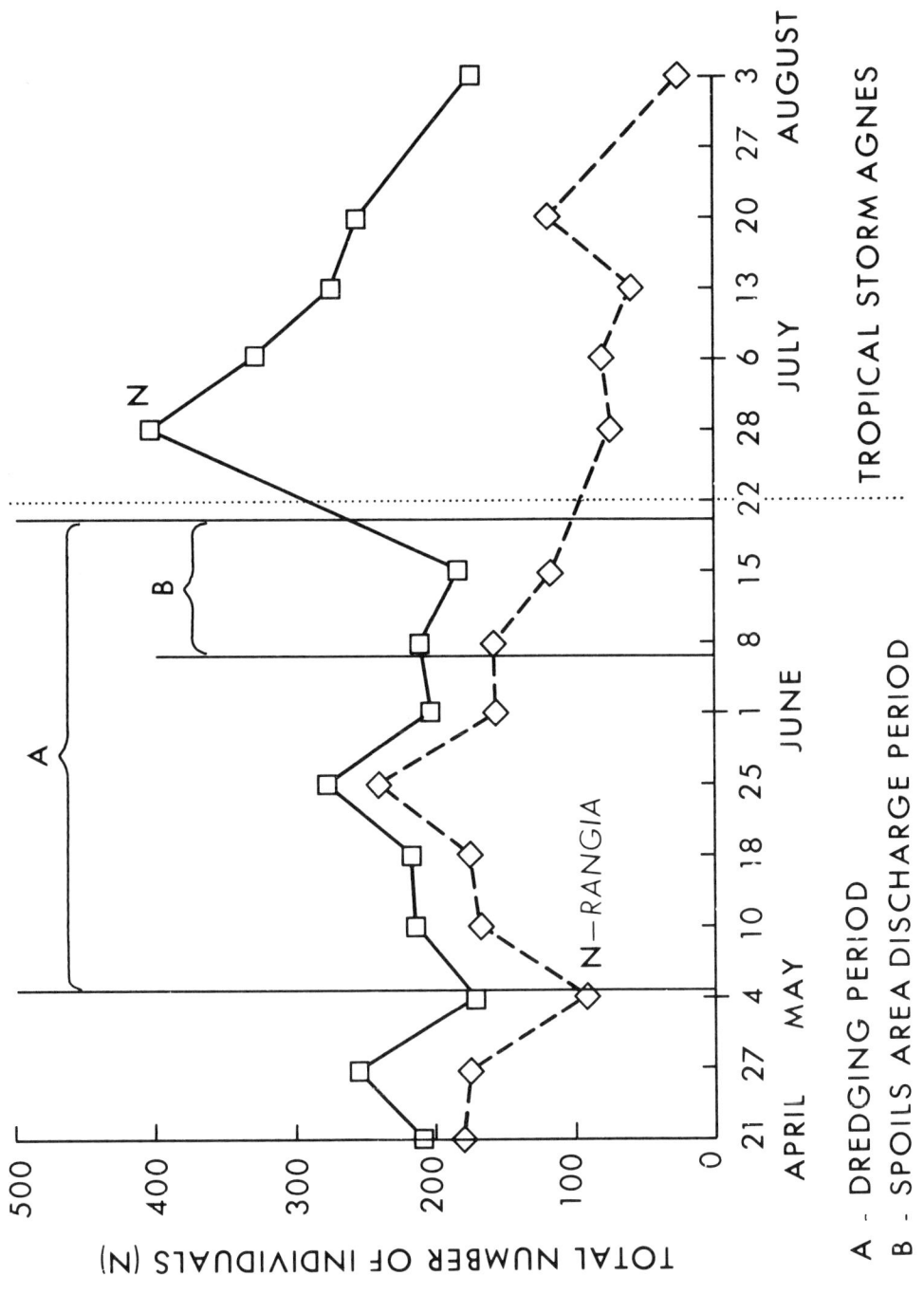

Figure 3. Number of species, total number of individuals, and total number of individuals excluding *Rangia cuneata* during the study period.

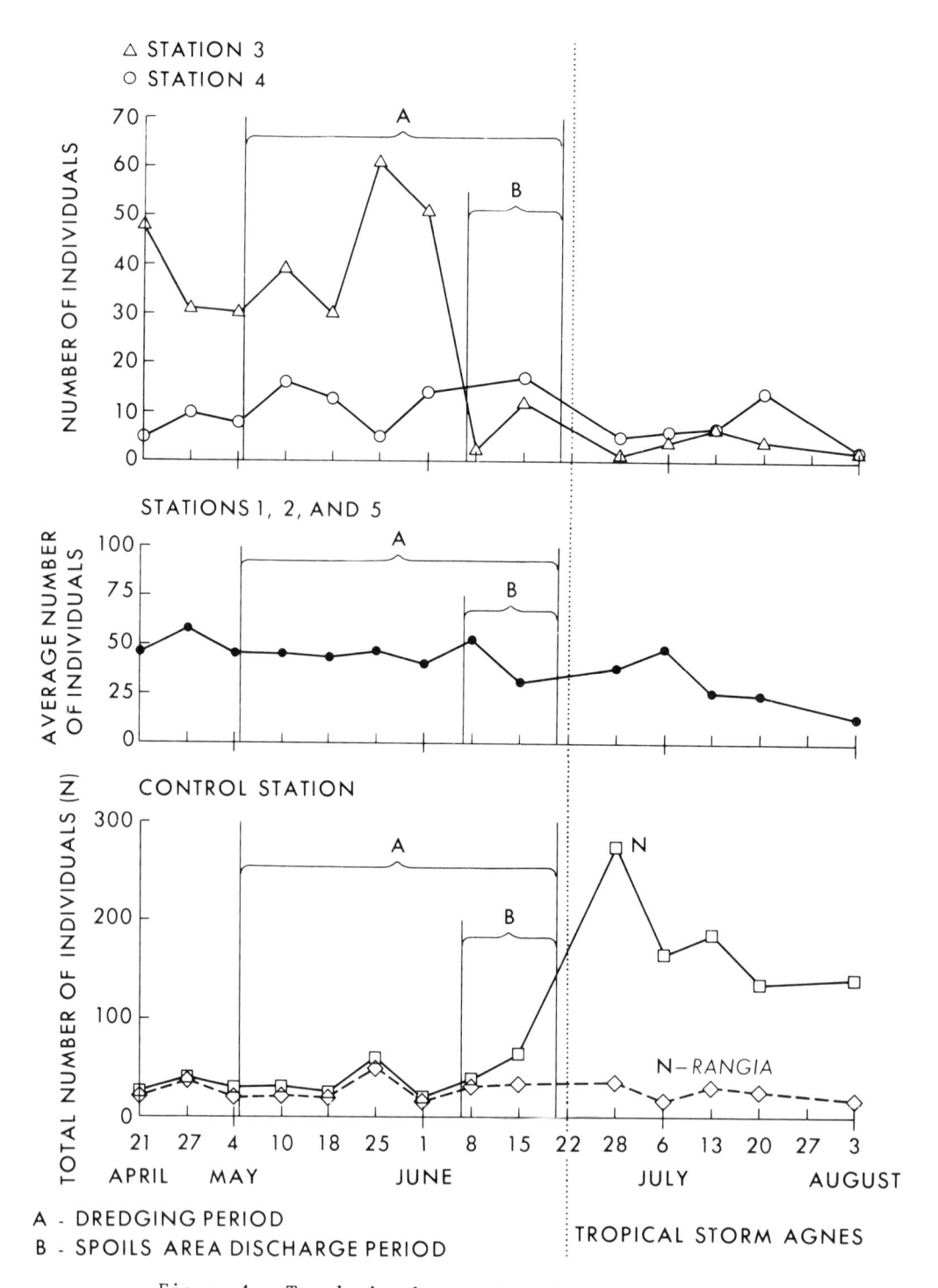

Figure 4. Trends in the total number of individuals during the study period.

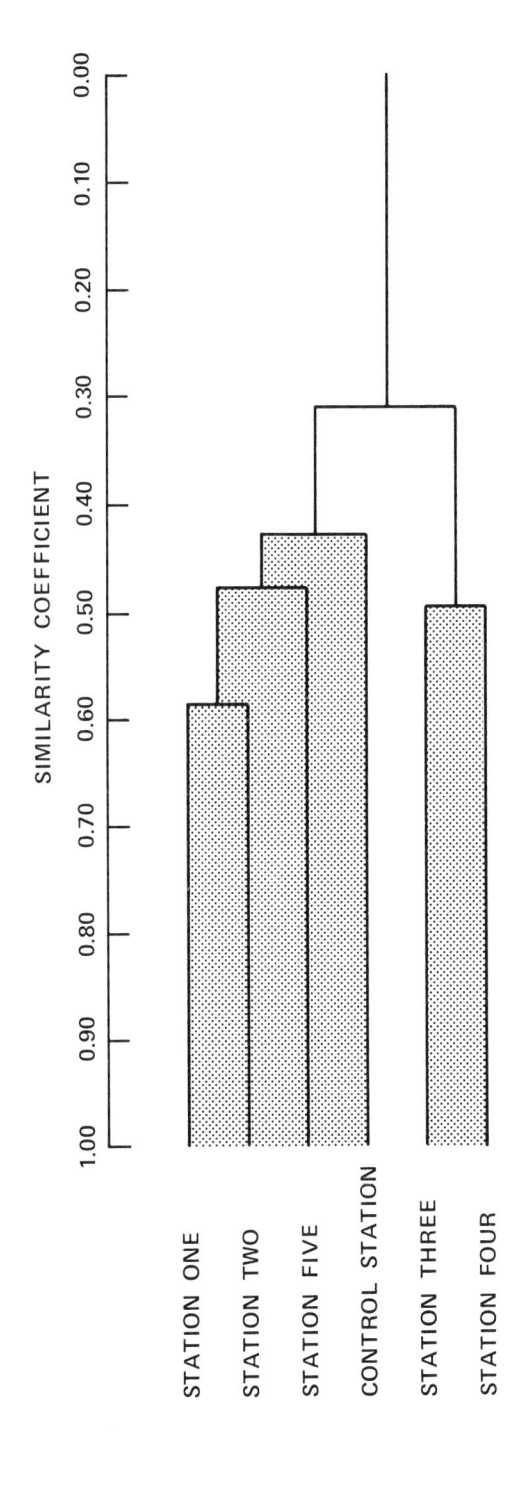

Figure 5. Dendrogram of stations based on species occurrence and abundance.

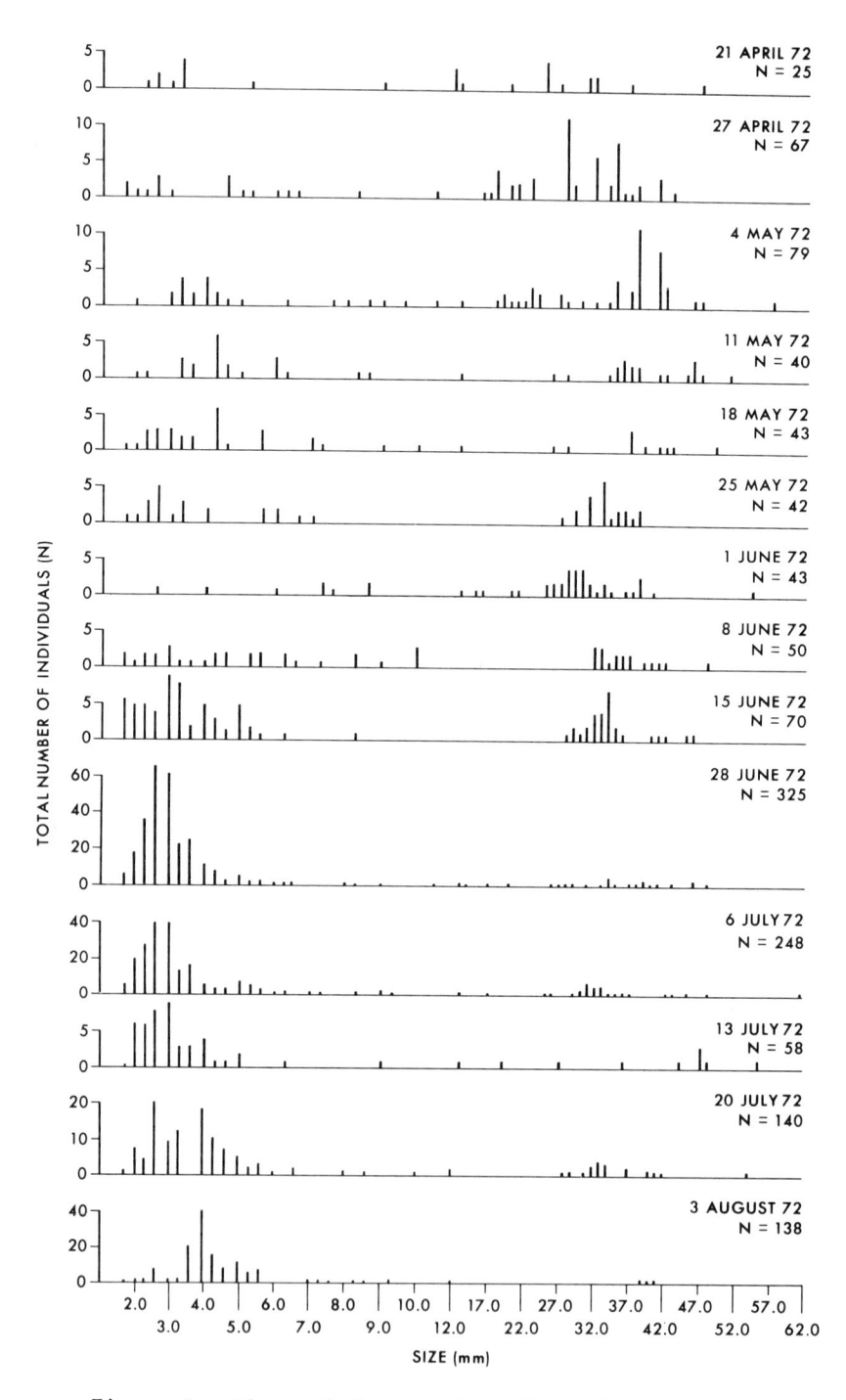

Figure 6. Size and frequencies of *Rangia cuneata* during the study period.

SOME EFFECTS OF TROPICAL STORM AGNES ON THE SEA NETTLE POPULATION IN THE CHESAPEAKE BAY[1]

David G. Cargo[2]

ABSTRACT

The incidence of scyphistomae and podocysts of the sea nettle, *Chrysaora quinquecirrha*, has been monitored in Maryland and Virginia waters since 1965. An as yet unexplained mortality of all stages was observed in the summer and fall of 1971. The effects of Tropical Storm Agnes compounded these drastically low levels to the extent that very few polyps were present in 1972 and then only in tidewater areas which were less seriously affected by the storm.

Despite experimental evidence showing that the re-emergence of polyps, which had encysted under the storm conditions, is directly related to a rise in salinity, recovery in the Bay has been slow. Minor recovery was noted late in 1972 but not until the spawning of medusae had taken place in the summer of 1973 was any real recovery apparent.

More customary numbers of polyps and cysts are now (1975) present throughout most of their range and a return to usual levels of infestation is anticipated.

INTRODUCTION

Agnes displayed a severe and long-lasting impact on the ecology of the Chesapeake Bay. A number of individuals have documented various aspects of this impact as summarized by CBRC (1973). One of the estuarine organisms most seriously affected was the sea nettle, *Chrysaora quinquecirrha*.

RESULTS AND OBSERVATIONS

The population of spawning medusae of the sea nettle *Chrysaora quinquecirrha* in 1971 was very low despite the normal numbers of polyps present in the fall of 1970 (Table 1). The reason for this remains unexplained. Very few planula larvae had set on the customary substrata during the summer of 1971 as samples of shells dredged in the fall of 1971 and on June 5-7, 1972 indicate (Table 1). For this reason, the drastic effects of the storm are not so dramatic when samples taken shortly after the storm are compared with earlier ones. However, other data show that a significant effect was being felt.

At Solomons near the mouth of the Patuxent River, a 24-hour sampling was in progress on June 27-28, 1972 designed to measure the influx and egress of *Chrysaora* ephyrae from Mill Creek at its mouth. Only a few ephyrae were taken, since the creek was already reflecting the effects of the storm. On the day before the sampling period (June 26), the salinity in Mill Creek, 3/4 mile upstream from the transect, was 7.3 ppt (more customary salinity would be 10-12 ppt). At 1400 on the 27th, the salinity had already dropped to 3.1 ppt and the young medusae and

---

[1] Contribution No. 570, U. M. Center for Environmental and Estuarine Studies.
[2] Chesapeake Biological Laboratory, Center for Environmental and Estuarine Studies, University of Maryland, Solomons, Md. 20688

418 Cargo

Table 1. Numbers of shells in sample showing sessile *C. quinquecirrha* as indicated by the following code:
a = blank   c = 11-50 polyps   e = podocysts only   N.O. = not occupied
b = 1-10 polyps   d = >50 polyps   * = bottom salinity (ppt)

Location	Nov 1970 a b c d e	Nov 1971 a b c d e	Jun 1972 a b c d e	10-13 Jul 1972 a b c d e	Jul-Aug 1972 a b c d e	Oct 1972 a b c d e	Jun 1973 a b c d e	Nov 1973 a b c d e
Hackett Pt. 38° 58'30"N - 76° 25'00"W	3 7 6 0 0 *13.5	13 0 0 0 0 *12.7	22 0 0 0 0 *6.8	N.O.	20 0 0 0 0 *3.3	16 0 0 0 0 *13.3	13 0 0 0 0 *4.8	13 0 0 0 0 *15.4
Prospect Bay/Eastern Bay 38° 44'05"N - 76° 14'05"W	2 4 4 3 0 *15.2	9 0 0 0 0 *13.8	20 0 0 0 0 *9.7	17 0 0 0 0 *5.7	N.O.	12 0 0 0 0 *12.8	12 0 0 0 0 *7.5	11 2 0 0 0 *13.7
Herring Bay 38° 45'35"N - 76° 31'35"W	11 2 3 1 0 *13.4	- 0 0 0 0 *12.4	29 0 0 0 0 *8.6	N.O.	17 0 0 0 0 *6.9	15 0 0 0 0 *13.1	11 0 0 0 0 *6.5	11 0 0 0 0 *15.3
Broad Cr./Choptank R. 38° 44'33"N - 76° 13'52"W	5 4 4 0 0 *14.5	N.O.	15 6 1 0 0 *9.4	13 0 0 0 0 *4.9	15 0 0 0 1 *--	15 0 0 0 0 *11.3	15 0 0 0 0 *9.1	8 7 0 0 1 *12.5
Ragged Pt/L. Choptank R. 38° 31'06"N - 76° 17'35"W	5 6 3 0 0 *15.3	7 0 0 0 0 *14.5	28 5 0 0 0 *9.1	21 0 0 0 2 *5.0	15 0 0 0 0 *6.4	12 0 0 0 0 *14.0	8 0 0 0 0 *8.8	11 5 3 0 0 *14.7
Solomons/Patuxent R. 38° 21'25"N - 76° 29'17"W	N.O.	N.O.	N.O.	20 0 0 0 0 *5.3	22 0 0 0 1 *4.9	13 0 0 0 0 *13.3	N.O.	4 7 4 0 0 *13.6
Mouth/Honga R. 38° 14'23"N - 76° 06'00"W	7 2 1 1 0 *16.8	5 2 1 0 0 *15.8	16 0 0 0 0 *9.5	20 0 0 0 0 *7.7	19 0 0 0 0 *7.9	13 1 0 0 0 *14.0	13 0 0 0 0 *--	4 6 2 4 3 *15.6
Haines Pt./Tangier Sd. 38° 12'00"N - 75° 57'18"W	12 2 0 0 0 *17.0	11 0 1 0 0 *15.5	17 1 0 0 0 *12.1	20 0 0 0 0 *8.4	20 0 0 0 0 *11.8	10 2 0 0 0 *13.8	15 0 0 0 0 *10.8	8 4 0 0 0 *14.8
Windmill Pt/St. Mary's R. 38° 09'18"N - 76° 19'48"W	N.O.	N.O.	N.O.	N.O.	16 0 0 0 2 *6.6	11 1 0 0 1 *14.2	11 0 0 0 0 *9.2	0 4 6 7 0 *14.6
Cornfield Pt./Potomac R. 38° 02'42"N - 76° 19'48"W	N.O.	N.O.	N.O.	N.O.	N.O.	10 0 0 0 0 *15.2	12 0 0 0 0 *9.9	8 4 1 0 0 *14.9
Marumsco Cr/Pocomoke Sd. 37° 57'00"N - 75° 44'10"W	N.O.	13 0 0 0 0 *15.0	N.O.	20 0 0 0 1 *6.8	17 1 0 0 0 *9.0	9 3 0 0 0 *13.8	11 1 0 1 2 *10.0	13 3 3 0 0 *15.9

ephyrae were absent and apparently being destroyed. Table 2 gives the field data for this period and a similar one for 1971.

Table 2. Number of ephyrae taken in hourly samples at Shipyard Point in St. John's Creek.

Time	June 23-24, 1971 Surf. Sal. 10.8 ppt Ephyrae	June 27-28, 1972 Surf. Sal. 2.5 ppt Ephyrae
1400	41	17
1500	35	0
1600	131	1
1700	155	0
1800	93	0
1900	63	0
2000	21	0
2100	5	0
2200	0	0
2300	6	0
2400	6	0
0100	15	0
0200	25	0
0300	22	0
0400	52	0
0500	64	No samples-Weather
0600	90	0
0700	136	0
0800	103	0
0900	300	0
1000	229	0
1100	213	0
1200	99	0
1300	41	0
1400	63	0

During this period, polyps and strobila were also being destroyed in our running-water aquaria so that by June 30, only encysted forms remained. These did not begin excysting until the last half of July when the salinity had reached 6-7 ppt. Two replicated experiments were initiated during July using a mixture of Instant Ocean salts and river water to test the actual time and salinity suitable to induce excystment. We found that starting at 5.5 and 6.7 ppt, initiation of excystment required 12 days at 6.7 ppt, 8 days at 7.0 ppt, 6 days at 8.8 ppt, 5 days at 9.3 ppt and 10.7 ppt, 3 days at 11.0 ppt, and 2 days at 12.3 ppt. These were conducted at 25°C. Strobilation was first observed from these excysted polyps in 12 days at 9.3 ppt and in 7 days at 14.0 ppt. No other strobilation activity was noted in these trials. Fig. 1 graphically shows this relationship.

The local river salinity reached 7 ppt on July 5 and fluctuated between 5 and 8 ppt for the next 20 days. Consequently, little excystment and no strobilation was possible before August 1 in this area. Our customary (5 minutes, 0.5 mm mesh aperture) ephyrae and medusae sampling in St. John's Creek with 0.5m plankton nets also showed the effects (Fig. 2). Starting on May 15, the data show that no ephyrae were taken until June 12 when one was secured. Seventeen were taken on June 19 when the salinity was 7-9 ppt. A few were taken on the 20th (5) and 26th (2). On July 3, (salinity: 3.1 ppt) no ephyrae were taken and they were absent in all subsequent samples during 1972. Only nine medusae were taken in September. Fig. 2 shows the five-year average for comparison.

Oyster dredge samples throughout the customary range (salinity between 5 and 25 ppt) of sea nettle polyps were begun on July 10 and continued on a regular basis until late fall. The samples for July 10-13 showed only cysts throughout the area samples, even in Tangier Sound where the storm's effects were minimal (Table 1). No medusae were observed at any time during this sampling. Later, on July 24-August 2, (Table 1) when a partial recovery of the Chesapeake Bay system had progressed for several weeks, the only localities where polyps were taken were the higher salinity stations where the effects of Agnes were the least marked. During this second survey, several large medusae were sighted near the Bay Bridge and in the Little Choptank River.

The third survey, in October (Table 1) showed continued recovery. Polyps were more numerous but still confined mostly to the Tangier and Pocomoke Sound area. One shell from the St. Mary's River also bore some polyps. Salinities at this time were higher, although still well below normal (25-35 ppt) for this time of the year. Swimming medusae were noted at a number of stations at this time. Individual sea nettles, mostly very large, were seen at a number of widely separated locations in Maryland. In the Bay off Parkers Creek about 20 miles north of Solomons, more than 25 medusae were seen during a 5-minute period. They were all large (6 in) and appeared healthy and robust.

A sampling of the oyster bottoms at 23 locations in Virginia on Oct. 30-Nov. 2 (Table 3) reiterated the shortage of sea nettle polyps. Only four samples, all taken in quiet sub-estuaries opening directly into the Bay, showed live polyps. In one of these samples (Piankatank River), the polyps were quite abundant.

Full recovery of the Bay to average ($\simeq$ 12.5 ppt) salinity values was not observed at Solomons until late March, due mainly to the higher than normal precipitation occurring during the winter of 1972-1973. In the summer of 1973, more normal environmental conditions returned to the area. However, our samplings of sea nettle populations indicated little improvement from what was found in the summer and fall of 1972. Table 1 shows salinities were still somewhat depressed and no sea nettle polyps or cysts were observed in Maryland samples except in Pocomoke Sound. Only one ephyra was taken in our weekly plankton samples from the Solomons area and medusae were not observed until early July (Fig. 2). Most of these medusae appeared late in the season, were very large and a large number of them were of the red phase, a form common in the southern part of the Bay but uncommon in Maryland waters. Most of these medusae probably originated in the southern part of the Bay.

The medusae present, however, apparently were quite successful in spawning in most areas of the Bay. Only in the northern half of Maryland's portion of the Bay were polyps absent during the fall of 1973 (Table 1). Gravid females retained in flowing water tanks at Solomons during late summer spawned heavily on clean oyster shells provided, and the Bay surveys in Maryland and Virginia in October and November indicate that spawning and subsequent stolon and cyst production were widespread (Tables 1 & 3).

Table 3. Numbers of shells in sample showing sessile *C. quinquecirrha* in Virginia as indicated by following code:
   a = blank
   b = 1-10 polyps
   c = 11-50 polyps
   d = >50 polyps
   e = podocysts only
   * = bottom salinity (ppt)

	Nov 1972 a b c d e	Dec 1973 a b c d e
Sandy Pt/Great Wicomico R. 37° 49'25"N - 76° 18'41"W	13 1 0 0 1 *14.2	3 1 3 3 5 *----
Parrott Is/Rappahannock R. 37° 36'10"N - 76° 24'05"W	10 0 0 0 0 *13.3	8 3 2 1 0 *----
Godfrey Bay/Piankatank R. 37° 30'59"N - 76° 20'53"W	6 8 6 0 0 *14.3	12 1 1 1 2 *23.4
Ware Neck/Mobjack Bay 37° 26'11"N - 76° 27'20"W	17 2 2 0 0 *17.5	7 0 0 0 0 *----
Clay Bank/York R. 37° 22'10"N - 76° 38'43"W	12 0 0 0 0 *17.2	11 1 0 0 0 *----
Blunt Pt/James R. 37° 02'45"N - 76° 32'25"W	14 0 0 0 0 *12.0	6 0 1 4 2 *----

From this evidence, it is obvious that a major temporary effect was impressed by Agnes upon the sea nettle population in Maryland waters of the Chesapeake Bay. Partial recovery has been observed and some reinfestation has occurred. It appears however, that the storm might have a far reaching effect for several years.

DISCUSSION AND CONCLUSIONS

These results indicate that in many areas of the Bay, conditions have returned to levels which allow extensive set, asexual budding and growth. This has happened to such an extent that the levels of infestation of sessile stages in much of the Bay now approach those observed in 1970. This suggests a widespread resurgence of the nettle in the near future.

It was obvious, as observations of the storm's impact upon many organisms became available, that the sea nettle was among those organisms markedly affected by the influx of fresh water. The timing of the storm played a major factor, occurring at the peak of strobilation when the release and growth of ephyrae and small post-ephyrae were at a maximum. Chemical changes were no doubt commonplace, thus affecting success of the development of strobila, as in the case of *Aurelia* (Spangenberg 1968). Prey organisms for the young nettles may have been less available; and, indeed, many chemical characteristics of the water masses were such that all stages were subjected to levels at or beyond the demonstrated tolerances for these factors. For these reasons, compounded by an already reduced population due to unknown causes, the tidewater areas of the Bay have enjoyed relatively little sea nettle infestation since 1970.

The physical influx of such a freshwater flow very likely exerted additional burdens upon the survival of the newly released medusae. The ephyrae are most abundant in the deeper, quiet sub-estuaries where the salinities average 15 ppt and there is a relatively small watershed. Siltation is usually slight in these areas and, indeed, silting itself has been shown to cause the complete demise of polyps (Schultz, Leonard P., personal communication). That the storm caused both of these conditions in addition to an actual sweeping out of the weak, swimming ephyrae from nursery areas is unquestioned. The results reported here and elsewhere are, for many reasons, quite understandable and fully corroborate previous assessments of the effects of such a storm.

ACKNOWLEDGEMENTS

The results reported herein were supported by the U. S. Army Corps of Engineers under Contract DACW 61-73-C-9348 and by the University of Maryland.

Special credit is due Mr. H. T. Pfitzenmeyer, M. J. Reber, C. J. Kucera, D. A. Neumann and Capt. M. E. O'Berry, all of the Chesapeake Biological Laboratory. Ms. Frances Younger prepared the figures. Ms. Sandra A. Riddle and Ms. Karen A. Cogan did the typing and helped with other necessary clerical tasks.

LITERATURE CITED

Chesapeake Bay Research Council. 1973. Effects of Hurricane Agnes on the environment and organisms of Chesapeake Bay. A report to the Philadelphia District U. S. Army Corps of Engrs. (Aven M. Andersen, Coordinator), Chesapeake Bay Inst. Contrib. 187, Nat. Resour. Inst. Contrib. 529, Va. Inst. Mar. Sci. Spec. Rep. Appl. Mar. Sci. & Ocean Engr. 29, 172 p.

Spangenberg, Dorothy B. 1968. Recent studies of strobilation in jellyfish. Oceangr. Mar. Biol. Ann. Rev. 6:231-247.

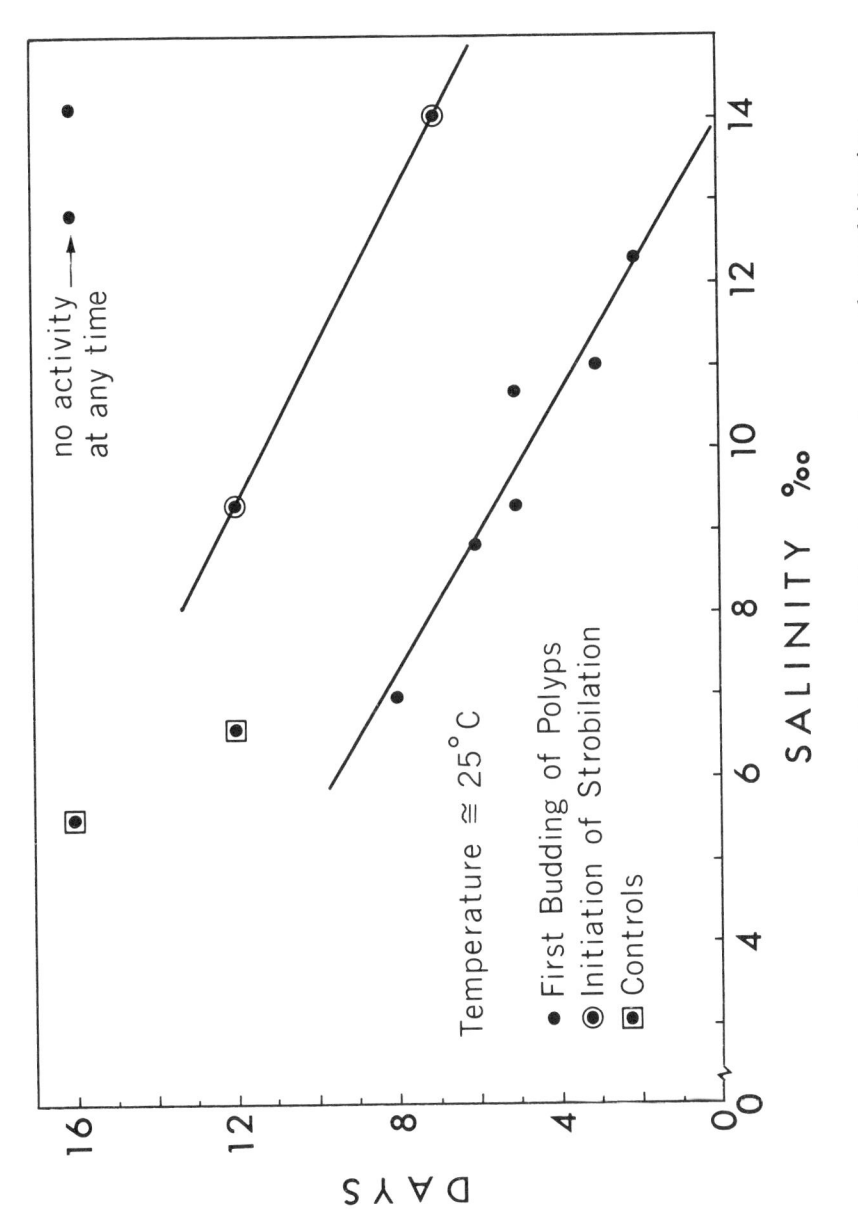

Figure 1. Relationship of rising salinity to excystment and strobilation among the scyphistomae of *Chrysaora quinquecirrha*.

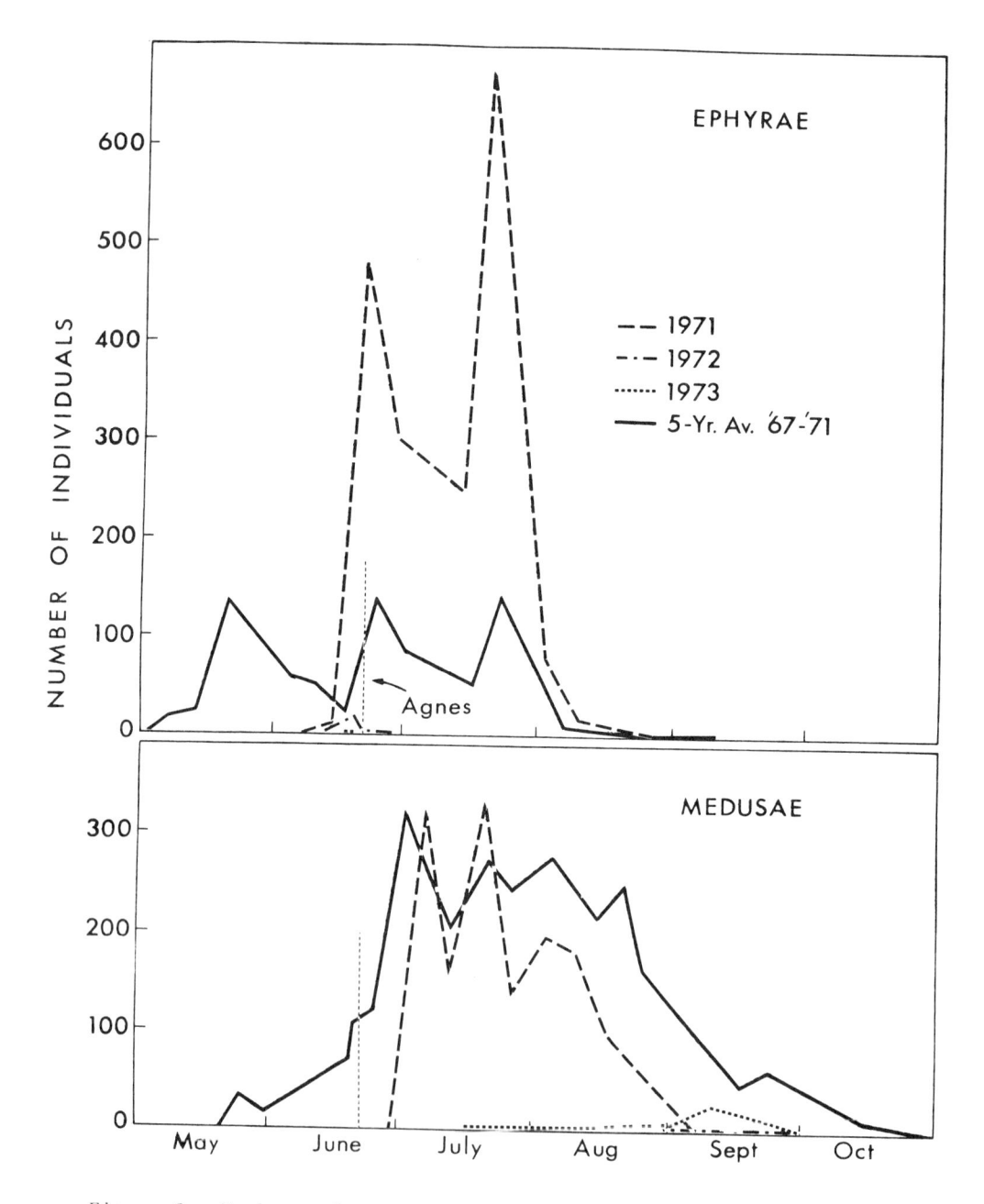

Figure 2. Numbers of sea nettles obtained in paired plankton hauls at 2 stations in St. John's Creek, near Solomons, Md.

EFFECTS OF TROPICAL STORM AGNES ON ZOOPLANKTON IN THE LOWER CHESAPEAKE BAY[1]

George C. Grant[2]
Burton B. Bryan[2]
Fred Jacobs[2]
John E. Olney[2]

ABSTRACT

Sampling techniques in use since August 1971 were employed to study effects of Tropical Storm Agnes on lower Chesapeake Bay zooplankton following the storm's passage on June 21, 1972. Mean catches of copepods, cladocerans, barnacle larvae, decapod larvae, chaetognaths, and fish eggs and larvae were calculated for the entire study area and six subareas from 8" bongo net collections. A single subarea was selected for specific identifications within major taxa of zooplankton.

Biomass, as estimated from settled volume and dry weight, was reduced following flooding in 1972. The average dry weight in August 1972 was 89 mg/m^3 compared with 269 mg/m^3 in 1971, a year of more nearly normal rainfall. Copepods, dominated by the euryhaline *Acartia tonsa*, were least affected by flooding, although the seasonal maximum may have been delayed one month. Cladocerans, however, were decimated following the flood; *Evadne* and *Penilia* were eliminated from most of the lower Bay. Most of the observed reduction in biomass is attributed to severe effects of freshwater runoff on polyhaline cladoceran populations, which had not fully recovered by the summer of 1973.

Other groups of zooplankton were more or less reduced in numbers, but within the extremes observed for copepods and cladocerans.

INTRODUCTION

Agnes reached Norfolk, Virginia on the night of June 21, 1972. Record rainfall over most of the Chesapeake drainage basin followed closely on the heels of heavy rain from frontal activity the preceding week. Preceding months of 1972 were also unusually wet, especially in late winter and early spring. Freshwater discharge from the Potomac and Susquehanna rivers, principal tributaries to the Chesapeake, exceeded by 50% mean flows for the 10-month period preceding Agnes. Salinities in Chesapeake Bay, therefore, were already well below normal (CBRC 1973), and intensified the ecological stress induced by Agnes.

Literature contains little information on effects of floods and hurricanes on zooplankton populations. Numerous authors have reported alterations in fish populations (Baughman 1949, Breder 1962, Robins 1957, Simmons & Hoese 1960, Tabb & Jones 1962). Effects on oysters and other benthic invertebrates include reports by CBRC 1973, Andrews 1973, Burbanck 1961, Butler 1949, 1952, Lunz 1956, May 1972, and Stone and Reish 1965.

---

[1] This research was supported, in part, by the Chesapeake Research Consortium, Inc., through funds supplied by the RANN program of the National Science Foundation. Contribution No. 760 from the Virginia Institute of Marine Science, Gloucester Pt., Va. 23062

[2] Virginia Institute of Marine Science, Gloucester Point, Va. 23062

Agnes interrupted the eleventh monthly survey in a 2-year investigation of zooplankton in lower Chesapeake Bay, begun in August 1971. We were therefore able to adapt an ongoing survey program to an impromptu study of flood effects and to compare flood data with before-and-after survey results. Quantitative zooplankton collections obtained with bongo samplers were analyzed by plankton groups. Summer surveys in 1971 and 1973 are compared with the flood period of 1972.

METHODS AND MATERIALS

Stations sampled monthly from August 1971 through July 1973 were randomly selected prior to each cruise from within each of eight subareas (A-H in Fig. 1). This technique was altered following Agnes for three weekly cruises beginning June 29-30, 1972, when those stations completed on June 19-20 (subareas A-F) were revisited. Random selection of stations was reinstituted with the July 24-27 regular monthly cruise. Data collected in subareas G and H are omitted from consideration in this paper, so that flood and non-flood periods are directly comparable, and are not included in any area means.

Hydrographic data reported herein were collected by means of a submersible pump, lowered to an even number of meters from the surface, but safely off the bottom. Temperatures were recorded and water samples collected for salinity and dissolved oxygen (DO) determinations at each 2-meter interval to the surface. Surface samples were taken within the upper half meter. A fuller description of sampling techniques may be found in Grant (1972).

Zooplankton was collected (Fig. 2) with 8-inch bongo samplers, fitted with 202µ mesh nets (Biological Methods Panel Committee on Oceanography 1969; McGowan & Brown 1966). General Oceanics, Inc. flowmeters were attached to the gear and used for volumetric estimates of water sampled. Tows were made from near the bottom to the surface in a stepped oblique technique, usually one minute per 2-meter interval.

Each tow yielded replicate samples, one from each of the bongo nets. The first was rinsed in distilled water and frozen over dry ice. After initial concentration in a 110µ sieve. The second was preserved in 5% formalin. Frozen samples were lyophilized in the laboratory for dry weights; preserved samples were settled for 24 hours in Imhoff cones for determination of settled volume, then split into successively smaller aliquots for counts and identification, first of the larger and rarer taxa, then of the smaller and numerically dominant zooplankters.

RESULTS

Hydrography of Lower Chesapeake Bay

Surface salinity in the lower Chesapeake Bay (below 37°23'N) fluctuates around mean values of from 19 ppt in the northwest portion to 25 ppt at Cape Charles during summer months (Stroup & Lynn 1963). This is the polyhaline region of the Bay. Mean summer temperatures at the surface over the years included in Stroup and Lynn (1963) range between 26 and 27.5°C, decreasing from west to east and toward the Bay mouth.

Mean surface temperatures and salinities within subareas A-F are shown in Fig. 3 for the period June 19-20 to September 1972. Grand means for comparable months in other years of our survey are also included. An initial rapid drop in salinity from June 19-20 to June 29-30 probably reflects local runoff and discharges from lower Chesapeake rivers such as the James, York, and Rappahannock. Surface salinity fluctuated during the next two weeks at abnormally low levels, before dropping to record lows in late July, one month after the passage of Agnes. Surface temperatures varied inversely with salinity, increasing rapidly from early to late July. Temperatures also elevated rapidly in this period at depth, increasing 2-5°C at 2 and 4 meters. At 6 meters, increases were from 1-3°C except in subareas C and E, where higher salinity water was encountered. The appearance of warm, low salinity surface waters in late July indicates a lag period of one month before full effects of flood waters from the Susquehanna were felt in the lowermost, normally polyhaline, region of Chesapeake Bay. Salinities were still below normal in September, but had recovered by October 1972. Except for the peak in late July, surface temperatures during summer 1972 were considerably below long-term averages.

Depletion of DO at deeper levels of the lower Chesapeake Bay, especially in deep holes found in the lower York and Rappahannock rivers, is a frequently observed summer phenomenon (VIMS, unpublished data). In the lower Bay proper in non-flood years, occasional measurements in the range of 3.4-4.3 mg/liter were taken from depths of 4-12 meters, from July through September. In the summer of 1972 DO was more severely reduced and over a greater area of the lower Bay (Table 1), especially after the arrival of flood waters from northern Chesapeake Bay. Low oxygen also was observed closer to the surface than in non-flood years, at depths of 4 meters.

Table 1. Occurrence of low dissolved oxygen measurements in lower Chesapeake Bay, June-September 1972.

Dates (1972)	Subarea*	Depth (m)	DO_2 (mg/liter)
June 19-20	E (2)	6-10	3.9-4.4
	F (1)	8-10	3.5-4.3
June 29-30 - no notably low values			
July 6-7	E (1)	10-11	3.5-3.7
July 13-14	E (1)	6-12	3.2-3.6
July 24-27	A (1)	6	1.6
	D (2)	6-8	1.1-2.1
	E (3)	4-12	1.0-3.1
	F (3)	4-16	2.1-4.9
Aug 15-21	A (1)	4-6	2.6-3.6
	D (3)	4-8	0.2-3.9
	E (2)	6-12	1.4-3.4
	F (1)	8-10	3.6-3.7
Sept 12-14	D (1)	6-8	4.0

*number of stations in parentheses

Zooplankton Biomass in Lower Chesapeake Bay

Two basic estimates of plankton biomass were routinely derived from bongo net samples: 24 hour settled volume of preserved samples in ml/m^3 and dry weight of frozen samples in mg/m^3. Agreement between the two estimates is generally good, except in periods of ctenophore abundance. *Mnemiopsis leidyi*, fully incorporated in dry weight estimates, disintegrates in formalin. Remaining tissues tend to float in preserved samples, so are not included in settled volume measurements. *Beroe ovata* presents similar difficulties, but of lesser magnitude. The resulting disagreement between estimates when *Mnemiopsis* is abundant is evident in June 1973 data shown on Fig. 4. Apparent agreement between estimates in June 1972 is deceptive since in that month most ctenophores were counted and discarded before freezing of samples. A slight divergence in estimates in late July 1972 may be due to the presence of *Beroe*.

The high biomass found in August 1971 was not attained in the flood year of 1972. Instead, biomass decreased to relatively low levels after the surge of *Mnemiopsis* in June and remained low (ca. 100 mg/m^3) throughout the summer (Fig. 4).

General Composition of Zooplankton

Data presented in Tables 2 and 3 were obtained from counts of major taxa separated in initial sorting of preserved bongo collections. Certain minor groups, such as mollusk larvae, polychaete larvae, amphipods, isopods and mysids are excluded from the tables.

Density of copepods, cladocerans, barnacle larvae, decapod larvae, chaetognaths, and fish eggs and larvae are given for the study area in Table 2 and for individual subareas in Table 3. Cruises included are those from immediately prior to Agnes through the summer of 1972 and, for comparison, August and September 1971 and June and July 1973. Copepods increased rapidly from an annual June low to a maximum in September 1972, when counts corresponded to those observed in August 1971. Cladocerans decreased drastically following Agnes and, except for a slight recovery in August, remained in very low numbers throughout the summer of 1972. In August 1971, cladocerans were dominant over copepods, accounting for a large part of the biomass peak. Their scarcity in 1972 may conversely account for the absence of a normal August peak in biomass. One year after the flood, June 25-27, 1973, the few cladocerans seen were taken in subareas B and C, near the mouth of the Bay.

Barnacle larvae decreased steadily from June through September 1972; counts in June and July 1973 were even lower. The peak in decapod larvae numbers occurred in August 1972, but at an order of magnitude lower than that of 1971. Chaetognaths normally peak in September, so escaped much of the flood's effect. Numbers of fish eggs and larvae, on the average, were not notably reduced in 1972.

Species Composition of Selected Groups in Subarea D

Subarea D, located in the shallower portions of the western side of lower Chesapeake Bay from Wolf Trap to Back Bay, was subjected to the greatest reduction in salinity among the six subareas considered here, and is selected for a detailed examination of zooplankton changes. The deviation from normal T-S relationships in this subarea produced by flooding from Agnes is evident in Fig. 5.

Diversity among copepods is naturally low in Subarea D during summer because of the overwhelming dominance by *Acartia tonsa*. However, after Agnes the diversity was decreased even further (e.g. H' in August 1971 was approximately 0.037, but only 0.008 in August 1972). Reappearance of certain species occurred in September 1972 (Table 4). Low diversity among summer copepods was repeated in 1973, with the nearly complete absence of species other than *A. tonsa* in June and July.

Table 5 gives catches within species of both cladocerans and chaetognaths. Three species of cladocerans are normally present in this subarea during summer.

Table 2. Density of major zooplankton taxa in the lower Chesapeake Bay, before and after Tropical Storm Agnes. Mean number per cubic meter for subareas A-F combined; based on stepped oblique tows with 8" bongo nets, mesh size 202μ.

	Primary Consumers			Secondary Consumers			
			Barnacle			Fish	
Dates	Copepods	Cladocera	Larvae	Decapods	Chaetognaths	Eggs	Larvae
PRE-FLOOD							
Aug 16-19, 1971	19,750	77,570	24	1,429	17	49	25
Sept 21-23, 1971	12,800	1,078	11	51	226	<1	<1
June 19-20, 1972	175	1,062	225	38	<1	69	1
FLOOD (1972)							
June 29-30	260	41	85	19	<1	27	4
July 6-7	1,510	78	48	11	<1	48	2
July 13-14	12,740	87	43	22	<1	83	9
July 24-27	13,720	8	9	12	2	72	24
Aug 15-21	12,690	2,015	4	101	15	32	13
Sept 12-14	20,400	1	4	26	102	0	<1
POST-FLOOD (1973)							
June 25-27	638	260*	3	22	<1	36	<1
July 23-24	2,542	1,655	34	63	1	115	12

*all at a few seaward stations

Table 3. Density of major zooplankton taxa within individual subareas A-F during flood period of 1972 and comparable months in 1971 and 1973.

Numbers/m³

Dates	Copepods	Cladocerans	Chaetognaths	Barnacle Larvae	Decapod Larvae	Fish Eggs	Fish Larvae
			SUBAREA A				
16-19 Aug 71	13,180	5,734	18	0	3,357	13	4.7
21-23 Sept 71	4,670	311	315	21	25	0	1.2
19-20 June 72	117	448	<1	139	64	74	0.6
29-30 June 72	791	51	<1	111	43	29	13.9
6-7 July 72	1,459	143	<1	77	6	68	2.5
13-14 July 72	12,000	8	<1	52	28	63	12.6
24-27 July 72	5,655	<1	<1	10	38	76	16.6
15-21 Aug 72	15,340	1,215	10	3	27	50	8.2
12-14 Sept 72	12,030	3	14	3	20	0	0.3
25-27 June 73	378	0	<1	1	6	17	0.4
23-24 July 73	4,381	838	1	13	70	174	7.6
			SUBAREA B				
Aug 71	13,040	153,500	20	81	1,558	52	13.0
Sept 71	2,913	3,221	238	13	46	<1	0.1
19-20 June 72	191	2,622	0	78	75	99	1.3
29-30 June 72	114	<1	<1	169	7	3	0.03
6-7 July 72	1,431	69	<1	19	13	15	2.3
13-14 July 72	16,560	110	<1	55	37	86	17.6
24-27 July 72	no samples						
15-21 Aug 72	9,886	757	3	9	91	40	19.0
12-14 Sept 72	32,780	<1	405	6	42	0	0.5
June 73	480	3	<1	0	19	40	0.7
July 73	2,065	1,091	4	30	80	290	35.1
			SUBAREA C				
Aug 71	5,734	53,090	5	0	2,937	116	46.4
Sept 71	3,764	431	358	12	34	<1	0.2
19-20 June 72	134	1,111	0	64	18	58	0.9
29-30 June 72	117	3	<1	31	11	13	1.7
6-7 July 72	1,410	255	0	15	4	115	4.3
13-14 July 72	19,920	340	1	27	22	125	10.2

Table 3. (Continued)

Dates	Copepods	Cladocerans	Chaetognaths	Barnacle Larvae	Decapod Larvae	Fish Eggs	Fish Larvae
			SUBAREA C (Continued)				
24-27 July 72	6,872	30	2	2	12	82	20.2
15-21 Aug 72	4,500	8,168	15	1	206	22	12.3
12-14 Sept 72	25,560	<1	102	3	23	0	1.1
June 73	1,204	1,225	1	10	61	62	0.3
July 73	3,385	4,042	3	21	166	136	24.8
			SUBAREA D				
Aug 71	38,120	16,960	8	35	467	5	42.7
Sept 71	57,530	2	113	7	74	0	0.8
19-20 June 72	131	20	0	1,256	17	16	0.5
29-30 June 72	74	10	0	65	11	48	0.2
6-7 July 72	3,453	1	0	43	31	47	1.2
13-14 July 72	8,203	0	0	17	10	19	2.9
24-27 July 72	15,290	0	<1	18	5	77	11.5
15-21 Aug 72	18,900	<1	<1	7	21	50	7.8
12-14 Sept 72	15,640	6	16	10	20	0	0.2
June 73	468	0	0	5	10	22	0.6
July 73	1,712	1,210	<1	51	16	89	4.9
			SUBAREA E				
Aug 71	18,880	42,890	15	1	512	61	30.0
Sept 71	19,320	615	50	12	58	0	0
19-20 June 72	185	967	0	163	12	90	2.0
29-30 June 72	40	255	0	14	21	16	0.7
6-7 July 72	119	0	0	11	7	11	0.4
13-14 July 72	10,900	<1	<1	54	13	44	4.0
24-27 July 72	17,140	2	3	9	4	92	27.7
15-21 Aug 72	12,090	79	21	3	54	33	16.2
12-14 Sept 72	14,920	<1	28	1	16	0	0
June 73	332	0	0	<1	15	15	0.2
July 73	1,471	1,503	<1	36	25	50	6.3

Table 3. (Continued)

Dates	Copepods	Cladocerans	Chaetognaths	Barnacle Larvae	Decapod Larvae	Fish Eggs	Fish Larvae
			SUBAREA F				
Aug 71	35,130	114,000	30	0	384	27	18.1
Sept 71	8,518	417	181	2	71	<1	0.2
19-20 June 72	265	531	0	123	16	57	1.6
29-30 June 72	53	0	0	100	7	47	1.3
6-7 July 72	803	8	0	88	4	39	1.7
13-14 July 72	4,577	29	<1	44	16	152	4.1
24-27 July 72	22,400	<1	1	10	5	37	36.9
15-21 Aug 72	18,310	176	26	4	136	17	12.9
12-14 Sept 72	18,990	<1	49	1	32	0	0.3
June 73	683	0	<1	<1	7	39	0.1
July 73	1,823	783	1	52	17	46	2.8

Numbers/m^3

Table 4. Species composition of copepod collections in Subarea D, summer months of 1972 and comparable months in non-flood years.

Numbers Per Cubic Meter

Dates	*Acartia tonsa*	*Labidocera aestiva*	*Pseudodiaptomus coronatus*	*Temora turbinata*	*Paracalanus crassirostris*	*Oithona* sp.	Unidentified harpacticoid	*Centropages hamatus*	*Eurytemora* sp.
PRE-FLOOD									
Aug 1971	36,740	35	29	24	16	--	8	--	--
Sept 1971	55,730	37	1,517	--	--	--	--	--	--
June 19-20, 1972	131	--	--	--	--	--	--	<1	--
FLOOD (1972)									
June 29-30	72	--	--	--	--	2	--	--	--
July 6-7	3,451	--	--	--	--	--	--	--	--
July 13-14	8,203	--	--	--	--	--	--	--	--
July 24-27	15,216	--	--	--	--	25	37	--	--
Aug 15-21	18,879	<1	--	--	82	10	--	--	11
Sept 12-14	6,607	10	13	--	--	--	--	--	--
POST-FLOOD (1973)									
June 25-27	534	--	--	--	--	--	2	--	--
July 23-24	1,803	--	--	--	--	--	--	--	--

Table 5. Species composition of cladoceran collections and abundance of the chaetognath *Sagitta tenuis* in Subarea D, summer months of 1972 and comparable months in non-flood years.

	Numbers Per Cubic Meter				Chaetognatha
	Cladocera				
Dates	*Podon polyphemoides*	*Evadne tergestina*	*Penilia avirostris*		*Sagitta tenuis*
PRE-FLOOD					
Aug 1971	399	14,580	2,193		6
Sept 1971	<1	2	0		110
June 19-20, 1972	20	0	0		0
FLOOD (1972)					
June 29-30	10	0	0		0
July 6-7	1	0	0		0
July 13-14	0	0	0		0
July 24-27	0	0	0		0
Aug 15-21	<<1	<1	0		<1
Sept 12-14	6	<<1	0		16
POST-FLOOD (1973)					
June 25-27	0	0	0		0
July 23-24	1,210	0	0		<<1
Aug (cladoceran cruise)	183	646	53		--

Podon polyphemoides decreased rapidly after Agnes to complete absence in mid- and late July; a slight recovery occurred in September. None were taken in June 1973; sizable catches occurred in July. *Evadne tergestina*, very abundant in August, 1971, was absent in June and July 1972, very rare in August and September, absent again in June and July 1973. *Penilia avirostris* was present in Subarea D only in August 1971. A post-survey cruise in August 1973, for the sole purpose of assessing cladoceran abundance, revealed the presence of all three species.

The single species of warm-water chaetognath taken in this subarea, *Sagitta tenuis*, was considerably reduced in numbers during August and September of 1972 compared with catches in the previous year.

Fish larvae were also considerably reduced in 1972, although this subarea is not normally very productive. Anchovies, most likely *Anchoa mitchilli*, were the most abundant larvae, reduced from $35/m^3$ in August 1971 to $6/m^3$ in August 1972. Two as yet unidentified species, one a goby and the other a blenny, were also present in low numbers both years. Several species, although rare, were present in August 1971, but absent in August 1972. They were *Cynoscion regalis*, *Gobiesox strumosus*, *Sphaeroides maculatus*, *Symphurus plagiusa*, *Trinectes maculatus* and *Syngnathus fuscus*.

Other zooplankton groups not incorporated in Tables 3-5 include gastropod larvae, polychaete larvae, amphipods, isopods, phoronid larvae and mysids. Among these, only gastropod larvae were numerically important. They were fairly abundant just prior to Agnes ($110/m^3$), then declined during flooding until the late July cruise. Numbers remained low thereafter. The following year, June and July 1973, found gastropod larvae very abundant and occasionally dominant over copepods (to $750/m^3$).

DISCUSSION

In attempting to evaluate effects of the flood on zooplankton populations, catastrophic changes induced by Agnes must be distinguished from normal annual and seasonal variations. The present exercise in evaluating a catastrophic perturbation on a particular segment of the biota has demonstrated to us the need for long-term monitoring of seasonal and annual changes in abundance. With data from only one complete and two partial summer periods, a description of normal zooplankton composition must be highly speculative. However, the data reveal distinct trends in zooplankton abundance and the storm significantly altered existing distributional and numerical zooplankton patterns during and directly following the flood period.

Normally, intensive predation by the ctenophore *Mnemiopsis leidyi* on crustacean populations results in very low numbers of copepods in May and June in the lower Bay (Burrell 1968). This occurred in both 1972 and 1973. As *Mnemiopsis* is reduced in July through predation by another ctenophore, *Beroe ovata*, the crustacean zooplankton, dominated by *Acartia tonsa*, builds to an August maximum. Our data indicate that the copepod maximum in 1972 was delayed until September (Table 2), even though the dominant *A. tonsa* is a widely distributed copepod occupying salinities as low as those encountered during the flood (Heinle 1972, Joseph & Van Engel 1968, Burrell 1972), and *Beroe* appeared throughout the lower Bay, as usual, in July 1972. However, no lasting effect on this euryhaline copepod was evident. Copepod abundance in September 1972 compared closely with that observed in August 1971, and we conclude that Agnes had little effect on total copepod population size.

Barnacle larvae and gastropod larvae were reduced in numbers following the flood, as were decapod larvae, chaetognaths, and to a lesser extent, fish eggs and larvae. The greatest effect by far was on the Cladocera. In August 1972, the mean number of cladocerans per cubic meter in the lower Bay was 2×10^3 as compared with 77×10^3 in August 1971. Collections from Subarea B in August 1971 showed counts of 153×10^3, compared with 0.8×10^3 in August 1972; other subareas were similar in reduction of cladoceran counts. *Evadne tergestina* and *Penilia avirostris*, two high salinity species, comprised the bulk of the cladoceran population in Subarea D in summer 1971 (Table 5). Bosch and Taylor (1968) found that the lowest salinities inhabited by these two species in Chesapeake Bay were 15.75 ppt and 18.12 ppt, respectively. In view of this, it is not surprising that *Evadne* and *Penilia* were almost completely eliminated from the lower Bay following Agnes.

The reduction in Cladocera may have been a relatively long-term one, extending through the summer of 1973. In June 1973, cladocerans were found only at a few seaward stations, and these consisted of *Podon polyphemoides* and *Evadne nordmanni*, remnants of a May peak. *Evadne tergestina* and *Penilia avirostris* did not reoccur in the lower Bay until July 1973 or in Subarea D until August 1973. It is quite likely that overwintering resting eggs deposited in the lower Bay were destroyed by the flood, so that repopulation had to await the influx of individuals from coastal water populations. Andrews (1973) has stated that recovery was slower among benthic organisms having crawling or sedentary larvae than among species with pelagic larvae. Planktonic organisms such as cladocerans, with a part of their life cycle tied to the bottom, might also be expected to recover slowly.

Cladocerans were the dominant zooplankton organisms in the lower Chesapeake Bay in August 1971. By August 1972, after flooding from Agnes, they were essentially absent. The severe reduction in cladocerans probably accounts for most of the observed loss in plankton biomass, down from 269 mg/m^3 dry weight in August 1971 to 89 mg/m^3 in August 1972.

LITERATURE CITED

Andrews, J. D. 1973. Effects of Tropical Storm Agnes on epifaunal invertebrates In Virginia estuaries. Chesapeake Sci. 14:223-234.

Baughman, J. L. 1949. Random notes on Texas fishes. Part I. Texas J. Sci. 2:117-138.

Biological Methods Panel Committee on Oceanography. 1969. Recommended procedures for measuring the productivity of plankton standing stock and related oceanic properties. National Academy of Sciences, Washington, D. C., 59 p.

Bosch, H. F. and W. R. Taylor. 1968. Marine cladocerans in the Chesapeake Bay estuary. Crustaceana 15:161-164.

Breder, C. M. 1962. Effects of a hurricane on the small fishes of a shallow bay. Copeia 1962(2):459-462.

Burbanck, W. D. 1961. The disappearance of *Cyathura polita* from the Ashepoo River, South Carolina, after hurricane "Gracie" in 1959. Ecology 42(3): 606-607.

Burrell, V. G., Jr. 1968. The ecological significance of a ctenophore, *Mnemiopsis leidyi* (A. Agassiz), in a fish nursery ground. M.A. Thesis, College of William and Mary, Williamsburg, Va., 61 p.

Burrell, V. G., Jr. 1972. Distribution and abundance of calanoid copepods in the York River estuary, Virginia, 1968 and 1969. Ph.D. Dissertation, College of William and Mary, Williamsburg, Va., 234 p.

Butler, P. A. 1949. An investigation of oyster producing areas in Louisiana and Mississippi damaged by flood-waters in 1945. U. S. Fish & Wildl. Serv. Spec. Sci. Rep. (8):1-29.

Butler, P. A. 1952. Effects of floodwaters on oysters in Mississippi Sound in 1950. U. S. Fish & Wildl. Serv. Res. Rep. (31):1-20.

Chesapeake Bay Research Council. 1973. Effects of Hurricane Agnes on the environment and organisms of Chesapeake Bay. A report to the Philadelphia District U. S. Army Corps of Engrs. (Aven M. Andersen, Coordinator), Chesapeake Bay Inst. Contrib. 187, Nat. Resour. Inst. Contrib. 529, Va. Inst. Mar. Sci. Spec. Rep. Appl. Mar. Sci. & Ocean Engr. 29, 172 p.

Grant, G. C. 1972. Plankton Program, I. Zooplankton. Pages 28-76 in Chesapeake Research Consortium, Inc., Ann. Rep. for June 1, 1971-May 31, 1972, 754 p.

Heinle, D. R. 1972. Tentative outline for inventory of zooplankton organisms: *Acartia tonsa* (copepod). Chesapeake Sci. 13(Suppl.):S176-178.

Joseph, E. B. and W. A. Van Engel. 1968. The characterization of coastal and estuarine fish nursery grounds as natural communities. Va. Inst. Mar. Sci. Final Rep., Nov. 1965-Aug. 1967, Contract 14-17-0007-531, Commercial Fish Res. and Development Act, 43 p.

Lunz, G. R. 1956. Oysters, diseases and downpours. Assoc. Southeast. Biol. Bull. 3:20.

May, E. B. 1972. The effect of floodwater on oysters in Mobile Bay. Proc. Natl. Shellfish Assoc. 62:67-71.

McGowan, J. A. and D. M. Brown. 1966. A new opening-closing paired zooplankton net. Univ. Calif. Scripps Inst. Oceanogr. (Ref. 66-23).

Robins, C. R. 1957. Effects of storms on the shallow water fish fauna of southern Florida with new records of fishes from Florida. Bull. Mar. Sci. Gulf Caribb. 7(3):266-275.

Simmons, E. G. and H. D. Hoese. 1960. Studies on the hydrography and fish migrations of Cedar Bayou, a natural tidal inlet on the central Texas coast. Publ. Inst. Mar. Sci. Texas 6:56-80.

Stone, A. N. and D. J. Reish. 1965. The effect of freshwater runoff on a population of estuarine polychaetous annelids. South. Calif. Acad. Sci. Bull. 64(3):111-119.

Stroup, E. O. and R. J. Lynn. 1963. Atlas of salinity and temperature distributions in Chesapeake Bay 1952-1961 and seasonal averages 1959-1961. Johns Hopkins Univ., Chesapeake Bay Inst., Graph. Summ. Rep. 2.

Tabb, D. C. and A. C. Jones. 1962. Effect of Hurricane Donna on the turtle grass beds of Biscayne Bay, Florida. Bull. Mar. Sci. Gulf Caribb. 11(2):191-197.

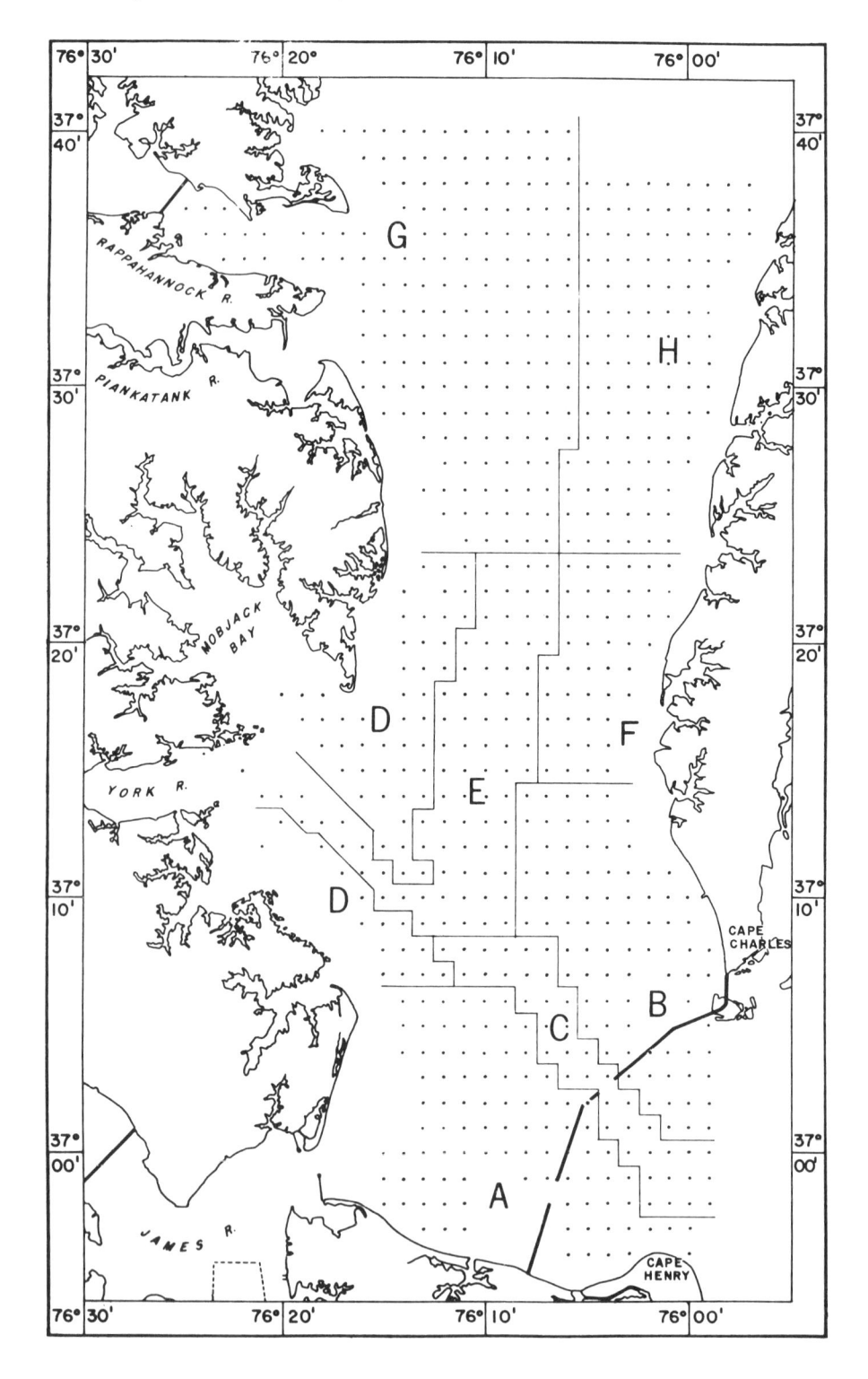

Figure 1. Area of study during two-year zooplankton survey of lower Chesapeake Bay where stations were chosen at random from within each of the eight subareas A-H each month. Flood studies consider only subareas A-F.

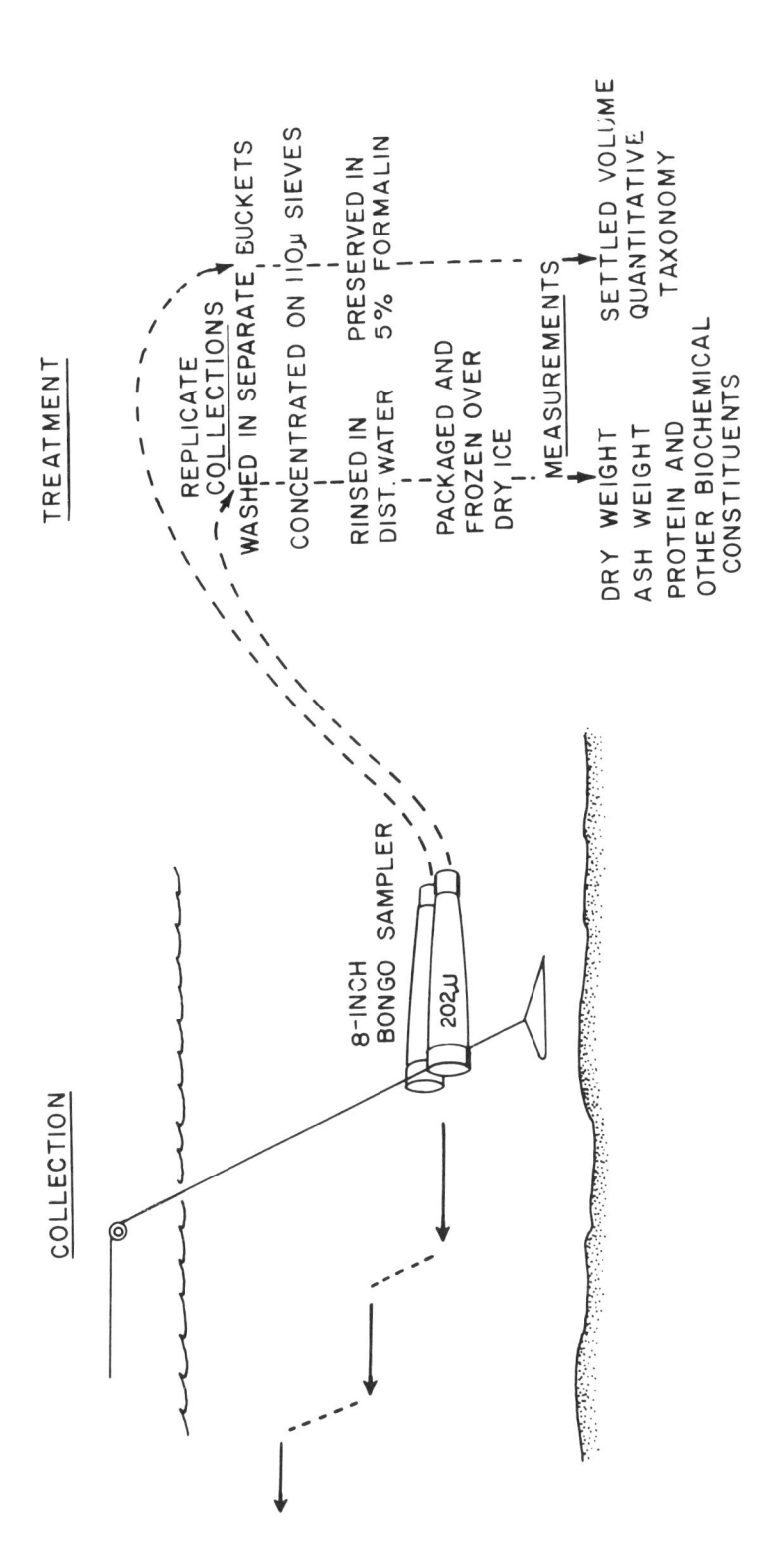

Figure 2. Diagram of zooplankton collection method and subsequent treatment of samples.

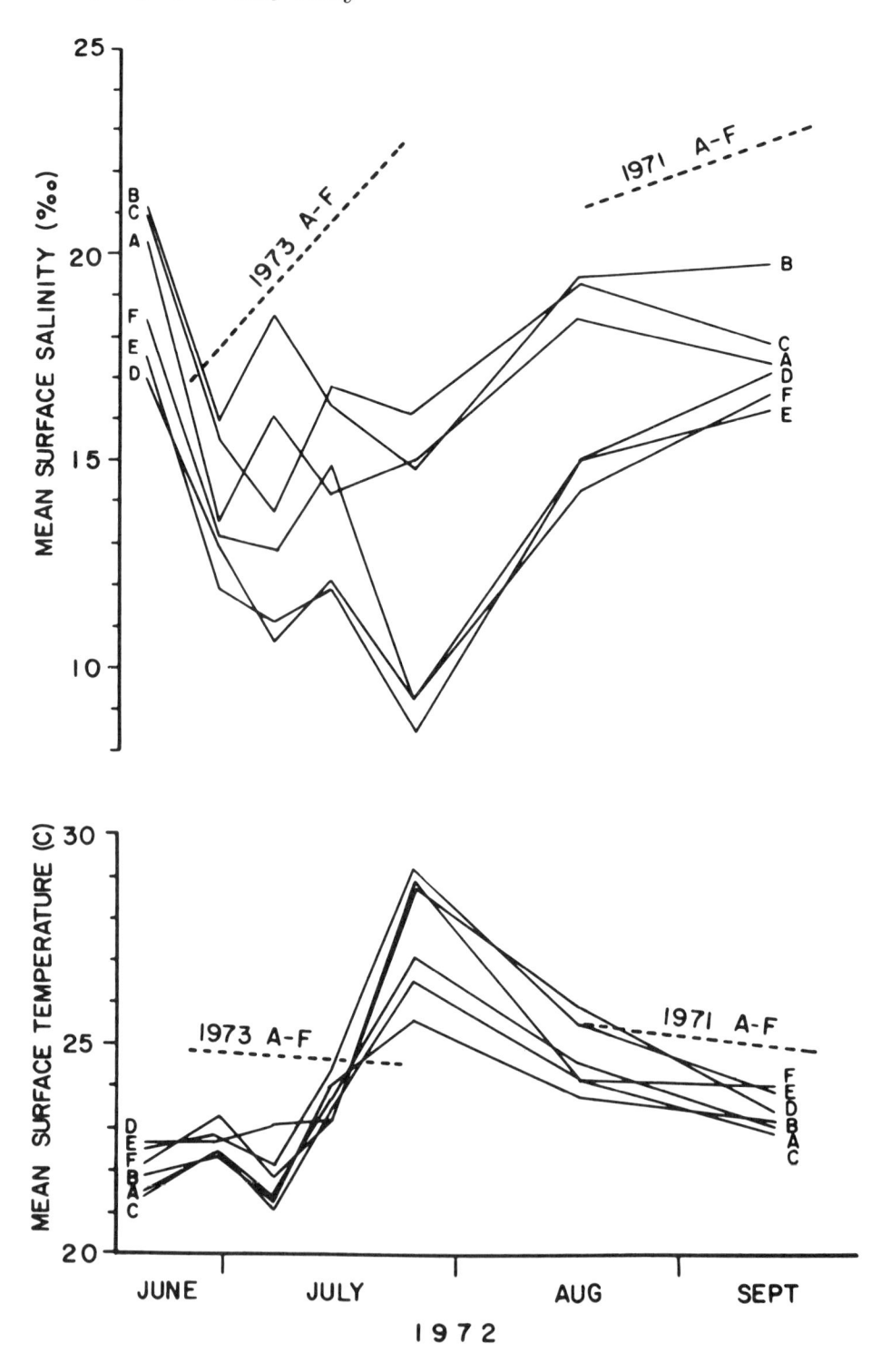

Figure 3. Mean surface salinity (top) and temperature (bottom) within subareas A-F, June 19-20 - September 12-14, 1972 (solid lines); total area means for comparable months in non-flood years connected by dashed lines.

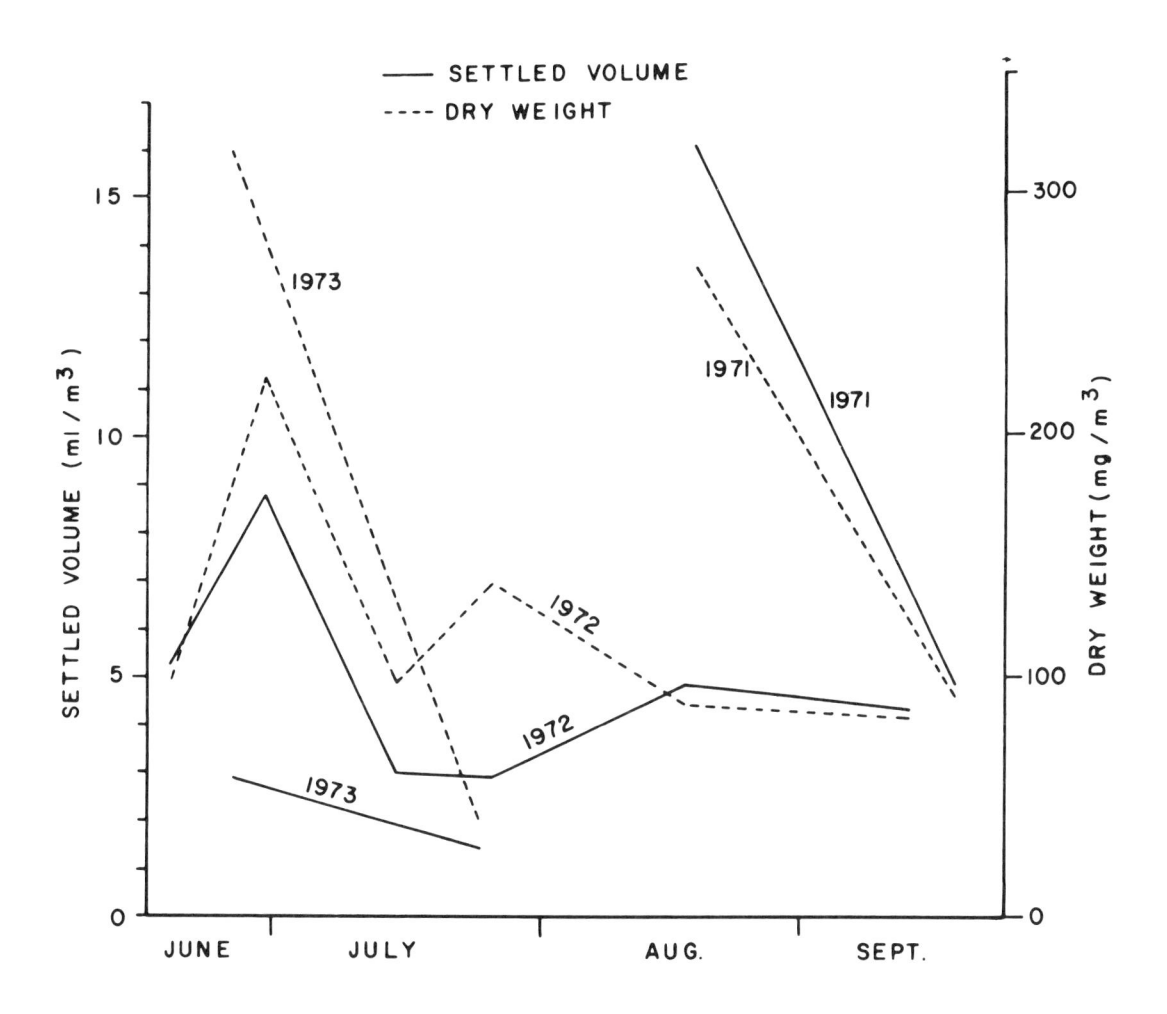

Figure 4. Estimates of zooplankton biomass in the lower Chesapeake Bay, dry weight in mg/m^3 and settled volume in ml/m^3, June 19-20 - September 12-14, 1972, and comparable months in 1971 and 1973; mean values for the entire study area (subareas A-F).

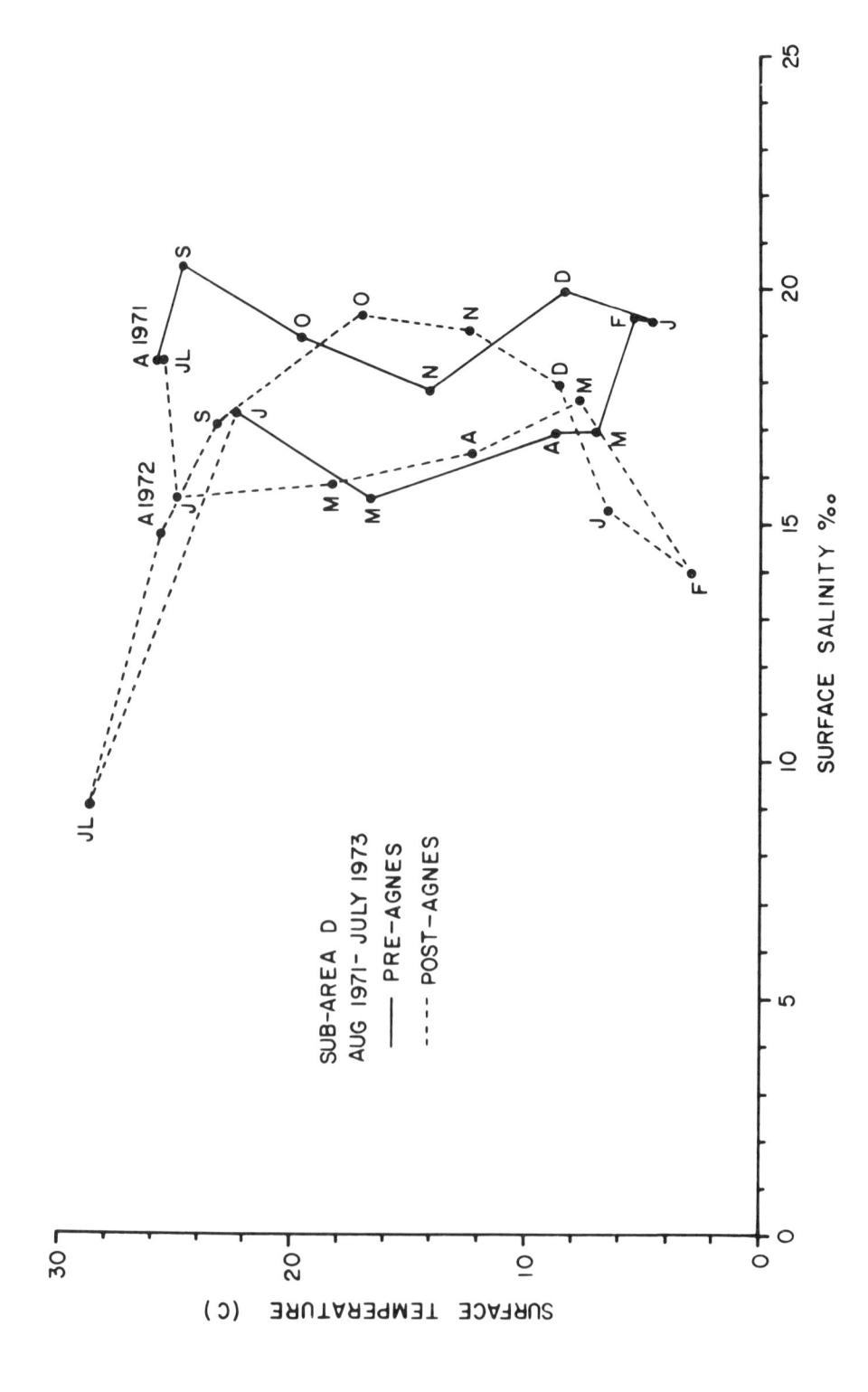

Figure 5. Mean surface temperature-salinity relationship in subarea D, August 1971-July 1973.

EFFECTS OF TROPICAL STORM AGNES ON STANDING CROPS AND AGE STRUCTURE OF ZOOPLANKTON IN MIDDLE CHESAPEAKE BAY[1]

D. R. Heinle[2]
H. S. Millsaps[2]
C. V. Millsaps[2]

ABSTRACT

General magnitude of densities and the age structures of populations of dominant summer species, particularly *Acartia tonsa*, were not affected by Agnes. Salinities at mid-Bay did not drop below the range tolerated by *A. tonsa*.

Species with requirements for higher salinities, e.g., *Oithona brevicornis*, disappeared after Agnes. They were detected again in late August or early September but failed to achieve densities recorded in other years.

INTRODUCTION

Our involvement in a survey of zooplankton in the vicinity of the Calvert Cliffs nuclear power plant provided an opportunity to examine the effects of the passage of Agnes on the seasonal distribution and age structure of the abundant species of zooplankton at that site. There have been no previous quantitative studies of zooplankton in mid-Chesapeake Bay that employed nets of sufficiently small sized mesh to retain all stages of the very abundant copepod nauplii. The 103 µm mesh used by Bosch and Taylor (1973a, b) probably retained most copepod nauplii but their papers dealt only with the cladoceran *Podon polyphemoides*. There are therefore no early historical data available for comparison with the seasons immediately following Agnes.

Data on seasonal abundance of several dominant species in the Patuxent estuary during 1969 and 1970 are presented in this paper to assist in the interpretation of the consequences of reduced salinity following Agnes.

METHODS

Seven stations off Calvert Cliffs (Fig. 1) were sampled for microzooplankton by towing a deep-well submersible pump at depths of 1 m, mid-depth, and near bottom. Separate samples were obtained from each depth. The bottom samples were generally collected within 1 to 4 m above bottom. Actual depths of the stations to the nearest meter were: 3-A, 10 m; 2-A, 9 m; 1-A, 7 m; 1-B, 10 m; 2-C, 35 m; 2-B, 13 m; and 3-B, 15 m. Towing speed was 1 to 3 m per second. Collection of each sample required 2 to 3 min. pumping. All three samples at each station were usually collected within a 10 to 15 min. interval. On deck, the flow from the pump passed to a collecting device with two "Y" valves in series. The first split provided flow through a standpipe containing a Yellow Springs Instruments oxygen probe. The second split allowed passing any portion desired of the remaining flow through a No. 20 (70-74 µm opening) plankton net mounted in a deck plankton collector, a water-filled Plexiglas tube with an overflow pipe. Water displaced through the net was metered by collecting it in a 19-liter carboy.

---

[1]Contribution No. 572, U. M. Center for Environmental and Estuarine Studies.
[2]Chesapeake Biological Laboratory, Center for Environmental and Estuarine Studies, University of Maryland, Solomons, Md. 20688

The zooplankton in the net were concentrated to about 100 ml and preserved with 10% formalin. Just prior to counting, samples were concentrated to approximately 30 ml. Four one-ml subsamples were withdrawn from each well-mixed sample and counted separately. All metazoans were counted by species. Copepods were further divided into the groups: nauplii, copepodites, and adults. Several species which have yet to be identified were provided numbers (e.g., Rotifer No. 1, 2, etc.). Tintinnid protozoans were counted as such. Polychaete larvae, clam larvae, and barnacle nauplii and cyprids were likewise not identified to species.

The subsampling routine was such that each organism counted represented 300 to 600 individuals per cubic meter. In converting counts to numbers per cubic meter, all calculations were rounded to the nearest 500 at the final step.

Salinity was measured in situ with either an Inter-Ocean CSTD probe (Model 500) or with a Beckman RS-5 salinometer.

Instantaneous death rates were calculated by the method described by Heinle (1966) with minor modifications. Average densities from all three depths were used in the estimates. Development times through naupliar and copepodid stages (Heinle 1969) were used to calculate death rates of nauplii (d_n) by assuming steady state, thus: numbers of copepodids = (numbers of nauplii) $e^{-d_n T_n}$; where T_n is the duration of the six naupliar instars. Death rates of copepodids (d_c) were calculated in the same manner assuming that $T_n = T_c$. The latter assumption would underestimate d_c when death rates were high and overestimate when death rates were low because of changes in the geometric mean age of adults under varying death rates.

RESULTS

The calanoid copepod *Acartia tonsa* (Dana) was the most abundant summer zooplankter. The variations in surface density at the seven stations were usually 3- to 5-fold (Fig. 2). The copepods were generally stratified vertically during the summer with the bulk of the population in the upper 10 to 15 m. Neither the abundance nor spatial heterogeneity of *A. tonsa* appear to have been affected by Agnes. In the Patuxent estuary, *A. tonsa* was found to the limits of salt intrusion during the summer (Fig. 3). Salinity at Calvert Cliffs did not fall much below 2 ppt after Agnes.

The widely distributed rotifer *Brachionus plicatilis* (Müller) was much less abundant during the summer of 1972 than during 1971 or 1973 (Fig. 4). Up to 360,000 *B. plicatilis* per m^3 were observed in August 1973. This species was most abundant during December 1971 and January 1972 when over 1×10^6 per m^3 were observed at some stations. Abundances were lower during the winter following Agnes.

The cyclopoid copepod *Oithona colcarva* (Bowman 1975) was observed at densities up to 87,000 per m^3 during August 1971 and was present during July (Fig. 5). During 1972, *O. colcarva* appeared in September but failed to achieve the maximum densities observed the preceding year. This species was less abundant during August 1973 also. The densities observed during September through December were similar in all years. This species commonly occurs from July through December in the Patuxent (Heinle 1969).

The harpactacoid copepod *Scottolana canadensis* (Willey) was a minor component in the mid-Bay plankton during 1971. Up to 100,000 per m^3 were found immediately after the flood waters reached Calvert Cliffs (Fig.6). Unlike during 1971, *S. canadensis* persisted through August 1972. Small numbers were present from September through December both years. In the Patuxent, *S. canadensis* was present from May through November at salinities of from 0 to 14 ppt (Fig. 7).

The two calanoid copepods most common during the winter, *Eurytemora affinis* (Poppe), a low salinity form, and *A. clausi* (Giesbrecht), a high salinity form, appear to have shifted their relative abundance during the year following Agnes. During the spring of 1972 up to 152,000 *A. clausi* per m^3 were observed (Fig. 8). The following spring the maximum abundance was 60,000 per m^3 but densities during April 1973 were much lower than during April 1972.

Densities of only 40,000 *E. affinis* per m^3 were observed during the spring of 1972, while up to 300,000 per m^3 were observed during 1973 (Fig. 9). This species was found into July 1972 at Calvert Cliffs. In the Patuxent, *E. affinis* was found only at salinities lower than 5 ppt during late summer but extended its distribution to salinities as high as 15 ppt during April, the period of maximum abundance (Fig. 10).

Surface salinities were not sufficiently different during the months of March through May to account for the shift from *A. clausi* to *E. affinis* as the winter dominant (Table 1). February was the only month during the spring of 1973 when the surface salinities were appreciably lower.

Some organisms, especially polychaete larvae, were essentially absent during the two months following Agnes in contrast to other years when they were a significant component of the plankton.

The apparent instantaneous death rates of nauplii (dn) and copepodids (dc) give a direct estimate of population turnover rate, and with correction for effects of temperatures on growth rate, provides a general impression of the age structure of the population. For example, relatively low values of dn and high values of dc indicate approximately equal numbers of nauplii and copepodids and fewer numbers of adults. When dn and dc are nearly equal, the age structure resembles a negative exponential function. High values of dn and low dc imply an age structure dominated by nauplii.

The apparent death rates of *A. tonsa* were high during the summer and fall suggesting turnover times of 1 to 2 days (turnover time = 1/d) (Fig. 11). The range of values was large during the summer and extremely large for copepodids during the fall when populations were declining. There were no apparent differences between years except for higher values during late April and early May 1973. The values of dn and dc were usually of the same magnitude during the summer, suggesting a negative exponential age structure and non-selective mortality during all years.

There were sufficient numbers of *E. affinis* during only three months each year to calculate dn and dc. The higher values of dn during 1973 probably resulted from the presence of that species later in the year when temperatures were higher and turnover times shorter (Fig. 12).

DISCUSSION AND SUMMARY

The only major changes in the surface plankton at mid-Bay following Agnes appear to have been a decrease of the rotifer *B. plicatilis* and the near absence of polychaete larvae and the cyclopoid copepod *O. colcarva* during the summer months. Salinities at Calvert Cliffs (Table 1) did not become sufficiently low to eliminate the normal summer dominant, *A. tonsa*. By fall, salinities had returned to near normal and the composition of the plankton had returned to that

Table 1. Salinities at Station 2C, Calvert Cliffs during 1972 and 1973. Surface (Sur) = 1 m, mid-depth (Mid) = 15 m, and bottom (Bot) = 30 m.

1972

	17. Jan.	19 Feb.	15 Mar.	6 Apr.	19 Apr.	3 May	15 May
Bot.	15.0	18.4	15.8	16.5	17.1	18.5	14.4
Mid.	11.0	16.7	11.3	16.3	15.6	14.7	12.4
Sur.	9.0	12.9	9.4	9.4	8.7	9.2	8.4

	31 May	14 June	28 June*	13 July	28 July	9 Aug.	23 Aug.
Bot.	18.8	19.1	-	10.3	20.3	21.0	18.8
Mid.	14.9	19.1	8.8	14.1	14.1	19.7	17.0
Sur.	8.1	9.1	1.1	5.1	5.1	7.9	11.7

	11 Sept.	22 Sept.	2 Oct.	16 Oct.	16 Nov.	14 Dec.
Bot.	18.8	19.6	20.9	16.2	19.2	16.0
Mid.	17.0	14.7	16.1	14.7	16.0	16.0
Sur.	11.7	14.0	13.8	14.7	15.7	15.4

1973

	11 Jan.	14 Feb.	15 Mar.	12 Apr.	25 Apr.	7 May	21 May
Bot.	16.1	12.6	20.7	20.7	14.0	18.4	16.1
Mid.	14.6	11.4	20.3	19.8	11.2	17.8	14.1
Sur.	9.2	6.4	9.2	9.8	9.3	8.6	9.7

	4 June	18 June	2 July	16 July	30 July	13 Aug.	27 Aug.
Bot.	16.0	-	17.5	16.9	19.1	22.9	24.9
Mid.	15.7	-	9.1	14.9	18.3	18.4	21.2
Sur.	7.5	-	8.7	7.5	10.2	11.5	12.4

	17 Sept.	29 Oct.	27 Nov.	19 Dec.
Bot.	22.2	18.0	17.2	12.0
Mid.	16.5	15.7	14.9	11.0
Sur.	16.5	15.7	14.2	10.0

*Salinities determined with an American Optical refractometer.

of preceding and following years. During the winter following Agnes *E. affinis* was more abundant than *A. clausi*, the dominant species the preceding winter. Changes in salinity were not sufficient to cause the shift.

The instantaneous death rates found (Figs. 11 & 12) suggest turnover times comparable to those reported by Heinle (1969) in the Patuxent estuary. Average summer rates are near or above maximum productivities suggested for *A. tonsa* from laboratory studies (Heinle 1969). As standing crops and turnover times were not apparently different during the summer after Agnes, it appears that production of zooplankton was not greatly stimulated.

Jeffries (1962a, b; 1964) has suggested that succession of zooplankton in estuaries is regulated by physical factors. Careful examination of Figs. 2-10 reveals features that support the hypothesis, especially during the spring. The early summer increase of *A. tonsa* occurred during April. The total density of zooplankton was much lower during May than during any other month. The narrow geographic scope of our sampling prevents us from knowing the extent of the phenomenon in the Bay. Substantially more temporal and spatial overlap of species occurred in the upper Patuxent (Heinle 1969).

ACKNOWLEDGEMENTS

This work was supported by the University of Maryland, the U. S. Atomic Energy Commission, and the U. S. Department of the Interior, Office of Water Resources Research.

We would like to thank the many people who assisted with the tedious counting of samples in this program.

LITERATURE CITED

Bosch, H. F. and R. W. Taylor. 1973a. Distribution of the cladoceran *Podon polyphemoides* in the Chesapeake Bay. Mar. Biol. 19:161-171.

Bosch, H. F. and R. W. Taylor. 1973b. Diurnal vertical migration of an estuarine cladoceran, *Podon polyphemoides*, in the Chesapeake Bay. Mar. Biol. 19:172-181.

Heinle, D. R. 1966. Production of a calanoid copepod, *Acartia tonsa*, in the Patuxent River estuary. Chesapeake Sci. 7:59-74.

Heinle, D. R. 1969. The effects of temperature on the population dynamics of estuarine copepods. Dissertation, Univ. of Maryland, 132 p.

Jeffries, H. P. 1962a. Succession of two *Acartia* species in estuaries. Limnol. Oceanogr. 7:354-364.

Jeffries, H. P. 1962b. Salinity-space distribution of the estuarine copepod genus *Eurytemora*. Int. Rev. Gesamten Hydrobiol. 47:291-300.

Jeffries, H. P. 1964. Comparative studies on estuarine zooplankton. Limnol. Oceanogr. 9:348-358.

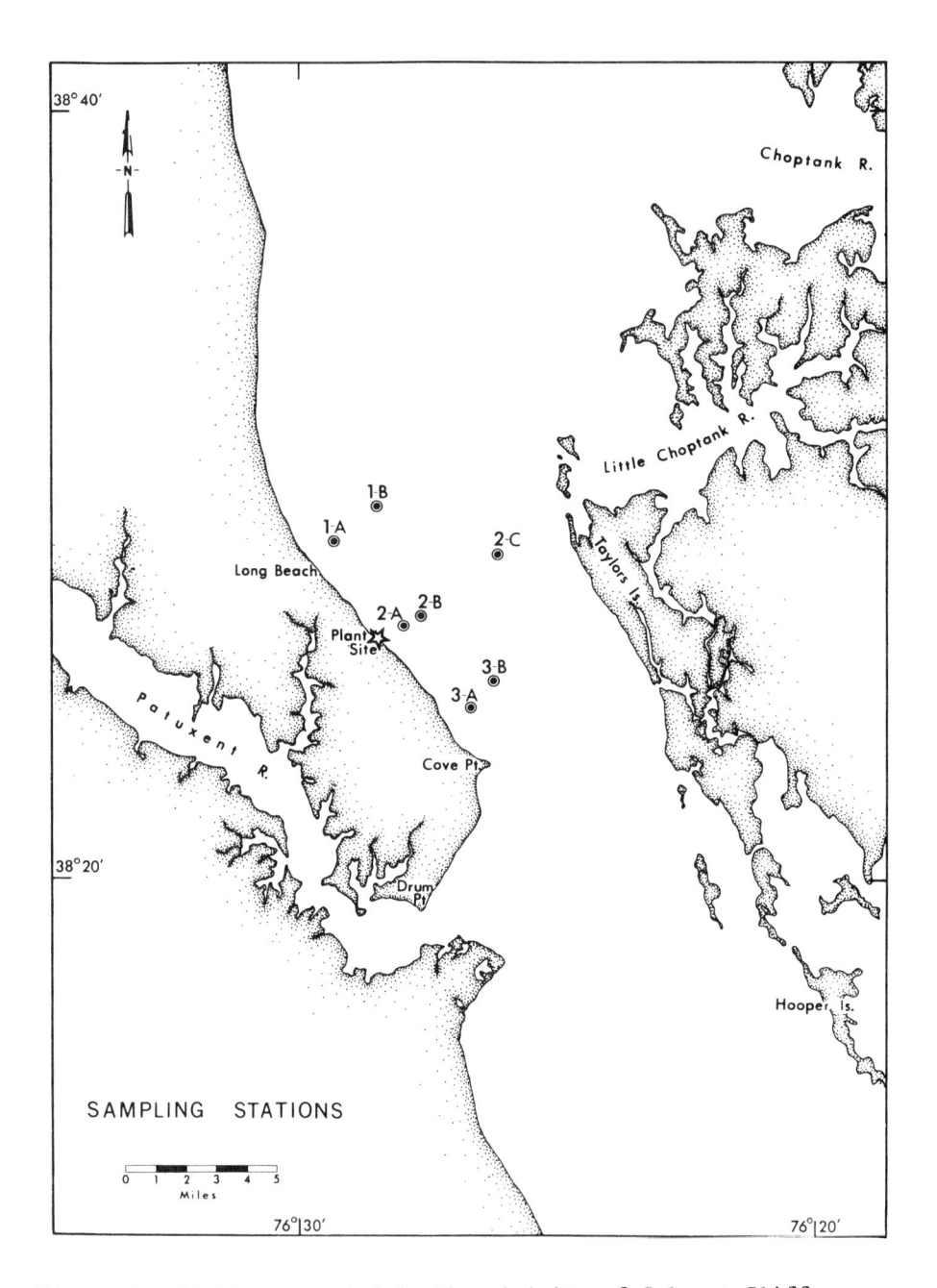

Figure 1. Stations sampled in the vicinity of Calvert Cliffs.

Station	Coordinates Latitude	Longitude	Station	Coordinates Latitude	Longitude
1-A	38°28'33"	76°28'15"	2-C	38°28'17"	76°23'27"
1-B	38°29'43"	76°26'53"	3-A	38°24'31"	76°23'33"
2-A	38°26'26"	76°26'09"	3-B	38°24'55"	76°23'00"
2-B	38°26'57"	76°25'27"			

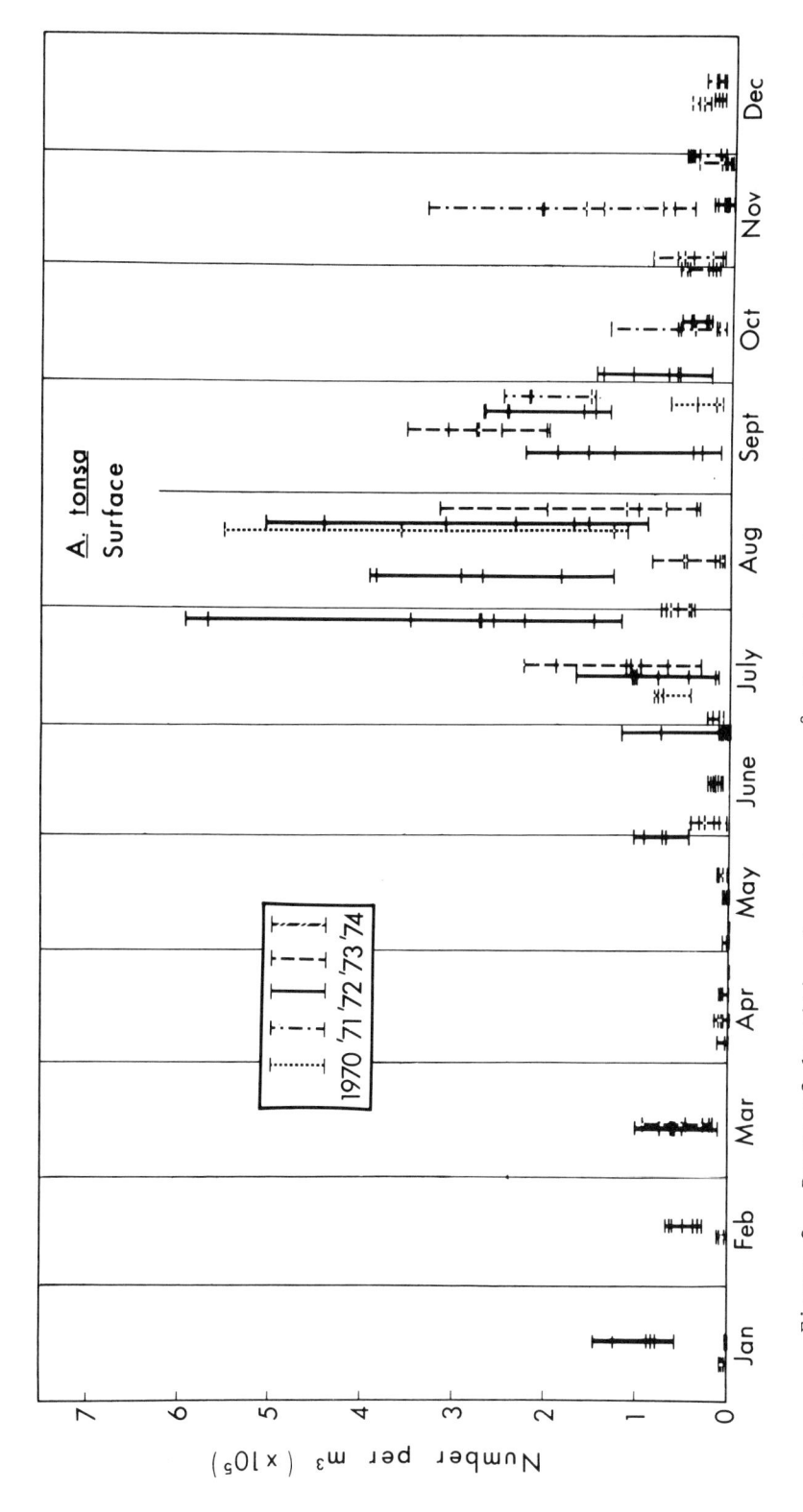

Figure 2. Range of densities in numbers per m³ of *A. tonsa* in surface (1 m) samples taken off Calvert Cliffs during 1970 through 1973. Seven stations were sampled on most occasions after September 1971. Four stations were sampled during 1970.

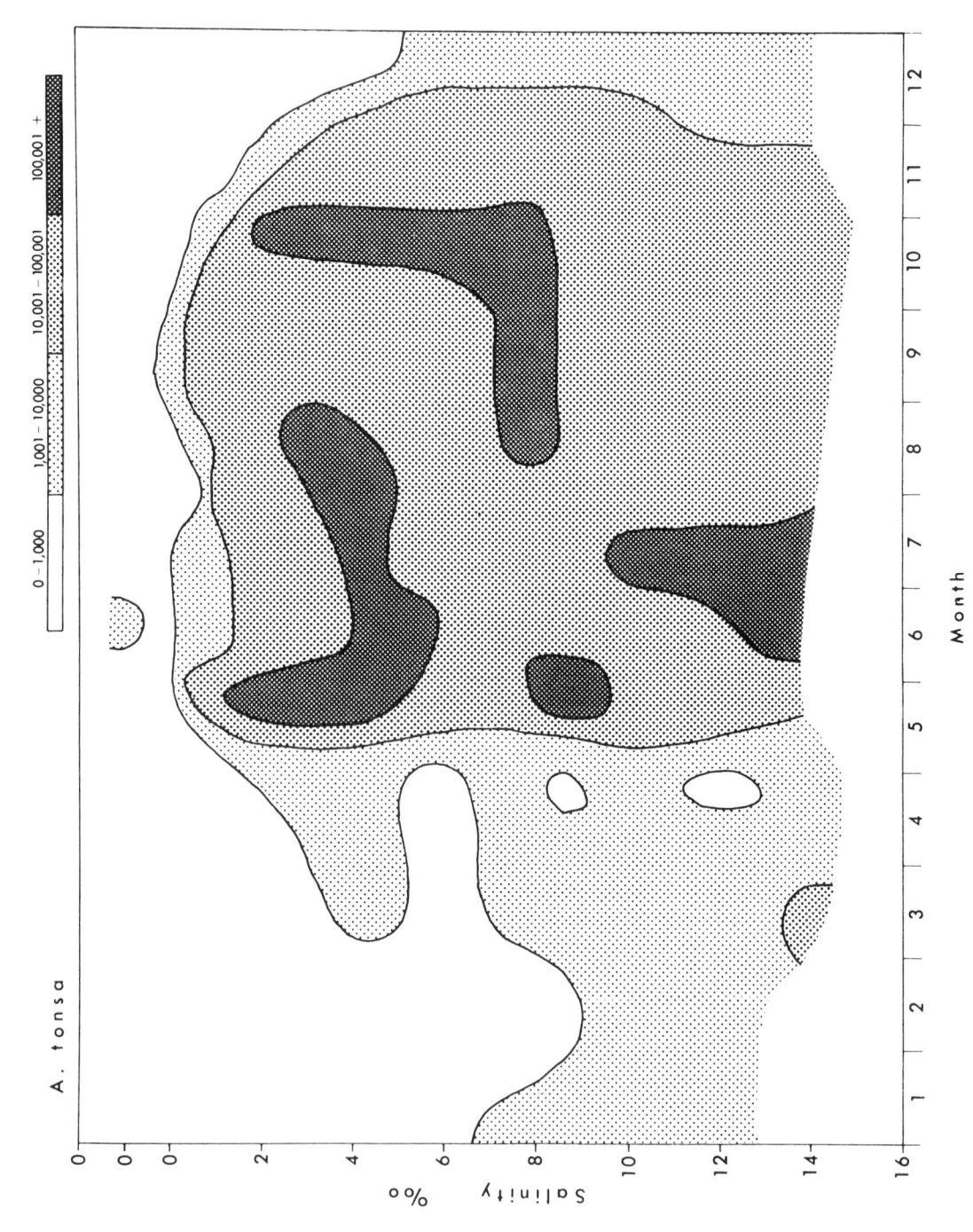

Figure 3. Time versus salinity distribution of *A. tonsa* in the Patuxent estuary during 1969.

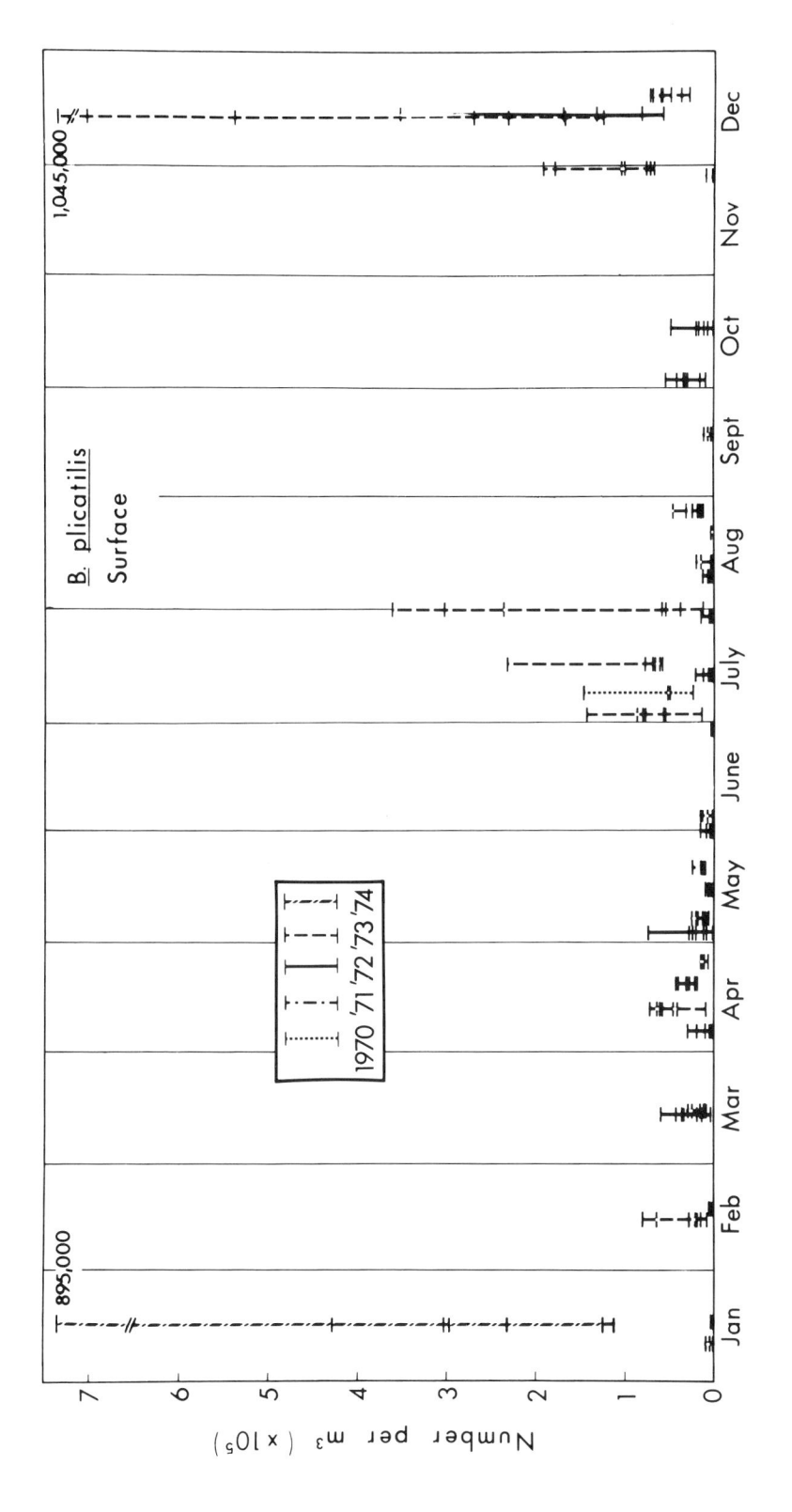

Figure 4. Range of densities in numbers per m³ of *B. plicatilis* in surface (1 m) samples.

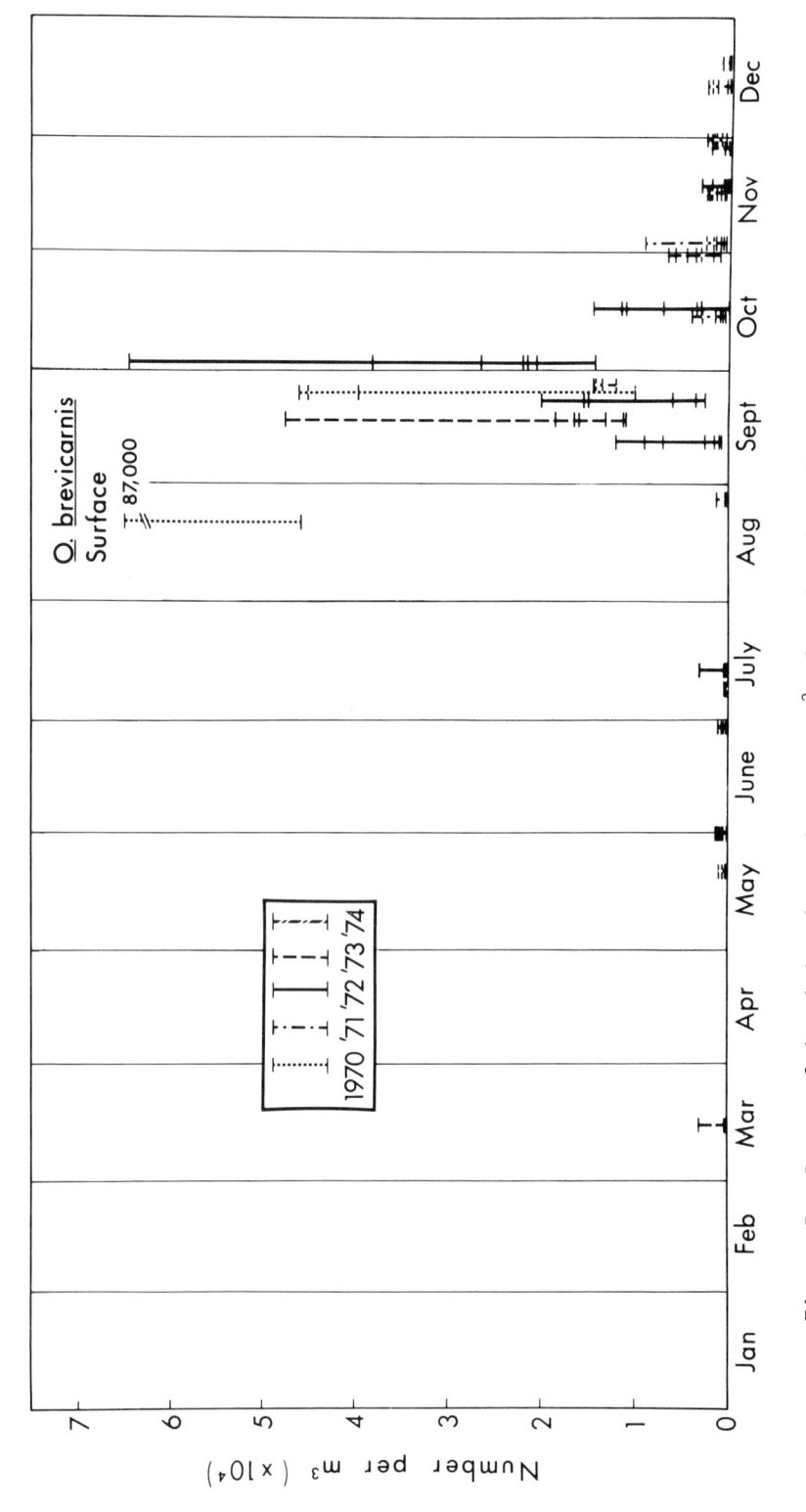

Figure 5. Range of densities in numbers per m³ of *O. brevicornis* in surface (1 m) samples.

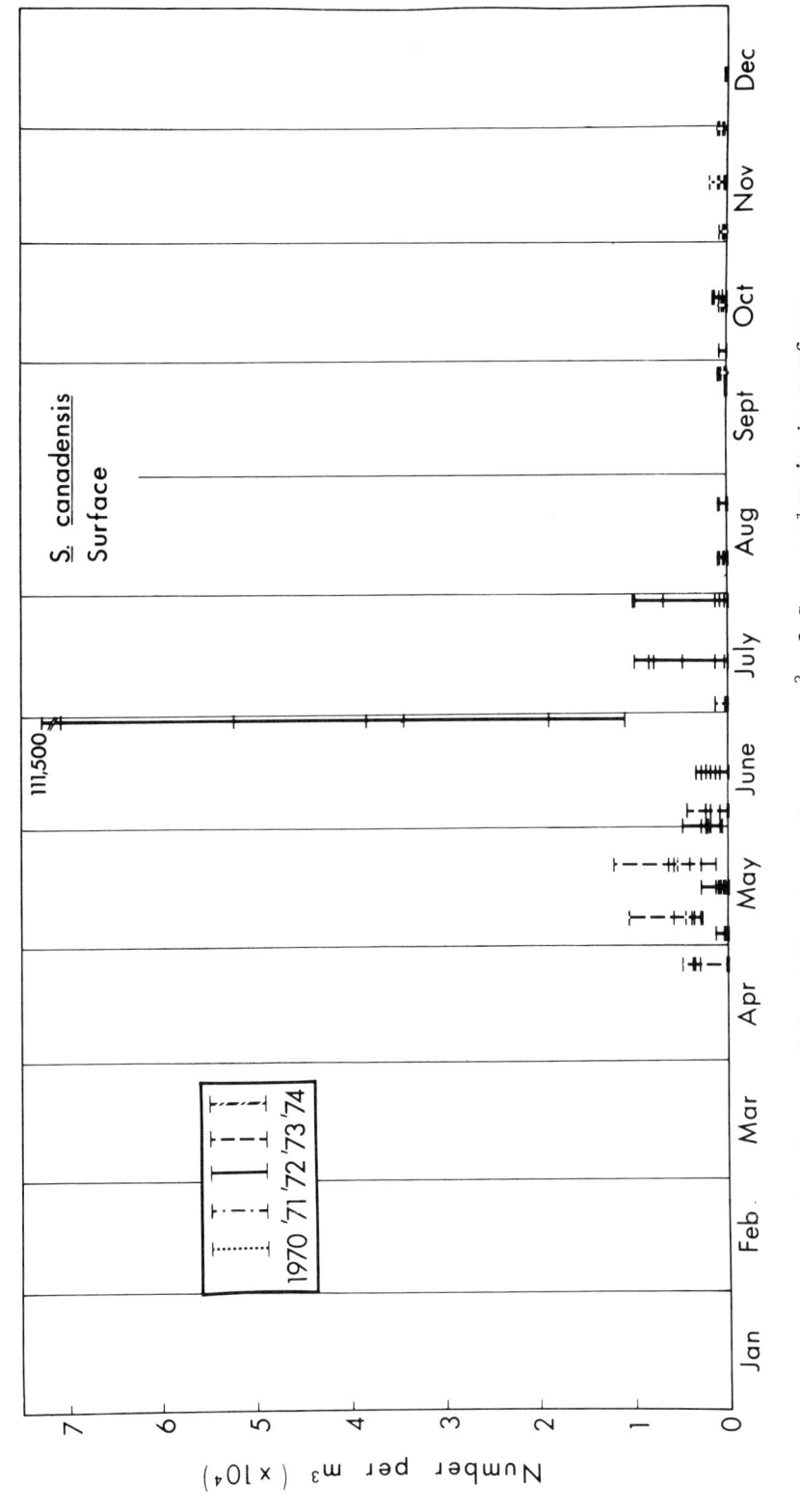

Figure 6. Range of densities in numbers per m³ of *S. canadensis* in surface (1 m) samples.

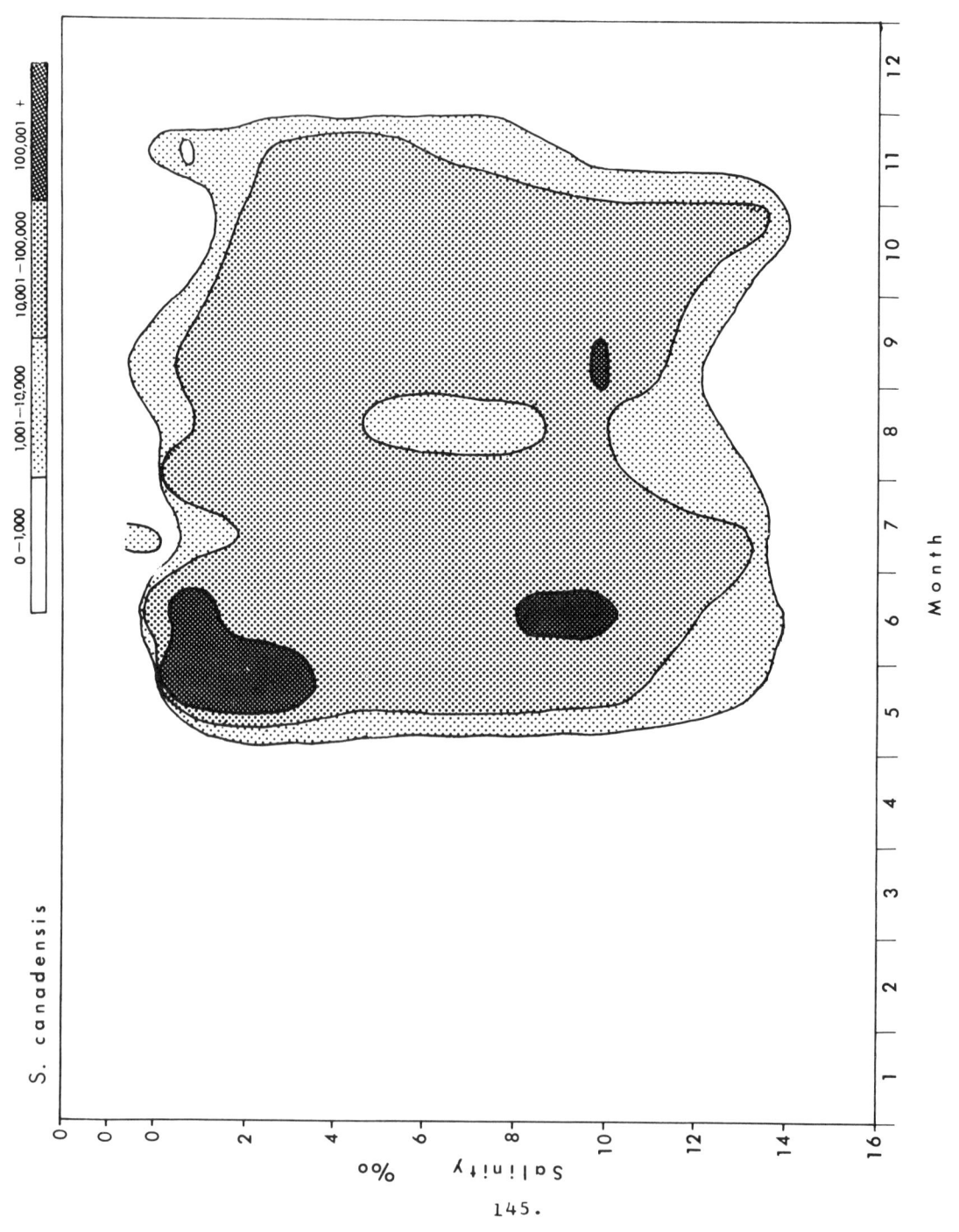

Figure 7. Time versus salinity distribution of *S. canadensis* in the Patuxent estuary during 1969.

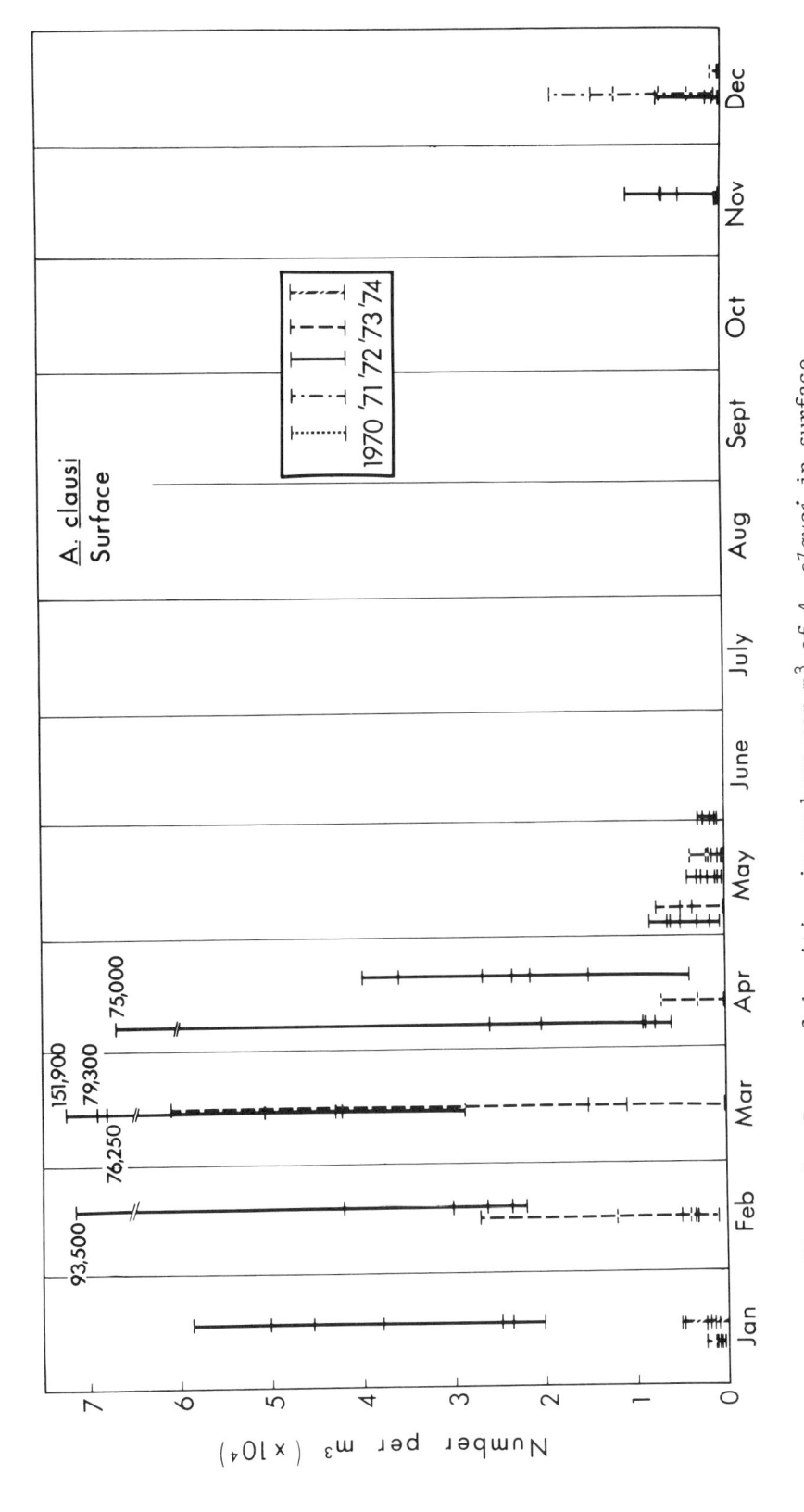

Figure 8. Range of densities in numbers per m³ of *A. clausi* in surface (1 m) samples.

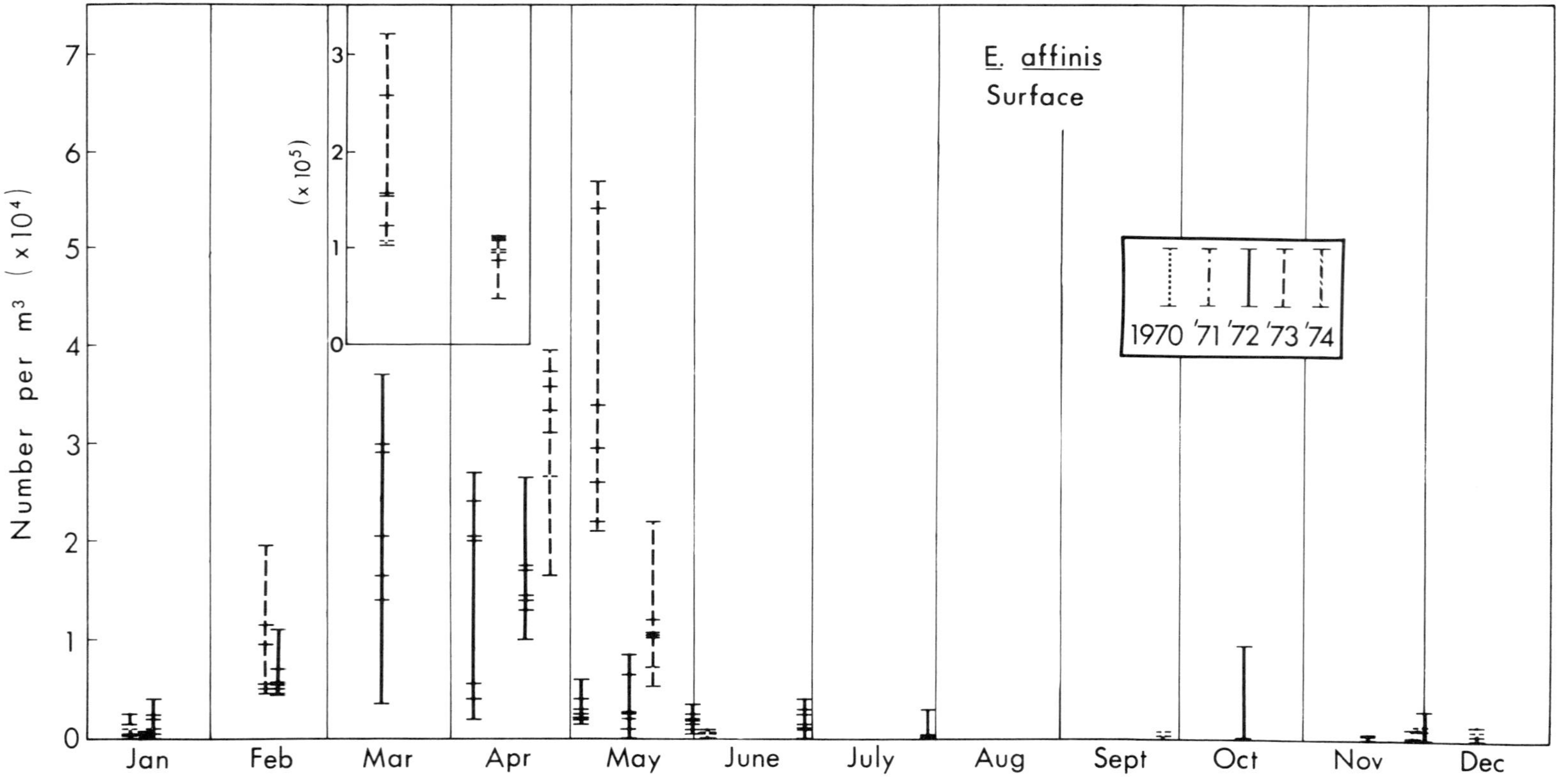

Figure 9. Range of densities in numbers per m³ of *E. affinis* in surface (1 m) samples.

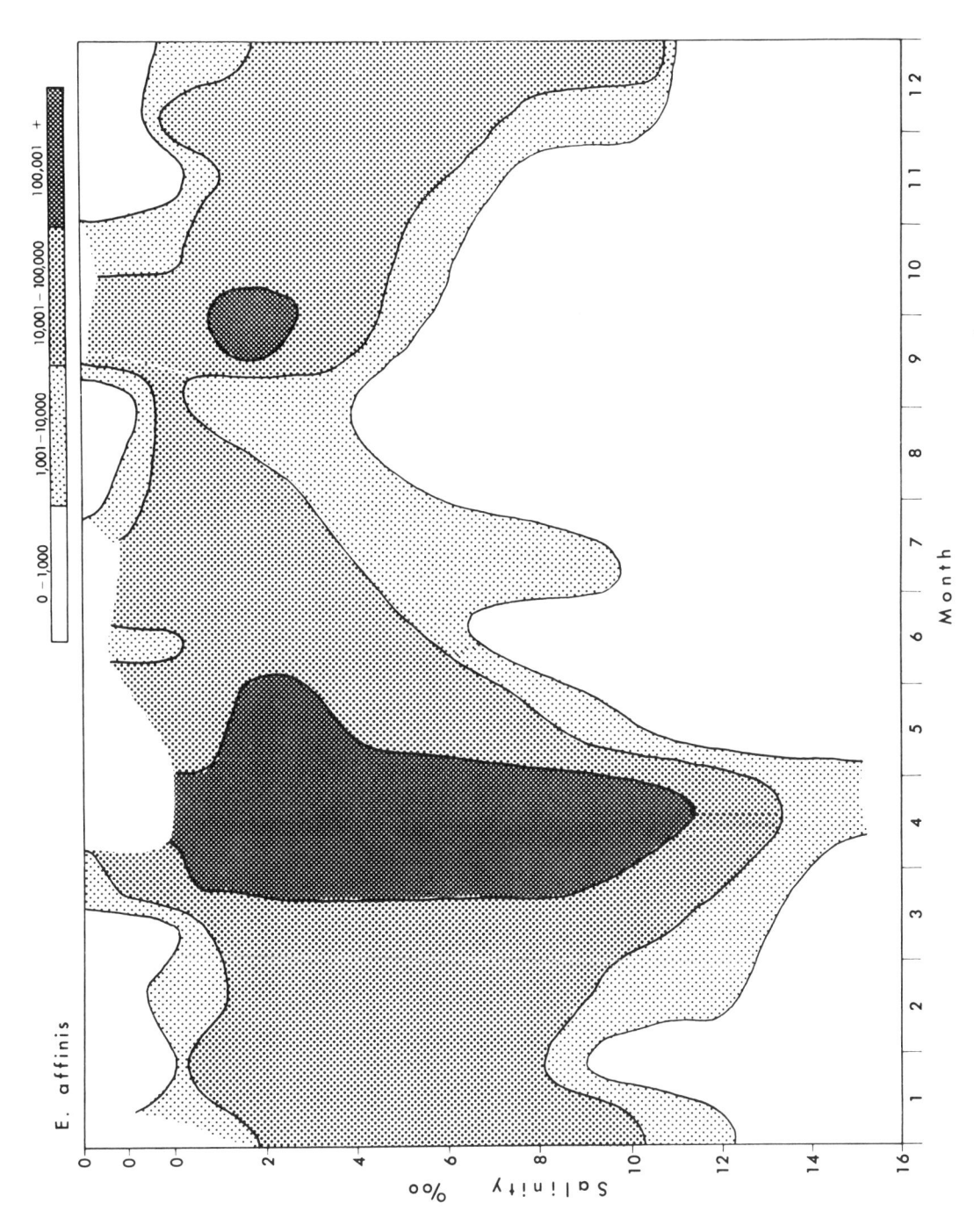

Figure 10. Time versus salinity distribution of *E. affinis* in the Patuxent estuary during 1969.

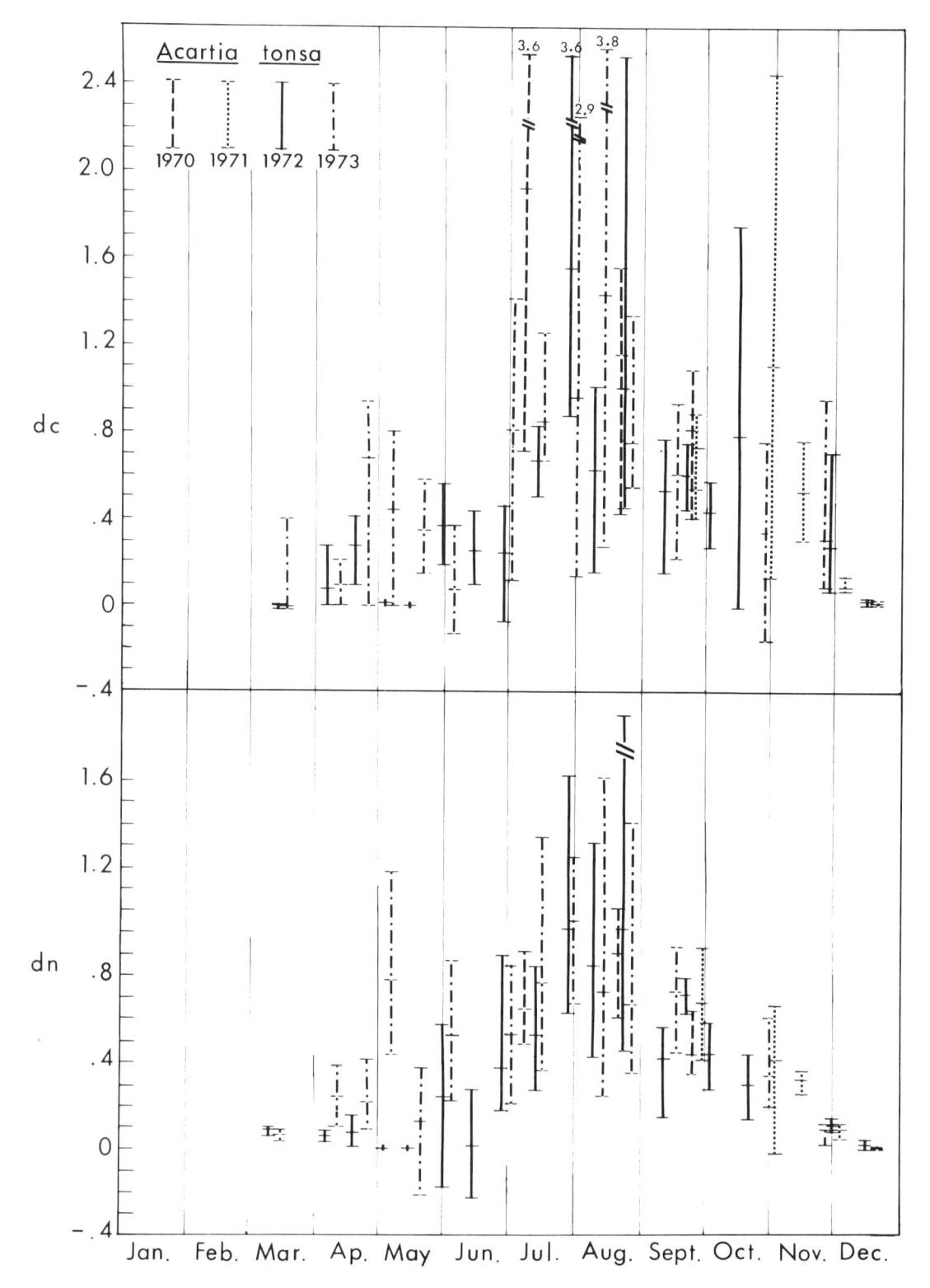

Figure 11. Apparent instantaneous death rates of nauplii (dn) and copepodids (dc) of *A. tonsa* calculated from age structures of mean populations at each station. Means and ranges are plotted.

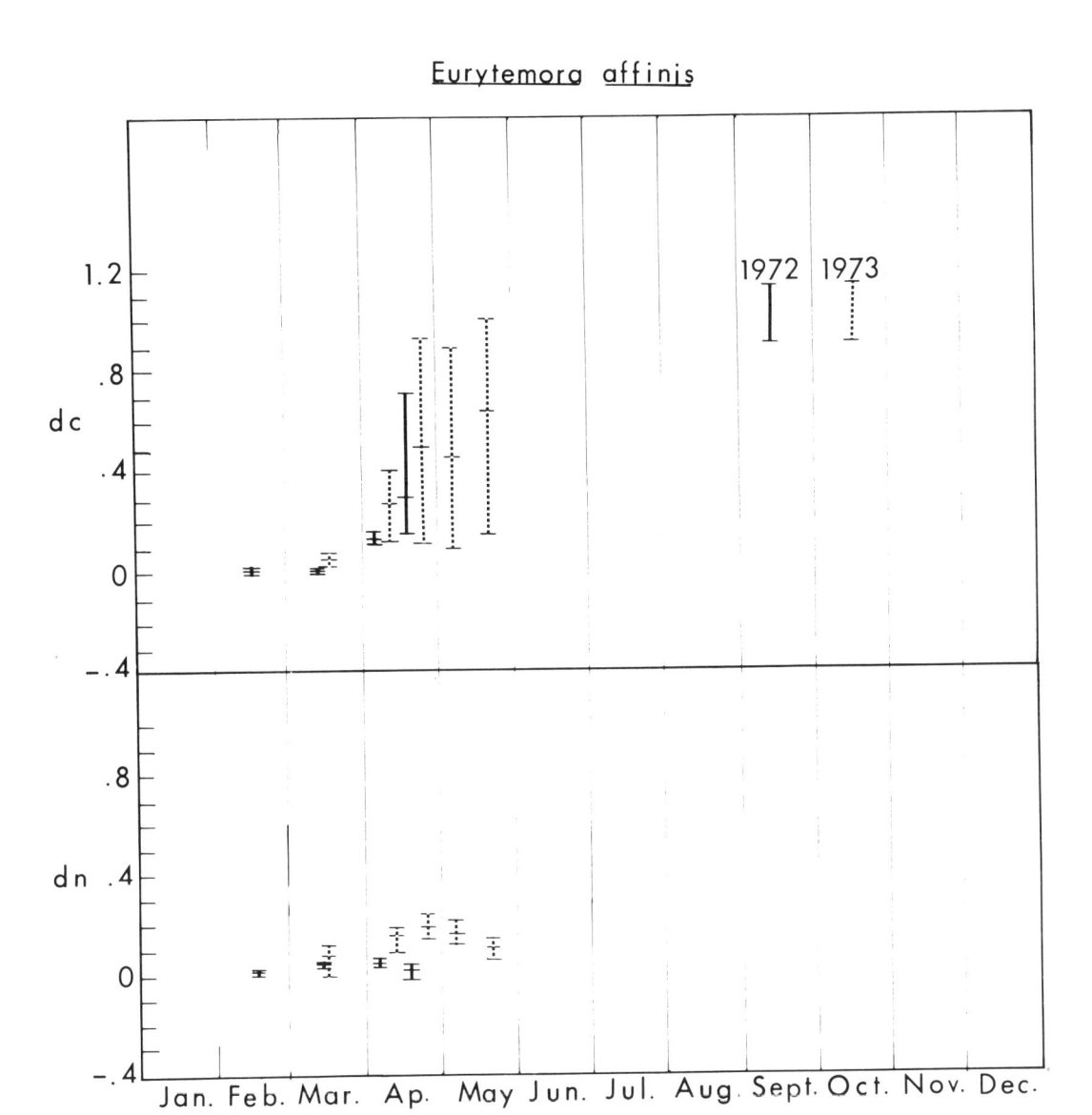

Figure 12. Apparent instantaneous death rages of nauplii (dn) and copepodids (dc) of *E. affinis* calculated from age structures of mean populations at each station. Means and ranges are plotted.

SHORT-TERM RESPONSE OF FISH TO TROPICAL STORM AGNES IN
MID-CHESAPEAKE BAY[1]

Douglas E. Ritchie, Jr.[2]

ABSTRACT

The only observed response of fishes to Tropical Storm Agnes in a mid-Chesapeake Bay study area was a one-month movement of bottom species from shallow, lowered-salinity areas to higher-salinity, deeper stations. By August, patterns of catches and salinity values were nearly equal to those before Agnes.

INTRODUCTION

The movements of fish in reaction to the freshwater input of Agnes (June 21-22, 1972) were revealed by a series of otter trawl catches off Calvert Cliffs Nuclear Power Plant (Fig. 1).

METHODS

Ten stations were occupied monthly, June through August 1972. They were paired to sample five different depths from 14 to 95 ft.

A 25 x 3-ft otter trawl with 1.5-in stretch mesh (S.M.) wings and body, 3/8-in S.M. cod inner liner, and equipped with a tickler chain, was towed on the bottom for 0.5 nautical mile, i.e., 15 minutes duration at 2 to 2.5 knots from *R. V. Aquarius*.

Each catch was processed immediately. Although several parameters were measured, only the numbers of each species will be covered here. Average depth was calculated from electronic depth recordings. Salinity was measured by an induction salinometer at surface and bottom at the end of each tow.

RESULTS

Salinity declined by as much as 6 ppt in July at most stations but by August it was similar to that of June 1972, pre-Agnes (Table 1). August salinity values for the mid-Bay are usually higher than in June (Beaven 1960; Stroup & Lynn 1963; Ritchie & Genys 1975).

Eleven species of fish were landed during these cruises (Table 1). In June, three species of fish came from two stations with depths down to 64 ft. A single hogchoker was the only bottom fish taken; other species were probably caught while the net was hauled back.

Only three stations produced white perch, a bottom species, and three mid-water species (again probably from near-surface waters) in July (Table 1). No bottom fish came from depths averaging less than 42 ft or in salinities less than 11.8 ppt.

---
[1]Contribution No. 571, U. M. Center for Environmental and Estuarine Studies.
[2]Chesapeake Biological Laboratory, Center for Environmental and Estuarine Studies, University of Maryland, Solomons, Md. 20688.

Table 1. Numbers of fish caught by fifteen-minute tows of 25-ft otter trawl during June, July, and August 1972 in mid-Chesapeake Bay off the Calvert Cliffs nuclear steam electric station. (Only stations where fish were caught are listed).

	PRE-AGNES								POST-AGNES						
	JUNE		JULY								AUGUST				
Station	112	101D	Total	101B	101C	104C	Total	112	113	101A	104A	104B	104C	104D	Total
Depth of Stations (Ft.) where fish were caught	31	64		42	68	64		27	14	20	42	47	64	95	
Salinity, ppt (Top)	9.9	10.0		5.3	5.4	5.7		8.9	8.8	8.6	9.1	8.9	9.0	8.9	
(Bottom)	11.9	17.6		5.7	11.8	11.2		9.9	9.1	10.5	13.2	15.6	18.3	19.9	
Species:															
Bay anchovy	625	72	697	6	5	1	12		1030			2			1032
Blueback herring	2		2												
Hogchoker	1		1					4	25	24				1	54
White perch					4		4		1	1					2
Alewife				1			1								
Atlantic menhaden				1			1								
Spot								3	7	43					53
American eel										2					2
Striped bass										1					1
Rough silverside											8	20			28
Oyster toadfish													1		1
TOTALS	628	72	700	8	9	1	18	7	1063	71	8	22	1	1	1173
TOTAL OF ALL SPECIES			700				18								1173
NUMBER OF SPECIES	3	1	3	3	2	1	4	2	4	5	1	2	1	1	8

In August, five demersal species and three pelagic species were taken at seven of the ten stations and over the entire range of depths. Salinity values were equal to those in June. Most catches were small in number. The species were hogchoker, spot, American eel, oyster toadfish, and white perch (demersals) and bay anchovy, rough silverside, and striped bass (pelagic species). The species present, their relative abundance, and their distribution with depth were nearly as would be expected for this area had Agnes not occurred.

SUMMARY

The effects of Agnes on fish in mid-Chesapeake Bay were of short duration. During July 1972, demersal fish apparently moved away from depths less than 40 ft. By August, when salinity returned to those of the pre-Agnes June, demersal species were taken from all depths sampled as one would expect them to be.

LITERATURE CITED

Beaven, G. F. 1960. Temperature and salinity of surface water at Solomons, Maryland. Chesapeake Sci. 1(1):2-11.

Ritchie, D. E., Jr. and J. B. Genys. 1975. Daily temperature and salinity of surface water of Patuxent River at Solomons, Maryland, based on 30 years of records (1938-1967). Chesapeake Sci. 16(2):127-133.

Stroup, E. D. and R. J. Lynn. 1963. Atlas of salinity and temperature distributions in Chesapeake Bay 1952-1961 and seasonal averages 1949-1961. Chesapeake Bay Inst. Graph. Summ. Rep. 2 (Ref. 63-1):X; 1-410.

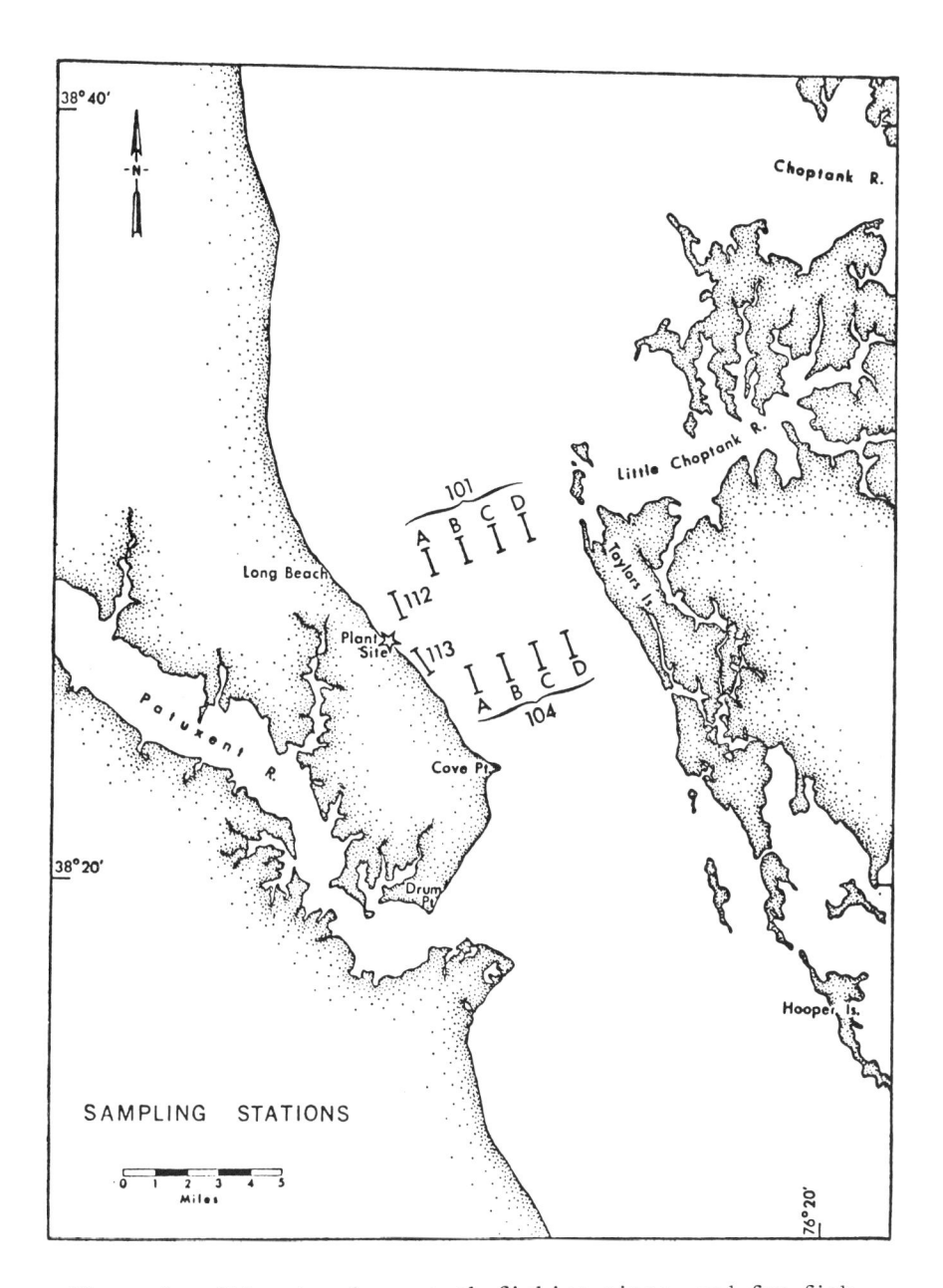

Figure 1. Otter trawl research fishing sites used for fish survey of offshore waters, Calvert Cliffs Nuclear Steam Electric Station.

THE EFFECTS OF TROPICAL STORM AGNES ON FISHES IN THE JAMES, YORK, AND RAPPAHANNOCK RIVERS OF VIRGINIA[1]

Walter J. Hoagman[2]
Woodrow L. Wilson[2]

ABSTRACT

Intensive trawl surveys during and after Tropical Storm Agnes were mounted on the James, York, and Rappahannock Rivers to measure the effects of the floodwaters on the distribution and abundance of fish. The direct effect of Agnes on the fish populations was minor and temporary. The normal zone was extended downriver. A substantial portion of the lower-river (marine) species was also displaced downstream and into Chesapeake Bay, but had returned by the follow-up surveys. No adult mortalities due to Agnes were detected. Although we know vast quantities of fish larvae and other plankton were swept into Chesapeake Bay, the overall impact on all fish appears to have been slight.

INTRODUCTION

The Department of Ichthyology of the Virginia Institute of Marine Science mounted intensive trawl surveys in the James, York, and Rappahannock Rivers during and after Agnes to determine the impact of the floodwaters on the resident and seasonal fishes.

METHODS

The initial survey took place between 28 June and 3 July 1972, the week after Agnes flooded the upper rivers and sent massive amounts of fresh water through the lower estuaries. The sampling scheme consisted of five replicate tows at three stations with a 30-ft, semi-balloon, bottom trawl (3/4-inch-mesh codend) for 7.5 minutes each. Six additional stations per river were sampled once. All stations were between the mouth of the river and just into the normal freshwater zone, with the replicate stations taken near the mouth, near the freshwater interface, and midway between. All surveys were conducted from the R/V *Langley*. River miles are given in Fig. 1.

Followup studies were made twice. Between 8 August and 7 September 1972 the sampling scheme of the initial survey was repeated to measure the recovery of the fish populations. In addition, five replicate samples were collected at Mile 39 in the James River. Between 30 October and 8 November 1972 another survey was undertaken with single tows at eight stations in the lower James, York, and Rappahannock Rivers up to Mile 36-50.

RESULTS

The results presented here represent conditions at the time of the three surveys. The entire isohaline movement during and after Agnes is covered in other sections of this volume. Fish were counted and measured individually but, for

---

[1] Contribution No. 761, Virginia Institute of Marine Science
[2] Virginia Institute of Marine Science, Gloucester Point, Va. 23062

simplicity, only average lengths are presented here. All trawling was performed in waters deeper than 7 m, thus we cannot provide data on changes in fish populations in the shoal communities. Because the fresh water ran out primarily along the surface (down to 3-6 m, mainly), we suspect the shoal fishes were affected first and most: they probably sought the deeper, more saline waters adjacent to their immediate locale. Conclusions in this report are therefore limited to the mainstream bottom community.

James River

The flood crest passed Richmond on 23 June 1972. It passed down the river as a surge, depressing salinities in the lower 30 miles within two days, with the lowest salinities reported on 28 June (Chesapeake Bay Research Council 1973). A sharp halocline was established by 28 June but the salinities rebounded in the bottom waters within 10 days. The stratified condition, with fresh water at the surface (0-8 m) and much higher salinity below, disappeared as the flow weakened. By 25 August (63 days after the crest) the salinity profile at a particular station (on low slack tide) was fairly uniform without the pronounced halocline of earlier dates. There was a net downstream displacement of approximately 6-10 miles in the bottom salinities between the worse case and the normal (Fig. 2). Oxygen was adequate for fish in the James River on every survey at the stations sampled ($O_2 \geq 4.7$ ppm).

The summer fish populations of the James River follow the typical pattern of estuarine migrants in the lower reaches (e.g. spot, *Leiostomus xanthurus*; Atlantic croaker, *Micropogon undulatus*; and weakfish, *Cynoscion regalis*) with a gradual transition to resident freshwater species in the middle to upper sections (e.g. channel and white catfish, *Ictalurus punctatus* and *I. catus*; American eel, *Anguilla rostrata*; and juvenile shad, *Alosa sapidissima*). White perch *(Morone americana)* have been partially absent from the James River since 1971 (St. Pierre & Hoagman 1974) and the striped bass *(Morone saxatilis)* has been at very low abundance (Merriner & Hoagman 1974).

The freshwater species did not move downstream appreciably with the 1 ppt isohaline. White and channel catfish were equally abundant at Mile 25 during and after Agnes; none were captured at Mile 10 during Agnes, even though the salinities had fallen to tolerable levels. Juveniles of blueback herring (*Alosa aestivalis*), American shad, and alewife *(Alosa pseudoharengus)*, which are pelagic and normally live in fresh water, were not captured at Mile 10-13 during Agnes or later. Carp *(Cyprinus carpio)* and brown bullhead *(Ictalurus nebulosus)* were not found any further downstream during Agnes than after.

Apparently the normal zones of residence for the freshwater species were maintained, even though the size of the zone had been extended downriver temporarily. The downstream extension of the freshwater zone was probably too rapid for these species to become aware of the expanded area and move into it. Since they normally live in fresh water, the additional flow provided no stimulus to leave their normal habitat.

Most species in the lower river can only tolerate particular minimum salinities. For these species, fresh water can be considered a pollutant that causes avoidance, or death if they are entrained. Being mobile, most would be expected to avoid falling salinities by moving out with the flow. Since most of these species commonly live in salinities 4 to 20 ppt in the lower rivers, the 0.5 to 5 ppt displacement (oligohaline zone) can be considered the avoidance zone.

Atlantic croaker and weakfish (grey trout) moved approximately 10 miles downstream during Agnes but had returned upriver two months later (Table 1).

Table 1. Catches and mean lengths (mm) of six major fishes captured in the James (J), York (Y), and Rappahannock (R) Rivers during and after Tropical Storm Agnes. Stations with asterisks (*) had five replicate tows made. A dash (-) indicates that no tow was made.

River and Mile	White Catfish 28 Jun - 3 Jul Number per tow	Mean length	8 Aug - 7 Sep Number per tow	Mean length	30 Oct - 8 Nov Number per tow	Mean length	Channel Catfish 28 Jun - 3 Jul Number per tow	Mean length	8 Aug - 7 Sep Number per tow	Mean length	30 Oct - 8 Nov Number per tow	Mean length
J-00	0		0		0		0		0		0	
05	0		0		0		0		0		0	
10*	0		0		-		0		0		-	
13	0		0		0		0		0		0	
19	2	95	6		1	175	1	232	2	275	1	437
25*	4	122	2	129	3	87	1	181	7	136	6	110
27	2	111	2	76	2	100	17	178	14	142	15	105
32	5	97	0		0		36	146	17	131	4	175
36	6	112	2	25	0		19	144	36	132	26	97
39*	-		8	90	-		-		48	167	-	
Y-00*	0		0		0		0		0		0	
05	0		0		0		0		0		0	
10	1	387	0		0		0		0		0	
15	3	230	0		0		0		0		0	
20	4	252	5	281	1	295	0		0		0	
25*	3	247	10	278	2	339	0		0		0	
30	2	116	8	200	51	215	0		0		0	
35	8	85	4	148	41	200	0		0		0	
40	3	214	38	86	13	13	0		3	128	5	177
50	0		2	232	8	151	0		4	166	3	153
R-00	0		0		0		0		0			
05*	0		0		0		0		0			
10	0		0		0		0		0			
15	0		0		0		0		0			
20*	1	276	0		0		0		0			
25	0		3	282	1	254	0		0			
30	3	171	27	180	0		0		0			
35*	8	89	4	136	7	153	0		0			
40	40	137	6	89	9	65	0		0			

Table 1. Cont'd.

River and Mile	Hogchoker 28 Jun - 3 Jul Number per tow	Hogchoker 28 Jun - 3 Jul Mean length	Hogchoker 8 Aug - 7 Sep Number per tow	Hogchoker 8 Aug - 7 Sep Mean length	Hogchoker 30 Oct - 8 Nov Number per tow	Hogchoker 30 Oct - 8 Nov Mean length	American Eel 28 Jun - 3 Jul Number per tow	American Eel 28 Jun - 3 Jul Mean length	American Eel 8 Aug - 7 Sep Number per tow	American Eel 8 Aug - 7 Sep Mean length	American Eel 30 Oct - 8 Nov Number per tow	American Eel 30 Oct - 8 Nov Mean length
J-00*	111	125	69	129	8	109	0		0		0	
05	339	115	54	126	94	134	0		0		0	
10	121	114	85	120	-		2	423	1	435	1	
13	85	120	38	117	49	130	3	424	4	485	3	289
19	17	96	39	90	103	120	4	321	14	290	4	439
25	2	64	6	89	56	105	1	313	8	305	0	
27	0		2	58	103	109	0		13	291	0	
32	1	51	0		6	111	0		6	302	0	
36	1	36	0		10	99	0		1	310	0	
39*	-		2	50	-		-		13	315	-	
Y-00	102	120	0		0		0		0		0	
05	52	131	1	140	7	131	0		0		0	
10	57	117	4	130	349	134	2	447	0		1	412
15	446	102	0		155	128	1	505	3	476	1	315
20	66	109	119	134	423	136	0		13	505	0	
25	30	135	98	91	54	125	2	405	4	497	6	353
30	7	95	26	72	183	117	0		1	470	5	391
35	11	53	9	77	238	116	0		0		1	410
40	6	43	27	64	520	119	2	300	0		1	420
50*	17	37	8	70	185	105	1	373	0		0	
R-00*	9	121	95	118	3	120	0		0		0	
05	42	108	52	122	1	123	1	489	2	525	0	
10	144	107	294	121	17	125	1	499	6	516	0	
15	119	105	41	122	1	110	1	461	0		0	
20	63	88	174	124	1	78	2	519	2	564	0	
25	7	90	80	111	33	118	2	578	20	417	2	431
30	7	86	76	101	352	113	3	575	36	498	3	419
35	27	42	78	77	514	109	1	593	1	467	2	638
40*	9	73	119	61	67	88	1	357	11	272	0	

Table 1. Cont'd.

River and Mile	Spot 28 Jun - 3 Jul Number per tow	Spot 28 Jun - 3 Jul Mean length	Spot 8 Aug - 7 Sep Number per tow	Spot 8 Aug - 7 Sep Mean length	Spot 30 Oct - 8 Nov Number per tow	Spot 30 Oct - 8 Nov Mean length	Atlantic Croaker 28 Jun - 3 Jul Number per tow	Atlantic Croaker 28 Jun - 3 Jul Mean length	Atlantic Croaker 8 Aug - 7 Sep Number per tow	Atlantic Croaker 8 Aug - 7 Sep Mean length	Atlantic Croaker 30 Oct - 8 Nov Number per tow	Atlantic Croaker 30 Oct - 8 Nov Mean length
J-00*	13	188	88	177	34	159	61	144	188	185	3	117
05	2	137	3	201	22	154	59	148	102	170	33	132
10*	8	161	33	170	-		25	152	57	190	-	
13	61	151	12	180	32	141	39	143	21	152	44	119
19	0		4	127	29	130	10	87	219	143	28	99
25*	0		0		0		1	79	16	124	9	68
27	0		0		0		0		7	97	1	22
32	0		0		0		1	66	1	75	2	27
36	0		0		0		0		14	53	7	39
39*	-		0		-		-		6	94	-	
Y-00*	73	173	2	176	2	151	129	139	1	142	1	97
05	7	167	9	186	11	140	68	136	0		4	86
10*	4	168	14	138	2	131	14	121	5	178	13	128
15	30	113	8	127	4	142	38	128	0		8	96
20	1	115	190	124	2	152	51	128	26	180	39	114
25*	27	89	31	120	1	124	607	112	10	166	22	105
30	0		5	112	9	134	0		55	80	60	90
35	0		0		0		0		23	60	12	62
40	0		0		0		0		0		0	
50	0		0		0		0		0		0	
R-00	84	154	112	139	39	133	0		26	133	35	87
05*	149	111	599	137	119	137	1	146	7	115	7	95
10	34	98	310	137	302	128	0		2	184	16	126
15	125	99	50	125	9	151	21	135	0		5	85
20*	29	99	112	127	36	131	5	123	9	150	10	74
25	7	93	9	137	251	132	0		10	161	15	108
30	28	94	2	144	115	139	5	103	22	178	26	134
35*	0		35	123	23	123	0		63	108	5	91
40	0		4	87	4	124	0		36	80	10	60

Table 1. Cont'd.

River and Mile	Weakfish 28 Jun - 3 Jul Number per tow	Mean length	8 Aug - 7 Sep Number per tow	Mean length	30 Oct - 8 Nov Number per tow	Mean length
J-00*	4	170	54	195	28	117
05	5	162	10	67	45	121
10*	2	164	21	85	0	
13	0		12	106	9	134
19	0		4	156	0	
25*	0		7	31	0	
27	0		1	27	0	
32			4	28	0	
36			3	28	0	
39			0			
Y-00*	6	193	1	76	0	
05	1	189	0		0	
10*	0		1	91	0	
15	1	200	0		3	174
20	0		7	89	1	105
25*	0		3	202	0	
30	0		27	88	0	
35	0		0	91	0	
40	0		0		0	
50	0		0		0	
R-00	14	206	71	99	2	122
05*	1	167	22	133	3	124
10	0		3	73	124	111
15	1	174	7	199	3	106
20*	1	38	15	107	1	147
25	0		3	137	8	151
30	0		12	126	28	123
35*	0		27	107	3	142
40	0		0		0	

At Mile 25 the Agnes survey captured none of these species, although two months later 80 croakers and 34 weakfish were taken in five replicate tows. Maximum abundance of these fish was within the lower 10 miles during both surveys, but the fresh water at Mile 19 on 18 and 29 June (Fig. 2) may account for their absence upriver. Hogchoker were displaced but not to the extent of croaker or weakfish. Silver perch were not captured at Mile 10 during Agnes but were present up to Mile 19 during the first follow-up survey. Between Mile 10 and the mouth (Mile 0) no detectable displacement of fishes occurred.

By 2 November 1972, the distribution of croaker, spot, hogchoker, silver perch, and bay anchovy was essentially the same as during the follow-up survey in August. The freshwater species were distributed as in previous surveys, with channel catfish and white catfish first appearing at Mile 19 and increasing in abundance upriver.

At Miles 0 and 10 on 28 and 29 June there were 107 spot, 431 croaker, and 27 weakfish captured with the 10 tows. On 8-10 August at the same stations there were 604 spot, 1,223 croaker and 376 weakfish captured. This represents a fourfold increase of these species after the flood in the lower river. Crowding in the lower river during Agnes would have shown a reverse pattern. Time of year and variability at stations may account for part of this increase, but it seems a substantial portion of the populations was actually "moved" out of the James during Agnes, and not merely crowded together downstream. We did not sample outside of the lower James River during Agnes, or the follow-ups, so we cannot demonstrate this effect except circumstantially. The mean lengths during Agnes, and from the follow-up surveys, indicate no substantial recruitment of the youngest year-classes to the river populations.

The flood waters, being of lesser density, overlaid the denser water of the lower James River. The sharp halocline (in vertical profile) continued to at least 19 July (Chesapeake Bay Research Council 1973). This fresher water affected the shoals first and would be expected to move the euryhaline fishes into deeper and saltier waters. The trawl catches during Agnes would have been greater at equivalent stations if this were the only displacement, because during the follow-up surveys the shoals had returned to normal salinity and the fishes would be redispersed. Since the Agnes catches were far less, and spot, croaker, and weakfish were downstream, it seems the flood water not only caused the fish to move downstream, but additionally caused a pronounced movement out of the river.

York River

The York River with its much smaller watershed did not have the equivalent freshwater input as the James River. However, being much smaller overall, the proportional impact was the same as or greater than in the James River. The initial flood surge displaced the 1 to 5 ppt isohaline to the maximum, eight days after the crest passed the fall line on 23-26 June 1972. We trawled the York on 29 and 30 June. At this time the fresh water had diluted the entire river and the normal bottom salinities were displaced approximately 12 miles downriver (Fig. 3). Vertical stratification, while evident, was not as intense as in the James River.

The fish populations in the York River reacted similarly to those in the James River. The freshwater species were captured somewhat lower in the river than normal, as shown by the distribution of white catfish and hogchoker (Table 1). There was no mass movement detectable. The euryhaline species in the lower salinity zones between Miles 15 and 28 moved downriver in response to the decreased salinities.

Large numbers of croaker and spot were captured at Mile 25 during Agnes and again during the follow-up of 21-23 August. Their average lengths, however, show that it was only the very young that remained in the lowered salinity after Agnes. At Mile 25 the spot were 89 mm, and the croaker 112 mm. Two months later the respective length were 120 mm and 166 mm, too much for growth alone. The average size for both croaker and spot increased downriver during Agnes (Table 1), and it is these fish that probably repopulated the middle York River over the following two months. Weakfish were not captured above Mile 15 during Agnes, but were fairly abundant between Miles 20 and 35 two months later.

Catches at Miles 0 and 10 during 21-23 August must be omitted from evidence used to demonstrate displacement because the oxygen was nearly depleted near bottom. On 3-8 November oxygen was normal at these stations for mid-fall and the fish were again present.

Rappahannock River

The "worst case" condition occurred on 10 July 1972, 18 days after Agnes crested at the fall line. The 1 ppt isohaline (at low slack water) intersected bottom at Mile 34, 5 ppt at Mile 23, and 8 ppt at Mile 6 (Chesapeake Bay Research Council 1973). The initial displacement occurred 2 days after the flood crested at the fall line on 22 June. A layer of extremely fresh water 5-10 m deep established a pronounced halocline in the lower river, with fresh water (surface readings) extending below Mile 15. Recovery was slowed by a 10 m sill which blocked the mouth to high salinity bay water.

Salinity data collected during bottom trawling show there was fresh water at Mile 35 several days after Agnes (Fig. 9). Before Agnes and during 6-7 September, the salinity was about 2.5 ppt at Mile 35, depending on tide. During the first follow-up, 1 ppt extended above Mile 40. Oxygen was low (2-3 ppm during Agnes and the first follow-up between Miles 5 and 25, but spot, hogchocker, and weakfish were still abundant.

The fish catches at Mile 35 during Agnes, contained few of the euryhaline varieties such as spot, croaker, and weakfish; but after 2 months they had repopulated the area up to Mile 40 (Table 1). As in the James River, the overall catch during the effects of Agnes was much less than during the follow-up surveys, suggesting that they were displaced out of the river temporarily. Average lengths indicated no tendency for a size selective migration to the lower river, as found in the York. Eels and hogchokers were found at most stations during the two major surveys.

The lower river contained fair quantities of all expected species during the 30-31 October survey (Table 1). These fish were distributed nearly uniformly up to Mile 30 where the bottom salinity was 5.3 ppt. The salinity fell to 0 ppt at mile 40 and the euryhaline species declined in abundance.

CONCLUSIONS AND DISCUSSION

In all rivers sampled, the yearling and older fishes that normally occupy salinities 3 ppt and higher were displaced downstream by the flood waters of Agnes. This displacement was 8-14 miles and extended throughout the river, apparently resulting in a portion of the population being "forced" out of the rivers. Increased concentration of fish in the lower portions of the rivers was not detected.

The direct effect of the Agnes flood waters on the fish populations seems to have been minor and temporary. Dead or sick fish were not captured and no fish kills were apparent from surface observations. Freshwater species moved downstream somewhat but not to the extent possible based on the changed salinity structure. The shoal-water fish probably moved first into deeper water, then downstream as the channel water became fresher.

The effects on larval fish have not yet been determined. A special study using suspended plankton nets in mainstream during Agnes (Hoagman, unpublished data) showed large quantities of fish larvae and other plankton were swept out of the James and Rappahannock Rivers into Chesapeake Bay. When these samples are processed, a better assessment of the indirect effects on the nurseries and direct effects on the larvae will be more clearly known.

When Hurricane Camille flooded the James River in late August 1969 the fish reacted as in this study. Camille caused the James to crest at 24.9 ft in Richmond, 11.6 ft less than Agnes. The maximum discharge was 282,000 cfs for Camille compared to 319,000 cfs for Agnes.

VIMS made trawl surveys of the James River before, during, and after Camille. The survey used surface, bottom, and midwater trawls from Mile 0 to Mile 80 near Richmond. Replicate bottom tows were not taken, and Miles 10, 15, and 20 were not sampled during the flood. The surface-trawl catches of juvenile alewife, blueback herring, and American shad showed a pronounced downstream movement of these pelagic species. Before the flood they were concentrated between Miles 46-65, but during the flood they were most abundant between Mile 25 and Mile 40 (St. Pierre et al. 1970).

Displacement of the lower river species between Miles 6 and 24 was unproven for Camille because these stations were not occupied. Three to four weeks after Camille, the fish populations of the central James returned to the pre-flood distribution, but the catches were less at 8 bottom and 16 midwater stations over a 50 mile stretch of river.

Of two other reports on the effects of hurricanes on fishes, Hubbs (1962) considered that the small pools he investigated in Texas were drastically changed; whereas Tabb and Jones (1962) believe no permanent damage was done to the fish populations in north Florida Bay by Hurricane Donna. The temporary effects of Donna were widespread and the changes in environment caused considerable fish movement.

For the freshwater and euryhaline species, it appears no real damage (i.e., reduction in abundance due to death) was done to the stocks because of Agnes. The temporary displacement had little effect on the commercial and sportfishing activities of the lower rivers. The estuarine nursery grounds were repopulated within several weeks. Unless the ichthyoplankton were seriously affected, the overall impact of Agnes on the fish populations in Virginia seems to have been negligible.

LITERATURE CITED

Chesapeake Bay Research Council. 1973. Effects of Hurricane Agnes on the environment and organisms of Chesapeake Bay. A report to the Philadelphia District U. S. Army Corps of Engrs. (Aven M. Andersen, Coordinator), Chesapeake Bay Inst. Contrib. 187, Nat. Resour. Inst. Contrib. 529, Va. Inst. Mar. Sci. Spec. Rep. Appl. Mar. Sci. & Ocean Engr. 29, 172 p.

Hubbs, C. 1962. Effects of a hurricane on the fish fauna of a coastal pool and drainage ditch. Texas J. Sci. 24(3):289-296.

Merriner, J. V. and W. J. Hoagman. 1974. Feasibility of increasing striped bass populations by stocking of underutilized nursery grounds. Final Rep., Virginia AFS-6-3, Anadromous Fish Act, Bass (To June 1973), 77 p.

St. Pierre, R., W. J. Davis, and E. J. Warinner. 1970. Effects of flooding on fish populations in the James River. VIMS Manuscript, 16 p.

St. Pierre, R. and W. J. Hoagman. 1974. Drastic reduction in the white perch (*Morone americana*) population in the James River, Virginia. Chesapeake Sci. 16(3):192-197.

Tabb, D. C. and A. C. Jones. 1962. Effect of Hurricane Donna on the aquatic fauna of North Florida Bay. Trans. Am. Fish. Soc. 91:375-378.

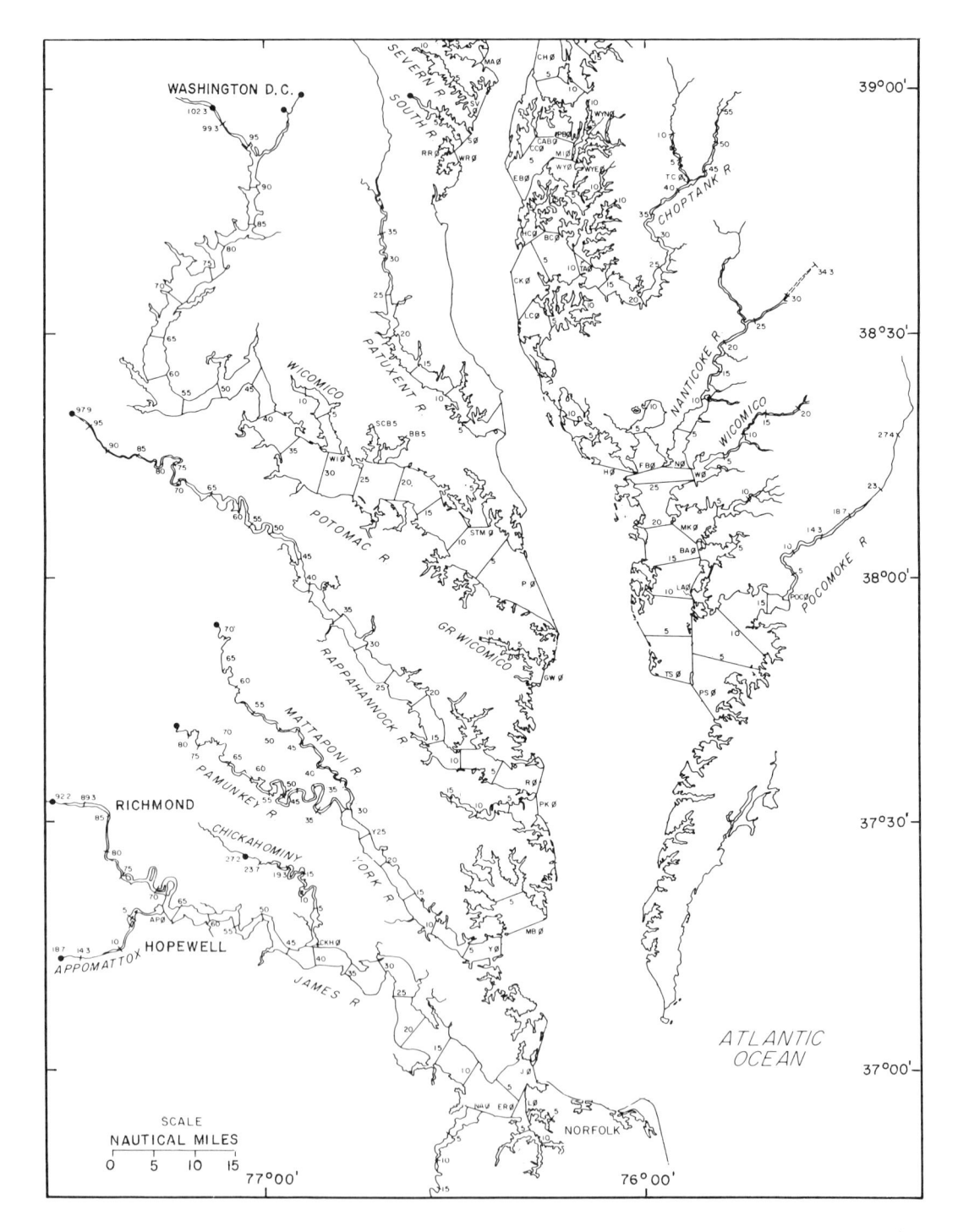

Figure 1. Locator map of study area with river miles indicated.

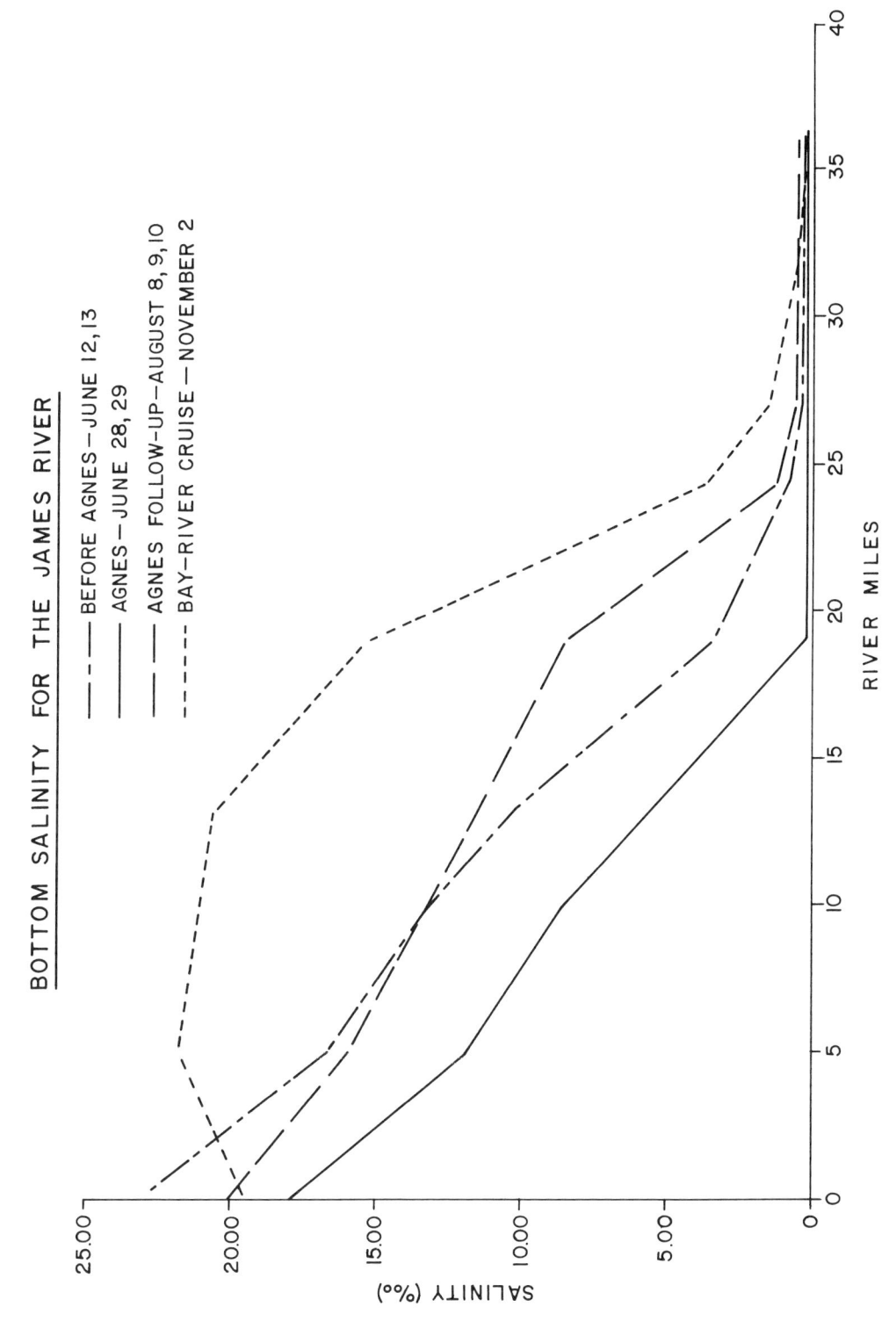

Figure 2. James River bottom salinity during trawling operations before, during, and after Agnes. Mainstream only.

Figure 3. York River bottom salinity during trawling operations before, during, and after Agnes. Mainstream only.

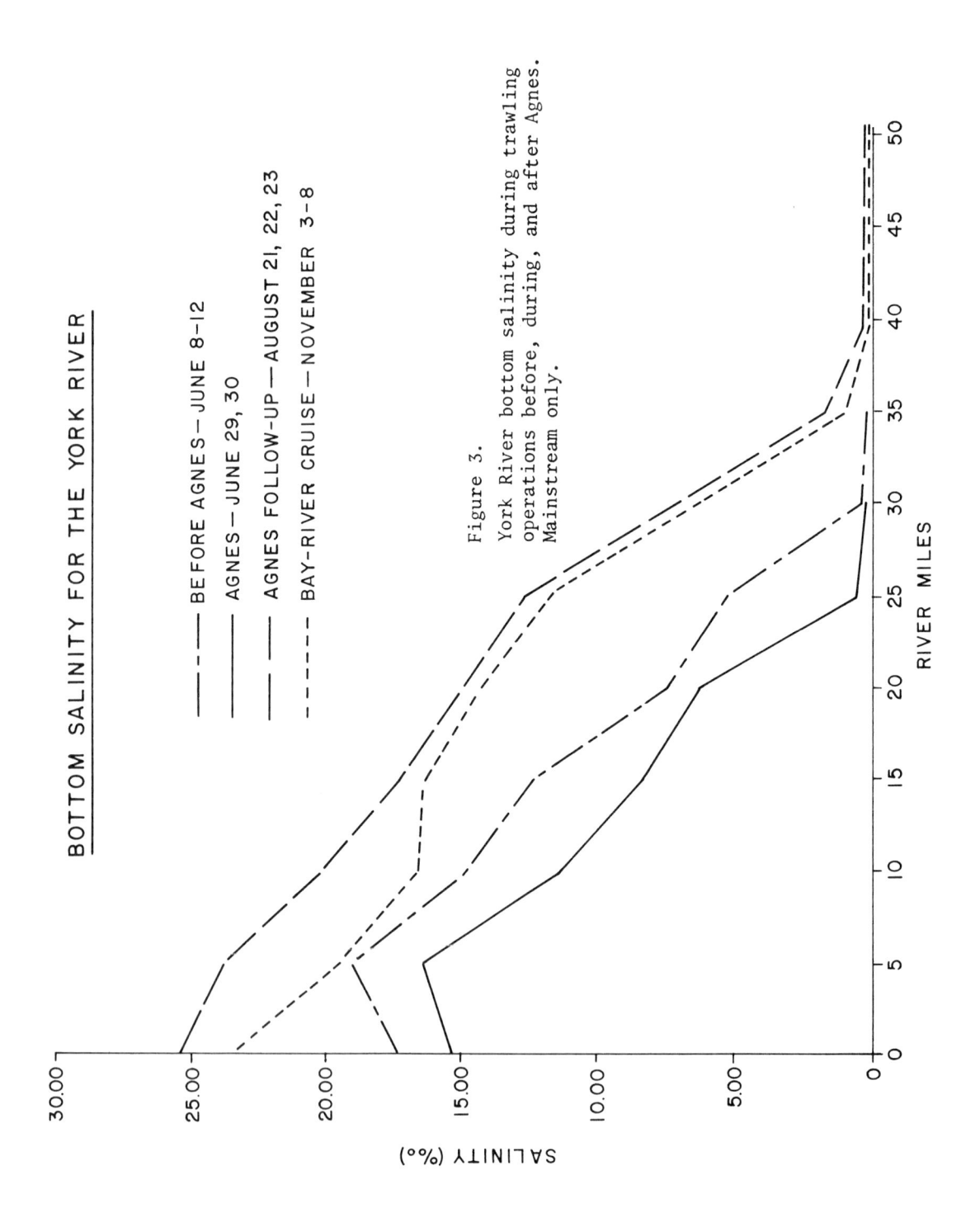

BOTTOM SALINITY FOR THE RAPPAHANNOCK RIVER

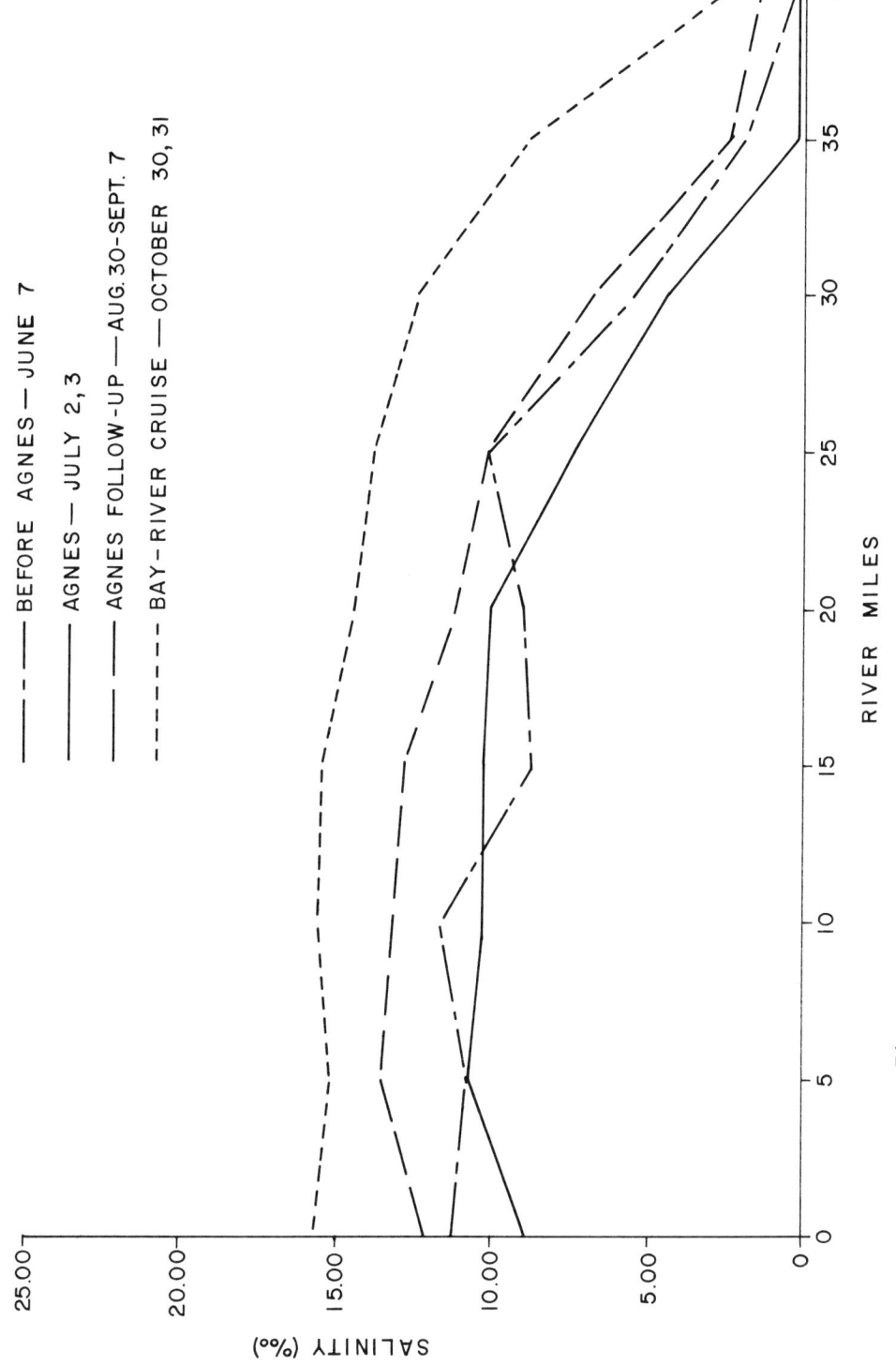

――――― BEFORE AGNES — JUNE 7
――― AGNES — JULY 2, 3
――――― AGNES FOLLOW-UP — AUG. 30-SEPT. 7
------- BAY-RIVER CRUISE — OCTOBER 30, 31

Figure 4. Rappahannock River bottom salinity during trawling operations before, during, and after Agnes. Mainstream only.

MORTALITIES CAUSED BY TROPICAL STORM AGNES
TO CLAMS AND OYSTERS IN THE RHODE RIVER
AREA OF CHESAPEAKE BAY

Robert L. Cory[1]
J. Michael Redding[2]

ABSTRACT

Post-Agnes surveys of clams and oysters in the Rhode, West, and South Rivers and an adjacent area of Chesapeake Bay showed that all species of mollusks that were sampled had suffered significant mortalities from the combined effects of high temperature and low salinities.
Mortalities ranged from 100% for soft-shell clams, *Mya arenaria*, to 6% for oysters, *Crassostrea virginica*. Other species sampled were *Brachidontes recurvus*, *Macoma balthica*, *M. phenas*, *Mulinia lateralis*, *Rangia cuneata*, and *Tagelus plebeius*.
Even though the samples indicated generally high mortalities, most species showed evidence of recovery one year later, and within two years commercial harvest of the soft-shell clams had resumed. Undoubtedly, the capability of the tributary systems to modify the influence of Chesapeake Bay was helpful in maintaining populations of mollusks in the Upper Bay.

INTRODUCTION

The sudden influx of fresh water into Chesapeake Bay, from Agnes in combination with a prolonged period of hot July weather, produced a period of stress on the Upper Bay's biota. Consequently, during late July and early August 1972 a reconnaissance survey was made of the clam population in the Rhode and West Rivers and adjacent bay and of the oyster populations in the same areas as well as in the South River. After finding no live specimens of the soft-shell clam, *Mya arenaria*, it was decided that a "follow up" survey should be conducted in 1973 to evaluate their degree of recovery. The oyster, *Crassostrea virginica*, appeared to suffer only slight mortality, so a second survey of this species was not conducted.

LOCATIONS AND METHODS

The Rhode, West, and South Rivers are located on the northwestern shore of Chesapeake Bay, about 11.2 km (7 miles) south of Annapolis, Md. Though called rivers they are better described as extensions of the Chesapeake Bay. Their water quality, particularly in the bayward portions, is chiefly influenced by exchange with adjacent bay waters.

---

[1] U. S. Geological Survey, Chesapeake Bay Center for Environmental Studies, Smithsonian Institution, Rt. 4., Box 622, Edgewater, Md. 21037
[2] Chesapeake Bay Center for Environmental Studies, Smithsonian Institution, Rt. 4, Box 622, Edgewater, Md. 21037

Clams

The soft clam investigation in 1972 was conducted with a commercial hydraulic clam dredge owned by Mr. Vernon Wilde, Galesville, Md. Samples were dredged from the 18 stations shown in Fig. 1. The dredging head cut an 86.0 cm (34") trench and the conveyor belt had a mesh size of 3.8 (1½"). The 1973 collections were made with the Chesapeake Biological Laboratory's clam dredge boat, *Venus*, with a dredge head of 91 cm (36") width and a belt mesh of 0.95 cm (3/8"). Both rigs were capable of sampling water depths from 0.6 to about 2.8 meters (2 to 7 ft.); the main difference between the two was in the size of the belt mesh; small clams passed through the mesh of the commercial dredge.

During the 1972 clam survey, a sample consisted of one-half bushel of shells, clams, etc. collected as it fell from the moving belt. The sample was not quantitative in terms of bottom area covered. In the 1973 survey, counts were based on quantitative collections from dredged areas estimated to be 5.55 square meters (60 sq. ft.) each. The area sampled was determined by driving a pole in the bottom near the bow of the moving vessel and then dredge and pole were removed after the boat had advanced 6.1 m (20 ft.). The distance covered was then multiplied by the width of the dredge head to determine the area sampled. Counts and length measurements were made on live and recently dead animals (termed "boxes"). The length measurements were made on the long axis of the shell regardless of hinge position.

The clam investigations were conducted primarily to assess the degree of a reported mortality to the soft-shell clam *M. arenaria*. The capture and recording of other species was a by-product of this effort.

In 1972, the Baltic clam, *Macoma balthica*, a common noncommercial clam in this area, was collected by sifting through 0.5 bushel samples of mud clumps taken from the soft bottoms found inside Rhode and West Rivers. On the harder, sandier bottoms this small clam fell through the mesh of the commercial dredge. In 1973 it was retained on the small-mesh dredge used.

Oysters

The 1972 oyster survey (14 and 15 August) sampled 21 stations in the Rhode, West, and South Rivers, and in the adjacent Chesapeake Bay (Fig. 2). Collections were made from a commerical dredge boat using a standard oyster dredge lined with a 13 mm (0.5 in) mesh net of Nytex nylon. Each station was sampled by taking as many "licks" with the dredge as necessary to fill a 0.5-bushel container with oysters and shells.

In addition to counts of live and dead oysters, a few live oysters from each station were opened for visual inspection of their general appearance and the state of their gonads.

RESULTS

Clam Surveys

The results of the July 1972 clam survey are shown in Table 1. The figures are qualitative and only indicate the degree of mortality suffered by the adults of the common mollusks of these waters.

Table 1. Ratios of live to dead mollusks collected on the 1972 clam survey.

Species	STATIONS										Total	Per Cent Dead
	1	2	3	4	5	6	7	8	9	10		
Mya arenaria[1]	0:0	0:0	0:162	0:21	0:87	0:13	0:88	0:47	0:155	0:26	599	100
Macoma balthica[2]	3:84	8:10	2:0	0:0	0:0	0:0	0:0	0:0	1:20	0:0	128	89
Brachiodontes recurvus[1]	0:0	0:0	0:0	0:0	0:0	0:0	0:0	36:16	0:0	18:1	71	23
Crassostrea virginica[1]	0:0	0:0	5:0	6:5	0:0	18:2	0:0	11:3	0:0	9:6	65	25
Tagelus plebeius[1]	0:0	0:1	7:1	0:0	10:20	0:0	3:10	1:0	2:6	0:0	61	62
Rangia cuneata[1]	0:0	0:0	0:0	0:0	0:0	0:0	0:0	1:0	0:0	0:0	1	0

[1]From 0.5 bushel sample taken with a commercial clam dredge.
[2]Collected from 0.5 bushel mud clumps.

Table 2. Number of live mollusks collected on the 1973 clam survey.

Species	STATIONS[1]																		Total
	1	2	3	4	5	7	8	9	10	11	12	13	14	15	16	17	18		
Macoma balthica	-	4	32	149	1	15	35	7	123	-	1	1	1	-	1	67	5		442
Macoma phenax	1	1	3	10	-	11	85	36	37	-	1	-	1	1	1	77	10		275
Mya arenaria	-	-	1	2	-	10	1	15	65	-	-	-	-	-	-	38	43		175
Rangia cuneata	-	-	-	1	1	1	5	7	51	-	-	-	-	-	-	6	1		73
Mulinia lateralis	-	-	-	-	-	-	-	-	1	-	-	-	-	-	-	1	1		3
Tagelus plebeius	-	-	-	-	-	-	-	-	-	-	-	-	-	-	-	1	-		1
TOTAL	1	5	36	162	2	37	126	65	277	2	1	2	1	2	190	60			969

[1]Station No. 6 was not sampled in 1973.

In 1972, not one live soft-shell clam was found. Recently dead soft-shell clams, most with decaying meats attached, ranged in length from 42 to 106 mm; peaks occurred at the 60, 70 and 80 mm lengths. The smallest collected were probably those spawned early in the autumn of 1971. According to Manning (1965) growth is rapid and the clams reach marketable size, 51 cm (2 inches)[1], in about 16 to 22 months. In 11 months they should have been in the 40 to 50 mm range.

Baltic clams were only collected at stations 1, 2, 3 and 9 where the bottom substrate was predominantly mud with small amounts of sand. At the other stations which were predominantly sand with small amounts of mud this species easily washed through the dredge mesh. The data on this species indicate two things: they suffered a high degree of mortality and the mortality was greatest at both ends of the area studied, i.e., in Chesapeake Bay and towards the head of Rhode River. Station 3 appeared to have the least mortality. The specimens collected in the 1972 survey ranged in length from 16 to 36 mm, with peak abundances at 28 and 30 mm.

Of the four other species collected in 1972, the short razor clam, *Tagelus plebeius*, with 62% dead, suffered the third greatest mortality. Though not conclusive, data from station 3, where 88% were alive, also indicate a lesser mortality in that area. The razor clams collected ranged in length from 58 to 80 mm, with most between 66 and 78 mm.

The bent mussels, *Brachidontes recurvus*, and American oysters, *Crassostrea virginica*, were collected together, with the mussels attached to the oyster shells. The mortalities observed, 23 and 25% respectively (Table 1), although probably caused by the stress were not considered serious. The mussels ranged in length from 14 to 42 mm, with 85% between 26 and 30 mm. The oysters ranged in length from 55 to 183 mm, with most between 100 and 155 mm.

Only one small, 30-mm-long, live brackish-water clam, *Rangia cuneata*, was collected on the 1972 survey. This species, taken at station 8, appears to be rather small in this area and was not retained by the commercial dredge.

The 1973 survey showed that although the mollusk populations were reduced because of Agnes, some individuals of all species survived and the populations were recovering (Table 2).

The populations of soft-shell clams had substantially recovered. They were found in greatest abundance at the most southerly station we sampled, station 10, where 65 were collected, an average of 12 per square meter. Lengths ranged from 22 to 65 mm, with a peak at 40 mm. Only two of the 175 were over 48 mm. A 65 mm clam collected at station 18 was considered to be two years old, and therefore had been there since the autumn of 1971, or before Agnes. The absence of any other specimens larger than 48 mm indicates that the mortality found by the 1972 survey was suffered by juveniles as well as by adults, otherwise, with a whole year to grow, many juveniles would have attained adult size and would have been collected.

Of the all the mollusks collected in 1973, *M. balthica* occurred most frequently. In the Rhode River stations it occurred in greatest abundance at station 4, with clams ranging in lengths from 9 to 38 mm, with peaks at 21 and 33 mm. The other stations in Rhode River contained fewer and generally smaller clams. Two stations in Chesapeake Bay (10 and 17) produced good numbers of baltic clams, with peak lengths about 25 mm at both places, about the average size for this clam.

---

[1] Market size now is 63 mm (2.5 in).

Table 3. Numbers of live and dead oysters collected during the August 1972 oyster survey.

Location	Station No.	Live	Recent Dead	Total	Percent Dead
Rhode & West Rivers	1	0	0	0	-
	2	0	0	0	-
	3	21	11	32	34
	4	75	24	99	24
	5	2	1	3	33
	6	49	12	61	20
	7	16	17	33	51
	8	61	6	67	10
	TOTAL	224	71	295	Avg. 24
Chesapeake Bay	9	0	0	0	-
	10	24	6	30	20
	11	88	6	94	6
	12	41	3	44	7
	13	10	0	10	0
	14	59	1	60	2
	TOTAL	222	16	238	Avg. 7
South River	15	7	1	8	12
	16	37	3	40	8
	17	73	2	75	3
	18	30	1	31	3
	19	29	0	29	0
	20	116	1	117	1
	21	38	12	50	24
	TOTAL	330	20	350	Avg. 6

The short razor clam, *Tagelus plebeius*, appears to have suffered most from Agnes. In contrast to 1972 when 23 live razor clams were collected from 5 of the 10 sampling stations (Table 1), in 1973 only one live razor clam was collected from 17 stations, and that one came from station 17, a Bay station. None was collected from any of the river stations.

Missing completely from the 1973 collections were oysters and mussels, but that was because the dredge encountered no oyster beds in 1973.

In another contrast, two species absent from the 1972 samples, *Macoma phenax* and *Mulinia lateralis*, were collected in 1973, although they were known to be present in this area before Agnes. The *M. phenax* ranged in length from 12 to 24 mm, with a peak at 18 mm.

Brackish-water clams, *Rangia cuneata*, were much more abundant in the 1973 samples than in those of 1972, a total of 73 versus 1. These were all small, though, with lengths ranging from 12 to 36 mm, with a peak at 24 mm.

Oyster Surveys

Surveys to determine the effect of Agnes on oysters were undertaken only in August 1972. In general, oysters suffered lower mortalities than the clams did. Mortalities were greatest in the Rhode and West Rivers, where the samples contained from 0 to 51% recently dead oysters, with an average of 24%; whereas samples from Chesapeake Bay and South River contained from 0 to 24% recently dead, with an average of about 7% (Table 3). The samples collected with the commercial oyster dredge, with an average mortality of 24%, agree well with the samples collected earlier with the clam dredge with an estimated 25% mortality.

Of the live oysters examined visually, a few were judged to have spawned recently and were rated poor, a frequent rating for oysters in the early autumn, but most appeared to be in good condition.

DISCUSSION

The results of the three surveys are only qualitative and not directly comparable, but they do provide a general picture of what effect Agnes had on the benthic mollusks in the Rhode River area.

Comparisons of the samples collected in 1972 and 1973 to determine the effects of Agnes must be tempered by previous knowledge of the molluscan community in the area and the effectiveness of the different gears. In 1972 small species such as *Macoma* and *Mulinia* and the young of *Mya* and *Rangia* passed through the large mesh of the commercial clam dredge, whereas in 1973 they were readily caught by the dredge of R/V *Venus*. The samples, however, did show what they were intended to show: that the mortalities of adult clams and oysters were high shortly after Agnes, that one year later many clam were still alive, and that some surviving *Mya* had spawned successfully after Agnes.

Added support to successful spawning by *Mya* has been provided by Mr. Frank Wilde (personal communication). He reports that numerous "seed clams" were present in the sediments of Jack Creek, a small embayment near our station No. 10, and that by mid-March 1974, commercial clammers were working offshore from Jack Creek. Thus, it appears that the set resulting from the autumn 1972 *Mya* spawning was sufficient to support a limited commercial harvest of market-sized clams about 16 months after the previous crop of adults was killed off.

The reason some mollusks died while others survived is probably related to the levels of temperature and salinity they were subjected to. Details of changes in water quality for this area are documented elsewhere in this symposium and in Cory's 1975 report on the water quality of the Rhode River collected by a U. S. G. S. continuous water quality monitor located near station 11 (Fig. 1).

Salinity at the monitor station at a depth of one meter increased from a spring low of 3.8 ppt on 1 June 1972 to a post-Agnes high of 7.0 ppt on 23 June. By 12 July salinity had dipped to the unseasonal low of 2.1 ppt, but then rose to 3.5 ppt by 5 August, and continued to increase to an annual high of 12.5 ppt in mid-November. Average monthly salinities for July and August of 1972 were about 6 and 3 ppt, respectively, lower than for other years observed.

Temperature at the monitor station decreased rapidly from 26° to 20°C during the week of 12 to 23 June, as cool, freshened water flowed in from Chesapeake Bay. Coincident with the period of depressed salinity, temperature at the one-meter level during a three-week "heat wave" increased to 32.6° by 19 July and remained above 30° until 26 July.

In general, during the period of freshening from the influx of Agnes-diluted Bay water, the Rhode and West Rivers had reverse salinity gradients with respect to the Bay. The circulation was such that nearly fresh Bay water (salinity of less than 1 ppt) flowed in at the surface and the saltier (salinity of 4 to 7 ppt) Rhode and West River water flowed out at the bottom. This condition persisted from mid-June until mid-July when the salt content of the Bay water began to increase and cooler, saltier Bay water flowed into the Rhode and West Rivers at the bottom. With exchange rates highest near the entrance to the Bay (Han 1975), this meant that the bottom organisms in that area were subjected to less of the environmental stresses (low salinity and high temperature) than those living in the shallow inshore areas or in the Bay proper. The capability of these tributary systems to modify the influence of Chesapeake Bay was undoubtedly helpful in maintaining the populations of mollusks in the Upper Bay.

CONCLUSIONS

The following conclusions can be made from the two clam surveys and the one oyster survey made in the Rhode, West, and South Rivers and vicinity after the passage of Agnes:

1) High temperatures together with low salinities during a six-week period from mid-June to the end of July 1972 fatally stressed many of the shallow-water, Upper Bay mollusks.

2) The soft-shell clam, *Mya arenaria*, suffered a drastic mortality which affected many of all size-classes; but juvenile *Mya* had repopulated the area by autumn 1973 and had grown to market size, 63 mm, by March 1974, when limited commercial operations resumed.

3) The baltic clam, *Macoma balthica*, and another small clam, *M. phenax*, both had high mortalities in the shallow water of the upper Rhode River. Both species were found in abundance near the mouths of the Rhode and West Rivers where their mortalities appeared to be minimal.

4) The short razor clam, *Tagelus plebeius*, suffered only moderate mortalities immediately after Agnes, but were almost completely absent from samples one year later. The cause of this absence is unclear but it may represent a delayed reaction to Agnes.

5) A survey of the oysters, *Crassostrea virginica*, in the Rhode, West, and South River estuaries in August 1972 indicated an average mortality for all stations of about 25%.

6) Residual salty water flowing out along the bottom of the tributaries during the period of freshening and followed by cooler and saltier bottom water flowing in from Chesapeake Bay during the subsequent summer increase in salinity probably reduced the stress on the benthic mollusks living near the mouths of the tributaries.

ACKNOWLEDGEMENTS

We are grateful for assistance from: Hayes Pfitzenmeyer, for advice and assistance on the clam collections and for reviewing the manuscript; Richard Younger, captain of the R/V *Venus*; Dr. Eugene Cronin, for making the *Venus* available (all personnel from the Chesapeake Biological Laboratory, Solomons, Md.); and Mr. Frank Wilde of Shadyside, Md. for loaning us his commercial oyster dredge and dredge boat.

LITERATURE CITED

Cory, R. L., J. M. Redding, and M. M. McCullough. 1975. Water quality in Rhode River at Smithsonian Institution Pier near Annapolis, Maryland, April 1970 through December 1973. Water Resour. Invest. 10-74, U. S. Geological Survey, pp. 1-67.

Han, G. 1975. Salt balance and exchange in the Rhode River, a tributary embayment to the Chesapeake Bay. Chesapeake Bay Inst., Johns Hopkins Univ. Tech. Rep. 89, Ref. 75-1, 143 p.

Manning, J. H. 1965. The Maryland soft shell clam industry and its effects on Tidewater resources. Maryland Bd. Nat. Resour. Rep. No. 11, 25 p.

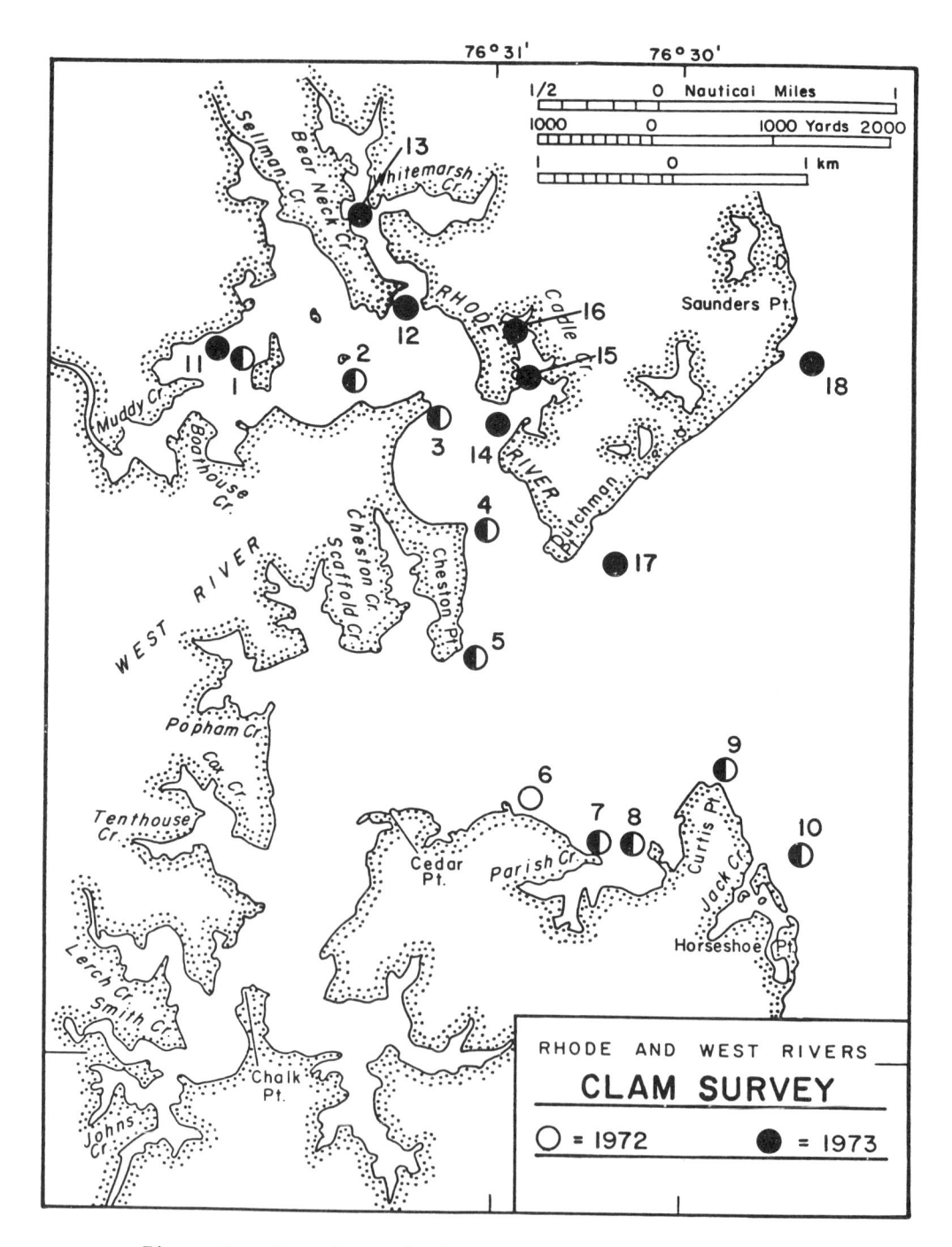

Figure 1. Locations of clam sampling stations in the Rhode and West Rivers.

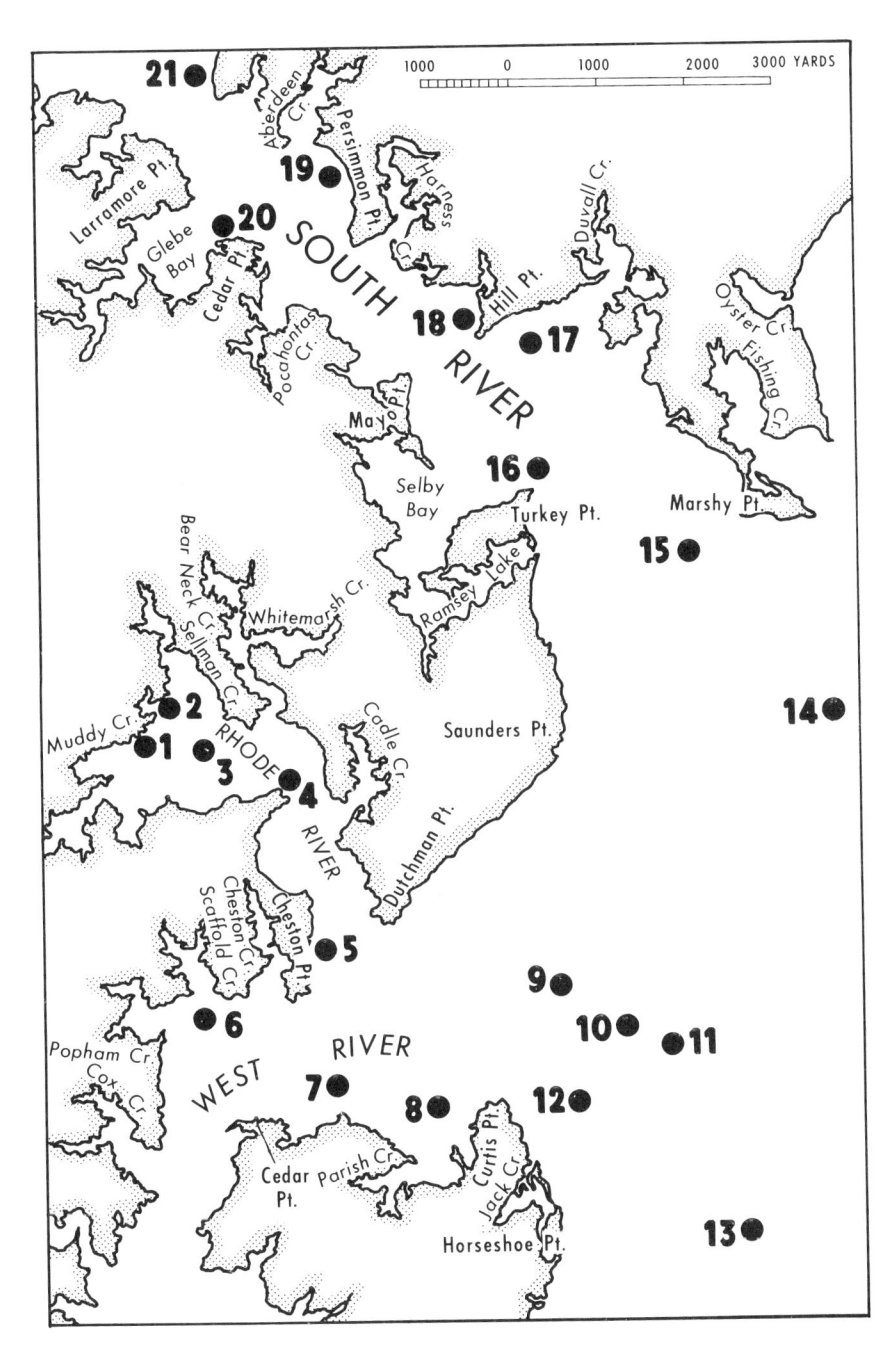

Figure 2. Locations of oyster sampling stations in the Rhode, West, and South Rivers.

THE EFFECT OF TROPICAL STORM AGNES ON OYSTERS, HARD CLAMS, SOFT CLAMS, AND OYSTER DRILLS IN VIRGINIA[1]

D. S. Haven [2]
W. J. Hargis, Jr.[2]
J. G. Loesch [2]
J. P. Whitcomb [2]

ABSTRACT

Tropical Storm Agnes had a major effect on the molluscan fisheries of Virginia. One effect was the direct mortality of oysters, *Crassostrea virginica*, in the upper parts of many estuaries. Typical losses on leased bottoms were: the James River, 10%; the York River, 2%; the Rappahannock River, 50%; and the Potomac River tributaries (Virginia) 70%. Economic loss was in excess of 7.9 million dollars. There was a nearly complete absence of oyster larvae attachment (setting) in 1972. Other effects of Agnes included a nearly complete loss of soft clams, *Mya arenaria*, in the Rappahannock River. Hard clams, *Mercenaria mercenaria*, were killed in the upper part of the York River. Oyster drills, *Urosalpinx cinerea*, were eliminated from the Rappahannock and reduced greatly in numbers in the York and James Rivers.

INTRODUCTION

Agnes entered the Chesapeake Bay area on 21 June 1972 where it released massive quantities of rain on the watersheds of many of the river systems which support oysters and clams in their lower reaches. This excessive rainfall followed a period when salinities in the principal estuaries from the Potomac to the James had already been depressed by unusually heavy rainfall during the preceding two months (Chesapeake Bay Research Council 1973).

As a direct result of this new influx of fresh water there was a displacement of the 0.5 ppt salinity isohaline line in certain systems as far as 18 miles down-river from its usual location. Sedentary animals such as mollusks could not escape these abnormal salinities. In many sections, a complete destruction of existing populations occurred. The severity of mortality depended on the region, species, salinity, length of exposure, temperature, and, in some locations, on dissolved oxygen (DO) levels.

Oysters (*Crassostrea virginica*) are active only when average salinity values exceed about 4 ppt. Below this they close their shells and become inactive (do not pump water or feed). The length of time oysters will survive in salinities less than 4 ppt depends in part on genetic factors and conditioning, but water temperature is by far the more important aspect. During the cooler months (<5.5C), oysters are inactive (Galtsoff 1964) and, therefore, salinity levels ranging from 0.0 to 4.0 ppt cause little mortality over periods of 2 to 3 months. But as the temperature rises, mortalities increase. At these salinities, 3 weeks is about the maximum period that oysters can survive at 21 to 27C (Andrews, Quayle, & Haven 1959).

---

[1]Contribution No. 762, Virginia Institute of Marine Science
[2]Virginia Institute of Marine Science, Gloucester Point, Va. 23062

Hard clams (*Mercenaria mercenaria*) are more sensitive to low salinities than oysters, but little information exists concerning their resistance to prolonged periods of stress. Available evidence suggests that during the warmer months salinity levels below 10 ppt for 1 to 2 weeks are lethal (Castagna & Chanley 1973).

Soft clams (*Mya arenaria*) are killed when salinities below 2.5 ppt persist for one to two weeks (Castagna & Chanley 1973).

Oyster drills (*Eupleura caudata* and *Urosalpinx cinerea*) are less tolerant than oysters to low salinities. Salinities of about 9 ppt are lethal in 2 or 3 weeks at summer temperatures (Carriker 1955, Zachary & Haven 1973).

Planktonic larvae of mollusks were also subject to abnormal conditions as a result of Agnes. Larvae were exposed to high turbidity and unfavorable currents which swept them from setting areas; in many areas salinities were below their tolerance limits. Many clams and oysters failed to spawn since they were inactivated by the low salinities.

Low oxygen was detected in many rivers after Agnes. There has been little research on chronic effects of low oxygen on survival of hard clams and oysters. Recent studies at the Virginia Institute of Marine Science (VIMS), however, indicate that if levels are below 1.5 mg/l, pumping and feeding efficiency of both species is impaired. Levels of about 0.8 mg/l cause nearly 100% mortality in oysters in about one week; hard clams show complete mortality in about 3 to 4 weeks. Larvae of both species survive less than one week at 0.8 mg/l (Haven, Walsh & Bendl 1973).

To estimate the impact of Agnes on Virginia's molluscan resources, on 22 June 1972, VIMS began a major program to evaluate its impact on the oyster, the hard clam, the soft clam, and the oyster drills. These species will be covered in four separate sections.

OYSTER MORTALITY AND ECONOMIC IMPACT

Methods

The objectives of this study were to estimate the quantity and value of oysters killed on public and private bottoms by Agnes. Ideally, such estimates should be based on the density of oysters (numbers or bushels per unit area) in a given area prior to the storm in relation to similar data collected after the mortality period. Quantitative data of this type were lacking, but on file were many years of information on the percentage of live to dead oysters on many representative public bottoms and private leases. This baseline information indicates what normal mortalities were in each region and, when compared with similar data collected after Agnes, made possible an estimate of the percentage killed in various areas. Later, the percentage mortality data were used, as will be discussed below, to estimate quantities of oysters killed on leased bottoms and in estimating the percentage of the total resources killed on public bottoms.

Estimates of the percentage of oysters killed were based upon an extensive series of weekly or biweekly samples of bottom material collected from public oyster rocks and private oyster leases beginning on 21 June 1972 and extending to December 1972. About 1,200 stations were visited in 110 separate field sampling trips between 21 June and December 1972. Frequent sampling in many areas established limits and gradients of mortality in respect to water depth and time. Bottom material was dredged or tonged from the bottom and numbers of live, dying, and recently dead oysters (recent boxes) in these samples were counted. The percentage mortality was calculated as:

$$\text{Percentage Mortality} = \frac{\text{No. recent boxes + No. dying}}{\text{No. alive + No. recent boxes + No. dying}} \times 100$$

Live oysters had tightly closed valves. The valves of dying oysters gaped open but their meats were intact. A recently dead oyster (recent box) had valves still attached, with inner surfaces still white and clean.

On leased bottoms the number of oysters killed by Agnes was determined by sampling only those areas known to contain oysters, information supplied principally by the lease holders. The total killed was then determined by multiplying the percentage dead with estimates of the number of bushels of oysters on the bottoms prior to Agnes. The number of bushels actually on the bottom was determined by multiplying the total volume of seed oysters planted in 1969-1970, 1970-1971, and 1971-1972 seasons (because it takes three seasons for seed to grow to market size) by a factor of one or two, depending upon the area (because Virginia growers obtain from one to two bushels of mature oysters from every bushel of seed they plant). The number of bushels lost was then multiplied by the expected price per bushel to arrive at an estimate of economic loss.

For public oyster grounds, it was possible to chart the areas in each system where public bars were located from records on file at the Virginia Marine Resources Commission (VMRC). On these public beds, almost all oysters originated from a natural set of oysters. Therefore, total volume present on the bottom could not be estimated as it was on leased bottoms. Also on these same bottoms there has never been a quantitative study of the density of oysters. However, data were obtained from the VMRC as to the approximate percentage of the State's total resource growing in river systems. Also, the productive regions were delimited (as they were for leased areas) by hydrographic factors and diseases. Data on annual production for public bottoms were obtained from the annual reports of the VMRC.

Hydrographic Data

Data on oxygen, salinity and temperature presented in this paper were obtained from a previous report on Agnes (Chesapeake Bay Research Council 1973). Additional information on the hydrography of the study area is presented earlier in this volume.

Oyster Mortality in the James River

Introduction. The James River is the major oyster producing region in Virginia and the source of over 75% of all seed planted in the state by private growers. Oysters grow in commercial quantities from Deep Water Shoals, the upriver limit of oyster production, down to the mouth of the Nansemond River (Fig. 1). In 1971 its public rocks produced 439,294 bushels of seed and 170,849 bushels of market oysters (Table 1). As outlined previously, accurate production data for private leases are not available.

Table 1. Oyster production in bushels from public grounds in Virginia and from the Potomac River, 1969, 1970, and 1971.

River	1969	1970	1971	Average
Potomac[1]				
Virginia tributaries	25,642	13,074	31,824	23,513
Potomac				
Landed in Md.[3]	91,113	117,818	210,989	
Landed in Va.[1]	220,022	266,275	172,403	
Total (Md. & Va.)	311,135	384,093	383,392	359,540
York [1]	204	360	716	427
James [1]				
Market [2]	157,669	143,778	170,847	157,431
Seed	486,536	264,203	439,294	430,011
Rappahannock [1]	29,402	23,698	65,949	39,683

[1] Source - Annual Reports Virginia Marine Resources Commission.
[2] Mostly 2½ inch oysters.
[3] Source - Fisheries Statistices of the USNMFS.

Hydrographic Data. Salinity levels in the James River fell rapidly in response to the influx of flood water into the estuarine portion of the system and by 28 June, 5 days after the flood waters crested, the surface water was fresh (0.05 ppt) just below the James River Bridge which is about 29.0 km below the point where the 0.05 ppt isohaline normally occurs. Recovery was relatively rapid, and by 19 July all beds except those at Deep Water Shoals showed from 1 to 5 ppt. By 25 August, salinities on all beds exceeded 5 ppt, except at Deep Water Shoals and Horsehead where they ranged from about 1 to 4 ppt which is not high enough to sustain activity. Subsequently, there was further recovery in all down-river areas except at Deep Water Shoals and Horsehead where salinities remained low for the rest of the year because of continued rains.

Water temperatures and DO were within tolerable levels for oysters.

Mortality - Public Grounds. Oysters began dying on the public rocks at Deep Water Shoals, Horsehead Bay, and Mulberry Point (1.6 km above Horsehead) between 10 and 14 July, about 14 days after the onset of the excessive freshwater flows associated with Agnes. By 24 July, when salinities again permitted oyster activity, the following mortalities were estimated: Deep Water Shoals, 57%; Mulberry Point, 25%; and Horsehead, 18% (Andrews, unpublished). Low salinity conditions (less than 1 ppt) returned in September 1972 due to excessive rainfall and by December 1972, mortality had increased to 95% at Deep Water Shoals and to 40% at Horsehead. Mulberry Point showed little change. Samples collected from downriver oyster rocks such as Point of Shoals, Wreck Shoals, and below, where about 85% of the oysters occur, indicated no significant mortalities due to low salinity.

Based on the preceding data, we estimate that about 5% of the oysters on public rocks in the James River System were killed by flood waters associated with Agnes.

Mortality - Private Leases. Monitoring oyster mortality on private leases did not begin until 8 September 1972. At that time it was reported that certain shallow-water beds (1.2-1.8 m) close to the shoreline had suffered moderate to heavy mortality; adjacent deep-water beds had not. This mortality probably coincided with that on the public bars up-river, i.e., between 10 and 14 July 1972. Significant mortalities along shore at depths less than about 1.8 m, extended almost 4 miles further downriver than they did in the deeper water toward the center of the river. The most probable reason for this pattern is the horizontally stratified nature of the salinity pattern. That is, oyster beds in the shoreline waters were subjected to lower salinities than those in the deeper waters.

Off the mouth of the Warwick River and off Mulberry Island in very shallow water (1.8 m) on the northern shore of the James, the percentage killed ranged from 20 to 96%.

Along the southern shore of the James significant mortalities were observed only in a small area of leased bottom at a depth of about 1.8 m in the upper river about 0.2 km below Point of Shoals. Down river from there, in the open river, mortalities were normal (<10%).

Oysters in three adjacent tributary creeks on the southern side of the lower James River suffered significant mortality due to fresh water from their head waters. In the Chuckatuck and Nansemond Rivers, oysters near the upper reaches of each system and in very shallow water (2-4 feet) showed an average mortality of about 24%. In the upper part of the Pagan River in very shallow water, mortality ranged from 20 to 23%. In the deeper water in all three systems, oysters showed less than 10% mortality, which is about normal for the region and season.

Interviews with VMRC and oyster growers indicated that about 20% of all the oysters growing in the James on leased bottoms were located in the mortality area. Estimates based on comparison of percentages killed in relation to this distribution indicate that about 10% of all oysters on leased bottoms in the James were destroyed as a result of Agnes.

Oyster Mortality in the York River

Introduction. York River oysters grow in commercial quantities from Bell Rock to about 3.2 km below Pages Rock (Fig. 1). In the last few years, public rocks production has been very low; in 1971 only 716 bushels were produced (Table 1). Oyster production is unknown from private leases in the York area.

Hydrographic Data. Salinities were strongly depressed in the York River after Agnes. On 5 July 1972, 11 days after the flood crest passed the fall line, the 2 ppt surface isohaline was at Bell Rock and the 10 ppt at Gloucester Point, significantly below a normal of about 10 ppt and 18 ppt, respectively for this period. Conditions slowly improved, but substantial vertical stratification was not evident until 24 July 1972.

Water temperatures stayed within tolerable levels for oysters, and although DO was not measured no evidence of anaerobic conditions was apparent from the dredged samples.

Mortality - Public Grounds. Oyster mortality on public grounds in the York River was negligible. At the furthest up-river productive public rock

(Bell Rock) salinities did not persist at critical levels long enough to kill significant numbers of oysters.

Mortality - Private Leases. Oysters killed on private leases were confined to shallow depths (0.6-1.2 m) on the west side of the river extending about 3.2 km downriver from Bell Rock. As in the James, oysters died prior to 11 July; the percentage killed on these beds was estimated at 33%. Because over 95% of all private plantings are located further downriver, only about 2% of the oysters on leased bottoms in the York were killed as a result of Agnes.

Oyster Mortality in the Rappahannock River

Introduction. Oysters are harvested commercially from Accaceek Point to about 3.5 km below the Norris Bridge (Fig. 1). Production from the public oyster rocks in 1971 was 65,949 bushels (Table 1). Production from leased areas in 1971 is unknown.

Hydrographic Data. Salinities were low for over a month in the upper Rappahannock River prior to the arrival of Agnes, with values at Bowlers Rock fluctuating from 1 to 5 ppt. No unusual oyster mortalities, however, were reported prior to 21 June 1972.

Five days after Agnes struck, the 1 ppt isohaline was located below Jones Point, where it remained, with minor fluctuations, through 11 July. During this same period, the 5 ppt isohaline extended from Jones Point at a depth of 6.1 to just below the Norris Bridge where it reached the surface. By 24 July conditions had partially recovered and at Accaceek Point, near the up-river limit of oyster culture, salinities ranged from 1 ppt at the surface to about 5 ppt at the bottom. By the second week in August, conditions had slowly improved and salinities were averaging about 5 ppt at Bowlers Rock.

Neither water temperature nor DO were critical in the areas where the major oyster mortality occurred. On 14 July, however, water on the bottom of the channel near Bowlers Rock contained only 1.0 mg/l oxygen, a level that will stress oysters if they are open and pumping.

Mortality - Public Grounds. Extensive sampling of the public beds in the Rappahannock from Bowlers Rock to the river mouth showed no mortality associated with Agnes. It is within this area that an estimated 98% of the oysters on public bottoms occur. The single exception to the preceding evaluation of mortality was at Russ Rock, a small isolated area, located 8 km above Accaceek Point. This area was not investigated, but verbal reports suggest that nearly all oysters died there. Based on these data, we were able to estimate that less than 2% of the oysters on the public rocks on the Rappahannock were killed as a direct result of Agnes. The absence of mortalities on public bottoms was due to their location at depths where salinities did not reach critical values for a sufficient time.

Mortality - Private Leases. Oysters began dying in the upper river on leased bottoms at depths less than 1.8 m about 30 June, and by 10 July oyster mortality was progressing at such a rapid rate that thousands of oyster meats were observed floating at the surface near Bowlers Rock where they were fed upon by seagulls. This period of peak mortality occurred after oysters had been subjected to salinities averaging 1 ppt or lower for about two weeks and to levels averaging from 1 to 5 ppt for at least a month before this. Although mortality decreased as salinities rose to higher levels, oysters continued to die until about the second week in August.

The area where oyster mortalities were highest on the south side of the river began at the up-river limit of oyster culture on leased bottoms, about 0.8 km above Bowlers Rock, and extended about 4.8 km down-river. Within this zone most of the oysters had been planted at depths ranging from 1.2 to 1.8 m, and here mortalities ranged from 69 to 100%. At slightly greater depths (1.8 to 3.1 m), where salinities were slightly higher, mortalities were lower and ranged from 19 to 74%. At depths exceeding about 3.0 m, mortalities were still lower, but few oysters are planted at those depths in the up-river section of the river because suitable bottoms are not available. Below the region of heavy mortality on the south side and extending down-river to McKans Bay, there were a few isolated pockets of mortality ranging from 22 to 50% in shallow water (0.9 m).

The area of maximum mortality on the north side on leased bottoms also occurred in shallow water at depths less than 1.8 m and it extended from Accaceek Point to about 1.6 km below Farnham Creek. In this zone mortality ranged from about 48 to 86%. Mortality was lower (19 to 74%) at depths between 1.8 m to 3.1 m. In deeper water, oyster mortality was lower, but again most oysters in this portion of the river are planted at depths less than about 3.0 to 4.4 m. Mortality caused by fresh water was light to zero below Farnham Creek.

We estimated, on the basis of interviews with oyster growers and officials of the VMRC, that prior to Agnes, about 70% of the oysters growing on leased bottoms in the Rappahannock existed in the area of heavy mortality extending from 0.8 km above Bowlers Rock to about 3.2 km below. Within this area the mean mortality based on 61 samples was 67%; therefore, about 47% of the resource was killed. Additionally, we estimated that about 3% of the oysters downstream were killed, largely those in McKans Bay. Therefore, we conclude that about half of the oysters on leased bottoms were killed by Agnes.

Oyster Mortality in the Corrotoman River

This river is a tributary on the north shore of the lower Rappahannock River (Fig. 1). Its public rocks in 1971 produced 1,911 bushels of market oysters (VMRC 1971), but the annual harvest from private leases is not known.

A single survey made on public and private oyster grounds on 22 September 1972 showed that mortality in the uppermost portion of the river on private leases ranged from 20% at the Otterman Wharf Ferry Landing to 22% off John Creek. The public rocks had less than 20% mortality.

Oyster Mortality in the Potomac River

Introduction. The Potomac River exclusive of its tributaries contains no leased bottoms. In 1971 the entire river produced over 383,392 bushels of market sized oysters (Table 1). Oyster mortality was nearly complete in the upstream reaches of this system as a direct result of Agnes.

Hydrographic Data. Salinities in the upper half of the oyster-producing region of the Potomac River had been abnormally low for about 3 months prior to Agnes. From mid-April to 16 June in the area extending from the Potomac River Bridge to Colonial Beach, surface salinities ranged from 1.0 ppt to 4.2 ppt (Chesapeake Bay Research Council 1973). On 28 June, 6 days after Agnes, salinity had dropped below the measurable level (0.5 ppt) there. Salinity levels during the next 16 days, except for minor effects due to tidal fluctuations, remained virtually unchanged, and on 14 July there was almost no measurable salt in the surface water as far downstream as Swan Point. By 29 August salinity levels had risen enough to permit normal physiological activity in oysters at Cedar Point where salinities were about 4 ppt at the surface and 6 ppt at the bottom.

Low DO was not a factor in mortalities of the upper Potomac River nor was termperature. DO in mid-channel from Cobb Island to the up-river limits of oysters ranged from 3.9 to 6.6 mg/1 on 28 June. Surface temperatures there ranged from 18.0 to 20.5C. By 20 July when oysters were dying rapidly, DO in the same region ranged at the bottom in mid-channel from 4.1 to 7.8 mg/1. Surface temperatures ranged from 25.4 to 27.5C.

Oyster Mortality. A few oysters began to die on the uppermost public bars in the Potomac River a week or two prior to Agnes due to low salinities which had prevailed there for about three months. Thereafter, with the influx of added fresh water due to the storm, mortality slowly increased. By 20 July, after exposure to fresh water for about three weeks, mortality increased sharply reaching 21 to 46% from Cobb Island to the up-river limit of oysters. By 1 August mortality climbed to nearly 100%. Salinities returned to normal at the end of August after which mortality among surviving oysters returned to typical levels (about 5% per year).

Estimates based on production records on file at the Potomac River Fisheries Commission indicated about half of the oysters existing in the Potomac River were located above Cobb Island, where mortality approached 100%.

The boundary line in the river which divided the zone of nearly complete mortality from that where mortalities were not significant was a line extending from Cobb Island in Maryland across the Potomac to Popes Creek in Virginia. Below this line, mortality was light and by mid-July less than 6% of the oysters had died on any of the downriver bars. By the end of August, 11% mortality had occurred among oysters further down the Potomac River at Coles Point.

Oysters on Bluff Point Bar, in St. Clements Bay, however, were all killed. By 1 September, oysters living in the lower Potomac River appeared in good condition as shown by surveys of VIMS and the Chesapeake Biological Laboratory.

Oyster Mortality in the Virginia Tributaries to the Lower Potomac

Introduction. Oyster mortality occurred on public and private bottoms in Virginia creeks tributary to the lower Potomac River, as well as in main stem of the system. Major mortality occurred in Lower Machodoc and Nomini Creeks, and scattered mortality was observed in Yeocomico River, Bonum, Jackson and Gardner Creeks (Fig. 1).

Lower Machodoc Creek. A survey on 20 July 1972 indicated that up to 40% of the oysters had died in the lower reaches of this system. In the upper half, few if any, died. Surface salinities in the creek varied from 1.9 to 2.3 ppt and DO in channel bottom water varied from 4.2 to 6.8 mg/1. By 28 July mortalities began to increase and in a section near the mouth (Glebe Creek) average 80%; an adjacent down-river location showed 17% dead. Upstream, mortality ranged from 43 to 75%. At this time DO in the channel ranged from 3.8 to 2.6 mg/l which is not considered critical. By 22 August, however, DO had declined to critical levels with values below 0.5 ppm. As a result, mortality of oysters reached nearly 100% over most of the system.

Nomini Creek. A survey on 20 July showed 12% of oysters on public rocks near the mouth dead; up-river mortality was higher, from 24 to 93% on some private beds. Channel salinities ranged from 6.1 to 7.7 ppt.

On 22 August 1972 anaerobic conditions developed in the mid-section of the system, with less than 0.5 mg/l DO in the bottom water in some sections. As a result, the oysters and oyster shells became black and smelled of hydrogen sulfide when lifted from the water. By 22 August nearly all oysters had died in Nomini Cut near the creek entrance and at many other locations up-river.

Yeocomico River. The Yeocomico River was not surveyed in the month following Agnes. A survey on 28 July showed oysters in 3.0-3.7 m near the mid-portion of the system apparently unaffected. Surface salinities near the mouth on 28 July averaged 3.3 ppt; marginal for oyster survival. Bottom water samples in mid-channel on 24 August showed 1.4 to 5.1 mg/l DO and salinity 6.1 to 8.4 ppt. Water temperatures were 29.8C to 28.2C at the surface and from 26.2C to 28.2C at the bottom.

Surveys in December 1972 indicated that several leased areas in shallow water in Shannon Creek (an arm of the Yeocomico) had experienced from 60 to 95% mortality. It was impossible to assign a specific cause to the mortality in Shannon Creek then, but depressed salinity and oxygen conditions associated with Agnes were probably responsible.

Bonum, Jackson, and Gardner Creeks. A survey in November 1972 showed that scattered kills had occurred in these three creeks tributary to the Potomac River below Lower Machodoc Creek, but inadequate sampling makes it impossible to generalize about the timing or cause of mortalities.

Conclusions on the Tributaries. In evaluating the percentage of the oyster populations killed in the Potomac River tributaries on the Virginia side of the system, it was difficult to generalize due to the scattered nature of the plantings and the wide variation in mortality. Interviews with oyster growers and VMRC officials plus data on percentage mortality caused us to conclude that about 70% of the oysters in these tributaries were killed by Agnes.

Regions Not Suffering Oyster Mortalities from Agnes

Watersheds associated with the Eastern Shore (Seaside and Bayside), the Piankatank and Great Wicomico Rivers, the Mobjack Bay Region and Lynnhaven Inlet received relatively little input of fresh water from Agnes. Thus, their oyster populations were less affected by the storm. However, these areas all were influenced by the storm in that salinities were lower than average and typical patterns of circulation of the water masses were influenced. This secondary effect while difficult to evaluate undoubtedly had an adverse effect especially in respect to setting of oyster larvae.

Estimates of the Economic Loss by Private Planters Due to Agnes

Introduction. One of the most pertinent problems was to determine where seed was planted in the 1969-1971 period. Based on information supplied by oyster growers and VMRC officials, it was determined that in this period the following percentages of the state's entire seed production were planted in the indicated rivers: James, 5%; York, 10%; Rappahannock, 50%; and Potomac tributaries, 25%. A summation of mortalities on leased bottoms in the same system, as shown in the preceding section, gave the following: James, 10%; York, 2%; Rappahannock, 50% and Potomac tributaries, 25%.

Using the preceding data plus additional information on total seed prediction for the state, values were calculated for economic loss due to Agnes (Table 2). A value of $5.10 per bushel (from records of the Potomac River Fisheries Commission) was assigned for 1969. This value was increased 20% each year to compensate for inflation and an increased demand.

Table 2. Losses of oysters in four estuaries of Virginia, in terms of volume and dollars.

Estuary	Year	Seed Planted Bushels	Oysters Killed[1] (Bushels)	Value Lost to[2] Growers
James (10%)[3]	1969	27,058	5,412	$ 27,601
	1970	18,308	3,662	22,411
	1971	28,284	5,657	41,522
	Sum	73,650	14,731	91,534
York (2%)	1969	54,116	2,165	11,042
	1970	36,616	1,465	8,966
	1971	56,568	2,263	16,610
	Sum	147,300	5,893	36,618
Rappahannock (50%)	1969	270,580	270,580	1,379,958
	1970	183,082	183,082	1,120,462
	1971	282,804	282,804	2,075,781
	Sum	736,466	736,466	4,576,201
Potomac Tributaries (70%)	1969	135,290	189,406	965,971
	1970	91,541	128,159	784,333
	1971	141,420	197,989	1,453,239
	Sum	368,251	515,554	3,203,543
Totals	1969	487,044	467,563	2,384,572
	1970	329,547	316,368	1,936,172
	1971	509,076	488,713	3,587,152
	Sum	1,325,667	1,272,644	7,907,896

[1] Estimated by multiplying the seed planted by the percentage killed and assuming that the oysters would have doubled in volume by harvest time.
[2] Estimated by multiplying bushels killed by a yearly market value ($5.10 for 1969, $6.12 for 1970, and $7.34 for 1971. See text).
[3] The parentheses contain the estimated percent mortality caused by Agnes.

Evaluation of these data shows the estimated minimal loss of oysters in Virginia was about 7.9 million dollars. We regard this estimate as minimal because there was no way to account for the loss of natural set. In addition, indirect economic damage was not assessed, for example, after a planting suffered oyster mortalities of 50% or more, the planter involved may have found it uneconomical to harvest the remaining low density populations which may die later from predators or diseases. Additional losses were sustained due to loss of markets and other similar economic factors.

THE EFFECTS OF AGNES ON OYSTER SETTING IN VIRGINIA IN 1972

Introduction

Oyster set (attachment of larvae to solid substrate such as shells) in Virginia River Systems dropped to the lowest level on record in 1972. Our studies indicate that this phenomenon was the direct result of the effects of Agnes. Low set occurred in all major oyster producing areas in the state; no area received sufficient set to be of any commercial value.

Methods

The virtual absence of a set in all river systems in Virginia in 1972 was established on the basis of an extensive program of monitoring conducted by VIMS since 1947. The data collected over the years by Andrews (unpublished) and later by Haven (Marine Science Bulletin 1969-1972) have been summarized. Three techniques have been used in the bay to monitor setting:
 1) Fall or winter surveys are made on representative public oyster bars. In this program, bushel samples of bottom material are dredged or tonged from the bottom. Later, numbers of spat, small oysters, market oysters, shells and boxes are counted. This technique measures the surviving spatfall as it occurs under natural conditions in the estuary. It is influenced by predators such as drills or crabs which may eat the small spat and by fouling organisms such as Bryozoa, Tunicata, or barnacles which may cover the shell and inhibit setting of the larvae. In this study, only data on numbers of spat per bushel are presented.
 2) Small wire bags containing about ¼ bushel of oyster shells are placed at representative locations in the estuary in June just prior to the onset of setting. They are recovered in October after the setting season has ended. This method shows the surviving set on shell and indicates what the set might be if shell is planted each year as it is often done in repletion activities and by private growers.
 3) Ten to fifteen oyster shells are strung on a piece of wire and suspended at representative stations from ropes off the bottom each week during the setting season. At the end of the week the strings are removed and replaced. Oyster spat setting on the smooth side of each shell are counted with the aid of a microscope and the results are tabulated for each station as the average numbers of spat per shell face per week. In this study the weekly set is summed to give the total set per season at each station. This latter parameter is termed the total theoretical set; it indicates what the set might be under optimum conditions where predators and fouling organisms had little effect.

Although information on setting has been collected since 1947, only that part collected since 1961 is used to show the impact of Agnes on setting. The reason is that beginning about 1960 there began a gradual decline in setting in the lower bay (Haven unpublished). Therefore, this recent period is regarded as more typical of conditions just prior to the storm.

To illustrate the impact of Agnes on setting, the average set for each of the three preceding techniques was tabulated for representative stations in each river system for the 1961 to 1971 period and the ranges compared to values for the 1972 set.

The stations selected for comparison are representative of the range of the productive natural rocks in each system. The stations selected for study in this paper (Fig. 1) are: the James, from Deep Water Shoals to the James River Bridge; the Rappahannock, from Bowlers Rock to 1 km above the Norris Bridge; and the York, from Bell Rock to Gloucester Point. In the Corrotoman, Great Wicomico, and Piankatank Rivers, the stations were located from the mouths of each system to the upper limit of oyster culture.

Results

In the James, York, and Rappahannock Rivers, with two minor exceptions, the 1972 set for all three methods of measuring was lower than for any single year during the preceding 10 years (Table 3).

In the Piankatank, Great Wicomico and Corrotoman Rivers, the 1972 set without exception was lower than the average for any of the years for which data exist within the 1961-1971 period.

Discussion

The absence of a set or the vey low set which occurred on shellbags or natural cultch at the end of the 1972 season might have been due to an excessive mortality associated with unfavorable environmental conditions. However, this was not the case, because few spat set on the shell strings. Significant setting simply did not occur in 1972.

There are several reasons why setting in Virginia reached such low levels following Agnes. Some of the reasons are apparent, others cannot be clearly documented.

Oysters spawn in Virginia from late June to the last of September, with peak spawning occurring during the months of July, August, and September. Therefore, it is evident that Agnes hit at the start of the spawning season and many oysters especially in the Potomac, Rappahannock, York, and James were inactivated by low salinities. Therefore, oysters did not spawn (if at all) until much later in the season.

Another factor was that salinities were low (5 ppt or below) far into the summer and oyster larvae do not develop or may die if exposed to 5 ppt or less for a week or two (Galtsoff 1964).

In river systems such as the Great Wicomico, the Piankatank, and in Mobjack Bay which typically received fair to moderate sets, the fresh water did not kill appreciable numbers of oysters and the salinity remained above levels that cause inactivity. Nevertheless, conditions associated with the storm are believed to have adversely influenced set. Possible reasons include the disruption of normal circulation patterns (Wood & Hargis 1971). Other reasons might include excessive silt levels, and low levels of DO.

Table 3. Effects of Agnes on oyster spatfall in Virginia. The post-MSX period (1961-1971) is compared with similar data for 1972.

River	Method	No. Stations	Period	Range
James	Shell bags (spat/shell)	3	1961-1971 1972	0.9 - 4.3 0.0 - 0.1
	Shell Strings (spat/shell face)	4	1961-1971 1972	3.8 - 9.8 0.7 - 3.3
	Natural Cultch (spat/bushel)	5	1961-1971 1972	62 - 216 0 - 26
Rappahannock	Shell bags (spat/shell)	5	1961-1971 1972	0.0 - 3.8 0.0 - 0.03
	Shell strings (spat/shell face)	8	1969-1971 1972	0.0 - 3.8 0.0 - 0.0
	Natural Cultch (spat/bushel)	5	1961-1971 1972	5 - 93 0 - 0
York	Shell bags (spat/shell)	4	1963-1971 1972	0.1 - 4.0 0.0 - 0.03
	Shell strings (spat/shell face)	1	1963-1971 1972	44.1 0.3
	Natural Cultch (spat/bushel)	4	1961-1971 1972	9 - 25 0 - 10
Piankatank	Shell bags (spat/shell)	6	1963-1971 1972	1.4 - 4.2 0.01- 0.14
	Shell Strings (spat/shell face)	4	1964-1971 1972	12.7 - 67.9 0.0 - 1.2
	Natural Cultch (spat/bushel)	5	1961-1971 1972	277 - 546 0 - 10
Great Wicomico	Shell bags (spat/shell)	6	1965-1971 1972	2.4 - 8.2 0.01- 0.05
	Shell strings (spat/shell face)	10	1964-1971 1972	19.0 - 2.34 0.0 - 3.1
	Natural Cultch (spat/bushel)	10	1968-1969 1972	2.0 - 200 0.0 - 0.01
Corrotoman	Natural Cultch (spat/bushel)	5	1961-1972 1972	183 - 256 0 - 2

In the James River seed-oyster area, the density of oysters on the bottom had been slowly declining since 1960 and the 1972 failure represented a further reduction in the number of 1 to 2 inch seed oysters available for harvest in 1973 and for 3 to 4 years thereafter. In the Piankatank and Great Wicomico Rivers, setting was below average in 1970 and 1971 and the failure in 1972 intensified the seed problem.

The availability of market oysters from the public bottom in Virginia has been slowly declining since the onset of MSX (Andrews 1967), and the absence of recruitment for 1972 will bring lower densities later in 1974 and 1975 when the oysters would normally reach maturity.

MORTALITY AND GROWTH OF HARD CLAM POPULATIONS ASSOCIATED WITH AGNES

Introduction and Methods

In Virginia, hard clam production is largely from public bottoms, and two of the most productive areas are the York and James Rivers. The latter area, however, is classed by the Virginia Bureau of Shellfish Sanitation as restricted for the harvest of shellfish, which means that clams harvested there may not be sold directly for human comsumption unless they are replanted (relaid for cleansing or depuration) on pollution-free bottoms for 15 days with water temperatures higher than 50 F. Such relaying of presumed polluted James River clams is widely practiced. Therefore, in evaluating losses of hard clams associated with Agnes it was necessary to consider: 1) losses on public bottoms in the James River and elsewhere; and 2) losses of relaid clams held on leased bottoms.

Mortality on leased bottoms was determined by collecting clams from the bottom with hand tongs. The number of recent boxes (valves still attached by a hinge) in the total sample was used to determine the percentage dead.

On public bottom, hard clam populations were sampled with a hydraulic tow dredge. Again, percentage mortalities were determined.

Areas referred to in this section are given by their local name and also as the number of km above the mouth of the river systems (Fig. 1).

York River

Public Bottoms. On 10 and 11 August 1972, 12 tows were made in the lower York River in an area extending from the Vepco Power Plant (4.0 km) to Queen's Creek (17.7 km). The study showed only a light mortality (5.3%) (Table 4).

Table 4. Numbers of live hard clams and boxes at various locations in the York River, Virginia, taken with a hydraulic tow dredge 10-11 August, 1972.

Station	Km Above Mouth	Number Live	Number Boxes	Water Depth (m)
Northside River				
Green Point	12.3	0	0	1.4
Mumfort Island	10.5	4	0	1.4
R-28	8.8	8	1	1.4
Waterview	6.4	12	0	1.2
Gaines Point	5.6	31	1	1.5
Perrin	3.2	62	2	1.4
Southside River				
Queen's Creek	17.7	0	0	1.4
Sandy Point	12.1	19	2	1.4
N-27	8.0	14	1	1.4
Yorktown	6.4	7	1	1.4
Wormley Creek	6.4	12	0	1.4
VEPCO	4.0	12	1	1.4

Private Beds. In the Perrin River near the mouth of the York River (2.4 km) there was a serious mortality of relaid hard clams on bottoms leased by one clam-processing firm. An investigation on 10 and 11 August 1972, showed that in one area 58% of the clams on a hard bottom had died, presumably during the past month. Clams stored in floats in an adjacent area suffered 82% mortality.

Off Timberneck Creek (12.8 km) clam beds belonging to a second company were sampled. On one plot in 1.2 m of water, 33% of the relaid clams died. Further inshore in shallower water, a bed containing 1,000 bushels of small hard clams experienced 100% mortality. These clams were valued by their owner when planted at $25 per bushel. Therefore, their loss was estimated at about $25,000.

Clams in very shallow water in the lower York River were exposed to salinities about 10 ppt or less for at least a month after Agnes. Studies have shown that hard clams cannot tolerate salinities ranging below about 10.0 to 12.5 ppt (Castagna & Chanley 1973). We conclude, therefore, that the fresh water associated with Agnes was the cause of the mortality. It is not clear why clams died on private leases and not on public bottoms. A probable reason may be that clams on the leased bottoms were subject to a salinity stress plus an added stress due to harvest, transport, and replanting.

In the month following Agnes, clam boxes washed up along shore in several locations in the York River from about 0 km to 19.3 km. The number was never large and ranged from about 5 to 20 per 50 m of shoreline. No estimate of percent mortality could be made because the density of the population these clams came from was unknown.

Poquoson River

Two companies with relaid clams on leased bottoms were visited on 11 August 1972; both sustained serious losses of hard clams originally harvested in the James River. Six separate sites belonging to one company were visited. Most of these clams had been transferred from the lower James River to the Poquoson River in the three week period after Agnes. Mortality ranged from 11 to 100%, averaging 44%. But even before Agnes clam mortality in this area was high. Clams which had been moved from the James River by a second company were sampled in early June 1972. The average mortality was 40% in four areas sampled. Salinity levels during the mortality period were not measured. Therefore, it is not possible to tell why these clams died.

James River

Ten stations sampled during August 1972 in the lower James River in the vicinity of Newport News (8.8 km) yielded no abnormal mortalities.

Adverse Effects on Growth

From 1968 to 1972 VIMS personnel conducted a study of hard clam growth in the lower Chesapeake Bay. A comparison of growth rates in years of normal salinity range with years of low salinities indicated a significant decline in the growth of clams in 1972 in the lower York and James Rivers as a result of Agnes (Loesch & Haven 1973).

Conclusions

The high mortality occurring among clams held in leased areas in the York and Poquoson Rivers probably resulted from the stress of moving and marginal salinities. A light mortality of undetermined magnitude occurred in the York River in very shallow water along shore from the mouth to about 19.3 km up-river.

MORTALITY OF SOFT CLAMS POPULATIONS ASSOCIATED WITH AGNES

Introduction and Methods

Mortality of the soft clam, *Mya arenaria*, was investigated in the Rappahannock River in March 1973, and near the mouth of the Piankatank River (off Gwynns Island) in October 1972, with a conventional Maryland-type, hydraulic, soft clam escalator. The R/V *Mar-Bel*, a VIMS vessel, was employed in the former area; at the latter site a privately owned commercial vessel was chartered by the VMRC. In making these surveys, the vessels were operated on public and private grounds where surveys completed in previous years showed soft clams to be present.

In addition to escalator samples, observations were made in the York River by washing clams from the bottom with a water jet produced by a portable gasoline powered pump.

Areas referred to are followed by station location expressed as km above the mouth of the river and may be located by reference to Fig. 1.

Rappahannock River

Fifteen stations were occupied in the river at seven separate locations on 12 and 13 March 1973. No living soft clams were found. However, shells were collected at many of these locations, indicating existence of previous populations.

Prior studies in the Rappahannock River by VIMS personnel and a private company indicated soft clams had formerly been abundant at several of the stations sampled in the study. An oyster grower (Mr. Ferguson) located at Remlick, Va., stated (personal communication) that about 25 to 30 bushels of soft clams per day could be harvested at the time he ceased commercial escalator harvesting on his grounds in the Morattico area (Fig. 1) in 1968. A similar abundance on these same bottoms were noted in a study by VIMS personnel when 38 bushels of soft clams were escalated from a ½ acre plot in approximately 8 hours in 1968. Again, in 1969, in the same location, 34 bushels of soft clams were harvested by VIMS in about 13 hours. Down-river at Parrott Island, 34 bushels of soft clams were escalated in 13 hours. Below Parrott Island to the mouth of the Rappahannock River, soft clams were represented by sparse populations during 1969 (Haven, Loesch & Whitcomb 1973).

The total absence of soft clams in the 1973 survey in relation to their former abundance indicated populations died in the Rappahannock River cataclysmicly. This probably occurred in the aftermath of Agnes due to the dramatic decrease in salinity, because *Mya arenaria* cannot tolerate salinities below 2.5 ppt for more than two weeks (Castagna & Chanley 1973). Salinities in the vicinity of the Norris Bridge in the lower river on 28 June 1972 were between 1 and 5 ppt. Therefore, we conclude that salinities were low enough to have produced the observed mortality.

York River

Samples collected with the jet sampler on 4, 10 and 15 July and 1 August 1972 indicated no unusual mortality of soft clams had occurred at Gloucester Point (8 km). Similar results were obtained from additional samples collected at Foxes Creek (26 km) from 10 to 31 July 1972. Salinity levels at the above sites never fell below 5 ppt.

Mouth of Piankatank River (Gwynns Island)

Eighteen stations were occupied in the study of soft clam populations off Gwynns Island on 2 October 1972. Shells of soft clams were abundant. Living clams were observed at 11 of the 18 stations but not in great densities; the best catch was about 10 clams per minute. The salinity levels soft clams were exposed to in this location after Agnes are not known.

MORTALITY OF OYSTER DRILLS (UROSALPINX CINEREA AND EUPLEURA CAUDATA) FOLLOWING AGNES

Introduction and Methods

The oyster drills *Urosalpinx cinerea* and *Eupleura caudata*, are the most destructive predators of oyster spat and small oysters in Chesapeake Bay where salinities exceed about 15 ppt (Andrews 1955). Salinities ranging from 7 to 10 ppt may kill oyster drills in 8 to 14 days (Zachary & Haven 1973). Agnes undoubtedly had a major influence on drill populations because of their sensitivity to low salinities. VIMS personnel evaluated the extent of this effect during August and September 1973 at locations where drills were a problem prior to Agnes.

The method used in detecting the presence of drills was commonly used in past studies (Andrews 1955). It consisted of constructing small wire bags (1.5 inch mesh) about 1 foot wide and 18 inches long which were filled with small freshly collected live oysters. Ten of these wire bags were attached by ropes 30 feet long to a 500-foot collecting line. At weekly intervals the wire bags were raised and shaken over a fine mesh wire screen which retained all drills attached to the live oyster "bait" during the past week. These data were used to show catch per trap per week.

Drill traps were placed in the lower parts of the James, York, Rappahannock, and Piankatank Rivers during August, September, and the first week of October 1973. The locations of specific rocks or areas mentioned in these sections are also given as km above the river mouth.

Results

The survey showed fresh waters associated with Agnes effectively killed drills in many areas where they were formely abundant (Table 5).

In the James River, prior to Agnes, the range of drills extended from about 1.6 km above the James River Bridge 14.5 km to the mouth of the river (Hargis 1966). After Agnes, drills were absent from the upper half of the range. None were collected on traps placed at the James River Bridge and at the mouth of the Nansemond River (8.8 km). At Hampton Flats (5 km) the catch was 3.0 per trap per week, indicating that salinities had not appreciably reduced population levels there.

Table 5. Drill trap catch of *Urosalpinx cinerea* and *Eupleura caudata* following Agnes at various locations during August and September, 1973.

Location	Upriver Location (km)	No. of Weeks Traps Exposed	Total Catch	Drills/Trap/Week
James River				
James River Bridge (Brown Shoal)	14.5	3	0	0
Nansemond River	8.8	3	0	0
Hampton Flats	5.5	3	52[1]	1.7
York River				
Gloucester Point	8.0	2	41[2]	2.1
Green Rock	14.5	2	2	0.1
Pages Rock	16.1	1	0	0
Rappahannock River				
Parrott Island	9.0	2	0	0
Mouth Corrotoman River (Towles Point)	18.0	3	0	0
Great Wicomico River - Gwynn I.				
Mouth of system	0	1	0	0

[1] 28 *Eupleura*
[2] 1 *Eupleura*

In the York River before Agnes, drills infested the area from off Aberdeen Creek (16 km) to the mouth of the river. Drills were still found at Gloucester Point (8 km) after Agnes, but the up-river range was greatly diminished by the fresh waters. Only two were captured at Green Rock (14.5 km) and none at Pages Rock (16.1 km), locations where drills caused extensive damage prior to Agnes.

The lower Rappahannock River, below Towles Point (18 km), has had a long history of drill infestation. In a study conducted in 1961 *Urosalpinx cinerea* were trapped at Towles Point. In addition, egg cases were obtained here indicating reproduction was occurring (Griffith & Engle 1961). No drills were collected in our study during the fall of 1973 from just south of Towles Point to Parrotts Rock (9 km) (Fig. 1).

Salinity levels in the Rappahannock River were below 10 ppt for about a month after the onset of Agnes in the lower river. This was sufficient to kill the drills.

No drills were obtained in traps set at the mouth of the Piankatank River in our study. Surveys made by laboratory personnel in previous years showed heavy drill infestation in this area prior to Agnes.

It is estimated that drills will not regain their former levels of abundance in the areas where they were killed for at least three or four years, thus oysters in this period will have a good chance for survival.

SUMMARY

Agnes had a major influence on all aspects of the molluscan fishery in Virginia waters of the Chesapeake Bay. Mortality of mature oysters on leased bottoms in the major systems was estimated as follows: Rappahannock: 50%; Potomac tributaries: 70%; James: 10%; and York: 2%. The economic loss on the private leases was estimated at 7.9 million dollars.

Oyster losses on public bottoms were low within the state and were estimated as follows: James River: 5%; York: 0%; and the Rappahannock: 2%. Losses on the public bottom in the Potomac were about 50%. As a direct result of the storm, the set of oyster larvae was the poorest on record. Hard clams were killed by Agnes in the York River by fresh water. Mortality, however, was heaviest in replanted populations. No effect was noted in the James River. There appeared to be a complete mortality of soft clams in the Rappahannock River from the Morattico region to the river entrance. In the York River and off Milford Haven, losses appeared minimal. Agnes greatly reduced oyster drill populations in the upper parts of their range in the James and York Rivers. In the Piankatank and Rappahannock Rivers it appears that most of the drills were killed.

LITERATURE CITED

Andrews, J. D. 1955. Trapping oyster drills in Virginia, Part I. The effects of migration and other factors on catch. Proc. Natl. Shellfish Assoc. 46:140-154.

Andrews, J. D. 1967. Interaction of two diseases of oysters in natural waters. Proc. Natl. Shellfish Assoc. 57:38-49.

Andrews, J. D., D. B. Quale, and D. S. Haven. 1959. Freshwater kill of oysters (*Crassostrea virginica*) in James River, Virginia, 1958. Proc. Natl. Shellfish Assoc. 49:29-49.

Carriker, M. B. 1955. Critical review of biology and control of oyster drills, *Urosalpinx* and *Eupleura*. U. S. Fish & Wildl. Serv. Spec. Sci. Rep. 148, 150 p.

Castagna, M. and P. Chanley. 1973. Salinity tolerance of some marine bivalves from inshore and estuarine environments in Virginia waters of the western mid-Atlantic coast. Malacologia 12(1):47-96.

Chesapeake Bay Research Council. 1973. Effects of Hurricane Agnes on the environment and organisms of Chesapeake Bay. A report to the Philadelphia District U. S. Army Corps of Engrs. (Aven M. Andersen, Coordinator), Chesapeake Bay Inst. Contrib. 187, Nat. Resour. Inst. Contrib. 529, Va. Inst. Mar. Sci. Spec. Rep. Appl. Mar. Sci. & Ocean Engr. 29, 172 p.

Galtsoff, P. S. 1964. The American oyster, *Crassostrea virginica*, Gmelin. Fish. Bull., U. S. Fish & Wildl. Serv. 64:1-480.

Griffith, W. and J. B. Engle. 1961. Role of salinity in the distribution of the oyster drill *Urosalpinx cinerea* in the Rappahannock River. Unpubl. MS.

Hargis, W. J., Jr. 1966. Final Report, Operation James River--An evaluation of physical and biological effects of the proposed James River navigation project. Va. Inst. Mar. Sci. Spec. Rep. Appl. Mar. Sci. & Ocean Engr. 7.

Haven, D. S., J. G. Loesch, and J. P. Whitcomb. 1973. An investigation into commercial aspects of the hard clam fishery and development of commercial gear for the harvest of molluscs. Final Rep. 3-124-R, U. S. Dept. of Commerce, Natl. Mar. Fish. Serv., Comm. Fish. Res. Develop. Act 88-309.

Haven, D. S., D. Walsh, and R. E. Bendl. 1973. Annual report to the National Science Foundation, 1973. Effects of low value of dissolved oxygen and high hydrogen sulfide on molluscs. Ann. Rep., Nat. Sci. Found. from Va. Inst. Mar. Sci.

Loesch, J. G. and D. S. Haven. 1973. Estimated growth functions and size-age relationships of the hard clam, *Mercenaria mercenaria*, in the York River, Virginia. The Veliger 16(1):76-81.

Virginia Marine Resources Commission. 1969-1972. Annual Reports of the Virginia Marine Resources Commission. Newport News, Va.

Wood, L. and W. J. Hargis, Jr. 1971. Transport of bivalve larvae in a tidal estuary. Fourth Marine Biology Symp. (D. J. Crisp, ed.), Cambridge Univ. Press, pp. 29-44.

Zachary, A. and D. S. Haven. 1973. Survival and activity of the oyster drill *Urosalpinx cinerea* under conditions of fluctuating salinity. Mar. Biol. 22(1):45-52.

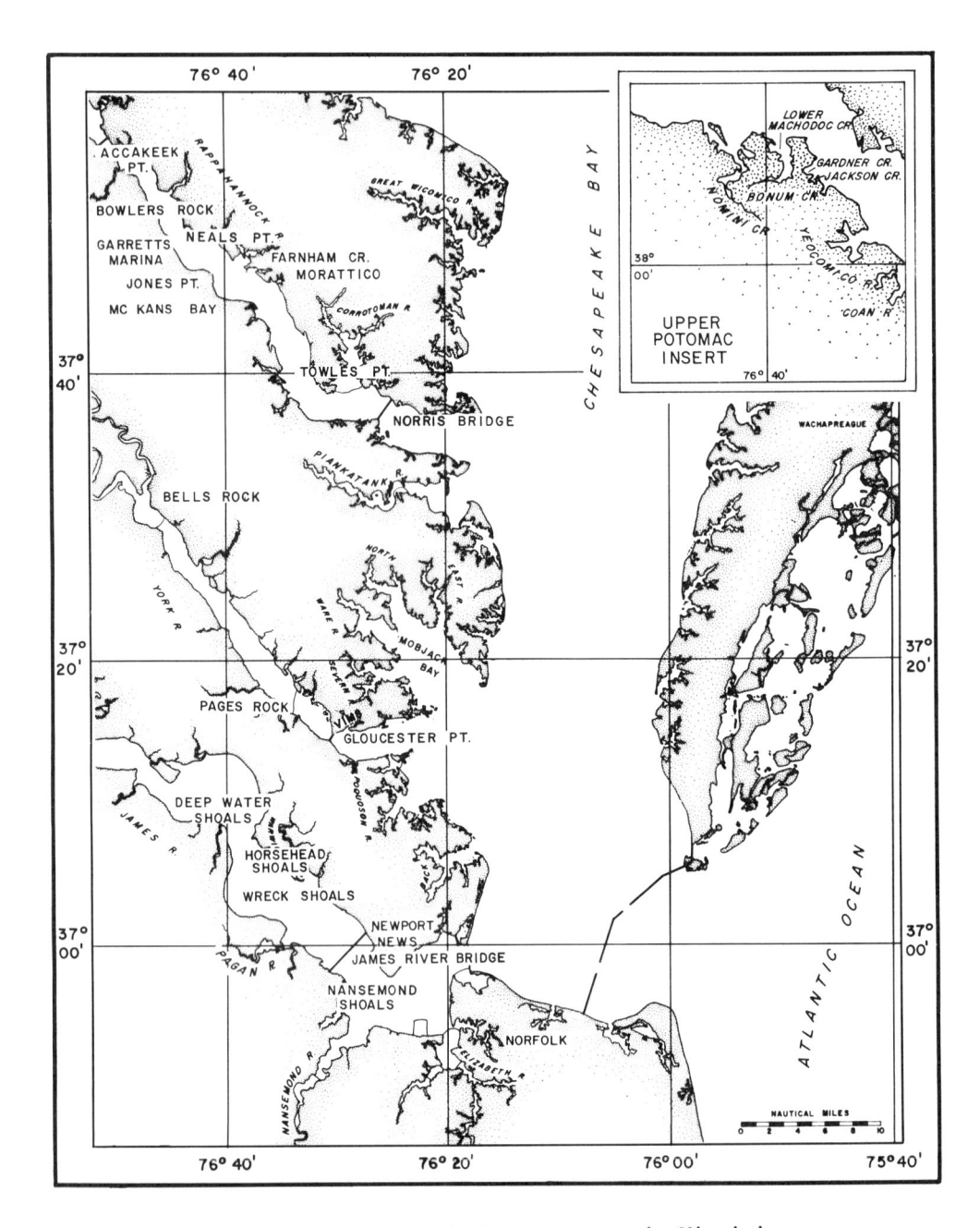

Figure 1. Location of river systems in Virginia.

A COMPARATIVE STUDY OF PRIMARY PRODUCTION AND STANDING CROPS
OF PHYTOPLANKTON IN A PORTION OF THE UPPER CHESAPEAKE BAY
SUBSEQUENT TO TROPICAL STORM AGNES[1]

M. E. Loftus[2]
H. H. Seliger[3]

ABSTRACT

The natural phytoplankton communities in Rhode River, West River, and an adjacent section of the Chesapeake Bay have been studied intensively since June 1969. Thus statistically significant comparisons of baseline data were made on both short term and long term effects of Tropical Storm Agnes on phytoplankton populations in the above area. Two major effects were 1) the immediate and dramatic change in relative species composition to favor the large dinoflagellates, and 2) a three-fold increase in mean total phytoplankton standing crops during 1973. Data are presented on production rates and efficiencies, nutrient levels, and standing crops of phytoplankton.

INTRODUCTION

In the aftermath of Agnes (June 21, 1972) the highly turbid, nutrient-rich floodwaters entered the upper regions of the Chesapeake Bay. These waters displaced and partially mixed with the surface waters and their endogenous crop of phytoplankton. In the following weeks the southward flow, the mixing of the lower density flood waters with the underlying bay water, and the upstream displacement of deeper mid-bay waters were coincident with the alteration of the relative composition and mass of the phytoplankton populations in the Rhode River. We had studied the phytoplankton crops in this subestuary for 2.5 years prior to Agnes and have 1.5 years of observations since the storm. These data on standing crops, species composition, primary productivity, nutrient concentrations, and salinity and temperature patterns in the surface waters of this subestuary, provide a means of evaluating the short and possibly long term effects of this storm on the phytoplankton community. We have compared seasonal mean values of the above parameters before and after Agnes for both Rhode River and the adjacent bay region over the time period 1970-1973.

METHODS

Figure 1 shows the Rhode River, Md., located approximately 5 miles south of Annapolis and the Severn River mouth. The river has a (mean low water) surface

---

[1]Contribution No. 864 from the McCollum-Pratt Institute and Department of Biology, The Johns Hopkins University. This work was supported by U. S. Atomic Energy Commission Contract No. AT(11-1) 3278 and National Science Foundation Grant No. GI 32110. M.E.L. received support under National Institutes of Health Training Grant GM-57.
[2]Current address: Ecological Analysts, Inc., York Building, 8600 LaSalle Road, Towson, Maryland 21204.
[3]Department of Biology, The John Hopkins University, Baltimore, Md. 21218

area of 5.9×10^6 m^2, a depth of 1.98 m, and a volume of approximately 11.46×10^6 m^3 (Pritchard & Han 1972). Its mouth joins with that of the West River resulting in common exchange with the bay. These rivers receive minimal freshwater inflow and therefore are only tributary-embayments. The dashed line (Fig. 1) indicates the location and extent of transects used to monitor the phytoplankton crop, its productivity, and associated nutrients beginning in 1970 and continuing through 1972. River data taken in 1973 are compared with data collected at Station 854-G (Fig. 1). At weekly to biweekly intervals, water from 0.5 meters was pumped continuously into a 20-liter carboy while traversing each transect. At the end of each transect vertical profiles of temperature and salinity were taken using a Beckman conductivity salinometer. Water samples were collected at several depths for pH, alkalinity titration, chlorophyll analysis (Loftus & Carpenter 1971), rates of primary productivity measured by ^{14}C uptake (Steeman-Nielsen 1952), and plankton enumeration. Nutrient samples were taken by filtration of averaged transect samples through pre-washed 0.45 μ Millipore or 984H Reeve Angel filters. The filtrates were stored at -20°C in acid-washed bottles until analyzed for inorganic phosphate (soluble reactive) (Murphy & Riley 1962), filter-passing dissolved organic phosphate (Strickland & Parsons 1968), nitrite (Bendschneider & Robinson 1952), nitrate (Morris & Riley 1963) and ammonia (Solórzano 1969). Phytoplankton were counted from freshly collected samples using Sedgwick-Rafter (1 ml), Palmer (0.1 ml) and hemocytometer (10^{-14} ml) counting chambers. Other samples were preserved with Lugol's iodine or Bouins' solution and counted in the above chambers. Dominant species or forms were counted without volume concentration.

Carbon assimilation rates were measured routinely by incubation in situ at 0.5 times the Secchi disc depth (~45% light). Beginning in 1972, productivity as a function of light intensity was measured using a) neutral density plastic screens to achieve 100, 61, 31, 18, and 4% of surface light intensities or b) incubation in situ at depths of 0.1, 0.5, 1.0, 1.5, and 2.0 meters. Incubation periods were either 1 or 2 hours duration between 1200 and 1430 hr. Duplicate light and dark bottles were filtered on prewetted 0.45 μ Millipore filters which, after rinsing with filtered bay water, were dissolved in dioxane-omnifluor solution in scintillation counting vials. For the calculation of photosynthetic carbon uptake rates inorganic carbon concentrations were determined from alkalinity titration, pH and the dissociation constants of the bicarbonate system (Buch 1945). The in situ and simulated in situ (neutral density screens) productivity/m^2 were calculated integrations over the photic zone. The simulated in situ productivity/m^2 was calculated assuming a) a constant attenuation coefficient calculated from Secchi depth (~20% surface light), b) the light available for photosynthesis was approximately one-half the Langley irradiance measure by a unfiltered pyroheliometer during the incubation period c) the fraction of light reflected from the water surface was 5% of that incident and d) chlorophyll $\underline{a}$ was uniformly distributed in the photic zone. Assimilation numbers were calculated for each screen transmission. The best regression rectangular hyberbola was calculated (Wilkinson 1961) to fit the saturation curve of hourly productivity per unit chlorophyll $\underline{a}$ (assimilation number) vs. hourly irradiance (Langleys/cm^2/hr^{-1}). These light response equations were integrated over the photic zone (from the surface to the depth of 1% surface light). On four occasions in situ and simulated in situ productivity integrals were compared directly. Analysis of variance showed that the two procedures were not significantly different (C.V. = 21%). In two separate experiments the midday hourly rates ranged between 1/8 and 1/10 of the total daily productivity; a factor of 9 was used to estimate the daily rates of productivity from midday values. Since these values were based on short term ^{14}C uptake the estimates were assumed to be more closely related to gross production.

RESULTS AND DISCUSSION

The comparison of means within the Rhode River (Transect 1) was used to evaluate the effects of Agnes on the phytoplankton crops using the Student's t statistic. A framework for overall comparison of standing crop chlorophyll $\underline{a}$ is essential. Seliger and Loftus (1974) have used a paired comparison analysis of standing crops to show that no significant differences were observable between the annual mean difference of chlorophyll $\underline{a}$ in the Rhode and West Rivers in 1970 and 1971. The West River crop was significantly higher than that of the Rhode River in 1972 by $\sim$4 µg/l. The annual Rhode River chlorophyll $\underline{a}$ mean exceeded that of the adjacent bay by $\sim$7 µg/l in 1970 and $\sim$10 µg/l in 1972. No significant difference was found in 1971. This summary illustrates the degrees of similarity among algal biomasses of the Rhode River, the West River and the adjacent bay region. In the remaining discussion we will consider only the comparison within the Rhode River (Transect 1) over time.

Annual values of surface water salinities in the Rhode River transect 1 (line o) and local rainfall (histogram) are shown in Fig. 2. The variations in the salinities reflect exchange with the bay. These fluctuations are not well correlated with local rainfall. The general pattern of the annual salinity variations indicates a winter decline to a spring minimum, coincident with maximum Susquehanna River flow occurring between April and June, followed by a regular increase to a fall maximum in October-December. In 1972 the delivery of the Susquehanna floodwaters of Agnes to the Rhode River occurred within 5 days of the storm (next sampling) when a decrease from 7.1 ppt to 2.5 ppt in salinity was observed. Approximately six weeks elapsed before river surface salinities returned to pre-Agnes values. The salinity patterns for the remainder of 1972 and for 1973 were typical of previous years.

In the region of the river sampled by transect 1, Han (this vol.) showed that the pre-Agnes water mass was displaced by flood waters. Coincident with this displacement we observed a marked increase in several nutrients. During the period from five days prior to Agnes to six days following there were increases in nitrate (0.9 to 149 µg at/l); nitrite (0.13 to 0.86 µg at/l), ammonia (2.0 to 33 µg at/l) and dissolved organic phosphate (0.4 to 1.2 µg at/l). No significant change in inorganic phosphate was observed. Two weeks following the storm the elevated levels were reduced to those characteristic of pre-Agnes water and normal summer levels with the notable exception of the dissolved organic phosphate which remained at high levels in the bay, West River and Station 7 throughout the summer.

Coincident with the pulse of nutrients delivered to the subestuary in the flood waters the algal (chlorophyll) standing crop increased by a factor of 5-10 over that observed for pre-storm periods. This change is documented in Fig. 3a where the extractable chlorophyll $\underline{a}$ concentrations are shown for transect 1 for 1972. The elevated levels of algal biomass continued through August but subsided in the fall and winter months. This increase was the result of blooms of *Gymnodinium nelsoni*, indicated by the reduction of chlorophyll passing a 10 µ Nitex net (Fig. 3b). Typically the size distribution of phytoplankton in the bay is dominated by nannoplankton (Seliger 1972; Loftus et al. 1972; McCarthy et al. 1974; Seliger & Loftus 1974) as is that of the rivers where 80-90% of the chlorophyll $\underline{a}$ passes through a 20 µ Nitex net. Fig. 3b shows that the relative distribution in the post-Agnes period has shifted from the normal nannoplankton dominance to 60-80% dominance by *G. nelsoni*. The single reversion to nannoplankton dominance is coincident with a second water mass exchange documented by Han (this vol.). The dinoflagellate bloom persisted until early September when chlorophyll $\underline{a}$ returned to pre-Agnes fall levels and the nannoplankton dominance returned. The effect of Agnes on the physical and biological parameters of the Rhode River apparently ended by the fall of 1972.

However, the chlorophyll a standing crop over 4 years of this survey (Fig. 4) shows the abnormally high chlorophyll standing crops in 1973. In all previous years a distinct seasonal procession of dominant dinoflagellates was observed in the river and adjacent bay regions. In winter, December through February, the small dinoflagellate *Katodinium rotundatum* is sometimes found in cell densities of $>10^5$/ml and chorophyll a concentrations exceeding 100 µg/l. These high densities are unusual as normal winter crops of chlorophyll a usually do not exceed 10 µg/l. During the period May-June, *Prorocentrum minimum* blooms are often evident in the western shore bay waters. *P. minimum* also occurs in high densities ($>10^5$/ml) when these blooms are carried into the rivers. The dominance of *P. minimum* gives way to *G. nelsoni* in late June to mid-July. *G. nelsoni* remains through September and is occasional in October sampling. The latter species declines when water temperatures decrease below $\sim$20°C. The order of appearance of dinoflagellates has been consistent throughout the four years of this survey; only the intensity and persistence were greater throughout 1973.

The statistical significance of the changes in chlorophyll standing crops are shown in Table 1. The mean values for chlorophyll a were calculated for all seasons of the four-year study. The Student's t statistic was used to determine the significance of the differences observed between seasons within each of the years and between like seasons of following years. The significance of the observed differences is indicated symbolically (see table legend) at each season boundary line in the table. Treating the data in this way tends to reduce the effects of gradual seasonal trends while showing significance when abrupt changes occurred. The three-fold increase in the 1972 summer standing crop following Agnes was extremely significant compared with spring and fall values for the same year as well as the summer levels of the preceding year but not different from the crop in the summer and fall of 1973. This testing procedure separates the short term effects of the tropical storm from its possible long term influence. The fall of 1972 and winter 1973 means are not significantly changed from preceding seasons while the increase in the warmer months of 1973 is apparent where the mean spring-summer crop exceeded that of pre-Agnes levels by nearly four-fold. An increase of this magnitude was not observed in the bay proper in 1973 suggesting that the tributary embayments might have been differentially influenced by the tropical storm in 1972.

No significant differences could be found in the productivity per square meter (100-1% of surface light) values from in situ and simulated in situ measurements made before and after Agnes in 1972. Similarly no difference in productivity was detectable between the summer values taken in 1972 as compared with 1973. When the 1972 and 1973 values were combined and productivity (gC/m^2 hr) was plotted as a function of ambient light intensities during the incubation period (Fig. 5) there appeared to be seasonal differences. The light saturation level for the colder months is lower than that for the warmer seasons. Taylor and Hughes (1967) made estimates of productivity over the photic zone in the upper Chesapeake Bay and its tributaries during the summer months of 1964. Primary productivity ranged from 1.5 to 3.5 gC/m^2-day as compared to our present values at light saturation of 1.5 to 2.4 gC/m^2-day. There is therefore no evidence to indicate that the changes in standing crops within the photic zone during 1972 and 1973 have increased the productivity of the upper bay region over the past decade. The gross annual production in 1972, 1973, based on average daily rates was 0.6 KgC/m^2. This value is similar to gross but far greater than net productivity values estimated by Flemer (1970) for the upper bay channel. Our value is also similar to gross annual production found by Flemer and Olmon (1971) in the Patuxent River. It is evident that no significant changes in gross productivity per square meter can be attributed to the effects of Agnes and that primary productivity in the Rhode River does not appear to be limited by light.

Table 1. Mean Values for Chlorophyll a (Rhode River) (µg/liter)

Year	Winter J 1-90 M	Spring-Summer A 91-240 A	Fall S 241-365 D
1970	8.2 n = 1	27.27 ± 10.85 ID n = 16 NS	(*) 11.25 ± 5.29 n = 6	NS
	—— ID ——	—(***)—	—** —
1971	8.87 ± 4.96 n = 7	NS 11.97 ± 4.43 NS n = 19 (*)	21.74 ± 8.29 n = 7	**
	—NS—	—(*)———(***)—	— NS —
1972	6.27 ± 3.4 n = 9	19.3+9.5 56.75+19.1 (***) n=9 (***) n=6 (***)	18.9 ± 7.6 n = 3	NS
	—NS—	—(***)———(NS)—	—**—
1973	10.36 ± 5.9 (***) n = 5	74.6 ± 34.7 (NS) n = 12 NS	59.8 ± 35 n = 6

ID: insufficient data
NS: not significant P>.05
*: significant .05 >P >.01
**: very significant 0.01> P.001
***: extremely significant .001>P

No statistical differences in the inorganic nutrient trends could be associated with Agnes as the variances were large relative to the means. Concentrations of NO_3^- in the winter months have always been significantly greater than those during summer and fall. No significant differences could be found in the nitrite or inorganic phosphate. However a significant increase was evident in the concentrations of dissolved organic phosphorus (DOP). As shown in Table 2 concentrations of DOP in 1973 were nearly four times greater than those during 1972. As was previously noted the pulse of Agnes' floodwater was coincident with elevated DOP concentrations in Rhode River (Transect 1). This declined within two weeks of the storm, and throughout the remainder of 1972 concentrations were comparable to those found in Transect 1 during 1971 (Table 2). The observations of high DOP throughout 1973 in Rhode River were coincident with high bacterial levels ($>10^6$/ml) in surface waters observed by Dr. M. Faust (personal communication) and with depletion of dissolved oxygen in bottom waters during June and August (D. Correll; R. Cory, personal communication).

Dissolved organic phosphate concentrations in surface waters (0-2 m) at Station 854-G were collected on bi-monthly cruises of the Chesapeake Bay Institute (Taft 1974) and supplements our data (Table 3). Pre-Agnes values are essentially the same. However, the high concentrations of DOP in the Rhode River and West River in 1973 could not have originated in the bay. Apparently the source of the elevated dissolved organic phosphorus was within the tributary embayments.

Following Agnes, the deposition of sediment in the upper bay was extensive (Schubel, this vol.). Suspended sediment as high as 200-300 mg/l reached the Rhode River area several weeks after the flood. Twenty to twenty-five percent of this

Table 2. Dissolved organic phosphorus µg atom/liter

Year	Winter J 1-90 M	Spring-Summer A 91-240 A	Fall S 240-365 D
1970	ND	1.93 ± 0.068 NS n = 2 (***)	0.77 ± 0.39 NS n = 6 NS	NS
1971	0.35 ± 0.57 NS n = 7 NS	0.43 ± 0.66 NS n = 16 NS	0.77 ± 1.08 NS n = 8 NS	NS
1972	0.51 ± 0.76 NS n = 9 (*)	0.39 ± 0.72 NS n = 13 (***)	0.34 ± 0.39 (*) (*)	
1973	1.98 ± 0.9 NS n = 4	1.92 ± 0.48 NS n = 7	1.84 ± 0.66 n = 3	

ID: insufficient data
NS: not significant P>.05
 *: significant .05>P>.01
 **: very significant 0.01>P>.001
 ***: extremely significant .001>P

Table 3. Dissolved organic phosphate (µg at./l): Yearly range.

Location:	Rhode River	Chesapeake Bay
Station:	Tran. IA	Tran IIIA (854-G)*
Year		
1972	(0.03 - 1.2**)	(0.035 - 2.54**)
1973	(1.09 - 3.30)	(0.30 - 0.55)*

**Post-Agnes values

material was organic. Since the sediment load returned to its normal range of 2-30 mg/l (Schubel et al. 1970) several months after the storm a considerable deposition must have occurred in the river channels and in the embayments.

CONCLUSION

The rapid increases in standing crops of phytoplankton associated with the influx of floodwaters of Agnes was the result of the growth of an algal species endemic to the area during the summer season. The blooming condition continued through the period of salinity recovery and nutrient depletion until the fall decline in temperature. Further effects of Agnes on the phytoplankton were not evident in the fall of 1972 or winter of 1972-1973. Productivity of the increased standing crop following Agnes was not light limited. It was neither significantly different from estimates of gross production per square meter made during the year prior to Agnes nor from those made a decade earlier (Taylor

& Hughes 1967).

The observations that 1) the standing crops of phytoplankton in 1973 equalled or surpassed the previous year's post-Agnes levels, 2) a 4-fold elevation of dissolved organic phosphate levels occurred in 1973 river samples compared with those observed in pre-Agnes samples and in post-Agnes bay water (Taft 1974) and 3) oxygen depletion in bottom waters of the river in 1973 was associated with high (>10^6/ml) bacterial levels, are consistent with the hypothesis that the amplification of the endogenous species concentrations in 1973 resulted from the utilization of the organic material deposited in the bottom sediments by Agnes in the preceding year.

ACKNOWLEDGEMENTS

We wish to express our thanks to Mr. Robert Cory who has made available his pyranometer records and for his helpful discussion of oxygen distributions in the Rhode River. We thank Dr. Maria Faust for information regarding the bacterial populations in the river and Dr. Jay Taft for supplying us with dissolved organic phosphate data in the bay waters adjacent to the rivers. Some data on standing crops were collected during Dr. W. R. Taylor's "Procon" cruises aboard the *Ridgely Warfield* supported by U. S. Atomic Energy Commission contract AT(11-1)3279 and National Science Foundation grant GA-33445. We especially acknowledge the diligence and technical skill of Mrs. Catherine Eisner without whom this work would not have been possible.

LITERATURE CITED

Bendschnieder, K. and R. J. Robinson. 1952. A new spectrophotometric method for the determination of nitrite in sea-water. J. Mar. Res. 11:87-91.

Buch, K. 1945. Kolsyrejamviktem i Baltiska Havet. Fennia 68:1-208.

Flemer, D. A. 1970. Primary production in the Chesapeake Bay. Chesapeake Sci. 11:117-129.

Flemer, D. A. and J. Olmon. 1971. Daylight incubator estimates of primary production in the mouth of the Patuxent River, Maryland. Chesapeake Sci. 12:105-110.

Loftus, M. E. and J. H. Carpenter. 1971. A fluorometric method for determining chlorophyll *a*, *b*, and *c*. J. Mar. Res. 29:319-338.

Loftus, M. E., D. V. SubbaRao, and H. H. Seliger. 1972. Growth and dissipation of phytoplankton in Chesapeake Bay. I. Response to a large pulse of rainfall. Chesapeake Sci. 13:282-299.

McCarthy, J. J., W. R. Taylor, and M. E. Loftus. 1974. Significance of nanoplankton in the Chesapeake Bay estuary and problem associated with the measurement of nanoplankton productivity. Mar. Biol. 24:7-16.

Morris, A. W. and J. P. Riley. 1963. The determination of nitrate in seawater. Anal. Chim. Acta 29:272-279.

Murphy, J. and J. P. Riley. 1962. A modified single solution method for the determination of phosphate in natural waters. Anal. Chim. Acta 27:31-36.

Pritchard, D. W. and G. Han. 1972. Physical hydrography of the Rhode River. Ann. Rep., Chesapeake Research Consortium, May 31, 1972, p. 446-457.

Schubel, J. R., C. H. Morrow, W. B. Cronin, and A. Mason. 1970. Suspended sediment data summary, August 1969-July 1970 (mouth of Bay to head of Bay). Chesapeake Bay Inst., Johns Hopkins Univ., Spec. Rep. 19, Ref. 70-10.

Seliger, H. H. 1972. Annual Report, Chesapeake Research Consortium, May 31, 1972, p. 458-551.

Seliger, H. H. and M. E. Loftus. 1974. Growth and dissipation of phytoplankton in Chesapeake Bay. II. A statustical analysis of phytoplankton standing crops in the Rhode and West Rivers and an adjacent section of the Chesapeake Bay. Chesapeake Sci. 15(4):185-204.

Solorzano, L. 1969. Determination of ammonia in natural waters by the phenol hypochlorite method. Limnol. Oceanogr. 14:799-801.

Steeman-Nielsen, E. 1952. The use of radioactive carbon (^{14}C) for measuring organic production in the sea. J. Cons. Perma. Int. Explor. Mar. 18: 117-140.

Strickland, J. D. H. and T. R. Parsons. 1968. A practical handbook of seawater analysis. Fish. Res. Bd. Canada Bull. No. 167, 311 p.

Taft, J. L., III. 1974. Phosphorus cycling in the plankton of Chesapeake Bay. Johns Hopkins Univ. Ph.D. Thesis.

Taylor, R. W. and J. E. Hughes. 1967. Primary productivity in the Chesapeake Bay during the summer of 1964. Johns Hopkins Univ. Tech. Rep. 34, Ref. 67-1.

Wilkinson, G. N. 1961. Statistical estimations in enzyme kinetics. Biochem. J. 80:324-332.

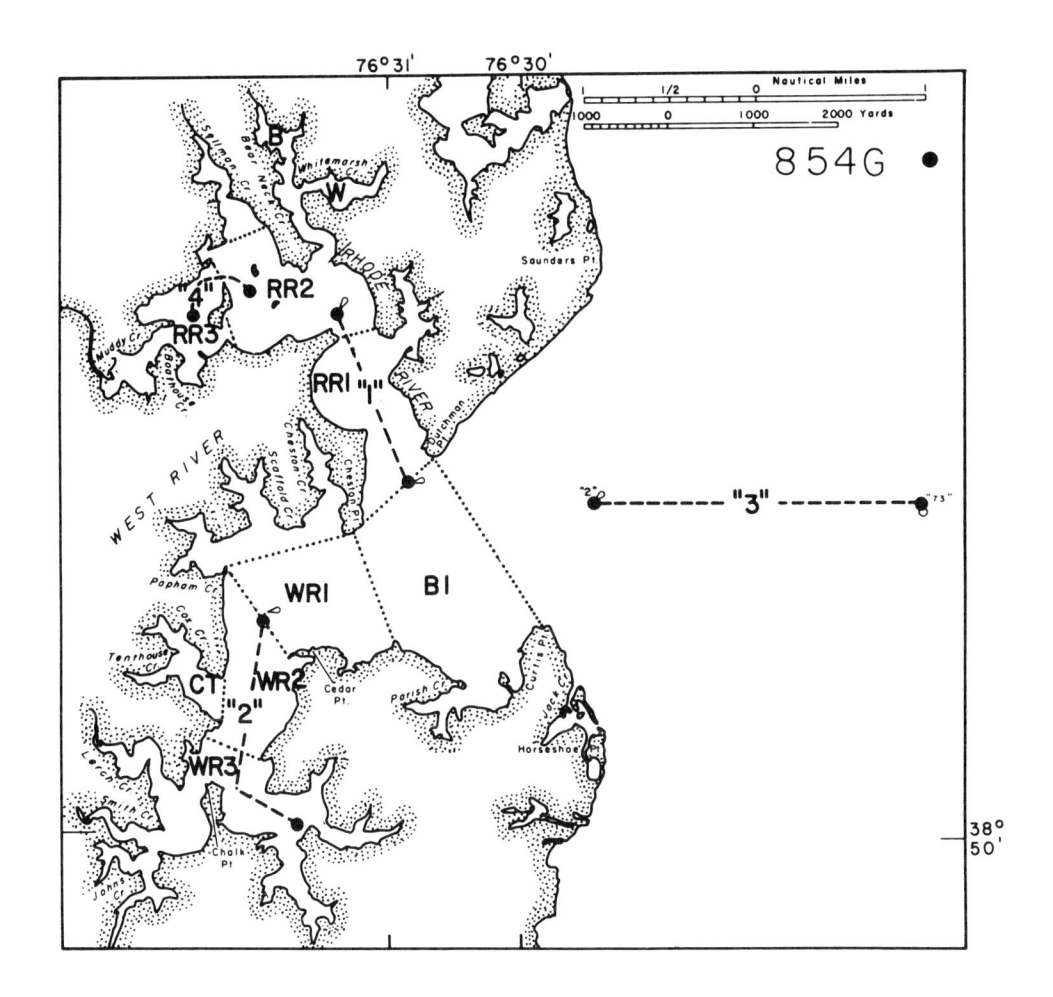

Figure 1. Sampling location in Rhode River, West River, and Chesapeake Bay, 1970-1973. Station 854-G (upper right) was used for 1973 dissolved organic phosphate collections by Dr. J. Taft.

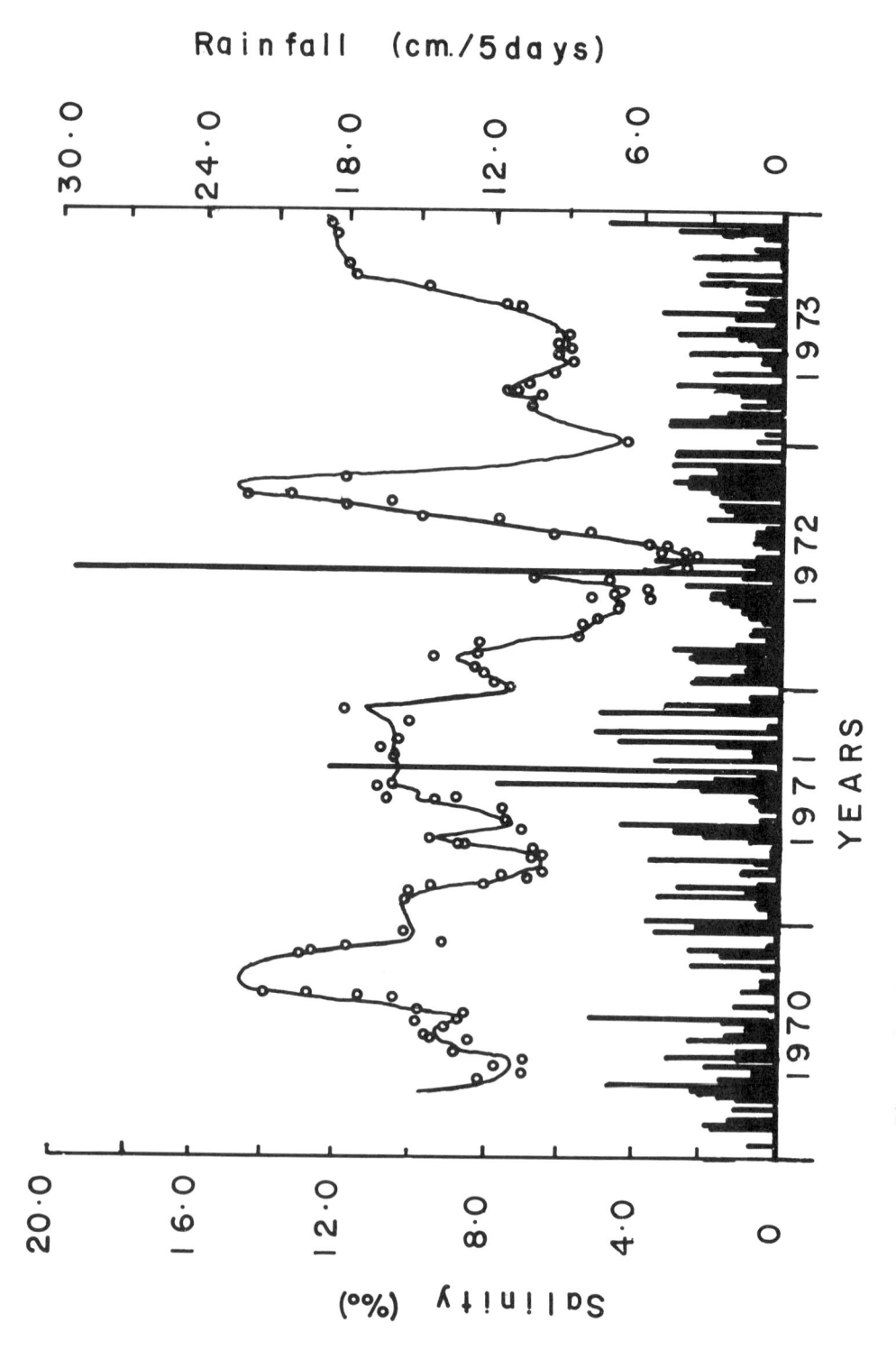

Figure 2. Four-year surface water salinity values measured at Station 7 in Rhode River with histogram of local rainfall (cm/5 days).

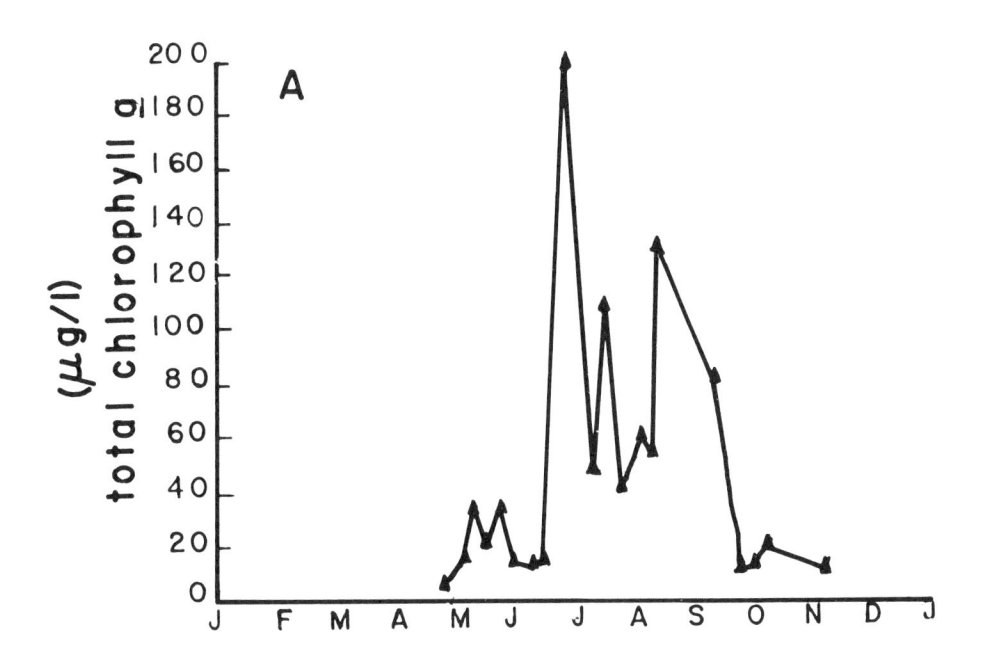

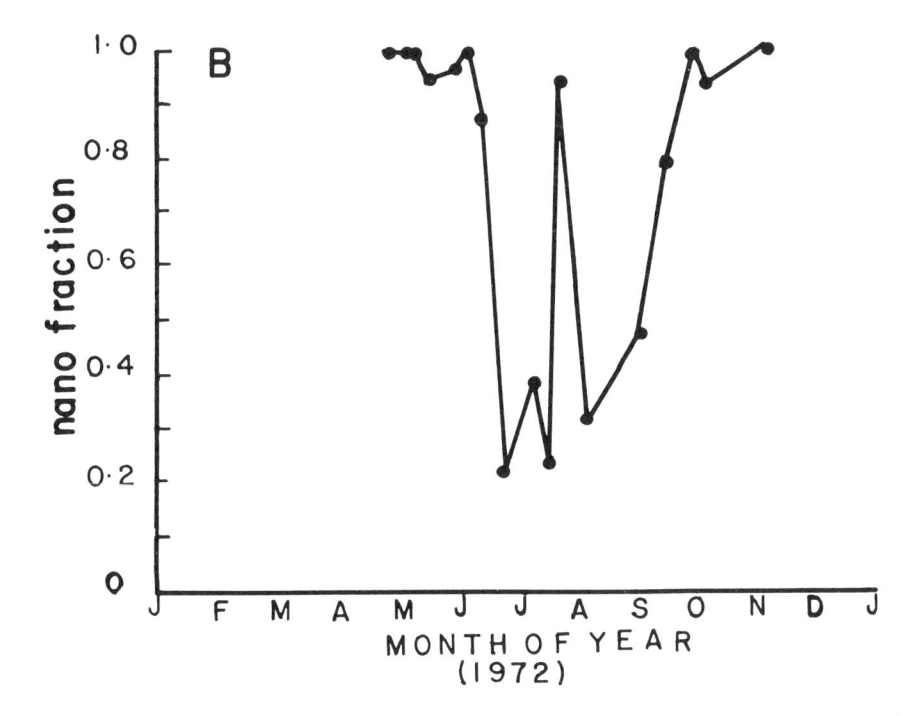

Figure 3. a) Chlorophyll a standing crop concentrations at Station 7 (beginning of Transect 1) in 1972.
b) Fractions of chlorophyll a which passed through a 10 μ Nitex net over the same period.

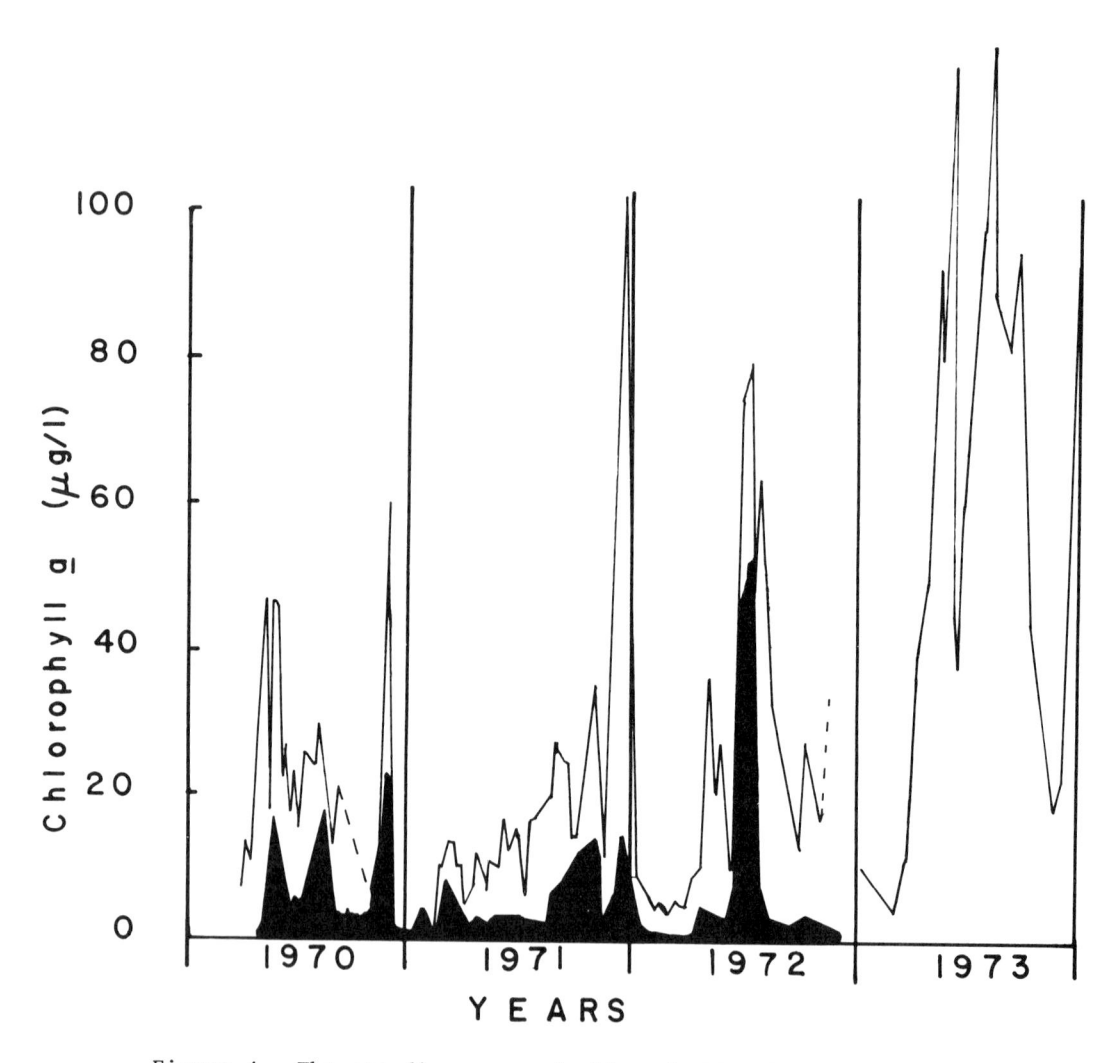

Figure 4. The standing crop of chlorophyll a in Transect 1 1970-1973 (line) and the portion of that crop retained on a 20 μ Nitex net (solid portion). Size distributions were not made in 1973.

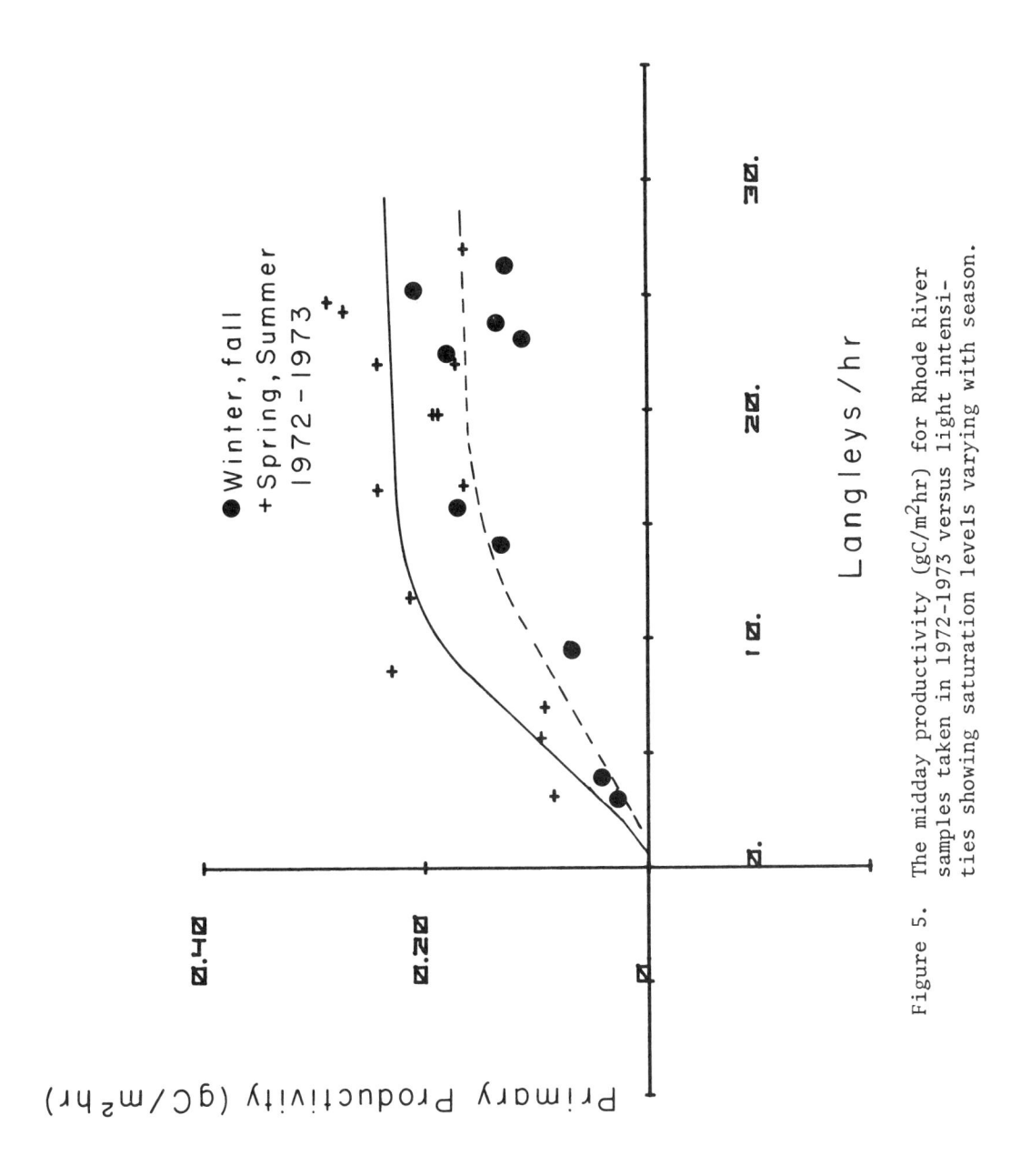

Figure 5. The midday productivity (gC/m²hr) for Rhode River samples taken in 1972-1973 versus light intensities showing saturation levels varying with season.

EFFECTS OF TROPICAL STORM AGNES ON THE BACTERIAL FLORA OF CHESAPEAKE BAY

J. D. Nelson, Jr.[1]
R. R. Colwell[1]

ABSTRACT

An investigation of the mercury-metabolizing bacterial microflora of Chesapeake Bay was undertaken during a period of time which encompassed the 9 months preceding and 11 months following Tropical Storm Agnes. Physical parameters, including ambient mercury concentrations, were measured, and the viable, aerobic, heterotrophic total and mercury-resistant bacterial populations were enumerated at stations located in Baltimore Harbor, off Matapeake, and in Eastern Bay. Comparison of data before and after June 1972 indicated increases in total and mercury-resistant populations, particularly in surface water, followed by a gradual return to levels observed prior to the storm.

INTRODUCTION

The influence of bacteria on the mobilization of mercury from sediments in Chesapeake Bay has been under study in our laboratory (see references). An ancillary goal of the studies was an assessment of the numbers of potentially Hg^0-evolving aerobic, heterotrophic bacteria. The proportion of $HgCl_2^-$ resistance among the total bacterial community, determined by plate-counting, has been shown to be a viable method for estimating in situ activity (Nelson & Colwell 1975). During the course of this investigation, in which extensive data relative to numbers of viable, aerobic, heterotrophic bacteria were accumulated, Chesapeake Bay was subjected to the effects of Agnes. Consequently, some information on the effect of the storm on the overall changes in the heterotrophic, aerobic, bacterial community and on the specific microbial population represented by the Hg-resistant bacteria, was fortuitously derived during the studies.

MATERIALS AND METHODS

Surface water (1 meter depth) samples were taken aseptically, using Niskin sterile bag samplers. Surface sediment samples were taken with a Petite Ponar grab sampler (Wildlife Supply Co., Saginaw, Mich.). Samples were diluted in artificial estuarine salts solution (10.0 gm NaCl, 2.3 gm $MgCl_2 \cdot 6H_2O$, and 0.3 gm KCl per liter) on an agar nutrient medium containing estuarine salts, casamino acids 0.5%, yeast extract 0.1% and glucose 0.2%. The plates were incubated at 20°C for 1 week, and colonies were counted. $HgCl_2$ was incorporated into the same medium at a concentration of 6 ppm, and mercury-resistant bacteria were enumerated in the same manner. Numbers of mercury-resistant bacteria were normalized and expressed as percent of the total viable count.

Dissolved oxygen (DO) was measured using a Y.S.I. (Yellow Springs Instrument Co., Yellow Springs, Ohio) oxygen electrode, and salinity was determined using a portable induction salinometer (Beckman Instruments Co., Baltimore, Maryland). Total mercury concentration in sediment was determined by flameless atomic ab-

---
[1]University of Maryland, College Park, Md. 20742.

sorption spectrophotometry at the National Bureau of Standards, Gaithersburg, Maryland, by W. E. Blair and F. Brinckman.

Details of the microbiological and analytical methods, media, and regents used are given elsewhere (see references).

RESULTS AND DISCUSSION

At approximately 6-week intervals, from Fall 1971 to Spring 1973, samples of surface water and surface sediment layers were plated on nutrient medium, with and without $HgCl_2$ supplementation. All physical and bacteriological data were separated into two groups: data preceding June 1972 ("before Agnes"), and those following ("after Agnes") (Table 1). With the exception of mercury determinations, all data were grouped by season, permitting comparisons based on samples taken within 2-week periods in consecutive years.

Table 1. Physical Parameters Measured Before and After Tropical Storm Agnes.

Season	Location	Before/After Agnes	Sediment (Hg Total)(ppm) (95% confidence interval)	Surface Temp(C)	Salinity (ppt)	Turbidity Secchi(m)	Dissolved O_2 (ppm)
Fall	B	B		17.5	7.25	0.60	9.6
		A	0.97 ± 0.04	20.0	8.80	1.30	9.3
	M	B		9.5	-	1.75	10.5
		A	0.20 ± 0.08	7.0	6.99	1.50	11.4
	EB	B		8.0	-	1.50	-
		A	0.04 ± 0.01	7.0	-	1.60	-
		B		10.0	-	2.00	-
		A	0.07 ± 0.01	9.0	-	2.00	-
Winter	B	B	0.82 ± 0.03	2.8	2.5	0.5	13.2
		A	0.67 ± 0.01	5.5	4.3	0.5	12.6
	M	B	0.29 ± 0.02	5.5	-	2.00	12.1
		A	0.20 ± 0.03	4.7	-	0.50	12.4
	EB	B	0.04 ± 0.01	2.4	10.6	2.00	13.2
		A	0.08 ± 0.02	4.7	9.2	1.50	12.4
Spring	B	B	0.82 ± 0.02	8.7	3.6	0.50	11.0
		A		11.2	4.8	0.40	11.0
	M	B	0.27 ± 0.03	6.1	3.8	0.80	11.9
		A		8.0	8.3	1.00	11.2
		B	0.37 ± 0.02	18.2	6.81	1.50	8.80
		A	0.14 ± 0.006	15.2	7.24	1.00	-
	EB	B	0.05 ± 0.01	16.0	7.06	1.75	10.0
		A	0.102± 0.008	16.3	9.13	2.00	-

B = Baltimore Harbor
M = Matapeake
EB = Eastern Bay

Changes resulting from Agnes were evaluated statistically by a t-test of the significance of differences in population means. The happenstance of this particular event, i.e., Tropical Storm Agnes, does not allow interpolation with great precision, but certain trends were quite evident.

The statistical results in Table 2 are based upon seasonally paired observations of physical parameters before and after Agnes (Table 1). Techniques of sufficient precision and accuracy for the detection of Hg in water were not available at the time, hence data for sediment appear in Table 1. For the sake of uniformity, all data were divided into "before" and "after" groups, and "t's" were calculated and two-tailed tests of significance were performed. There was no apparent significant, permanent change in mercury concentration in sediment, water temperature, turbidity, or DO concentration, whereas, salinity appears to have increased slightly. The mercury data indicated that extensive deposition of mercury did not occur after the storm, i.e., as a result of the storm.

Table 2. Comparative physical characteristics of Chesapeake Bay before and after Tropical Storm Agnes.

Comparison	t^1 (difference of means)	Degrees of freedom	Significance level[2]
Sediment Hg total	+0.6182	28	n.s.
Surface water temperature	+0.1500	20	n.s.
Surface salinity	+1.1018	12	n.s.
Turbidity (Secchi disc)	-0.5657	20	n.s.
Dissolved oxygen	-0.3697	12	n.s.

[1] treated as unpaired comparisons.
[2] n.s. = not significant at 0.20 level or less (two-tailed test).

However, comparison of total population levels of bacteria before and after the storm indicated that temporary, but measurable, increases occurred (Tables 3 and 4) in both total and mercury-resistant bacteria. The data were examined after grouping by location and season. Numbers of bacteria in the water increased significantly overall after the storm. An overall increase in total numbers of bacteria in the sediment was also noted, but it was found to be less significant. In addition, significant increases were observed at all three sampling locations, i.e., in Baltimore Harbor, off Matapeake, and in Eastern Bay. A breakdown of the data by season indicates that by Spring 1973, the population levels returned to "normal."

Population levels of mercury-resistant bacteria exhibited a seasonal periodicity that was observed prior and subsequent to the storm. Thus, Agnes did not produce drastic ecological effects on this physiological group of bacteria. A reproducible Spring peak in total, viable numbers of mercury-resistant bacteria was observed, both in 1972 and 1973 for water and sediment (Figs. 1 and 2). A secondary peak in Fall 1972 may have been a result of the storm, but this hypothesis could not be confirmed because sampling was terminated prior to Fall 1973 because of termination of the project in which this study was included. Statistical analysis of the data indicated similar changes in a specific segment of the bacterial community. Mercury-resistant bacteria in the water increased after the storm (Table 4). In contrast, mercury-resistant bacteria in sediment decreased.

Table 3. Numbers of bacteria in water and sediment before and after Tropical Storm Agnes.

Season	Location	Before/After Agnes	TVC Water (x10²)	Sediment (x10⁵)	% of TVC Resistant (6 ppm Hg Cl₂) Water	Sediment
Fall	B	B	65.0	1.3	4.6	5.6
		A	4100.0	16.0	92.5	5.7
	M	B	8.7	1.0	<1.2	2.0
		A	75.0	7.4	6.0	1.5
	EB	B	6.9	35.0	<1.7	<.04
		A	14.0	5.5	0.7	0.5
	EB	B	6.9	8.7	1.5	<.11
		A	16.0	11.0	0.3	0.19
Winter	B	B	2200.0	20.2	0.3	8.4
		A	3400.0	23.0	8.9	3.3
	M	B	40.0	1.0	5.8	3.9
		A	86.0	7.4	9.0	0.9
	EB	B	84.0	20.0	<0.2	0.08
		A	110.0	13.9	0.2	0.60
Spring	B	B	-	16.0	-	29.8
		A	-	16.0	-	10.2
	M	B	120.0	9.4	6.5	7.0
		A	108.0	19.9	11.1	6.1
	M	B	130.0	2.7	0.30	1.50
		A	980.0	8.3	0.01	2.54
	EB	B	220.0	8.0	1.30	1.30
		A	260.0	3.6	0.08	0.31

B = Baltimore Harbor
M = Matapeake
EB = Eastern Bay

Figure 3 shows the relative magnitude of the changes in total and mercury-resistant bacterial populations for each of the seasons following Agnes. A general tapering off of the ratio, after-before, from fall to spring was evident for all three stations. Generally, Baltimore Harbor water showed the greatest relative changes in bacterial populations, and Eastern Bay showed the least changes.

More extensive data would have been helpful in establishing a firm baseline for the comparison of seasonal data. However, the results do point out that quantitative changes, albeit transient, predicated by Agnes, extended to the microbial community. Whether or not these changes can be attributed to direct input of terrestrial run-off or to indirect affects of the input from Agnes is an important question which cannot yet be answered unequivocally.

Table 4. Comparative population levels of bacteria in water and sediment of Chesapeake Bay before and after Tropical Storm Agnes.

Comparison	t[1] (Difference of means)	Degrees of freedom	Significance level[2]
Total heterotrophic population			
Water	+1.5303	9	0.10
Sediment[3]	+0.6298	10	n.s.
Baltimore Harbor			
Water + sediment	+1.3448	4	0.15
Matapeake			
Water + sediment	+1.1732	7	0.15
Eastern Bay			
Water	+2.653	3	0.05
Sediment	-1.3682	3	n.s.
Fall[4]			
Water + sediment	+1.0215	7	0.20
Winter[5]			
Water	+1.0927	2	0.20
Sediment	+0.2558	2	n.s.
Spring[6]			
Water	+1.0487	2	n.s.
Sediment	+0.9000	3	n.s.
Mercury-resistant population[7]			
Water	+1.2715	12	0.15
Sediment	-0.9373	18	n.s.

[1]Treated as paired comparisons. (After Agnes-before Agnes). One-tailed test of significance.
[2]n.s. no significant increase (less than .20 level).
[3]Paired water samples and paired sediment samples were combined because of insufficient data for separate analysis.
[4]1971 vs. 1972.
[5]1972 vs. 1973
[6]1972 vs. 1973.
[7]Expressed as percent of total population resistant to 6 ppm $HgCl_2$.

LITERATURE CITED

Colwell, R. R. and J. D. Nelson, Jr. 1974. Bacterial mobilization of mercury in Chesapeake Bay. Proc. Internat. Conf. on Transport of Persistent Chemicals in Aquatic Ecosystems, Ottawa, Canada, Sect. 3, pp. 1-10.

Colwell, R. R., S. Berk, G. S. Sayler, J. D. Nelson, Jr., and J. Esser. 1975. Mobilization of mercury by aquatic microorganisms. Proc. Internat. Conf. on Heavy Metals in the Environment, Oct. 27-31, 1975, Toronto, Ontario, Canada (in press).

Colwell, R. R., G. S. Sayler, and J. D. Nelson, Jr. 1975. Microbial mobilization of mercury in the aquatic environment. Environmental Biochemistry Proc. Symp., sponsored by Canada Centre for Inland Waters, Burlington, Ontario, April 8-12, 1975 (in press).

Nelson, J. D., Jr. and R. R. Colwell. 1972. Metabolism of mercury compounds by bacteria in Chesapeake Bay. Proc. 3rd Internat. Congr. Marine Corrosion and Fouling, Oct. 2-6, 1972, Nat. Bur. Standards, Gaithersburg, Md., Northwestern Univ. Press, Evanston, Ill., pp. 767-777.

Nelson, J. D., Jr., H. L. McClam, and R. R. Colwell. 1972. The ecology of mercury resistant bacteria in Chesapeake Bay. Marine Techn. Soc., Proc. 8th Ann. Conf., Washington, D. C., pp. 303-312.

Nelson, J. D., W. Blair, F. E. Brinckman, R. R. Colwell, and W. P. Iverson. 1973. Biodegradation of phenylmercuric acetate by mercury-resistant bacteria. Appl. Microbiol. 26:321-326.

Nelson, J. D., Jr. and R. R. Colwell. 1975. The ecology of mercury-resistant bacteria in Chesapeake Bay. J. Microbial Ecol. 1:191-218.

Sayler, G. S., J. D. Nelson, Jr., and R. R. Colwell. 1975. Role of bacteria in the bioaccumulation of mercury in the oyster, *Crassostrea virginica*. Appl. Microbiol. 30:91-86.

Vaituzis, Z., J. D. Nelson, Jr., L. W. Wan, and R. R. Colwell. 1975. Effects of mercuric chloride on growth and morphology of selected stains of mercury-resistant bacteria. Appl. Microbiol. 29:275-286.

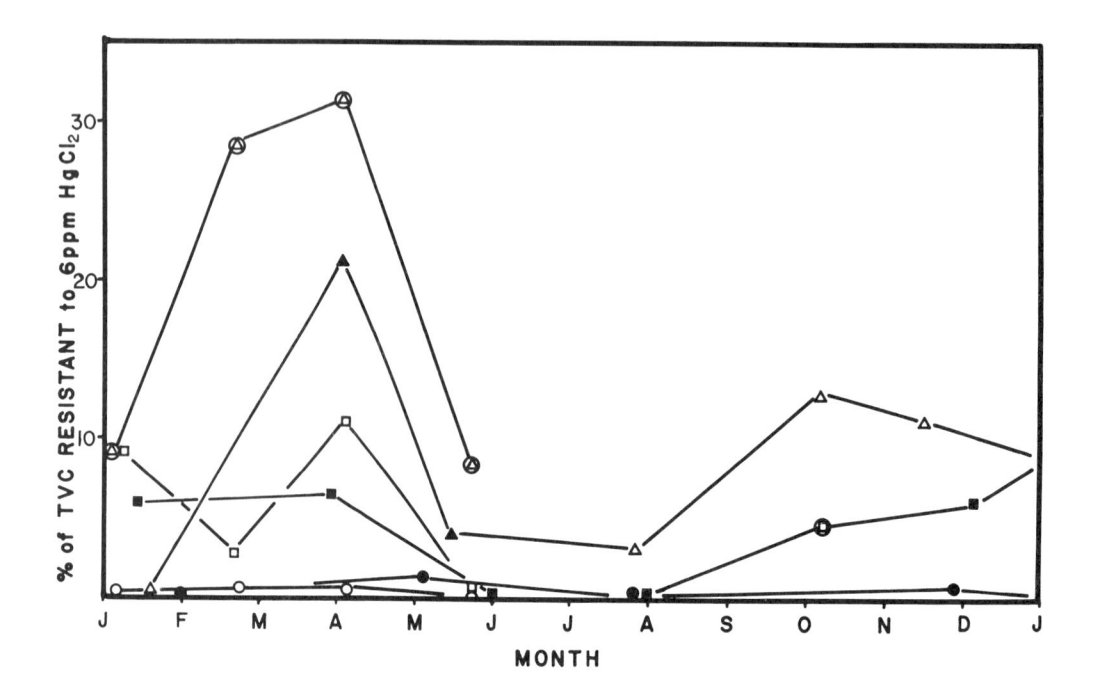

Figure 1. Percentage of the total, viable, heterotrophic, aerobic bacterial population in surface water resistant to 6 ppm of mercuric chloride during the period 1972-1973. Symbols: Baltimore Harbor, Colgate Creek - 1972 (△) and 1973 (⊕); Baltimore Harbor, Sparrow's Point - 1972 (▲); off Kent Island at Matapeake - 1972 and 1973, respectively (■, □); Sandy Point - 1972 (⊡); and Eastern Bay off Parson's Island - 1972 and 1973, respectively (●, ○).

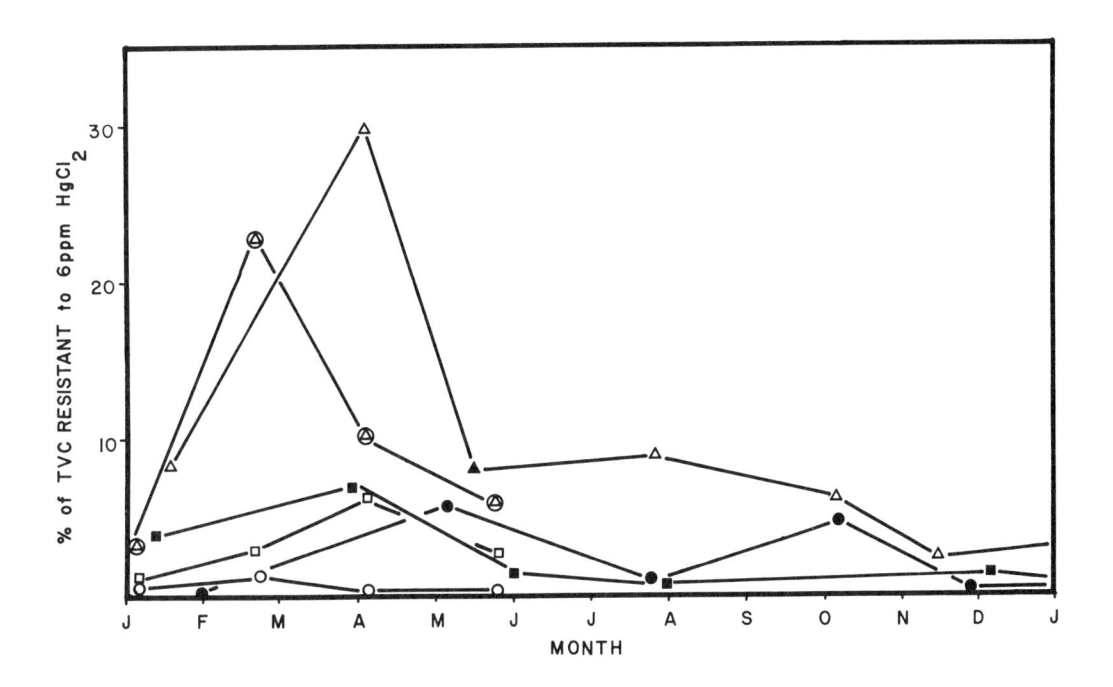

Figure 2. Percentage of the total, viable, heterotrophic, aerobic bacterial population in surface sediment resistant to 6 ppm of mercuric chloride during the period 1972-1973. Symbols: Baltimore Harbor, Colgate Creek - 1972 and 1973, respectively (△,Ⓐ), and Sparrow's Point - 1972 (▲); off Kent Island at Matapeake - 1972 and 1973, respectively (■,□); and Eastern Bay off Parson's Island - 1972 and 1973, respectively (●,○).

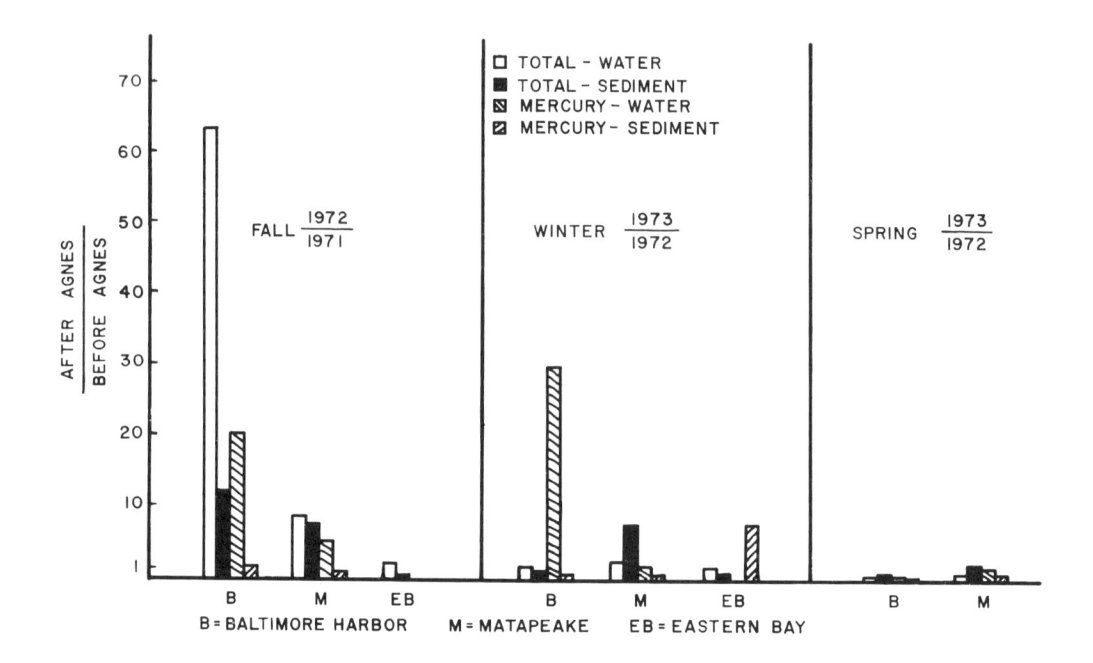

Figure 3. Relative changes in numbers of viable, aerobic, heterotrophic total and mercuric chloride resistant bacteria in surface water and surface sediment layers resulting from Tropical Storm Agnes. Total viable bacterial count in water and sediment, respectively (▫, ▪). HgCl$_2$ resistant bacterial count in water and sediment, respectively (▨, ▨).

SPECIES DIVERSITY AMONG SARCODINE PROTOZOA FROM
RHODE RIVER, MARYLAND FOLLOWING TROPICAL STORM AGNES

Thomas K. Sawyer[1], Sharon A. Maclean[1],
Wayne Coats[2], Mary Hilfiker[2],
Pat Riordan[2], Eugene B. Small[2]

ABSTRACT

Protargol-impregnated sarcodine protozoa attached in situ to glass coverslips submerged in both shallow marshes and in water column stations of the Muddy Creek-Rhode River subestuary in upper Chesapeake Bay, Maryland were studied during the months of May through August 1972 and 1973. Generic identifications were made where possible of 11 types of Amoebida, 2 of Proteomyxida, 3 of Heliozoa, and 3 of Testacea, and the relative abundance of each type was recorded for the different stations sampled each year. Semi-quantitative estimates were made of the density of other protozoa, algae, and diatoms occupying the substrate and competing for space. Ciliates, diatoms, and detritus were the dominant components of the total biomass while amoebae of the genus *Cochliopodium*, *Mayorella*, and *Thecamoeba*, and the heliozoan, *Acanthocystis*, were the only sarcodines which periodically attained a high population density. Amoeba were more numerous and species diversity was greater after Tropical Storm Agnes in 1972 than at any comparable period in 1973. *Acanthocystis* was present periodically in large numbers in late June and July of both years. Although specific differences were noted during the months that followed Agnes in 1972 and comparable periods in 1973, the current lack of knowledge on the role of other ecological factors on estuarine protozoa precludes any interpretation of the cause and effect relationships introduced by the storm.

INTRODUCTION

Ecological relationships among aquatic ciliate protozoa in the Muddy Creek-Rhode River subestuary have been under investigation since late 1971 (Small 1972a) and preliminary findings have been summarized. Permanent slide preparations of protozoa have been summarized. Permanent slide preparations of protozoa collected from the river have been re-examined for sarcodine protozoa and new data have been acquired which include the amoebae, heliozoa, testacea, and proteomyxa. Documentation of the diversity and relative abundance of the principal protozoa of the study area should provide valuable historical data on aquatic microzoa and should yield useful baseline data for measuring future environmental changes and their effect on the structure of microscopic communities.

The impact of Agnes in late June 1972 on economically important fish and shellfish of the Bay already has been assessed by state and federal agencies, and commercial losses are documented in the records. Similar assessment of changes in the protozoan community is now in progress but several years of intensive study still are required for a complete appraisal. Findings to be summarized here represent the periods May through August 1972, and May through August 1973,

---
[1] U. S. Department of Commerce, Middle Atlantic Coastal Fisheries Center, Oxford, Md. 21654
[2] Department of Zoology, University of Maryland, College Park, Md. 20740

and are presented to provide a comparison of species diversity during spring and early summer of the two study years. Studies on the protozoa were not started until late 1971 and, unfortunately, data are not available for comparable seasons prior to Agnes. Further studies are in progress to compare the fall and winter months of 1972 and 1973 and all seasons of 1974.

METHODS

Collection sites were established at the head of Muddy Creek and at several points along its course eastward to the Rhode River. Stations included areas in the marsh and adjacent mud flats in Muddy Creek and open water in both the creek and the Rhode River. Microscopic studies on sarcodines were made from material collected in Muddy Creek at Stations G and H, at Station $A16$ situated in the Rhode River at the end of a laboratory dock at Java Farm, and at Station I, a channel marker in the River. Plastic containers, each containing 8 clean coverslips, were positioned at each station and allowed to remain for 5 to 7 days before removal to the laboratory for processing. Coverslips and attached microorganisms were fixed in Bouin's solution and stained by protargol silver impregnation. One slide from each group of 8 was examined by scanning all fields with a 25X bright-field objective and two others were scanned with a 40X bright-field objective in which every third optical field was studied. Occasionally, coverslips from a single collection were overgrown and covered with detritus, so that fewer than 3 of them were suitable for examination. Stations, dates, and numbers of preparations examined are summarized in Table 1.

Protozoa on each coverslip were identified to genus or to a group-specific common name. Sarcodine protozoa usually were all within recognized genera, while other groups, such as ciliates, dinoflagellates, choanoflagellates, and diatoms, were grouped non-specifically for purposes of recording their contribution to the total biomass. Relative numbers were scored on a 4 plus basis: 4 plus indicating highest abundance or a "bloom"; 3 plus, a significantly high population without overcrowding; 2 plus, a moderate population found in about 50% of the field examined; 1 plus, a low population found in about 25% of the fields; and plus/minus, an organism found as rarely as once or twice per coverslip. Occasionally a rare organism was recorded from only 1 of 3 coverslips from the same collection. Protozoa representing all groups were subdivided into 27 categories (Table 2), and the number present in each collection was expressed as a fraction of the total. Data from each collection were then recorded and analyzed to determine both diversity and relative abundance by date and by station. A large number of coverslips (104) were examined in 1972 in order to estimate the effect of decreased salinity after Agnes on species diversity and abundance, while a smaller number (25 coverslips) were examined in 1973 to compare equivalent time periods in 1972. Further examinations are in progress to increase both the number of stations and dates, and the number of months studied in 1972 and 1973.

RESULTS

Morphologic characteristics of 11 different types of Amoebida on protargol-stained coverslips were, with one exception, adequate for placing them in established genera. The exception, small lobose amoebae of the limax-type, requires a knowledge of mitotic patterns to separate species of *Hartmannella* and *Vahlkampfia*, and for purposes of survey-type study, they are best placed in the general category of "limax" amoebae. Specific identification to species was not attempted in the present study since living material was not examined and mitotic patterns were not determined. Information from historical literature and from a recent study on amoebae of Chesapeake Bay (Sawyer 1971) was adequate for making most

Table 1. Stations, Dates, and Number of Coverslips Examined for Sarcodine Protozoa in Muddy Creek and Rhodes River, Maryland.

Station	Date (No. Examined)							Total

1972

	5/12	5/19	6/30	7/7	7/21	7/28	8/25	
G0	0	0	6	1	0	3	0	10
G1	0	2	2	3	1	1	0	9
G4	0	3	8	7	0	3	0	21
H	8	7	5	4	3	3	3	33
A16	0	3	2	3	3	1	0	12
I	4	4	3	1	3	4	0	19
								104

1973

	5/21	6/19	6/29	7/20	7/27	8/22	Total
G0	1	0	0	0	0	0	1
G1	1	0	0	0	0	0	1
G2	0	1	0	1	1	1	4
G3	1	0	0	0	0	0	1
G4	1	0	0	0	0	0	1
H	2	2	0	1	2	0	7
A16	2	0	3	0	2	0	7
I	1	0	0	0	1	1	3
							25

Table 2. General Classification of Organisms Recognized on Coverslip Preparations from Muddy Creek and Rhodes River, Maryland.*

GROUP I - AMOEBIDA	GROUP II - PROTEOMYXIDA	GROUP III - HELIOZOIDA
Cochliopodium sp. #1	Proteomyxan sp. (?)	Acanthocystis (?)
Cochliopodium sp. #2	Microgromia sp. (?)	Actinophrys (?)
Mayorella sp. #1		Raphidocystis (?)
Mayorella sp. #2		
Thecamoeba sp. #1	GROUP IV - TESTACEA	GROUP V - MASTIGOPHORA
Thecamoeba sp. #2		
Hartmannella sp. #1	Difflugia (?)	Dinoflagellates
Hartmannella sp. #2	Centropyxis sp. (?)	Choanoflagellates
Vannella	Trinema (?)	Flagellates, various
Hyalodiscus		
Vahlkampfia		
	GROUP VI - CILIATA	GROUP VII - ALGAE & DIATOMS
	Ciliates, sessile	
	Ciliates, free	Not identified; free or attached
	Tintinnids	

*Preliminary identifications made from protargol stained preparations, living material not examined.

generic distinctions and for recognizing several different species within the same genus. The different genera and species (designated sp. #1, sp. #2, etc.) are summarized in Table 2. The second group, designated proteomyxans, contains some of the most diverse and poorly understood taxa in protozoan literature; two probable proteomyxans were recognized. The first type produced a branched network with most branches perpendicular to each other and extended outward for considerable distances. The second resembled testate forms of *Microgromia*. In the third group, the heliozoa, three species tentatively were identified as belonging to the genera *Acanthocystis*, *Actinophrys*, and *Raphidocystis*. Stained specimens of both sexual and asexual stages of heliozoa were observed and in some instances progressive stages in their life cycles were detected on slides collected at weekly intervals. Species belonging to the first two genera occasionally were present in moderate-to-large numbers while *Raphidocystis* was found infrequently and in small numbers. The fourth group, the testaceans, were represented by three distinctly different species of *Difflugia*, *Centropyxis*, and *Trinema*. *Difflugia* sp. was characterized by the extrusion of filose or spiny protoplasmic strands, *Centropyxis* by the presence of spines on the external test, and *Trinema* by its very small size and a wreath-like halo which formed the test. The fifth group, the flagellates, could not be identified with certainty but three different morpho-types were noted. Flagellates stained with protargol usually had a large complex nucleus and flagella which were visualized best with phase contrast microscopy. Dinoflagellates were characterized best by clear representation of their scales or plates, and choanoflagellates were small, bell-shaped with a thread-like stalk, and similar to species within the genus *Salpingoeca* Clarkes, 1867. Flagellates on protargol-treated preparations, particularly dinoflagellates and choanoflagellates, were not studied for taxonomic determinations but were noted in order to estimate their seasonal appearance and contribution to the total population of organisms per coverslip. The sixth group, the ciliates, were not identified but were grouped into sessile forms and free forms for purposes of recording population densities and seasonal appearance. The seventh group, the diatoms and microalgae, were not identified but were noted for the same purposes as flagellates and ciliates. Certain species of diatoms were noted when sarcodines were observed to have an apparent feeding preference for a particular type. Diatoms were abundant at all stations throughout the seasons studied except at Station *I* in 1973. High sarcodine abundance was associated with large numbers of diatoms at certain times but not at others, indicating that the species of diatoms present as food were more important than were the numbers alone. Summary data for all groups showed that in 1972 diatom abundance and protozoan diversity were uniformly larger than they were in 1973. Differences could not be attributed solely to the effects of Agnes since the influence of food organisms and other natural factors on sarcodines is largely unknown. Observations on each group of organisms and specific differences noted during 1972 and 1973 are summarized below:

Group I - Amoebida

Genus Cochliopodium: *Cochliopodium* (2 species) was found in moderate to large numbers at all shallow water mud flat stations sampled in 1972, and in smaller numbers at water column stations, with greater numbers present in July then in May and June. The larger species (Fig. 1) was noted to feed primarily on large diatoms rather than on more abundant small types, and the second species, being smaller, probably fed on bacteria. Both types of *Cochliopodium* behaved differently in 1973, small numbers of the large form were present only on June 19 coverslips from the mud flat, and moderate numbers of the small form were present only on coverslips from dockside Station *A16*.

Genus Mayorella: *Mayorella* (2 species) was found in small numbers at all mud flat and water column stations sampled in 1972. Numbers of amoebae similarly were small in 1973 and coverslips frequently were negative for mayorellids. Feeding preferences were not observed with either species (Fig. 2).

Genus Thecamoeba: *Thecamoeba* (2 species) was found frequently in small to moderate numbers at all stations sampled in 1972. The larger species (Fig. 3) fed extensively on diatoms and probably bacteria. Both species fed voraciously and often were observed in large masses of plant material and detritus where they were easily overlooked and difficult to recognize. Both species were present in small numbers at all stations in 1973 except at Stations *A16* and *I*.

Genus Hartmannella: *Hartmannella* (2 species) was found rarely or in small numbers at all stations sampled in 1972. Food preferences of the larger species (Fig. 4) and smaller species were not detected. Both species were found in approximately the same numbers and with the same frequency in 1973 as they were in 1972.

Genus Vannella: *Vannella* (1 species) was found in small numbers at all stations sampled in 1972. Amoebae within the genus are recognized bactivores but probably feed on fungi and yeasts as well. Specimens on stained coverslips frequently were wrinkled and distorted and were not photographed; one well-known species in Chesapeake Bay has been illustrated (Sawyer 1971). *Vannella* was found in very small numbers only at station *G4* in 1973.

Genus Hyalodiscus: *Hyalodiscus* (1 species) was found in very small numbers and only at Station *G1* sampled in 1972. The amoebae were small and overstained with protargol and were not suitable for photomicrography; food preference was not detected. In 1973 *Hyalodiscus* was found in very small numbers at mud flat stations *G3* and *G4*.

Limax-type amoebae: Small lobose amoebae with blunt, rounded pseudopods were found at all stations except *A16* sampled in 1972 and were present in small to moderate numbers. Limax amoebae are recognized bactivores and are known to attain high population densities in the presence of abundant food organisms. On protargol stained coverslips the amoebae were small, densely stained, and not suitable for photomicrography; limax amoebae from Chesapeake Bay were illustrated earlier (Sawyer 1971). Essentially similar findings were found at stations sampled in 1973.

Group II - Proteomyxa

Two organisms of uncertain taxonomic status were placed tentatively in the Proteomyxida, a poorly defined group which often is considered to be intermediate between the Amoebida and the Mycetozoa. The first organisms produced a branched network of protoplasm which resembled fungal hyphae. Shrinkage and dark overstaining on coverslips prohibited a specific identification but the presence of isolated organisms and the lack of connected anastomosing filaments suggested that it should be included in the sarcodine group. Until living material is available for study, the first organism is considered provisionally to be related to *Biomyxa*. *Biomyxa*-like organisms, usually only one per coverslip, were found at least once per station in 1972 but only at Station *H* on one occasion in 1973.

The second organism (Fig. 5) produced long thin beaded anastomosing filopodia and resembled certain delicate *Microgromia* species of testaceans. Since the test or shell characteristic for the Testacea was not detected on protargol-stained specimens, this organism is included (only provisionally) in the proteomyxan group. The species was present in small to moderate numbers at all stations sampled in 1972 with the smallest number occurring at Station *I*. In 1973 specimens were rare at all stations except Station *I*, where it was absent. Food preferences for the two organisms in this group were not detected.

Group III - Heliozoa

Genus Acanthocystis: One species of *Acanthocystis* (Fig. 6) was found in numbers ranging from negative to bloom proportions at all stations sampled in 1972. Stages ranging from cysts to sexual and asexual forms were found at different sampling times and sequential growth stages were found on coverslips collected in successive weeks. Surface scales viewed with phase contrast optics were especially distinct after staining with protargol. Food preferences were not determined. In 1973 a bloom was detected at Station *H* late in July while only small numbers were found at the *G* Station and Rhode River Station *I*; dockside Station *A16* was consistently negative.

Genus Actinophrys: One species of *Actinophrys* (Figs. 7a, b, c) was found in small to moderate numbers at all stations except *G1* in 1972. Several different life cycle stages ranging from probable encysted zygotes to gamete stages (Figs. 7b, c) were noted. In 1973 numbers ranged from negative to moderate and blooms were not detected; large populations were found only at Station *A16* on June 29, and Stations *G* and *I* were negative. Food preferences were not detected.

Genus Raphidocystis: One species of an extremely small heliozoan probably *Raphidocystis* (Fig. 8) was found in numbers ranging from negative to moderate in 1972 with the largest numbers occurring in late July at water column Stations *A16* and *I*. Only the filose floating form was detected on stained coverslip preparations, probably because the extremely small adult stage would produce even smaller intermediate growth stages. In 1973 all stations were negative for this genus.

Group IV - Testacea

Genus Difflugia: *Difflugia* (1 species) was found in small to moderate numbers at all stations sampled in 1972. Stained specimens had rounded tests without surface ornamentation and the protoplasm, when extended to the outside, had a broad filose appearance (Fig. 9) with broad-based spines. Most specimens were found in areas of the coverslip which were densely packed or matted with plant material and detritus. In 1973 the organism was present at all stations except Station *I* and was most numerous at Station *H*. Food preference was not detected.

Genus Centropyxis: *Centropyxis* (1 species) was found at all stations except Station *I* in 1972, and numbers per slide ranged from negative to few. Stained specimens in this genus were larger than specimens of the other genera and several spiny processes (Fig. 10) extended from the test. Most specimens were present in mats of plant material and detritus and often were difficult to identify with certainty. In 1973 specimens were negative to moderate in numbers and populations usually were similar to those observed in 1972. Food preference was not detected.

Genus Trinema: *Trinema* (1 species) was found in negative to moderate numbers at all stations sampled in 1972. Well-fed stained specimens were small, round, and packed with diatoms and unfed specimens had clear round protoplasm slightly contracted within the test (Fig. 11a). One specimen (Fig. 11b) probably *Trinema*, was found with the protoplasm extended through the operculum. Moderate populations were found at all stations except Stations *I* and *A16* in 1973. Food preference was a single species of diatom that was slightly shorter than the internal diameter of the test.

Group V - Mastigophora

Three basic types of flagellates were found on stained coverslips: flagellates resembling elongate *Euglena* and *Paranema* and rounded *Ochramonas;* armored dinoflagellates; and stalked choanoflagellates. The first group was present in few to moderate numbers at all stations in 1972 and in smaller numbers in 1973. The second group, the dinoflagellates, was observed in very large numbers at Station $G4$ and in negative to moderate numbers at other stations in 1972. In 1973 dinoflagellates were numerous on July 20 and August 22 at Station $G2$ and negative to rare at the other stations. Small flagellates were difficult to detect on preparations with dense populations of ciliates, diatoms, or detritus and probably were frequently overlooked. Organisms representing the *Mastigophora* (Figs. 12 & 13) were enumerated but not identified for the purpose of estimating their contribution to biomass on each coverslip.

Group VI - Ciliata

Ciliate protozoa were recorded as sessile or free-living but were not identified to family or genus. In 1972 the relative abundance of both groups ranged from negative to moderate at the mud flat stations and from moderation to bloom proportions in the river stations. Conceivably, coverslip traps at mud flat G Stations may have experienced periods of exposure at low tide which did not affect traps submerged in the deeper water column stations. Results for 1973 were similar to those from 1972 with the exception that, on rare occasions, ciliates were very few in number or absent, and on May 21 there was a peak number of tintinnids at water column Station $A16$. Data summarized from all collections showed that sessile ciliates and accompanying diatoms made up the bulk of the total biomass per coverslip preparation, and the free-swimming ciliates were the second most abundant organisms.

Group VII - Algae & Diatoms

Phytoplankton were recorded only in terms of their relative abundance because, in the company of abundant organic detritus, they occupied most of the substrate provided by glass coverslips. This group occasionally settled in numbers sufficient to exclude most of the sarcodine species. Although phytoplankton species were not identified, evidence was found which suggested that a large sarcodine species diversity often was associated with a large phytoplankton diversity. Conversely, a narrow phytoplankton diversity, independent of numbers, was associated with a narrow sarcodine diversity.

The lack of data from the study area prior to 1972 seriously limited the extent to which the influence of Agnes on sarcodine protozoa could be estimated and the extent to which studies in 1972 and 1973 could be compared. Data derived from protargol-impregnated sarcodines were remarkably useful in view of the fact that such treatment is not ideally suited for the specific protoplasmic staining of nonciliated protozoa, especially for identifications to the species level. Microscopic examination of a large number of coverslip preparations, collected specifically for studies on the ciliate protozoa, has provided a biological inventory of the sarcodine protozoa of a low salinity area of the Chesapeake Bay that is without historical precedent.

DISCUSSION

The opportunity to examine large numbers of stained slides prepared by investigators in the Department of Zoology, University of Maryland as part of ongoing research on the ciliate protozoa of the Muddy Creek-Rhode River subestuary, has yielded new and valuable data on sarcodine protozoa from a low-salinity brackish area of the Chesapeake Bay system. The value of the newly acquired data is greatly enhanced by the fact that coverslip traps submerged for periods of 5-7 days were collected at weekly intervals throughout all seasons of a several year period. Studies currently in progress at the University were initiated late in the summer of 1971, and, consequently, historical data on protozoa prior to summer of 1972, when Agnes reduced the salinity of the Bay to levels approaching that of fresh water, are not available in published literature. Information is available, however, for several months preceding Agnes, and a comparison of those data with new data from ongoing work will be available for future studies. For the present, findings made in 1972, the year of Agnes may be used as a basis for comparing findings in subsequent years and recording significant changes in species diversity and abundance. Indications already are emerging which suggest that the drop in salinity to 2-3 ppt after Agnes may have encouraged certain emergent species which thrive best in an oligohaline (<5 ppt) environment. Similarly, heliozoa, amoebae which transform to a radiate floating stage, and certain sediment or bottom-dwelling species may have been distributed or redistributed throughout the water column. It was of interest to note that among the sarcodine protozoa from both marsh or mud flat stations and water column stations, approximately twice as many species were found in 1972 as in 1973. Similarly, the large species of *Cochliopodium*, *Mayorella*, and *Thecamoeba* were present in greater numbers and were found more frequently in 1972 than they were in 1973. Further studies in 1974 should reveal whether 1972 indeed was unusual in terms of the diversity and numbers which were recorded. Furthermore, it will be interesting to follow the heliozoa and testacea which historically are recognized to have a larger species diversity in fresh water than in saline habitats, yet were encountered frequently in 1972 and 1973.

One very important indirect effect of Agnes has been to call attention to the paucity of reports on ecological inter-relationships among brackish and marine aquatic protozoa. Attempts to measure the influence of major environmental change on aquatic species are hindered by the lack of knowledge concerning normal changes in nutrient cycling, temperature, components of the food web, competition, etc. Recent studies on marine ciliates and bacteria have added substantially to our knowledge of aquatic interactions but virtually nothing has been published on similar interactions with marine sarcodines. Beers and Stewart (1967) pointed out that microplankton (20-200µ in size) and ultramicroplankton (2-20µ) are not retained by No. 3 netting (333µ) commonly used for plankton collections. Collections in the California current (Beers and Stewart 1967) showed that protozoa made up 59-83% of all the animals collected per station and were represented by radiolaria, foraminifera, and ciliates. Lighthart (1969) enumerated both bacteria and protozoa from Puget Sound and the adjacent Pacific Ocean and reported that numbers decreased horizontally from shore to the open sea, and vertically from the sediment towards the surface of the water column. On the basis of such observations, marshlands and shallow rivers as found in the Muddy Creek-Rhode River complex would be expected to have high population densities of both bacteria and protozoa. Preliminary summarizations (Small 1972a) of studies with ciliate protozoa confirm the findings of Beers and Stewart (1967) and Lighthart (1969), and suggest that diversity and biomass are more complex than heretofore expected. Lackey (1961) reported even earlier that ciliate protozoa were the largest group that he collected, both in terms of numbers and biomass, and stated that complex interactions between protozoa, algae, and bacteria reqquire long-term study.

Bovee (1965) conducted a detailed long-term study in freshwater streams in Florida and reported the direct influence of natural factors (pH, temperature, B. O. D., etc.), and pollutant additions, on species of aquatic amoebae; counterpart studies on marine amoebae have not yet been accomplished. Bovee (1965) found species of heliozoa in April and May when the water temperature was over 21°C and pH was 7.9, and pointed out that the average water temperature over an extended length of time is more important than the temperature at a given sampling interval. Temperature, therefore, would be expected to vary significantly in shallow marshes and creeks as opposed to the open ocean, and should receive attention in future studies in Chesapeake Bay. Bovee (1961, unpublished) examined 230 water samples from freshwater, brackish, and marine sources in Florida and found amoebae belonging to 11 families and 28 genera; 4 of these 22 species were marine, 2 were from salt marshes, and 1 was in brackish water. He reported that amoebae were abundant in water ranging from 20-25°C, and most of them lived in association with algae, diatoms, molds, bacteria, yeasts, and nematodes, associations which commonly were noted on coverslips examined in the study presented here. Sawyer (1971) reported 8 genera of amoebae from the Tred Avon River, upper Chesapeake Bay, where seasonal salinities ranged from 9-16 ppt prior to Agnes, and also reviewed characteristics and problems related to species identifications.

Current studies on marine amoebae by both Sawyer and Bovee (unpublished and in press) during the period 1971 to 1974 have added new genera and species, and have extended the geographic range of known species. Thirty-five species were found in Chincoteague Bay of which 20 were new (Sawyer, in press), and 26 subsequently were found in control and dump sites in the New York Bight (unpublished) at stations ranging from 30 to 140 meters in depth. During this period Bovee (unpublished, 1973) made collections in Chincoteague Bay, Va., Sandy Hook Bay, N. J., New York Bight dump sites, and Woods Hole, Mass., and collectively found 18 genera and 53 species of which 26 probably are new. Studies mentioned briefly in the foregoing discussion are designed to estimate the species composition of marine habitats and to provide specific descriptions of identifying characteristics for each genus and species.

Future studies are needed to explain the naturally occurring factors which account for differences in both diversity and population density. When such information is available, it should be possible to interpret the effects of events such as periods of dryness, unseasonal rainfall, sediment disturbance, and pollutant additions on diversity and abundance of recognized protozoa in structured communities. Small (1972b) has briefly summarized the present status of research on ciliate protozoa in Chesapeake Bay, and current studies are providing valuable information on certain of the considerations summarized above. Fenaux (1973) stated that the first serious descriptions of planktonic animals were published in 1775, and quantitative studies with plankton nets date from about 1888. Progress in protozoan research since the late 1700's and 1800's has been significant but is seldom sufficient to permit a detailed analysis to be made of the effects of major environmental alterations on established marine and estuarine communities.

LITERATURE CITED

Beers, T. R. and G. L. Stewart. 1967. Micro-zooplankton in the euphotic zone at five locations across the California Current. J. Fish. Res. Bd. Canada 24: 2053-2068.

Bovee, E. C. 1961. Studies on the naked, lobose, free-living Amobida. Progress report to the National Institutes of Health, Bethesda, Md. on Research Grant E-1158 (C, C1, C2, C3, unpublished), 22 p.

Bovee, E. C. 1965. An ecological study of amebas from a small stream in northern Florida. Hydrobiologia 25:69-87.

Bovee, E. C. 1973. A preliminary report on the amoebae of the Atlantic Coast from Virginia to Massachusetts, including the New York Bight. Report to National Marine Fisheries Service, Sandy Hook Laboratory, Highlands, N. J., unpubl., 19 p.

Fenaux, R. P. O. 1973. Current trends in Mediterranean planktology. South African National Oceanographic Symp., Cape Town, 6-10 August 1973 (abstracts). Sea Fish. Br., Dep. Ind., Sea Point, Cape Town, South Africa, p. 3-4.

Lackey, J. B. 1961. Bottom sampling and environmental niches. Limnol. Oceanogr. 6:271-279.

Lighthart, B. 1969. Planktonic and benthic bacteriovorous protozoa at eleven stations in Puget Sound and adjacent Pacific Ocean. J. Fish. Res. Bd. Canada 26:299-304.

Sawyer, T. K. 1971. Isolation and identification of free-living amoebae from upper Chesapeake Bay, Maryland. Trans. Am. Microsc. Soc. 90:43-51.

Small, E. B. 1972a. Studies of estuarine ciliate protozoa as a function of environmental change in the Rhode River. Chesapeake Research Consortium, Inc., Ann. Rep. June 1, 1971-May 31, 1972, p. 318-330.

Small, E. B. 1972b. Free-living protozoa of the Chesapeake Bay exclusive of Foraminifera and the flagellates. Chesapeake Sci. 13(Suppl.):S96-S97.

LEGEND

Figures 1-13. Aquatic protozoa from the Muddy Creek-Rhode River subestuary of upper Chesapeake Bay, Maryland. Figs. 1-4. Amoebida. Fig. 5. Proteomyxan-like organism. Figs. 6-8. Heliozoa. Figs. 9-11. Testacea. Fig. 12. Dinoflagellate. Fig. 13. Flagellate.

1. *Cochliopodium* with ingested diatoms, bright-field X560. Note deeply stained hull units (arrows).

2. *Mayorella* with broad-based digitate pseudopods, phase contrast X560.

3. *Hartmannella* with blunt, rounded anterior pseudopod, phase contrast X560.

4. *Thecamoeba* with blunt anterior pseudopod and longitudinal ridges or folds, phase contrast X560.

5. Proteomyxan-like species resembling the testacean genus *Microgromia*, bright-field X560. Note extensive beaded protoplasmic strands (arrows).

6. *Acanthocystis* division stage, phase contrast X560. Note refractile surface scales and large ingested food organism (arrows).

7. a. *Actinophrys* division stage, phase contrast X560. Note nuclei of 4 developing trophozoites and ingested food organism (arrows).

   b.& c. Probable pre-gamete stages of *Actinophrys*, bright-field X560. Note empty capsule and lid in b. (arrow).

8. Small heliozoan, probably *Raphidocystis*, phase contrast X560.

9. *Difflugia* with spine-like extensions of protoplasm, bright-field X560.

10. *Centropyxis* with lateral spines originating from the test, bright-field X560. Note large operculum (arrow).

11. a. Resting forms of a small testacean, probably *Trinema*, phase contrast X560.

    b. Feeding form, probably *Trinema*, with extruded protoplasm, bright-field X560.

12. Unknown dinoflagellate often present in large numbers on coverslip preparations, phase contrast X560.

13. Unknown large flagellate, phase contrast X560. Note flagella (arrows).

All photographs made from protargol-impregnated specimens attached to glass coverslips and fixed with Bouin's solution.

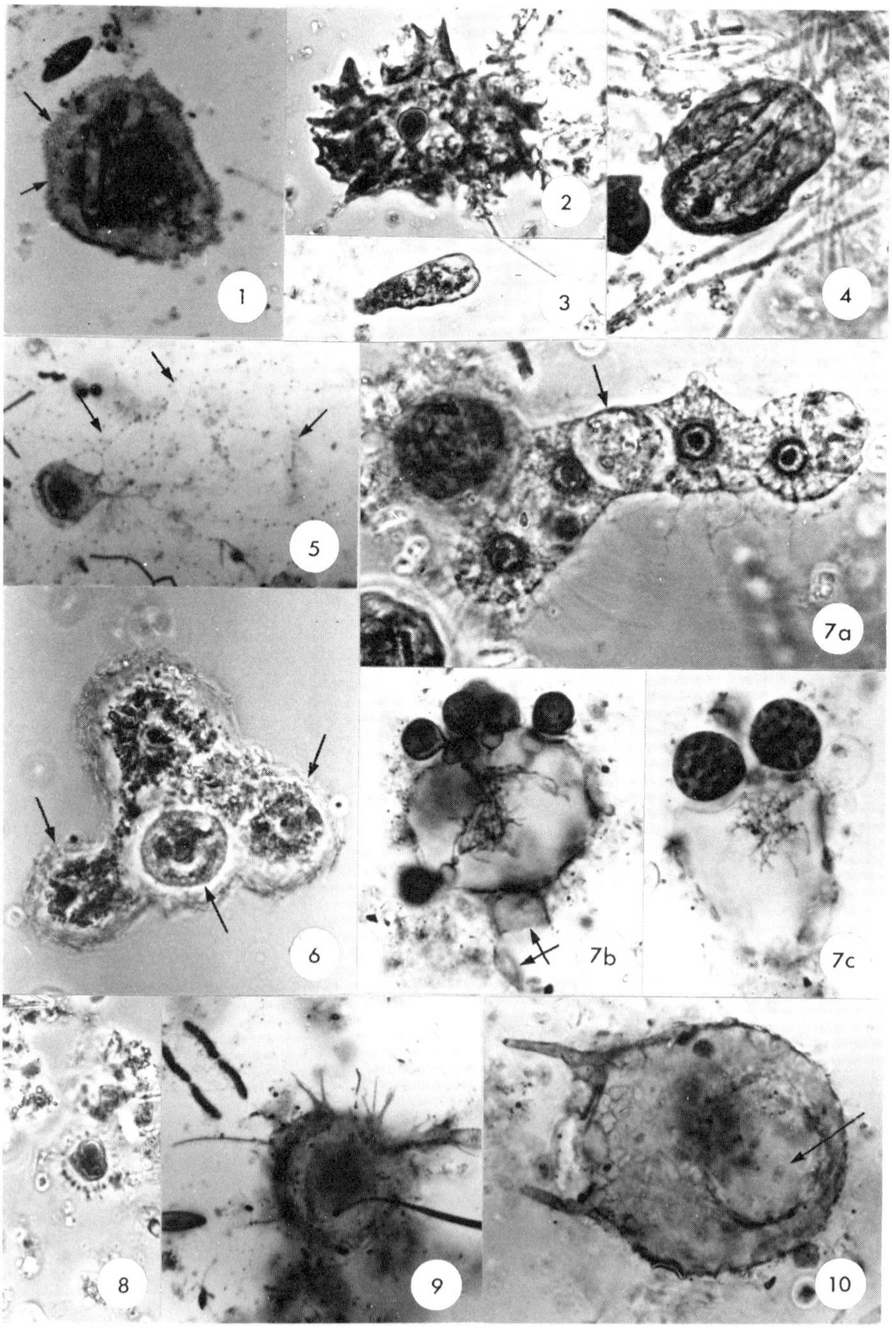
1
2
3
4
5
7a
6
7b
7c
8
9
10

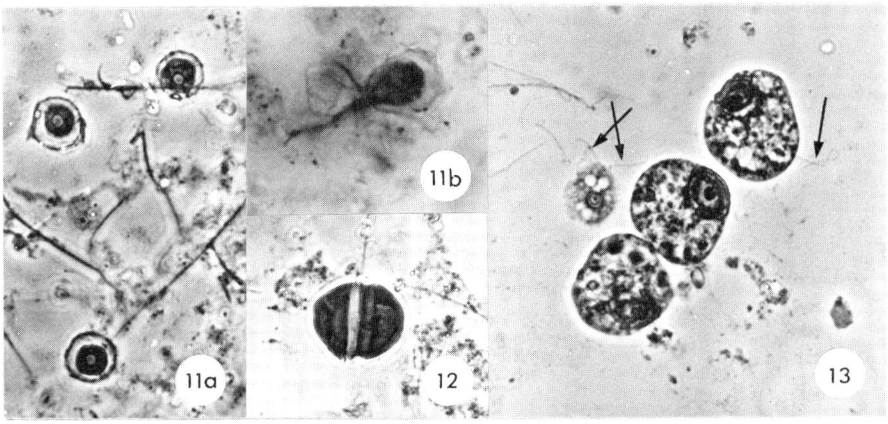
11a
11b
12
13

THE IMPACT OF TROPICAL STORM AGNES
ON MID-BAY INFAUNA[1]

D. Heyward Hamilton, Jr.[2]

ABSTRACT

A continuing intensive study of the macroscopic infaunal community at a location near Cove Point in mid-Chesapeake Bay had been under way for one year prior to the advent of Tropical Storm Agnes. Data on species abundance for two major biotopes show a significant impact associated with the storm, including complete mortality of a residual population of *Gemma gemma*, the most abundant (10^5 m^{-2}) infaunal species. Repopulation has been rapid, and normal biomass quantities in the sediments were maintained despite continuous changes in the composition of the infaunal community.

The arrival of Agnes in the Chesapeake Region was in one sense fortuitous, since it occurred over a year after a detailed study of the infauna at a location near Cove Point had been initiated. More than a year of information is thus available before the advent of the storm as a basis for comparison of pre- and post-storm conditions.

Details of sampling methodology and design have been previously considered (Hamilton & LaPlante 1972). A 1mm mesh screen was used to separate organisms from sediment samples taken with an anchor dredge (Holme 1971) at quarterly intervals. The samples averaged 14ℓ in volume, representing an average of .18m^2 of surface. Paired samples were taken at 25 locations during the first year of the program and at 11 stations after this; locations are shown in Fig. 1. Two major sediment types were studied: an inshore sand bottom at a depth of 3-6m and an offshore silt-clay zone at a depth of 8-12m.

The reliability and precision of the dredge are indicated in Tables 1 and 2. Given in Table 1 are paired biomass (as ash-free dry weight) values for every station which has been under continuous study since the beginning of the program. Sample size is too small to permit statistical evaluation, but the data show that a high precision was routinely achieved, regardless of sediment type or the quantity of organisms present.

As an additional check on the validity of the sampling methodology, Kendall's rank correlation coefficient was calculated for a group of 10 randomly selected stations. Only first year results were considered and only the following six species included: *Macoma balthica*, *Gemma gemma*, *Mulinia lateralis*, *Cyathura polita*, *Edotea triloba*, and *Nereis succinea*. Criteria for the choice of these species were: positive identification, presence throughout the year, and quantitative sampling. Some samples lacked even these species, preventing calculation of the correlations for some sample pairs, identified as N. D. in Table 2. With some exceptions, the values are high, indicating generally strong correlation in the ranked abundance of the six species.

Tenore (1972) reported on a seasonal pattern of estuarine benthos, in the Pamlico River, N. C. The data for number of organisms caught per sample at Cove

---
[1]Contribution No. 684, Center for Environmental and Estuarine Studies, Univ. of Maryland.
[2]Division of Biomedical and Environmental Research, U. S. Energy Research & Development Administration, Washington, D. C. 20545.

Table 1. Precision of anchor dredge infaunal samples. Data are g ash-free dry wt m^{-2}, S = Sand, O = Mud Biotope.

STATION	SEDIMENT	MAY 71	AUG. 71	NOV. 71	FEB. 72	MAY 72	AUG. 72	NOV. 72	FEB. 73
3A	S	20.0	1.57	1.83	3.98	6.9	1.72	2.54	3.27
B	S	21.5	0.64	.81	5.08	8.1	1.36	2.82	3.53
4A	S	30.0	2.96	4.17	1.64	2.4	.47	1.86	3.20
B	S	28.0	2.03	4.04	2.58	2.6	.44	3.13	2.94
13A	S	0.5	2.17	.44	4.69	3.1	1.14	2.22	2.50
B	S	2.5	0.60	1.16	2.97	4.1	1.14	2.90	1.51
14A	S	40.0	4.32	4.86	11.99	55.0	2.56	2.88	0.64
B	S	28.0	8.64	4.04	11.50	64.1	1.17	2.88	0.96
15A	S	24.75	2.50	1.00	0.79	3.5	1.39	1.75	1.04
B	S	47.5	2.90	1.52	2.63	4.0	1.11	1.04	1.34
24A	S	9.0	1.95	1.95	5.45	7.5	1.01	1.14	1.98
B	S	15.0	1.88	1.78	5.94	29.3	0.86	2.55	1.89
11A	O	11.2	0.08	.79	2.07	4.6	.03	.20	1.08
B	O	5.5	0.10	.99	2.17	5.6	.00	.30	1.16
12A	O	7.0	0.29	1.34	3.89	8.1	.28	.45	2.79
B	O	5.5	0.43	2.40	5.01	7.2	.00	.29	2.97
21A	O	6.0	0.08	.33	2.34	3.7	.17	.25	0.95
B	O	9.0	0.05	.47	2.37	3.6	.00	.30	0.85
23A	O	20.5	3.13	1.40	3.09	18.9	9.89	3.67	2.61
B	O	18.1	2.74	1.16	4.20	24.7	9.94	5.55	3.78

Table 2. Kendall's rank correlation coefficent for randomly selected anchor dredge samples (replicated) from Cove Point. See text for further details.

STATION	MAY 71	AUG 71	NOV 71	FEB 72	$\bar{x}$
1	N.D.	N.D.	N.D.	.33	.33
2	N.D.	1.00	1.00	.80	.93
5	N.D.	1.00	1.00	.33	.78
8	1.00	1.00	1.00	1.00	1.00
9	1.00	1.00	1.00	N.D.	1.00
14	1.00	1.00	.66	.80	.87
17	.33	N.D.	1.00	1.00	.78
18	.66	1.00	1.00	-.33	.58
22	1.00	1.00	1.00	.33	.83
25	.73	1.00	1.00	.33	.77
$\bar{x}$	.82	1.00	.96	.51	.82

Point also show a seasonal pattern (Fig. 2). Although trends are indicated, the pattern is somewhat erratic; the seasonality is much clearer when changes in biomass with time are examined (Figs. 3 & 4). Within limits the biomass of macroscopic infauna in both of the major sediment types is constant when a comparison by season is made. The interval between May and August, a period of transition from spring to summer conditions of temperature and dissolved oxygen, is a period of declining infaunal standing crop. The factors responsible for this are complex. Probably predation and low oxygen concentrations are major influences. The storm, which arrived during this period of natural mortality in the last week of June 1972, produced significant changes in total infaunal standing crop. The figures show that abundance of organisms was returning to pre-Agnes conditions by fall of that year. The vertical lines, which bracket the mean plus or minus one standard error, are indicative of the patchiness of the infaunal distributions, particularly in spring following larval settlement.

The passage of the storm was not without impact. Table 3 shows that a substantial, although declining, population of the lamellibranch *Gemma gemma*, present at Cove Point prior to the storm, disappeared in August 1972. On July 12, two samples were taken, each of which contained large quantities (approximately 20,000 individuals) of recently dead *Gemma*. The shells taken were unlike any previously observed. The valves were still attached, retaining the luster characteristic of living shells, but no tissue was present. *Mulinia lateralis* and *Mya arenaria* apparently experienced similar mortalities although shells of recently dead animals were not observed. *Macoma balthica* and *Macoma phenax* appeared more resistant to storm-induced stresses because significant numbers of each persisted through February of 1973. Subsequent data from the study area (Cargo, personal communication) indicate that both *Mulinia* and *Mya* have become re-established and continue to be major components of the molluscan fauna at Cove Point. Only one *Gemma* has been taken since the storm.

Although the difference in number of samples taken reduces the strength of the comparison shown in Table 4, a trend toward fewer species and fewer individuals of those species present in both biotopes after the passage of Agnes is apparent. Of 18 species taken in the mud biotope, 7 were present in both 1971 and 1972. Eight species present in 1971 were not present in 1972 while three species not present in 1971 were present in 1972. Because of the homogeneity of the sand and mud biotopes in terms of sediment characteristics (Hamilton & LaPlante 1972, 1973), it is unlikely that a zero catch in either of the two years was due to insufficient sampling. With respect to the 22 species taken in the sand biotope, 14 were present

Table 3. Summary of anchor dredge catch, Cove Point, Md., 1971-1973. Data are total unadjusted values for the number of samples (in parentheses) in the collection, separated by biotope.

	May 71 (20) Mud	May 71 (25) Sand	Aug 71 (20) Mud	Aug 71 (28) Sand	Nov 71 (20) Mud	Nov 71 (28) Sand	Feb 72 (20) Mud	Feb 72 (28) Sand	May 72 (8) Mud	May 72 (12) Sand	Aug 72 (8) Mud	Aug 72 (12) Sand	Nov 72 (6) Mud	Nov 72 (12) Sand	Feb 73 (6) Mud	Feb 73 (12) Sand
Coelenterates																
Diadumene leucolena	2	3	0	1	0	2	0	1	2	2	0	0	0	0	0	0
Crustaceans																
Callinectes sapidus	0	1	0	0	0	1	0	0	0	0	0	0	0	0	0	0
Chiridotea caeca	0	0	0	4	0	5	0	7	0	0	0	6	0	4	0	3
Crangon septemspinosa	0	0	0	0	0	1	0	0	0	94	0	1	0	2	0	0
Cyathura polita	2	100	2	27	0	3	0	7	1	0	1	2	0	1	2	14
Edotea triloba	52	294	0	36	0	27	4	33	28	19	1	13	2	1	0	0
Leucon americanus	154	22	0	1	1	1	8	0	47	2	0	0	2	0	0	0
Oxyurostylis smithi	3	0	0	0	0	0	0	0	0	0	0	0	0	0	0	0
Palaemonetes pugio	0	0	0	1	0	0	0	0	0	1	0	0	0	0	0	0
Sphaeroma quadridentatum	0	0	0	0	0	5	0	12	0	0	0	0	0	0	0	0
Unidentified amphipods	144	4741	1	6	5	70	4	391	14	290	0	42	208	48	132	123
Polychaetes																
Eteone lactea	10	22	2	2	1	26	1	37	34	67	0	0	23	6	1	2
Glycera dibranchiata	20	65	3	19	0	51	4	30	3	7	0	0	0	1	0	0
Glycinde solitaria	0	0	0	18	163	132	40	25	15	1	1	0	4	1	9	0
Heteromastus filiformis	35	36	18	87	2	124	31	113	46	44	7	60	25	66	57	43
Nereis succinea	1077	16	43	96	470	52	571	12	388	31	3	77	306	36	120	6
Pectinaria gouldii	239	15	0	11	6	6	0	4	8	0	0	0	1	0	0	0
Prionospio pinnata	0	0	0	2	291	17	180	277	214	707	3	111	41	784	671	504
Pseudeurythoe paucibranchiata	0	0	6	3	0	0	0	38	3	0	0	0	0	0	0	0
Scoloplos fragilis	1114	154	570	183	10	34	7	29	47	28	0	11	4	43	0	1
Scoloplos robustus	23	3	473	214	6	205	9	162	51	54	0	1	5	56	10	89
Unidentified	0	13	0	24	0	1	1	1	0	1	0	0	2	2	0	1
Flatworms																
Stylochus ellipticus	0	4	0	1	1	147	0	12	0	5	0	1	1	7	0	1
Nemerteans																
Micrura leidyi	15	287	4	71	3	172	9	166	21	37	15	35	11	36	12	36
Mollusks																
Acteocina canaliculata	55	23	3	9	5	86	5	31	4	1	0	0	0	0	0	0
Acteon punctostriatus	0	4	0	0	1	147	1	4	0	1	0	0	0	0	0	0
Brachidontes recurvus	0	4	0	0	0	0	0	0	0	0	0	0	0	0	0	0
Gemma gemma	984	232113	90	59019	2	2696	15	1025	34	2229	0	0	0	1	0	0
Haminoea solitaria	0	0	0	5	0	315	6	73	0	0	0	0	0	0	0	0
Laevicardium mortoni	0	0	0	0	0	0	0	0	0	0	0	0	0	0	0	0
Macoma balthica	222	21	81	18	21	7	329	243	738	368	275	23	15	36	49	7
Macoma phenax	22	120	9	43	55	465	57	302	24	70	5	59	12	61	9	40
Mulinia lateralis	5	199	78	910	68	682	699	981	598	1994	8	3	5	8	3	1
Mya arenaria	2	0	0	1	12	6991	185	12936	376	2928	0	1	0	35	100	335
Tagelus plebius	0	2	0	0	0	0	0	1	0	0	0	0	0	0	0	0

Table 4. August catches in a normal year compared to those following Agnes for two biotopes at Cove Point.

SPECIES	MUD BIOTOPE-AUGUST 1971 (20)	1972 (8)	SAND BIOTOPE-AUGUST 1971 (28)	1972 (12)
C. polita	2	1	27	2
C. caeca	-	-	4	6
E. triloba	0	1	36	13
E. lactea	2	0	2	0
L. americanus	-	-	1	0
G. dibranchiata	3	0	19	0
G. solitaria	0	1	18	0
H. filiformis	18	7	87	60
N. succinea	43	3	96	77
P. gouldii	-	-	11	0
P. pinnata	0	3	3	111
P. paucibranchiata	6	0	3	0
S. fragilis	570	0	183	11
S. robustus	473	0	214	1
S. ellipticus	-	-	1	1
M. leidyi	4	15	71	35
A. canaliculata	3	0	9	0
G. gemma	90	0	59019	0
H. solitaria	5	0	-	-
M. balthica	81	275	18	23
M. phenax	9	5	43	59
M. lateralis	78	8	910	3
M. arenaria	-	-	1	1

in both years while 8 were present in 1971 but not in 1972. None of these present in 1972 were absent in 1971.

Of the many water quality characteristics influenced by the discharges associated with Agnes, probably none was of greater significance for the sedentary biota than salinity. No salinity data were taken at the Cove Point location, but data for nearby locations for long term averages (Stroup & Lynn 1963) and depressions induced by Agnes (CBRC 1973) show that a significant salinity change occurred. These data (summarized in Table 5) show a comparison of salinity values for Chesapeake Bay Institute stations north (834 F) and south (181) of the Cove Point area with long term averages read from isohalines for spring (lower values) and summer (higher values) in this segment of the Bay. A sharp drop of 7-8 ppt was associated with the storm. By the middle of August, roughly 6 weeks later, salinities had begun to return to near normal values, but were still depressed. The long term summer average for salinity in this area is 13 ppt. Salinities in August after Agnes were still 5.5-6 ppt below normal. All but the most euryhaline organisms were subjected to an acute salinity stress in the days immediately after Agnes, as well as chronic salinity stress which persisted beyond mid-summer, an appreciable duration in terms of the short life span of most of these species.

Wass (1972) has summarized salinity ranges for many Chesapeake Bay macroinvertebrates; the data for those occurring at Cove Point are summarized in Table 6. This information strongly suggests that the osmotic problems caused by Agnes exceeded the adjustment capacity of *Mulinia lateralis* and *Mya arenaria*, while the greater probability of survival in *Macoma balthica* and *Macoma phenax* might have been predicted. The data suggest that *Gemma* might have tolerated the lower

Table 5. Surface salinities at Chesapeake Bay Institute stations north and south of Cove Point during early summer, 1972, as compared with long term averages. (CBRC 1973)

DATE	834 F (NORTH)	818 (CEDAR PT.) (SOUTH)
6-7 June	8.5	8.5
26-29 June	1.0	2.5
7 July	2.5	3.0
14-15 July	4.0	5.0
22-24 July	2.0	4.0
5 August	6.0	5.5
16-17 August	7.5	7.0
Long Term (1952-1961) Averages (Stroup & Lynn 1963)	9-13	9-13

Table 6. Salinity ranges of Cove Point species in terms of Venice classification system. Adapted from Wass (1972).

ORGANISM	CATEGORY	SALINITY RANGE (ppt)
Diadumene leucolena	Upper Meso- and Polyhaline	12-30
Callinectes sapidus	Euryhaline	-
Chiridotea caeca	Meso- to Euhaline	5-40
Crangon septemspinosa	Euryhaline	-
Cyathura polita	Oligo- and Mesohaline	0.5-18
Edotea triloba	Near Euryhaline	-
Leucon americanus	Meso- to Euhaline	5-40
Oxyurostylis smithi	Upper Meso- to Euhaline	12-40
Paleomonetes pugio	Oligo- to Polyhaline	0.5-30
Sphaeroma quadridentatum	Oligo- to Polyhaline	0.5-30
Eteone lactea	Meso- and Polyhaline	5-30
Glycera dibranchiata	Polyhaline	18-30
Glycinde solitaria	Upper Meso- and Polyhaline	12-30
Heteromastus filiformis	Oligo - Euhaline	-
Nephtys bucera	Euhaline	-
Pectinaria gouldii	Upper Meso- and Polyhaline	12-30
Polydora ligni	Oligo - Euhaline	-
Prionospio pinnata	Euryhaline	-
Pseudeurythoe paucibranchiata	Upper Meso- and Polyhaline	12-30
Scoloplos robustus	Polyhaline	18-30
Stylochus ellipticus	Upper Meso- and Polyhaline	12-30
Micrura leidyi	Upper Meso- and Polyhaline	12-30
Acteocina canaliculata	Polyhaline	18-30
Acteon punctostriatus	Polyhaline	18-30
Brachidontes recurvus	Meso- to Euhaline	5-40
Gemma gemma	Meso- and Polyhaline	5-30
Haminoea solitaria	Polyhaline	18-30
Laevicardium mortoni	Polyhaline	18-30
Macoma balthica	Lower Mesohaline	5-12
Macoma phenax	Mesohaline	5-18
Mulinia lateralis	Upper Meso- and Polyhaline	12-30
Mya arenaria	Upper Meso- and Polyhaline	12-30
Tagelus plebius	Upper Meso- and Lower Polyhaline	12-24

salinities, although evidence already presented indicates otherwise. This suggests that salinity was only the foremost of an array of stresses associated with the influx of fresh water, or that some of the tolerance limits are inexact.

The impact of Agnes on the infauna at Cove Point was distinct although not so dramatic as reported elsewhere in this symposium. Only one abundant species (*G. gemma*) seems to have essentially disappeared from the area, although several forms of moderate to low abundance (e.g. *Leucon americanus, Acteocina canaliculata, Acteon punctostriatus*) were not taken during the post-storm sampling. The general eurytopy of temperate estuarine forms apparently enabled the survival of a typical abundance of invertebrates at Cove Point, despite the magnitude of a perturbation with a frequency measured in centuries.

ACKNOWLEDGEMENTS

We gratefully acknowledge the financial support of the Columbia LNG Corporation throughout the course of this program. We also thank Dick LaPlante and Julie Hunt, who gave valuable technical help.

LITERATURE CITED

Chesapeake Bay Research Council. 1973. Effects of Hurricane Agnes on the environment and organisms of Chesapeake Bay. A report to the Philadelphia District U. S. Army Corps of Engrs. (Aven M. Andersen, Coordinator), Chesapeake Bay Inst. Contrib. 187, Nat. Resour. Inst. Contrib. 529, Va. Inst. Mar. Sci. Spec. Rep. Appl. Mar. Sci. & Ocean Engr. 29, 172 p.

Hamilton, D. H. and R. S. LaPlante. 1972. Cove Point benthic study. Annual Rep., Nat. Resour. Inst., Univ. of Maryland, Ref. 72-36, 66 p.

Hamilton, D. H. and R. S. LaPlante. 1973. Cove Point benthic study. Annual Rep., Nat. Resour. Inst., Univ. of Maryland, Ref. 73-53, 98 p.

Holme, N. A. 1971. Macrofauna sampling. Chapter 6 in Methods for the Study of Marine Benthos (N. A. Holme, A. D. McIntyre, eds.), IBP Handbook No. 16, Blackwell Scientific Publications.

Stroup, E. D. and R. J. Lynn. 1963. Atlas of salinity and temperature distributions in Chesapeake Bay, 1952-1961, and seasonal averages, 1949-1961. Chesapeake Bay Inst. Johns Hopkins Univ. Graph. Summ. Rep. 2, 409 p.

Tenore, K. R. 1972. Macrobenthos of the Pamlico River estuary, North Carolina. Ecol. Monogr. 42(1):51-69.

Wass, M. L., et al. 1972. A check list of the biota of lower Chesapeake Bay. Va. Inst. Mar. Sci. Spec. Sci. Rep. 65, 290 p.

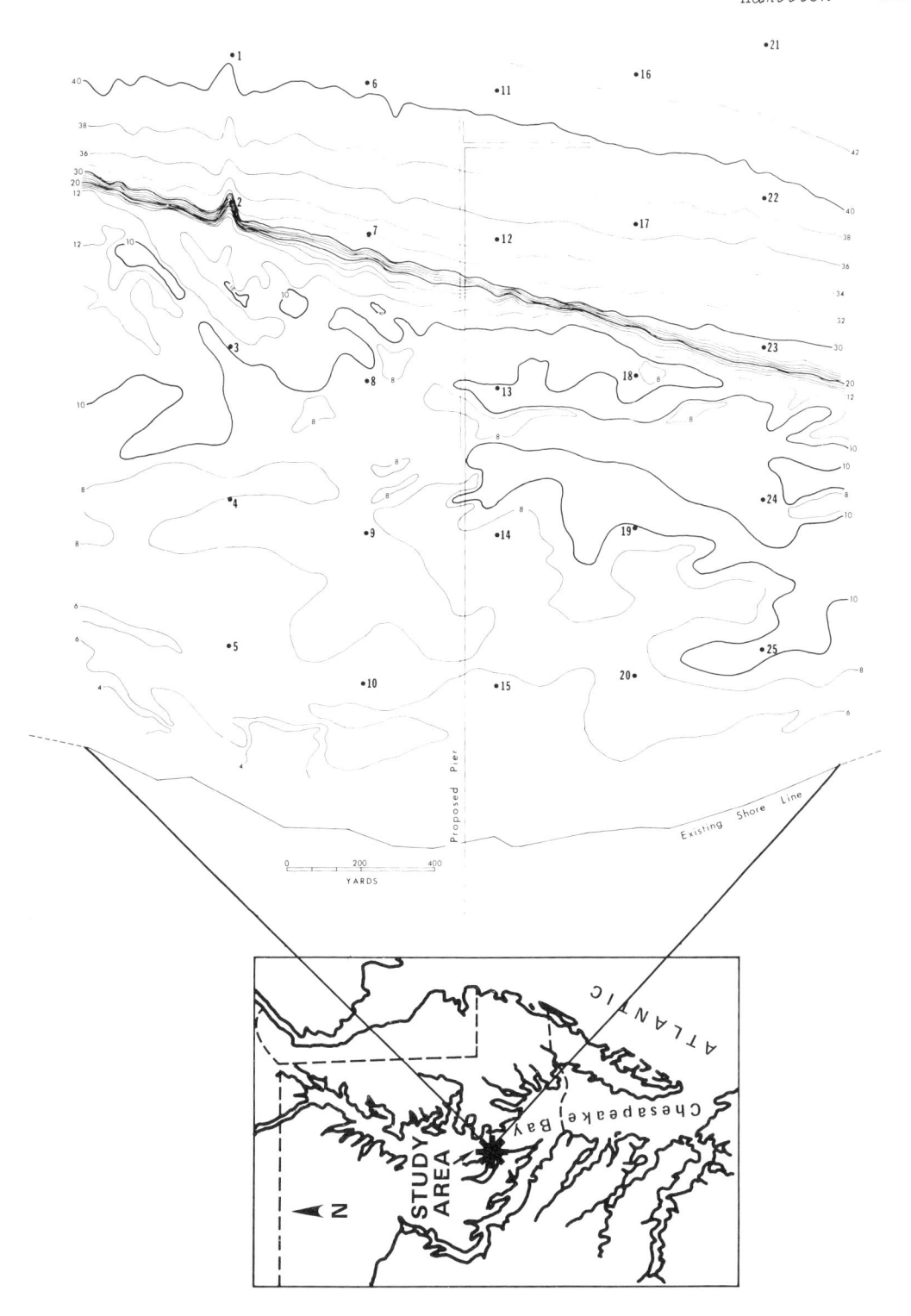

Figure 1. Location of study area and arrangement of sampling grid.

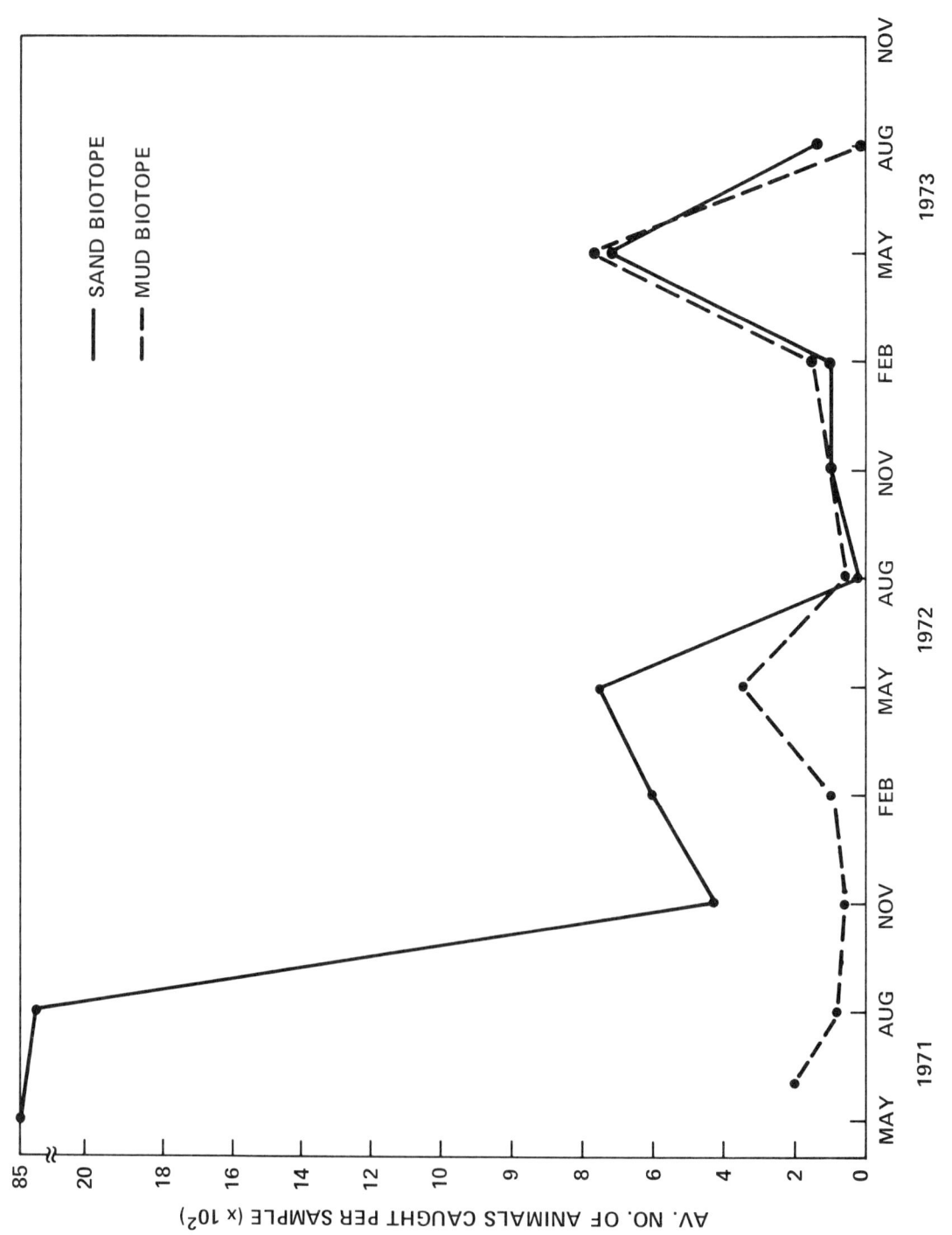

Figure 2. Temporal pattern of total mud and sand infaunal catch.

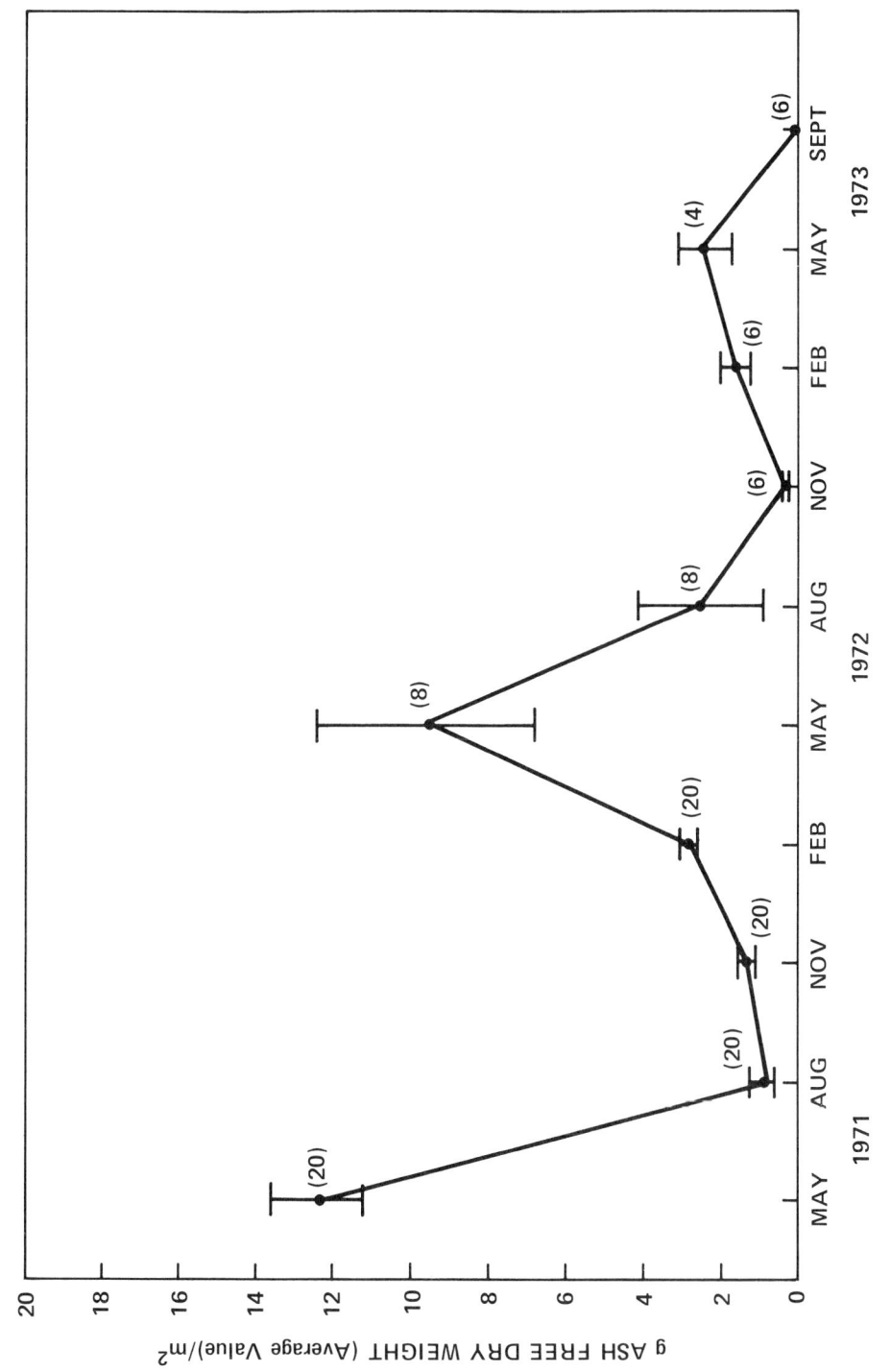

Figure 3. Temporal pattern of mud biotope standing crop.

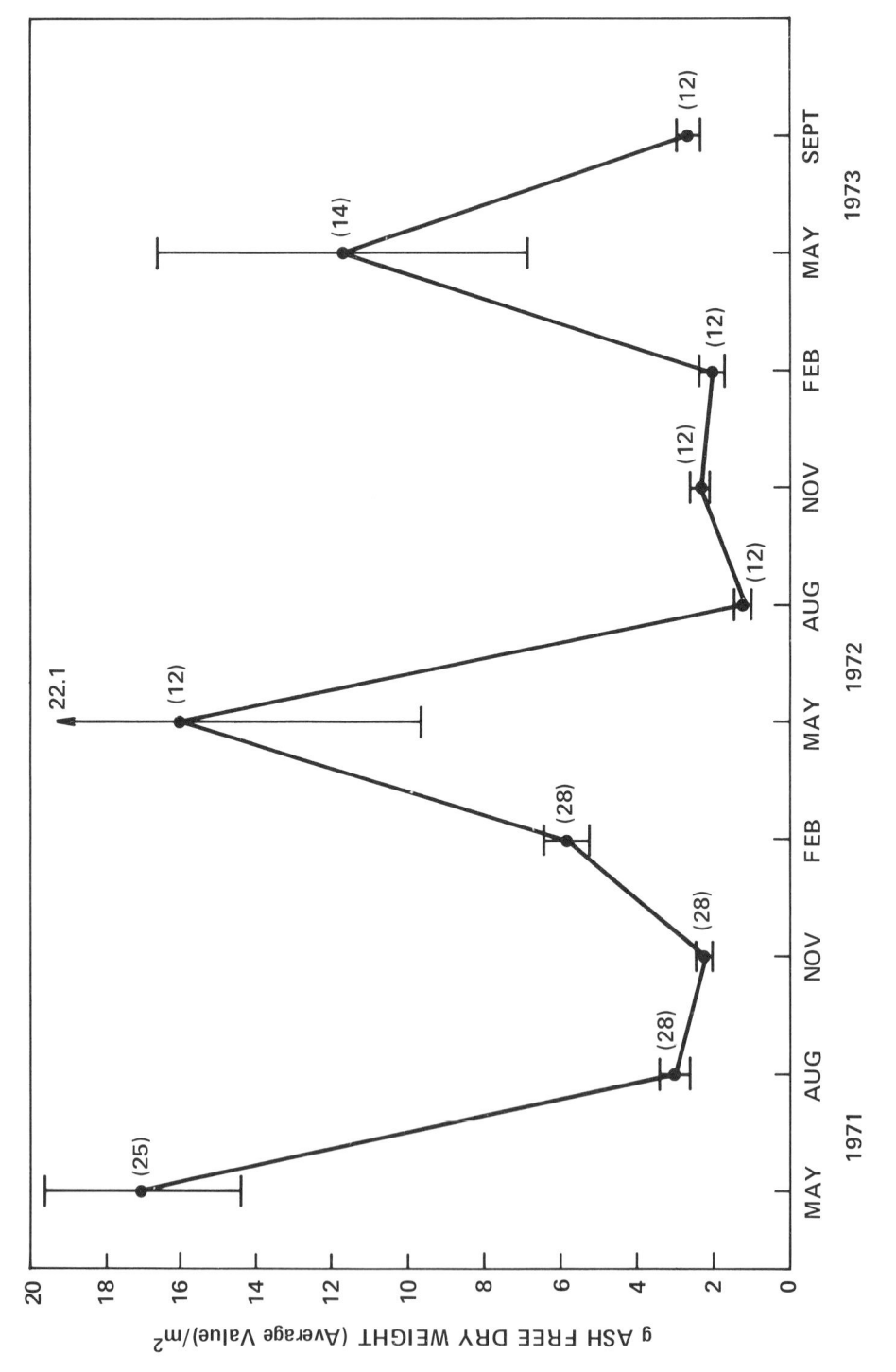

Figure 4. Temporal pattern of sand biotope standing crop.

PATTERNS OF DISTRIBUTION OF ESTUARINE ORGANISMS AND THEIR
RESPONSE TO A CATASTROPHIC DECREASE IN SALINITY[1]

Peter F. Larsen[2]

ABSTRACT

The occurrence of Tropical Storm Agnes during an ongoing study on the community structure of the macrobenthos associated with the James River oyster reefs provided a unique opportunity to document the responses of this assemblage to such a disturbance. The spatial and temporal patterns of abundance of 18 important taxa are examined in this paper. Eight species exhibited limited upestuary penetration, six were most successful in the upper part of the estuarine segment studied, two were most abundant in the mid-section of the study area, and two were ubiquitous. In the post-Agnes period, six species exhibited reduced population levels, three experienced population increases, three became relatively more abundant at the downestuary sites and reduced at the upestuary sites, three became relatively more abundant at the upestuary sites and reduced at the downestuary sites, while no significant response was shown by three others.

Hypothetical response categories are advanced to explain these responses. The freshet arrived and removed stenotopic species (category 1 response) which allowed others to fill the void in abundance (category 2 response). Other species essentially extended their range downstream where conditions were not optimal but were reduced in their original range because of the physiological stress caused by very low salinities (category 3 response). With the return of higher salinities the larvae of more stenohaline species settled where there was open space, i.e., at the upestuary sites (category 4 response). Some species showed no significant changes in abundance (category 5 response).

INTRODUCTION

On June 21, 1972, during a year study of the benthic macrofaunal community associated with the James River oyster reefs, Agnes crossed the Chesapeake Bay drainage basin. The torrential rains associated with this storm caused flooding, which on the James River was exceeded only by the flood of 1771. Salinity, below normal before the freshet, dropped to zero throughout the normally oligohaline and mesohaline estuarine area under consideration. Salinity increased steadily throughout the summer but normal seasonal values were not attained for the remainder of 1972. A complete description of the flood and its immediate effects can be found in the Chesapeake Bay Research Council (1973) report.

The Agnes flood occurred immediately after the completion of a season's sampling and allowed documentation of faunal responses to a major natural disturbance. Impacts of natural catastrophes of this sort are well recorded in the scientific literature (cf. Brongersma-Sanders 1957) but usually the reports do not consider the characteristic dynamics of natural systems because of a lack of pre-catastrophe data. Recent accounts of the effects of abnormally low salinities on estuarine fauna include Thomas and White (1969), May (1972) and Andrews (1973).

---
[1]Contribution No.763, Virginia Institute of Marine Science.
[2]Fisheries Research Laboratory, Department of Marine Resources, West Boothbay Harbor, Me. 04575.

The spatial and temporal patterns of abundance of several important species from the oyster reef assemblage are examined in this paper, and based on these distributions, hypothetical response categories are advanced to explain the general responses of estuarine organisms to freshets.

MATERIALS AND METHODS

Eight study sites were selected along the estuarine area in the lower James River, the southernmost tributary of the Chesapeake Bay system (Fig. 1). These sites encompassed the entire range of physical and biological conditions of the productive natural oyster reefs.

Samples were taken in December 1971 and March-April, June and September 1972. The June sampling period ended five days before the passage of Agnes and the September samples were taken about 11 weeks after it. At each site, six 0.0126 m^2 samples were obtained from randomly selected points around a fixed landmark. The samples from a surface operated suction sampler (Larsen 1974) were retained in a 0.5 mm mesh bag. The bags were removed from the sampler, labeled, and placed in a MgSO$_4$ solution to relax the organisms. Full strength formalin was added after a few hours to produce approximately a 5% solution.

The preserved samples were separated into three size groups, using 9.5, 4.0 and 1.0 mm screens, to facilitate handling. All organisms were removed from the samples, identified and enumerated. Shell-boring animals were recovered by carefully fragmenting each shell with pliers.

Some physical characteristics of these sites, observed during sampling, are contained in Table 1, and long term average and post-Agnes salinity regimes are presented in Figs. 2 and 3 respectively. As 1971 and 1972 were wet years, the isohalines in Fig. 2 are slightly further upestuary than they were in the immediate pre-Agnes period.

Table 1. Physical characteristics of the sampling sites in the James River estuary, Virginia.

Site	Depth Range (m)	Mean Depth	Observed Salinity Range (ppt)	Kilometers from mouth
1	2.1-4.1	3.1	13.8-17.6	18.9
2	1.8-4.0	2.9	11.5-13.4	21.4
3	1.5-2.7	1.9	10.4-13.7	25.2
4	2.7-3.7	3.2	9.5-12.3	29.2
5	1.2-3.4	2.4	2.0-9.3	35.0
6	2.7-5.5	3.9	2.1-9.6	35.3
7	1.8-3.7	2.7	1.6-8.8	37.1
8	1.4-5.1	2.8	0.7-6.2	42.4

A preview of the data was undertaken to identify species whose abundance and frequency of occurrence were sufficient enough upon which to base conclusions about the significance of changes in these parameters.

RESULTS

The 192 samples yielded 142 recognizable taxa. Of the 124 non-colonial taxa, 18 had the criteria of abundance and frequency of occurrence to justify analysis of their distributions. Together they accounted for over 93% of the individuals collected. These species were: the coelenterate *Diadumene leucolena*; the flatworm *Stylochus ellipticus*; the mollucks *Mya arenaria*, *Brachidontes recurvus*, *Congeria leucophaeta*, and *Odostomia impressa*; the annelids *Streblospio benedicti*, *Polydora ligni*, *P. websteri*, *Boccardia hamata*, *Heteromastus filiformis*, *Nereis succinea*, and *Peloscolex* spp. (includes *P. heterochaetus* and *P. gabriellae*); the crustaceans *Cassidinidea lunifrons*, *Corophium lacustre*, *Melita nitida*, and *Balanus improvisus*; and the tunicate *Molgula manhattensis*. The spatial and temporal abundances of these species are tabulated in Table 2a-r.

These species exhibited four principal spatial patterns of abundance. Eight species, the polychaetes *S. benedicti*, *P. ligni*, and *H. filiformis*, the mollusks *M. arenaria* and *O. impressa*, the polyclad *S. ellipticus*, the anemone *D. leucolena*, and the tunicate *M. manhattensis*, were most abundant at the downestuary sites and were absent, or present in reduced numbers, at the upestuary sites. Six species, the polychaete *B. hamata*, the crustaceans *C. lunifrons*, *C. lacustre*, and *B. improvisus*, and the mollusks *B. recurvus* and *C. leucophaeta* were most abundant in the middle or upper section of the estuarine segment studied and exhibited reduced populations toward the downestuary end of the sampling area. The polychaete *P. websteri* and the amphipod *M. nitida* had centers of abundance at sites 2-4 with reduced population levels throughout the rest of the estuarine segment. The annelids *N. succinea* and *Peloscolex* spp. were ubiquitous over the study area.

Several patterns of changes in population levels were noted between June and September. Six taxa, *S. benedicti*, *H. filiformis*, *Peloscolex* spp., *M. arenaria*, *O. impressa*, and *D. leucolena* all experience definite population declines. *M. arenaria* and *D. leucolena* also suffered range contractions. Remarkable population gains were registered between June and September by *B. hamata*, *C. leucophaeta*, and *M. manhattensis*, and the latter two also experienced range extensions. *C. lacustre* and *B. recurvus* exhibited increases in abundance at the four downestuary sites in September while maintaining population levels comparable to the levels of previous sampling periods at the upestuary sites. Fewer individuals of *B. improvisus* were found at the four upestuary sites in September than in any previous sampling period while the second highest population level of the study year was found at the downestuary sites in this month. In September, both species of *Polydora* exhibited yearly low population levels at the downestuary sites and yearly highs at the upestuary sites. *S. ellipticus* showed reduced numbers but a wider distribution in September relative to June. The abundance and distributions of *N. succinea*, *M. nitida*, and *C. lunifrons* were not significantly changed between June and September.

DISCUSSION

An attempt is made here to categorize the responses of these 18 estuarine species to the sudden and severe physiological stress caused by the Agnes freshet. Such categorization is difficult in biological systems (Korringa 1957) as one is usually dealing with gradients of response and not a finite number of discrete alternatives. There will always be exceptions and borderline cases and the more specific a categorization, the more exceptions there will be. Sampling, in this case, covered only one yearly cycle and thus allowed no information on the normal year to year fluctuations in population levels. Some of the species considered in this paper ordinarily exhibit population fluctuations in the summer, but it is assumed that in the summer of 1972 these fluctuations were accentuated by the Agnes freshet. A significant amount of distributional information has been presented on

Table 2. Spatial and temporal abundance of select species in the James River oyster assemblage. September samples are post-Agnes. Values are the sum of six replicates.

a) *Diadumene leucolena*

site	D	M	J	S
1	90	431	255	54
2	516	219	205	0
3	882	240	317	23
4	483	220	195	4
5	0	0	0	0
6	0	0	0	0
7	0	0	0	0
8	0	0	0	0

b) *Stylochus ellipticus*

site	D	M	J	S
1	2	0	2	2
2	7	1	24	0
3	3	0	178	4
4	6	2	67	1
5	23	0	0	1
6	0	0	0	2
7	4	0	0	2
8	1	0	0	2

c) *Mya arenaria*

site	D	M	J	S
1	32	45	62	4
2	39	242	133	1
3	6	134	28	1
4	15	170	60	0
5	0	0	0	0
6	0	0	0	0
7	0	0	0	0
8	0	0	0	0

d) *Brachidontes recurvus*

site	D	M	J	S
1	0	0	0	1
2	31	38	22	179
3	61	29	24	151
4	57	65	59	274
5	197	34	140	89
6	94	4	6	110
7	321	836	268	186
8	123	14	56	97

e) *Congeria leucophaeta*

site	D	M	J	S
1	0	0	0	0
2	0	0	0	0
3	0	0	0	0
4	0	0	0	6
5	0	0	0	149
6	0	0	0	19
7	0	1	0	237
8	4	0	3	793

f) *Odostomia impressa*

site	D	M	J	S
1	179	23	51	10
2	226	160	123	10
3	87	38	70	0
4	97	134	128	0
5	2	1	0	0
6	0	0	0	0
7	3	14	2	0
8	0	0	0	0

g) *Streblospio benedicti*

site	D	M	J	S
1	16	242	1087	7
2	24	256	799	12
3	123	101	462	27
4	57	121	210	14
5	3	0	0	0
6	1	0	0	0
7	0	0	0	0
8	0	0	0	0

h) *Polydora ligni*

site	D	M	J	S
1	45	417	330	6
2	80	238	588	117
3	100	80	826	57
4	47	86	528	24
5	38	1	9	35
6	5	1	0	26
7	27	64	17	26
8	3	0	0	51

Table 2. (Continued)

i) *Polydora websteri*

site	D	M	J	S
1	7	28	47	0
2	55	43	87	30
3	153	39	107	35
4	59	53	63	12
5	10	2	5	39
6	6	3	0	6
7	6	50	9	22
8	2	0	1	3

j) *Boccardia hamata*

site	D	M	J	S
1	0	19	11	7
2	3	1	0	361
3	128	27	44	277
4	21	20	25	166
5	121	17	15	97
6	86	16	2	45
7	92	101	85	106
8	16	2	15	26

k) *Heteromastus filiformis*

site	D	M	J	S
1	86	151	146	100
2	216	247	184	57
3	109	111	108	67
4	149	169	84	67
5	39	6	13	9
6	17	11	2	9
7	61	43	38	13
8	31	8	9	1

l) *Nereis succinea*

site	D	M	J	S
1	36	252	122	28
2	96	63	28	151
3	254	226	107	329
4	186	163	91	214
5	259	13	57	88
6	23	7	1	32
7	76	203	63	88
8	37	8	30	33

m) *Peloscolex* spp.

site	D	M	J	S
1	203	584	385	25
2	100	280	408	62
3	1005	853	605	181
4	397	564	266	64
5	622	57	68	36
6	89	68	2	48
7	381	702	77	63
8	134	27	104	63

n) *Cassidinidea lunifrons*

site	D	M	J	S
1	6	19	5	7
2	129	53	37	73
3	276	96	52	145
4	74	56	43	73
5	495	12	33	90
6	27	4	1	2
7	214	310	55	39
8	209	3	72	54

o) *Corophium lacustre*

site	D	M	J	S
1	1	2	2	17
2	4	0	0	145
3	18	1	49	133
4	5	1	4	24
5	153	9	202	174
6	43	19	59	37
7	197	177	700	108
8	66	13	640	301

p) *Melita nitida*

site	D	M	J	S
1	12	33	33	11
2	55	25	102	106
3	281	100	733	178
4	85	89	297	117
5	298	3	23	40
6	13	3	0	13
7	124	36	21	44
8	65	1	30	88

Table 2 (Continued)

q) *Balanus improvisus*

site	D	M	J	S
1	0	0	0	39
2	56	12	293	173
3	43	16	5567	363
4	25	33	3964	695
5	3445	76	2148	163
6	341	69	58	102
7	2016	4330	3254	187
8	779	93	452	109

r) *Molgula manhattensis*

site	D	M	J	S
1	0	1	1	422
2	0	0	0	428
3	0	0	0	21
4	0	0	0	7
5	0	0	0	0
6	0	0	0	0
7	0	0	0	0
8	0	0	0	0

these 18 taxa and, despite the above limitations, generalizations derived from the data may have far-reaching implications and predictive value in similar future situations.

The first faunal response to the freshet, and the one most often reported in the literature, was the decrease in abundance of six species. A summer decline in abundance of the soft clam *M. arenaria* is normal in the Chesapeake Bay due to predatory and/or physiological stress (J. Lucy, personal communication). The population reductions experienced by the other species showing this type of response, however, were probably related more directly to the very low salinities of the immediate post-Agnes period, as four of them had previously exhibited limited upestuary penetration. The numerical decrease of *H. filiformis* between June and September is noteworthy since Tenore (1972) reported this species as most abundant and in the summer in the Pamlico River estuary, North Carolina.

The second response category was population increases in the wake of the freshet. The increase in numbers of *B. hamata*, *C. leucophaeta*, and *M. manhattensis* was caused by the coincidence of their reproductive periods with the recovery of salinity levels and the space left open by the decline of many shell-fouling species.

B. improvisus, *C. lacustre*, and *B. recurvus* exhibited a third response category characterized by an atypical, relatively high abundance at the four downestuary sites, coupled with a relatively low abundance at the four upestuary sites. A high proportion of the spring set of *B. improvisus* survived at downestuary sites evidently due to reduced biotic stress caused by the decrease in salinity. At the upestuary sites salinity was limiting for much of the summer, resulting in abnormally low population levels. On the other hand, *C. lacustre* and *B. recurvus* probably set after the freshet peaked. They opportunistically flourished in the open space on the downestuary sites whereas upestuary suboptimal salinities limited recruitment.

The two species of *Polydora* exhibited a fourth response pattern to the freshet, characterized by a relatively high abundance at the upestuary sites with a below normal population at the downestuary sites. Larvae of *P. websteri* are found in the water column throughout the year in Louisiana (Hopkins 1958) and from April to August in Maine (Blake 1969) so it appears safe to assume that *P. websteri* reproduces throughout much of the year in the Chesapeake Bay region. Orth (1971) demonstrated that the peak of *P. ligni* reproduction occurs in the spring, but significant numbers of larvae were still found at the end of May. Larvae of this species occur in Maine waters through the end of September (Blake 1969). *P. ligni* may reproduce throughout the warm period of the year as was found for the European species *P.*

ciliata (Daro & Polk 1973) which Rasmussen (1973) equates with *P. ligni*. Daro and Polk (1973) found that three to four generations of *P. ciliata* are normally produced each year; the generations after the first do not show heavy settlement because of larval mortality and interspecific competition. These data indicate that both common species of *Polydora* found in this study have a potential for recruitment that lasts through the summer months.

The occurrence of the freshet interrupted recruitment of *Polydora* in the study area. This may have been caused by the cessation of adult spawning, increased larval mortality, or a combination of both. Before successful recruitment was reestablished much of the space on the lower four reefs was utilized by more tolerant species, e.g. *B. recurvus* and *C. lacustre*, or by species with well-timed reproductive cycles, e.g. *M. manhattensis* and *B. hamata*. Salinities at all sites were at yearly highs by late summer. This put the upestuary sites within the optimal salinity range of the *Polydora* species. As *Polydora* recruitment resumed, the larvae were relatively more successful in settling at the upestuary sites than at the more heavily fouled downestuary sites.

The flatworm *S. ellipticus* suffered a population loss while experiencing a range extension between June and September, which can be considered a modified category 4 response.

A fifth category of response was no significant change. If the species which exhibited this response, *N. succinea*, *M. nitida* and *C. lunifrons*, were affected by the freshet, the sampling design was not adequate to demonstrate it.

ACKNOWLEDGEMENTS

The support of D. S. Haven throughout the course of the field study is gratefully acknowledged. The contribution of R. J. Orth, D. S. Haven and J. D. Andrews, who reviewed an earlier form of this paper, are appreciated.

LITERATURE CITED

Andrews, J. D. 1973. Effects of Tropical Storm Agnes on epifaunal invertebrates in Virginia estuaries. Chesapeake Sci. 14:223-234.

Blake, J. A. 1969. Reproduction and larval development of *Polydora* from northern New England (Polychaeta: Spionidae). Ophelia 7:1-63.

Brongersma-Sanders, J. 1957. Mass mortality in the sea. Pages 941-1010 in Hedgpeth, J. W. Treatise on marine ecology and paleoecology. Vol. 1, Ecology. Geol. Soc. Am. Mem. 67.

Chesapeake Bay Research Council. 1973. Effects of Hurricane Agnes on the environment and organisms of Chesapeake Bay. A report to the Philadelphia District, U. S. Army Corps of Engrs. (Aven M. Andersen, Coordinator), Chesapeake Bay Inst. Contrib. 187, Nat. Resour. Inst. Contrib. 529, Va. Inst. Mar. Sci. Spec. Rep. Appl. Mar. Sci. & Ocean Engr. 29, 172 p.

Daro, M. H. and R. Polk. 1973. The autecology of *Polydora ciliata* along the Belgian coast. Neth. J. Sea Res. 6:130-140.

Hopkins, S. H. 1958. The planktonic larvae of *Polydora websteri* Hartman (Annelida, Polychaeta) and their setting on oysters. Bull. Mar. Sci. Gulf Caribb. 8:268-277.

Korringa, P. 1957. Lunar periodicity. Pages 917-934 in Hedgpeth, J. W. Treatise on marine ecology and paleoecology. Vol. 1, Ecology. Geol. Soc. Am. Mem. 67.

Larsen, P. F. 1974. A remotely operated shallow water benthic suction sampler. Chesapeake Sci. 15:176-178.

May, E. B. 1972. The effect of floodwater on oysters in Mobile Bay. Proc. Nat. Shellf. Assn. 62:67-71.

Orth, R. J. 1971. Observations on the planktonic larvae of *Polydora ligni* Webster (Polychaeta: Spionidae) in the York River, Virginia. Chesapeake Sci. 12:121-124.

Rasmussen, E. 1973. Systematics and ecology of the Isefjord marine fauna (Denmark). Ophelia 11:1-507.

Tenore, K. R. 1972. Macrobenthos of the Pamlico River Estuary, North Carolina. Ecol. Monogr. 42:51-69.

Thomas, M. L. H. and G. N. White. 1969. Mass mortality of estuarine fauna at Bideford, P.E.I., associated with abnormally low salinities. J. Fish. Res. Bd. Canada 26:701-704.

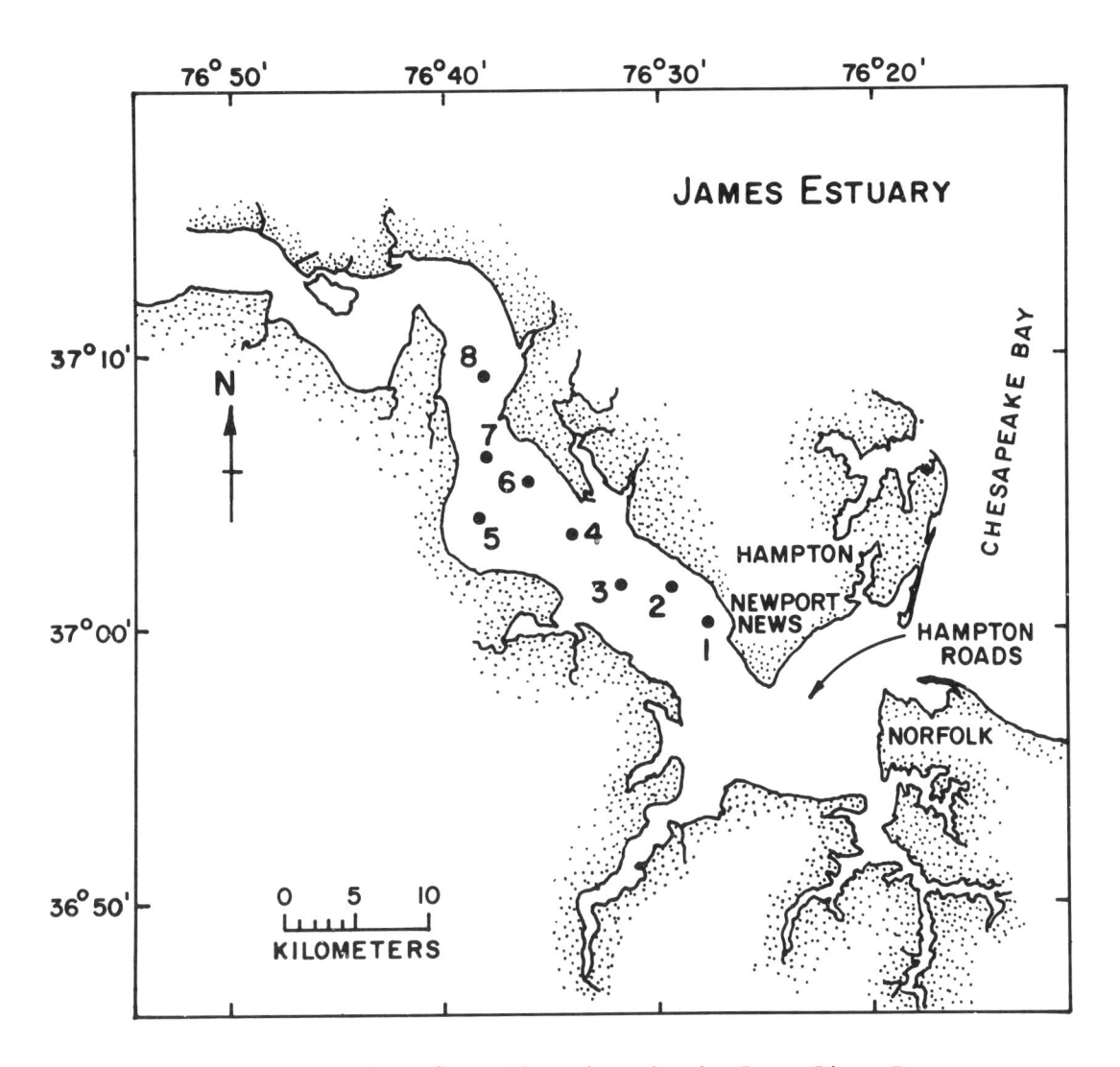

Figure 1. Location of sampling sites in the James River Estuary.
Common names for these sites are:
Site 1 - Brown Shoal
Site 2 - Thomas' Rock
Site 3 - White Shoals
Site 4 - Wreck Shoals
Site 5 - Point of Shoals
Site 6 - The Swash
Site 7 - Horsehead Shoal
Site 8 - Deepwater Shoal

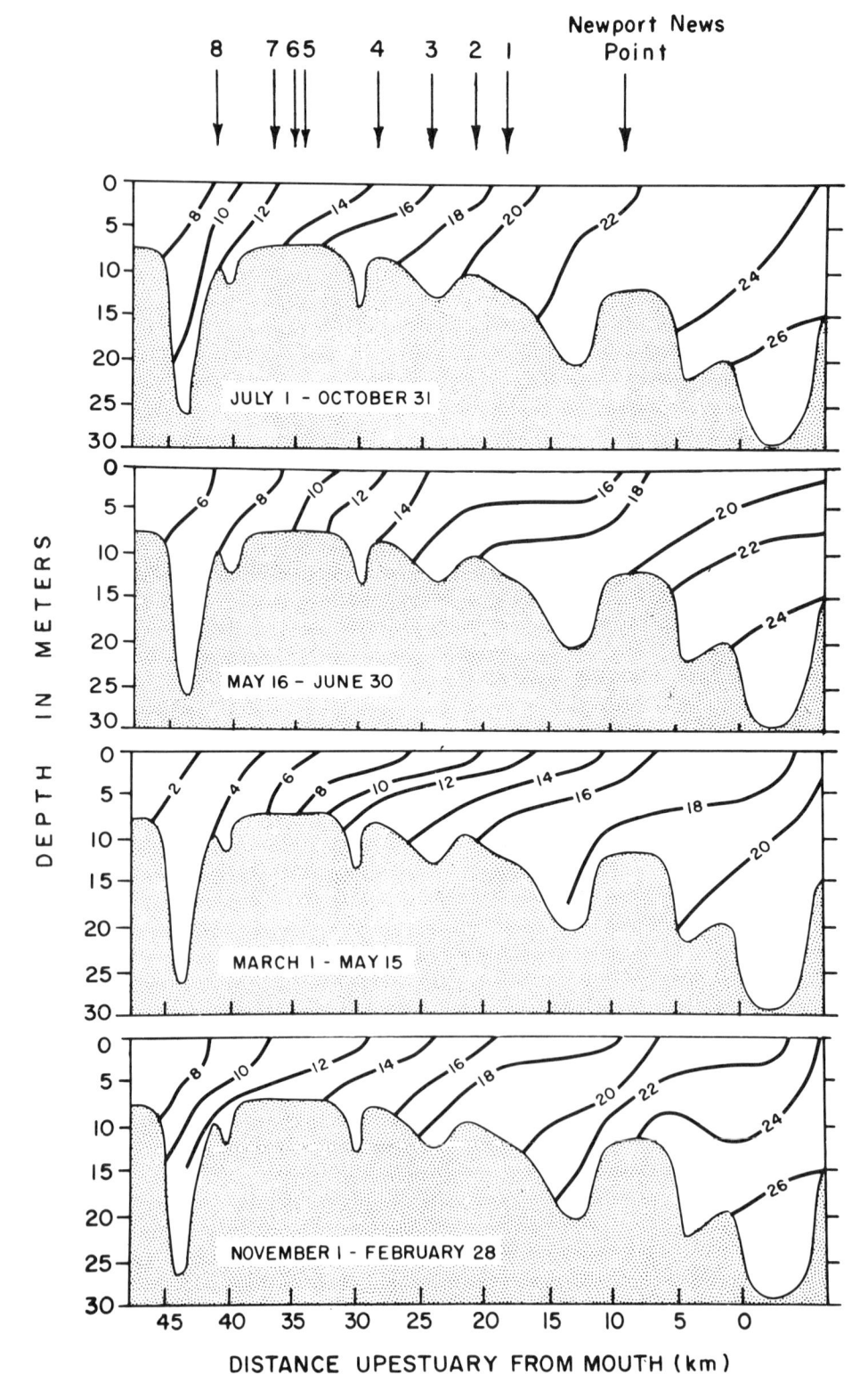

Figure 2. Normal seasonal salinity distribution of the James River Estuary. Average from 1944 to 1965 (modified from Nichols 1972).

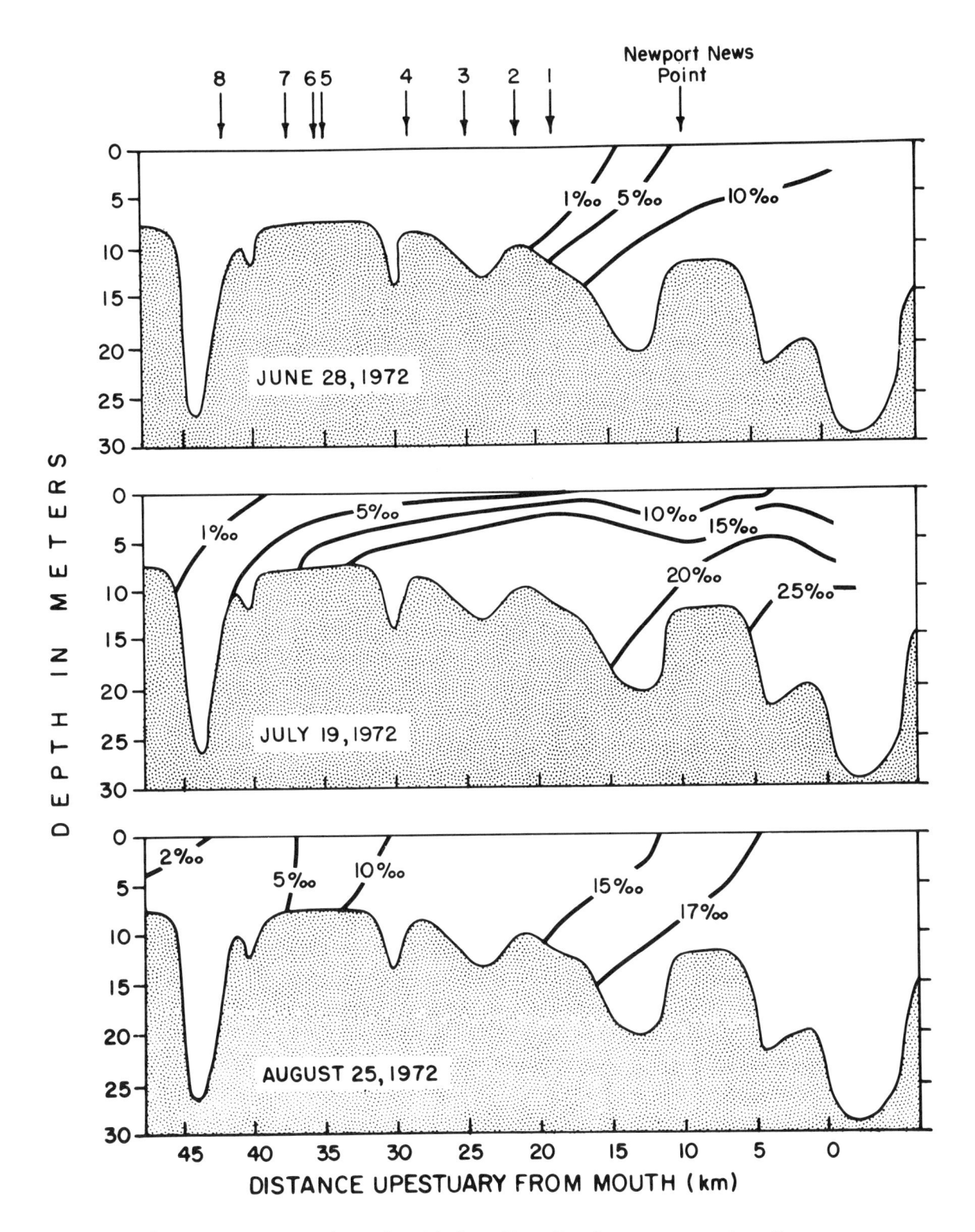

Figure 3. Alteration of salinity distribution as a result of Hurricane Agnes (modified from Chesapeake Bay Research Council 1973).

THE EFFECT OF TROPICAL STORM AGNES ON THE
BENTHIC FAUNA OF EELGRASS, ZOSTERA MARINA,
IN THE LOWER CHESAPEAKE BAY[1]

Robert J. Orth[2]

ABSTRACT

Tropical Storm Agnes caused major changes in the macroinvertebrate assemblages of both epifauna and infauna in eelgrass, *Zostera marina*, beds. Species abundance and density of infauna declined by one-third to one-half of values found prior to Agnes. Typical members of the infaunal community such as the amphipods, *Ampelisca* spp. and *Lysianassa alba*, the polychaetes *Sabella microphthalma* and *Exogone dispar*, ostracods and gastropods were either absent or rare following Agnes. Epifaunal density was much higher than that recorded before Agnes but the number of species was reduced. This high density was attributed to several species, e.g. *Molgula manhattensis*, which appeared to occupy space left open by the absence of typical members of this community, e.g. *Paracereceis caudata* and *Bittium varium*. The abnormally low salinities following Agnes affected various species in different ways. Some species were totally eliminated, severely reduced in abundance or, in a few euryhaline species, not affected at all. In some populations it appeared that adults survived but juveniles suffered high mortalities. Recovery and reestablishment by many species will be complicated by the disappearance of eelgrass in some portions of the Bay.

INTRODUCTION

The eelgrass, *Zostera marina*, ecosystem in the Chesapeake Bay is highly productive, furnishing shelter and food for young fishes and blue crabs and providing an essential habitat for many invertebrate species. The invertebrate fauna normally associated with *Zostera* beds is among the most diverse and dense in the Chesapeake Bay (Marsh 1973; Orth 1973).

In June 1972 freshwater runoff from Tropical Storm Agnes caused a drastic reduction in salinity in the shallows where *Zostera* grows throughout the lower Bay. The *Zostera* beds and associated fauna experienced a salinity reduction of as much as 10% or 50% of their normal values. Salinity remained low for periods ranging from one to two months (see papers in this symposium concerning salinity changes in the lower Bay).

The objective of this report is to document the major changes in the *Zostera* community related to the apparent salinity stress following Agnes.

MATERIALS AND METHODS

Two study sites (Fig. 1) in the York River were chosen at which baseline data already existed. Mumfort Island was the site of a study of epibiota by Marsh

---
[1] Contribution No.764, Virginia Institute of Marine Science.
[2] Virginia Institute of Marine Science, Gloucester Point, Va. 23062 and Department of Zoology, University of Maryland, College Park, Md.

(1970) during 1967-1968 and a study of infauna by Orth (1971) in 1970. Sandy Point at the mouth of the river was also sampled during the infaunal study in 1970.

Infaunal samples were taken approximately every two months beginning in August 1972 and continuing through July 1973 at Sandy Point. Epifaunal samples were taken in March and July 1973 at Sandy Point and in July, October and December 1973 at Mumfort Island. A large epifaunal collection was taken from artificial *Zostera* in August 1972 at Sandy Point and from Brown's Bay in the Mobjack Bay in November and December 1973 and January 1974. Artificial eelgrass, consisting of polypropylene plastic strands, was placed in the York River immediately after Agnes. Fauna associated with artificial *Zostera* resembled that from real *Zostera*.

Methods of sampling epifauna and infauna were identical to those employed by Marsh (1970) and Orth (1971). Ten infaunal samples were taken randomly with a Plexiglas corer (area of 0.007 m^2); epifaunal samples were taken by clipping plants at the base and placing them in cloth bags with a 0.5 mm nylon mesh bottom. Where epifaunal and infaunal samples were taken together, the epifaunal sample was taken first, followed by the infaunal sample. The latter were washed through a 1.0 mm mesh screen. Epifauna and sediment was stripped from blades of eelgrass and washed through a 0.5 mm mesh screen. Cleansed plants were oven-dried at 80°C for 48 hours and weighed. Abundance of non-colonial species was expressed as numbers of individuals per gram dry weight of *Zostera*.

RESULTS

Infauna

Post-Agnes infaunal samples show some changes in numerically dominant species (Fig. 2, Table 1). Epifaunal species found in the sediments in winter when *Zostera* is scarce (Orth 1973) are excluded from this analysis. Samples taken after Agnes at Sandy Point were compared with those taken at Sandy Point and Back River in 1970. Samples at Sandy Point (B) and Back River (A) were very similar in species composition in 1970 (Orth 1973).

Table 1. Species recorded in infaunal collections in 1970 but not present or rare after Agnes at Sandy Point.

Prionospio heterobranchia (Polychaeta)
Exogone dispar (Polychaeta)
Lysianassa alba (Amphipoda)
Cylindroleberis mariae (Ostracoda)
Sarsiella zostericola (Ostracoda)
Anadara transversa (Bivalvia)
Acteocina canaliculata (Gastropoda)
Triphora perversa (Gastropoda)
Mangelia plicosa (Gastropoda)
Acteon punctostriatus (Gastropoda)

The eurytopic species, *Heteromastus filiformis*, *Scoloplos robustus*, *Streblospio benedicti*, *Eteone heteropoda*, *Edotea triloba*, *Polydora ligni*, *Nereis succinea*, *Spiochaetopterus oculatus*, *Glycinde solitaria*, and *Peloscolex gabriellae* (Fig. 2), seemed least affected by Agnes. Their abundances were similar to those recorded in 1970. These species are widely distributed throughout the estuary (Boesch 1971) in a wide range of sediment types. Their euryhalinity presumably allowed them to tolerate salinity stress imposed by Agnes.

Species apparently most affected by Agnes were those which brood their young or are more stenohaline (Table 1). The amphipods *Ampelisca abdita*, *A. vadorum* and *Lysianassa alba*, the isopod *Cyathura burbancki*, and the polychaetes *Sabella microphthalma*, *Exogone dispar* and *Prionospio heterobranchia*, numerically dominant species in 1970, were rare or absent following Agnes (Table 1, Fig. 2). *Macoma balthica* and *M. mitchelli*, normally restricted to the mesohaline and oligohaline zones (Wass 1972), were found at the mouth of the York River in 1973 for the first time (Orth 1971). At Mumfort Island, *M. balthica* had densities up to 14/m^2 in 1970; in July 1973 they reached 514/m^2.

At Sandy Point numbers of species and faunal densities were reduced following Agnes (Fig. 3). The high density in March 1973 was similar to that of March 1970 (at Station B, *Polydora ligni* made up 47% of the total in the Post-Agnes data and 36% prior to Agnes).

Recovery of species associated with *Zostera* was interrupted by the dramatic decline of *Zostera* in the York River and other areas of the lower Bay in August 1973. This was apparently due to the influx of the migratory cownose ray *Rhinoptera bonasus*, which uprooted vast areas of *Zostera* (Orth 1975).

Infaunal samples collected in November and December 1973 and January 1974 in Mobjack Bay, one of the few places where *Zostera* was still dense, revealed densities of *Ampelisca vadorum*, *Cyathura burbancki*, *Acteocina canaliculata* and *Sabella microphthalma* similar to those recorded in 1970. Other species (Table 1) have been found only rarely if at all.

Epifauna

Although quantitative sampling did not commence until March 1973, qualitative observations made during July and August 1972, a large qualitative sample taken from a plot of artificial *Zostera* in August 1972, and the absence of epifauna usually taken in infaunal samples during the winter months confirmed the patterns observed in the quantitative data. Changes in epifauna after Agnes are represented by taxonomic groups.

Anthozoa. *Aiptasiomorpha luciae*, the only anemone collected in 1968, was present from late March to early December, with peak densities in late summer and fall. This species was present on the artificial *Zostera* in August 1972 and appeared unaffected by Agnes. One year after Agnes, *A. luciae* was rare, with densities (0.3-1.8 ind./g *Zostera*) far below those recorded by Marsh for these same times of year. Adults apparently survived the Agnes freshet but subsequent recruitment was poor, perhaps because of poor survival of juveniles or reproductive failure. This pattern was common to several species discussed below.

Turbellaria. *Euplana gracilis* and *Stylochus ellipticus* were the most abundant flatworms in 1968. They were very abundant in August 1972 on artificial *Zostera* and again in 1973. As suggested by Andrews (1973) these species appeared unaffected by Agnes.

Rhynchocoela. *Zygonemertes virescens* and *Tetrastemma elegans* were common previously and both species were found in August 1972. In July through December 1973, densities were similar or higher than those recorded by Marsh in 1968.

Polychaeta. *Nereis succinea*, found throughout the year, was most abundant from June to October during Marsh's study. After Agnes, *N. succinea* was extremely abundant on the artificial *Zostera*, and was found throughout 1973, being most abundant during the summer with densities higher than previously recorded (Table 2).

Platynereis dumerilii was found on artificial *Zostera* after Agnes. In 1973 it was absent in summer but occurred in high densities in the fall. Marsh reported a summer increase in 1968.

Polydora ligni was also present on artificial *Zostera* and was dense at several locations during 1973 (Table 2). The seasonal abundance pattern differed from that found by Marsh. However, *Polydora* populations vary greatly in space and time. Large numbers were found on *Zostera* in Mobjack Bay in January 1974. Marsh found none in January 1968.

No other polychaete species were found on artificial *Zostera* after Agnes and only three other species have since been collected on *Zostera*. *Podarke obscura* and *Hydroides dianthus* were recorded frequently by Marsh but have not occurred since Agnes. *Sabella microphthalma* was found throughout the year in 1968. None were recorded after Agnes on artificial *Zostera*, but densities in July and October 1973 were similar to those of Marsh for this same period in 1968. *Brania clavata*, another polychaete abundant from April to October 1968 was infrequent in very low densities and not until December 1973 was it recorded in high densities. This apparently reflected gradual recovery of the populations.

Less abundant species found by Marsh, but not recorded after Agnes, were *Nereiphylla fragilis*, *Exogone dispar*, *Pista palmata*, *Odontosyllis fulgurans*, and *Lepidonotus variabilis*.

Prosobranchia. *Bittium varium* was one of the most numerically dominant species in the *Zostera* community prior to Agnes. It was abundant on artificial *Zostera* after Agnes and was observed frequently on live eelgrass during July and August 1972. In 1973, few *Bittium* were recorded from samples taken in March and July but they increased in abundance in October and November 1973 (Table 3) the decline in *Bittium* populations during late 1972 and early 1973 probably resulted from juvenile or larval mortality during the Agnes freshet. Egg capsules are deposited in May and June (Marsh 1970), with the new year class constituting an increasing proportion of the population from June through fall (Fig. 4). A random sample of *Bittium* taken in August 1972 revealed the almost complete absence of young, suggesting they were largely killed by the salinity reduction. Adults die in the fall, and the few surviving juvenile *Bittium* comprised the small spring and summer 1973 populations. Infaunal samples taken in the winter (1972-1973), when *Bittium* is normally found in the sediment, contained few live snails, another indication that young had suffered high mortalities in the summer. Repopulation during the summer of 1973 appeared successful. In July 1973, 18% of all *Bittium* were 1.5-2.0 mm and in October 1973 the entire sample (592 individuals) consisted of individuals less than 3.5 mm. Densities on *Zostera* were similar to those recorded for the same periods by Marsh. Infaunal samples taken January 1974 in Mobjack Bay revealed an abundance of *Bittium* in the sediments.

Crepidula convexa was abundant throughout 1968 and on artificial eelgrass after Agnes. In the summer and fall 1973, *Crepidula* reached a density, in some samples, 200 times that found during the same period in 1969 (Table 3). Blades of *Zostera* were covered with *Crepidula*. *Crepidula*, apparently unaffected by Agnes, was able to exploit space made vacant by the demise of other dominant species.

Table 2. Number of individuals per gram dry weight of *Zostera* for three species of polychaetes found on *Zostera* before and after Agnes. Mean and range for each sampling period for Marsh's three stations (pre-Agnes) and for replicates taken after Agnes at each station listed (SP - Sandy Point; MI - Mumfort Island; BB - Brown's Bay).

Species	Post-Agnes				Pre-Agnes			
Nereis succinea	1972	Aug	(SP)	abundant	1967	Nov	0.8	(0.02 - 2.3)
	1973	March	(SP)	0.08		Dec	0.3	(0.09 - 0.7)
		July	(SP)	10.1 (0 - 0.5)	1968	Jan	0.2	(0.02 - 2.3)
		July	(MI)	9.1 (7.8 - 12.6)		March[1]	0.1	(0.1 - 0.2)
		Oct	(MI)	11.1 (8.3 - 9.9)		June[2]	4.7	(1.2 - 11.0)
		Nov	(BB)	0.1 (8.2 - 13.9)		July	2.9	(1.7 - 4.7)
		Dec	(MI)	0.6 (0 - 0.2)		Aug	1.2	(0.4 - 2.0)
		Dec	(BB)	0.3 (0.1 - 0.5)		Oct	6.0	(3.3 - 7.7)
	1974	Jan	(BB)	0.03 (0 - 0.08)		Nov	0.5	(0.07 - 1.0)
						Dec	0.2	(0.08 - 0.4)
Platynereis dumerilii	1972	Aug	(SP)	present	1967	Nov	2.0	(0.2 - 5.7)
	1973	March	(SP)	0		Dec	0.7	(0.2 - 2.0)
		July	(SP)	0.2 (0 - 0.4)	1968	Jan	0.8	(0.1 - 2.4)
		July	(MI)	0		March[1]	0.03	(0 - 0.1)
		Oct	(MI)	4.2 (2.8 - 5.6)		June[2]	4.7	(1.2 - 11.0)
		Nov	(BB)	2.4 (1.8 - 2.9)		July	2.9	(1.7 - 4.7)
		Dec	(MI)	1.5		Aug	1.2	(0.4 - 2.0)
		Dec	(BB)	3.8 (3.4 - 4.1)		Oct	7.1	(4.5 - 10.1)
	1974	Jan	(BB)	2.8 (2.2 - 3.4)		Nov	0.5	(0.07 - 1.0)
						Dec	0.2	(0.08 - 0.4)
Polydora ligni	1972	Aug	(SP)	present	1967	Nov	0	
	1973	March	(SP)	2.2 (1.7 - 3.2)		Dec	0	
		July	(SP)	77.6 (46.8 - 96.1)	1968	Jan	0	
		July	(MI)	5.4 (4.4 - 6.3)		March[1]	0.2	(0.07 - 0.4)
		Oct	(MI)	9.3 (7.9 - 10.7)		April	56.4	(13.0 - 86.5)
		Nov	(BB)	3.5 (0.9 - 6.0)		May	48.6	(0.3 - 144.9)
		Dec	(MI)	0		June[2]	0.3	(0.05 - 0.9)
		Dec	(BB)	16.9 (14.5 - 19.3)		July	0.02	(0 - 0.05)
	1974	Jan	(BB)	11.8 (9.9 - 13.7)		Aug	0	
						Oct	7.9	(0.8 - 13.7)
						Nov	0.02	(0 - 0.07)
						Dec	0	

[1] represents first sampling period for that month.
[2] represents second sampling period for that month (see Marsh 1970 for sampling dates)

Table 3. Number of individuals per gram dry weight of Zostera for Bittium varium and Crepidula convexa. Mean and range of each sampling period given.

Species	Post-Agnes					Pre-Agnes			
Bittium varium	1972	Aug	(SP)	abundant		1967	Nov	3.7	(1.9- 6.5)
	1973	March	(SP)	0.08	(0 - 0.5)		Dec	5.5	(3.5- 9.3)
		July	(SP)	9.0	(4.7- 14.7)	1968	Jan	3.5	(0.8- 8.2)
		July	(MI)	1.7	(1.5- 1.8)		March[1]	0.8	(0.3- 1.4)
		Oct	(MI)	61.6	(58.8- 64.4)		June[2]	23.6	[12.1- 40.0]
		Nov	(BB)	66.5	(62.0- 71.0)		July	60.1	[14.4-121.2]
		Dec	(MI)	1.1			Aug	66.2	[27.1-124.7]
		Dec	(BB)	11.1	(10.4- 12.7)		Oct	96.3	[39.3-203.1]
	1974	Jan	(BB)	7.0	(5.2- 8.7)		Nov	43.8	[33.2- 60.1]
							Dec	57.8	[46.3- 71.8]
Crepidula convexa	1972	Aug	(SP)	abundant		1967	Nov	14.1	(8.3- 22.9)
	1973	March	(SP)	4.3	(0.7- 6.8)		Dec	10.6	(6.1- 13.1)
		July	(SP)	119.5	(72.8-164.7)	1968	Jan	11.6	(4.9- 22.0)
		July	(MI)	108.5	(96.0-121.0)		March[1]	3.0	(1.3- 5.7)
		Oct	(MI)	851.8	[817.6-886.0]		June[2]	8.1	(4.2- 13.2)
		Nov	(BB)	164.3	[108.2-220.4]		July	18.2	[11.3- 24.8]
		Dec	(MI)	40.0			Aug	38.4	[10.0- 46.1]
		Dec	(BB)	40.0	(39.6- 40.3)		Oct	12.3	(6.6- 21.8)
	1974	Jan	(BB)	34.1	(23.5- 44.6)		Nov	10.5	(6.7- 17.6)
							Dec	9.1	(5.7- 15.5)

[1] represents first sampling period for that month.
[2] represents second sampling period for that month (see Marsh 1970 for sampling dates).

Only two other prosobranch species were collected after Agnes but their densities on *Zostera* were low compared to *Bittium* and *Crepidula*. Populations of adult *Mitrella lunata* were found in August 1972. Few were found in the sediments in the winter where they had occurred previously, and low densities occurred in summer 1973. Densities remained depressed, at least to January 1974. This suggests heavy juvenile mortality after Agnes.

Epifaunal populations of *Urosalpinx cinerea* were apparently decimated after Agnes (Andrews 1973). Few individuals were seen in the 1972-1973 collections made from *Zostera* blades; however, the density of *Urosalpinx* in infaunal samples was similar to those recorded previously.

Opisthobranchia. The abundance of most opisthobranchs was severely reduced after Agnes. Four pyramidellids were recorded before Agnes, with *Odostomia impressa* the most common. Only one specimen of *O. bisuturalis* was found after Agnes and *O. impressa* was rare (only nine specimens have been recorded in all epifaunal collections since Agnes).

Elysia catula, a saccoglossan occurring throughout the year in Marsh's 1968 study, has not been found since Agnes. Collections by R. Vogel (personal communication) verify that not only *E. catula* but also *Doris verrucosa*, a dorid nudibranch previously less common than *Elysia*, have not been found in the York River since Agnes. *Tenellia fuscata*, *Ercolania vanellus*, and *Cratena pilata* appeared unaffected by Agnes (R. Vogel, personal communication). *Doridella obscura*, another dorid nudibranch, was very abundant on the artificial *Zostera* after Agnes and occurred at densities of 11-18 ind./g eelgrass during the summer of 1973. *Doridella*, present in Marsh's study at lower densities, is very euryhaline and might have responded to the increase in available space, as did *Crepidula convexa*.

Bivalvia. *Anadara transversa* occurred in almost every collection by Marsh, but after Agnes only one specimen was found at Mumfort Island in October 1973 and at Sandy Point in July 1973. *A. transversa* was also absent from infaunal samples where it occurred previously. Andrews (1973) also found this species rare after Agnes.

Amygdalum papyria, a relatively euryhaline species not recorded by Marsh, was found on artificial *Zostera* and at Sandy Point in July 1973 and also at Mumfort Island in October and December 1973.

Urochordata. *Molgula manhattensis* was one of the few extremely abundant species after Agnes (Table 4). After Agnes, recruitment of *Molgula* was suppressed for two months by fluctuating salinities, after which it recovered quickly as salinities became more favorable (Andrews 1973). The cause for this increase in abundance on *Zostera* is unknown, but may be a result of increased survival of juveniles because of reduced predation or reduced competition from other sessile fauna.

Botryllus schlosseri and *Perophora viridis*, two colonial ascidians found by Marsh, have not been seen since Agnes.

Cirripedia. *Balanus improvisus*, a very euryhaline species, was found throughout 1968, with peak abundances in May and early June. Competition for space and predation appear to be factors that lead to reduction in their numbers under normal conditions. After Agnes, *Balanus* was present on artificial *Zostera* and more abundant one year after Agnes than in 1968 (Table 4). Whether this was because of reduced predation (*Stylochus* was still abundant and found in many empty barnacle tests), relaxed competition for space, or simply unusually successful recruitment is unknown.

Isopoda. *Paracerceis caudata*, previously one of the numerically dominant members of the *Zostera* community, has been absent since Agnes. Present throughout the year in Marsh's study, numerous *Paracerceis* were observed during 1970 and 1971 on *Zostera* blades. It appears that *Paracerceis* populations were eliminated, at least in the York River-Mobjack Bay area.

Two other isopod species, *Erichsonella attenuata* and *Idotea baltica*, were present in densities similar to those found by Marsh. These species seemed unaffected by Agnes. Abundant *Idotea*, including ovigerous females and juveniles, were observed just after Agnes.

Amphipoda. Amphipods, normally abundant and diverse on *Zostera*, were comparatively depauperate after Agnes. Only three species were found on artificial *Zostera* in August 1972, and only ten species have been recorded from August 1972 to October 1973, compared with 23 species recorded by Marsh. The group is now less diverse (Fig. 5); data on number of species in November and December 1973 are similar to 1968 data, but the species composition in these two periods of 1973 compared to 1968 are different. The dominant species, *Cymadusa compta*, *Ampithoe valida*, *Elasmopus laevis*, *Caprella penantis*, and *Paracaprella tenuis*, were present before and after Agnes, but the rarer species, such as *Colomastix halichondriae*, *Rudilemboides nageli*, *Melita* spp., and *Stenothoe* spp. were absent from all collections.

Cymadusa compta and *Ampithoe valida* [reported by Marsh as *A. longimana*, which is found in higher salinities (Bousfield 1973)], were the most abundant species recorded by Marsh in 1968 and both were found after Agnes in similar densities (Table 5). *Cymadusa compta* was extremely abundant on the artificial *Zostera* in August 1972 with many ovigerous females present.

Elasmopus laevis and *Gammarus mucronatus* were also present during Marsh's study. *Elasmopus* was found in York River collections made in July 1972 but was absent from artificial *Zostera* in August 1972 and has not re-occurred there. It was found in the Mobjack Bay in November and December 1973 and January 1974. *Gammarus mucronatus* was very abundant just after Agnes and in March 1973, but has since been rare. *Caprella penantis* and *Paracaprella tenuis* were rare after Agnes in the York River but found in Mobjack Bay from November 1973 to January 1974 at moderate to high densities.

DISCUSSION

The macroinvertebrate community associated with *Zostera* was severely affected by the salinity stress following Agnes. Though quantitative data are not available for the periods just prior to and immediately after Agnes, qualitative observations together with existing background quantitative data reflected a major disturbance during this period. Later quantitative samples also support this conclusion. Many of the numerically dominant species, as well as less abundant species, were either rare or absent immediately after Agnes. A few species, e.g. *Crepidula* and *Molgula*, became more abundant than prior to the influx of Agnes. Of course, the degree to which the effects of Agnes were responsible for observed population changes is questionable. Little is known about physiological tolerances of Chesapeake Bay organisms. Even less is known about importance of biological interactions and func-

Table 4. Number of individuals per gram dry weight of *Zostera* for *Balanus improvisus* and *Molgula manhattensis*. Mean and range of each sampling period given.

Species		Post-Agnes				Pre-Agnes		
Balanus improvisus	1972	Aug	(SP)	abundant		1967	Nov	0.3 (0.02- 0.5)
	1973	March	(SP)	0.05	(0 - 0.3)		Dec	1.4 (0.1 - 3.9)
		July	(SP)	30.0	(15.3- 46.5)	1968	Jan	0.7 (0.3 - 1.8)
		July	(MI)	4.6	(1.4- 7.8)		March[1]	0.07 (0 - 0.2)
		Oct	(MI)	24.5	(22.6- 26.4)		May	18.5 (0.7 -50.0)
		Nov	(BB)	26.5	(14.5- 38.4)		June[1]	9.2 (0.8 -23.0)
		Dec	(MI)	6.3			July	0.04 (0 - 0.1)
		Dec	(BB)	17.7	(13.8- 21.5)		Oct	0.2 (0.04- 0.7)
	1974	Jan	(BB)	1.3	(0.8- 1.7)		Nov	0.1 (0 - 0.2)
							Dec	0.03 (0 - 0.08)
Molgula manhattensis	1972	Aug	(SP)	0		1967	Nov	0.4 (0 - 1.2)
	1973	March	(SP)	0			Dec	0
		July	(SP)	154.8	(87.2-232.8)	1968	Jan	0
		July	(MI)	158.2	(127.8-188.5)		March[1]	0
		Oct	(MI)	37.1	(33.8- 40.3)		June[2]	1.3 (0.03- 3.9)
		Nov	(BB)	0.9	(0.1- 1.6)		July	7.7 (0.2 -22.1)
		Dec	(MI)	0			Aug	0.5 (0.1 - 1.3)
		Dec	(BB)	0.7	(0.5- 0.)		Oct	2.6 (1.6 - 3.5)
	1974	Jan	(BB)	0.03	(0 - 0.08)		Nov	0.06 (0 - 0.2)
							Dec	0.03 (0 - 0.08)

[1]represents first sampling period for that month.
[2]represents second sampling period for that month (see Marsh 1970 for sampling dates).

Table 5. Number of individuals per gram dry weight of *Zostera* for *Cymadusa compta* and *Ampithoe valida* (and *longimana*). Mean and range for each sampling period given.

Species		Post-Agnes			Pre-Agnes		
Cymadusa compta	1972	Aug	(SP)	very abundant	1967	Nov	3.9 (0.5 - 10.7)
	1973	March	(SP)	0.9 (0.7- 2.0)		Dec	1.0 (0.4 - 1.8)
		July	(SP)	1.9 (1.4- 2.3)	1968	Jan	2.7 (1.8 - 4.0)
		July	(MI)	5.2 (5.0- 5.3)		March[1]	0.8 (0.3 - 1.5)
		Oct	(MI)	4.5 (4.4- 4.5)		June[2]	0.3 (0.04- 0.5)
		Nov	(BB)	9.9 (6.2-13.6)		July	3.4 (1.5 - 4.8)
		Dec	(MI)	6.1		Aug	2.1 (0.3 - 3.0)
		Dec	(BB)	23.7 (23.4-24.0)		Sept	5.2 (0.3 - 13.1)
	1974	Jan	(BB)	13.0 (8.5-17.4)		Oct	39.1 (2.7 - 79.5)
						Nov	2.5 (0.2 - 5.0)
						Dec	1.4 (0.2 - 2.3)
Ampithoe valida & *longimana* (*valida* at Mumfort Island only)	1972	Aug	(SP)	Present	1967	Nov	1.1 (0.6 - 2.8)
	1973	March	(SP)	0.2 (0 - 1.0)		Dec	2.2 (1.4 - 3.9)
		July	(SP)	0.3 (0 - 0.6)	1968	Jan	2.5 (1.4 - 3.1)
		July	(MI)	18.1 (16.5-19.6)		March[1]	1.2 (0.5 - 2.6)
		Oct	(MI)	4.3 (3.6- 4.9)		June[2]	1.2 (0.3 - 2.2)
		Nov	(BB)	0.6 (0.2- 0.9)		July	16.9 (7.0 - 35.6)
		Dec	(MI)	3.1		Aug	5.6 (1.4 - 12.4)
		Dec	(BB)	5.9 (3.4- 8.3)		Sept	3.2 (0.2 - 7.6)
	1974	Jan	(BB)	4.6 (2.2- 7.0)		Oct	69.7 (15.0 -150.2)
						Nov	3.7 (2.0 - 6.3)
						Dec	0.8 (0.4 - 1.2)

[1] represents first sampling period for that month.
[2] represents second sampling period for that month (see Marsh 1970 for sampling dates).

tional roles of species in communities.

The effects of Agnes on species in the *Zostera* community were of several types:
1) Species which may have suffered some mortality but became more abundant after Agnes, e.g. *Molgula manhattensis, Crepidula convexa.*
2) Species totally eliminated, severely reduced, or with poor recovery, e.g. *Ampelisca* spp., *Lysianassa alba, Paracerceis caudata, Podarke obscura, Elysia catula.*
3) Species whose adult populations survived but the juveniles suffered high mortalities resulting in reduced populations later, e.g. *Bittium varium.*
4) Species unaffected, e.g. *Heteromastus filiformis, Polydora ligni, Cymadusa compta, Erichsonella attenuata.*

The *Zostera* community in the Chesapeake Bay is a diverse assemblage of species in a wide variety of taxonomic groups. However, after Agnes both the epifauna (Fig. 6) and infauna (Fig. 3) were less diverse than previously. This was a result of the loss of a few numerically dominant species and many of the less abundant ones. No comparably catastrophic environmental stresses on the *Zostera* community have occurred for at least 10-15 years, thus allowing the development of high species diversity.

The biotic interactions (intra- and interspecific competition, predation, etc.) on the species associated with *Zostera* are not known but experimental manipulation of dominant species may provide an insight to some of the biological control mechanisms operating in the *Zostera* community. Agnes provided a natural experiment on community, selectively removing some dominant members and reducing the abundance of others. The observations, keeping in mind the noted limitations, suggest some interesting interrelationships. However, these are mostly correlative and further experimentation needs to be carried out for verification.

Crepidula convexa and *Molgula manhattensis* were found in far greater abundance than recorded prior to Agnes (Tables 4 & 5). *Crepidula* broods its young and *Molgula* has short-lived planktonic larvae, yet both were recruited successfully after Agnes. This may have been due to increased larval survival resulting from a depression of planktonic predators, a relaxation of predation on newly settled juveniles, or to space left open by eliminated competition.

Bittium varium, a usually abundant herbivorous grazer reduced by Agnes, may play a major role of keeping space open on the *Zostera* blade by grazing on space-consuming algae and eliminating newly settled larvae by its grazing action.

Whether and when the *Zostera* community returns to pre-Agnes conditions of high species diversity, depends on potential recruitment from refuge populations of species almost totally eliminated (in this context, species which disperse via planktonic larvae, or swimming juvenile or adult stages, would be expected to reestablish faster than those which brood their young and disperse by crawling), the ability of those species reduced in abundance to increase their populations and the absence of any major disturbance, either man-made or natural. Reestablishment of these species in *Zostera* will be complicated by the absence of *Zostera* in areas where it normally occurs. The mass destruction of *Zostera* by the cownose ray, *Rhinoptera bonasus*, has eliminated much of the *Zostera* in the lower Bay, thus removing, at least temporarily, an important habitat for many species.

LITERATURE CITED

Andrews, J. D. 1973. Effects of Tropical Storm Agnes on epifaunal invertebrates in Virginia estuaries. Chesapeake Sci. 14:223-234.

Boesch, D. F. 1971. Distribution and structure of benthic communities in a gradient estuary. Ph.D. Dissertation, College of William and Mary, Williamsburg, Va., 120 p.

Bousfield, E. L. 1973. Shallow-water gammaridean amphipoda of New England. Cornell Univ. Press, 312 p.

Marsh, G. A. 1970. A seasonal study of *Zostera* epibiota in the York River, Virginia. Ph.D. Dissertation, College of William and Mary, Williamsburg, Ga., 155 p.

Marsh, G. A. 1973. The *Zostera* epifaunal community in the York River, Virginia. Chesapeake Sci. 14:87-97.

Orth, R. J. 1971. Benthic infauna of eelgrass, *Zostera marina*, beds. M. S. Thesis, University of Virginia, Charlottesville, Va., 79 p.

Orth, R. J. 1973. Benthic infauna of eelgrass, *Zostera marina*, beds. Chesapeake Sci. 14:258-269.

Orth, R. J. 1975. Destruction of eelgrass, *Zostera marina*, by the cownose ray, *Rhinoptera bonasus*, in the Chesapeake Bay. Chesapeake Sci. 16:205-208.

Wass, M. L. 1972. A check list of the biota of lower Chesapeake Bay. Va. Inst. Mar. Sci. Spec. Sci. Rep. No. 65, 290 p.

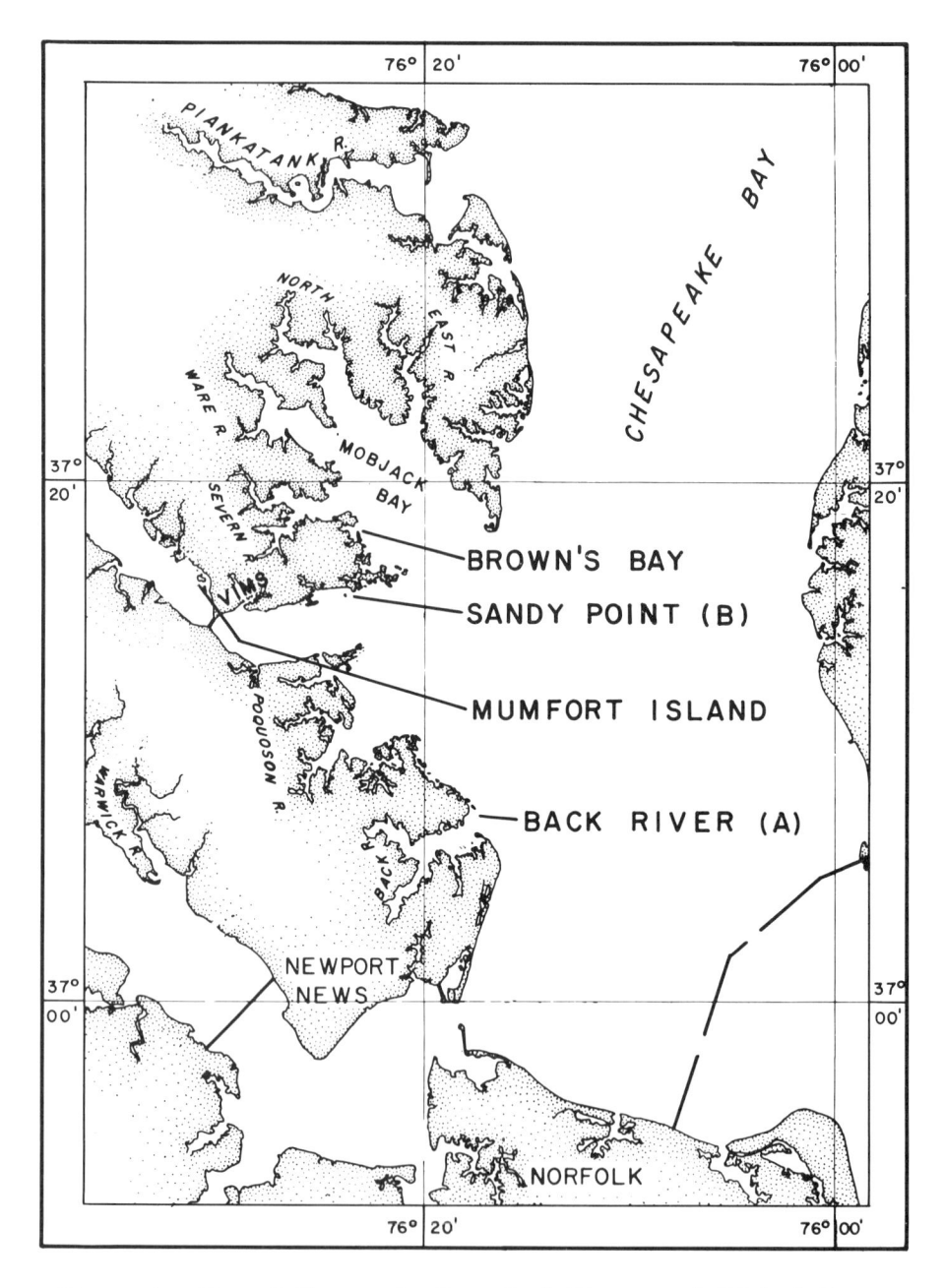

Figure 1. Map of Chesapeake Bay, York River, and Mobjack Bay showing sampling locations.

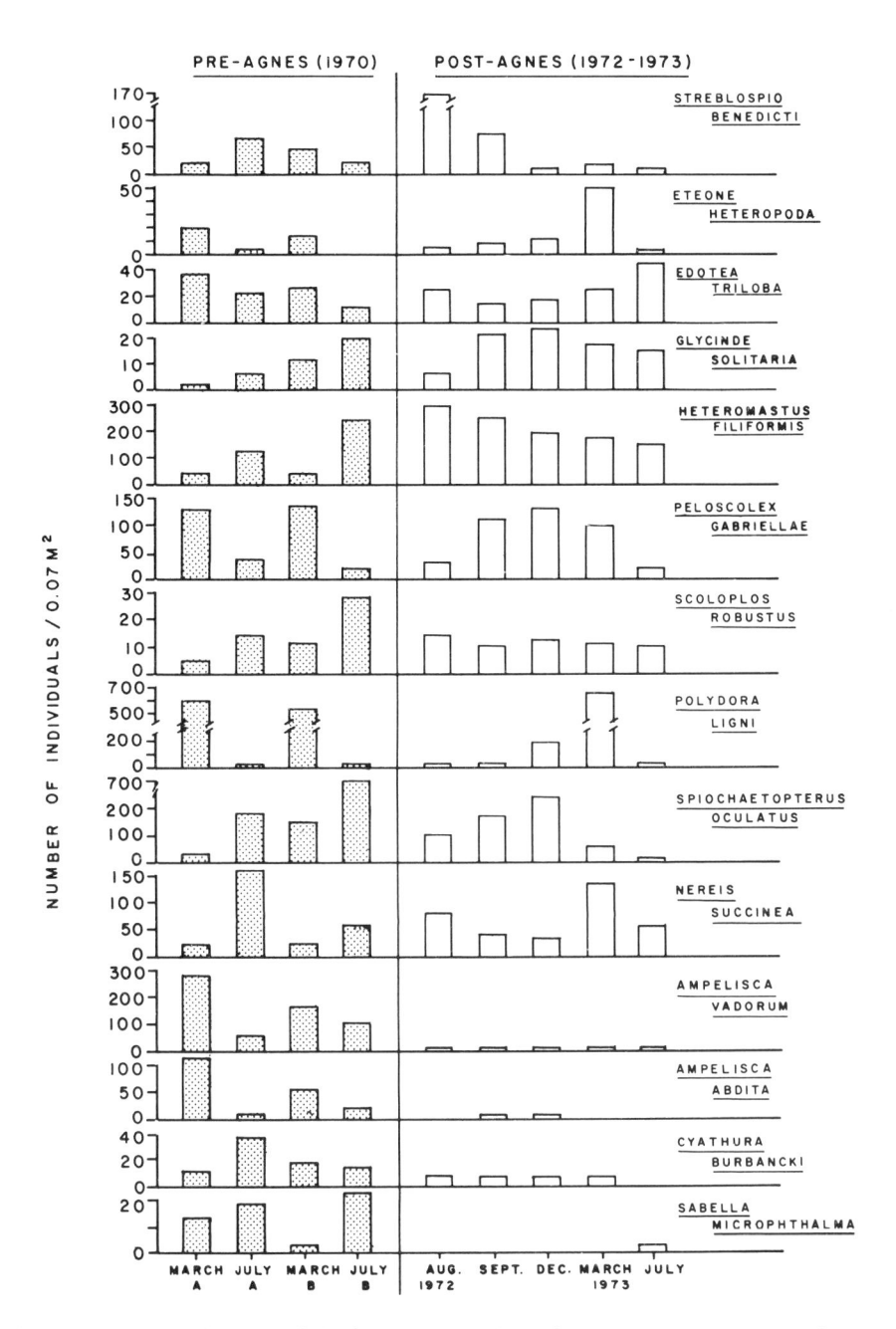

Figure 2. Densities of infaunal species found prior to and after Agnes. Pre-Agnes data from Orth (1971) for two selected stations A and B. Post-Agnes data collected at station B (Sandy Point).

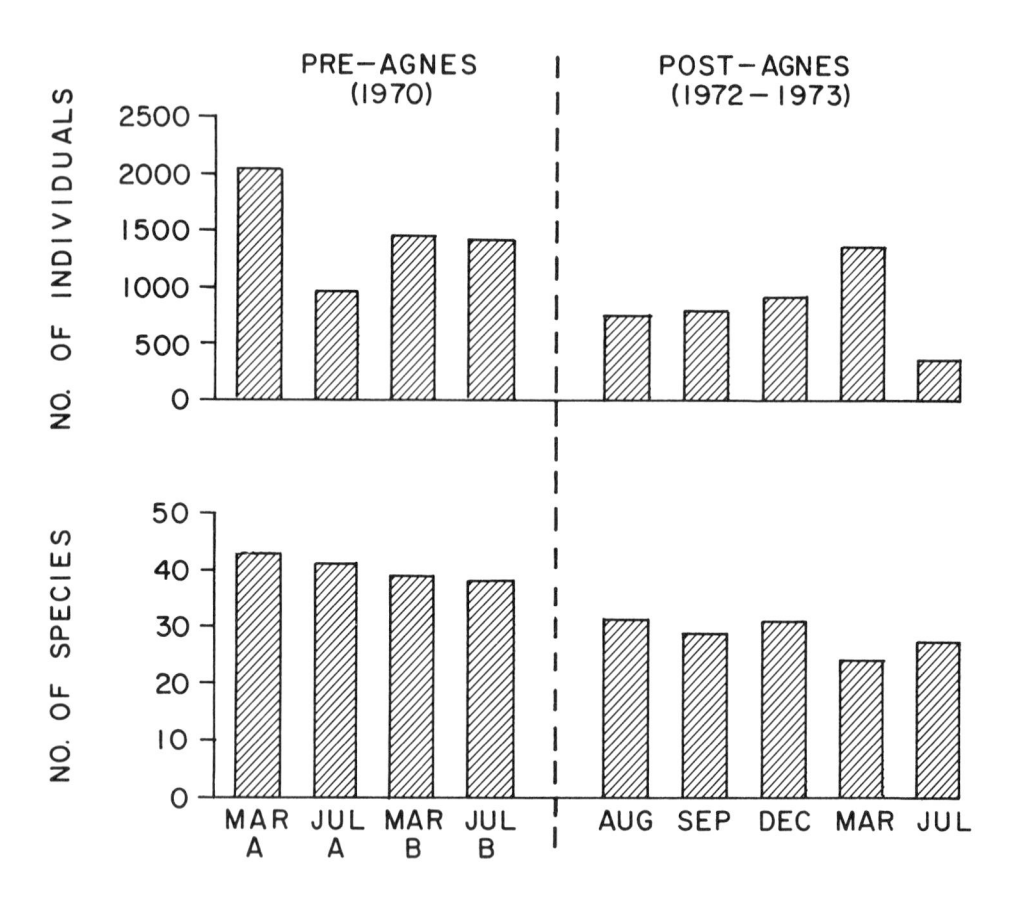

Figure 3. Total number of species and individuals found in 10 cores (0.07 m^2) at stations A and B in 1970 and at station B in 1972-1973.

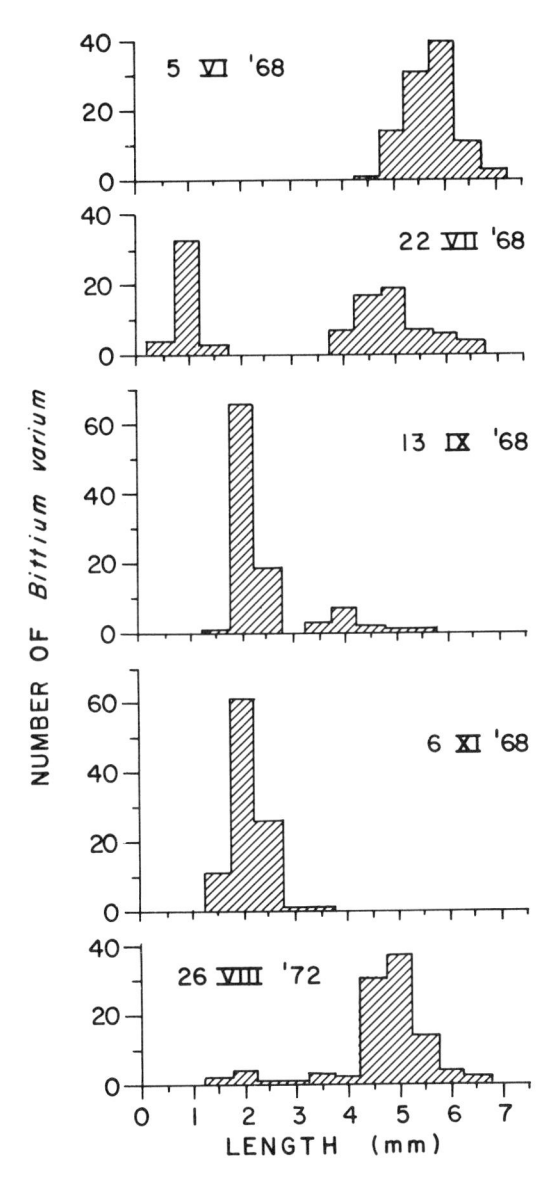

Figure 4. Length x frequency histograms of 100 *Bittium varium* selected randomly from Mumfort Island (Marsh's station C) on each of four sampling dates in 1968 and similar sample taken from artificial eelgrass in August, 1972, at Sandy Point.

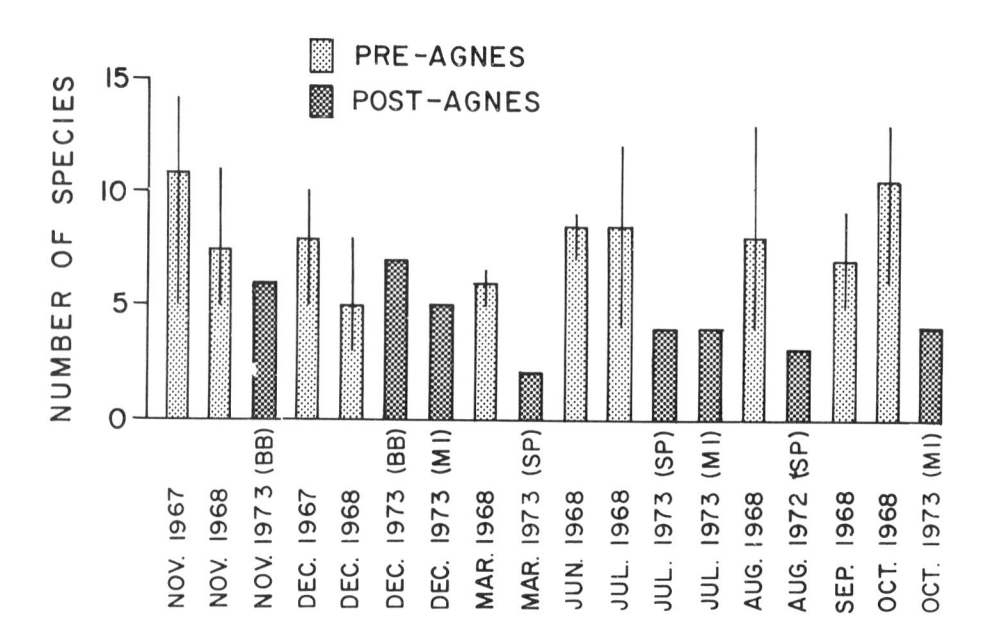

Figure 5. Number of amphipod species found before and after Agnes. Pre-Agnes data from Marsh at Mumfort Island. Mean and range of three stations sampled by Marsh in 1967 and 1968 for number of epifaunal amphipod species and total number of amphipod species found after Agnes at selected areas (BB-Browns Bay; MI-Mumfort Island; SP-Sandy Point).

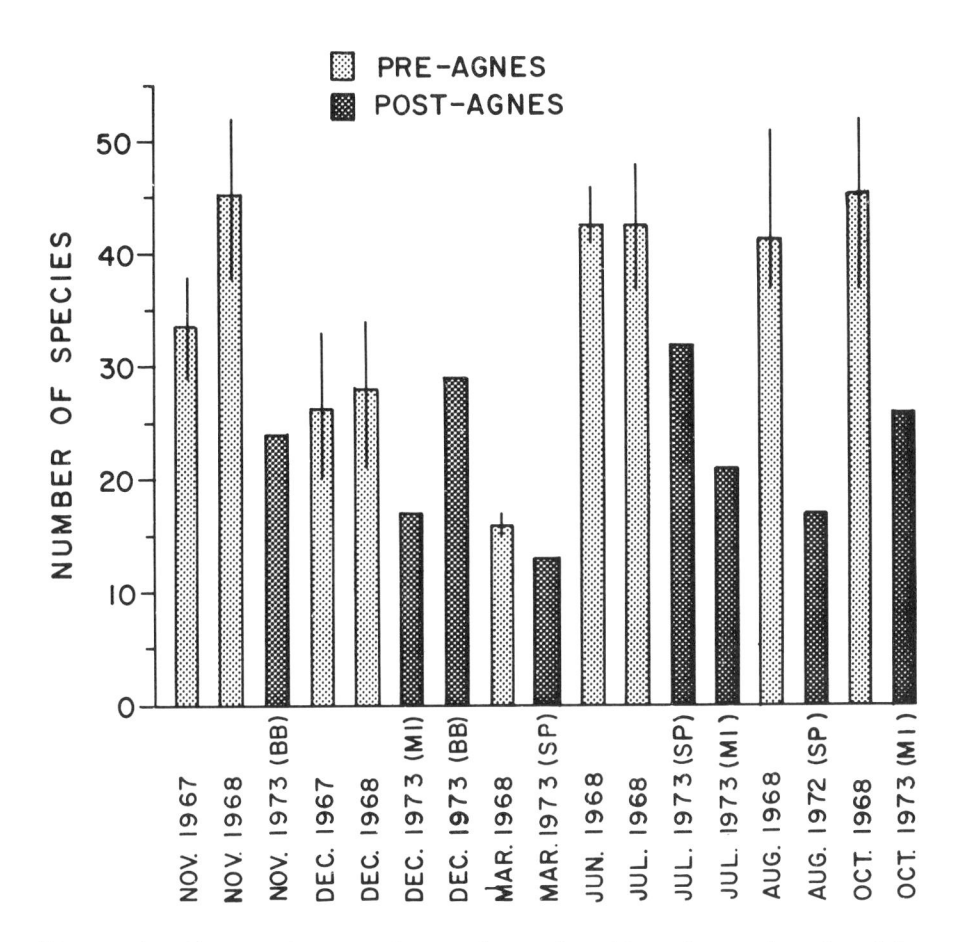

Figure 6. Mean and range for number of epifaunal species found by Marsh at three stations at Mumfort Island, 1967, 1968, and total number species found at several locations after Agnes.

EFFECT OF TROPICAL STORM AGNES ON SETTING OF
SHIPWORMS AT GLOUCESTER POINT, VIRGINIA[1]

Marvin L. Wass[2]

ABSTRACT

Surveillance of shipworm infestation at Gloucester Point, Va., began in 1958. Borer attack by *Bankia gouldi* occurred in July to early October each year until the passage of Agnes greatly reduced setting. Populations returned to near normal in 1975. Salinity was shown to vary with watershed rainfall.

INTRODUCTION

Structural damage to wooden vessels and piers by shipworms has been a continuing problem in estuaries above 5 ppt salinity (Turner 1974). Since the shipworm, *Bankia gouldi*, the only teredinid mollusk that occurs commonly in Chesapeake Bay (Scheltema & Truitt 1954), is limited by about 10 ppt, wooden structures are relatively safe in oligohaline waters. Salinity at Gloucester Point is usually between 15 and 20 ppt, but has ranged from 7.2 to 25.2 ppt.

METHODS

Panels of clear pine have been exposed monthly at the Virginia Institute of Marine Science (VIMS) pier since September 1958, to determine the annual period and magnitude of infestation. Initially, panels were exposed at the request of the Clapp Laboratories, Duxbury, Mass., the exposed panels being sent to the laboratory for identification of boring and fouling organisms until VIMS assumed responsibility in 1966. Long-term "test panels" were first exposed for 8 months, while "controls" were down for 1 month. Test panels were changed to a 6-month rotation in 1969. All dates given are for the month in which the panel was last out. Panels were considered riddled if borer holes occurred in every inch of a split panel.

Test panels were originally 3/4 x 6 x 12-inch loblolly pine panels. Recently, "inch" boards have been only 5/8 inch in thickness. Formerly attached to a vertically hung pipe, they were later tethered to a horizontal galvanized pipe hung about 2 feet off the bottom. The latter system, using nylon cord through two holes in the board to tie to the pipe, has largely prevented further panel loss. Prior to then, numerous panels were lost, partly due to heavy infestations. Panels were usually changed on the first day of each month. After prominent fouling organisms are recorded, the panels are split finely enough to reveal all borer holes. If pallets of borers are evident, counts are made before the panels are split.

RESULTS

Setting of larval borers at Gloucester Point occurs from at least the first week in July through the first week in October. Intensive peak setting occurred during months of warmest water temperatures. Setting data may be compared by

---
[1] Contribution No. 778, Virginia Institute of Marine Science.
[2] Virginia Institute of Marine Science, Gloucester Point, Va. 23062

years and by months. Over the 18-year period, sets occurred in July in 73% of the recovered panels; August, 80%; September, 73% and October, 44%. On a monthly basis, average sets of shipworms for the 4 months were: July, 3.5 (6.8%); August, 12.5 (24.3%); September, 27 (52.5%) and October, 8.4 (16.4%).

From the inception of the project through December 1972, test panels removed from September through February were virtually always riddled. The heaviest borer set in a control was 92 individuals in September 1971 (Table 1). Furthermore, 1971 was the only year in which the test panel retrieved at the end of July was riddled as were all test panels through February 1972 (Table 2). Although Agnes occurred before the known setting time, test panels held 16 borers in July and 6 in August, whereas only 2 *Bankia* occurred in controls, 1 each in July and September. Yet, all test panels retrieved in the last 4 months of 1972 were riddled.

Table 1. Numbers of *Bankia gouldi* in control panels before and after Agnes.

	June	July	August	September	October
1971	0	1	3	92	14
1972	0	1	0	1	0
1973	0	0	2	0	0
1974	0	1	0	0	0
1975	0	7	1	0	0

Table 2. Riddled or numbers of *Bankia* in test panels put out 6 months earlier.

	June	July	August	September	October
1971	0	riddled	riddled	riddled	riddled
1972	0	16	6	riddled	riddled
1973	0	pl*	few	11	pl*
1974	0	0	6 med.	7 large	3 large
1975	0	3	30 med.	35 large	25 large

*Panel lost

Continuing wet years kept subsequent test panels from becoming riddled until September 1975. In 1973, test panels held 12 borers in February and 2 in March. Although controls continued to have low sets through 1975, test panels nearly always held some borers and in 1975 were virtually back to normal.

Climate

In this decade, the climate has been increasingly more maritime, i.e., warmer in winter, cooler in summer. Not only have temperatures been well above average each winter, but rainfall has also been high. The average precipitation for the York River drainage basin towns of Ashland, Partlow, Walkerton, West Point, and Williamsburg combined was 37.15 inches for the 6 dry years 1963-1968; whereas in 1971 through 1975, it averaged 52.05 inches, a mean difference of 14.9 inches. In 1975, rainfall exceeded records of the previous 15 years by at least 4 inches, except at Ashland. Williamsburg had 17.9 more inches than it had in 1960, due largely to 18.45 inches in September, about 2/3 of which fell on September 1.

During the same period, salinity for the York River at Gloucester Point averaged 19.5 ppt, while average annual rainfall was 44.85 inches. A linear regression analysis (Fig. 1) gave a 74% correlation, using yearly mean salinities and mean annual rainfall for the five localities above. Following Agnes, surface salinity dropped to as low as 7.2 ppt at Gloucester Point and cross-sections below the York River bridge ranged below 13.8 ppt through July (Andrews 1973). Since 1970, salinity has continued below average; whereas for 9 years, 1962 through 1970, it averaged 20.3 ppt.

DISCUSSION

The effect of Tropical Storm Agnes on *Bankia gouldi* illustrates the plight of many species disrupted by the low salinity (Andrews 1973; Boesch, Diaz & Virnstein, in press). Adult *Bankia* survived but were largely unable to recolonize new panels until 1975 in significant numbers, whereas many other invertebrates, mainly epifauna, have still not returned to Gloucester Point.

The evidence indicates that the panels longest in the water are most likely to be penetrated by *Bankia* larvae. Culliney (1973) has shown that *Teredo navalis* is extremely sensitive to humic acids (Gelbstoff) and *Bankia* somewhat less so. One might conjecture that new, well-dried panels may not exude as much humic acid as would panels submerged longer. Or perhaps larvae are better able to survive where fouling organisms are present. There is some indication that panels only partly fouled are optimal for larval penetration. The July and August panels (down 6 months) had only 16 and 6 *Bankia* respectively, whereas the last 4 months of the year had riddled test panels.

The negative correlation of rainfall with salinity was best when data from all five gauge stations were combined, probably because precipitation often varied greatly between stations during the warmer months. The finding of only two larvae in control boards in 1972 could be due to one or more reasons: 1) low salinity, 2) low oxygen, 3) unfavorable habitat due to lack of certain fouling organisms, or 4) excess turbidity. A combination of low salinity and low oxygen would seem most likely. More perplexing is the riddling of test panels during the last 4 months of 1972. Since there is no record of shipworms setting in June over the entire study, the presence of riddled test panels through December would seem to indicate a set in early July before the effect of Agnes was felt. A few shipworms did set after salinity rose again, since test panels put down in September and October 1972, held 12 and 2 borers respectively when retrieved in February and March 1973.

Since 1970, late summer and autumn have trended to wet weather. The resultant lowered salinity, in addition to a dearth of adult shipworms, apparently resulted in the lack of a late summer set adequate to produce riddled panels until 1975. So great was their recovery, that stakes put out to suspend oyster trays broke off at the base from *Bankia* attacks, worse in 1975 than for many years (Jay D. Andrews, pers. comm.). As further testimony to the possible effect of fouling, or perhaps "wood leaching", as an aid to setting, nearly riddled panels were retrieved in January and February 1976. Placed in August and September these boards obtained a good set while the controls had one borer set in August and none later.

While salinity had begun dropping in 1971 before the onslaught of Agnes, fouling organisms of many species still competed for space on the panels. In winter, the hydroid, *Gonothyraea loveni*, quickly spread a network over new panels. Barnacles often set heavily in spring and less so in autumn. By April heavy fouling by *Polydora ligni* occurred (Orth 1971). This small polychaete captures fine particles and sediments them around its tube, the multitude of adjoining tubes covering the upper panel surface with a heavy layer of fine dark clay. Little of this soft encrustation survives until mid-May. The serpulid, *Hydroides dianthus*, has returned to the panels, but in far lower numbers than in the 1960's.

Three tunicates have been associated with the panels. Most prominent of all the fouling organisms is the solitary sea squirt *Molgula manhattensis*. This large species has a phenomenal reproductive potential, and although it briefly disappeared following the storm, it quickly reappeared in greater numbers. Now back to normal, it heavily fouls panels down over 2 months. The colonial ascidians, *Perophora viridis* and *Botryllus schlosseri*, have not occurred on the panels since the storm, and it seems unlikely that they will until a sustained salinity near 20 ppt occurs.

Newly placed panels (Fig. 2) are usually covered with a layer of eggs deposited by the toadfish, *Opsanus tau*, or more often, the skilletfish, *Gobiesox strumosus*. The tightly placed ova and parental care preclude setting of invertebrates until the eggs hatch. However, fouling is normally much less on the undersides of panels where these fish eggs are deposited.

In spite of competition from fouling animals, wherever salinity and temperature conditions are favorable, shipworm infestation of wood seems inevitable. From the view of man's activities, it would seem that the decreasing use of wooden boats and the trend toward using salt-treated or creosoted piles and bulk heading would tend to reduce the shipworm habitat in the future. Countering this trend is the placement of many new pound net poles and oyster-ground marking stakes annually.

ACKNOWLEDGEMENTS

The manuscript was reviewed by J. D. Andrews and D. F. Boesch. M. H. Roberts aided the rainfall-salinity correlation analysis. D. F. Boesch, R. J. Orth, R. Vogel and other VIMS students assisted in changing panels and in identification and enumeration.

LITERATURE CITED

Andrews, J. D. 1973. Effect of Tropical Storm Agnes on epifaunal invertebrates in Virginia estuaries. Chesapeake Sci. 14:223-234.

Boesch, D. F., R. J. Diaz, and R. W. Virnstein. In press. Effects of Tropical Storm Agnes on soft-bottom macrobenthic communities of the James and York estuaries and the lower Chesapeake Bay. Chesapeake Sci.

Culliney, J. L. 1973. Settling of larval shipworms, *Teredo navalis* L. and *Bankia gouldi* Bartsch, stimulated by humic material (Gelbstoff). 3rd Int. Congress on Marine Corrosion and Fouling, Natl. Bur. Standards, Gaithersburg, Md., 1973.

Orth, R. J. 1971. Observation on the planktonic larvae of *Polydora ligni* Webster (Polychaeta: Spionidae) in the York River, Virginia. Chesapeake Sci. 12: 121-124.

Scheltema, R. S. and R. V. Truitt. 1954. Ecological factors related to the distribution of *Bankia gouldi* Bartsch in Chesapeake Bay. Chesapeake Biol. Lab., Pub. 100, Solomons, Maryland, 31 p.

Turner, R. D. 1974. In the path of a warm, saline effluent. Bull. Am. Malacological Union, Inc., 1973: 36-41.

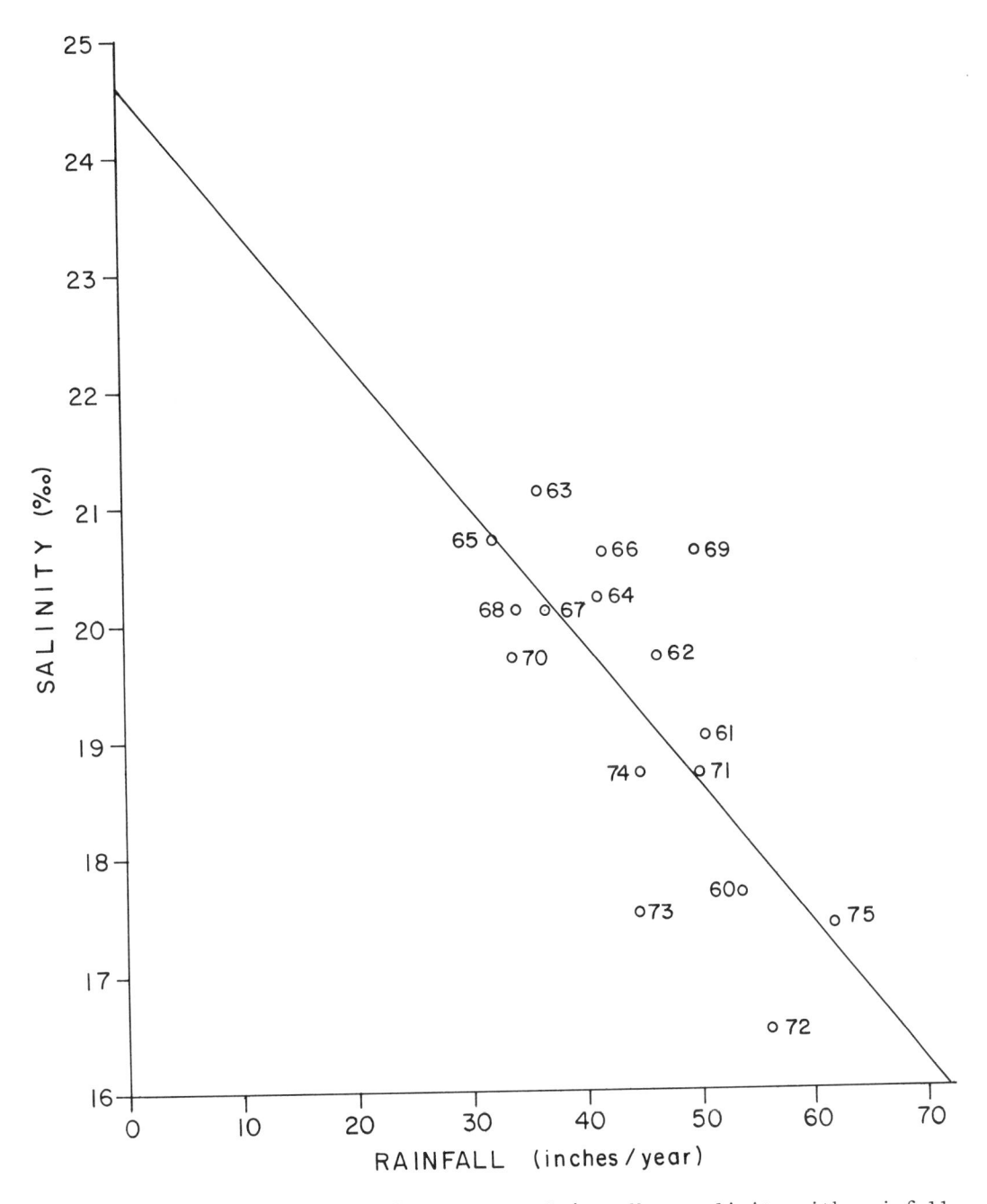

Figure 1. Correlation of Gloucester Point, Va., salinity with rainfall in the York River drainage basin, 1960-1975.

Figure 2. Eggs of oyster toadfish on underside of shipworm panel.

THE DISPLACEMENT AND LOSS OF LARVAL FISHES
FROM THE RAPPAHANNOCK AND JAMES RIVERS, VIRGINIA,
FOLLOWING A MAJOR TROPICAL STORM[1]

Walter J. Hoagman[2]
John V. Merriner[2]

EXPANDED ABSTRACT

Methods

Two days after Tropical Storm Agnes, the Virginia Institute of Marine Science (VIMS) established an anchor station at Mile 15 in the Rappahannock and Mile 10 in the James River. Both stations were in mainstream, manned constantly for 10 days, and took continuous current data from meters placed at 0, 6, 8, and 15 m in the Rappahannock and 0, 4, 5, 8, and 14 m in the James. Concurrently, 0.85 m diameter plankton nets of No. 1 nylon mesh were hung in the flowing surface water for 10 minutes hourly. A small collection of midwater (4 m) plankton samples was obtained from the James station. The shoal areas were not sampled for larval fish. The ichthyoplankton and zooplankton captured were preserved and later identified to species. Both rivers experienced constant ebb tide for several days because of the freshwater layer sweeping to Chesapeake Bay. Below 4 m, the currents were often strongly opposed to the surface currents.

The number of larvae swept to Chesapeake Bay was computed by using a moving average of three of the 10 minute counts and the current measurements; then expanding to hourly estimates, with the flood tide periods subtracted from the loss on ebb tide to obtain daily totals. Outflow was nearly constant from 0-4 m over the first 6 days of sampling, but to allow for shoal areas where currents may have been slower and larvae less affected, only the upper meter in each river was used to calculate total volume containing fish larvae passing the river mile. The program began June 24, 1972 on each river and was discontinued on July 7, 1972, providing 184 surface samples from the Rappahannock, and 61 surface samples and 23 midwater samples from the James.

Results

Eighteen species of fish larvae were captured from the Rappahannock and 22 species from the James. *Gobiosoma bosci* typically made up 75-99% of the daily catches, followed by *Anchoa mitchilli* with 2-20%. The next most abundant were *Syngnathus fuscus* and *Menidia menidia*.

All of the *G. bosci* were young, ranging 4.2 to 10.1 mm total length with an average of 6.2 mm. *A. mitchilli* ranged 3.0 to 18 mm but the vast majority were near 10 mm. *S. fuscus* were below 50 mm average length and *M. menidia* were typically close to 20 mm. *Alosa sapidissima* and *Alosa* sp. (river herring) were few in numbers and very small (6.0-11 mm).

Catches of larval fish were highest at the beginning, then tapered off rapidly after six days. On June 24, 65 million *G. bosci* and 0.7 million *A. mitchilli* were swept past Mile 15 on the Rappahannock. Seven days later, strong flood tides began to reverse the flow and millions of fish larvae were carried

---

[1]Contribution No. 799, Virginia Institute of Marine Science.
[2]Virginia Institute of Marine Science, Gloucester Point, Va. 23062.

back into the river. These estimates were subtracted from the estimated losses. In the Rappahannock, the grand total estimated loss from June 24 to July 7, 1972, using only the upper 1 m for calculation, was 93 million *G. bosci*, 3.6 million *A. mitchilli*, 1.8 million *S. fuscus*, and 2.2 million *M. menidia*. All other species had estimated losses of less than 0.5 million. The estimates are additionally conservative, because Agnes produced strong outflows on June 21, 22, and 23 and these days were not sampled or included in the total estimates.

The James River, using to 1 m only, lost 429 million *G. bosci*, 26 million *A. mitchilli*, 0.9 million *M. menidia*, 0.4 million *S. fuscus*, 13.1 million *Alosa aestivalis*, and 1.6 million *Dorosoma cepedianum*. June 25 and 28 were the days of greatest loss.

Midwater catches (4 m) of fish larvae from the strongly outflowing water demonstrate a high proportion of the total loss (unknown) was below 1 m. Between 0630 on June 25 and 1040 on June 28, 13 sets of surface and midwater samples were taken concurrently from the James station. The actual number of larvae captures was:

	G. bosci	*A. mitchilli*	*A. aestivalis*
Surface	627	62	56
Midwater	1,035	111	16
Midwater as % of Total Captured	62%	64%	22%

Based on these data, the 1 m estimate of total loss could be low by a factor of 5 to 10. The midwater data were not complete enough to allow volume estimates of flow or estimate total loss of fish larvae with catch data below 1 m.

Conclusions

The loss of fish larvae from both rivers was high for 4 species in the Rappahannock and 6 species in the James. The James lost 5 times the *G. bosci*, 7 times the *A. mitchilli*, and 30 times the *A. aestivalis* as the Rappahannock. The Rappahannock lost more *M. menidia* and *S. fuscus*, however.

In 1972, the juvenile *A. aestivalis* population in the James on September 1 was estimated to be 264 million (Hoagman & Kriete 1975). This was only 23% of the 1971 year class and 66% of the 1973 year class for the James.

Turner and Chadwich (1972) using 11 years of c/f sampling, gave a daily Z of 0.053 for *Morone saxatilis* over the sizes of 20 to 51 mm. Their Z may not apply to *A. aestivalis*, but without Z for these, an approximation of stock size on June 22 can be computed by substitution in the general population formula:

$$N_t = N_0 e^{(Zt)}$$

$$264 = N_0 e^{(-0.53 \times 71 \text{ days})}$$

$$N_0 = 613 \text{ mill.}$$

An estimated loss of 13 million *A. aestivalis* because of Agnes, therefore, represents a 2% loss of the larvae. This would be the low estimate. More likely, the loss was above 5%, but probably less than 20%.

We do not know the actual impact on the stocks which had substantial larval losses. Agnes did not wipe out any river or stock it seems. The fish populations with high losses were abundant the year after Agnes, but quantitative measures of the relationship are not available.

LITERATURE CITED

Hoagman, W. J. and W. H. Kriete. 1975. Biology and management of river herring and shad in Virginia. VIMS Adadromous Fish Project, Annual Report, AFC-8-2, 105 p.

Turner, J. L. and H. K. Chadwick. 1972. Distribution and abundance of young-of-the-year striped bass, *Morone saxatilis*, in relation to river flow in the Sacramento-San Joaquin estuary. Trans. Am. Fish. Soc. 101(3):442-452.

EFFECT OF AGNES ON JELLYFISH IN SOUTHERN CHESAPEAKE BAY

R. Morales-Alamo[1]
D. S. Haven[1]

ABSTRACT

Medusa populations of the stinging nettle *Chrysaora quinquecirrha* were lower than usual in Southern Chesapeake Bay in the summer of 1972. Mean daily number for a 5-day period was never higher than seven that year. In 1968, 1969, and in 1973 mean daily number peaked at around 100. Very low counts recorded in 1970 and 1971 were widespread in the region, while low counts in 1972 were confined to areas affected by low salinities due to Agnes. Small numbers of ephyrae were observed in most of the minor tributaries of the lower bay and in tributary creeks of the James and York Rivers near their mouths during July and August 1972. A severe decrease in number of polyps and podocysts on bottom shells was evident in the upper and middle estuaries of the James, York, and Rappahannock Rivers. Populations of polyps and podocysts were not seriously affected in the Great Wicomico and Piankatank Rivers, Mobjack Bay tributaries and Eastern Shore creeks. They also appeared unaffected in tributary creeks of the James and York Rivers near their mouths. Numbers of polyps and podocysts near the mouth of the Rappahannock River (including tributary creeks) were much lower than usual. Abundant numbers of medusae in the summer of 1973 indicated complete recovery from the storm's effects.

---

[1] Virginia Institute of Marine Science, Gloucester Point, Va. 23062

ECONOMIC IMPACTS

Richard J. Marasco, Editor

ECONOMIC IMPACTS OF TROPICAL STORM AGNES IN VIRGINIA

M. A. Garrett, Jr.[1]
L. G. Schifrin[1]

ABSTRACT

This study identifies and estimates the direct economic losses to Virginia businesses in the Chesapeake Bay area resulting from the ecological damages and adverse weather attributable to and associated with Tropical Storm Agnes. These losses are concentrated primarily in three areas: the shell- and finfishing and packing industries; the resort, tourist, sportfishing, and other recreational industries; and costs of restoring or repairing environmental or equipment damages. In determining the monetary totals of the damages incurred, the following approaches are utilized: a) losses in the fishing and packing industries are estimated by determining the amount of actual destruction resulting from the ecological effects of the storm, valued at existing market prices, and the reduced volume of processed and packaged shellfish and finfish. Additional inclusions are made for shell- and finfish displaced, though not destroyed, by the storm; b) losses to the recreation and related industries are reckoned by compilations and year-to-year comparisons of sales revenues in a wide variety of categories that reflect variations in tourist and related visitation and expenditure; and c) actual outlays are estimated for such things as channel dredging, debris removal, marine repairs, and other such expenditures made necessary by the storm. As noted, the analysis is concerned only with these direct economic losses, and it estimates these losses in approximate, but reasonably accurate, dollar magnitudes.

Tropical Storm Agnes created both immediate and far reaching economic impacts in Virginia, with major revenue losses occurring in the shellfish industry and tourist and recreation related industries, and minor losses to individuals and to the public as a result of various environmental effects of the storm. As will be shown, the largest economic impact, both immediate and long range, resulted from the decline in oyster production (and processing) caused by low water salinity levels created in the aftermath of Agnes.

Further substantial economic losses occurred in the tourism and recreation industries as one might expect; the Bay is a major tourist attraction and recreational resource, and thus an important revenue base. Tourism losses, although confined within a relatively brief period, were nonetheless large, and recreational use of the Bay, particularly for sportfishing, was seriously affected for most of the summer of 1972. Together, these losses were exceeded only by those in the shellfish industry.

In regard to other impacts, while damages of other sorts were relatively light, some costly dredging was made necessary by siltation, and some losses

---

[1]Department of Economics, College of William and Mary, Williamsburg, Va. 23185

to boat owners through storm-related damage were incurred.

Accordingly, the following analysis indicated that Agnes produced a surprisingly large and long-lasting impact on the economy of Virginia.

SHELLFISH AND FINFISH INDUSTRIES

Oyster Production

Determination of the economic impact of Agnes on Virginia oyster landings is made difficult by two factors. First, the effect of Agnes on oyster production must be assessed in part by a comparison of actual landings in 1972, 1973, and even 1974, with estimates of what landings would have been had Agnes not occurred; yet Agnes occurred at a time when Virginia oyster production appeared to be rebounding from several years of declining production attributable to disease and predators. Therefore, it is not at all clear as to what the "normal" volume of landings would have been in 1972, 1973, and 1974, given the uncertainties experienced in the past 15 years. Second, the impact of changes in the supply of oysters normally tends to cause prices to move in an opposite (and therefore offsetting) direction, depending on demand conditions. But further complicating this picture is the fact that Virginia production and also the demand for Virginia oysters are parts of the total supply and demand for oysters, and thus the supplies of Virginia and Maryland oysters each influence the prices for both. Thus, some explanation must be offered for price movements for Virginia oysters that do not agree with the total supply picture.

The recent effects of disease and predators on Virginia oyster production are well known: production declines from 21 million pounds (not including Potomac oysters) representing 33% of national production in 1959 to 7.5 million pounds of 12% of national production in 1969. Virginia oyster production began to recover after 1969 and by 1971 production had increased to 8.5 million pounds, representing a 13% improvement in just two years. But the impact of Agnes is clearly evidenced in subsequent production data. Oyster landings in 1972 and 1973 were just about 5.0 million pounds each year, decidedly lower than what would have been the case otherwise. How much lower depends, though, on whether we project the continued improvement in Virginia oyster landings as happening, or whether we take the 1971 level as a normal "recovered" level, which would have been maintained in 1972 and beyond.

In the period 1966 through 1973, the varying effects of disease, predators, and Agnes caused output to fluctuate between 5 million to 9 million pounds annually; on the other hand, prices showed a steady but modest decline from $.679 to $.638 per pound. Thus, before proceeding, we must take into account total supply factors as well as demand conditions in the market as a whole in order to reconcile the decline in Virginia oyster production with the prices that emerged in the markets for oysters.

The production decline in Virginia oysters attributable to Agnes was not itself sufficient to cause a noticeable change in oyster prices; however, the combined Maryland-Virginia loss in oyster landings may have created a significant increase in price. Yet, the average Virginia price fell from $.647 per pound to $.638 per pound between 1971 and 1972 (Table 1). This, combined with the fact that Virginia prices are consistently higher per pound than Maryland prices, which suggests a somewhat unique demand for Virginia oysters, indicates an understandable shift in demand between 1971 and 1972, a reduced demand for oysters attributable to the pollution created by Agnes. Partially offsetting shifts in both the demand and supply schedules between 1971 and 1972 are therefore responsible for the negligible decline in price.

Table 1. Dollar Loss in Virginia Oyster Landings Estimated from Natural Growth from Seed Planted (3 year cycle)

	Natural Growth from Seed Planted, Pounds		
	1969	1970	1971
James	35,175 lbs.	23,800 lbs.	36,769 lbs.
York	14,069	9,520	14,707
Rappahannock	1,758,770	1,190,033	1,838,463
Potomac tributaries	1,231,139	833,023	1,286,796
Totals	3,039,153	2,056,376	3,176,735
Price in Harvest Year	$.646	$.638	$.630
Dollar Value	$1,963,292	$1,311,967	$2,001,343

Total Dollar Value $5,276,602

Source: The Effects of Hurricane Agnes on the Environment and Organisms of Chesapeake Bay, prepared by the Chesapeake Bay Research Council, January 1973; Seventy-Fourth and Seventy-Fifth Annual Reports of the Marine Resources Commission, 1972 and 1973.

We must therefore take into consideration each of these effects in analyzing the economic impact of Agnes on Virginia oyster production. If we assume that oyster landings level off with the 1971 pre-Agnes production of 8.5 million pounds and that the supply and demand for all oysters would not have created a price in 1972 and 1973 that differed significantly from the actual prices, total value of Virginia landings lost in 1972 and 1973 equals $4.5 million, and, if we include 1974 values, the total loss approached $6.9 million (Table 2). We now compare these data with the loss that occurred from seed planted oysters in Virginia waters. These data are taken from The Effects of Hurricane Agnes on the Environment and Organisms of Chesapeake Bay: Early Findings and Recommendations, prepared by the Chesapeake Bay Research Council. This study provides excellent data on the absolute effect of Agnes on oyster mortality in Virginia waters by river. The mortalities are: James 10%, York 2%, Rappahannock 50%, and Potomac tributaries 70%. Using a growth factor of 2 for each bushel seeded and a conversion factor of 6.5 for bushels to pounds, we estimate losses solely from seeded oysters for the 3 year period of $5,308,369, or approximately $1 million less than our estimate had production equaled pre-Agnes output.

Two factors suggest these are realistic, if somewhat conservative estimates of Virginia losses. Given the production of oysters following Agnes, the market prices in 1972 and 1973 suggest that the decline in demand for oysters attributable to pollution in 1972 carried over to 1973; and, since pollution should not have been a factor in 1973, a semi-permanent decrease in demand seems to have occurred. Consequently, we would not expect the market price

Table 2. Dollar Loss in Virginia Oyster Landings Attributable to Tropical Storm Agnes Assuming Production Remained Constant at Pre-Agnes Level

Year	Actual Value Virginia Landings	Expected Value Virginia Landings[1]	Loss
1971	$5,509,965	$5,509,965	
1972	3,240,723	5,482,500*	$2,241,777
1973	3,176,750	5,397,500**	2,220,750
1974	3,000,000 (est.)	5,440,000***	2,440,000
		Total	$6,902,527

Source: Actual values and prices, Seventy-Fourth and Seventy-Fifth Annual Reports of the Marine Resources Commission, pp. 20-23.

[1] Expected landings of 8.5 million pounds.
*Actual price $.646 per pound. Data computed using $.645 per pound.
**Actual price $.638 per pound. Data computed using $.635 per pound.
***Data computed using $.640 per pound.

of oysters to have been substantially different from the actual prices, reflecting the validity of our estimates. Even if we assume that the production of Bay oysters, had Agnes not occurred, would have been sufficient to create a decline of as much as $.025 per pound and if we also assume Virginia production would have maintained a 6% rate of recovery from disease and predators, Virginia losses would exceed $8 million for the 3 year period (Table 3). We therefore conservatively estimate a $6 million loss in Virginia oyster landing attributable to Agnes for the period 1972 through 1974.

Table 3. Dollar Loss in Virginia Oyster Landings Attributable to Tropical Storm Agnes Assuming Production Rate of Growth of 6% per Year

Year	Actual Value Virginia Landings	Expected Value Virginia Landings[1]	Loss
1971	$5,509,965	$5,509,965	
1972	3,240,723	5,771,479*	$2,530,756
1973	3,176,750	6,022,919**	2,846,169
1974	3,000,000 (est.)	6,434,564***	3,434,564
		Total	$8,811,489

Source: Actual values and prices, Seventy-Fourth and Seventy-Fifth Annual Reports of the Marine Resources Commission, pp. 20-23.

[1] 6% rate of growth.
*Actual price $.646 per pound. Data computed using $.645 per pound.
**Actual price $.638 per pound. Data computed using $.635 per pound.
***Data computed using $.640 per pound.

A major effect of Agnes on oyster production may not, however, be felt for several years and probably can never be accurately documented. The aftermath of Agnes created water conditions that virtually eliminated oyster spawn and set during 1972 in both Maryland and Virginia waters. While the short run reduction in oyster production that will occur in 1975 and to a lesser extent in 1976 represents a direct effect, a more serious long range effect can result from a semi-permanent reduction in the stock of oysters producing spawn and set. Failure to replenish the oyster stock needed to provide seed oysters throughout the state, especially precisely at the time when the total oyster supply is at the lowest ebb in recorded history, could cause a much longer recovery period to reach some "normal" or "desired" level of production. This, in turn, could result in a reduction in total revenues over a period of several years compared with the revenues that would accrue under an optimum plan of recovery for oyster production. The fiscal consequence of a reduced stock of seed oysters deserve serious consideration.

Indirect Impact in the Oyster Industry

The inter-relationships among industries suggest that a reduction in the output of any good, especially a natural resource such as oysters, will have far reaching intermediate effects. Projecting the total indirect effects of the decline in oyster production does not, however, appear to be warranted. While determination of the indirect effects are theoretically possible, substitutability of the final product would substantially reduce the impact in each stage of production beyond initial processing, thus the majority of the indirect effects will be captured in the local impact of the oyster processing sector.

Data on the absolute dollar value of oyster processing in Virginia are not available, but realistic estimates can be made with existing information. Virginia oyster processing establishments process the total production of Virginia oysters and a substantial portion of Maryland oyster landings. A study conducted at the University of Maryland indicated that 60% of the Maryland oyster landings are processed in Virginia. Multiplying the absolute dollar value of oyster landings processed in Virginia by the value added in the processing stage thus provides a measure of the economic impact of the Virginia oyster processing sector. In order to determine the economic impact of Agnes, we simply calculate the pounds lost in oyster landings multiplied by $.50.

The estimated dollar loss in value added in oyster processing (Table 4) exceeds $4.7 million for 1972 and 1973, and because 1974 oyster landings in Maryland are expected to decline sharply, the three year loss exceeds $7.2 million. These estimates presume the Virginia oyster production would have leveled off at the pre-Agnes output; thus again, we consider this to be a most conservative estimate. For example, had the rate of recovery experienced in 1969-70 in oyster production in Virginia continued through 1973 with Maryland production remaining relatively stable, the two year loss in value added attributable to Agnes would exceed the conservative estimate of $4.7 million.

Clam, Crab, and Finfish Production

Agnes produced a relatively minor economic impact on the clam industry in Virginia compared with the production of oysters. While soft shell clams are extremely vulnerable to reduced salinity levels, Virginia waters do not produce them in sufficient quantities for commercial harvesting. Hard shell clams

Table 4. Estimated Dollar Loss in Value Added in Virginia in the Oyster Processing Industry Attributable to Tropical Storm Agnes.

	Pounds Lost in Oyster Landings[1]	Value Added Loss in Processing-Virginia (current dollars)
Maryland, 1972	1,723,500	$ 861,800
Virginia, 1972	3,483,400	1,741,700
Maryland, 1973	849,000	424,500
Virginia, 1973	3,520,800	1,760,400
	Subtotal, 1972-1973	$ 4,788,400
Maryland, 1974	1,063,800	531,900
Virginia, 1974	3,812,500	1,906,300
	Total	$ 7,226,600

[1]Virginia data obtained from Table 1. Assumes 60% of Maryland oyster landings are processed in Virginia. Maryland data obtailed from a comparable study conducted for Maryland included in this volume.

survived the fresh water of Agnes with one prominent exception. Clams that had been relocated from the James River to the York River and its tributaries in order to undergo a 3 week cleansing period suffered almost 100% mortality. Normally, minimal loss is observed in relocated clams, and, since only relocated clams were affected, the combination of stress and low salinity levels was apparently responsible. Using data from the Marine Resources Commission, we estimate 7,035 bushels of relocated clams were affected by Agnes producing a direct loss of $42,410[1].

Revenues from the production of crabs and finfish indicate that Agnes did not reduce catches in 1972 or 1973. The crab catch continued to conform to biologists' expectations and catches of menhaden, the most significant contributor to the dollar value of Virginia finfish, increased approximately one-third in 1972. Biologists' field samples indicate that both crabs and finfish remained in waters with acceptable salinity levels according to their species and were therefore pushed out of the bay and its tributaries by the infusion of fresh water but returned as salinity levels returned to normal. There is the possibility, however, that Agnes affected the spawning of certain finfish which could affect future catches, e.g., the recent decline in shad production, but this may be extremely difficult to determine.

---

[1]11,725 bushels of clams were relocated from the James River in June 1972. Approximately 60% are relocated in the York River and its tributaries and 40% to the eastern shore. Since the minimum relocation period is 15 days, we assume all clams relocated during June to the York River area were affected by Agnes, although obviously a portion of the clams relocated in May were also affected.

TOURIST AND RECREATION RELATED INDUSTRIES

Tourism

Revenue from travel business in Virginia exceeds one billion dollars annually and represents 2.1% of the national travel market. Consequently, a disaster such as Agnes would obviously have a noticeable effect on tourist-oriented industries. The monthly Travel Barometer maintained by the Travel Development Department of the Virginia State Chamber of Commerce indicated tourism for the State as a whole during June of 1972 was down 6.9% from its projected level. An analysis of the Barometer data indicated that tourist attractions from Fredericksburg west were not adversely affected by Agnes. Thus, not unexpectedly, the tourism decline in June 1972 seems fully attributable to Agnes since it was limited to the Tidewater area and the Eastern Shore. While precise data are not available on tourism expenditures *per se*, we compared the absolute dollar expenditures on three tourist-oriented industries, by quarter, from 1970 through 1973 for counties and cities in Tidewater and Eastern Shore. The data were obtained from Department of Taxation, Taxable Sales Annual Reports, and cover expenditures for "motels and hotels", "restaurants", and "sports and hobbies".

The seasonal characteristics of tourism in eastern Virginia are easily discernible. The revenue from tourist-oriented industries during the summer quarter ranges from 2 to 4 times the revenue of the winter quarter. We can, therefore, legitimately treat any significant year-to-year changes in expenditures in these industries during the spring and summer quarters to changes in the volume of tourism from year-to-year. But the data as noted are in quarterly form, and they indicate that if the effects of Agnes carried over into July, they were so small as to be masked by the data for the entire summer quarter. (Obviously, if data were available by month, it might be possible to detect some effect from Agnes during July.) Moreover, it is possible that the effects of Agnes are not masked by the data for the spring quarter of 1972 because tourism during June accounts for a relatively large part of all tourism during the entire spring quarter.

In order to determine the direct effects of Agnes on tourism, we calculate the expected revenue for the spring quarter of 1972 for the three industry groups, and subtract from it the actual spring quarter revenue. Expected revenue is determined by a simple average of the revenue for the spring quarters of 1971 and 1973 and checked with the growth rate for 1970-73 for consistency. We have also checked for inordinately large increases between 1972-73 during the summer quarter in order to ensure that the increase is not influenced inordinately by additions in facilities between 1972 and 1973.

We estimate a total direct loss in tourism revenues attributed to Agnes of $3.5 million, with $2.5 million of that loss in the Norfolk and Virginia Beach areas (Table 5). This is a most conservative estimate since we have included only those industries primarily tourist oriented; i.e., purchases of all other goods and services by tourists, such as gasoline, groceries, theater admissions, etc., have not been included. Moreover, we do not believe we have been able to capture the total loss even in the tourist oriented industries. For example, using the monthly index of tourism prepared by the Bureau of Business Research at the College of William and Mary, expenditures in Colonial Williamsburg during June of 1972 were 6-9% below the expected level; yet, the simple straight line method we have employed indicated no discernible actual loss. In order to be consistent among geographical areas, however, our data do not include the decline in anticipated tourism for Colonial Williamsburg attributable to Agnes.

The largest charter and head-boat fleets in Virginia are either already ocean-based, such as those at Wachapreague on the Eastern Shore or at Rudee Inlet in Virginia Beach, or have easy access to the ocean, such as those based in the Norfolk area from Willoughby Bay to Little Creek, and on the North Shore of Virginia in Lynnhaven Inlet.

Table 5. Revenue Loss Attributable to Tropical Storm Agnes in Selected Tourist Oriented Industries in Virginia Cities and Counties.

	Actual 1972	Expected 1972	Loss
Cities			
Hampton	$ 4,996,210	$ 5,225,120	$ 228,910
Newport News	5,331,229	5,751,171	419,942
Norfolk	15,061,768	16,230,514	1,168,756
Williamsburg/ James City County		No Discernible Loss	
Virginia Beach	10,763,851	12,160,545	1,396,694
Counties			
Accomack	875,727	913,944	38,267
Gloucester	230,455	259,052	28,597
Lancaster	837,893	979,273	141,380
Mathews	81,944	86,255	4,311
Middlesex	183,586	186,598	3,012
Northampton		No Discernible Loss	
Northumberland	76,582	96,928	20,346
		Total	$3,450,215

Source: See Appendix II.

Thus, to whatever extent charter and head-boat fishing revenues were affected by Agnes, losses were confined to the inshore boats, which are fairly few in number, usually lower priced, and see less intensive normal utilization than the offshore boats. Estimates of even these losses are difficult to come by, because these boats are largely decentralized in location and do not use central booking services. However, a total loss to them within the $25,000 to $50,000 range is estimated to have occurred, and the mid-point of that range, $37,500, is accepted as a reasonably accurate loss figure.

While charter boat fishing is by far the dominant fraction of the commercial sportfishing industry in Virginia and losses to it related to Agnes were not large, there are other elements of commercial sportfishing which, though much smaller in size, suffered proportionately more damages. In this regard, particularly, are the fishing pier and small-boat rental businesses.

There are nine commercial fishing piers in Virginia, only two of which extend into the ocean; the remaining seven are all located at the mouth of or inside the Bay. While precise damage figures to these seven are not available on the individual basis, our survey of pier owners and operations indicates that the average losses to the four larger piers in the lower Bay attributable to Agnes were approximately $15,000, representing more the continuing impact throughout the summer than the absence of business during the storm period. Losses to other piers totalled no more than $10,000. Thus, Agnes-related revenue losses to fishing pier operators are estimated at $70,000.

Another facet of this economic sector is small boat rentals. The Virginia State Travel Service cites 21 listings of small boat rentals for saltwater fishing; for the most part, these are small businesses or relatively small ancillary operations to larger businesses. A survey of these firms indicates revenue losses of $1,000 or more to have been very infrequent, and that total losses to them from Agnes probably come only to the $10,000 level.

Thus, the total Agnes-related losses to all elements of the commercial sportfishing industry in Virginia total approximately $117,500, which when viewed in the full perspective of all losses, represents a small and quite insignificant, perhaps surprisingly so, dollar figure.

Private Sportfishing. But if commercial sportfishing losses were small, private sportfishing revenue losses were not. The analysis indicates that losses in this sector are the product of two factors: the total number of boating days (average boats per day multiplied by the number of days affected) lost to Agnes and the typical monetary outlay per boat per day.

In regard to lost boating days, this study, on the basis of extensive interviews with marina operators, boat owners, sportfishermen, and local radio and TV boating-weather-and-conditions reporters, has determined that Bay conditions were such that there was virtually no pleasure boating for a period of three weeks after the storm. Beyond that period, boating was still affected; various estimates indicate that for several reasons boating was down by as much as 50% for another three weeks, and by roughly 25% for the "rest of the summer," at a minimum taken to mean through August and probably through September as well. The reasons given include both "real" and "psychological" factors, with the demarcation between the two being unclear. Among the cited factors are fear of personal injury and property losses from striking floating or submerged tree trunks and other debris brought downriver, particularly in the James River, by the storm-swollen waters, fear of contamination directly or through eating contaminated fish, from sewage in the waters, or because the sediment-laden waters made navigation in general difficult and rendered boating and its concomitant water contact sports of skiing and swimming less attractive than otherwise.

Of course, the three week period after the storm and the periods beyond are not clearly separable. The return of boating from a zero level to a 50% level and then to a 75% level occurred at a rate of change other than a simple stair-step fashion. However, the use of stair-step demarcations is simply an averaging method for the time within the respective periods, and does no violence either to the logic or accuracy of the findings and figures.

But percent figures, although somewhat meaningful in themselves, must be translated into absolute terms. Thus, the relevance of some sort of measure of the number of boats typically on the Bay during the months of June, July, and August is clear. To determine these boats-in-use figures, numerous marina, launch-ramp, and charter boat operators were interviewed, and among them there was virtual agreement on these important points:

1) On a typical good-weather weekend-day or holiday, the boats in the lower Bay are likely to number between 3,000 and 3,000 at any one time, and total around 5,000 for the day as a whole;

2) The weekday volume of boat use is no lower than 10 or 15% of the weekend or holiday volume, and probably approaches 20%, the latter figure emerging

because of the effect of vacationers in July and August, and of good fishing in June. However, because the estimates range from 10 to 20%, this study employs the midpoint, 15%, as a plausible relationship between weekday and weekend boating traffic in the Bay;

3) The average daily expenditure for the typical boat, a 21 or 23 footer occupied by 3 to 4 adult, usually male, fishermen, although now somewhat higher, was about $20 in 1972. This outlay covers food and beverages, fuel, bait, tackle, and ramp charges;

4) Normal boating days lost, in part or totally, to weather in June and July is no more than the equivalent to five days each month. It is probably the case that weather losses are a bit higher in the later summer, as the prevailing winds shift from the Southwest;

5) Of the total private boat launchings into Chesapeake Bay at least 85% are in the lower Bay area; thus the middle and upper Bay boat "population" are roughly 15% of the lower Bay population, and raising the lower Bay figures by 15% provides a reasonable estimate of the total private boat launchings on the Bay.

Accordingly, this study estimates the economic impacts of Agnes on the private sportfishing-related industries in this summer as follows. Beginning with June 22, 1972, four time spans are defined, the first running for three weeks until July 12, the second for three more weeks until August 2, and the third covering the remainder of August, and a fourth encompassing the third and extending beyond it, covering the remainder of August and all of September. These periods are labeled A, B, C_1, and C_2 respectively in Table 6, in which these economic losses are detailed. Next, these time periods are divided into "weekend and holiday" days and "mid-week" days, the first being labeled "1", and the other "2". Thus, 1A comprises weekends or holidays within the first three-week period; and $2C_1$ comprises mid-week days between August 3 and 31. In Column 2, the boats-per-day are provided, 5,000 plus 750 (15% of 5,750) per mid-week day.

Column 3 of Table 6 indicates the effects of Agnes on private boating: a decline of 100% in period A, 50% in period B, and 25% in period C.

Column 4 shows the absolute number of boat-days lost because of Agnes, and these figures represent the product of Column 1 (days), Column 2 (normal boat launchings), and Column 3 (percentage decline in launchings). These figures, multiplied by an average outlay of $20 (Column 5), provide the dollar losses shown in Column 6. Against the sum of these losses, we have deducted the estimated "normal weather loss", to arrive at a net loss figure. Note, though, that because two alternative time periods for period C are employed, the final loss figures are given in terms of a range rather than a precise amount. However, for computational purposes, we have again chosen the range mid-point as a single representative loss figure.

OTHER IMPACTS

Tropical Storm Agnes caused economic losses other than those relating to commercial and private sportfishing. Among the other losses are those that concern channel maintenance, debris removal, and boat and marine equipment damages. However, of these, only the first looms large as an out-of-pocket loss. Reports from the U. S. Army Corps of Engineers, Norfolk District, indicate that thusfar the flooding of 1972 in the tidewater estuaries of Chesapeake Bay has caused these expenses, above normal maintenance costs: $575,000 (including survey and other costs) for dredging in the James River; and $120,000

Table 6. Dollar Losses to Private-Boating Industries from Agnes.

(1)	(2) Normal-Boats-Per Day	(3) Decline (%)	(4) Boat-Day Losses	(5) Avg. Expenditure/Boat-Day	(6) Loss ($)
1. Weekend and Holiday Days					
A (June 22-July 12) 8	5,750	100	46,000	20	920,000
B (July 13-Aug 2) 6	5,750	50	17,250	20	345,000
C_1 (Aug 3-31) 8	5,750	25	11,500	20	230,000
C_2 (Aug 3-Sept 30) 18	5,750	25	25,875	20	517,500
2. Mid-Week Days					
A 13	860	100	11,180	20	223,600
B 15	860	50	6,450	20	129,000
C_1 21	860	25	4,515	20	90,300
C_2 41	860	25	8,815	20	176,300
		Gross total loss	$(A+B+C_1)$		1,937,900
			$(A+B+C_2)$		2,311,400
Normal Weather Loss					
15% June/July (1A, 1B, 2A, 2B)					242,640
20% Aug/Sept ($1C_1$, $1C_2$, $2C_1$, $2C_2$)					
$1C_1 + 2C_1$					64,060
$1C_2 + 2C_2$					138,760
		Total Normal Weather Loss	$(A+B+C_1)$		306,700
			$(A+B+C_2)$		381,400
		Net Loss	$(A+B+C_1)$		1,631,200
			$(A+B+C_2)$		1,930,000
		Mid-Range (approx.)			1,780,000

(including survey costs) for dredging of the Rappahannock River channel, for a total of $695,000.

Our extensive interviews support the opinion that debris removal was largely left to nature itself, or was done by marina and waterfront property owners and boaters themselves on a non-commercial basis. Furthermore, it also seems that many boat damages were to the small boat owner and to a considerable extent were of small individual magnitude. Many of these damages were hull scrapes and gouges that were repaired by the owners themselves. Occasional more severe damages to private boats occurred, but these, although once in a while quite sizable in dollar value, seem to have been the exception and did not loom large in total. It is estimated from survey data that small-boat damages commercially repaired totalled slightly under $10,000 in the Norfolk, Virginia Beach area.

In regard to damages to larger boats which tend to be more costly per incident, data provided by the United States Coast Guard Station in Norfolk indicate that there were several Agnes-connected incidents of damage to five barges and one Navy vessel, also totalling an estimated $10,000 in damages.

The reporting requirements for marine casualty damages specify at least a $1,500 damage level, and thus all losses above that amount presumably were reported.

Accordingly, the economic damages from Agnes relating to channel clearance, debris removal, and boat damages come to an estimated $715,000.

CONCLUSIONS

As a result of our widespread survey of the economic losses resulting from Tropical Storm Agnes, the salient conclusions are these:
1) The amount of damage, conservatively estimated, approaches $20 million;
2) The impact of damages was uneven, very severe in certain sectors of the shellfishing, tourism, and recreation industries, and less so in other sectors of these industries and in other industries;
3) The timing of damages includes both immediate and long term damages, which though not equally severe, are separately of a severe magnitude.

Table 7 shows in summary form the economic losses from Agnes in Virginia as ascertained in this study.

Table 7. Total Economic Losses in Virginia from Tropical Storm Agnes

Area	Amount	
Shellfish and Finfish Industries		
Oysters	$6,900,000	
Oyster Processing	7,200,000	
Clams	42,410	
Total		$14,142,410
Tourism and Recreation		
Tourism	3,500,000	
Recreation		
Commercial Sportfishing	117,500	
Private Sportfishing	1,780,000	
(Recreation Subtotal)	(1,897,500)	
Total		5,397,500
Other Impacts		
Channel Dredging	695,000	
Boat and Ship Damages	20,000	
Total		715,000
Grand Total		$20,254,910

Appendix I. Virginia Oyster Landings, Total Value of Virginia Landings and Price of Oysters per Pound, 1969-73

Year	Virginia Landings (Pounds of Meat)	Total Value Virginia Landings	Price Per Pound
1973	4,977,400	$3,176,750	$.6382
1972	5,013,268	3,240,723	.6464
1971	8,441,538	5,509,965	.6527
1970	8,043,274	5,426,940	.674
1969	7,446,827	5,019,310	.674

Source: <u>Marine Resources Commission</u>, Annual Reports for the Fiscal Years Ending June 30, 1972 and June 30, 1973.

Appendix II. Dollar Expenditure on Tourist Related Industries[1], Spring and Summer Quarters 1970-73, Virginia Counties and Cities by Tropical Storm Agnes

	April-June 1970	July-Sept 1970	April-June 1971	July-Sept 1971	April-June 1972	July-Sept 1972	April-June 1973	July-Sept 1973
Counties								
Accomack	745,264	1,183,667	899,366	1,460,972	875,727	1,752,107	928,622	2,307,438
Gloucester	173,620	227,741	209,278	260,351	230,455	298,251	308,827	490,708
James City	781,001	1,372,741	1,165,074	1,553,121	1,158,348	1,657,552	1,136,299	1,964,145
Lancaster	704,746	880,540	1,021,120	1,062,069	837,893	1,287,956	896,477	1,529,280
Mathews	94,630	131,838	73,049	108,427	81,944	159,171	99,461	194,493
Middlesex	176,702	276,583	174,927	323,405	183,586	341,703	198,270	378,837
Northampton	454,494	762,718	517,641	751,862	576,914	844,926	551,543	989,041
Northumberland	84,099	97,963	101,868	97,557	76,582	119,511	95,894	112,441
York	808,085	923,477	991,265	1,811,191	1,509,943	1,960,622	1,484,294	2,086,143
Cities								
Hampton	4,666,391	5,012,766	4,678,151	5,164,259	4,996,210	6,163,088	5,772,090	7,589,106
Newport News	4,763,025	4,969,563	5,347,206	5,702,814	5,331,239	6,421,468	6,155,137	7,202,734
Norfolk	14,105,652	14,903,400	14,561,981	16,471,984	15,061,768	18,367,405	17,899,047	22,091,767
Virginia Beach	9,914,769	16,981,189	11,243,314	18,603,080	10,763,851	22,017,347	13,077,777	25,618,938
Williamsburg	7,410,871	8,349,959	8,583,185	9,352,395	8,787,409	9,491,435	8,356,764	12,120,307

[1]Tourist related industries include motels and hotels, restaurants, and sports and hobbies.
Source: Virginia Department of Taxation, Taxable Sales Annual Reports, 1970-73.

THE MARYLAND COMMERCIAL AND RECREATIONAL
FISHING INDUSTRIES: AN ASSESSMENT OF THE ECONOMIC
IMPACT OF TROPICAL STORM AGNES[1]

Lee Smith[2]
Richard J. Marasco[2]

ABSTRACT

This study was designed to assess the economic losses suffered by the commercial fisheries, the recreation industry, and recreators, and to determine the costs of debris removal and repair to navigation works. The oyster and clam industries bore the heaviest burden of loss among the fisheries, and their losses will continue for at least several years. Bay area motels, restaurants, and marinas incurred decreases in business particularly in areas where boating was banned or difficult. Although there was no monetary record of it, a large loss was also felt by Bay users, sport fishermen, boaters and swimmers, who were not able to participate in these recreation activities in the Bay for 1 to 8 weeks.

INTRODUCTION

Tropical Storm Agnes set records for physical and economic damage to the Eastern Seaboard. The major tributaries of the Chesapeake Bay reached record flood levels. The impact of this freshwater flooding had major economic repercussions on the fisheries, the sportsmen, and the recreation industries which depend upon the Bay.

Maryland commercial fishermen harvest numerous species of finfish and shellfish from the Chesapeake Bay annually. In 1971, landings of fishery products from the Bay by Maryland watermen totaled 61 million pounds and were valued at $16.5 million, dockside (National Marine Fisheries Service 1973).

The Bay has historically been used by large numbers of people for recreation. In the State of Maryland alone, there were about 80,000 recreational boats registered in 1972. In addition, motels and restaurants in Bay adjacent counties did a $210 million taxable business in fiscal 1973, much of which was related to Chesapeake Bay recreation.

This study was aimed at assessing the economic losses suffered by the commercial fisheries, the recreation industry, recreators, and determining the cost of debris removal and repair to navigation works. Concern was limited to first round or primary effects. No attempt was made to isolate multiplier effects; that is, economic repercussions experienced by other sectors of the Maryland economy.

---
[1]Contribution No. 5194, Scientific Article No. A2213, Dept. of Agricultural and Resource Economics, University of Maryland.
[2]Department of Agricultural and Resource Economics, University of Maryland, College Park, Md. 20742

THE MARYLAND COMMERCIAL FISHING AND FISH PROCESSING INDUSTRIES

Oysters

Large oyster mortalities were experienced in the Chesapeake Bay during the summer of 1972 due to Agnes. Frequent samplings of the oyster bar located in the Bay indicated that the low salinity associated with Agnes was the primary cause of oyster deaths (Chesapeake Bay Research Council 1973). It was estimated that 699,000 bushels of market size oysters were killed in the Maryland section of the Bay and its tributaries.[1]

In order to determine the economic impact of Agnes, for 1972 and 1973, it was necessary to estimate what landings and revenues would have been without Agnes. Estimation of losses for 1974 required estimates of both actual landings and revenues given the occurrence of Agnes and potential landings and revenues which would have been realized if Agnes had not affected the Bay. The results which would have occurred without Agnes were referred to as the "nonoccurrence case".

Oyster landings in Maryland for 1972 totaled 16,006,821 lbs., or 2.4 million bushels (National Marine Fisheries). If in addition the oystermen had harvested all the dead market-size oysters, the 1972 harvest would have been 3,075,582 bushels. This was considered a highly unlikely result. A small percentage of the oysters killed by Agnes would have died due to natural causes. It was suggested that 5% would be a good estimate of the natural mortality.[2] In addition, it was felt that the probability that all the dead market size oysters would have been harvested in the same year was small. Even if all the oysters which were market size in the summer of 1972 had been harvested the following season, they would not all have been landed in the calendar year 1972. Therefore, it was assumed that 60% of the dead market size oysters would have been harvested in 1972 and about 30% the following year. Given this set of assumptions, the 1972 "nonoccurrence" oyster harvest was estimated to be 2.8 million bushels or 18,879,292 pounds.[3] A demand equation was developed and used to estimate price given the nonoccurrence quantity. Total revenue in the absence of Agnes was estimated to be $11,022,000 given the expected harvest price of 58.3¢/lb. Actual 1972 total revenue was $9,420,178 for a loss of about $1,602,000.

If Agnes had not occurred, oyster landings in 1973 would have been higher by a percentage of the Agnes-killed oysters. Assuming that 30% of the dead oysters would have been harvested in 1973, nonoccurrence landings were estimated to be about 20.1 million pounds. Actual landings were 18.7 million pounds. The calculated non-occurrence revenue was not significantly different from occurrence revenue, so the loss for 1973 was taken to be zero.

---

[1] The Maryland Department of Natural Resources estimated that 611,000 bushels of oysters in the Bay and its tributaries were killed by Agnes. Potomac mortalities were estimated to be about equal to the amount landed in the Potomac in the 1972-1973 and 1973-1974 seasons, since it was believed that this catch comprised most of the surviving adult oysters and sampling determined that average mortality on the Potomac had been 50%.

[2] Personal communication, Aven Anderson, University of Maryland, Natural Resource Institute.

[3] It must be assumed further, that Maryland oystermen would have been capable of harvesting 2.8 million bushels, a larger quantity than has been landed in recent years. This assumption seems reasonable. Editor.

In addition to the total loss to oystermen during 1972 and 1973, $1,602,000, there were also losses at the processors' level. Historically, a large percentage of the oysters harvested in Maryland have been shipped out of the state to be shucked. Sources estimate that 60% of the oysters harvested in Maryland waters have historically been sold to processors located elsewhere. Assuming that the remaining 40% was shucked by Maryland processors, 1,715,000 lbs. of oysters did not reach the processor level because of Agnes. Research on seafood processing indicated that the normal markup by oyster processors was about 78% of the ex-vessel price, which suggested that the processors' markup was about 50¢/lb. in 1971. Assuming the processors' markup remained at 50¢/lb., the 1972-1973 loss to Maryland processors was estimated to be $857,500.

Oyster production has been very high in 1974. The state's as yet incomplete figures show 8.7 million pounds of oysters landed. If production could continue at this rate, the 1974 catch would be around 21,750,000 lbs. However, most of the available evidence indicates that harvesting will slow down drastically when the fall season opens. According to reports from the State Department of Natural Resources, Anne Arundel County has already experienced a shortage of oysters. Oystermen who were bringing in 40 bushels per day last September were bringing in only 10 bushels per day when the season closed.

The state planted 997,876 bushels of seed in 1972 in the Bay and in 1971 in the Potomac, which should have been the basis for the 1974 harvest. A bushel of seed produces, roughly, 2 bushels of oysters under Bay conditions. Converting this to pounds indicated a possible 13,471,316 lbs. of oyster meats. Natural reproduction could increase this potential nonoccurrence catch but the amount was unknown. The price that corresponds with this quantity was estimated to be about $1/lb., therefore total nonoccurrence revenues would have been about $13.5 million. It was estimated that 131,300 bushels of seed oysters were killed by Agnes, decreasing the potential harvest in 1974 by 1,773,000 lbs.

However, it was determined that the greatest revenue loss to oystermen was due to the timing of the oyster harvest. The high monthly catches in the spring of 1974 have brought low prices. Even if the buyers had perfect foresight and realized that the year's catch will be small, the market price would be low because most oysters are consumed fresh. Revenue at the oystermen's level has amounted to $5.2 million from the spring catch. If the crop remaining to be harvested in the fall is, as was predicted, only 3 to 4 million lbs., the price should increase then. A price of $1 per pound for 3 million pounds could increase the year's revenue to only $8.2 million, or $5.3 million less than the nonoccurrence revenue. In addition, processors will have lost about $354,600 in 1974 on the seed which was killed in 1972.

Soft Clam

Agnes had a two-fold effect on soft-shell clams. The first effect was to pollute the Bay with sewage overflow, forcing the Health Department to prohibit clamming. The second effect was on the condition of the clams. Low salinity, silt, and increased water temperature caused grave damage to the clams. Biological surveys indicated the market-sized clams suffered 90-100% mortality and that 95% of the surviving commercially harvestable soft-shell clams were concentrated in less than 5,000 acres (Maryland State Department of Health and Mental Hygiene). This decimation of adult soft clams was the reason the State Department of Natural Resources imposed a ban on clamming in September 1972, when the health ban was lifted.

The maturation of surviving spawn led the State Department of Natural Resources to lift its conservation ban on June 1, 1973, allowing clamming in Talbot

County, with a reduced quota of 15 bushels a day. On June 23, the clam bars were closed again by the Health Department. Although the water was of acceptable purity levels, the clams tested on the docks exhibited high bacterial counts. There was and still is controversy between the clammers and the state as to whether this second closing was justified. As an outgrowth of this controversy, the Maryland Department of Health and Mental Hygiene formed a study group in June 1973 to attempt to determine the extent and the cause of the excessive bacterial counts that led to the ban on harvesting of soft clams. The results of the study indicated that in the two study areas, the soft clam populations were in poor physical condition with over 90% of the clams exhibiting signs of stress. Further, a direct correlation was established between bacterial levels and the degree of physical deterioration of the clams. After continued testing, Talbot County was opened up again for clamming in late August.

The clam fishery has gone through five stages between the occurrence of Agnes and the present:
1) Clamming was closed because of pollution in water;
2) Clamming was closed because so few live clams were left;
3) Clamming was open in limited areas;
4) Clamming was closed due to high bacterial levels in harvested clams; and
5) Clamming was open again in some areas.

The catch losses incurred during stages 1 and 2 were clearly caused by Agnes. The basic cause of the second health ban was difficult to establish although it was probably Agnes. Separating the impact of stage 4 from the other closures appeared impossible, so the entire loss has been attributed to Agnes.

Over the period 1962-1971, clam landings averaged 6,735,696 lbs. per year. Landings varied from a low of 6 million to 8.1 million and there was no clear trend or cycle. It was assumed that the 10 year average was a "normal" landings figure. In 1972, with clamming allowed for less than half the year, landings were only 1,949,520 lbs. Revenue to the clammers was $1,014,782. If Agnes had not occurred, and landings had been "normal," revenue would have been $3,391,554. The clammers' lost revenue then amounted to about $2,376,000.

In 1973, clamming was allowed for almost 5 months, but on only 5,000 acres. Landings totaled 632,808 lbs. The actual price of 83.3¢/lb produced a revenue of $527,340. The normal catch receiving the predicted price would have yielded $3,607,622, for a loss of $3,080,282.

The ex-vessel price for soft-shell clams increased 30¢/lb. to 83¢/lb in 1973. This increase was considerably less than the price elasticity would have suggested. However, the increase was limited by competition with New England producers. Most of the soft-shell clam catch was traditionally shipped to New England, primarily for steaming. Since the New England States also produce soft-shell clams, and have the advantage of lower shipping costs, the price received for Chesapeake clams cannot rise above the New England price. The upper limit in 1973 appeared to be about $10-11/bushel, or 83¢/lb. In addition, there probably was a downward shift in demand for Bay soft clams. Since Maryland clams were unavailable for 14 months, many users probably switched to New England suppliers or discontinued the sale of soft-shell clams. The shrinkage of the market was reflected in the fact that before Agnes 40-50% of the soft clams were processed to some degree before leaving the state, but since Agnes they have all been shipped live.

Soft clam landings for 1974 were estimated from landings in the fall of 1973. Although biologists at the Maryland State Department of Natural Resources estimated that an additional 10,000 acres may be "harvestable bottom," by late 1974, as of May 1974 no additional areas had been opened to clamming. It was therefore assumed that the 1974 soft clam harvest would be landed from the areas which had been opened in 1973. The experience of several past years indicated that around 32% of the total year's catch was normally harvested between September and the end of December. Applying this to the actual fall catch in 1973 yielded an estimate of 1,980,700 lbs. a year from these clam beds. According to the demand equation for clams, predicted price would be above the maximum; therefore it was assumed that prices in 1974 will again be about 83¢/lb. The resultant revenue to the clammers was estimated to be $1,644,000.

If deflated personal income does not increase in 1974, the predicted price for the normal catch would be 55.5¢/lb. The nonoccurrence clammer revenue of $3,738,311 would have been $2,094,317 larger than that which will be realized in 1974 in the occurrence case.

The total loss through 1974 in revenues to the clammers, given the recovery assumptions, was estimated to be $7,550,600. Additional losses have been and will be experienced by the processing sector. A telephone canvass of packing houses after Agnes by the state DNR found that 1,200 people were unemployed due to the lack of shellfish. Processing margins for clams have not been well researched, but the effort and expense involved in processing clams, per pound, should be similar to that involved in processing oysters. Therefore, a processor's markup of 78% was applied to the portion of clams which had been processed, pre-Agnes. If 50% of the nonoccurrence 1972-1974 catch had been processed, revenue to the processors would have been $4,188,000. Since Agnes, soft clams have not been processed because of the loss of the New England market, so there was no revenue to packing houses from soft clams, except the $395,765 which was realized in 1972 before Agnes struck. This revenue loss of $3,791,855 was added to the clammer's loss, for an industry loss in Maryland of about $11.3 million.

Blue Crabs

There was considerable disagreement, at least until this summer, about the effect of Agnes on the blue crab population. In a report prepared by the Chesapeake Bay Research Council it was surmised that the crab population and reproduction were not seriously harmed, although there was not much evidence one way or the other. On the other hand, another report for the Corps of Engineers stated that the blue crab population was "seriously depleted as a result of the storm" (Seiling 1973). More recent scientific sampling indicates that this was not the case (Chesapeake Biology Laboratory, Maryland DNR). Last fall the DNR found the "normal" size distribution of crabs in its fall sampling program (William Outten).

There is great yearly variation in crab population and landings and it had appeared that 1972 was going to be a low year even before Agnes. Moreover, the catch had been very high -- over 4 million pounds during each summer month. Table 1 shows that while June landings were below the average of the past 4 years, July and August were above, the latter by a large amount. It was concluded on the basis of this and the current information from the biologists that there was no significant loss of blue crabs.

Finfish

No evidence was located which indicated that Agnes caused mortality among adult finfish. Their mobility allowed them to move to saltier water. Biologists suspected that the flushing of the Bay during the peak spawning period could have been destructive to eggs and larvae of some finfish; however, spawning also takes place in the fall. The opinion of experts was divided as to the long run impact of Agnes on fish populations. In the absence of biological evidence, it was assumed that there will not be a significant decrease in future finfish catch due

Table 1. Hard and Soft Crabs Landed in Maryland for Selected Months, 1968-1971 (Thousand lbs.).*

Year	June	July	August
1968	1640	2850	2334
1969	687	4195	5747
1970	3590	547	4352
1971	4035	4366	5055
Average 1968-1971	2488	2739	3283
Actual 1972	2100	3753	5385

*National Marine Fisheries, Maryland Landings

to Agnes.

Even though fish populations were not harmed, catch could have been down in 1972 due to the week of bad weather and the displacement of fish. In order to determine if the catch was lower, landings statistics for 10 years were examined. Although the 1972 catch of all finfish, 13.9 million lbs., was less than the 10 year average, it was within one standard deviation of the average. Thus, the statistical evidence did not support the contention that Agnes lowered the finfish catch.

ECONOMIC IMPACT ON RECREATION INDUSTRIES AND USERS

The recreation activities most predominant on the Chesapeake Bay historically have been fishing, hunting, crabbing, boating, swimming and sightseeing. Many of these activities were curtailed by Agnes. Boating and therefore fishing were made difficult by floating debris in many areas of the Bay. Beaches were closed because of pollution for varying periods of time. Sportcrabbing and sportfishing were less productive than normal because finfish and crabs moved down the Bay after Agnes for at least a month.

Businesses which served the recreationists - motels, restaurants, marinas, bait and boat sales places - lost revenue because of the Agnes-caused decrease in recreation activity.

Motels

In order to measure the economic impact of Agnes on the motel industry, motels in all of the counties which touch on the Bay were surveyed. Some of these were on major highways or at some distance from the water. Almost 50% of the responding motels indicated their customers were all either on business or transient, rather than recreating in the area. According to survey response, the total decrease in business which motel operators attributed to Agnes' effect on the Bay amounted to around $180,788.[1] For some motels, business decreased only during the immediate period of the storm, but for others, business was poor all summer.

When the average number of visitors per summer week was compared to the total decrease in customer days, by county, it was evident that there was considerable

---

[1] From the sample it was estimated that 7,726 visitor days were lost. This was multiplied by a per person room rate of $9 and inflated by 2.6 to represent all motels.

economic impact on certain areas. In Baltimore, Cecil, Kent, Queen Anne's and St. Mary's, the loss to the average motel was most of a week's business. Since most of the motels involved were summer-oriented, this loss represented a significant part of the year's business.

Marinas

In 1972 there were approximately 320 marinas (places which charged for rental of 10 or more slips and moorings) along the Chesapeake Bay and its estuarine tributaries. Sources used to estimate the impact of Agnes on marinas were An Economic Analysis of Marinas in Maryland, The Boating Almanac, 1973, and a small telephone survey conducted in March 1974.

The primary impact of Agnes on the marinas operated through the 4-6 week curtailment of pleasure boating caused by the weather and debris. This lowered gas and oil sales, which averaged 25% of marina gross volume. It also affected subsidiary businesses such as transient slip rentals, food and hardware sales and attached or nearby restaurants and lodging. The impact on marine associated restaurants and lodging was not estimated because of lack of information. Slips and mooring rentals, the mainstay of the business, were held at the normal level in most marinas, but were lower in others. A loss in rentals was experienced by marinas at considerable distance from the large cities whose slips were rented by the week or month rather than normally for the whole season, and by those where silting closed channels or otherwise hampered navigation.

A few places suffered physical damage but the amount was seldom large. Marinas reported some increase in repair work due to collisions involving debris or propeller or impeller failure due to silt, but this work did not come close to outweighing the loss in other revenues.

Had Agnes not occurred, the marina business would have grossed close to $9 million, not including restaurant business. This figure was derived by updating the average revenue per marina in Economic Analysis of Marinas in Maryland by use of the average increase in slip rental price, in gas prices, and in the consumer price index, and multiplying by the number of marinas operating on the Bay in 1972. Marina operators on average, according to the survey, believed their loss of revenue was about 13% of their business, or a gross dollar loss of $1.1 million.

The counties in which most marinas did incur losses were Harford, Queen Anne's, Charles, St. Mary's. As in the case of motels, the losses in these areas were considerably higher than the average for the Bay area. In the upper and midsections of the Bay, marina operators attribute their loss to a decrease in boat usage due to debris in the Bay and the ban on water contact sports and to a lesser extent on silt impeding their channels. In the lower section of the Bay, the major problem was a decrease in transient boat traffic, probably caused by fear and the problems in the upper Bay.

Other Recreation Industries

There was no direct information on the impact of Agnes on the restaurant business. However, the motel survey found that probably 7,726 motel customer days were lost because of Agnes. Motel customers make heavy use of the restaurant industry. An A. D. Little study of tourism in Maryland in 1972 found that motel-staying boaters spend about 80% as much at eating and drinking places as at motels. This figure was applied to the motel loss to obtain an estimate of the loss to Bay area restaurants of $144,631. This was considered a minimum estimate of losses because it only measures the loss of customers who stayed in motels. Restaurant customers who lived in the area and made day trips to boat, fish or sightsee, could also have been discouraged from eating in restaurants because of Agnes and its aftermath.

The A. D. Little study also reported that a motel-staying recreationist would probably spend an additional $6 to $8 per day in the area. Using the same method as above, at least $54,000 in gross revenue was lost to retail establishments, such as gas stations, around the Bay because of Agnes.

State parks also serve the recreationist. Four state parks along the Bay attributed to Agnes decreases in income amounting to $70,000.

Impact on Recreation Users

The use of the Bay for swimming, fishing, boating, and other recreation activities affords utility to large numbers of recreators. The travel-cost method was used to estimate the value of the Bay to sportfishermen. There have been extensive creel censuses of the Bay which provided information on how many anglers came from different distant zones in order to fish on the Chesapeake Bay in the summer.

These data along with information from the motel survey were used to construct the numbers per thousand population traveling from increasing distance zones to fish. The miles traveled were transformed into 1972 costs, to yield a curve relating total cost per visit and the number of participants per thousand population (Fig. 1). This was defined as the demand for the whole recreation experience.

It was considered possible that the costs of the trip may yield some utility of their own. For instance, eating in a restaurant on the way home from a fishing trip may be enjoyable. The cost of the meal then does not indicate that the Bay user would have been willing to spend that much in order to use the Bay. Required was an estimate of what sportfishermen would be willing to pay for the use of the Bay. This was accomplished by hypothesizing different fees for use of the Bay, and estimating how many people would fish on the Bay at various fees. The resulting curve was considered equivalent to a true demand for the resource-fishing in the Chesapeake Bay (Fig. 2). The total value to fishermen of their use of the Bay proper for the summer season, the area under the resource demand curve, was estimated to be approximately $1,060,000. This estimate was based on the number of angler days in 1962. National surveys in 1960 and 1970 found that the number of Atlantic saltwater angler days increased by 30% over ten years. Assuming this rate of growth in sportfishing also held for Chesapeake, total value to the Bay users in Maryland was $1,378,000 in 1972. Since the time costs to users were not included, this estimate was thought to be low.

After Agnes, portions of the Bay were closed to boating. Even when boating was allowed, it was more hazardous than normal because of floating debris. It seems likely that pier fishing was also interrupted, due to inclement weather and lack of fish. In any case, pier fishing accounts for less than 1% of the angler days on the Bay proper. On average, fishing from boats was not possible for 3 weeks of usually prime fishing time. According to the 1962 Elser creel census, 16% of the season's angler-days would have occurred during this 3-week period. The value lost then was about $220,500. It was assumed that these recreational experiences were lost and not merely postponed. This seems valid for the large number of backyard boaters who normally go out almost every weekend during the summer. There was probably no realistic substitute for this experience. It also seems to be true for vacationers, according to the records of motel owners. They had not experienced additional business later in the season that might have been delayed vacations. This was considered a validate of the assumption that about 3 weeks worth of recreational experiences were irretrievably lost.

The loss estimated above was the value lost to fishermen only. Non-fishing boat owners were similarly affected. National and state boat owner surveys have found that only 40% of all boat owners use their boats for fishing. This indicates that there were 1.5 times as many non-fishing boat owners as fishing ones. Assuming that they use their boats for pleasure cruising or water skiing with the same frequency as fishermen, and their demand curve for the Bay was similar to that of the sportfishing boat owners, and additional $330,700 in recreational value was estimated to have been lost by non-fishing recreationists because of Agnes.

The closure of the Bay to water contact sports, which lasted until August 12, 1972, in the upper part of the Bay, caused a loss to large numbers of people. However, because of the lack of information, it was impossible to estimate the loss to recreationists who would have participated in water contact sports.

CLEANUP COST

Opening up the Bay and its tributaries for navigation required debris and silt removal. The Army Corps of Engineers undertook most of this work. In some cases Agnes merely forced an acceleration of the normal dredging schedule. The Corp's expenditure on seven major cleanup projects in the Bay or saltwater portions of its tributaries has amounted to about $900,000. Some projects remain undone. Additionally private groups along the shore, the State of Maryland and the Coast Guard also did some cleanup work.

CONCLUSION OF ECONOMIC ASPECTS IN MARYLAND

On the Chesapeake Bay, Agnes had its major economic impact on the oyster and clam industries. Oystermen and the oyster processors lost revenue because of the storm. Their losses will continue and may even increase.

Oyster mortalities in 1972 have caused losses to oystermen and oyster processors for 3 years. Data available indicated that the oyster industry incurred losses during 1972 and 1973 amounting to approximately $2.4 million; it appeared that the 1974 loss will be about $5.6 million. The complete lack of set in 1972, attributed to Agnes, and the poor set in 1973, partially a result of Agnes, will mean that the oyster population will be low in 1975 and 1976. The resulting losses to the oyster industry will be large.

The clam population was affected much more drastically than the oyster. The soft clam industry's 1972-1973 loss amounted to $7.8 million while the probable 1974 loss amounted to $3.6 million. Although the losses so far have been greater than in the oyster industry, the resource appears to be recovering well.

Bay-oriented recreation businesses lost money, because normal recreation activities were curtailed by Agnes and its aftermath. Motels, marinas, and restaurants in specific areas along the Bay experienced considerable decreases in volume. The total amount of the loss, $1.6 million, was not large but the impact was concentrated in several small counties.

Bay users suffered nonmonetary losses because of Agnes. Many people were unable to participate in their favorite free recreation activities for many days. This short-lived impact was valued at $551,200.

Cleanup work made necessary by Agnes imposed costs on state and federal governments and private groups. The amount spent, $900,000, was small compared to inland cleanup costs, but there is still work which needs to be done.

The total monetary figure in current dollars arrived at was $22.5 million.

LITERATURE CITED

Boating Almanac. 1973. G. W. Bromley and Co., Inc., New York.

Chesapeake Bay Research Council. 1973. Effects of Hurricane Agnes on the environment and organisms of Chesapeake Bay. A report to the Philadelphia District U. S. Army Corps of Engrs. (Aven M. Andersen, Coordinator), Chesapeake Bay Inst. Contrib. 187, Nat. Resour. Inst. Contrib. 529, Va. Inst. Mar. Sci. Spec. Rep. Appl. Mar. Sci. & Ocean Engr. 29, 172 p.

Elser, H. J. 1965. Chesapeake Bay Creel Census, 1952. Dept. of Chesapeake Bay Affairs and Natural Resources Inst., Univ. Maryland.

Little, A. D. 1972. Tourism in Maryland. Md. Dept. Economic & Community Development, Annapolis, Md.

Lyon, G. H., D. F. Tuthill, and W. B. Matthews, Jr. 1969. Economic analysis of marinas in Maryland. Agricultural Experiment Sta., Univ. Maryland, College Park, Md.

Maryland State Department of Health and Mental Hygiene. Cooperative study to determine cause and extent of high bacterial counts found in *Mya arenaria* in 1973. Unpublished report.

National Marine Fisheries Service. Maryland Landings. U. S. Dept. of Commerce.

National Marine Fisheries Service. 1973. 1970 Salt-water angling survey. U. S. Dept. of Commerce, Washington, D. C.

Seiling, F. 1973. Updated evaluation of damage to Maryland commercial fisheries due to Hurricane "Agnes". Maryland Dept. of Natural Resources. Unpublished report.

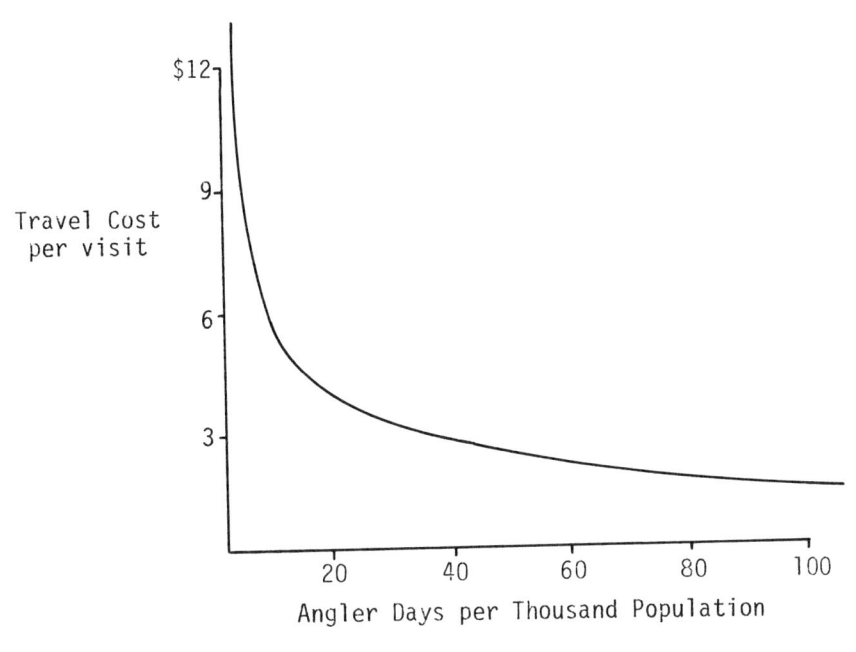

Figure 1. Demand for the recreation experience. Source: Elser (1965) and motel survey conducted by author.

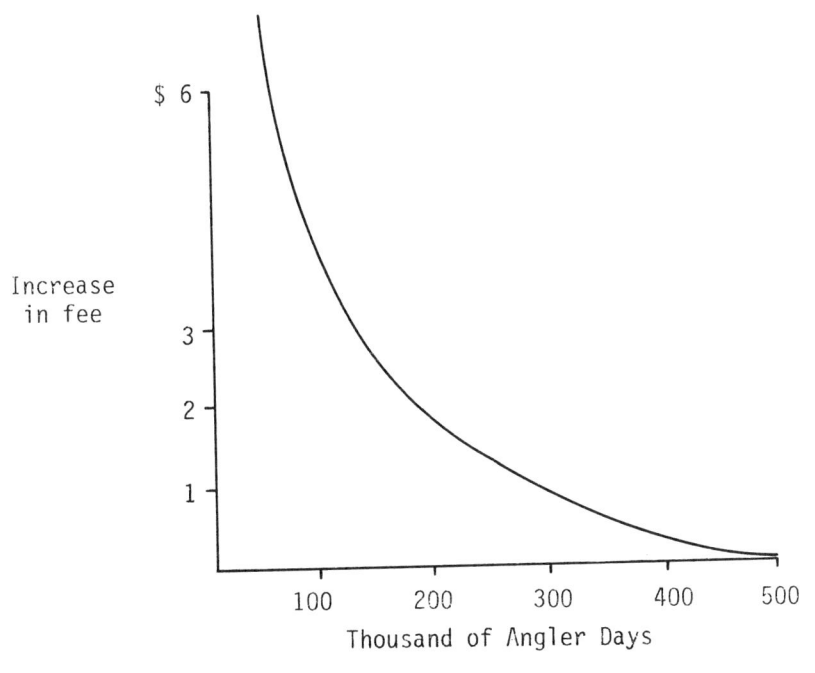

Figure 2. Demand for use of the bay for sportfishing.

PUBLIC HEALTH IMPACTS
M. P. Lynch, Editor

PUBLIC HEALTH ASPECTS OF TROPICAL STORM AGNES
IN VIRGINIA'S PORTION OF
CHESAPEAKE BAY AND ITS TRIBUTARIES[1]

M. P. Lynch[2]
J. Claiborne Jones[2]

ABSTRACT

All Virginia waters within Chesapeake Bay were closed for the taking of shellfish for direct consumption on 23 June 1972. This initial closing was in anticipation of high microorganism levels accompanying flood waters moving downstream. Various areas beginning with the lower portion of Chesapeake Bay were reopened beginning on 20 July 1972. By 5 October 1972 all areas closed as a result of Tropical Storm Agnes were reopened. No increased incidents of infectious diseases caused by waterborne microorganisms were noted in Virginia which could be attributed to Tropical Storm Agnes. In contrast to the Maryland situation, no Virginia waters were closed for water contact as a result of Agnes. No incidents of injuries or health impairment due to hazardous substances entering the estuarine portion of Virginia's waters were reported. Accumulations of pesticides and heavy metals in shellfish were insufficient to cause a public health hazard. Prompt action by the Virginia Health Department in closing the waters of the Bay and its tributaries to shellfish harvesting for human consumption undoubtedly prevented marketing of contaminated shellfish. The higher coliform counts in Virginia's estuaries in late June 1972 appeared to be primarily associated with runoff from the initial rains. The overall public health impact of Agnes on the Virginia waters of Chesapeake Bay and its tributaries was minimal with the exception of economic dislocations caused by shellfish and water contact recreation closings.

SHELLFISH AREA CLOSINGS

On 23 June 1972, the Virginia Department of Health closed Virginia waters in Chesapeake Bay and its tributaries for the taking of shellfish for direct human consumption. This action was a precautionary measure, taken in anticipation of high coliform levels of flood waters entering the shellfish harvesting areas of the Commonwealth.

The Virginia Bureau of Shellfish Sanitation immediately initiated an extensive sampling program to monitor coliform levels throughout the Bay and its tributaries. The results of this sampling indicated that anticipated high levels of coliforms did not occur throughout the Bay. Portions of the Bay and tributaries were reopened commencing 20 July 1972. By 5 October 1972, all areas closed as a result of Agnes were reopened. Fig. 1 recapitulates the timing of the reopening of Virginia waters to shellfish harvesting.

---
[1]Contribution No.765, Virginia Institute of Marine Science
[2]Virginia Institute of Marine Science, Gloucester Point, Va. 23062

WATER-CONTACT RECREATION BANS

No water contact recreation bans in Virginia waters were imposed by the Virginia Department of Health. The 30 June 1972 ban on water contact recreation imposed in all Maryland waters in Chesapeake Bay and tributaries had an impact on those Virginia counties with Potomac River shorelines since Maryland has jurisdiction in the Potomac to the Virginia shore.

On 14 July 1972, the City of Norfolk, Virginia, Health Department closed the Sarah Constant Shrine Beach and recommended against use of other city beaches in the Ocean View area (Fig. 2). On 19 July the U. S. Navy closed two beaches at Little Creek, Virginia and its one beach along Virginia Beach's oceanfront. The City of Norfolk action was triggered by high fecal coliforms at the Sarah Constant Beach (up to 6000/100 ml) and Community and Ocean View beaches (between 240-400/100 ml). The U. S. Navy action was taken on the basis of total coliform counts of more than 1000/100 ml (fecal coliforms at the Navy beaches measured only 4/100 ml).

The higher than normal coliform and fecal coliform readings in the Norfolk area were initially attributed to Agnes, but subsequent investigation disclosed that the source was a broken sewer main in Norfolk which overflowed into a storm drain entering Chesapeake Bay near the Sarah Constant Beach.

WATERBORNE PATHOGENS

No incidents of increased disease caused by waterborne pathogens were reported in Virginia. Infectious hepatitis cases reported in Virginia during the months of June-August 1972 (Table 1) were 56% fewer than in the comparable period of 1971. In 1973, there were 52% fewer cases of infectious hepatitis reported than in 1972. It is not known whether this is part of a downward trend in hepatitis infections or due in part to reduced shellfish consumption during the period after Agnes.

Table 1. Cases of infectious hepatitis reported in Virginia during the months of May - October 1971 through 1973. These figures were obtained from Virginia Department of Health.

Month	1971	1972	1973
May	184	103	40
June	140	72	67
July	228	107	37
August	111	88	36
September	90	73	75
October	118	55	43

MISCELLANEOUS HAZARDS

Analysis of pesticide and heavy metal levels in shellfish (Huggett & Bender, this vol.) indicated that accumulations of these materials did not reach levels to constitute health hazards as a result of Agnes.

Estimates of quantities of hazardous material entering the estuaries with flood waters were incomplete. The City of Richmond indicated 1,000 lbs. of liquid chlorine were lost in the flood, but no injuries were reported attributable to this loss. No incidents of injuries or health impairments due to hazardous materials entering the estuarine portion of Virginia were reported.

COLIFORM LEVELS

Introduction

Preliminary assessment of coliform levels in Virginia's estuarine waters (Chesapeake Bay Research Council 1973) indicated a general rise in coliform levels during the aftermath of Agnes. At the time of the initial assessment (September-October 1972) it was not possible to assign cause for all the rise to direct or indirect effects of Agnes. Initial reports of Agnes effects indicated many sewage treatment plants were inundated with resultant additional amounts of sewage to the waters entering the study area.

Rough approximations of sewage bypassed in various Virginia tributaries were developed from memoranda reports from wastewater plant operators in the files of the Virginia Department of Health (Table 2). These approximations are very crude and only indicate that a large amount of sewage was bypassed as a result of Agnes.

Table 2. Approximate estimates of sewage bypassed in Virginia streams as a result of Tropical Storm Agnes (developed from memoranda in Virginia Department of Health files).

Tributary	Estimated Bypass (Million Gallons)
James River	400
Potomac River[1]	40
Rappahannock River	Undetermined
York River	None

[1]Does not include estimates of bypassed sewage from Washington, D. C. or Suburban Maryland.

The only body of data of sufficient completeness to provide insight into the relation of Agnes and coliform levels in Virginia's estuarine waters is that of Virginia Department of Health's Bureau of Shellfish Sanitation (BSS). The BSS conducted intensive monitoring of Virginia's waters at stations that had been sampled at periodic intervals in previous years. The BSS made all data obtained from coliform monitoring available to us for analysis in relation to Agnes.

The BSS utilizes an MPN technique for coliforms. Lactose broth medium is used for presumptive coliforms with total coliforms confirmed in brilliant green bile broth and fecal coliforms confirmed in EC medium (APHA 1971).

Certain deficiencies inherent in the data base prevent definite analysis of the interrelationship between coliform levels and Agnes. These inherent deficiencies are:
1) Prior to the summer of 1972 the BSS was staffed at a very low level. The frequency of sampling at the established station was very low. Fortunately, the increased staffing of the BSS to a more realistic level occurred coincidently with Agnes so that sampling during Agnes' aftermath and subsequent periods (particularly the summer of 1973) does provide a data base of some utility.

2) The BSS is charged with monitoring coliform levels in terms of public health significance. Their laboratory methodology does not normally enable discrete enumeration of coliform levels greater than 1100/100 ml. Increased levels could not therefore be determined in areas which normally have coliform levels of 1100/100 ml or greater.

3) The sampling sites for routine BSS sampling were chosen with a view of monitoring areas of shellfish production, i.e., natural oyster rock, active private beds and areas in proximity to known sewage outfalls. Few of the sampling stations are located in mid-channel areas of the Bay or major tributaries or in the vicinity of the mouths of these tributaries.

Selection of Stations

For the purposes of this attempt to analyze available coliform data in terms of Agnes, the data available from the stations located at the points indicated in Fig. 3 were evaluated for the three year period 1971 through 1973. These stations were selected using the following criteria:

1) Location along the salinity gradient: An attempt was made to select stations along the major tributaries at the uppermost site sampled, the mouth of the tributary and approximately evenly spaced sites along the tributary. In the Bay proper, sites were selected which would provide good geographical coverage.

2) Proximity to mid-channel: Where possible sites closest to the channel were chosen for detailed analysis. Because of the siting factors mentioned previously, it was not possible to find sites in mid-stream of the major tributaries or the Bay proper.

3) Normal condition for the area: Those sites which routinely had coliform levels in excess of 1100/100 ml were eliminated from the detailed analysis.

4) Completeness of data: After the previous criteria were used to select possible sites for analysis, the completeness of data, particularly presence of data in all three years (1971-1973) during the months of May through October was used for the final criterion.

It must be added at this point that few stations proved to be ideal for analysis and that many compromises in selection had to be made.

RESULTS

At most stations selected for analyses, highest coliform counts coincided with periods of heaviest rainfall. A typical rainfall-coliform relationship is shown in Fig. 4. Total coliform levels were already relatively high in May and early June (prior to Agnes) at many stations throughout the Bay. For the most part, these relatively high levels can be attributed to runoff associated with the heavy rainfall during these months. Although total coliform levels were relatively high, fecal coliform levels did not appear markedly elevated (Fig. 5).

At most of the stations analyzed which were not usually considered polluted, total coliform levels during the summer of 1972 exceeded total coliform levels during the summer of 1971 or 1973 (Fig. 6). This was not, however, true of fecal coliforms. With few exceptions, fecal coliforms did not reach excessively high levels at these stations during the aftermath of Agnes.

CONCLUSION

The runoff associated with local rainfall from Agnes resulted in increased levels of total coliforms throughout the Virginia portion of Chesapeake Bay and its tributaries. With few exceptions these high levels of total coliforms were not accompanied by a rise in fecal coliforms. In local areas, where wastewater treatment plants were taxed to or beyond capacity fecal coliform counts reached very high levels. These high fecal coliform levels did not appear to extend beyond local waters. There did not appear to be any excessively high coliform levels (either fecal or total) developed in the shellfish harvesting areas as a result of high coliform levels in flood waters from above the fall line.

SUMMARY

Prompt action by the Virginia Department of Health in closing the waters of the Bay and its tributaries to shellfish harvesting for direct human consumption, undoubtedly prevented marketing of contaminated shellfish. The higher than usual coliform levels in Virginia waters appeared to be primarily associated with run-off from the initial rains. Overall, with the exception of economic dislocations in the shellfish industry due to closings and the recreation industry because of Maryland's water contact recreation ban, the public health impacts of Agnes in Tidewater, Virginia were minimal.

ACKNOWLEDGEMENTS

We wish to thank Mr. Cloyde Wiley for making the Bureau of Shellfish Sanitation data available to us and for his encouragement during this effort. Particular thanks must also go to Mr. Steve Shaw of the BSS who patiently explained the files to us and suffered graciously through the many interruptions while we screened the data. We wish to emphasize, however, that all interpretation and evaluation of the data in terms of Agnes are those of the authors and not the BSS. This evaluation was supported by contract DACW-31-73-C-0189 from the U. S. Army Corps of Engineers, Baltimore District to the Chesapeake Research Consortium, Inc.

LITERATURE CITED

APHA. 1971. Standard methods for the examination of water and sewage. Am. Public Health Assoc. and Am. Water Works Assoc., Ninth Edition.

Chesapeake Bay Research Council. 1973. Effects of Hurricane Agnes on the environment and organisms of Chesapeake Bay. A report to the Philadelphia District U. S. Army Corps of Engrs. (Aven M. Andersen, Coordinator), Chesapeake Bay Inst., Contrib. 187, Nat. Resour. Inst. Contrib. 529, Va. Inst. Mar. Sci. Spec. Rep. Appl. Mar. Sci. & Ocean Engr. 29, 172 p.

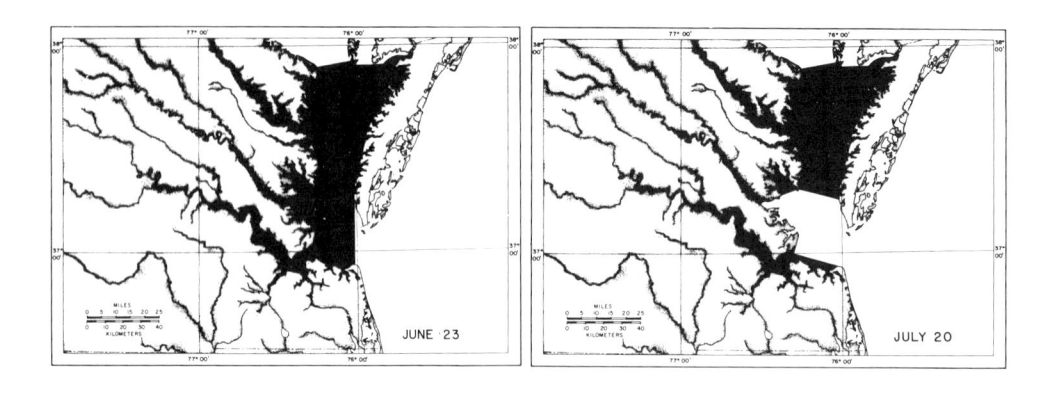

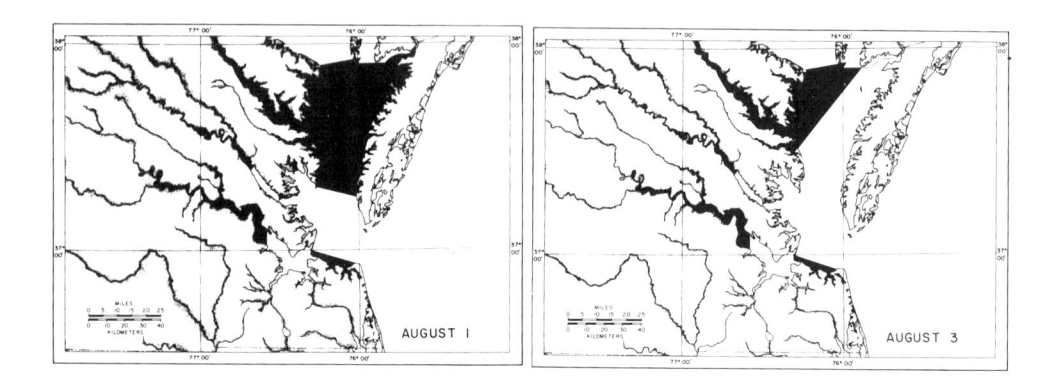

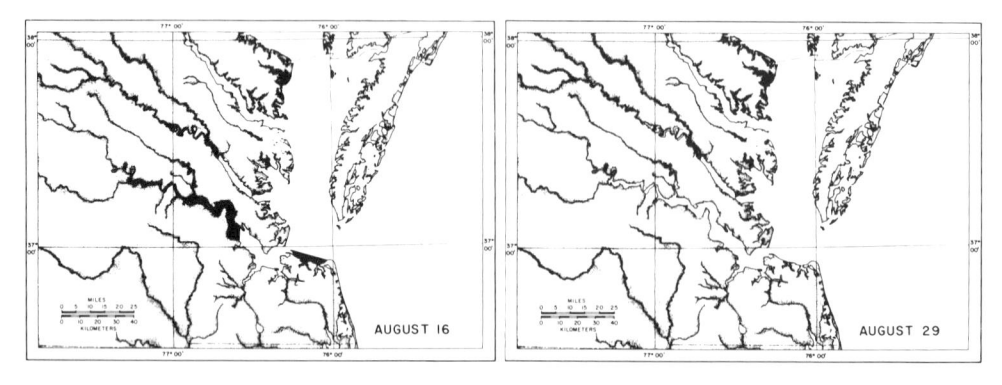

Figure 1. Areas of the Commonwealth of Virginia closed for the harvesting of shellfish for direct human consumption in the aftermath of Tropical Storm Agnes, summer of 1972. By 5 October 1972, all areas closed as a result of Agnes were reopened. Closings normally in effect, remained in effect.

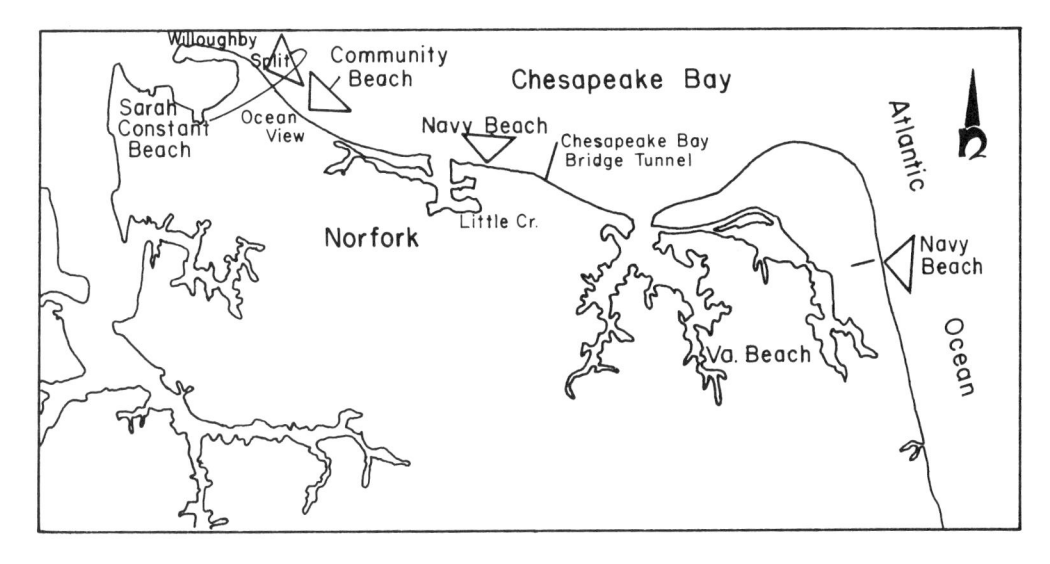

Figure 2. Beaches in the Norfolk and Virginia Beach area closed for water contact recreation during parts of July 1972.

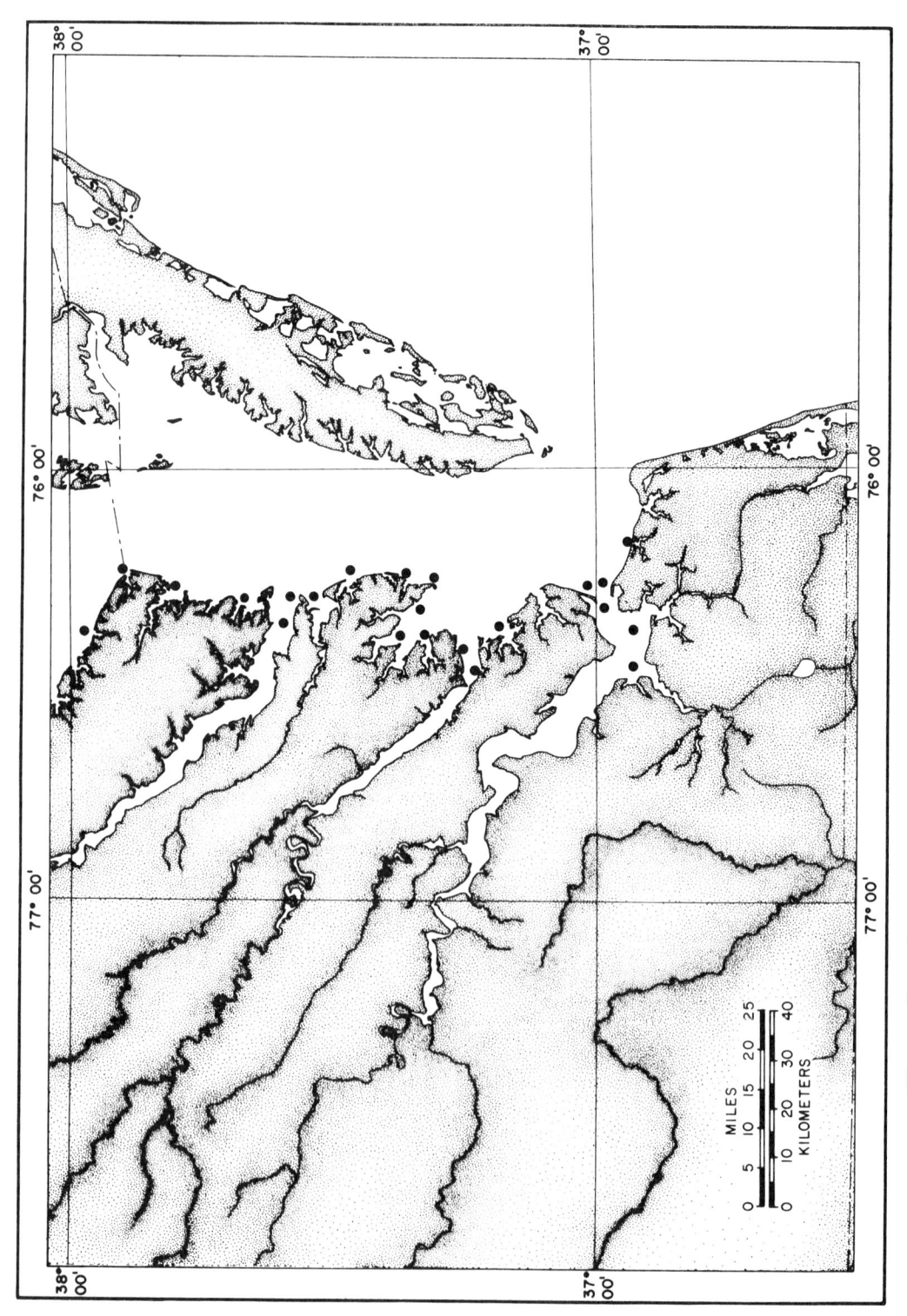

Figure 3. Virginia stations selected for analysis of Agnes effects on estuarine bacterial levels.

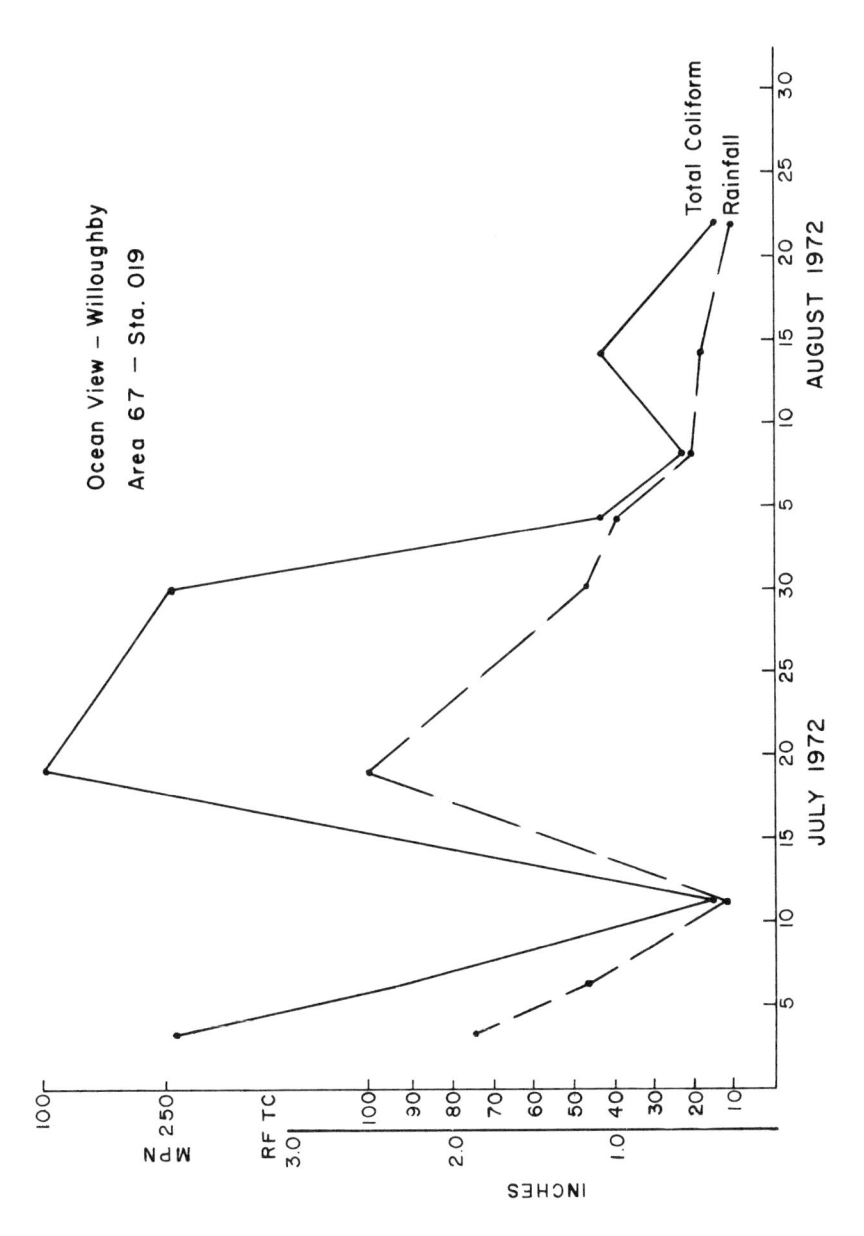

Figure 4. Typical rainfall - total coliform relationship seen in Bureau of Shellfish Sanitation data.

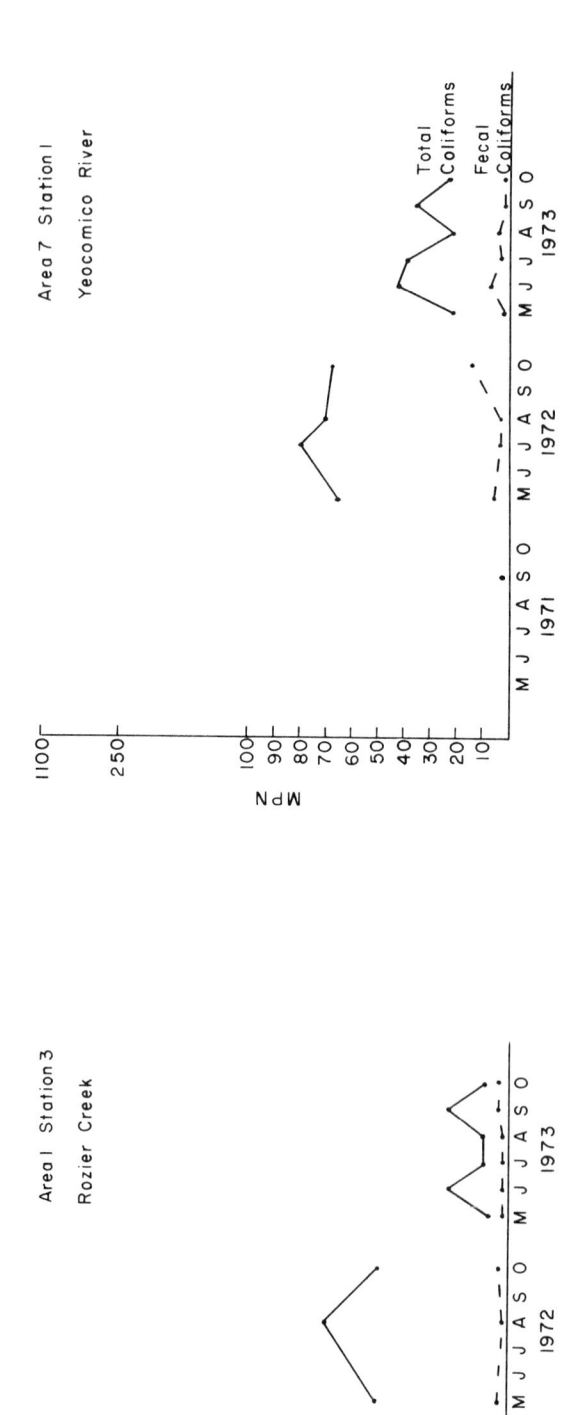

Figure 5. Examples of elevated total coliform levels in Virginia waters prior to Agnes.

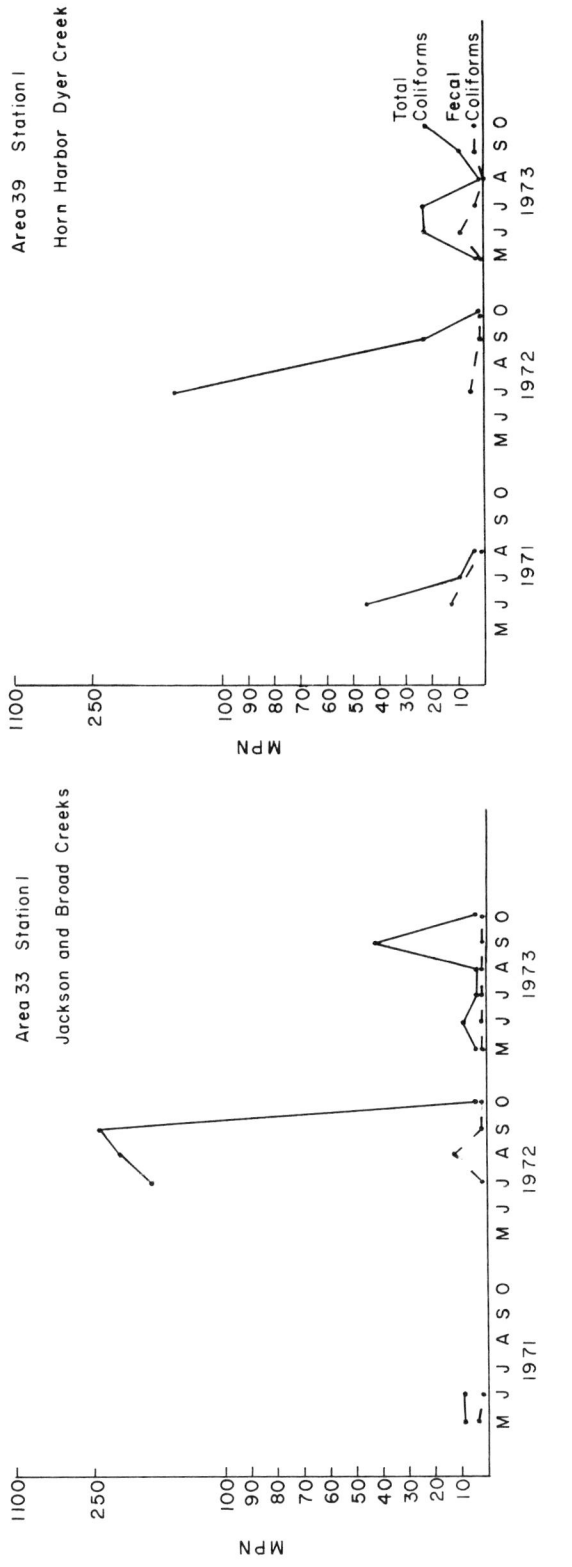

Figure 6. Examples of typical total coliform and fecal coliform patterns during the aftermath of Tropical Storm Agnes.

PUBLIC HEALTH ASPECTS OF TROPICAL STORM AGNES
IN MARYLAND'S PORTION OF CHESAPEAKE BAY[1]

Aven M. Andersen[2]

ABSTRACT

The floods of Tropical Storm Agnes carried raw sewage, hazardous chemicals, silt, and debris into Chesapeake Bay. These substances presented many threats to public health. In the interest of public health, the Maryland Department of Health and Mental Hygiene closed the bay to shellfish harvesting on 23 June 1972 and to water-contact recreation on 30 June 1972. As the levels of coliforms in shellfish and water in the various parts of the bay dropped to acceptable levels, MDHMH reopened these areas. By 28 July 1972, four weeks after Agnes, all areas of the bay had been reopened to water-contact recreation, and by 1 August 1972 most of the shellfish growing areas had been reopened. Apparently the closings instituted by MDHMH prevented any significant public health problems.

INTRODUCTION

The Agnes floods created the potential for a number of public health problems. Raw sewage from inundated sewage treatment plants and broken sewer lines as well as numerous hazardous chemicals from flooded industries, warehouses, and railroad cars washed into the bay with the storm runoff. A number of illnesses could have resulted from ingesting the bay water and serious skin irritations could have resulted merely from contact with the water. Moreover, because shellfish like clams and oysters concentrate water-borne microorganisms, the ingestion of contaminated shellfish could have produced illnesses. The massive debris carried into the bay was still another threat to water skiers and bathers. Therefore, in the interest of public health, the Maryland Department of Health and Mental Hygiene closed the bay to shellfish harvesting and water-contact recreation.

SHELLFISH HARVESTING

On 23 June 1972 the Maryland Department of Health and Mental Hygiene closed all the shellfish growing waters within the boundaries of the State of Maryland to shellfish harvesting (Chaney 1972). This action was authorized under Article 43, Section 228B Annotated Code of Maryland, which states; "whenever. . . any area of waters of the State devoted to the production or storage of shellfish is so polluted so that shellfish produced or stored in such area is a hazard to the public health, (MDHMH) . . . shall restrict such area for the taking of shellfish."

Subsequent sampling showed extremely high levels of coliforms in clams, oysters, and bay water. Although detailed analyses of the 5,000 or so water and shellfish samples is incomplete, a few examples will illustrate the high coliform levels encountered.[3]

Oysters collected from Teagues oyster bar in the Patuxent River on 10 July 1972, 17 days after Agnes, gave a coliform MPN count of 240,000/100g and a stan-

---

[1]Contribution No. 582, U. M. Center for Environmental and Estuarine Studies.
[2]Chesapeake Biological Laboratory, Center for Environmental and Estuarine Studies, University of Maryland, Solomons, Md. 20688.
[3]Preliminary information provided by Mr. William C. King, Maryland Department of Health and Mental Hygiene.

dard plate count of 810,000/g.

The highest counts were obtained on 11 July 1972 from oysters and clams collected off Franklin Manor, an area about 12 nautical miles south of the Chesapeake Bay Bridge. The clams had a coliform MPN count of 240,000≥/100 g and a standard plate count of 12,000,000/g, whereas the oysters had a coliform MPN count of 240,000≥/100 g a fecal coliform MPN count of 240,000≥/100 g and a standard plate count of 63,000,000/g.

Samples collected about three nautical miles above the Bay Bridge, near Love Point, on 12 July 1972 had coliform MPN counts of 240,000/100g for clams and oysters and standard plate counts of 1,100,000/g for clams, and 340,000/g for oysters.

High counts remained for some time in the upper bay, as clams collected near Gibson Island (above the Bay Bridge) on 27 July 1972 had coliform MPN counts of 240,000/100g and a standard plate count of 1,600,000/g.

As the levels of coliform bacteria in the water and shellfish dropped to acceptable levels, MDHMH permitted the various areas to be reopened to shellfish harvesting (Table 1).

Although the coliform levels dropped and the clam beds were clean from a public-health point-of-view, on 1 September 1972 the Maryland Department of Natural Resources instituted its own ban on soft-shell clam harvesting to prevent overharvesting of the surviving clams.

Table 1. Dates of closing and reopening Maryland waters to shellfish harvesting following Tropical Storm Agnes.

Date	Action taken by the Maryland Department of Health and Mental Hygiene (MDHMH)
23 June 1972	MDHMH closed all shellfish growing waters within the boundaries of the state of Maryland to shellfish harvesting.
6 July 1972	Sinepuxent, Assawoman, and Chincoteague Bays reopened.
10 July 1972	Big Annemessex River, Manokin River, Monie Bay, Choptank River (sections 3, 4, and 5), Harris Creek and Broad Creek reopened.
14 July 1972	Fishing Bay reopened.
1 Aug. 1972	Nanticoke River, Pocomoke Sound, Tangier Sound, and Honga River reopened.
1 Sept. 1972	Maryland Department of Natural Resources (MDNR) institutes a ban on all soft-shell clam harvesting to protect the remaining clam populations.
18 Sept. 1972	MDNR opens the oyster season.

WATER-CONTACT RECREATION

Because of the large amounts of raw sewage washed into the bay, as well as the discoloration, debris, and strong currents, the Maryland Department of Health and Mental Hygiene warned residents on 27 June 1972, five days after Agnes passed over the bay, that they should not participate in water-contact recreation in Maryland waters.

On 30 June 1972, eight days after Agnes, MDHMH closed the Chesapeake Bay and all its tributaries within the boundaries of the State of Maryland for water contact recreation (Solomon 1972).

Three days later Dr. Solomon continued the ban. He stated: "The overall picture at this time indicates considerable organic and inorganic contaminants, and this fact, combined with a weather forecast for more rain, makes it mandatory that the ban continue." With the eleven state laboratories constantly monitoring the condition of the bay, Dr. Solomon promised ". . . that as soon as the waters become safe for swimming, I will lift the ban." (MDHMH 1972).

The department began lifting the ban on 7 July 1972 when it approved the waters of Anne Arundel, Baltimore, and Calvert counties. By 28 July 1972 the last of the waters closed by Agnes were reopened (Table 2). Thus, from 30 June 1972 until 28 July 1972, almost one month, parts of the bay were closed to water-contact recreation.

Table 2. Dates of closing and reopening the Maryland portion of Chesapeake Bay to water-contact recreation following Tropical Storm Agnes.

Date	Action taken by the Maryland Department of Health and Mental Hygiene (MDHMH).
27 June 1972	MDHMH warned that residents should not participate in water recreational activities for the next few days.
30 June 1972	MDHMH announced a ban on swimming, water skiing, skin diving, and other body contact activities for the Chesapeake Bay and all its tributaries within the State of Maryland.
3 July 1972	Ban continued on water-contact recreation.
7 July 1972	Ban lifted for Baltimore, Anne Arundel, and Calvert counties.
12 July 1972	Ban lifted for Queen Annes, Talbot, Wicomico, and Somerset counties.
13 July 1972	Ban lifted for Dorchester County.
15 July 1972	Ban lifted for all but the Maryland portion of the Susquehanna River, the upper Choptank River, the Patuxent River, and the Potomac River above the U. S. Route 301 bridge.
28 July 1972	Ban lifted for the remaining areas, the Susquehanna River, Upper Choptank River, Patuxent River, and St. Marys County and Charles County portions of the Potomac River.

CONCLUSIONS

Although much raw sewage and many hazardous chemicals were washed into Chesapeake Bay by the Agnes floods, no serious health problems arose. The effects were largely economic: shellfish harvesters and processors were put out of work and the seashore recreation industry suffered because of the ban on water-contact sports, especially since it occurred over the Fourth-of-July weekend. The lack of serious health problems can probably be credited to the rapid preventative action taken by the Maryland Department of Health and Mental Hygiene.

LITERATURE CITED

Chaney, Howard E. 1972. Letter of 23 June 1972 notifying Mr. Fred Seling, Maryland Fish and Wildlife Administration, that all shellfish growing areas of Maryland are closed. Md. Dept. Health & Mental Hygiene, Environmental Health Administration, 1 p.

Maryland Department of Health and Mental Hygiene. 1972. "Ban to continue on water contact in Chesapeake and tributaries." Md. Dept. Health & Mental Hygiene, Public Information Serv., News Release, 3 July 1972, 2 p.

Solomon, Niel. 1972. "Order banning water contact recreation in Chesapeake Bay and all its tributaries." Md. Dept. Health & Mental Hygiene, Memorandum to all Deputy State Health Officials and all Local Health Officers, dated 30 June 1972, 1 p.

LIBRARY OF CONGRESS CATALOGING IN PUBLICATION DATA

Chesapeake Research Consortium.
   The effects of tropical storm Agnes on the Chesapeake Bay estuarine system.

   (CRC publication; no. 54)
   Appendices include papers which were presented at a symposium held May 6-7, 1974 at College Park, Md. under the sponsorship of the Chesapeake Research Consortium.
   1. Oceanography--Chesapeake Bay. 2. Atlantic States--Hurricane, 1972. I. Davis, Jackson. II. Title. III. Series: Chesapeake Research Consortium. CRC publication; no. 54.
GC511.C46 1977   551.4'614'7   76-47392
ISBN 0-8018-1945-8